CONVERSION FACTORS

LENGTH

1 meter (m) = 39.37 in. = 3.281 ft = 6.214×10^{-4} mile = 10^{10} Å = 10^{15} fermis

1 in. = 0.02540000 m; 1 ft = 0.3048 m; 1 mile = 1,609 m

1 nautical mile = 1,852 m = 1.1508 miles = 6,076.10 ft

1 angstrom (Å) = 10^{-10} m; 1 mil = 10^{-3} in.; 1 rod = 16.5 ft; 1 fathom = 6 ft

AREA

1 m^2 = 10.76 ft^2 = 1,550 $in.^2$; 1 hectare = 10^4 m^2 = 2.471 acres

1 ft^2 = 929 cm^2; 1 $in.^2$ = 6.452 cm^2 = 1.273×10^6 circular mils; 1 acre = 43,560 ft^2

VOLUME

1 m^3 = 35.31 ft^3 = 6.102×10^4 $in.^3$

1 ft^3 = 0.02832 m^3; 1 U.S. gallon = 231 $in.^3$; 1 liter = 1.000028×10^{-3} m^3 = 61.02 $in.^3$

TIME AND FREQUENCY

1 year = 365.2422 days = 8.766×10^3 hr = 5.259×10^5 min = 3.156×10^7 sec

1 sidereal day (period of earth's revolution) = 86,164 sec

1 hertz (Hz) = 1 cycle/sec

SPEED

1 m/sec = 3.281 ft/sec = 3.6 km/hr = 2.237 mi/hr = 1.944 knots

1 km/hr = 0.2778 m/sec = 0.9113 ft/sec = 0.6214 mi/hr

1 mi/hr = 1.467 ft/sec = 1.609 km/hr = 0.8689 knot

MASS

1 kg = 2.205 lb mass = 0.06852 slug

1 lb mass = 0.4536 kg = 0.03108 slug; 1 slug = 32.17 lb mass = 14.59 kg

DENSITY

1 g/cm^3 = 1,000 kg/m^3 = 62.43 lb mass/ft^3 = 1.940 slug/ft^3

1 lb mass/ft^3 = 0.03108 slug/ft^3 = 16.02 kg/m^3 = 0.01602 g/cm^3

FORCE

1 newton (N) = 10^5 dynes = 0.1020 kg wt = 0.2248 lb

1 lb (force) = 4.448 newtons = 0.4536 kg wt = 32.17 poundals

PRESSURE

1 newton/m^2 = 9.869×10^{-6} atm = 1.450×10^{-4} lb/$in.^2$ = 0.02089 lb/ft^2

$\qquad$ = 7.501×10^{-4} cm Hg = 4.015×10^{-3} in. of water = 10^{-5} bar

1 lb/$in.^2$ = 144 lb/ft^2 = 6,895 newtons/m^2 = 5.171 cm Hg = 27.68 in. of water

1 atm = 406.8 in. of water = 76 cm Hg = 1.013×10^5 newtons/m^2

$\qquad$ = 10,330 kg wt/m^2 = 2,116 lb/ft^2 = 14.70 lb/$in.^2$ = 760 torr

WORK, ENERGY, HEAT

1 joule (J) = 0.2389 cal = 9.481×10^{-4} Btu = 0.7376 ft-lb = 10^7 ergs = 6.242×10^{18} eV

1 kcal = 4,186 joules = 3.968 Btu = 3,087 ft-lb

1 eV = 1.602×10^{-19} joule; 1 unified amu = 931.48 MeV

1 kW-hr = 3.6×10^6 joules = 3,413 Btu = 860.1 kcal = 1.341 hp-hr

POWER

1 hp = 2,545 Btu/hr = 550 ft-lb/sec = 745.7 watts = 0.1782 kcal/sec

1 watt (W) = 2.389×10^{-4} kcal/sec = 1.341×10^{-3} hp = 0.7376 ft-lb/sec

ELECTRIC CHARGE

1 coulomb (C) = 0.1 abcoulomb = 2.998×10^9 statcoulombs (esu)

1 faraday = 96,487 coulombs; 1 electronic charge = 1.602×10^{-19} coulomb

ELECTRIC CURRENT

1 ampere (A) = 0.1 abampere (emu) = 2.998×10^9 statamperes (esu)

ELECTRIC POTENTIAL DIFFERENCE

1 abvolt (emu) = 10^{-8} volt (V); 1 statvolt (esu) = 299.8 volts

CAPACITANCE

1 statfarad (esu) = 1 cm of capacitance = 1.113×10^{-12} farad (F)

MAGNETIC FLUX

1 weber (Wb) = 10^8 maxwells = 10^5 kilolines

MAGNETIC INTENSITY B

1 tesla (T) = 1 newton/amp-m = 1 weber/m^2 = 10,000 gauss = 10^9 gamma

MAGNETIZING FORCE H

1 amp-turn/m = 0.01257 oersted = 0.01257 gilbert/cm

Final July 10th

David Harrison
621 N Rogers
Irving
254-7972

**ELEMENTS
OF PHYSICS**

ELEMENTS
OF PHYSICS

ALPHEUS W. SMITH
Late Professor of Physics
Ohio State University

and

JOHN N. COOPER
Professor of Physics
U.S. Naval Postgraduate School

McGRAW-HILL BOOK COMPANY
New York St. Louis San Francisco Düsseldorf
Johannesburg Kuala Lumpur London Mexico Montreal
New Delhi Panama Rio de Janeiro Singapore Sydney
Toronto

EIGHTH EDITION

This book was set in Vega Light with Century Expanded Italic by York Graphic Services, Inc., printed by the Murray Printing Company, and bound by Rand McNally & Company. The designer was Barbara Ellwood; new drawings were done by York Graphic Services, Inc. The editors were Jack L. Farnsworth and Laura Warner. Peter D. Guilmette supervised production.

**ELEMENTS
OF PHYSICS**

Library of Congress Catalog Card Number 70-38998

07-058631-4

4567890 MURM 79876543

Preface   vii

CONTENTS

The role of physics in modern life continues to expand. Since publication in 1964 of the seventh edition of "Elements of Physics," space satellites have been launched by the score, man has set foot on the moon, and innumerable applications of physics have been made in communications, in transportation, and in industry. A knowledge of the principles of physics is increasingly imperative if one wishes to keep abreast of what is happening in the broad technological segment of human activity. It is the purpose of this book to introduce the student to the fundamental laws of physics and to give him some feeling of their power and beauty. Toward this end, a wealth of applications and examples is introduced, with particular attention to the physics of everyday experience. It is our belief that an understanding of physical phenomena adds a new dimension to the beauty of the rainbow and an enriched appreciation of such scientific advances as those which permit us to view in full color events taking place many hundreds of kilometers away.

This text is written for students who have not studied calculus. The treatment of physical laws and their applications has been handled as quantitatively and rigorously as the mathematical sophistication of the prospective student permits. While there are many situations which can be treated better by use of calculus, there are others in which careful physical reasoning can give a more penetrating understanding of a resulting equation than the glib application of calculus would.

The rapid increase in the use of the International System of Units throughout the world has encouraged further emphasis on mks units, although the British engineering system is employed in mechanics and heat. Definitions of key quantities of the cgs system are presented, but no problems are introduced in either cgs electrostatic or electromagnetic units.

Although the time-honored division of the book into parts on mechanics, heat, wave motion and sound, light, electricity, and modern physics has been retained, there has been considerable reorganization in an effort to make the text more useful to the student and his professor. In several areas material has been added, including a section on the Coriolis acceleration in Chap. 19, one on lasers and masers in Chap. 46, and one on the Hall effect in Chap. 48.

There is ample material in this text for three full semesters of study. The classical physics sections alone require a full academic year for reasonably thorough coverage. For a two-semester course which includes considerable modern physics, it is possible to omit many selected chapters without seriously neglecting the background knowledge required for most subsequent chapters. Among the chapters which are sometimes omitted are those on rotational motion, fluids, elastic properties of matter, sound, atmospheric physics, optical instruments, spectra and color, chemical and thermal electromotive forces, generators and motors, and alternating currents.

It is with sorrow that we report the death of Alpheus W. Smith in April, 1968, some three months after his ninety-second birthday. A man of outstanding character and ability, he had a distinguished career at Ohio State University as Professor, Head of the Department of Physics and Astronomy, and Dean of the Graduate School. In recognition of his great contributions, the physics building at Ohio State University

PREFACE

has been named in his honor. Although Dean Smith was not able to participate directly in the preparation of recent editions of "Elements of Physics," a sincere effort has been made to retain the spirit and vision with which he imbued the first five editions.

The junior author is deeply grateful to many persons who have offered constructive criticism of the previous edition and made suggestions for improvement. It would be impossible to acknowledge by name all the professors and students who have reported errors, proposed changes, or otherwise contributed to the development of this book. However, he wishes to acknowledge his special indebtedness to Professor G. N. Koehl, of George Washington University, and to Professors E. C. Crittenden, Jr., A. R. Frey, H. E. Handler, S. H. Kalmbach, L. O. Olsen, J. D. Riggin, J. R. Neighbours, J. V. Sanders, and W. B. Zeleny, all colleagues at the Naval Postgraduate School. Finally, special thanks are due the author's wife Elaine, who typed and helped prepare the entire manuscript.

Corrections, criticisms, and suggestions from those who examine or use this book will be greatly appreciated.

JOHN N. COOPER

**ELEMENTS
OF PHYSICS**

MECHANICS

CHAPTER ONE
PHYSICS AND MEASUREMENT

EVERY DAY we move about, we see, we hear, we interact with our environment in myriad ways. The science of physics is concerned with the inanimate aspects of the universe and our interactions with it. How are sound and light produced, propagated, and detected? How does an automobile engine, an electric motor, or a jet engine operate? How can the motions of billiard balls, space capsules, and planets be predicted? These are all questions which can be answered in terms of the principles of physics. No one who lives a normal, active life can avoid knowing and using physics in a *qualitative* way. It is one purpose of this book to help you achieve a *quantitative* knowledge of many of the topics of physics and to broaden the range of physical phenomena with which you are acquainted.

PROBABLY NO ASPECT of physics is more familiar than the motions of objects such as balls or automobiles. Therefore we start our study with mechanics, the branch of physics dealing with the motions of bodies and with the influence of forces on these motions. To obtain precise relationships, we require quantitative measurements in terms of carefully established standards. In this first chapter we define the fundamental units of length (the *meter*), of time (the *second*), and of mass (the *kilogram*), in terms of which we can express any quantity we meet in mechanics.

In a broad sense physics is the branch of knowledge which describes and explains the material world and its phenomena and which uses the resulting understanding to create new areas of human experience. In terms of this sweeping definition, other physical sciences—such as chemistry, geology, astronomy, and the engineering disciplines—are branches of the basic science *physics*, which a century ago was known as *natural philosophy*.

At present, it is customary to use *physics* in a more restricted sense. Aspects of nature which are ordinarily regarded as clearly in the domain of physics are mechanics, sound, heat, electricity and magnetism, atomic and nuclear structure, and the properties of solids, liquids, gases, and plasmas. Broad areas of knowledge in which physics overlaps related sciences are described by names such as astrophysics, biophysics, chemical physics, and geophysics. Of course, these fields are not distinct; they merge into one another. Sharp distinctions between the sciences are neither necessary nor desirable.

Throughout the ages, men of intelligence have endeavored to explain what went on in the world about them. There has been a never ending struggle to formulate a systematic set of concepts about the world we live in. In part, men have been driven by curiosity—an urge to learn ''why.'' In part, they have been motivated by the conviction that their efforts to understand natural phenomena would lead to the possibility of controlling these phenomena. Thousands upon thousands of men have contributed to the vast store of knowledge and information which constitutes the physics of today. Thousands more are actively working to add new facts to this body of knowledge and to organize facts already known into logical patterns. Physics is growing at an unprecedented rate. It is a science full of life and vigor.

Physics is as old as man's curiosity about his environment and the wealth of processes which it manifests. In the phenomena which impinge on our senses we find many regularities and a great orderliness, suggesting that these phenomena are subject to certain laws. Physical scientists seek to discover these laws of nature, to organize them into a logically elegant structure, and to use them for the benefit of man.

Everyone is familiar with at least some of the regularities of nature. A small child soon learns that released objects fall to the floor; a more mature observer may be impressed by the regularity of the movement of the sun, moon, and stars. To find the relationship between these apparently unrelated processes required the genius of Newton. Yet many an important contribution to physics has been made by men of modest intellectual achievements. In the words of Aristotle,

> The search for Truth is in one way hard and in another easy. For it is evident that no one can master it fully nor miss it wholly. But each adds a little to our knowledge of Nature, and from all the facts assembled there arises a certain grandeur.

Physics is a rapidly evolving science with many branches in widely varying stages of development. Typically the first stage consists of the observation of some physical phenomenon. Then measurements are made, and there are efforts to formulate ''laws'' in terms of which the observations can be related quantitatively. As knowledge grows, a creative facet of physics appears; men learn how to apply the laws in ways which may revolutionize our mode of life. For example, in the eighteenth century scientists made great progress in understanding heat. This knowledge led to the development of heat engines, and their use opened the way for steamships, railroads, automobiles, and aircraft in our transportation system and for tractors on our farms. In the nineteenth century physicists investigated and began to understand electrical phenomena. Creative synthesis of this knowledge brought about the widespread use of electric energy for light, for heat, for electric motors, and for radio and television. Investigation into the discharge of electricity through gases led to the discovery of x-rays, which not only provided a vital tool for the study of crystals and atomic structure but also paved the way for revolutionary advances in the diagnosis and treatment of diseases. In the twentieth century research in physics has expanded to unprecedented levels and has given us insight into the properties of solids and into the structure of atoms and nuclei. This knowledge in turn has been utilized to develop solid-state circuits, computers, and nuclear power sources.

Progress in physics brings with it progress in commerce and industry, in medicine, and in the related sciences. Much of the difference between the way we now live in the United States and the way the Indians lived 500 years ago is associated with our greater knowledge of physical phenomena and how they can be controlled.

1.2 STANDARDS OF LENGTH AND MASS

In physics and related sciences, facts and laws are expressed in precise language. For the most part the terms used in physics are defined in an unambiguous way, so that scientists all over the world mean exactly the same thing by the same term. In order to formulate valid relationships between physical quantities, accurate measurements must be made. Thus, physics is a quantitative science. In 1883, Lord Kelvin emphasized this quantitative aspect in the following words:

> I often say that when you can measure what you are speaking about, and express it in numbers, you know something about it; but when you cannot measure it, when you cannot express it in numbers, your knowledge is of a meager and unsatisfactory kind. It may be the beginning of knowledge, but you have scarcely, in your thoughts, advanced to the stage of science, whatever the matter may be.

In making measurements of physical quantities, we ordinarily perform a series of operations which involve the comparison of an unknown quantity with an accepted standard of the same kind. Thus when we say that a road is 10 meters (m) wide, we mean that a standard of length called a *meter* must be applied to it 10 times in succession in order to cover its entire width. Similarly, a body is said to have a mass of 25 kilograms (kg) if its mass is 25 times as great as the mass of a standard kilogram.

At one time the length of a king's foot or the span of his hand served as an adequate standard of length, and a selected stone as a standard of mass. However, as commerce between countries increased, and as scientists began to make more and more accurate measurements, established standards of length and mass became of increasing importance. In 1791 the French National Assembly adopted the *metric system*, in which the standard of length was the *Meter of the Archives*, a platinum bar fabricated to have a length one ten-millionth of the distance from the North Pole to the equator as then measured on the quadrant through Paris. As a standard mass, French scientists prepared a platinum cylinder, known as the *Kilogram of the Archives*, which had a mass as near as they could make it to the mass of 1,000 cubic centimeters (cm^3) of pure water at maximum density. (Subsequent measurements have shown that a volume of 1,000.028 cm^3 of water at maximum density has a mass of 1 kg. This volume is known as the *liter*.) The French system, slightly modified, is now used almost universally in sc ic work.

the United States a chaotic set of standards had
a bushel basket in South Carolina held 68 in.^{3} more
ket in New York, while a pound of meat in Massa-
ighter than a pound of meat in Maine. These facts
the Congress in 1821 by John Quincy Adams, then
In his "Report on Weights and Measures" Adams

easures may be ranked among the necessaries of
dividual of human society. They enter into the
angements and daily concerns of every family.
ary to every occupation of human industry; to the
security of every species of property; to every
de and commerce; to the labors of the husband-
uity of the artificer; to the studies of the philoso-
arches of the antiquarian; to the navigation of
he marches of the soldier; to all the exchanges
he operations of war. The knowledge of them,
se, is among the first elements of education,
d by those who learn nothing else, not even
This knowledge is riveted in the memory by
on of it to the employments of men throughout

In ablished a bureau which later became the
Nati ds, an invaluable national asset.

s from many civilized countries met in Paris
to di idards of measurement. It was generally
agree were most suitable and that copies should
be ma shed to adopt the *metric system* of units.
Thirty ue of an alloy of platinum (90 percent) and iridium (10 percent). On these bars were marked fine transverse lines separated as nearly as possible by a distance equal to the length of the Meter of the Archives. These new standards were made in 1880, and in 1889 they were distributed to the governments participating in the convention of 1875. The United States obtained two of these standard "meters," of which Bar No. 27 was the primary standard of length for the United

Figure 1.1

United States national standard of mass. A standard kilogram is a platinum-iridium cylinder about 39 mm in diameter and 39 mm in height.

States until 1960. Bar No. 6 was most nearly equal to the Meter of the Archives, and it was declared the *International Standard Meter*. It is kept at the International Bureau of Weights and Measures at Sèvres, near Paris.

At the same time, 40 standard kilograms, with masses as nearly as possible equal to that of the Kilogram of the Archives, were constructed from the platinum-iridium alloy. The new mass which agreed most closely with the Kilogram of the Archives is now the *International Mass Standard*. The primary mass standard of the United States is Kilogram No. 20 (Fig. 1.1), one of the two delivered to this country in 1889. With the kilogram the legal unit of mass, the pound mass is defined to be 0.45359237 kg.

By the middle of the twentieth century the prototype meter bar was no longer good enough as a standard of length. Secondary standards could be compared with it only to an accuracy of roughly 1 part in 10 million, while technological requirements for modern industry called for tolerances of the order of 10^{-7} m. In general, it is desirable that a master standard be at least 10 times more accurate than the scientific and industrial measuring systems which are derived from it. As a consequence, at the 1960 International Conference on Weights and Measures the meter was redefined in terms of the wavelength (Fig. 1.2) of the orange light emitted by the krypton isotope of mass 86. Krypton is found in the air, and a standard based on the light from krypton 86 has the advantage that it can never be destroyed, stolen, damaged, or lost. With this standard of length it is possible to make measurements with an accuracy of the order of 1 part in 100 million. Further, anyone with the appropriate equipment can reproduce the present standard of length, defined as follows:

The meter is a length 1,650,763.73 times the wavelength of the orange-red light given off by the pure krypton isotope of mass 86 when it is excited in an electric discharge.

The meter is the legal standard of length in the United States; however, most everyday measurements are made in British units, which are related to the meter by definition:

One inch is 0.0254 m exactly.

1.3 UNIT OF TIME

In both the metric and British systems, the second is the fundamental unit of time. This unit originated from the rotation of the earth in the following manner: The time from the instant the center of the sun is on the meridian plane (the meridian plane of an observer is a plane determined by the axis of rotation of the earth and the point at which

Figure 1.2

The wavelength λ is the distance from crest to crest (or trough to trough) of a wave.

the observer is located on the earth's surface) until the sun is centered on the meridian the following day is called a *solar day*. For several reasons solar days are not all of identical duration; the average over 1 year is called the *mean solar day*. This period is divided into 24 hr, each hour into 60 min, and each minute into 60 sec. Thus, a solar day contains 86,400 sec. For many years the second was defined as 1/86,400 of a mean solar day.

However, the earth's rotation is subject to small variations which are not thoroughly understood. As a result, the mean solar day is not an entirely satisfactory quantity for defining a standard of time. The 1960 International Conference on Weights and Measures redefined the second as 1/31,556,925.9747 of the solar year 1900. In 1968 a new definition was adopted:

The second is the duration of 9,192,631,770 periods of the radiation corresponding to the transition between the two hyperfine levels of the fundamental state of the atom of cesium 133.

1.4 THE INTERNATIONAL SYSTEM OF UNITS

The 1960 General Conference of the International Bureau of Weights and Measures established the Système International d'Unités (abbreviated SI) by selecting the *meter* as the basic unit for *length*, the *kilogram* for *mass*, the *second* for *time*, the *ampere* (Sec. 36.8) for *current*, the *kelvin* (Sec. 15.7) for *temperature*, and the *candela* (Sec. 29.1) for *luminous intensity*. Larger and smaller units are related to the standard units in terms of powers of 10. Thus the kilometer is 1,000 m, while the millimeter is 0.001 m. Table 1.1 lists the prefixes for fractions and multiples together with their abbreviations. In most scientific work metric units are used, and the International System is widely accepted. However, the cgs system, based on the centimeter, the gram, and the second (together with other electrical units), retains an important following.

Many other systems have been devised. The *British absolute system* has as its fundamental units the *foot*, the *pound mass*, and the *second*. It is not necessary that the fundamental units of system involve length, mass, and time. Indeed, as we shall see, the *British engineering system* is based on the *foot*, the *second*, and the *pound force*.

An overwhelming number of countries, and 90 percent of the world's population, now use the metric system, including virtually all the nations of Europe, Asia, and South America and most of those of Africa. The United Kingdom (Fig. 1.3), South Africa, and Australia are actively converting to the metric system. As of 1971 the United States was one of the few countries neither presently metric nor in the process of conversion, although a change to the metric system had been officially proposed.

1.5 DERIVED UNITS

When a physical observable is expressed in terms of a combination of the fundamental quantities, its units are said to be *derived*. For example, the units of area may be written as the square of units of length. Thus,

Table 1.1
Fractions and Multiples (SI)

Factor	Prefix	Abbreviation
*10^{-1}	deci	d
*10^{-2}	centi	c
10^{-3}	milli	m
10^{-6}	micro	μ
10^{-9}	nano	n
10^{-12}	pico	p
10^{-15}	femto	f
10^{-18}	atto	a
*10	deka	da
*10^{2}	hecto	h
10^{3}	kilo	k
10^{6}	mega	M
10^{9}	giga	G
10^{12}	tera	T

*The conference recommended that these factors be used only when there is a strongly felt need and that compound prefixes, for example, mμ, be avoided.

Figure 1.3
Miss Metric poster used in the United
Kingdom to help prepare citizens for
the shift from British to metric units.
[*Construction Industry Training
Board.*]

we may express areas in square meters or square feet. Similarly, the
dimensions of volume are lengths cubed—cubic meters, cubic feet, etc.
Speed and density are examples of the many quantities in mechanics
which are written in terms of derived units.

Speed The average speed of a body is defined as the ratio of the
distance it travels to the time required to pass over that distance. Thus,
speed is obtained by dividing a length by a time. Appropriate units of
speed are meters per second, feet per second, miles per hour, etc.
The speeds of ships and aircraft are often measured in *knots*. One knot
is one nautical mile (1,852 m or 6,076 ft) per hour.

Example

A track man runs a measured kilometer in 2 min 27 sec. Find his
average speed in meters per second.

$$\text{Average speed} = \frac{\text{distance}}{\text{time}} = \frac{1{,}000 \text{ m}}{147 \text{ sec}}$$

$$= 6.80 \text{ m/sec} \qquad \text{////}$$

Density The density of a body is defined as the ratio of its mass to its volume. Thus

$$d = \frac{m}{V} \tag{1.1}$$

where d is the density, m the mass, and V the volume. Among the familiar units for density are kilograms per cubic meter, slugs (Sec. 6.5) per cubic foot, and grams per cubic centimeter.

Example

A quantity of mercury has a mass of 2.05 kg and occupies a volume of 151 cm³. What is its density?

$$d = \frac{m}{V} = \frac{2.05 \text{ kg}}{151 \text{ cm}^3} = 0.0136 \text{ kg/cm}^3$$

$$= 13 \text{ g/cm}^3 = 13,600 \text{ kg/m}^3 \qquad \text{////}$$

Example

Find the mass of air in a room which is 3.00 by 8.00 by 6.00 m. The density of the air is 1.29 kg/m³.

$$m = dV = 1.29 \text{ kg/m}^3 \times 144 \text{ m}^3 = 186 \text{ kg} \qquad \text{////}$$

The term "density" is often used in conjunction with certain other words to suggest the amount of some quantity per given amount of some other quantity. For example, we use the term *population density* to describe number of people per unit area. In electricity the term *charge density* means the ratio of electric charge to volume, and *surface charge density* is the ratio of electric charge to surface area. The ratio of the weight of an object to its volume is called its *weight density*. The ratio of the mass of a rope to its length is sometimes called its *linear density*. Whenever the word density appears in this text in any sense other than that of mass per unit volume, it will be qualified by an additional word.

1.6 CONVERSION OF UNITS AND SIGNIFICANT FIGURES

In the measurement of any physical quantity, the choice of units is ordinarily dictated by convenience and by the habits of the observer. In reporting any measurement, it is important to list not only the number of times the standard unit was included but also to state what this standard unit is. For example, the height of a man might be 6.08 ft, 73.0 in., or 1.85 m. The number which describes the height is meaningless unless the accompanying units are known.

We live in a country where most everyday measurements are made in the British system, but much scientific work involves metric units.

Since anyone who has to deal with scientific phenomena in the United States is essentially forced to use both British and metric units, it is important to be able to transform from one system to another. To facilitate this operation an extensive list of conversion factors is presented just inside the front cover.

It is convenient to convert a quantity from one unit to another in the form of the examples below. Note that one may cancel units just as though they were algebraic quantities. Note further that the factors by which one multiplies are all unity.

Example

Convert 60.0 mi/hr to feet per second and meters per second.

$$60.0\frac{\text{mi}}{\text{hr}} = 60.0\frac{\text{mi}}{\text{hr}} \times \frac{1\ \text{hr}}{3,600\ \text{sec}} \times \frac{5,280\ \text{ft}}{1\ \text{mi}}$$

$$= \frac{60.0 \times 5,280\ \text{mi}\ \text{hr}\ \text{ft}}{3,600\ \text{hr}\ \text{sec}\ \text{mi}} = 88.0\frac{\text{ft}}{\text{sec}}$$

$$= 88.0\frac{\text{ft}}{\text{sec}} \times \frac{0.3048\ \text{m}}{1\ \text{ft}} = 26.8\frac{\text{m}}{\text{sec}} \qquad \text{////}$$

Example

The density of aluminum is 2,699 kg/m³. Convert this density to grams per cubic centimeter and to pounds mass per cubic foot.

$$2,699\frac{\text{kg}}{\text{m}^3} = 2,699\frac{\text{kg}}{\text{m}^3} \times \frac{1\ \text{m}^3}{(100\ \text{cm})^3} \times \frac{1,000\ \text{g}}{1\ \text{kg}}$$

$$= 2.699\frac{\text{g}}{\text{cm}^3}$$

$$2,699\frac{\text{kg}}{\text{m}^3} = 2,699\frac{\text{kg}}{\text{m}^3} \times \frac{1\ \text{lb}_m}{0.4536\ \text{kg}} \times \frac{(0.3048\ \text{m})^3}{1\ \text{ft}^3}$$

$$= 168.5\frac{\text{lb}_m}{\text{ft}^3} \qquad \text{////}$$

In making conversions and in solving problems it is improper and wastes time to carry out calculations with more figures than are meaningful. For example, if we measure the thickness of a brick at six places and obtain the values 6.85, 6.88, 6.87, 6.86, 6.84, and 6.88 cm, the average of these values is 6.8633333 . . . cm, but obviously the numbers after 6.86 overstate the reliability of the average. Clearly, even the last 6 is not certain; all numbers following it have *no physical significance*. It is not only pointless to retain numbers which have no physical meaning, but it may be misleading as well. *All the figures in the numeric of a measurement are significant if they do not overstate the reliability.* In the first example above, the initial datum and the solutions are given to three significant figures, while in the second four significant figures are involved. Usually we shall treat all numbers as though we knew them to three significant figures and calculate answers accordingly.

In tables and inside the front cover we often state more than three significant figures for one of two reasons: (1) Sometimes for some special reason you may need to know the value of the quantity to more significant figures. For example, the speed of light in vacuum, an important physical constant, is 3.00×10^8 m/sec to three significant figures, but its value is known to be less than that by about 0.07 percent. (2) Whenever an answer depends on the *difference* of two nearly equal numbers, we must know the latter to several significant figures to get the difference to three significant figures. For example, $2{,}539.2 - 2{,}517.8 = 21.4$; in this case we need five significant figures in the original numbers to have three significant figures in the difference.

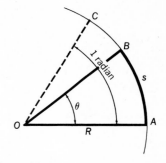

Figure 1.4
The angle θ in radians is the ratio of the arc length AB to the radius OA. The angle AOC is 1 radian ($= 57.2958°$).

1.7 MEASUREMENT OF ANGLES

Degrees, minutes, and seconds are familiar units in which angles are measured. A degree is the angle subtended at the center of a circle by an arc of length $\frac{1}{360}$ of the circumference of the circle. A minute is $\frac{1}{60}$ of a degree, and a second is $\frac{1}{60}$ of a minute.

    To measure the angle turned through by a rotating wheel, we are likely to use the *revolution*. The angle subtended by the full circumference is one revolution (rev).

    Still another unit which is often convenient for measuring angles is the *radian*. *A radian is the angle subtended at the center of a circle by an arc of length equal to the radius*. Thus, the angle AOC of Fig. 1.4 is 1 radian (rad), since the arc length AC is equal in length to radius OA.

    The angle AOB of Fig. 1.4 is represented by θ; in radians it is s (the length of arc AB) divided by the radius R:

$$\theta = \frac{s}{R} \tag{1.2}$$

The angle θ is the ratio of two lengths; consequently it is *dimensionless*. Since the circumference of a circle is 2π times the radius, it follows that there are 2π rad in a revolution. Hence,

$$360° = 2\pi \text{ rad} = 1 \text{ rev}$$

1.8 FORCE

One of the important concepts in physics is that of force. For the moment, we may think of a force as a push or a pull exerted upon some object. The most familiar forces are those which we experience through our own muscular activities. More generally, a force is an action, exerted by one body on another, that tends to change the state of motion of the body acted upon. Thus, when a man throws a baseball, he exerts a force upon it which changes its state of motion. We shall later define force explicitly in terms of effects on the motions of objects.

    We know that every body near the surface of the earth is pulled toward the earth by gravitational attraction, which we shall now use to define a unit of force. *The pound force (lb) is the net gravitational pull of the earth on a standard one-pound mass (lb_m) at a specified location* (essentially at mean sea level, 45° north latitude). The pound

Spring balance

Weight, mg

Figure 1.5

The weight of a body is equal and opposite to the force required to hold the body at rest against the net gravitational pull on it.

force is called a *gravitational* unit because it is defined in terms of the *pull of the earth on a standard mass at a specified location.*

The basic force unit for the mks system is the *newton* (N), which is properly defined (Sec. 6.4) in terms of its effect on the motion of a one-kilogram mass in a manner which is *independent of the location of this mass.* For this reason it is called an *absolute* unit of force. For the moment it is sufficient to observe that 1 N = 0.2248 lb.

The net pull of the earth, at our specified location, on a standard kilogram mass is known as the *kilogram weight* (1 kg wt = 9.807 N). Although the kilogram weight is sometimes a convenient force unit, it is not basic for any widely adopted system of units.

1.9 WEIGHT AND MASS

The weight of an object is defined as the net gravitational force of attraction acting on the object (Fig. 1.5). Its value depends on the location. If a good spring balance were calibrated at 45° north latitude and mean sea level and then transported to various points on the earth's surface, it would be found that the pull of the earth on a 1-lb mass is more than 1 lb force at the North Pole but less than 1 lb at the equator or on top of a high mountain. The weight of a body is slightly different on top of Pike's Peak than in New York City. These changes in weight are not large, but they can be detected readily with sensitive instruments. If an object which weighs 6 lb on the earth were taken to the moon, its weight there would be only 1 lb. In principle, we could reduce the weight of a body almost to zero by taking it sufficiently far off in space, as we shall see in Chap. 9.

On the other hand, the mass of a body at rest is the same at all points in the universe. This is one reason that the concept of mass is so important in physics. It is sometimes said that the mass of a body is the ''quantity of matter in the body,'' a property of the body which does not change as long as its speed is negligible compared with the speed of light. An object has mass whether or not it is near enough to the earth to have observable weight. The mass of a body reveals itself as *inertia*—the tendency of matter to resist any change in its state of motion. This important attribute of mass is discussed further in Chap. 6.

The larger the mass of a body, the harder the earth pulls on it. Indeed, the weight of a body at any place is directly proportional to its mass, and the mass of a body is often determined by comparing the pull of the earth on it with the pull on a standard mass at the same place. An unfortunate consequence is that the word ''weight'' is frequently used improperly in place of mass. For example, an analytical balance typically determines the mass of the unknown in terms of standard masses, yet we refer to this operation as ''weighing'' and call the result the ''weight'' of the unknown.

1 What is your weight in pounds and in newtons? Your mass in kilograms? Your height in inches and in meters?

2 What are some of the more important characteristics of a physical standard unit?

3 What repetitive phenomena could serve as reasonable time standards?

4 If you had a ruler with a smallest division of 1 mm, how accurately could you measure the length of this page? The thickness of this page alone? If you wished to know the thickness of the page more precisely, what could you do with just the ruler?

5 What advantages does the new international standard of length have over the platinum-iridium bar it replaced? What advantages did the meter bar have over the new standard?

6 A simple pendulum 1 m in length has a period very close to 2 sec. What would be the disadvantages of defining the time unit as one-half the period of a pendulum 1 m in length?

7 What advantages would accrue to the nation if the United States were to adopt the metric system for general commercial and government use? What are the disadvantages of abandoning the use of British units of length and force?

8 How could you determine the height of a tree or building from the level ground around it? What apparatus would you need, and how would you use it?

9 In what sense is it true that the earth makes one rotation each 24 hr? In what sense is it false? The sidereal day, widely used by astronomers, is the time interval between two successive passages of a star across the meridian. How many sidereal days are there in 1 year? Show that the length of a sidereal day is 86,164 sec.

10 If someone told you that the dimensions of every object in the universe, including you, had decreased by one-third last night while you slept, how could you refute him?

11 The density of a cylinder is to be determined by measuring its mass, length, and diameter. How big an error would be introduced in the final result by a 1 percent error in (*a*) the mass, (*b*) the length, and (*c*) the diameter? Why does the 1 percent error in the diameter introduce twice as large an error in the density as the corresponding errors in mass and length?

12 What are the dimensions of speed? Of density? List at least two kinds of units in which each may be measured. In general, what is the distinction between dimensions and units?

PROBLEMS

Unless otherwise specified, the numerics stated in the problems should be regarded as valid to three significant figures. In general, slide-rule accuracy is all that is desired; answers should be given to three (or at most four) significant figures.

1 Mount Logan is the highest mountain in Canada, with an elevation of 19,850 ft. Find its altitude in meters. *Ans.* 6,050 m

2 The highest mountain in the world is Everest at 8,848 m. Find its altitude in feet.

3 The highest peaks in South and North America are respectively Aconcagua at 6,960 m and McKinley at 6,194 m. Compute their altitudes in feet. *Ans.* 22,830 and 20,320 ft

4 Mount Kilimanjaro in Tanzania has an altitude of 19,340 ft, greatest in Africa, while Mount Cook has an elevation of 12,349 ft, greatest in New Zealand. Find these heights in meters.

5 Airline distances between New York and London and New York and Capetown are respectively 5,589 and 12,552 km. What are these distances in miles? *Ans.* 3,473 and 7,801 mi

6 From San Francisco it is 2,393 mi to Honolulu and 7,416 mi to Sydney by air. Find these distances in kilometers.

7 The Washington monument is reported to be 555 ft $5\frac{1}{8}$ in. high. To how many significant figures is the height given? How meaningful do you think the last significant figure is? Express the height in meters to three significant figures. *Ans.* 169 m

8 The meter was selected originally as 1/10,000,000 of the distance from the North Pole to the equator along the quadrant of the earth's surface through Paris. Assuming that the earth is spherical, find the radius and the circumference of the earth in kilometers and in miles.

9 The 800 pages of a book have a total thickness of 4 cm. Find the average thickness of a single sheet in micrometers (formerly called microns) and in mils (one mil = 0.001 in.).
Ans. 100 μm; 3.94 mils

10 On a certain day the speed of sound in air is 340 m/sec. Find the speed in kilometers per hour, in feet per second, and in miles per hour.

11 A supersonic aircraft has a speed of 2,520 km/hr. Find its speed in meters per second, in miles per hour, and in feet per second.
Ans. 700 m/sec; 1,570 mi/hr; 2,300 ft/sec

12 A jet airplane is moving at a speed of 600 mi/hr. Find its speed in feet per second, in meters per second, and in kilometers per hour.

13 The effective radius of the wheel of an automobile is 26 cm. Find the angle in revolutions and in radians through which the wheel rotates in going a distance of 1 km.
Ans. 612 rev; 3,850 rad

14 In going around a sharp curve, the driver of an automobile turns the steering wheel through an angle of 5 rad. Through how many degrees did it turn? Through how many revolutions? If the wheel has a radius of 16 cm, through what distance did a point on the outer edge of the wheel move?

15 The gasoline (petrol) tank of an automobile holds 60 liters. Find its volume in U.S. gallons (1 U.S. gal = 231 in.^{3}) and in British imperial gallons (277.4 in.^{3} = volume of 10 lb of water at 62°F).
Ans. 15.8 U.S. gal; 13.2 B.I. gal

16 Solid carbon dioxide has a density of 1,530 kg/m^3. What volume would be occupied by 30 cm^3 after evaporation to gas with a density of 1.9 kg/m^3?

17 A curve on a scenic highway is 250 m long and subtends an angle of 40° at the center of the circle of which it is a part. What is the radius of the curve? *Ans.* 358 m

18 The angle subtended by the sun's diameter at the surface of the earth is 0.5°. Find the approximate radius of the sun if it is at a distance of 1.5×10^8 km from the earth. The moon subtends essentially the same angle at the earth. What is the approximate radius of the moon if it is 3.8×10^5 km away?

19 There are 640 acres in 1 mi^2; a hectare is 10^4 m^2. Find the area of a rectangular plot of ground 50 m long and 40 m wide in (*a*) hectares, (*b*) square feet, and (*c*) acres.
Ans. (*a*) 0.2 hectare; (*b*) 21,500 ft^2; (*c*) 0.494 acre

20 A mountain peak 20 km (measured horizontally) away from the observer is sighted from a telescope at an elevation of 500 m above sea level. If the line of sight makes an angle of 10° with the horizontal, find the height of the peak above sea level.

21 An aquarium has dimensions of 90 by 60 by 40 cm. What mass of water does the aquarium hold when full? What is the volume in cubic feet?
Ans. 216 kg; 7.63 ft^3

22 The circumference of the circular magnet ring of the Brookhaven 33-GeV accelerator is approximately $\frac{1}{2}$ mi. Find the area circumscribed by the magnet ring in square feet and in square meters.

23 A lake has an area of 8×10^7 m^2. A layer of water 2.5 mm thick evaporates from this lake over a hot weekend. What mass of water was evaporated in this period? *Ans.* 2×10^8 kg

24 If 1 gal is 231 in.^{3}, find the volume in gallons of an aquarium of dimensions 2.5 by 2 by 1.2 ft. If 1 ft^3 of water weighs 62.4 lb, what weight of water does the tank hold when full?

25 What mass of water falls on 1 km^2 of land during a 2.54-cm (1-in.) rain? *Ans.* 2.54×10^7 kg

26 A rectangular lot has sides 125 and 100 ft long. Find the area in acres and in square meters.

27 An irregular crystal of quartz has a mass of 42.5 g. When it is submerged in water in a test tube of radius 1.5 cm, the water level is observed to rise 2.26 cm. Find the density of the quartz crystal. *Ans.* 2,660 kg/m^3

28 To determine the width of a canyon to be spanned by a horizontal bridge, a distance AB of 120 m is marked off parallel to one edge of the canyon. Point C on the opposite side of the canyon is sighted from points A and B. If AC is perpendicular to AB and BC makes an angle of 25° with AB, what is the width of the canyon?

29 In surveying for a straight road to traverse rough terrain between points A and B by a series of cuts and fills, a horizontal base line AC 80 m long is laid out. A transit shows that AB makes an angle of 34° with AC while CB makes an angle of 70° with CA. Determine the distance AB. *Ans.* 77.5 m

30 In astronomy distances are so huge that it is convenient to have some very large units in which to measure them. Among them are the *astronomical unit* (the average distance from the earth to the sun, about 1.496×10^{11} m), the *light-year* (distance traversed by light in vacuum in a time of 1 year), and the *parsec* (the distance at which one astronomical unit subtends an angle of 1 second of arc). Show that 1 parsec is 206,265 astronomical units, or 3.26 light-years. (The speed of light in vacuum is given along with other important constants inside the front cover.)

31 About 250 B.C. Eratosthenes measured the circumference and hence the radius of the earth. He observed that at noon on the first day of summer sunlight struck the bottom of a vertical well near Aswan, Egypt. At Alexandria, 5,000 stadia to the north, at the corresponding time the sun was $\frac{1}{50}$ of a circle south of the zenith (straight up). Calculate the radius of the earth if his stadium was 160 m (no one knows for sure). *Ans.* 6.37×10^6 m

CHAPTER TWO
VECTOR QUANTITIES

QUANTITATIVE EXPRESSIONS for all the measurable quantities of mechanics can be written in terms of fundamental units of length, time, and mass. For many physical observables, such as displacement, velocity, and force, the *direction*, as well as the magnitude, is important. To add such *vector* quantities we must take appropriate account of the directions, thereby using a form of geometrical addition which we develop and discuss in this chapter.

2.1 DISPLACEMENT

When a body moves from one location to another, it is said to undergo a *displacement*. Suppose, for example, that a helicopter takes off, flies due east 12 kilometers, and lands. It has undergone a displacement of 12 km east. If the helicopter takes off again and flies 25 km in a straight line before landing, how far is it from its original starting point? A specific answer to this question cannot be given until we have one *additional* piece of information: the direction of this second displacement. From the information now available we can only be sure that the helicopter lies somewhere on a circle with center at B and a radius of 25 km (Fig. 2.1). *A displacement is completely specified only when we state both its magnitude and direction.* The magnitude consists of a numeral and a unit.

Figure 2.1

Successive displacements from point A of 12 km east and 25 km in an arbitrary direction could bring a helicopter to *any* point on the circle *DEC*.

If the second displacement is in the same direction as the first, the helicopter is at point C and the total displacement is 37 km east. If the second displacement is in the direction opposite to the displacement AB, the helicopter is at D and the displacement is 13 km west. If the displacement is due north, the helicopter is at E and its distance from the starting point is $R = \sqrt{(25)^2 + (12)^2}$ km $= 28$ km.

Let us now specify the 25-km displacement to be at an angle of 53°N of E (Fig. 2.2). In our figure we represent each displacement as a directed line segment (arrow) pointed in the direction of the displacement with length proportional to the displacement. One way of determining the distance from the starting point to the final position is by making a scaled drawing and measuring the distance with a ruler. However, in most cases it is easier to solve a problem of this type analytically. We observe that the displacement BP of 25 km at an angle of 53°N of E takes the helicopter a distance BG to the east and a distance GP to the north. By definition (see Sec. A.1, Appendix A) we know that *in any right triangle the sine of an angle is the ratio of the side opposite the angle to the hypotenuse, while the cosine is the ratio of the side adjacent to the angle to the hypotenuse.* Therefore $BG/25 = \cos 53° = 0.6$ and $BG = 15$ km. Similarly $GP/25 = \sin$

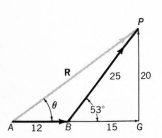

Figure 2.2

The resultant of the displacements 12 km east and 25 km 53°N of E is the vector **R**.

53° = 0.8, from which $GP = 20$ km. Thus, a displacement of 25 km at an angle of 53°N of E is equivalent to a displacement of 15 km east plus a displacement of 20 km north. We can, if we like, replace our 25-km displacement with the displacements BG and GP. We observe from Fig. 2.2 that the helicopter is now 27 km east of its original starting point and 20 km north. The distance AP is therefore given by

$$AP = \sqrt{(27)^2 + (20)^2} \text{ km} = 34 \text{ km}$$

From the right triangle AGP we see that $\sin \theta = \frac{20}{34} = 0.59$, from which $\theta = 36°$.

2.2 VECTORS AND SCALARS

Physical quantities which have both magnitude and direction and which add like displacements are called vector quantities. A vector quantity can be represented graphically by a directed line segment (arrow) or line vector. A knowledge of both the magnitude and the direction of a vector quantity is required for its complete description. Vectors in different directions cannot be added, subtracted, or multiplied by the methods of ordinary arithmetic; they are treated by geometrical techniques. In this chapter we deal with three types of classical† vector quantities: displacements, velocities, and forces. In later chapters, many more vector quantities such as acceleration, momentum, torque, angular velocity, and angular acceleration are introduced.

Other quantities in physics, such as mass, time, and volume, have only magnitude and can be completely specified by a numeral and a dimension. They do not involve any idea of direction. Such quantities are called *scalars*. They obey the ordinary laws of addition, subtraction, multiplication, and division. If 30 liters of gasoline is added to a tank which originally contained 7 liters, the tank contains 37 liters. If a 10-g mass is removed from the pan of a balance which holds 80 g, the mass remaining on the pan is 70 g. These are illustrations of the addition and subtraction of *scalar quantities*.

Vectors are usually denoted in print by boldface type, such as **F** for force, **v** for velocity, etc. Whenever a boldface symbol is used in text or in a formula, it indicates that direction as well as magnitude is of importance. Italic type is used for scalar quantities and for the magnitudes of vectors.

2.3 THE GRAPHICAL ADDITION OF VECTORS

A vector quantity can be represented graphically by a directed line segment, or arrow, in the direction of the vector with length proportional to the magnitude of the vector. To add two vectors **A** and **B**, we may represent each of them by an arrow of scaled length and find their sum (Figs. 2.3 to 2.5) by placing the tail of the second arrow at the point of the first. Alternatively, we might start with the second vector and

---

† Many quantities which behave like vectors in everyday situations fail to satisfy the definition above when speeds comparable to the speed of light are involved. For example, very high velocities do not add like displacements even though they do have magnitudes and directions.

Figure 2.3
The resultant of two displacements (*a*) in the same direction and (*b*) in opposite directions.

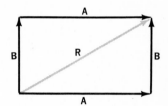

Figure 2.4
Resultant of two vectors at right angles to each other. Note that
R = A + B = B + A.

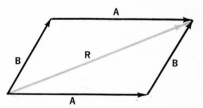

Figure 2.5
Resultant of two vectors not at right angles to each other; as always for vectors, **R = A + B = B + A**.

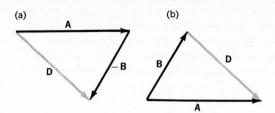

Figure 2.6
Subtraction of vector **B** from vector **A**
can be accomplished (*a*) by adding
−**B** to **A** or (*b*) by finding the vector
which must be added to **B** to give **A**.

add the first, as shown in Figs. 2.4 and 2.5. This gives rise to the *vector parallelogram*. The vector sum, known as the *resultant,* is given by the diagonal **R**. If the graphical representation of the vectors **A** and **B** is performed carefully, the resultant vector **R** can be measured with considerable accuracy.

To subtract one vector from another, we need only reverse the direction of the vector we wish to subtract and add this negative vector to the first. This is shown schematically in Fig. 2.6*a*, where **D** = **A** − **B**. An alternative way of subtracting **B** from **A** is to find that vector **D** which must be added to **B** to give **A** (Fig. 2.6*b*), since **B** + **D** = **A** if **D** = **A** − **B**.

There is no limit to the number of vectors which may be combined graphically. If we wish to determine the resultant of four vectors, we may add the third vector to the resultant of the first two, and the fourth vector to the resultant of the first three. For example, suppose that an object has four forces acting upon it: 8 lb east, 5 lb 30°N of E, 7 lb 37°W of N, and 3 lb due south, as shown in Fig. 2.7. The resultant is given by the vector **R**. The order in which the vectors are added makes no difference in determining the resultant. We shall calculate the value of **R** in Sec. 2.7.

2.4 VELOCITY

Suppose you take an automobile ride. The distance you travel, as read by the odometer, is a scalar quantity. If at the end of 3 hr you have traveled a distance of 240 km, your average speed has been 80 km/hr. Suppose that at this point you are 75 km due east of your starting point (Fig. 2.8). Then your displacement is 75 km east. *Displacement* is a vector quantity. It is not necessarily related in any simple way to the

Figure 2.7
(*a*) Four forces acting upon a body at
O. (*b*) The resultant of these four
forces is independent of the order in
which they are added.

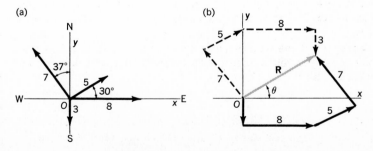

total *distance* traversed. The *average velocity* during the trip is *the ratio of the displacement to the time*. In this case the average velocity is 25 km/hr east. Note that *velocity is a vector quantity*, while *speed is a scalar*. When you return to your starting point, the average *velocity* is zero since the *displacement* is zero; the average speed is very different from zero.

Example

On a transcontinental trip a family begins to drive at 6 A.M. and stops at 5 P.M. In this period the car goes 495 mi and is 350 airline miles southwest of its starting point. Find the average speed and average velocity for the day.

$$\text{Average speed} = \frac{\text{distance}}{\text{time}} = \frac{495 \text{ mi}}{11.0 \text{ hr}} = 45 \text{ mi/hr}$$

$$\mathbf{v}_{av} = \frac{\text{displacement}}{\text{time}} = \frac{350 \text{ mi southwest}}{11.0 \text{ hr}} = 31.8 \text{ mi/hr southwest} \quad ////$$

In this chapter we shall add velocities, confining ourselves to cases in which the velocity is constant, or *uniform*, i.e., in which equal displacements are traversed in equal times. In later chapters we shall consider types of motion in which the velocity varies with time.

Figure 2.8
Displacement is a vector quantity; distance is a scalar quantity.

2.5 FRAME OF REFERENCE

Whenever we refer to the displacement or velocity of a body, we are always, by implication, referring these measurements to some other body or framework of lines which is regarded as fixed. Such a framework is called a *frame of reference*. Some problems can be greatly simplified by the choice of a suitable frame of reference.

With few exceptions we refer quantities to a system of axes fixed on the earth. When we say an object is at rest, we ordinarily mean it is at rest relative to the earth. Relative to the sun it is moving through space at a speed of 30 km/sec by virtue of the orbital motion of the earth (and at a speed which depends on latitude because of the spin of the earth on its axis). A passenger seated in a train traveling 100 km/hr is at rest relative to the car and the other passengers. If he walks forward in the car with a speed of 6 km/hr, his speed relative to the earth is 106 km/hr.

It is meaningless to speak of the "absolute" position of a body in space or to assign to the body an "absolute" velocity. Any position vector or velocity vector must be specified relative to some frame of reference.

In general, if the velocity of a body B relative to one frame of reference is $\mathbf{v}_{B1}$, its velocity relative to a second frame $\mathbf{v}_{B2}$ is given by the vector equation

$$\mathbf{v}_{B2} = \mathbf{v}_{B1} + \mathbf{v}_{12} \tag{2.1}$$

where $\mathbf{v}_{12}$ is the velocity of the first frame relative to the second. For example, if an airplane is flying with a velocity $\mathbf{v}_{PA}$ relative to the air and the air has a velocity $\mathbf{v}_{AG}$ relative to the ground, the velocity $\mathbf{v}_{PG}$ of the plane relative to the ground is given by

$$\mathbf{v}_{PG} = \mathbf{v}_{PA} + \mathbf{v}_{AG} \qquad (2.1a)$$

If $\mathbf{v}_{PA}$ is 500 km/hr east and $\mathbf{v}_{AG}$ is 80 km/hr east, $\mathbf{v}_{PG} = 580$ km/hr east. If $\mathbf{v}_{AG}$ is 80 km/hr west, $\mathbf{v}_{PG}$ is 420 km/hr east.

In general, if we have three bodies A, B, and C and we let $\mathbf{v}_{AB}$ represent the velocity of A relative to B, and so forth, then $\mathbf{v}_{AC} = \mathbf{v}_{AB} + \mathbf{v}_{BC}$. (Note that we add the vectors with the two B's adjoining. This is an example of what is sometimes called the *domino rule* of vector addition.) If we want the velocity of C relative to B, we observe that $\mathbf{v}_{CB} = \mathbf{v}_{CA} + \mathbf{v}_{AB}$. The same rule applies to other vector quantities.

Figure 2.9

Resolution of a vector **v** into its rectangular components V_x and V_y.

2.6 RECTANGULAR COMPONENTS OF A VECTOR

Consider a vector **V**, 5 units in length and directed at an angle of 37°N of E (Fig. 2.9). This displacement is equivalent to displacements of 4 units east and 3 units north. In any problem in which this 5-unit vector occurs, we may replace it with vectors of 4 units east and 3 units north. These two vectors, which are perpendicular to each other, are called the *rectangular components* of the 5-unit vector. By replacing a vector with its rectangular components, we can always reduce any problem in vector addition to a situation in which we are dealing with component vectors, all of which lie either along one line or perpendicular to this line. Thus, the replacement of a vector by suitable components results in the possibility of handling vector additions with a minimum of mathematical complexity.

If we wish to obtain the rectangular components of a vector in any mutually perpendicular directions, we proceed as follows. We place the origin of a rectangular coordinate system at the tail of the vector which we wish to resolve and then drop perpendiculars from the point of the vector to each of the coordinate axes. *The magnitude of the rectangular component in any direction is given by the magnitude of the vector multiplied by the cosine of the angle between the vector and the axis along which we seek the component.*

The vector **V** of Fig. 2.9 is the resultant of a vector in the x direction of magnitude V_x and one in the y direction of magnitude V_y. If we let **i**, **j**, and **k** represent vectors of unit length in the x, y, and z directions, respectively, we may write the vector equation

$$\mathbf{V} = \mathbf{i}V_x + \mathbf{j}V_y + \mathbf{k}V_z \qquad (2.2)$$

where $V_x = V \cos \alpha$, $V_y = V \cos \beta$, and $V_z = V \cos \gamma$, γ being the angle between **V** and the z axis. In this text we shall ordinarily confine our attention to problems in which we are concerned with only two components.

Figure 2.10

Resolution of the weight of an object into components parallel to and perpendicular to an inclined plane.

Example

A 50-newton (N) weight rests on an inclined plane (Fig. 2.10) making an angle of 30° with the horizontal. Find the components of the weight parallel and perpendicular to the plane.

Figure 2.11
Components BA and BD of the force
BC exerted on a canalboat by a cable.

We place the origin of a system of axes at the body, with the x axis parallel to the plane. Next we drop perpendiculars from the end of the vector **w** to find the lengths of X and Y.

$$X = w \cos \alpha = w \sin \theta = (50 \text{ N})(\sin 30°) = 25 \text{ N}$$
$$Y = w \cos \theta = (50 \text{ N})(\cos 30°) = 43.3 \text{ N} \qquad ////$$

Consider a canalboat as indicated in Fig. 2.11, and let a cable pull on it in the direction BC. This force produces two distinct effects on the boat. It moves the boat forward, and it pulls it to the bank. Two separate forces BA and BD might have been applied with the same result. The two forces that would produce the same effect as BC are the components of BC. Similarly, the velocity of the wind propelling a sailboat may be resolved into components perpendicular and parallel to the sail; the force perpendicular to the sail is then resolved into components perpendicular and parallel to the axis of the boat. The component perpendicular to the axis produces little effect because of the shape of the boat.

2.7 THE ANALYTICAL ADDITION OF VECTORS

In adding vectors analytically, it is desirable first to make a rough sketch of the vectors to be added. Then we choose a convenient set of mutually perpendicular axes. Usually, this involves a choice of the east-west and north-south axes or vertical and horizontal axes. However, in some cases it may be more convenient to choose a set of mutually perpendicular axes oriented in some special way, as in Fig. 2.10. Next, we find the components of all the vectors along the axes chosen. We may then ignore the vectors and work only with the components. We can readily add together all the vector components whose arrows are in the direction of each of the axes and then find their resultant by the use of the pythagorean theorem.

In general, if we are adding vectors **A**, **B**, and **C**, where $\mathbf{A} = \mathbf{i}A_x + \mathbf{j}A_y$, and so forth, the resultant **R** is given by

$$\mathbf{R} = \mathbf{A} + \mathbf{B} + \mathbf{C}$$
$$= \mathbf{i}A_x + \mathbf{j}A_y + \mathbf{i}B_x + \mathbf{j}B_y + \mathbf{i}C_x + \mathbf{j}C_y$$
$$= \mathbf{i}(A_x + B_x + C_x) + \mathbf{j}(A_y + B_y + C_y)$$

Thus each component of **R** has magnitude equal to the sum of the corresponding components of the vectors added. In dealing with the components of vectors along any given axis, we select one direction as positive; a component in the opposite direction is then assigned a negative sign. Thus, if we have a north-south axis and elect to call north the positive direction, a component of 5 km southward is written as −5 km north.

Example

Find the resultant of the four forces shown in Fig. 2.7.

First we find the x (east) and y (north) components of each force. It is convenient to list them in a table as shown. For the 5-lb force the x component is (5 lb)(cos 30°) = 4.33 lb, and the y component in pounds is 5 sin 30° = 2.50. For the 7-lb force the x (east) component is (−7 lb)(sin 37°) = −4.20 lb, and the y component is (7 lb)(cos 37°) = 5.60 lb. (In calculating these components we have tacitly assumed that the forces involved were 5.00 and 7.00 lb, i.e., that they were known to three significant figures at least.) From the table we see that the components of the resultant are 8.13 lb east and 5.10 lb north. The resultant R is given by

$$R = \sqrt{(8.13)^2 + (5.10)^2} \text{ lb} = \sqrt{92.1} \text{ lb} = 9.60 \text{ lb}$$

and the angle between R and the x axis is given by

$$\sin \theta = \frac{5.10}{9.60} = 0.531$$

so that $\theta = 32.1°$N of E. ////

	Components	
Force, lb	*x (east)*	*y (north)*
8 E	8.00	0
5 30°N of E	4.33	2.50
7 37°W of N	−4.20	5.60
3 S	0	−3.00
R	8.13	5.10

Example

In a football game two linemen are assigned to open a hole by blocking a guard. One of them exerts a force of 150 lb straight toward the goal line (Fig. 2.12), and the other exerts a force of 100 lb at an angle of 60° with the goal line. Find the resultant force.

The x component of the 100-lb force is (100 lb)(sin 60°) = 86.6 lb. The y component is (100 lb)(cos 60°) = 50.0 lb. The net force toward the goal line is then 236.6 lb, while the new force parallel to the goal line is 50.0 lb. The resultant is

$$R = \sqrt{(237)^2 + (50.0)^2} \text{ lb} = \sqrt{58,500} \text{ lb} = 242 \text{ lb}$$

and the angle θ is given by $\sin \theta = 50.0/242 = 0.207$, so that $\theta = 12°$. ////

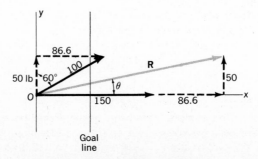

Figure 2.12
Finding the resultant force **R** by determining its components parallel to and perpendicular to the goal line.

Example

23

VECTOR QUANTITIES

An airplane cruises at a speed of 300 km/hr relative to the surrounding air. If the wind is blowing from the southwest at 40 km/hr, find the direction in which the pilot must point the airplane if he wishes to go due east. What is the velocity of the airplane relative to the ground?

Figure 2.13

Flying due east relative to the ground when the wind is blowing from the southwest is accomplished by flying at an angle θ south of east relative to the air.

The pilot must fly at an angle θ (Fig. 2.13) south of east such that the southward component of $\mathbf{V}_{PA}$ (velocity of plane relative to air) is exactly equal to the northward component of $\mathbf{V}_{AG}$ (velocity of air relative to ground).

$$300 \sin \theta = 40 \cos 45° = 28.3$$

$$\sin \theta = \frac{28.3}{300} = 0.0943$$

$$\theta = 5°25'S \text{ of E}$$

The velocity $\mathbf{V}_{PG}$ of the airplane relative to the ground is

$$\mathbf{V}_{PG} = 300 \cos \theta + 40 \sin 45° \text{ km/hr east}$$
$$= 299 + 28.3 \text{ km/hr east} = 327 \text{ km/hr east} \qquad ////$$

QUESTIONS

1 Can the vector sum of two unequal forces be zero? Of three unequal forces?

2 How should three vectors of equal magnitude be arranged if they are to have a resultant of zero?

3 When two vectors **A** and **B** are drawn from a common origin to form sides of a vector parallelogram (Fig. 2.5), one diagonal is the resultant. What is the other diagonal?

4 Can a 0.5-m steel tube be packed in a rectangular box not over 0.3 m long? Explain.

5 Does the odometer of an automobile measure displacement? If not, what does it measure?

6 Why does an airplane land and take off into the wind? When a plane is landing on an aircraft carrier, what is the most favorable direction of carrier and plane relative to the wind and to each other? Why?

7 An airplane has an airspeed of 300 mi/hr, and the wind is blowing with a speed of 50 mi/hr. What is the maximum possible speed of the plane relative to the ground? The minimum possible speed? Could the speed relative to the ground also be 300 mi/hr? If so, explain how.

8 What can you conclude about two vectors **A** and **B** if:

a **A** + **B** = **R** and $R^2 = A^2 + B^2$

b **A** + **B** = **A** − **B**

c **A** + **B** = **R** and $A + B = R$

9 With the aid of a figure explain how an iceboat can sail faster than 15 mi/hr in a 15 mi/hr breeze. What limits the speed which the boat can attain?

10 If the resultant of three vectors is zero, is it necessary that all three lie in the same plane? If so, prove it; if not, propose an example.

11 What justification can you offer for the assertion that force is a vector quantity?

PROBLEMS

1 An automobile is driven 40 km east and then 9 km north in going from one city to another. Write the displacement vector in the component form of Eq. (2.2). How far apart are the two cities? *Ans.* 40**i** + 9**j** km; 41 km

2 Find the magnitude of the single displacement that is equivalent to successive displacements of 40 and 60 m, the direction of the second

displacement being perpendicular to that of the first.

3 Find the resultant of the following four displacements: 9 m east, 13 m south, 2 m west, and 11 m south. *Ans.* 25 m 73.7°S of E

4 Find the magnitude and direction of the resultant of the three displacements 5 mi west, 3 mi north, and 10 mi 35°N of W.

5 A helicopter carrying mail flies 10 km due north and lands. It then flies 30 km in a direction 30°N of E. Find (a) how far north the helicopter is from its original starting point and (b) the magnitude and direction of its resultant displacement. *Ans.* 25 km; 36.1 km; 43.9°N of E

6 An airplane flies 200 miles in a direction 25°S of W. Find the components of its displacement toward the south and toward the west.

7 A helicopter undergoes three displacements, all in kilometers, in the xy plane. If the displacements are $r_1 = 6i - 5j$, $r_2 = 2i + 10j$, and $r_3 = -i + 7j$, write the resultant displacement r in the ij notation. What is the magnitude of r? *Ans.* $7i + 12j$ km; 13.9 km

8 A body is displaced 12 ft west and 5 ft north. What are the magnitude and direction of the resultant displacement?

9 A man pushes along the handle of a lawn mower which makes an angle of 35° with the horizontal ground. One component of the 50-N force he exerts drives the lawn mower along; another component pushes the lawn mower against the ground. Calculate the force components parallel to and perpendicular to the ground. *Ans.* 41.0 and 28.7 N

10 Calculate the horizontal and vertical components of a vector which has a magnitude of 40 lb and makes an angle of 25° with the horizontal.

11 An automobile is driven 30 km east, then 12 km south, and then 14 km west in 40 min. Determine the distance traveled, the displacement, the average speed, and the average velocity. *Ans.* 56 km; 20 km 37°S of E; 84 km/hr; 30 km/hr 37°S of E

12 A boy walks 2,500 m west, 1,000 m south, and then 500 m east in 1 hr. Determine the distance traveled, the displacement, the average speed, and the average velocity.

13 Let **i**, **j**, and **k** be unit vectors of 1-cm length in the x, y, z directions. A cube with 3-cm sides is placed with one corner at the origin and three edges of the cube lying along the x, y, and z axes. Write the position vector for each of the eight corners of the cube in terms of **i**, **j**, and **k**. *Partial answer:* $r = 3i + 0j + 3k$

14 Three football players participating simultaneously in a tackle exert the following forces on the ball carrier: 100 lb north, 120 lb 20°N of E, and 80 lb 35°W of N. Find the resultant of these forces.

15 Find the resultant of the following three forces: a force of 50 N making an angle of 30° with the $+x$ axis, a force of 80 N making an angle of 135° with the $+x$ axis, and a force of 30 N making an angle of 240° with the $+x$ axis. *Ans.* 62.4 N at 153°

16 Two helicopters take off from the same landing area. One proceeds 10 km 30°S of W; the second flies 6 km 45°S of E. Find the displacement of the second helicopter relative to the first.

17 A ferryboat goes straight across a river in which there is a current of 3 km/hr. If the speed of the boat relative to the water is 10 km/hr, find the direction in which it is pointed. What is its velocity relative to the earth? *Ans.* 17.5° upstream; 9.54 km/hr

18 A boat which has a speed of 9 mi/hr crosses a river in which the current is 2 mi/hr. At what angle must the boat be headed to land at a point directly opposite its starting point? What is the speed of the boat relative to the ground?

19 An airplane flies a compass course due south with speed relative to the air of 270 km/hr. After 1 hr the pilot finds himself over an airport 240 km south and 70 km east of his starting point. Calculate (a) the speed of the wind and (b) the direction in which he should set his course to fly due south with this wind. *Ans.* 76.2 km/hr; 15°W of S

20 The current in a river flows south at a rate of 1 mi/hr. A man rows a boat west at a rate of 3 mi/hr relative to the water. Find the speed of the boat relative to the land.

21 A ball is thrown from a moving car with a velocity relative to the car of 14 m/sec perpendicular to the direction of the car's motion. If the speed of the car is 20 m/sec, find the velocity of the ball relative to the ground. *Ans.* 24.4 m/sec at 35° to car's velocity

22 An airplane with an airspeed of 150 mi/hr is to

fly due east. The wind has a speed of 40 mi/hr from 30°N of E. In what direction should the aircraft be pointed? What is the groundspeed?

23 A helicopter has a velocity relative to the air of 40 knots 35°N of W. The velocity of the wind is 20 knots from west to east. Find the velocity of the helicopter relative to the ground.
Ans. 26.3 knots 29.1°W of N

24 A rope attached to a sled makes an angle of 40° with the ground. With what force must the rope be pulled to produce a horizontal component of 60 N? What will then be the vertical component of the force?

25 An airplane has a velocity of 500 knots 20°S of E relative to the air. If the wind has a velocity of 80 knots 60°N of E find the velocity of the airplane relative to the ground.
Ans. 520 knots at 11.3°S of E

26 A steamboat (see accompanying figure) is moving eastward with a speed of 30 km/hr relative to the earth. The wind is blowing from north to south with a speed of 10 km/hr. Find the velocity of the smoke relative to the boat.

Problems 26 to 28

27 A motor launch (see accompanying figure) cruises at 20 knots due west while the wind has a velocity of 15 knots due south. In what direction will a flag point if hung from the mast of the boat? *Ans.* 36.9°S of E

28 A steamboat (see accompanying figure) is moving eastward with a speed of 25 mi/hr. The wind is blowing from the north with a speed of 10 mi/hr. Find the velocity of the smoke relative to the boat.

29 A boat moves east at a speed of 25 km/hr. The wind is blowing from the northwest at 10 km/hr. To an observer on the boat in what direction does the trail of smoke from the boat lie? What is the speed of the smoke relative to the boat? *Ans.* 21.5°S of W; 19.3 km/hr

30 The pilot of an airplane is trying to keep on a beam toward an airport due west. The airspeed of this plane is 600 mi/hr. If the wind is blowing

from the southwest at 100 mi/hr, in what direction must he head the plane? Find the speed relative to the ground.

31 An aircraft has a speed of 250 km/hr relative to the air. On a day when the wind is 50 km/hr, a pilot flies due east relative to the ground above an east-west railroad track. In 30 min the plane is 125 km from his starting point. Calculate the wind direction and the heading of the plane (angle between its axis and the railroad) if the heading is north of east.
Ans. 5.7°E of S; 11.5°

32 The velocity of an aircraft relative to the air is 360 mi/hr north. The wind is blowing 100 mi/hr from the south and west with the wind velocity making an angle of 36.9° with east. Find the velocity of the aircraft relative to the ground.

33 Two trains traveling toward each other leave stations at different times. One train travels at a speed of 90 km/hr, the other at a speed of 80 km/hr. The stations are 120 km apart. How much later should the faster train start in order for the trains to meet halfway between the stations? *Ans.* 5 min

34 The pilot of an aircraft maintains an airspeed of 400 knots in flying from city A to city B, which is 1,000 nautical miles due east. There is a 60-knot wind from 30°S of W. What course does the pilot maintain? Find his groundspeed and flying time.

35 Ship A is steaming due north at 20 knots and observes ship B 14 nautical miles dead ahead. After 30 min B is 3 nautical miles due east of A. Both ships maintain constant velocity. (*a*) Write the velocity of B in the form $\mathbf{v}_B = a\mathbf{i} + b\mathbf{j}$, where $\mathbf{i}$ and $\mathbf{j}$ are unit vectors east and north, respectively. (*b*) Write an expression for the distance between the ships as a function of the time in hours.
Ans. (*a*) $6\mathbf{i} - 8\mathbf{j}$ knots
(*b*) $(196 - 784\,t + 820\,t^2)^{1/2}$ nau mi

36 Show that if ϕ is the angle between two vectors **A** and **B** when they are placed tail to tail, the magnitude of their resultant is given by $R = \sqrt{A^2 + B^2 + 2AB\cos\phi}$. Show also that the angle θ between **A** and **R** is given by $\tan\theta = (B\sin\phi)/(A + B\cos\phi)$.

NOW THAT WE KNOW how to find the resultant of two or more vector quantities, we are ready to consider the simultaneous action of several forces on a body. Our ultimate objective is to learn how to predict the future motion of any body subject to a known system of forces, provided we know the present position and velocity of the body. At first we consider only the case in which the body remains at rest or moves with constant velocity, reserving cases in which the velocity changes for Chap. 6.

CHAPTER THREE
SYSTEMS OF FORCES

3.1 FORCE AND MOTION

The idea of a force as a push or pull is a familiar one. From our everyday experience we know that we must exert forces on bodies whenever we wish to change their state of motion. It is our objective to obtain an understanding of the quantitative relationships between forces and motions, a subject which is called *dynamics*. But first we examine the conditions which must be satisfied in order that a body be at *rest*. This study is known as *statics*, and it is here that we begin our formal development of mechanics.

If a body at rest is to remain at rest, the resultant of all the forces which act on the body must be zero. Similarly, if a body is already in motion and the resultant force acting on it is zero, the velocity of the body remains constant. These statements follow directly from a great generalization known as *Newton's first law* of motion (see Chap. 6). In this chapter and the next we apply the condition that the resultant force acting on a body at rest is zero, but before doing so we must distinguish between the forces exerted *on a body* and the forces exerted *by the body*.

3.2 NEWTON'S THIRD LAW

Consider a cord of negligible weight from which a ball is suspended (Fig. 3.1*a*). If the cord is at rest, the upward pull of the ceiling is equal and opposite to the downward pull of the ball. The force exerted on the cord at either end is known as the *tension* in the cord. The cord, subjected to pulls at each end, is said to be "in tension."

We now direct our attention to the ball of Fig. 3.1*b*. When the ball hangs at rest, there are two forces acting *on it:* its weight **w**, which is the downward pull of the earth, and the pull **T** of the cord upward. By Newton's first law, the resultant of these two forces must be zero, so they must be equal in magnitude and opposite in direction.

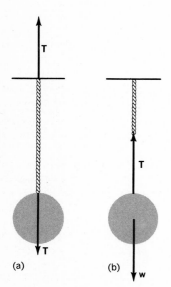

(a)   (b)

Figure 3.1

Equilibrium occurs (*a*) for the cord of negligible weight when the upward pull of the ceiling is equal and opposite to the downward pull of the ball and (*b*) for the ball when the upward pull of the cord is equal and opposite to its weight.

Note that the cord pulls upward on the ball with the force **T**, while the ball pulls downward on the cord with an equal and opposite force. This is an example of the application of Newton's *third law*, which states: *If body A exerts a force on body B, body B exerts an equal and opposite force on body A.* Forces in nature always occur in pairs which are equal and opposite. However, the two members of the pair always act on *different* bodies. In Fig. 3.1*a* the ceiling pulls up on the cord, which in turn pulls down on the ceiling with an equal and opposite force.

When you hold this book in your hand, you exert an upward force *on the book;* at the same time the book exerts an equal downward force *on your hand*. If a man pulls *on a rope* with a force of 40 newtons, the rope pulls back *on the man* with a force of 40 N.

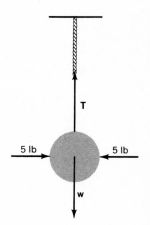

Figure 3.2
If the ball is to remain at rest when a force to the right is exerted on it, an equal force to the left is required.

3.3 CONCURRENT FORCES

When two or more forces act upon a body and the lines of action of these forces pass through a common point, the forces are said to be *concurrent*. By the "line of action" of a force we mean simply the line along which the force acts, extended indefinitely in both directions.

The two forces **w** and **T** acting on the ball of Fig. 3.1*b* are concurrent; both act along a line which passes through the center of the ball. Now, suppose that someone were to push on the ball with a force of 5 lb to the right, and you wish to keep the ball at rest. What must you do? Clearly, you must push with a force of 5 lb in the opposite direction (Fig. 3.2). In general, if we wish to keep the ball at rest, we must balance any force applied in any given direction with an equal force in the opposite direction. (Of course, we may replace any force with its rectangular components acting at the same point.)

A body that remains at rest or moves with constant velocity is said to be in a state of *equilibrium* (equal balance). For a body to be in equilibrium under the influence of any number of concurrent forces, *the vector sum of the forces acting on the body must be zero*. If we draw a vector diagram to represent the forces, the condition that the resultant be zero is equivalent to the condition that the force polygon must close (Fig. 3.3).

In treating the equilibrium of any body we must be sure to include *all* the forces which act *on that body* and *only* those forces. The first question to be resolved in working any equilibrium problem is: Of what body is the equilibrium being considered? Then *all* the forces exerted *on this body,* and no other forces, are taken into account. A figure

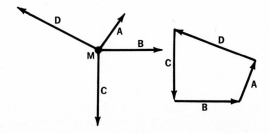

Figure 3.3
For a point mass *M* to remain at rest under the influence of forces, the resultant force must be zero so that the force polygon closes.

showing the body in question with *all* forces which act upon it is known as a *free-body diagram*. The drawing of such a figure is a valuable aid in solving problems involving the equilibrium or acceleration of any object.

3.4 EQUILIBRIUM UNDER CONCURRENT FORCES

Clearly, if the resultant of all the forces acting on a body is zero, the sum of the rectangular components of these forces along any axis must be zero. Thus, when we resolve the forces into components in mutually perpendicular x and y (and z if necessary) directions, for a body in equilibrium the sum of the x components is zero, or the sum of the components in the $+x$ direction is equal to the sum of the components in the $-x$ direction. Similarly, the sum of the y (and z) components is zero.

Usually, but not always, it is desirable to choose vertical and horizontal axes. In this case:

1 *The sum of all upward force components is equal to the sum of all downward force components.*

2 *The sum of all force components to the right is equal to the sum of all force components to the left.*

In mathematical shorthand these conditions may be written in the form

$$\Sigma F_{\text{up}} = \Sigma F_{\text{down}} \tag{3.1}$$
$$\Sigma F_{\text{right}} = \Sigma F_{\text{left}} \tag{3.2}$$

where Σ (Greek capital sigma) is a standard abbreviation for "the sum of."

Consider next the case in which the ball of Fig. 3.1 is pulled to the left by a horizontal force **F**, as indicated in Fig. 3.4. The ball is acted upon by three concurrent forces **F**, **T**, and **w**, the vector sum of which must be zero.

In treating a problem of this kind it is convenient to resolve the tension **T** into a vertical component $T \cos \theta$ and a horizontal component $T \sin \theta$. If we apply Eqs. (3.1) and (3.2) above, we obtain immediately

$$T \cos \theta = w$$
$$T \sin \theta = F$$

Figure 3.4
A ball in equilibrium.

Example

If $w = 20.0$ N and $\theta = 35°$, find F and T.

$$T \cos \theta = w \qquad\qquad F = T \sin \theta$$
$$0.819T = 20.0 \text{ N} \qquad\qquad = 24.4 \text{ N} \times 0.574$$
$$T = 24.4 \text{ N} \qquad\qquad = 14.0 \text{ N} \qquad\qquad ////$$

Example

A weight of 40 N is supported as shown in Fig. 3.5. Find the tension in each rope.

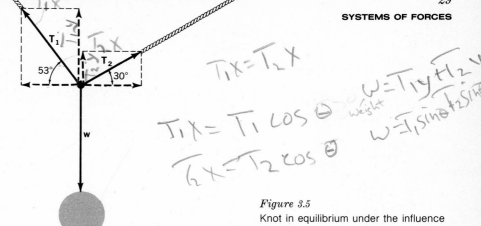

Figure 3.5
Knot in equilibrium under the influence of three forces **T**$_1$, **T**$_2$, and **w**.

The lower rope supports the 40-N weight, so the tension in it is 40.0 N. Now let us consider that the knot where the ropes meet is the body in equilibrium. It is acted upon by three forces w, T_1, and T_2.

The vertical component of T_1 is $T_1 \sin 53° = 0.800 T_1$, and the horizontal component is $T_1 \cos 53° = 0.600 T_1$. The vertical component of T_2 is $T_2 \sin 30° = 0.500 T_2$, and the horizontal component is $T_2 \cos 30° = 0.866 T_2$. Applying the force conditions (3.1) and (3.2) yields

$$0.800 T_1 + 0.500 T_2 = 40.0 \text{ N}$$
$$0.866 T_2 = 0.600 T_1$$

from which $T_1 = 34.9$ N and $T_2 = 24.2$ N. (Note that the larger force is exerted by the shorter rope in this case.) ////

A force which is equal and opposite to the resultant of two or more forces is called the *equilibrant*. Figure 3.6 shows the resultant and equilibrant of two forces **P** and **Q** which are not mutually perpendicular.

Figure 3.6
Resultant and equilibrant of two forces **P** and **Q**.

3.5 FRICTION

Frictional forces play an important role in mechanics, both for moving bodies and for those at rest. Unlike the weight of a body, which is constant at any particular location, we find that the frictional force between a body and a surface depends on many factors. For example,, when a heavy block of wood is pushed along the top of a table, the frictional resistance depends on the surface of the table, the surface of the block, how clean the surfaces are, the speed of the block, and the force pressing the surfaces together. When a body moves on a surface, there is always such resistance to the motion. In many situations friction is undesirable, and elaborate means are employed to make it as small as possible.

Friction has advantages as well as disadvantages. If it were not for friction between his shoes and the floor, a person would have great difficulty in moving about. When the pavement is covered with ice, friction is small and driving is difficult. Because of friction, belts cling to pulleys and drive machinery. Screws and nails stay in place in objects into which they are driven by means of friction.

3.6 KINETIC FRICTION

When an object is sliding over a surface, the direction of the frictional force is parallel to the surface and opposite to the direction of motion (Fig. 3.7). Let us denote by f the force which is just necessary to overcome the frictional force and keep a body moving with *constant speed* across the surface. It is found experimentally that the force necessary to overcome friction is proportional to the force pressing the surfaces against one another. This fact may be expressed by the equation

$$f = \mu_k N \tag{3.3}$$

where N is the force pressing the surfaces together and μ_k is a constant called the *coefficient of kinetic* (or sliding) *friction. The coefficient of kinetic friction between two surfaces is the ratio of the force required to overcome friction to the normal force pressing the surfaces together when one surface is sliding over the other at constant speed.* (Here *normal* means *perpendicular;* the normal force is at right angles to the surfaces.)

For ordinary surfaces the frictional force does not depend on the gross area of the rubbing surfaces. For our rough calculations we assume that friction is independent of velocity, although this is only a crude approximation. At very low speeds μ_k sometimes increases with speed, but not always; for moderate speeds (an inch per second to a few feet per second) it is often roughly independent of speed; at higher speeds μ_k typically becomes smaller as the speed goes up.

It should be noted that these statements about friction are worded rather guardedly. This is desirable because the rules which govern friction are only good working approximations. Friction is a very complex phenomenon. Even the smoothest surface is composed of hills and valleys from the microscopic point of view. The actual microscopic area

Figure 3.7

The coefficient of friction is the ratio of the force f required to overcome friction to the force N pressing the surfaces together.

of contact is generally only a tiny fraction of the macroscopic areas of the surfaces. A substantial body of evidence supports the idea that the frictional force is proportional to the area of contact. Ordinarily the frictional force is proportional to the load because the actual contact area is proportional to the load. Apparently, friction arises primarily from adhesion of molecules of the two surfaces which are in contact. Indeed, if we carefully polish and degas two steel surfaces, the friction is tremendously increased over that for rougher surfaces of the same material. Very thin contamination layers can result in large changes in observed friction. Some substances melt and flow under the influence of frictional forces, particularly when high speeds are involved. Such behavior, called *plastic flow*, often lubricates the surface.

(a)

(b)

Figure 3.8
The force pressing the surfaces together is (a) reduced to $w - V$ by the upward component V of the applied force **F** and (b) increased to $W + V'$ by the downward component V' of the applied force **F'**.

It is important to observe that the force N pressing two surfaces together is not necessarily the weight of the sliding object. To move a heavy trunk across a rough floor, one possibility is to pull on the handle as in Fig. 3.8a. We resolve the force **F** into two components; the horizontal component f is useful in overcoming the friction, while the vertical component V reduces the force pressing the surfaces together. In this case, $N = w - V$, where w is the weight of the trunk. To move the trunk by pushing downward (Fig. 3.8b) requires a larger force **F'**, since the force N pressing the surfaces together becomes $w + V'$, where V' is the downward component of **F'**.

3.7   THE COEFFICIENT OF STATIC FRICTION

If a body is at rest on a surface, a larger force is required to overcome the friction and put the body in motion than is required to keep the body in motion at constant speed. If we let f_s represent the force necessary to overcome friction in starting the body, we find that this force is also proportional to the force N pressing the surfaces together. Thus,

$$f_s = \mu_s N \qquad (3.4)$$

where μ_s is called the *coefficient of static friction. The coefficient of static friction between two surfaces is the ratio of the force required to overcome friction to the normal force pressing the surfaces together when the surfaces are at rest relative to one another.* Since the force necessary to overcome static friction is greater than the force necessary to overcome sliding friction, the coefficient of static friction is greater than the coefficient of kinetic friction. Table 3.1 lists the coefficients of kinetic and static friction for several types of surfaces.

Because static friction is always greater than kinetic friction, a good automobile driver is careful not to keep his wheels locked when he is braking to a stop. Once the wheels stop rotating, the tires slide over the pavement and the coefficient of kinetic friction is in effect. When the wheels are rolling without sliding, the point of contact between tire and pavement is at rest and the coefficient of static friction is applicable.

If a body is at rest on a surface, the frictional force is just large enough to permit the conditions for equilibrium to be satisfied. If the body is resting on a horizontal surface and no horizontal forces act, the frictional force is zero. If the surface is tilted slightly, the frictional force becomes just great enough to prevent the body from sliding. Thus

Table 3.1
Approximate Coefficients of Friction

Surface	μ_k	μ_s
Steel on:		
Steel (clean)	0.50	0.75
Steel (greasy)	0.15	0.20
Ice	0.01	0.02
Oak on oak	0.4	0.5
Greased surfaces	0.05	0.06
Rubber tire on:		
Dry concrete,		
low speed	0.7	0.9
Dry concrete,		
high speed	0.35	0.6
Wet concrete,		
low speed	0.5	0.7

the force of friction can be computed by the relation $f_s = \mu_s N$ only when the body is just at the point of sliding. The product $\mu_s N$ gives the *maximum* value of the friction when the body is at rest. The frictional force may be substantially less than this.

3.8 FRICTION ON AN INCLINED PLANE

In Fig. 3.9 a block of weight w rests on a plane inclined at an angle θ with the horizontal. Let the angle θ be varied until the block just slides down the plane with uniform velocity when once started. Represent this angle by θ_k. Let X represent the component of the weight down the plane and Y the component of the weight normal to the plane. The force of friction between the block and plane is parallel to and up the plane. Since the block experiences no acceleration, the frictional force up the plane is equal to the component of the weight parallel to the plane; thus $X = f_k = w \sin \theta_k$. The force pressing the surfaces together is just the component of the weight perpendicular to the plane; $N = Y = w \cos \theta_k$. By definition, the coefficient of kinetic friction is

Figure 3.9

The frictional force **f** has the same magnitude as the component X when the block slides down with constant speed; in this case the normal force **N** exerted on the block by the plane is equal and opposite to Y, the component of the weight perpendicular to the surface.

$$\mu_k = \frac{f_k}{N} = \frac{w \sin \theta_k}{w \cos \theta_k} = \tan \theta_k \qquad (3.5)$$

The tangent of the angle at which the block slides down the plane with uniform velocity is equal to the coefficient of kinetic friction.

If the angle θ is varied until it is the largest angle at which the block remains at rest on the plane, the component of the weight parallel to the plane ($w \sin \theta_s$) is equal to the force f_s necessary to overcome static friction, while the force pressing the surfaces together is equal to $w \cos \theta_s$, the component of the weight perpendicular to the plane. The angle θ_s at which we have satisfied these conditions is known as the *angle of repose*, and

$$\mu_s = \frac{f_s}{N} = \frac{w \sin \theta_s}{w \cos \theta_s} = \tan \theta_s \qquad (3.6)$$

3.9 REDUCING FRICTION

If a layer of liquid is introduced between two surfaces, the liquid flows over the surfaces and adheres to them. Instead of friction between two solids, there is friction between layers of the liquid. When oil is poured into the bearings of a machine, the sliding takes place primarily between layers of oil. Since friction between oil layers is much less than between metal surfaces, the frictional force is decreased. Oil is usually preferred to water as a lubricant, not because there is less friction between oil layers than between water layers, but because oil films have the property of staying between the metal surfaces, while water layers are readily squeezed out.

When frictional forces cannot be reduced sufficiently by lubrication, it is customary to substitute rolling friction for sliding friction. The friction of a solid rolling on a surface is far less than the friction of a solid sliding over the surface. For this reason, automobiles and railroad trains operate on wheels rather than on runners and often use ball bearings or roller bearings at the axles. When a car wheel rolls on a level track, it makes a slight depression in the track, and the wheel is somewhat flattened. As the wheel rolls, it is forced continually to climb out of this depression. This is one reason that even rolling wheels require a force to overcome friction. The amount of depression depends on the nature and area of the surfaces in contact.

QUESTIONS

1 A book rests on a horizontal desk top. What two forces act on the book? What are the reactions to these forces, and on what body is each reaction exerted?

2 Under what circumstances will the frictional force on a block on a rough inclined plane be directed down the plane?

3 A weight is hung from the middle of a rope. Will it ever be possible to get the rope to be horizontal by pulling on the ends? Discuss with a vector diagram.

4 If frictional forces are independent of area, why are automobile brake shoes of large area desirable?

5 One end of a rod is held up by a string while the other end rests on frictionless ice (or on a floating cork). Why does the string take up a vertical line in the absence of any additional forces regardless of the angle made by the rod with the horizontal?

6 What is the smallest number of forces not all acting in the same plane which can put a point mass in equilibrium?

7 A man lies in a hammock stretched between two trees. Are the supporting ropes more likely to break if the hammock is strung tightly or loosely? Explain using a vector diagram.

8 Show that for a body at rest on a rough horizontal plane, no force making an angle with the vertical smaller than the angle of repose can slide the body along the plane.

9 Prove that if a body is in equilibrium under the action of a set of forces, any one of the forces is equal and opposite to the resultant of all the other forces.

10 If the pulleys in the accompanying figure are of negligible mass, why can't the system as shown in the accompanying figure be in equilibrium? What must happen before equilibrium can be achieved for this pulley system? If all the pulleys

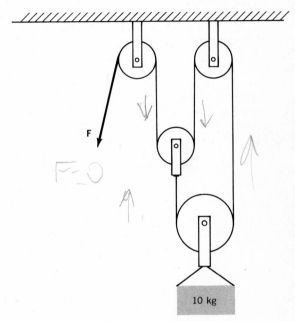

Question 10

had the same mass, what would be the smallest value which would allow the system as shown to be in equilibrium?

PROBLEMS

Draw an appropriate sketch for each problem.

1 A pendulum bob with a weight of 20 N hangs from a cord. A horizontal force sufficient to bring the cord to an angle of 30° with the vertical is applied to the bob. Find the horizontal force and the tension in the cord. What is the smallest force which will hold the bob in equilibrium? What is the tension in this case?

Ans. 11.5 N; 23.1 N; 10.0 N; 17.3 N

2 The ball of Fig. 3.4 has a weight of 6 lb. Find the horizontal force F required to hold the ball

at rest with $\theta = 25°$. What is the tension in the cord?

3 A block weighing 50 N rests on a frictionless plane inclined 25° to the horizontal. It is connected to a weight w by a rope of negligible weight. The rope runs up parallel to the plane and passes over a pulley so that the weight w hangs vertical. What must be the magnitude of the weight if the system is in equilibrium? What is the normal force exerted on the block by the plane? Draw an appropriate vector diagram showing all forces acting on the block.

Ans. 21.1 N; 45.3 N

4 A frictionless cart standing on an inclined plane that makes an angle of 15° with the horizontal is kept from rolling downhill by a force of 12 N applied in a direction parallel to the plane. What is the weight of the cart? What is the normal force exerted on the cart by the plane?

5 A carton weighing 100 lb rests on a horizontal loading platform. A horizontal force of 42 lb is required to start the carton in motion, but once it is moving, a force of 35 lb is adequate to keep it in motion at constant speed. Find the coefficients of static and kinetic friction.

Ans. 0.42; 0.35

6 A brake shoe is pressed against the rim of a moving wheel with a force of 80 N. If the coefficient of kinetic friction between the surfaces is 0.45, how much frictional force is developed?

7 A box weighing 60 N is on a horizontal floor. The coefficient of static friction between surfaces is 0.30, and that of sliding friction is 0.25. Find the frictional force under each of the following conditions: (*a*) no additional force is applied; (*b*) a force of 10 N is applied toward the east; (*c*) a force of 17.9 N is applied toward the west; (*d*) a force of 20 N to the south is applied steadily to the box.

Ans. (*a*) 0; (*b*) 10 N west; (*c*) 17.9 N east if block at rest, 15 N east if in motion; (*d*) 15 N north

8 (*a*) If block A of the accompanying figure weighs 5 lb and block B weighs 20 lb, what force F is required to pull them along at constant speed if $\mu_k = 0.24$? (*b*) If a string is fastened between the two hooks so A does not move,

Problems 8 and 9

what is F, assuming the coefficient of kinetic friction is 0.40 between the two blocks?

9 (*a*) If blocks A and B of the accompanying figure have masses of 2 and 10 kg, respectively, find μ_k between B and the table if a force of 25 N is required to pull the blocks at constant speed. (*b*) What force would be required to pull B if A were held fixed by a string between the hooks? Take $\mu_k = 0.40$ for the two blocks.

Ans. (*a*) 0.21; (*b*) 32.8 N

10 What force acting at an angle of 30° above the horizontal is necessary to move a mass of 10 kg with uniform velocity along a horizontal surface, the coefficient of friction being 0.30?

11 The coefficient of sliding friction between a block and the floor is 0.35. If a force of 20 N acting 37° above the horizontal gives the block a constant velocity, find the weight of the block.

Ans. 57.7 N

12 If the angle of repose for a 20-lb block of metal on an incline is 40°, what force parallel to the incline is necessary to cause the body to begin to move up the incline?

13 A body is in equilibrium under the influence of three forces, two of which are equal in magnitude and the third has a magnitude half as great. Find the angle between the two equal forces.

Ans. 29.0°

14 The end of a horizontal electric transmission line in which there is a tension of 2,000 N is fastened to the top of a vertical pole and to a guy wire which can sustain a tension of 6,000 N. What is the smallest angle which the guy wire can make with the pole if there is to be no resultant lateral force on the pole?

15 A 30-kg traffic light is supported by two wires, one of which makes an angle of 20° with the horizontal, while the second makes an angle of 10°. Find the tension in each.

Ans. 579 N; 553 N

16 A picture is supported by two wires fastened to the ends of the upper edge of the picture frame, which is horizontal. Each of the wires makes an angle of 65° with the vertical. What is the tension in each wire if the picture weighs 8 lb?

17 A bird weighing 4 N alights on a weightless telephone wire midway between two poles which are 60 m apart. Assume that the wire between the bird and each of the poles forms a straight line. If the center of the wire is 0.18 m below its level at the poles, and if the weight of the wire is neglected, calculate the tension in the wire due to the weight of the bird.
Ans. 333 N

18 A bird weighing 2 lb alights on a weightless wire midway between two poles 180 ft apart. Because of the weight of the bird, the wire, which was initially horizontal, sags 4 in. What is the tension in the wire due to the weight of the bird?

19 An inextensible tow rope 10 m long is stretched taut between a tree and an automobile. A force of 250 N at the center of the rope moves it 0.6 m perpendicular to the original line of the rope. Find the force exerted on the tree and on the automobile.
Ans. 1,050 N

20 Find the tension in each rope of the accompanying figure if the weight w is 80 lb.

21 Find the tension in each rope of the accompanying figure if the weight w is 50 N.
Ans. (*a*) 35.7, 40.4; (*b*) 62.5, 37.5; (*c*) 4.17, 31.3, 14.6 (all in newtons)

22 A man is pushing an 80-lb box along a level floor with a uniform velocity of 5 ft/sec by exerting a force on the box at an angle of 30° downward from the horizontal. If the coefficient of friction between sliding surfaces is 0.40, what is the magnitude of the force?

23 A block rests on a plane that is inclined at an angle of 30° to the horizontal. The coefficient of kinetic friction between the block and the plane is 0.20. If the block has a weight of 100 N, what force parallel to the incline is necessary to keep it sliding up the plane? What horizontal force would be required? *Ans.* 67.3 N; 87.9 N

24 A 20-lb frictionless car rests on an inclined plane that makes an angle of 35° with the horizontal. What force must be exerted to keep it from sliding down the plane if the force is applied so that it makes an angle of 30° downward with the horizontal?

25 In the accompanying figure block A has a mass of 5 kg, and block B a mass of 10 kg. The coefficient of kinetic friction between the blocks and the surface is 0.40. Force F is applied to pull B up the plane at constant speed. (*a*) Draw free-body diagrams for A and B, showing *all* forces acting on each. (*b*) Find the tension in the cord between A and B. (*c*) Find the magnitude of F. *Ans.* (*b*) 19.6 N; (*c*) 110 N

Problems 25 and 26

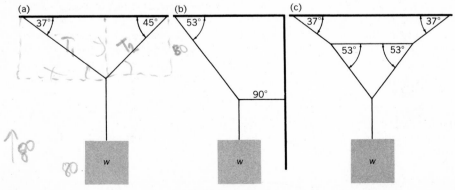

Problems 20 and 21

26 If A and B in the accompanying figure each weigh 20 lb and rest on surfaces such that the coefficient of friction is 0.30, what is the tension in the cord between A and B? Find what force F is needed to pull B up the plane.

27 Two cylindrical lengths of pipe, each weighing 80 lb, lie against one another on a smooth sidewalk. What horizontal force on each pipe is needed to keep them together when an identical length is placed in the groove between them?

Ans. 23.1 lb

28 A force of 60 lb exerted at an angle of 30° above the horizontal is needed to keep a 200-lb block moving along a cement floor. Find the coefficient of kinetic friction.

29 Find the compression in the strut and the tension in the cable of the accompanying figure if the strut is weightless and w is 400 N. The strut is free to rotate about a pin at the wall, so the strut exerts a force only along its length.

Ans. (*a*) 667 N, 533 N; (*b*) 693 N, 800 N; (*c*) 1,140 N, 857 N

30 Find the compression in the strut and the tension in the cable of the accompanying figure if the strut is weightless and w is 300 lb. The strut is free to rotate about a pin at the wall, so the strut exerts a force only along its length.

31 A uniform flexible cable of weight w is supported by hooks at the same height on neighboring telephone poles. If the cable makes an angle θ with the horizontal at the hooks, find (*a*) the force exerted on the cable by the hook on the right and (*b*) the tension in the cable at its lowest point.

Ans. (*a*) $w/(2 \sin \theta)$ at angle θ with horizontal; (*b*) $(w \cos \theta)/(2 \sin \theta)$

32 A tightrope walker of weight w stands one-third of the way along a horizontal tightrope of length L, which sags a maximum distance s below the horizontal line. If s is very small compared with L and the rope has negligible weight, show that the tension T is given by $2wL/9s$.

33 A block of mass m rests on a rough plane inclined at an angle θ with the horizontal. Show that if a gradually increasing horizontal force F is exerted on the block, the block begins to slide up the plane when $F = mg(\sin \theta + \mu_s \cos \theta)/(\cos \theta - \mu_s \sin \theta)$, where μ_s is the coefficient of static friction.

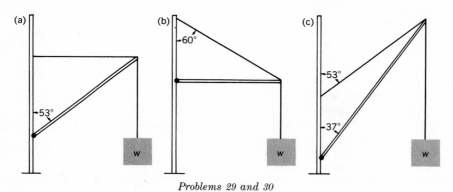

Problems 29 and 30

IN CHAP. 3 we learned that if the resultant force on a body is zero, the body maintains a constant velocity (which is zero if the body is initially at rest). However, the body may still have a rotational acceleration about its center of mass. We next consider what condition must be satisfied if there is to be no rotational acceleration, and we develop the general conditions for equilibrium which must be satisfied if a body is to remain at rest or move with constant linear and angular velocities.

CHAPTER FOUR
TORQUE AND EQUILIBRIUM

4.1 TORQUE

If two boys who weigh 50 and 60 lb, respectively, sit on opposite ends of a horizontal seesaw 12 ft long (Fig. 4.1), we know that the seesaw does not remain horizontal; the 60-lb boy moves downward, and the 50-lb boy upward. This occurs even though the upward force exerted by the support is equal to 110 lb (plus the weight of the seesaw, which we assume for the moment is negligible). A glance at the figure reveals that the forces are not concurrent and that the conditions required for a body to be at rest under the influence of concurrent forces are not enough to keep the seesaw at rest. The weight of the 60-lb boy acts to produce a clockwise rotation about the fulcrum, while the weight of the 50-lb boy acts to produce a rotation in the opposite direction. When both boys are at the ends of the seesaw, the two rotational tendencies do not balance each other and there is a resultant clockwise rotation. We all know from experience that the two boys can sit on the seesaw and keep it at rest, provided the heavier boy sits closer to the fulcrum. This suggests that the tendency to produce rotation depends not only on the force acting but also on a distance.

A simple experiment shows that if the 50-lb boy sits 6 ft from the fulcrum and the 60-lb boy 5 ft from the fulcrum, the seesaw is balanced. We observe that $60 \times 5 = 50 \times 6$. This suggests that the tendency to produce rotation depends on the product of force and distance. We must, however, inquire more closely into how this distance factor is to be determined. Suppose that our seesaw is elevated and that the 50-lb boy is suspended from a rope 8 ft long hanging from the end of the seesaw (Fig. 4.2). His distance from the fulcrum is then 10 ft. Will the seesaw still be balanced? Yes, indeed. An experiment would show that moving the 50-lb boy to this position had no effect on the balance. Thus, it is clear that it is *not* the distance from the fulcrum to the point at which the force is applied which is important; rather, it is the *perpendicular distance from the fulcrum to the line along which the force acts*. This distance is known as the *lever arm* (or *moment arm*). The

Figure 4.1
Rotational equilibrium of the seesaw occurs when the 60-lb boy at A moves to B.

Figure 4.2
The torque about the fulcrum depends on the product of the force and the distance from the fulcrum to the line along which the force acts.

Figure 4.3
Torque is a vector quantity. By
convention its direction is that in which
a right-handed screw would advance if
it were turned by the torque.
Alternatively, the direction is pointed by
the outstretched right thumb when the
fingers of the right hand curve in the
direction in which the screw would
rotate.

lever arm for any axis is the perpendicular distance from the axis to the line along which the force acts.

The product of the lever arm and the force (6 ft and 50 lb for the 50-lb boy on the seesaw) is called the *torque* (or the *moment* of the force). If the seesaw is to be in equilibrium, the clockwise torque about the fulcrum must be balanced by the counterclockwise torque. Not only is the clockwise torque equal to the counterclockwise torque about the fulcrum, but the net counterclockwise torque *about any axis* is equal to the net clockwise torque *about the same axis*, provided the resultant of the forces is zero. For example, if the weight of the seesaw itself is negligible, Eq. (3.1) requires that the fulcrum exert an upward force of 110 lb. If we choose an axis through the left end of the seesaw, we have a counterclockwise torque of 110 lb × 6 ft = 660 lb-ft and a clockwise torque of 60 lb × 11 ft = 660 lb-ft.

The torque τ exerted by a force about a fixed axis is the product of the lever arm and the force, where the lever arm is the distance from the axis to the line along which the force acts. Torque is a vector quantity with direction chosen by convention to be the direction in which a right-handed screw would advance if it were turned by the torque (Fig. 4.3). Alternatively, the torque has the direction in which the right thumb points when the right fingers curve to indicate the path along which points of the body would move under the influence of the torque. For the case shown in Fig. 4.4 the direction of the torque† is out of the

† In general, the torque τ exerted about any point is given by $\tau = \mathbf{r} \times \mathbf{F}$, where $\mathbf{r}$ is the vector from the point in question to the point where the force $\mathbf{F}$ is applied. The product $\mathbf{r} \times \mathbf{F}$ is known as the *vector* or *cross* product of $\mathbf{r}$ and $\mathbf{F}$. It has magnitude $rF \sin \theta$ and the direction described in Fig. 4.3.

Figure 4.4
The torque about any axis is the
product of the force and the distance
from the axis to the line along which
the force acts, called the *lever arm*.

page, and its magnitude is

$$\tau = rF \sin \theta \qquad (4.1)$$

where θ is the angle between **r** and **F**. Clearly, $F \sin \theta$ is the component of **F** perpendicular to the line joining the axis to the point of application of the force. Since $r(F \sin \theta) = F(r \sin \theta)$, the torque is the product of the force F and the lever arm $r \sin \theta$.

We shall be concerned primarily with forces confined to a single plane. In this case the torques are all perpendicular to the plane containing the forces. In describing the torques, we shall refer to them as clockwise if they act to produce a rotation in the direction in which the hands of a clock move and counterclockwise if they act in the opposite direction. The conventional direction for a clockwise torque is into the plane and for a counterclockwise torque out of the plane.

As an illustration of the importance of the lever arm in computing the torque, consider the operation of a bicycle. If a boy pushes down on the pedal with his full weight (Fig. 4.5), the turning moment produced depends on the location of the pedal. If it is straight down, no rotational effect is produced. The torque in this case is zero, because the lever arm is zero. The maximum torque occurs when the pedal is in the position marked A; in this position the lever arm is R. When the pedal is in position B, the lever arm is reduced from R to X.

4.2 CENTER OF GRAVITY

Every particle of an extended body possesses weight, so that the pull of the earth on the body is composed of a large number of forces directed toward the center of the earth. For a body of ordinary size these forces are essentially parallel to one another. The body can, however, be supported in equilibrium by a single upward force, provided its line of action passes through a point called the *center of gravity* of the body. This point can be located as follows.

If we suspend the body from a single point, it comes to rest with a definite orientation. Let us determine a line through the body directed vertically downward from the point of suspension (Fig. 4.6). Next, let us suspend the body from several other points on its surface and, in each case, determine the position of the line through the point of support and directed vertically downward. We find that all the lines determined in this way intersect at a point which is the *center of gravity* of the body.

The center of gravity of a uniform sphere, cube, or rod is at its geometrical center. The center of gravity of an axe or hammer is nearer the head than the handle. The center of gravity of a telephone pole or a baseball bat lies on the axis, but is closer to the thicker end. The center of gravity of a ring or of a tire is at the geometrical center, which, of course, does not lie in the material of the object. Although the center of gravity is a point fixed relative to the body, it does not necessarily lie within the body.

Since an extended body can be at rest when supported by a single upward force whose line of action passes through the center of gravity, the sum of the clockwise torques due to the individual forces on the

Figure 4.5
The vertical force produces no torque when the pedal is straight down; maximum torque is exerted when the pedal is at A.

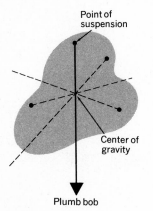

Point of suspension

Center of gravity

Plumb bob

Figure 4.6
The center of gravity of a body can be found by the experiment shown. If the only forces acting are the weight and an equal upward force exerted at a point of suspension, a vertical line through the suspension point passes through the center of gravity.

particles of the body must be equal to the sum of the counterclockwise torques about an axis through the center of gravity. For any axis the torque due to the weight of the body acting at the center of gravity is equal to the sum of the torques due to the forces of gravity on all the various parts of the body. When we are dealing with the extended body, we may consider all the weight of the body as concentrated at the center of gravity, and we may thus replace the individual forces acting on the many particles of the body as a single force equal to the weight. This is true not only for an axis determined by the center of gravity, but for any axis. *For the purpose of computing torques the entire weight of an object can be considered to act on the center of gravity.*

Example

Find the center of gravity of two spheres A and B of weights 0.1 and 0.4 N, respectively, connected by a light (negligible mass) rod 20 cm long (Fig. 4.7).

If the center of gravity is at a point C, the clockwise torque due to the larger weight w_2 about an axis through C is just equal to the counterclockwise torque due to the smaller weight w_1. Then

$$w_1 x_1 = w_2 x_2$$

Since $x_1 + x_2 = 20$ cm $= 0.20$ m,

$$(0.1 \text{ N})x_1 = (0.4 \text{ N})(0.2 \text{ m} - x_1)$$
$$x_1 = 0.16 \text{ m} = 16 \text{ cm} \qquad ////$$

4.3 EQUILIBRIUM

Thus far we have been considering the conditions which must be satisfied in order that a body remain *at rest*. Exactly the same conditions apply to a body which is in motion with *constant velocity*. When we ride in an automobile over a smooth road at a constant velocity of 40 km/hr south, the resultant of the forces which act on us is the same as when we are at rest. As we shall see in Chap. 6, if the resultant force acting on a body is not zero, the body is accelerated.

The condition that the vector sum of the torques on a body be zero in order that there be no rotation for a body at rest can be extended to include bodies rotating about a fixed axis as follows: *If the vector sum of the torques acting on a rigid body is zero, the rotational velocity of the body is constant.* When a resultant torque acts on the body, there is a change in its angular velocity, as we shall see in Chap. 11. *A body which has a constant linear velocity and a constant rotational velocity is in a state of equilibrium.* Statics involves a special case of equilibrium, namely, the case in which both the linear velocity and the rotational velocity have the constant value *zero.*

4.4 THE CONDITIONS FOR EQUILIBRIUM

In order that a rigid body be in equilibrium, two conditions must be satisfied:

x_1 — x_2

C

A

B

w_1

w_2

Figure 4.7
Center of gravity of two spheres connected by a rod of negligible mass.

1   *The vector sum (resultant) of all forces acting on the body must be zero;* $\Sigma\mathbf{F} = 0$.

2   *The vector sum (resultant) of all torques about any axis must be zero;* $\Sigma\tau = 0$.

The first condition for equilibrium guarantees that the center of gravity (more accurately the center of mass) of the body moves with constant velocity. The second condition guarantees that the rotational velocity of the body remains constant. In applying the second condition, there is no restriction as to what axis is chosen. It may be through the center of gravity of the body, through one end of the body, or at any convenient place.

In practice, it is usually convenient to use the conditions for equilibrium in a slightly different form. If we replace forces with their vertical and horizontal components, we may write the conditions† for equilibrium in the form:

1   *The sum of all upward forces is equal to the sum of all downward forces.*

   The sum of all forces to the right is equal to the sum of all forces to the left.

2   *The sum of all clockwise torques about any axis is equal to the sum of all counterclockwise torques about the same axis.*

$$\Sigma F_{\text{up}} = \Sigma F_{\text{down}} \tag{4.2}$$

$$\Sigma F_{\text{right}} = \Sigma F_{\text{left}} \tag{4.3}$$

$$\Sigma \text{ torques}_{\text{clockwise}} = \Sigma \text{ torques}_{\text{counterclockwise}} \tag{4.4}$$

In working equilibrium problems it is imperative that *all* forces acting on the body in question be considered, and *only* those forces. The first thing one must do is decide what body is to be dealt with. Then it is usually helpful to make a rough sketch of this body, representing all forces acting on it by suitably placed arrows. Next, one resolves all those forces which are neither horizontal nor vertical and replaces them with their components. At this point one is ready to apply the conditions for equilibrium. It is not necessary that we choose vertical and horizontal axes for applying the conditions for equilibrium, but in the majority of cases this is the most convenient choice.

Example

It is found by weighing that the front wheels of a truck support 15,000 N and the rear wheels 27,000 N. If the wheelbase (distance between axles) is 6 m, find the weight of the truck and the location of the center of gravity (Fig. 4.8).

By Eq. (4.2),

$$15,000 \text{ N} + 27,000 \text{ N} = w$$

$$w = 42,000 \text{ N}$$

† These conditions are written for the case of forces acting in a plane. The extension to three-dimensional systems is not difficult; all we need do is (1) add to the first condition for equilibrium that the sum of all components perpendicular to the plane determined by the first two axes equals zero and (2) apply the torque condition to three noncoplanar axes.

Figure 4.8
Center of gravity of a truck.

7.2 m

800 N

240 N

2.1 m

F

V

H

Figure 4.9

Equilibrium of a ladder leaning against a smooth wall.

Let d be the distance from the front wheels to the center of gravity, and find torques about the front axle as axis. By Eq. (4.4),

$$wd = 27{,}000 \text{ N} \times 6 \text{ m}$$
$$(42{,}000 \text{ N})d = 27{,}000 \text{ N} \times 6 \text{ m}$$
$$d = 3.86 \text{ m} \qquad \text{////}$$

Example

A uniform ladder weighs 240 N and is 7.5 m long. It leans against a smooth (frictionless) wall at a point 7.2 m above a cement driveway with its base 2.1 m from the wall. An 800-N man stands six-tenths of the way up the ladder. (*a*) Find the horizontal and vertical components of the force exerted on the ladder by the driveway. (*b*) If the coefficient of static friction between the ladder and the driveway is 0.2, what fraction of the way to the top can the man climb before the ladder slips?

The object in equilibrium is the ladder, and the forces exerted on it are shown in Fig. 4.9. H and V are the horizontal and vertical components of the force exerted on the ladder by the driveway.

a By Eq. (4.2): $V = 800 \text{ N} + 240 \text{ N} = 1{,}040 \text{ N}$

   By Eq. (4.3): $H = F$

If we choose the base of the ladder as axis, we obtain, by Eq. (4.4),

$$(7.2 \text{ m})F = (1.26 \text{ m} \times 800 \text{ N}) + (1.05 \text{ m} \times 240 \text{ N})$$
$$F = 175 \text{ N} \qquad \text{and} \qquad H = 175 \text{ N}$$

Note that we may use any axis for applying Eq. (4.4). It may be instructive to try other axes (such as one at the point where the ladder touches the wall or one through the center of the ladder) to assure yourself that we may indeed choose any axis.

b As the man climbs the ladder, F and H increase, but since H is provided by friction, it cannot exceed $0.2V = 208 \text{ N}$. Let q represent the fraction of the length of the ladder which the man has climbed. When $H = 208 \text{ N}$, F is also 208 N and applying Eq. (4.4) about the bottom of the ladder yields

$$7.2 \text{ m} \times 208 \text{ N} = (1.05 \text{ m} \times 240 \text{ N}) + (2.1q \times 800)\text{N-m}$$

from which $q = 0.74$. The ladder slips just before the man has climbed three-fourths of the way to the top! ////

Example

A derrick (Fig. 4.10) has a uniform boom 30 ft long which weighs 1 ton. A load of 3 tons is suspended from the end. The boom is hinged to a vertical mast and held up by a cable which makes an angle of 60° with the mast and which is fastened 10 ft from the end of the boom. If the boom makes an angle of 37° with the mast, find the tension in the cable and the force exerted on the boom by the hinge.

10'

3 tons

0.866T

T

0.5T

60°

1 ton

16'

37°

H

V

9'

12'

18'

Figure 4.10

Equilibrium of a derrick boom.

First, we resolve the tension into its horizontal and vertical components, which are $T \sin 60° = 0.866T$ and $T \cos 60° = 0.5T$, respectively.

Next let H and V represent the components of the force exerted on the boom by the hinge.

By Eq. (4.2): $\qquad V = 3 + 1 + 0.5T$

By Eq. (4.3): $\qquad H = 0.866T$

If we choose the hinge as the axis, Eq. (4.4) yields

$$(9 \times 1) + (0.5T \times 12) + (3 \times 18) = (0.866T \times 16)$$

$$T = 8.02 \text{ tons}$$

whence $V = 8.01$ tons, and $H = 6.95$ tons.

Note that we have replaced the force T with its components and then computed the torque due to each component. The resultant of the torques due to the components is equal to the torque due to the original force. ////

4.5 TYPES OF EQUILIBRIUM

The equilibrium of a body may be *stable*, *unstable*, or *neutral* (Fig. 4.11). When a body returns to its original position after being slightly disturbed, the equilibrium is said to be *stable*. A cone standing on its base is an illustration of this type of equilibrium. When this cone is tilted slightly and released, it returns to its original position.

If the cone rests on its vertex, it can be in equilibrium only when its center of gravity lies directly above the vertex. If it is slightly displaced, the cone falls over; it is in *unstable* equilibrium. A body in unstable equilibrium does not return to the original equilibrium position when slightly displaced but moves farther away. Lastly, a billiard ball resting on a horizontal table is in *neutral* equilibrium. When it is slightly displaced, it neither returns to its former position nor does it go farther away from the initial position. It remains in any position in which it finds itself. A cylinder or a cone lying on its side on a horizontal surface is also in neutral equilibrium.

The position of the center of gravity is of paramount importance in determining the stability of a body. The lower the center of gravity, the greater the stability of the body and the more difficult it is to overturn it. The body becomes unstable as soon as the vertical line through its center of gravity falls outside its base. The leaning tower of Pisa remains in stable equilibrium because, in spite of its leaning, the line of action of the weight falls inside the base. When an effort is made to turn a body over by pushing on it, the center of gravity rises for a body in stable equilibrium, descends for a body in unstable equilibrium, and remains at the same level for a body in neutral equilibrium.

Figure 4.11
Three kinds of equilibrium are *stable*, *unstable*, and *neutral*.

(a)

(b)

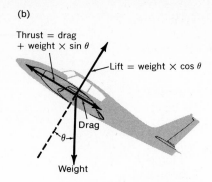

Figure 4.12

An aircraft in stable equilibrium. The velocity must be constant.

4.6 THE STABILITY OF AIRCRAFT

An airplane flying horizontally at constant speed is in equilibrium under the influence of four generalized forces: the lift, the weight, the drag, and the thrust (Fig. 4.12). The weight of the airplane may, of course, be regarded as concentrated at the center of gravity. The lift forces, which arise from pressure differences on the wing, fuselage, and tail sections of the aircraft, may be replaced in a similar way by a resultant lift vector. The friction resulting from the passage of air over the surface of the aircraft gives rise to a net retarding force known as the *drag*, while the aircraft engine, whether propeller or jet, provides *thrust* to overcome this drag.

For the aircraft to be longitudinally stable, the resultant of the torques about a transverse axis through the center of gravity must be zero. The airplane must be stable both longitudinally and laterally. Note that the conditions for equilibrium must be satisfied over a wide range of speeds and loadings. This can be accomplished by moving aileron, wing, or elevator surfaces in such a way as to keep the lift equal to the weight and the center of lift appropriately placed relative to the center of gravity. When the speed of an aircraft approaches the speed of sound, there may be radical changes in the way in which air flows over the aircraft surfaces. Such changes introduce sudden shifts in the position of the center of lift and thereby introduce large and sudden torques on the aircraft. For this and other reasons, it is dangerous to fly any type of aircraft above a maximum speed which depends upon the design.

QUESTIONS

1 Where are the positions of stable, neutral, and unstable equilibrium for a right circular cylinder?

2 Why does a person lean forward when he is pushing a heavy cart?

3 How could a body estimated to weigh 10 lb be weighed using a meterstick and a spring balance calibrated to read only up to 4 lb?

4 Why is a wagon loaded with hay more likely to turn over on a hillside than one loaded with an equal weight of gravel?

5 Is a ladder resting on a sidewalk and leaning against a wall any more likely to slip as a man climbs toward the top? Explain.

6 One end of a meterstick (or rod) rests on a pencil held by one hand and the other end on a horizontal eraser held by the other hand. If the two hands are slowly moved closer together until they touch, will the meterstick eventually roll off? Explain why it will not even though the coefficients of friction are different.

7 Can a body be in equilibrium under the influence of three forces which are neither concurrent nor parallel?

8 Many cathedrals have large buttresses outside and against their masonry walls. What is the purpose of the buttresses? Why don't ordinary brick houses need buttresses?

9 In the prototypes of the Concorde supersonic aircraft provision was made to pump 12,000 lb of fuel to a rear fuselage tank to give the desired handling and control characteristics. What does this fuel transfer do to the center of gravity? What does this provision imply about the net lift vector at supersonic speed as compared to the same vector at subsonic speed?

10 A brick rests on a horizontal sidewalk. An identical brick is placed on top of the first with three-tenths of its length overhanging. Can a third brick be added with three-tenths of its length overhanging the second? What is the limiting number which can be assembled in this way and still remain in equilibrium?

11 Consider the compressive forces in the backbone and the tension in the muscles of the back when one bends over to lift a heavy load. Why is a back injury much more likely to occur if a person bends his back to pick up a heavy load than if he bends his knees, keeping his back nearly vertical?

PROBLEMS

Draw an appropriate sketch for each problem.

1 When the front wheels of an automobile are run onto a platform scale, the scale balances at 1,800 lb; when the rear wheels are run onto the scale, it balances at 1,400 lb. What is the weight of the automobile and how far is its center of gravity behind the front axle? The distance between axles is 110 in. *Ans.* 3,200 lb; 48.1 in.

2 Weights of 10, 4, 6, and 8 N are located, respectively, at 25-, 50-, 75-, and 100-cm marks on a meterstick whose weight is negligible. What are the magnitude and location of the single upward force which will balance the system?

3 A uniform beam 3 m long weighs 200 N and is

supported at its ends by two walls. Find the reaction of the walls against the beam when a man weighing 750 N stands on the beam at a distance of 2 m from one end.
Ans. 600 and 350 N

4 A tapered pole 8 ft long weighs 30 lb. It balances at its midpoint when a 5-lb weight hangs from the slimmer end. Find the center of gravity.

5 A 700-N bricklayer is 2 m from one end of a uniform 500-N scaffold which is 8 m long. A pile of bricks weighing 400 N is 3 m from the same end. If the scaffold is supported at the two ends, calculate the force on each support.
Ans. 1,025 and 575 N

6 A 60-lb uniform beam 10 ft long is supported at its ends by two walls. Find the forces exerted on the supports when a 180-lb man stands on the beam at a distance of 3 ft from one end.

7 A 72-kg man stands on the ball of one foot. His shinbone exerts a downward force at the ankle joint roughly 10 cm back of the ball of the foot while the Achilles tendon pulls (almost) upward 4 cm further back. Using the crude approximation that the forces involved are all vertical, find the force exerted by the Achilles tendon and that exerted by the shinbone.
Ans. 2,470 N; 1,764 N

8 A uniform bar 2 yd long has a weight of 5 lb fastened to it at one end and a weight of 8 lb at the other end. The bar itself weighs 6 lb. Where could a single force be applied to balance the system, and how great would the force have to be?

9 A square is acted upon by the forces of 4, 6, 8, and 12 N, respectively, along the four sides. The forces all act to produce rotation in the same direction. If the length of a side of the square is 0.5 m, what is the resultant torque acting to rotate the square about an axis through its center? *Ans.* 7.5 N-m

10 A wheel of 20-cm diameter has an axle of 2-cm radius. If a force of 300 N is exerted along the rim of the wheel, what is the smallest force exerted on the outside of the axle which will result in zero net torque?

11 A bar of uniform cross section is carried by two

boys, one at either end of the bar. If the bar weighs 40 lb and is 11 ft long, where must a load of 50 lb be hung from the bar so that one boy will carry twice as much as the other?

Ans. 2.2 ft from stronger boy

12 A 100-lb boy and a 75-lb boy sit at the ends of a 12-ft seesaw. At what point should a 50-lb boy sit for the seesaw to be in equilibrium?

13 A uniform horizontal bar 6 ft long weighs 40 lb. Upward forces of 25 and 15 lb are exerted at the ends. Is the bar in equilibrium? Calculate the torque about each end, about the center, and about a point 5 ft from one end of the bar.

Ans. 30 ft-lb in each case

14 A uniform rod 120 cm long has a mass of 2 kg. It has a mass of 4 kg attached to it at one end, a mass of 6 kg at the other end, and a mass of 5 kg in the middle. Find the position of the center of gravity.

15 A telegraph pole is placed on a two-wheeled dolly located 3 m from the thicker end, and an upward force of 1,000 N at the thinner end is required to keep it horizontal. The pole is 12 m long and weighs 7,500 N. Where is the center of gravity? *Ans.* 4.2 m from the thicker end

16 A tapered pole which is 15 ft long and weighs 90 lb can be balanced at a point 6 ft from the thicker end. If it were to be supported at its ends, how much force would be needed at each end?

17 A cubical box 1.2 m on a side weighs 600 N and rests on a rough floor for which the coefficient of static friction is 0.80. If the center of gravity of the box is at its center, what is the greatest height at which a horizontal force can be applied to slide the box without tipping it? What is the lowest height at which a horizontal 400-N force can be applied to tip the box?

Ans. 0.75 m; 0.90 m

18 To a thin circular disk whose radius is 15 cm there is attached another circular disk of the same material whose radius is 5 cm. Find the center of gravity of the combination if the smaller disk has one point on its circumference at the center of the larger disk.

19 From a circular disk of 25 cm radius there is cut out a circle of 20 cm diameter. Find the center

of gravity of the remainder of the disk if the circumference of the hole passes through the center of the disk.

Ans. 1.90 cm from center of disk on the extension of the line between the center of the disk and the center of the hole

20 From a circular disk whose radius is 30 cm there is cut a circular hole whose radius is 12 cm. Find the center of gravity of the remainder of the disk if the hole has one point of its circumference at the center of the disk.

21 The horizontal bar AB in the accompanying figure is nonuniform, being 5 m long with its center of gravity 2 m from the wall. The bar weighs 300 N. A 900-N weight is suspended at the end B of the bar. What is the tension in the supporting cord if θ is 37°? What are the horizontal and vertical components of the force exerted on the bar by the pin at A?

Ans. 1,700 N; 1,360 N; 180 N

22 The horizontal bar AB in the accompanying figure is supported by the cord BC at one end and by a pin at the wall AC at the other end. The bar is of uniform cross section, weighs 25 lb, and is 5 ft long. If θ is 40°, what are the tension in the cord and the horizontal and vertical components of the force exerted on the bar by the pin at A?

23 In the accompanying figure AB represents a uniform sign 12 ft long and weighing 400 lb. If a 200-lb man stands 4 ft from a pin at A and θ is 37°, find the tension in the cable and the vertical and horizontal components of the force exerted on the sign by the pin.

Ans. 444 lb; 333 lb; 356 lb

24 The 500-N horizontal bar AB in the accompanying figure is nonuniform, being 4 m long with its center of gravity 1.6 m from the wall. A 750-N

Problems 21 to 24

weight is suspended on the bar 3 m from the wall. The angle θ is 60°. What is the tension in the cord BC? Find the horizontal and vertical components of the force exerted on the bar by the pin at A.

25 A uniform 50-lb gate is 4 ft high and 3 ft wide. It is supported by two hinges, the upper of which is 4 in. from the top and bears three-fifths the weight. The lower hinge is 4 in. from the bottom. Find the horizontal and vertical components of the force exerted on the gate by the lower hinge.   *Ans.* 22.5 lb; 20 lb

26 A 200-N door 2.5 m high and 0.9 m wide is hung by two hinges, each of which supports half the weight. If the hinges are 25 cm from the top and 25 cm from the bottom of the door, find the horizontal components of the forces exerted on the door by the hinges. Assume that the center of gravity of the door is at the geometrical center.

27 Two vehicles are crossing a bridge 20 m long. A passenger car weighing 12,000 N is 4 m from one end. A truck weighing 40,000 N is 8 m from the same end. If the bridge is symmetrical with respect to the center and weighs 8×10^5 N, what are the forces on the two supports at the ends of the bridge?
          Ans. 4.336×10^5 N; 4.184×10^5 N

28 If weights A, B, C, D, and E are, respectively, located at distances a, b, c, d, and e from one end of a light bar, use the two conditions for equilibrium to show that the center of gravity is located a distance y from that end of the bar, where

$$y = \frac{Aa + Bb + Cc + Dd + Ee}{A + B + C + D + E}$$

29 A uniform ladder 25 ft long weighs 60 lb. It leans against a vertical frictionless wall with its lower end 7 ft from the wall and its upper end 24 ft from the ground. Find the force exerted on the ladder by the wall and the horizontal and vertical components of the force exerted on the ladder by the ground.   *Ans.* 9.75, 9.75, and 60 lb

30 A nonuniform ladder is 10 m long. It weighs 160 N, and its center of gravity is 4 m from the bottom. The ladder leans against a smooth wall with its base on a sidewalk 6 m from the base of the wall. Find the force exerted on the ladder by the wall and the horizontal and vertical components of the force exerted on the ladder by the sidewalk. If the coefficient of static friction between ladder and sidewalk is 0.4, at what

angle would the ladder be on the verge of slipping?

31 A ladder is 8 m long, has its center of gravity 3 m from the bottom, and weighs 240 N. A man weighing 400 N stands halfway up the ladder, which makes an angle of 20° with the vertical. Find the force exerted on the ladder by the smooth wall and the horizontal and vertical components of the force exerted on the ladder by the ground.   *Ans.* 106 N; 106 N; 640 N

32 A uniform 20-ft ladder weighing 40 lb rests against the smooth side of a house so that it makes an angle of 60° with the ground. A 160-lb man stands three-quarters of the way to the top. What is the horizontal force exerted on the ladder by the house? What are the horizontal and vertical components of the force exerted on the ladder by the ground?

33 A uniform ladder 5 m long weighs 200 N. It leans against a smooth wall, making an angle of 60° with the horizontal ground. How far up the ladder can a man weighing 800 N climb before the ladder slips if the maximum frictional force the ground can provide is 400 N?
          Ans. 3.71 m

34 A uniform 26-ft ladder weighing 60 lb leans against a smooth wall with its base 10 ft from the wall. How far up the ladder can a 156-lb man climb if the maximum horizontal force available at the base is 60 lb?

35 If a light triangular frame ABC which is held together by pins (see accompanying figure) is equilateral with sides 2 m in length, and if w is 200 N, find the tension in member AC and the compressional force exerted by member AB.
          Ans. 57.7 N; 115 N

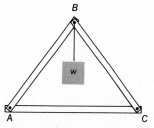

Problems 35 and 36

36 A simple light triangular frame ABC (see accompanying figure) is held together by pins. If sides AB and BC are both 5 ft long while AC is 6 ft, find the horizontal and vertical components of the force exerted on the pin at A by member BC when w is 60 lb. What is the tension in member AC?

37 A horizontal uniform steel beam weighs 30,000 N and is 6 m long. It is supported at a vertical wall by a pin at one end and by a cable making an angle of 37° with the horizontal, attached 1 m from the other end and running to a clamping point on the wall. A load of 80,000 N is suspended from the end of the beam away from the wall. Find the tension in the cable and the horizontal and vertical components of the force exerted on the bar by the pin. What is the direction of the vertical component?
Ans. 190,000 N; 152,000 N; 4,000 N; downward

38 If the load hanging from the end of the boom of Fig. 4.10 were increased from 3 to 4 tons, what would be the new magnitudes of H, V, and T?

39 A derrick similar to that of part (c) of the figure for Probs. 29 and 30 in Chap. 3 has a uniform boom 8 m long which weighs 8,000 N. A load w of 15,000 N is suspended from one end. Find the tension in the cable and the horizontal and vertical components of the force exerted on the boom by the pin at the base.
Ans. 40,700 N; 32,600 N; 47,400 N

40 A derrick similar to that of part (a) in the figure for Probs. 29 and 30 in Chap. 3 has a uniform boom 20 ft long which weighs 2,000 lb. A load w of 4,000 lb is suspended from the end. Find the tension in the cable and the horizontal and vertical components of the force exerted on the boom by the pin at the base.

41 A stepladder is 2 m high. The step side is 2.5 m long, uniform, and weighs 60 N. The other side is also uniform, is 2.5 m long, and weighs 20 N. The sides are hinged at the top, and there is a massless tie rod 1.5 m long which connects the midpoints of the sides. A 640-N man stands three-fourths the way to the top. Find the forces exerted on the two sides of the ladder by the floor and the tension in the tie rod. (*Hint:* Assume the floor is smooth initially; for the tension apply the equilibrium conditions to one or both sides.)
Ans. 450 N; 270 N; 390 N

42 Prove that a uniform cube of edge L will be in stable equilibrium if it is placed on top of a rough cylindrical pipe of radius R provided the diameter of the pipe is greater than L.

43 A homogeneous cylinder of radius R and mass m slides along a table without rotating when acted upon by a horizontal force **F** applied at an appropriate height y. (*a*) If μ_k is the coefficient of sliding friction, find y. (*b*) Prove that the cylinder is not in equilibrium unless $y = 0$ and then only if $F = \mu_k mg$.
Ans. (*a*) $y = R(1 - \mu_k mg/F)$

WE KNOW that if the resultant force on a body is zero, the body has a constant linear velocity. What happens if the resultant force is not zero? Then the velocity changes, and the motion is said to be *accelerated*. Before we relate the resultant force on the body to the rate of change of velocity in Chap. 6, we introduce the concept of *acceleration* and consider the motion of a body subject to a constant acceleration. A familiar example is a freely falling ball or projectile.

CHAPTER FIVE
TRANSLATIONAL MOTION

5.1 TYPES OF MOTION

The linear velocity of a body is constant when the vector sum of the forces acting on the body is zero. If the resultant force is not zero, the velocity of the body changes. Before we relate the time rate of change in velocity to the force, we develop some of the equations of *kinematics*, the science of motion. In this chapter we deal with the simpler but nevertheless important problems of linear motion. (In order to treat advanced problems in kinematics, it is necessary to use *calculus*, a branch of mathematics developed by Newton and Leibnitz to solve such problems.)

In the motion of an automobile chassis along a straight road or of an elevator up and down in its shaft, all points of the body move along parallel lines. Any object which moves in this way is said to undergo a motion of pure *translation*. On the other hand, a merry-go-round and the flywheel of a stationary engine turn about stationary axes. Such a body is said to undergo a motion of *pure rotation*, or an angular motion. All points in the object describe concentric circles about the axis. We defer further consideration of rotational motion to Chap. 11 and confine our attention in this chapter to several types of translational motion. When we deal with a rigid body, we can always resolve its motion, regardless of how complex it may be, into a translation of the center of mass and a rotation about the center of mass. The motion of the wheel of a train or of a boomerang flying through the air may be regarded as a combination of translational and rotational motions.

5.2 INSTANTANEOUS VELOCITY

In dealing with the motion of an object, we are concerned with how its displacement and velocity change in time. Consider the motion of an automobile. Let $\mathbf{r}_0$ be the radius vector from the origin to the center of gravity of the automobile at the instant we start our stopwatch, which

Figure 5.1

The average velocity over time interval t is the ratio of the displacement $\mathbf{s} = \mathbf{r} - \mathbf{r}_0$ to the time; the instantaneous velocity is the limit of the ratio $\Delta\mathbf{s}/\Delta t$ as Δt approaches zero.

hereafter reads the time t. (A subscript zero indicates the value of the quantity at the time $t = 0$.) Let $\mathbf{r}$ be the radius vector (Fig. 5.1) at time t sec later. During this period the average velocity (Sec. 2.4) is given by

$$\mathbf{v}_{\text{av}} = \frac{\mathbf{r} - \mathbf{r}_0}{t} = \frac{\mathbf{s}}{t} \tag{5.1}$$

In general, the average velocity over any time interval Δt is

$$\mathbf{v}_{\text{av}} = \frac{\Delta\mathbf{s}}{\Delta t} \tag{5.2}$$

where $\Delta\mathbf{s}$ is the displacement occurring in the time Δt. (Here and later, we use the Greek letter Δ as mathematical shorthand to mean "the change in.") Thus, if in Fig. 5.1 the car is located by $\mathbf{r}_1$ at time t_1 and by $\mathbf{r}_2$ at t_2, the average velocity between t_1 and t_2 is given by $(\mathbf{r}_2 - \mathbf{r}_1)/(t_2 - t_1)$, which is $\Delta\mathbf{s}/\Delta t$.

The velocity of the automobile may well be changing. If so, we may be interested in how fast (and in what direction) it is moving at a given instant—a quantity which we call the *instantaneous velocity*. If we want to know the velocity as the car passes a telephone pole, we may proceed as follows. We measure the average velocity of the car in the block containing the telephone pole by dividing the length of the block by the time required to cover the block. This will probably be closer to the velocity at the instant of passing than the average velocity over some longer period would be. We might come closer to the instantaneous velocity by determining where the car was 1 sec before it reached the pole and 1 sec after it passed the pole, but this still gives us an average velocity over a 2-sec interval. To get closer to the instantaneous velocity, we measure the displacement over shorter and shorter time intervals. As we make Δt smaller, this average velocity comes closer to the instantaneous velocity. We define the instantaneous velocity $\mathbf{v}$ as the limit approached by $\Delta\mathbf{s}/\Delta t$ as Δt decreases, approaching zero as a limit. We write this

$$\mathbf{v} = \lim_{\Delta t \to 0} \frac{\Delta\mathbf{s}}{\Delta t} \left(= \frac{d\mathbf{s}}{dt} \text{ in calculus notation} \right) \tag{5.2a}$$

The instantaneous speed is the magnitude of the instantaneous velocity. Ideally, the speedometer of an automobile reads instantaneous speed.

5.3 ACCELERATION

When an automobile is driven in city traffic, its velocity is continually changing. It is zero while the car is waiting for a red light to change; when the traffic light goes green, the velocity increases for a while and then varies up and down as traffic conditions require until the car is again brought to rest. Figure 5.2 shows a possible pattern for such a motion. Notice that after the first stoplight the car picked up speed rapidly, while after the second it gained velocity slowly.

In treating motion in which the velocity is changing, we define a new term: *acceleration. Acceleration is the time rate of change of*

Figure 5.2
Velocity plotted as a function of time
for an automobile.

velocity. The average acceleration over any period of time t is the change in velocity divided by the corresponding change in time.

$$\mathbf{a}_{\text{av}} = \frac{\mathbf{v} - \mathbf{v}_0}{t} \tag{5.3}$$

In the rather complex motion shown in Fig. 5.2 the acceleration varies, and we define instantaneous acceleration $\mathbf{a}$ as

$$\mathbf{a} = \lim_{\Delta t \to 0} \frac{\Delta \mathbf{v}}{\Delta t} \left(= \frac{d\mathbf{v}}{dt} = \frac{d^2\mathbf{s}}{dt^2} \text{ in calculus notation} \right) \tag{5.4}$$

From its definition acceleration has the units of velocity divided by time, since it tells us how much the velocity changes per unit of time. For example, acceleration may be expressed in meters per second per second (m/sec^2). In discussing the motion of an automobile, we might measure the change in velocity in miles per hour and the change in time in seconds. Then the acceleration would be in miles per hour per second. If an automobile decelerates from 80 to 60 km/hr in 10 sec, the average acceleration is -20 km/hr divided by 10 sec, or -2 km/hr-sec. In metric units we shall most often measure acceleration in meters per second per second.

Example

An automobile manufacturer advertises that his product can start from rest and reach 100 km/hr (62 mi/hr) in less than 10 sec. What minimum average acceleration is required?

$$a_{\text{av}} = \frac{v - v_0}{t} = \frac{100 \text{ km/hr}}{10 \text{ sec}} = 10 \text{ km/hr-sec}$$

$$= 2.78 \text{ m/sec}^2 = 9.11 \text{ ft/sec}^2 \qquad ////$$

The mathematical treatment of motions in which the acceleration varies irregularly is beyond the scope of this text. We shall confine our discussion to several very important types of motion in which the acceleration is either constant or varies in some relatively simple way. We begin by considering the case in which the acceleration is constant and the motion takes place along a straight line.

(a)

(b)

Figure 5.3

Velocity and displacement of a freely falling body as functions of time in (*a*) metric units and (*b*) British engineering units.

Figure 5.4

A feather and a steel ball experience equal accelerations when they fall freely in an evacuated tube.

5.4 UNIFORMLY ACCELERATED RECTILINEAR MOTION

If a ball is dropped and its velocity measured as a function of time, it is found that the velocity is 0.98 m/sec 0.1 sec after release, 1.96 after 0.2 sec, 2.94 after 0.3 sec, and so forth. The velocity and displacement vary with time after release as shown in Fig. 5.3. Each tenth of a second the velocity increases 0.98 m/sec; each second it increases 9.8 m/sec (or 32 ft/sec).

The acceleration of a freely falling body at mean sea level and 45° latitude is 9.80665 m/sec², or 32.17398 ft/sec². By a freely falling body we mean one for which the air resistance is negligible. A feather does not fall through the air with this large an acceleration because of the substantial air friction, but if it is released in an evacuated tube, it has the same acceleration as a steel ball (Fig. 5.4). When a body falls through the air, the effect of air friction depends on the velocity, size, shape, density, and surface of the falling object. This friction naturally reduces the acceleration somewhat for a falling body, but if a ball is thrown upward, air friction increases the downward acceleration on the way up. We represent the acceleration due to gravity by g. It varies slightly from place to place for reasons discussed in Sec. 9.9. In our problems on falling bodies we shall take $g = 9.80$ m/sec² or 32.0 ft/sec², thereby making the arithmetic of the problems a little simpler.

When the acceleration is constant, the average velocity during any time interval is given by

$$v_{av} = \frac{v + v_0}{2} \tag{5.5}$$

If a ball is released from rest and falls for 3 sec, its initial velocity is 0 and its final velocity is 29.4 m/sec. During these 3 sec the average velocity is $(29.4 + 0)/2 = 14.7$ m/sec. Similarly, if an automobile is *uniformly accelerated* from 20 to 60 km/hr, the average velocity during the acceleration is 40 km/hr. Equation (5.5) together with the equations defining average velocity and acceleration are the three fundamental equations for uniformly accelerated motion. When we measure displacement from the position of the body at $t = 0$,

$$v_{av} = \frac{s}{t} \tag{5.1a}$$

$$a = \frac{v - v_0}{t} \tag{5.3}$$

$$v_{av} = \frac{v + v_0}{2} \tag{5.5}$$

These basic equations may be combined in many ways, two of which are particularly useful, namely,

$$s = v_0 t + \tfrac{1}{2}at^2 \tag{5.6}$$
$$v^2 = v_0{}^2 + 2as \tag{5.7}$$

Equation (5.6) may be obtained by combining Eqs. (5.1a), (5.3), and (5.5) as follows:

$$s = v_{av}t \qquad \text{by Eq. (5.1a)}$$

$$= \frac{v + v_0}{2}t \qquad \text{by use of Eq. (5.5)}$$

$$= \frac{v_0 + at + v_0}{2}t \qquad \text{by use of Eq. (5.3)}$$

$$= v_0 t + \tfrac{1}{2}at^2$$

Equation (5.7) may be derived by multiplying $s = [(v + v_0)/2]t$ by $a = (v - v_0)/t$, which gives $as = (v^2 - v_0{}^2)/2$ or $2as = v^2 - v_0{}^2$.

In the relations developed above, it is important to observe that s is the *displacement* of the body and not the total distance traversed. For example, if we throw a ball up into the air with a speed of 96 ft/sec and ask what the displacement is at the end of 5 sec, the relations above will provide the answer, which is 80 ft. Figure 5.5 shows the displacement measured upward as functions of time and velocity. During the 5 sec the ball traverses a much greater distance. Indeed, at the end of 3 sec it is at a height of 144 ft. Further, the displacement s is

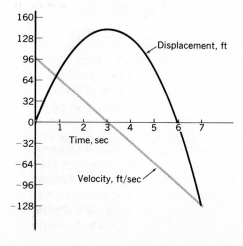

Figure 5.5
Displacement and velocity of a ball thrown upward at $t = 0$ with a speed of 96 ft/sec as functions of time.

measured *from the position at* $t = 0$. If we wish to measure the displacement from some other point, Eq. (5.6) must be replaced by

$$s = s_0 + v_0 t + \tfrac{1}{2}at^2 \tag{5.6a}$$

where s_0 is the displacement of the body at $t = 0$ from the new reference point.

A second caution has to do with the fact that displacement, velocity, and acceleration are *vector* quantities, and due regard must be given to direction. Which direction one elects to choose as positive is one's own free choice, but once it is decided that upward is positive, any vector quantity which is downward must be given a negative sign. If we throw a ball upward with a speed of 30 m/sec, we may decide to choose upward as the positive direction. If we do, $v_0 = 30$ m/sec, but $a = -9.8$ m/sec². An alternative choice would be to call downward positive. Then v_0 becomes -30 m/sec, and $a = +9.8$ m/sec².

Equations (5.5) to (5.7) have been developed *only* for the case in which **s**, **v**, **v$_0$**, and **a** all lie along the same straight line. If these vectors have different directions but **a** is constant, the x components of the vectors satisfy these equations—and so do the y and z components. Indeed, Eqs. (5.5) and (5.6) are valid equations when the quantities are written as vectors, but Eq. (5.7) is not.

Example

Galileo is reputed to have shown that a heavy ball and a light one fall together by dropping them from the leaning tower of Pisa, which is 54 m high. How long does it take a ball to fall this distance? What is the speed of the ball just before it strikes the ground?

Let us take downward as positive. Then $a = 9.8$ m/sec², $s = 54$ m, and $v_0 = 0$.

By Eq. (5.6),

$$s = v_0 t + \tfrac{1}{2}at^2$$
$$54 \text{ m} = 0 + \tfrac{1}{2}(9.8 \text{ m/sec}^2)t^2$$
$$t^2 = 11.0 \text{ sec}^2$$
$$t = 3.32 \text{ sec}$$

By Eq. (5.7),

$$v^2 = v_0{}^2 + 2as$$
$$= 0 + 2 \times 9.8 \text{ m/sec}^2 \times 54 \text{ m}$$
$$= 1,060 \text{ m}^2/\text{sec}^2$$
$$v = 32.6 \text{ m/sec}$$

Check:

By Eq. (5.5), $$v_{\text{av}} = \frac{v_0 + v}{2} = \frac{0 + 32.6}{2}$$
$$= 16.3 \text{ m/sec}$$

By Eq. (5.1), $$s = v_{\text{av}}t$$
$$= 16.3 \text{ m/sec} \times 3.32 \text{ sec}$$
$$= 54 \text{ m} \qquad\qquad ////$$

Example

55

TRANSLATIONAL MOTION

A ball is thrown upward from a bridge with a speed of 48 ft/sec. It misses the bridge on the way down and lands in the water 160 ft below. Find how long the ball rises, how high it goes, how long it is in the air, and its velocity when it strikes the water.

Let us take upward as positive. Then $v_0 = 48$ ft/sec, and $a = -32$ ft/sec². The ball rises until $v = 0$.

By Eq. (5.3),

$$a = \frac{v - v_0}{t}$$

$$-32 \text{ ft/sec}^2 = \frac{0 - 48 \text{ ft/sec}}{t}$$

$$t = \frac{48 \text{ ft/sec}}{32 \text{ ft/sec}^2} = 1.5 \text{ sec}$$

During this 1.5 sec,

$$v_{av} = \frac{v + v_0}{2} = \frac{0 + 48 \text{ ft/sec}}{2} = 24 \text{ ft/sec}$$

and

$$s = v_{av}t = 24 \text{ ft/sec} \times 1.5 \text{ sec} = 36 \text{ ft up}$$

By Eq. (5.6),

$$s = v_0 t + \tfrac{1}{2}at^2$$

$$-160 = 48t + \tfrac{1}{2}(-32)t^2 \qquad \text{(Note that } s \text{ is negative.)}$$

$$t^2 - 3t - 10 = 0$$

$$t = +5 \text{ sec (or } -2)$$

(Clearly we want the positive answer here. The negative answer tells us how much before we threw the ball we would have had to fire it upward from the water to have it pass the bridge going 48 ft/sec at $t = 0$.)

By Eq. (5.3),

$$a = \frac{v - v_0}{t}$$

$$-32 \text{ ft/sec}^2 = \frac{v - 48 \text{ ft/sec}}{5 \text{ sec}}$$

$$v = -112 \text{ ft/sec} \qquad \text{(The minus sign means } \textit{downward.}\text{)}$$

Check:

$$s = v_{av}t = \frac{v + v_0}{2}t$$

$$-160 \text{ ft} = \frac{-112 + 48}{2} \text{ ft/sec} \times 5 \text{ sec} = -160 \text{ ft} \qquad ////$$

5.5 PATH OF A PROJECTILE FIRED HORIZONTALLY

If a body is projected horizontally from the top of a tower of height h (Fig. 5.6) with a velocity v_x, it continues to move with the same horizontal velocity it had at the beginning of its path (any decrease caused by

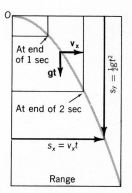

Figure 5.6
Path of a body projected horizontally. The downward component of the displacement at time t is $\tfrac{1}{2}gt^2$, while the horizontal component is $v_x t$.

the resistance of the air is neglected). At the same time the body falls because of the attraction of the earth. Hence, at any instant the velocity of the projectile has two components: a horizontal component which remains constant and a downward component which increases with the time. The horizontal component of the velocity and the horizontal displacement are independent of whether or not the body is falling. Similarly, the vertical components of the displacement, velocity, and acceleration are independent of whether or not the body is moving horizontally.

If a bullet is fired horizontally with a velocity of 700 m/sec, it falls 4.9 m in 1 sec, 19.6 m in 2 sec, and so forth, just exactly like a bullet which was simply dropped from the end of the gun. In this case the vertical velocity of the body after t sec is given by $v_y = gt$. The horizontal component of the velocity remains constant, so that the resulting velocity, which is tangent to the path of the body, has a magnitude $v = \sqrt{v_x{}^2 + v_y{}^2} = \sqrt{v_x{}^2 + g^2t^2}$. The horizontal displacement s_x of the body in time t is equal to $v_x t$, while the distance the body falls in the same time is $s_y = \frac{1}{2}gt^2$.

When an aircraft flying straight and level releases a bomb, the bomb continues to move forward with constant horizontal velocity until it strikes the ground (if air friction is neglected). The downward velocity component, which was zero at release, increases linearly with time because of the acceleration of gravity. In order to hit the target, the bomb must be released some time before the aircraft is directly above the target.

Example

A ball is thrown horizontally with a velocity of 20 m/sec from a tower 40 m high. Find the time of flight, the horizontal range, and the speed of the ball just before it strikes the ground.

Let us choose downward as positive. For the vertical motion, $a_y = 9.8$ m/sec^2, $v_{0y} = 0$, and $s_y = 40$ m. By Eq. (5.6),

$$s_y = v_{0y}t + \tfrac{1}{2}a_y t^2$$
$$40 \text{ m} = 0 + \tfrac{1}{2}(9.8 \text{ m/sec}^2)t^2$$
$$t = \tfrac{20}{7} = 2.86 \text{ sec}$$

The horizontal range s_x is the horizontal velocity (which is constant at 20 m/sec) multiplied by the time of flight.

$$s_x = v_x t = 20 \text{ m/sec} \times 2.86 \text{ sec} = 57.2 \text{ m}$$

Just before hitting, $v_x = 20$ m/sec, and, by Eq. (5.3),

$$v_y = a_y t = 9.8 \text{ m/sec}^2 \times \tfrac{20}{7} \text{ sec} = 28 \text{ m/sec}$$
$$v = \sqrt{v_x{}^2 + v_y{}^2} = \sqrt{(20)^2 + (28^2)} \text{ m/sec} = 34.4 \text{ m/sec} \qquad ////$$

5.6 PROJECTILE FIRED AT AN ANGLE WITH THE HORIZONTAL

If an object is given a velocity **V** at an angle θ with the horizontal, it follows a parabolic path (Fig. 5.7). If there were no gravity, this projectile would move along the line **V**t. In time t the effect of gravity is to bring

Figure 5.7
Path of a body projected at an angle θ
above the horizontal axis.

the projectile a distance $\frac{1}{2}gt^2$ below this line. In treating projectile problems we shall assume that air friction is negligible, that the horizontal component of the velocity remains constant, and that the vertical motion has uniform acceleration **g**. These are reasonable approximations for our purposes, but they are not entirely justified in many practical situations.

The first step in analyzing the motion of a projectile fired at an angle with the horizontal is to resolve the velocity of projection into vertical and horizontal components. If the velocity of projection **V** makes an angle θ with the horizontal, the horizontal component is $v_x = V \cos \theta$, the vertical component $v_y = V \sin \theta$. The horizontal velocity remains constant, but the vertical component changes. After time t the vertical component is $v_y = V \sin \theta - gt$. The projectile rises until the vertical component of the velocity becomes zero. At this instant the height is maximum, and the time of rise t_r is given by

$$0 = V \sin \theta - gt_r \qquad \text{or} \qquad t_r = \frac{V \sin \theta}{g}$$

The total time the projectile is in the air is the sum of the time of rise t_r and the time of fall t_f. Throughout the time of flight the horizontal component of the velocity remains constant at $V \cos \theta$. The horizontal range of the projectile is therefore $V \cos \theta \, (t_r + t_f)$. If the projectile falls to the same height from which it was projected, the time of fall is the same as the time of rise, and the time of flight is given by $t_r + t_f = (2V \sin \theta)/g$. In this case maximum range is attained when $\theta = 45°$ (Fig. 5.8) if air friction is negligible. Note that under these conditions the range is the same for any two complementary angles.

Figure 5.8
Paths of balls of the same initial speed (100 ft/sec) projected at different angles showing the positions of the balls after 1, 2, and 3 sec.

Example

A punter kicks a football with a velocity of 80 ft/sec at an angle of 37° with the horizon. Find the time during which the ball rises, how high it goes, and the length of the kick (from punter's toe to receiver's hands).

The vertical component of the initial velocity is (80 ft/sec)(sin 37°) = 48 ft/sec, and the horizontal component (80 ft/sec)(cos 37°) = 64 ft/sec. If we choose upward as positive for the vertical motion, $v_{0y} = 48$ ft/sec, and $a_y = -32$ ft/sec². At the top of the path $v_y = 0$; thus, by Eq. (5.3),

$$-32 \text{ ft/sec}^2 = \frac{0 - 48 \text{ ft/sec}}{t}$$

$$t = 1.5 \text{ sec}$$

On the way up, the average velocity is [by (Eq. 5.5)] 24 ft/sec. Since $s_y = v_{av}t$, we have $s_y = 24$ ft/sec × 1.5 sec = 36 ft, which is how high the ball goes.

It takes as long for the ball to come down as it did to go up, so that the total time in the air is 3.0 sec. In this time the horizontal velocity is constant at 64 ft/sec, so that

$$s_x = 64 \text{ ft/sec} \times 3.0 \text{ sec} = 192 \text{ ft or } 64 \text{ yd} \qquad \text{////}$$

Example

A stone is thrown from the edge of a cliff with a velocity of 16 m/sec at an angle of 30° above the horizontal. Four seconds later it strikes the ocean below. Find the height of the cliff above the ocean and the horizontal range of the stone.

The vertical component of the initial velocity is 16 sin 30° = 8 m/sec upward. If we choose upward as positive, $v_{0y} = 8$ m/sec, and $a_y = -9.8$ m/sec². The vertical distance down to the ocean is given by

$$\begin{aligned} s_y &= v_{0t}t + \tfrac{1}{2}at^2 \\ &= (8 \text{ m/sec} \times 4 \text{ sec}) + \tfrac{1}{2}(-9.8 \text{ m/sec}^2 \times 16 \text{ sec}^2) \\ &= 32 \text{ m} - 78.4 \text{ m} \\ &= -46.4 \text{ m} \end{aligned}$$

The horizontal range s_x is

$$s_x = v_x t$$

since v_x is constant.

$$v_x = 16 \text{ m/sec} \times \cos 30° = 13.9 \text{ m/sec}$$

$$s_x = 13.9 \text{ m/sec} \times 4 \text{ sec} = 55.6 \text{ m} \qquad \text{////}$$

1 Does a negative acceleration necessarily imply that a body is slowing down? What (or who) determines the sign of the acceleration in a particular problem?

2 Discuss why it is not possible in general to take account of the effects of air on the motion of a ball by simply reducing the value of g.

3 If a body has a uniform velocity, what is the relationship among its initial, final, and average velocities? If a body has a uniform acceleration, what is the relationship among its initial, final, and average velocities?

4 Under what circumstances may a body have (*a*) an acceleration but no velocity, (*b*) an acceleration but constant speed, (*c*) a velocity but no acceleration, and (*d*) an acceleration which is constantly changing?

5 Why are the sights on a rifle adjusted when the distance to the target is changed?

6 What is the effect of air friction on the vertical motion of a baseball? On the horizontal motion? Compare the path of the baseball assuming air friction with the ideal parabolic path by means of a sketch.

7 For a given initial speed a football kicked at 35° has the same range as one kicked at 55°. Why would the lower trajectory be better for a quick kick when the receiving safety man is close to the line of scrimmage? Why would the higher trajectory be preferred if the receiver is back to catch the ball?

8 If a body is thrown vertically upward and air resistance is not negligible, is the time of rise equal to the time of fall? Which is greater? Explain.

9 Why does an outfielder trying to cut off a run in a baseball game ordinarily try to get the ball to the catcher on the first bounce rather than on the fly?

10 Draw the path of a stone thrown backward with a speed of 40 ft/sec from the rear of a train traveling 50 ft/sec (*a*) from the point of view of the thrower and (*b*) from the frame of an observer on the ground.

11 What are the effects of air friction and the curvature of the earth on the motion of a long-range ballistic missile? What other factors must be taken into consideration?

12 Show that in the absence of air friction, the speed of a ball is the same at any two points on its trajectory which are at the same height. How do the velocities at these two points compare?

PROBLEMS

Unless otherwise specified, assume that air resistance is negligible and that g is 9.8 m/sec² or 32 ft/sec².

1 An airplane starting from rest has a uniform acceleration of 1.5 m/sec² south. What is its velocity at the end of 30 sec if this acceleration is maintained? *Ans.* 45 m/sec south

2 A uniformly accelerated body is moving with a velocity of 8 m/sec south; 5 sec later it has a velocity of 7 m/sec north. What is the acceleration?

3 A train has an acceleration of 0.75 m/sec² in a direction opposite that of its motion. How long a time will the train require to stop if it is initially going 30 m/sec (67 mi/hr)? How far will it travel in this time? *Ans.* 40 sec; 600 m

4 A freight train starts from rest, is uniformly accelerated, and travels 800 ft in 40 sec. Find the acceleration, the average speed, and the final speed.

5 An object passes the origin at $t = 0$ with a velocity of 21 m/sec to the right. It has a constant acceleration, and 12 sec later it is moving to the left with a speed of 15 m/sec. Calculate (*a*) the acceleration, (*b*) the position of the particle at $t = 12$ sec, and (*c*) the maximum displacement from the origin during the 12-sec period. *Ans.* (*a*) −3 m/sec²; (*b*) 36 m to right; (*c*) 73.5 m

6 A ball is dropped from a bridge. If it requires 4 sec for the ball to strike the ground below, how high is the bridge in meters? How fast is the ball traveling just before it reaches the ground?

7 A man has a 180-mi trip to make, of which the first third is over congested highways while the remainder is over a good freeway. If he averages 40 mi/hr over the first 60 mi, at what average speed must he drive the remaining 120 mi to average 55 mi/hr for the entire trip? *Ans.* 67.7 mi/hr

8   With what initial velocity must a ball be thrown upward to reach a maximum height of 20 m? How long is the ball in the air, assuming it is caught at the same height at which it was released?

9   A particle moves in a straight line with an acceleration of 3 m/sec². If it starts from rest, how far does it go in 4 sec? What is its velocity after 4 sec? How long a time is required for the particle to go 54 m? What is its speed as it passes the 54-m point?
Ans. 24 m; 12 m/sec; 6 sec; 18 m/sec

10   If a car going 60 km/hr (36 m/hr) can be stopped in 30 m, how much distance is required to stop the car when it is traveling 120 km/hr, assuming that the same constant acceleration is involved?

11   The takeoff speed of an airplane is 30 m/sec. What uniform acceleration must this plane have if it is to become airborne after a 150-m run? How long does the takeoff require?
Ans. 3 m/sec²; 10 sec

12   The maximum deceleration which the tires of an automobile can produce on a certain wet pavement is 15 ft/sec². Find the minimum distance in which a car traveling 100 ft/sec (68 mi/hr) can be stopped, measuring from the point at which the brakes are first applied. How long does it take?

13   An automobile traveling 30 m/sec can be stopped on a dry highway in a distance of 75 m, while on a wet pavement it requires 120 m, and on an icy pavement 600 m. (*a*) Calculate the average deceleration in each case. (*b*) Estimate the corresponding stopping distances for an initial speed of 15 m/sec. (*c*) If the reaction time required for a typical driver to set his brakes is 0.75 sec, calculate the total distance traversed after a driver traveling 30 m/sec on a wet pavement sees something requiring an emergency stop.
Ans. (*a*) 6, 3.75, and 0.75 m/sec²; (*b*) 18.8, 30, and 150 m; (*c*) 143 m

14   A naval aircraft is to be launched by a catapult from a carrier. If the catapult acts over a dis-

tance of 30 m, find the minimum average acceleration which will give the aircraft a speed of 50 m/sec at launching.

15   In March 1954 Col. J. P. Stapp rode a rocket-propelled sled down a track at 632 mi/hr. He blacked out but suffered no serious injury when he and the sled were stopped in 1.4 sec. Calculate the average acceleration in feet per second per second and in *g*'s.
Ans. −662 ft/sec²; 20.7 *g*'s

16   In a football game a passer throws the ball with a speed of 50 ft/sec. If he produces speed by moving the ball over a distance of 5 ft along a straight line, find the average acceleration during the throwing operation.

17   The speed of a baseball pitched by Bob Feller was once determined to be 145 ft/sec. What average acceleration must Feller have given the ball if this speed was produced over a distance of 9 ft?
Ans. 1,170 ft/sec²

18   A ball is batted straight up and returns to the level of the batter in 6 sec. How high does the ball travel? What are its velocity and acceleration at the top of its flight? What are its initial and final velocities?

19   A ball is thrown upward from the roof of a building. It has an initial velocity of 20 m/sec. How high is the building if on its downward flight it just misses the thrower and falls to the ground 5 sec after it is thrown? With what speed does it strike the ground?   *Ans.* 22.5 m; 29 m/sec

20   If an aircraft lands at a speed of 120 mi/hr (176 ft/sec) and the maximum deceleration which the braking system can produce is 12 ft/sec², find the minimum distance in which the plane can be stopped on level ground. How long does it take to bring the aircraft to rest?

21   A woman is reported to have survived a 144-ft fall from the seventeenth floor of a building when she landed on a metal ventilator box, which she crushed to a depth of 18 in. What was her average deceleration in feet per second per second and in *g*'s? (No doubt air friction was important here.)   *Ans.* 3,070 ft/sec²; 96 *g*'s

22   When an aircraft lands on a carrier, it is arrested by a restraining cable which is engaged by a tail hook as the aircraft passes over the cable. If the cable produces an average deceleration of 2.5 *g*'s (80 ft/sec²), how far does the aircraft move after it engages the cable at a speed of 120 ft/sec?

23   A basketball player can reach 8 ft when stand-

ing. With what speed must he jump to be able to reach 12 ft? *Ans.* 16 ft/sec

24 An automobile starts from rest, maintains a constant acceleration of 5 ft/sec² until its speed is 50 ft/sec, and then continues at this speed. Find the time required for the car to go 1,500 ft.

25 A rifle fires a bullet with a muzzle speed of 400 m/sec. If the rifle sights are aligned with the barrel, find how far above a bull's-eye 100 m away the rifle should be aimed to hit the center. (The usual practice is to adjust the sights to allow for the fall of the bullet for some prescribed target distance.) *Ans.* 0.306 m

26 A bullet has a speed of 2,000 ft/sec as it leaves a rifle. If it is fired horizontally from a cliff 24 ft above a lake, what is the horizontal range of the bullet?

27 A ski jumper takes off from a horizontal jump-off. He lands 11.0 m below the point of takeoff and 30 m from a vertical line through this point. Find the speed of the skier as he takes off.

Ans. 20 m/sec

28 A golf ball is driven horizontally with a speed of 90 ft/sec from a cliff 100 ft high, and it lands on a level plain below. How long is the ball is the air? How far from the base of the cliff does it strike the ground?

29 A boy on a bridge throws a stone horizontally with a speed of 35 m/sec, releasing the stone from a point 19.6 m above the surface of a river. How far from a point directly below the boy will the stone strike the water? *Ans.* 70 m

30 A baseball is batted, and 6 sec later it is caught by the center fielder a distance of 400 ft from the batter. (Assume the ball was hit and caught 4 ft above the ground; neglect air friction.) Find (*a*) the horizontal component of the ball's velocity and (*b*) the maximum height *above the ground* reached by the ball.

31 A boy kicks a soccer ball off the ground, giving it a speed of 15 m/sec at an angle of 53° with the horizontal. Find how long the ball is in the air, how high it goes, and the horizontal distance it traverses before it strikes the ground.

Ans. 2.45 sec; 7.35 m; 22.0 m

32 At the kickoff a football is given a velocity of 80 ft/sec at an angle of 53° with the horizontal. Find how long the ball is in the air, how high the ball goes, how far from the kickoff point the ball lands, and the magnitude of the velocity of the ball 3 sec after it is kicked. Compare your results with those of the first example in Sec. 5.6.

33 A golfer chips a ball over a sand trap to an elevated green, and 3 sec after he hits the ball it lands on the green 150 ft east and 36 ft above his initial lie. (*a*) Find the components of the initial speed of the ball as it left the club. (*b*) Find the speed of the ball when it strikes the green.

Ans. (*a*) 50 and 60 ft/sec; (*b*) 61.6 ft/sec

34 A golf ball is hit with a speed of 144 ft/sec at an angle of 30° with the horizontal from an elevated tee 40 ft above the fairway. How long is the ball in the air? At what horizontal distance from the tee does the ball strike the ground?

35 A baseball is batted with a speed of 120 ft/sec at an angle of 37° with the ground. Find how long the ball is in the air, how high the ball goes, how far from the plate it is caught, and the components of the velocity of the ball 4 sec after it is hit.

Ans. 4.5 sec; 81 ft; 432 ft; 96 and 56 ft/sec downward

36 A golf ball on a level fairway is given an initial velocity of 50 m/sec at an angle of 35° with the horizontal. If air friction is neglected, find (*a*) how long the ball is in the air, (*b*) how high it goes, and (*c*) how far away it lands.

37 A boy throws a stone toward the ocean with an initial velocity of 25 m/sec 37° above the horizontal. The stone just clears a fence at the edge of a cliff; the height of the fence is luckily the same as the height at which the stone is released. The stone lands in the ocean 50 m below the top of the fence. Find (*a*) the distance from the boy to the fence, (*b*) the horizontal distance from the fence to the point at which the stone hits the ocean, and (*c*) the speed of the stone the instant before it hits.

Ans. (*a*) 61.2 m; (*b*) 40.2 m; (*c*) 40.1 m/sec

38 A basketball player releases a ball 7 ft above the floor when he is 24 ft from the basket. The ball goes through the rim of the basket 10 ft above the floor 1.5 sec later. Find the horizontal component of the initial velocity, the vertical component of the initial velocity, and the maximum height above the floor reached by the ball.

39 A baseball is given a velocity of 100 ft/sec at an angle of 53° above the horizontal. It leaves the bat at a height of 4 ft above the ground. What are its height and distance from home plate (a) after 2 sec and (b) after 4 sec? (c) Will the ball clear a 20-ft fence 280 ft from the plate for a home run? By how much?
Ans. (a) 100 ft, 120 ft; (b) 68 ft, 240 ft; (c) yes, about 8 ft

40 Find the minimum speed with which a baseball must be hit at an angle of 53° with the horizontal to leave the bat at a height of 4 ft and clear a 40-ft screen at a distance of 320 ft from home plate. Assume air friction is negligible.

41 A projectile fired with initial velocity v_0 at an angle θ above the horizontal has only a downward acceleration of magnitude g in the flat-earth, frictionless approximation. (a) Write the vector equation giving the displacement **r** as a function of time. (b) Show that the equation of the trajectory is

$$y = x \tan \theta - \frac{gx^2}{2v_0^2 \cos^2 \theta}$$

Ans. (a) $\mathbf{r} = \mathbf{i}v_0 t \cos \theta + \mathbf{j}(v_0 t \sin \theta - gt^2/2)$

42 A golfer chips a ball high over a sand trap. If the ball leaves his club at an angle of 53° with the horizontal, what velocity must it be given to land 60 m distant? (Assume the green is at the same level as the initial lie of the ball.) How high does the ball go? How long is it in the air?

43 A boy riding a bicycle at a speed of 5 m/sec passes a parked car; 15 sec later the car starts and is uniformly accelerated for 10 sec, at which time it passes the cyclist. What was the acceleration of the car? *Ans.* 2.5 m/sec²

44 A ball is projected upward from the bottom of a tower that is 98 m high, and, at the same instant, another ball is dropped from the top of the same tower. If the balls meet at a point halfway between the top and the bottom of the tower, with what initial velocity was the ball projected upward? How high will this ball rise?

45 An automobile traveling 60 mi/hr in a 30 mi/hr zone passes a parked patrol car. If the automo-

bile maintains a constant speed and the patrol car a constant acceleration of 8 ft/sec², how long will it take the patrol car to catch the speeding vehicle, assuming the former starts as the latter passes by? *Ans.* 22 sec

46 Balls rolling on an inclined plane have an acceleration of 0.5 m/sec². One ball is released from rest 25 m up the plane and at the same instant a second ball at the bottom is rolled upward along the plane. If the second ball returns to the bottom at the same moment as the first, find the initial speed of the second ball and how far it rolled up the plane.

47 A ball is dropped from a window and is caught by a boy 44.1 m below. A second ball is tossed upward by another boy at the instant the first was dropped, and it is caught at the same instant the first ball is caught. How high did the second ball rise above the point from which it was tossed and caught? *Ans.* 11.0 m

48 A station wagon starts from rest and has a linear acceleration of 1.5 m/sec². A police car parked 100 m behind starts at the same instant and accelerates at 2 m/sec². Calculate (a) how long it took before the two vehicles were abreast and (b) the distance traveled by the station wagon.

49 An elevator moves upward at a speed of 10 ft/sec along the side of a building under construction. When the elevator is 20 ft above the ground, a boy on the ground throws a ball vertically upward with a speed of 60 ft/sec, releasing the ball 4 ft above the ground. Write expressions (numerical so far as possible) for the displacement, velocity, and acceleration as a function of time (a) for a reference frame with its origin on the ground and (b) a reference frame riding with the elevator.
Ans. (a) $y = 4 + 60t - 16t^2$ ft; $v = 60 - 32t$ ft/sec; $a = -32$ ft/sec²; (b) $y = -16 + 50t - 16t^2$ ft; $v = 50 - 32t$ ft/sec; $a = -32$ ft/sec²

50 Show that in the flat-earth, frictionless-atmosphere approximation the horizontal range R of a mass projected with initial speed V at an angle θ above the horizontal is given by $R = (2V^2 \sin \theta \cos \theta)/g$ and that the ratio of the maximum height reached to R is $(\tan \theta)/4$.

51 An object of mass m initially at rest at the top of a frictionless inclined plane of length L slides to the bottom in a time t. With what velocity must the body be started up the plane if it is to slide $0.81L$ before it starts back down? *Ans.* $1.8L/t$

WE WISH to be able to predict the motion of any body subjected to a known system of forces. We have learned how to determine the resultant of a system of forces, and we know that the acceleration of the body is zero if the force resultant is zero. But what happens when the resultant is not zero? The answer to this question was formulated in the seventeenth century by Isaac Newton as the second of his three celebrated laws of motion.

CHAPTER SIX
FORCE AND MOTION

6.1 NEWTON'S FIRST LAW OF MOTION

It is a familiar fact that the motion of a body is intimately related to the forces which act upon it. However, for centuries the way in which force and motion are related eluded the philosophers who speculated about the relationship. Aristotle expressed the idea that a body in motion comes to rest unless it has a force acting upon it continuously. For centuries men wondered what it was that pushed the planets around in the sky and what made the moon go around the earth. Although brilliant minds worked on the problem, no consistent and acceptable relationship between force and motion was established until the seventeenth century.

Galileo (1564–1642) performed a number of careful experiments on the motions of bodies; it is largely to his work that we owe the equations of kinematics in the preceding chapter. Galileo studied accelerated motion both by dropping bodies and by rolling balls down inclined planes. He observed that when friction was very small, a ball would roll for a great distance on a horizontal plane without stopping. Eventually, he became convinced that a ball on a perfectly frictionless horizontal plane would persist forever in its motion at constant speed. This revolutionary idea was one which was to prove most fruitful in developing an understanding of motion. Galileo discovered that bodies do not fall with constant speed; rather they have a *constant acceleration*. This suggested that the pull of the earth produced not the motion itself but the change in the state of motion.

Isaac Newton (1642–1727) accepted Galileo's conclusions and formulated this idea as his first law of motion, which may be stated as follows:

Every body continues in its state of rest or of uniform velocity in a straight line unless it is compelled to change that state by the application of some resultant external force.

No actual body is every completely free from external forces, but there are situations in which it is possible to make the resultant force

Figure 6.1
Newton's first law states that, unless acted upon by some resultant force, a body at rest remains at rest and a body in motion remains in motion with constant velocity.

Figure 6.2
Application of Newton's first law of motion.

approximately zero. In those cases we find that the body behaves in accordance with the first law of motion. Since we can never eliminate friction completely in our experiments, and since our efforts to compensate for it are imperfect, we must recognize that Newton's first law is an idealization. However, it is an idealization which provided the key for building a consistent and understandable theory of motion. There are, of course, many ways in which this great principle may be enunciated. Another statement is the following:

A body at rest remains at rest, and a body in motion remains in motion with constant velocity unless acted upon by some resultant force.

When a very small force acts on a reasonably massive body for a very short time, the resulting change in the velocity of the body is very small, and the zero-force requirement in Newton's first law is *almost* satisfied. Under these conditions we can snap a card from under a coin (Fig. 6.1) or jerk a tablecloth from under a glass. Newton's first law may also be invoked to explain how we put the head of a hammer onto the handle (Fig. 6.2).

6.2 INERTIA

In Newton's first law of motion there appears an important property of matter known as *inertia*—that property of matter by which it maintains a constant velocity in the absence of an unbalanced external force. When an automobile is suddenly stopped, the passengers obey Newton's first law and continue in motion with constant velocity until some external force changes their state of motion. Seat belts in an automobile can provide such an external force—one much preferred to that exerted by the windshield or dashboard. A man running on an icy sidewalk finds it difficult to stop suddenly because friction is inadequate to provide the necessary external force. When a baseball leaves the pitcher's hand, it continues to move with essentially constant velocity until it reaches the catcher's glove. No force is required to keep it moving. Of course, the ball is slowed down slightly by air resistance and is pulled toward the earth by its weight.

If a heavy ball is suspended from a string (Fig. 6.3) and an identical string is fastened below, the upper string breaks under a slowly increasing steady pull from below. But if a sudden jerk is applied to the lower string, it breaks. The inertia of the ball is so great that it isolates the upper string from the jerk.

All matter has inertia. The concept of mass was introduced by Newton as a measure of inertia. At any point on the earth the weight of any body is proportional to its mass, but if the body were in interplanetary space where it no longer had observable weight, it would still have inertia. It would still take the same force of toe on ball to give a football a specified acceleration. If the football were filled with mercury, its inertia would be greatly increased. To kick such a football would be just as painful in a rocket ship as on the earth, even though the mercury-filled ball might have little or no weight in interplanetary space.

If no external unbalanced force acts upon a body, it maintains a constant velocity. What happens if there is an external unbalanced force? To answer this question quantitatively, let us consider a set of idealized experiments.

1   Suppose that we have a perfectly level and frictionless table along which we can accelerate a mass of several kilograms. (We could make the friction very small by using small wheels with roller bearings.) If we now take an accurately calibrated spring balance and exert a force **F** on the mass, a certain acceleration is produced. Let us measure this acceleration. Next, let us exert exactly twice as great a force and again measure the acceleration (Fig. 6.4). We find the acceleration to be exactly double the first acceleration. If we again double the resultant force, the acceleration doubles once more and is thus four times the first acceleration. By measuring the acceleration for a large number of different resultant forces, we find that (within experimental error) the *acceleration* of our chosen mass *is directly proportional to the resultant force* **F** and its direction is that of the *resultant* force.

2   Next suppose that we choose a resultant force **F** and measure the acceleration it produces on a mass of 1 kg. Then let us keep the force constant but increase the mass accelerated to 2 kg (Fig. 6.5). We find the acceleration is half as great as it was the first time. If we increase the mass accelerated to 3 kg and measure the acceleration, we find it is one-third as great. If this experiment is performed for a number of different masses, we find that the data are consistent with the idea that *the acceleration* **a** *is inversely proportional to the mass accelerated.*

If we combine the results of these two series of experiments, we conclude that

$$\mathbf{a} = \frac{k\mathbf{F}}{m} \qquad (6.1)$$

This equation is a statement (in restricted form) of Newton's second law of motion. *The acceleration of a body is directly proportional to the resultant force acting upon it and is inversely proportional to the mass of the body.*

Figure 6.3
Inertia is one aspect of mass. A slow continuous increase in the downward force breaks the string above the heavy ball, but a sudden increase breaks the lower string.

Figure 6.4
The acceleration of a body is directly proportional to the resultant force acting on the body.

Figure 6.5
The acceleration due to a given resultant force is inversely proportional to the mass accelerated.

The proportionality constant k in Eq. (6.1) depends upon the units in which the force, the mass, and the acceleration are measured. By a suitable choice of these units one can make the constant k equal to unity. Such a choice results in convenience in handling the great structure of mechanics which is built around Newton's second law of motion. Unfortunately, many different schemes for making $k = 1$ have been devised. Each has advantages and disadvantages. In our work with Newton's second law and its consequences, we shall use two of the many possible choices—one for working in the metric system and a very different one for the English system.

6.4 THE NEWTON

In the metric system we have defined the fundamental units of mass (the kilogram), length (the meter), and time (the second). In order to write Newton's second law in the form

$$\mathbf{F} = m\mathbf{a} \tag{6.2}$$

we define a new unit of force called the *newton. One newton is that resultant force which produces an acceleration of one meter per second per second in a mass of one kilogram.*

The newton is the basic force unit in the mks system of units; 1 newton (N) = 0.2248 lb. The newton is a so-called "absolute" unit of force because its definition is made without any reference to the earth or its pull. From $\mathbf{F} = m\mathbf{a}$, we see that 1 N is equivalent to 1 kg-m/sec². The weight of 1 kg at sea level and 45° north latitude is 9.80665 N.

In dealing with tiny forces in the metric system of units, the *dyne* is sometimes used (1 dyne = 10^{-5} N). The dyne is the force required to produce an acceleration of one centimeter per second per second in a mass of one gram.

Example

An automobile has a mass of 1,600 kg. What force is required to give this automobile an acceleration of 1.20 m/sec² if there are frictional retarding forces totaling 200 N?

By Newton's second law, the unbalanced force is given by

$$F = ma$$
$$= 1{,}600 \text{ kg} \times 1.20 \text{ m/sec}^2$$
$$= 1{,}920 \text{ N}$$

This is the resultant, or unbalanced, force on the automobile. The total force is

$$1,920 \text{ N} + 200 \text{ N} = 2,120 \text{ N}$$ ////

6.5 GRAVITATIONAL UNITS OF FORCE

For the *British engineering system* of units an entirely different approach is adopted. The constant k in Newton's second law of motion is made equal to unity by choosing not a new unit of force but a new unit of mass.† In this system the basic unit chosen for acceleration is the foot per second per second, and the basic force unit is the pound, which is the gravitational attraction of the earth for a 1-lb mass at a place where the acceleration due to gravity has the value 32.17398 ft/sec². This force unit is known as a *gravitational unit,* since it is defined in terms of the pull of gravity on a standard object. Once we have chosen such a force unit and an acceleration unit, we make k equal to unity by choosing a new unit of mass. This unit, called the *slug* (*slug*gishness = inertia), is defined as follows: *One slug is the mass of a body which experiences an acceleration of one foot per second per second when acted upon by an unbalanced external force of one pound.*

A force of 1 lb gives a 1-lb mass an acceleration of 32.17 ft/sec² and gives a 1-slug mass an acceleration of 1 ft/sec². Therefore, the slug has a mass slightly more than 32 times that of the standard pound mass. In the British engineering system, forces are measured in pounds, masses in slugs, and accelerations in feet per second.

When the unbalanced force acting on a body is its weight $\mathbf{w}$, the resulting acceleration is $\mathbf{g}$, the acceleration due to gravity. For this case $\mathbf{F} = m\mathbf{a}$ becomes

$$\mathbf{w} = m\mathbf{g} \qquad (6.3)$$

If we divide Eq. (6.2) by Eq. (6.3), we obtain $\mathbf{F}/\mathbf{w} = \mathbf{a}/\mathbf{g}$, or

$$\mathbf{F} = \frac{w}{g}\mathbf{a} \qquad (6.4)$$

a form in which it is often convenient to put Newton's second law, especially when using the British engineering system of units.

Example

A 160-lb man stands in an elevator. Find the force exerted on the man by the floor of the elevator when the elevator (*a*) has an upward acceleration of 4.0 ft/sec², (*b*) has a downward acceleration of 4.0 ft/sec², and (*c*) has a constant downward velocity of 4 ft/sec.

a
$$F = ma = \frac{w}{g}a$$
$$= \frac{160 \text{ lb}}{32 \text{ ft/sec}^2} \times 4.0 \text{ ft/sec}^2 = 20 \text{ lb}$$

† An alternative giving rise to the British *absolute* system of units is to retain the pound as a unit of mass and to define a new unit of force called the *poundal.* One poundal is that force which produces an acceleration of one foot per second per second in a mass of one pound.

Figure 6.6
The action (force exerted on the trailer by the car) is equal and opposite to the reaction (force exerted on the car by the trailer).

This is the *unbalanced force* on the man. If E is the force exerted by the elevator, $F = E - w$, $20 = E - 160$, or $E = 180$ lb upward.

b   The unbalanced force is now 20 lb downward so that E is 20 lb less than w.

$$E = 160 - 20 = 140 \text{ lb}$$

c   If the velocity is constant, $a = 0$, and the unbalanced force is zero. The elevator exerts a force just equal to the man's weight.

$$E = w = 160 \text{ lb} \qquad\qquad ////$$

6.6   NEWTON'S THIRD LAW OF MOTION

In Sec. 3.2 we introduced Newton's third law in treating a system of bodies at rest. This law is equally valid in dealing with bodies in motion, either uniform or accelerated. The wheels of an automobile in motion push backward on the road, but the road pushes forward on the wheels with an equal force. *Whenever one body exerts a force upon a second body, the second body exerts an equal and opposite force on the first.*

Whenever a force acts upon a body, there is always an equal and opposite force exerted by the body (Figs. 6.6 and 6.7). A train pulls the locomotive back with a force which is exactly as great as the forward force which the locomotive exerts on the train. A helicopter pushes downward on the air with a force equal to that which the air exerts upward on the helicopter. The sun pulls on the earth and the earth pulls on the sun with equal and opposite forces. Newton stated his third law in the form: *To every action there is an equal and contrary reaction.* Here the term "action" is used to imply "force."

Direction of motion

Figure 6.7
The force of the water pushing on the nozzle is exactly equal and opposite to the force the nozzle exerts on the water.

6.7   APPLICATION OF NEWTON'S SECOND LAW

In a problem involving Newton's second law, the first question which we must ask ourselves is: To what body do we wish to apply $\mathbf{F} = m\mathbf{a}$? Once this question is answered, we must find *all the forces* which act on this body and determine their resultant $\mathbf{F}$. In many situations it is desirable to isolate the body and draw a diagram showing every force which acts on the body. Note that $\mathbf{F}$ is the resultant force, while m is the mass accelerated by $\mathbf{F}$.

Newton's second law can be applied to a system of bodies as well as to individual bodies. For example, it applies equally well to a system composed of an elevator, several people, and their luggage as to any one individual in the elevator.

Consider two masses m_1 and m_2 ($m_2 > m_1$) suspended by an inextensible string which passes over a frictionless pulley (Fig. 6.8). Such an arrangement is known as an *Atwood's machine*. We wish to find the tension T in the string and the acceleration a of the system. Since the string is inextensible, the accelerations of both masses have the same magnitude. We now apply Newton's second law to each mass separately. For m_2 there is a downward force m_2g and an upward force T; the resultant force is $m_2g - T$. For m_1, which rises, the unbalanced force is $T - m_1g$. Hence,

For m_2:
$$m_2g - T = m_2a$$

For m_1:
$$T - m_1g = m_1a$$

If we add these equations, we obtain

$$m_2g - m_1g = (m_1 + m_2)a$$

which is just what we obtain from applying Newton's second law to the system composed of m_1 plus m_2.

If $m_1 = 1.20$ kg and $m_2 = 1.80$ kg, we have

$$(1.80 \text{ kg} \times 9.80 \text{ m/sec}^2) - T = 1.80 \text{ kg} \times a$$
$$T - (1.20 \text{ kg} \times 9.80 \text{ m/sec}^2) = 1.20 \text{ kg} \times a$$

If we add these equations, we obtain

$$(17.64 - 11.76) \text{ kg-m/sec}^2 = 3.00 \text{ kg} \times a$$
$$a = 1.96 \text{ m/sec}^2$$

whence

$$T = 14.1 \text{ kg-m/sec}^2 = 14.1 \text{ N}$$

Figure 6.8
An Atwood's machine.

As a second illustration of the application of Newton's second law of motion, consider a mass M sliding along a smooth horizontal table (Fig. 6.9) with a second mass m fastened to it by means of a string that passes over a frictionless pulley.

Let T be the tension in the string, and a the acceleration of the system. For M the resultant force is T, since the weight Mg is exactly balanced by the upward force exerted by the table. Applying Newton's second law of motion to each mass separately, we obtain

For M:
$$T = Ma$$

For m:
$$mg - T = ma$$

If M weighs 8.0 lb, and m 2.0 lb, these equations become

$$T = \frac{8.0}{32} \text{ slug} \times a$$

$$2.0 \text{ lb} - T = \frac{2.0}{32} \text{ slug} \times a$$

from which $a = 6.4$ ft/sec^2, and $T = 1.6$ lb.

Next, let us suppose that the coefficient of kinetic friction between M and the table is 0.20. Then there is a frictional force (0.20×8.0) lb acting to the left, and the resultant force on M becomes $T - 1.6$ lb. Applying Newton's second law to the two masses now yields

Figure 6.9
Newton's second law of motion applies to the system as a whole and to each of the masses independently.

For M: $\qquad\qquad\qquad T - 1.6 \text{ lb} = \dfrac{8.0}{32} \text{ slug} \times a$

For m: $\qquad\qquad\qquad 2.0 \text{ lb} - T = \dfrac{2.0}{32} \text{ slug} \times a$

from which $a = 1.3 \text{ ft/sec}^2$, and $T = 1.9 \text{ lb}$.

6.8 MASS: ANOTHER POINT OF VIEW

Ordinarily, when we wish to determine the mass of a body, we do so by comparing its mass with that of standard masses on an equal-arm balance or by some equivalent method. The question may arise: Given a standard mass m_s, can we measure the mass of some other body by a method which is completely independent of the earth's gravitational pull? We can see that the answer is affirmative by considering an imaginary but fundamentally straightforward experiment.

Consider a spaceship located in a region in which there is no observable gravitational pull. In this ship imagine several bodies, one of which we arbitrarily agree is to be our standard mass, to which we assign a value m_s. To determine the mass m_1 of one of the other bodies, we allow it to interact with the standard body in the several different ways suggested by Fig. 6.10. In each case the interaction produces accelerations of magnitudes a_s and a_1 in opposite directions. We measure both these accelerations at the same instant and find that, for any interaction, the ratio a_1/a_s is always the same. We now define the ratio of the masses to be

$$\frac{m_s}{m_1} = \frac{a_1}{a_s} \tag{6.5}$$

so that $m_1 = (a_s/a_1)m_s$, or $m_1 a_1 = m_s a_s$, which is just what Newton's second and third laws predict.

(1) Compressed spring

(2) Electrical attraction

Figure 6.10
Equal but opposite forces are exerted on m_1 and m_s by one another when they interact. In the absence of any other net force on either body, $m_1 \mathbf{a}_1 = -m_s \mathbf{a}_s$, regardless of the particular way in which the bodies interact.

(3) Collision

In a similar way the mass m_2 of a second body can be determined by having it interact with the standard mass m_s. In this case $m_2 a_2 = m_s a_s$, where a_2 and a_s are the magnitudes of the accelerations of m_2 and m_s, respectively, when they are interacting only with each other. If m_2 and m_1 interact, we find $m_1 a_1 = m_2 a_2$. Thus, a consistent set of mass values can be assigned to any number of particles, even in a region in which none of the particles has any weight.

6.9 MOMENTUM AND NEWTON'S SECOND LAW

In Sec. 6.3 we introduced Newton's second law in a somewhat restricted form. In order to express this law in a more general form, we introduce linear *momentum*. The linear *momentum of a body is the product of the mass and the linear velocity.* Momentum is a *vector* quantity which has the direction of the velocity.

Example

A fullback with a mass of 90 kg is running with a velocity of 8 m/sec due north. Find his momentum.

Momentum = mass × velocity
$$= 90 \text{ kg} \times 8 \text{ m/sec north} = 720 \text{ kg-m/sec north} \quad ////$$

A force changes the momentum of a free mass upon which it acts. The rate at which the momentum of the body changes is directly proportional to the force. Indeed, Newton's second law may be written in the form: *The time rate of change of the momentum of a body is proportional to the net force acting upon the body and is in the direction of this resultant force.* As an equation, we may write

$$\mathbf{F} = \frac{\Delta(m\mathbf{v})}{\Delta t}\left(= \frac{d(m\mathbf{v})}{dt} \text{ in calculus notation}\right) \qquad (6.6)$$

When we deal with a single simple mass and a velocity small compared with that of light, the mass is constant, and we may write Eq. (6.6) as $\mathbf{F} = m\,\Delta\mathbf{v}/\Delta t = m\mathbf{a}$. However, there are several situations in which we must use Newton's second law in the form of Eq. (6.6). For example, at a speed approaching the speed of light, the mass of a body increases as the speed increases (Sec. 41.9). Similarly, if a sailor pulls a rope along a deck from a large coil, the mass in motion constantly increases. In rocket propulsion (Sec. 8.4) the mass of the rocket decreases as the hot gases are ejected. In each of these cases the mass being accelerated is not constant, and to find the velocity and displacement as functions of time we use Newton's second law in the form $\mathbf{F} = \Delta(m\mathbf{v})/\Delta t$.

QUESTIONS

1 Why is it that a piece of paper slips out from under a glass of water when the paper is jerked suddenly, yet the glass moves with it when the paper is pulled slowly?

2 Newton's third law says that action and reaction are always equal and opposite. If this is true, why don't they always cancel one another and leave no unbalanced force acting on any body?

3 If the same retarding force is available to stop a car regardless of its speed, how much is the stopping distance increased by doubling the speed?

4 Describe and discuss the problems of a man standing in a bus (*a*) when the brakes are suddenly applied and (*b*) when the bus has a large forward acceleration.

5 A book rests on a board which makes an angle of 30° with the horizontal. What forces act on the book? By what bodies are these forces exerted? What are the reactions to these forces, and upon what bodies are they exerted?

6 A serious neck injury is sometimes incurred by the driver of a stopped car which is struck from behind. Discuss the underlying physics.

7 Aristotle argued that if one stone lies on another, the upper one pushes down on the lower; as a consequence two stones fall faster than one, and a heavy stone falls faster than a light one. What is wrong with this argument?

8 One 0.5-kg mass is suspended from a spring balance, and an identical 0.5-kg mass is balanced on one pan of an equal-arm balance. If both masses are in an elevator, what happens in each case when the elevator is given an upward acceleration?

9 If a man and his parachute have a total mass *M* and he descends through the air at constant speed, what is the retarding force exerted by the air? How does the retarding force as the parachute opens compare with *Mg*?

10 What are the dimensions of mass in the British engineering system for which length, time, and force are the fundamental quantities?

11 A man stands on the platform of a large, sensitive scale. Discuss the reading as a function of time if the man tosses a heavy ball upward and catches the ball on its way down. How would things differ if he tossed the ball horizontally to a person standing nearby?

12 In city driving an automobile starts from rest, has a uniform acceleration until its speed is 15 m/sec, proceeds at constant speed for two blocks, and then decelerates uniformly to rest at a stop sign. Is the force required to stop the car equal in magnitude to the one which brought it up to speed? Justify your answer.

13 Show that if two bodies *A* and *B* initially at rest are acted upon only by their mutual interaction, the distance moved by *A* is to the distance moved by *B* as m_B is to m_A.

14 What is the fallacy in the assertion that you cannot throw a baseball because (*a*) by Newton's third law the ball exerts as big a force on you as you can exert on it, (*b*) the sum of these forces is zero, and (*c*) therefore, by Newton's second law, there is no acceleration?

15 Discuss the possibility of driving a sailboat by means of a blower which directs a stream of air against the sails. Which way would the boat be most likely to move? Why?

16 A frictionless plane in an elevator makes an angle θ with the horizontal. What is the acceleration of a block on the plane when the elevator has (*a*) constant velocity *v* upward; (*b*) constant velocity *v* downward; (*c*) constant acceleration *a* upward; (*d*) constant acceleration *a* (less than or equal to *g*) downward?

PROBLEMS

Use the approximate value $g = 9.8\ m/sec^2 = 32\ ft/sec^2$.

1 A Boeing 707 jet has a takeoff mass of 1.1×10^5 kg. Each of its four engines has a thrust of 77,000 N. Calculate the acceleration and the length of the ground run if the takeoff speed is 73 m/sec (140 knots).

Ans. 2.8 m/sec²; 952 m

2 What resultant force is required to give a 3,000-lb automobile an acceleration of 8 ft/sec²?

3 A football has a mass of 0.42 kg. What unbalanced force is required to give this football an acceleration of 400 m/sec²? If this force acts over a distance of 2 m, with what speed does the ball leave the passer?

Ans. 168 N; 40 m/sec

4 A man weighing 160 lb slides down a rope that can sustain only 135 lb. What is the smallest acceleration the man can have without breaking the rope?

5 A 70-kg man slides down a rope that serves as a fire escape. The maximum force that can be applied to the rope without breaking it is 600 N. Find the least acceleration that the man can have without breaking the rope. *Ans.* 1.23 m/sec²

6 A baseball has a mass of 0.145 kg. Find the average resultant force required to give this baseball a speed of 40 m/sec over a distance of 2.5 m.

7 A book rests on a horizontal seat in an automobile traveling 30 m/sec. If the coefficient of static friction is 0.25, calculate the shortest distance in which the car can be stopped without the book's sliding on the seat. *Ans.* 184 m

8 A catapult for launching aircraft from a carrier produces an acceleration of 3 g's (96 ft/sec²) on a 25,000-lb aircraft. Find the unbalanced force required to produce this acceleration. Through what distance must this unbalanced force act to give the plane a speed of 120 ft/sec?

9 A man exerts a constant force of 50 N for 5 sec on a 20-kg sled initially at rest on a frozen pond. If the coefficient of sliding friction is 0.1, find the acceleration of the sled while the man is pushing on it. What is the acceleration after he stops pushing? What is the maximum speed of the sled? How far does the sled move?
Ans. 1.52 m/sec²; −0.98 m/sec²; 7.6 m/sec; 48.5 m

10 A 5-kg block pulled along a level surface by a horizontal force of 24 N has an acceleration of 2 m/sec². Find the coefficient of friction. Draw a diagram showing all forces acting on the block.

11 A 192-lb man stands on a spring balance in an elevator. What does the balance read when the elevator has (a) a constant upward speed of 8 ft/sec, (b) a constant upward acceleration of 8 ft/sec², and (c) a constant downward acceleration of 8 ft/sec²?

Ans. (a) 192 lb; (b) 240 lb; (c) 144 lb

12 A man stands on a spring scale in an elevator. When the elevator is at rest, the scale reads

176 lb. What does it read when the elevator has an upward acceleration of 4 ft/sec²? A downward acceleration of 4 ft/sec²? What is the acceleration when the scale reads 187 lb?

13 A porter carries a 20-kg bag into an elevator. What force must he exert on the bag in order to hold it when the car is started with an upward acceleration of 2 m/sec²? When the car has a constant upward velocity of 2 m/sec? When the car has a downward acceleration of 2 m/sec²?

Ans. 236 N; 196 N; 156 N

14 The car of a frictionless elevator and its contents exert a force of 2 tons on the cables when at rest. How great is the force when an upward acceleration of 5 ft/sec² is being given to the elevator? When the acceleration is numerically the same but downward?

15 Two carts are in contact on a horizontal frictionless air track; cart A has a mass of 3 kg, and cart B a mass of 2 kg. (a) Find the force on A needed to give the two carts an acceleration of 1.2 m/sec². What is then the force exerted on B by A? (b) If the carts are given an equal acceleration in the opposite direction by pushing on B, find the force exerted on A by B. Explain why the force exerted on B by A in part (a) is not equal to the force exerted on A by B in part (b). *Ans.* (a) 6 N, 2.4 N; (b) 3.6 N

16 A string that can sustain a tension of 50 N is fastened to a mass of 4 kg lying on a horizontal table. What is the largest vertical acceleration that can be imparted to the mass without breaking the string?

17 The brakes of a 1,000-kg automobile can exert a retarding force of 3,000 N. Find the distance the car will move before stopping if it is traveling at the rate of 24 m/sec when the brakes are applied. *Ans.* 96 m

18 If the frictional forces on a 3,200-lb automobile add up to 180 lb retarding force, what would be the deceleration of the car coasting to a stop on a level road? How far would it coast from an initial speed of 60 ft/sec?

19 A 2-kg block rests on top of a 5-kg block which is on a table. If $\mu_k = 0.20$ and $\mu_s = 0.25$ for all

surfaces involved, find the largest horizontal force which can be applied to the lower block without having the upper block slide.

Ans. 30.9 N

20 A block of mass m slides down a plane that is inclined at 30° with the horizontal. If the coefficient of friction between the block and the surface of the incline is 0.2, find the acceleration of the block.

21 A 2-kg mass is placed on one of two identical carts initially at rest. The carts are then driven apart by an ideal massless spring and receive velocities of 6 and −10 m/sec. (*a*) Find the mass of the carts. (*b*) An unknown mass is added to the second cart, and the experiment is repeated; the observed velocities are now 4 and −2.5 m/sec, respectively. What mass was added? *Ans.* (*a*) 3 kg; (*b*) 5 kg

22 An 8-kg mass interacts with a 5-kg mass through a light stretched spring. If the 8-kg mass has an acceleration at some instant of 1.2 m/sec², find the net force on the 5-kg mass and its acceleration at that instant.

23 A 16,000-lb aircraft lands on a level field at a speed of 100 ft/sec. If the greatest acceleration which its brakes can produce is −8 ft/sec², find the minimum distance in which the aircraft can stop, the time required to stop, and the retarding force available.

Ans. 625 ft; 12.5 sec; 4,000 lb

24 A helicopter weighs 9,600 lb. Find the resultant force exerted on the helicopter by the air when it has an upward acceleration of 2 ft/sec². What is the force when the helicopter has a constant upward velocity of 2 ft/sec?

25 Each car of a toy train weighs 2 lb. In a train containing four cars and the engine, the engine pulls on the first car with a force of 3 lb. If the acceleration of the train is 5 ft/sec², with what force does the first car pull on the engine? What is the unbalanced force on the first car? What is the total frictional force on the four cars?

Ans. 3; 0.313; 1.75 lb

26 A 24,000-lb airplane starts from rest and has a uniform acceleration of 8 ft/sec². If it requires a speed of 90 ft/sec to take off, find (*a*) the minimum run for a takeoff, (*b*) the minimum time from rest to takeoff, and (*c*) the net thrust from the engine available for acceleration.

27 The three bodies A, B, and C of the accompanying figure have masses of 5, 2, and 3 kg, respectively. If they are given a vertical acceleration of 2 m/sec², find the tensions T_1, T_2, and T_3. *Ans.* 118, 59, and 35.4 N

28 If A, B, and C of the accompanying figure weigh 10, 6, and 8 lb, respectively, find the vertical acceleration, T_1, and T_3 if T_2 is 21 lb.

Problems 27 and 28

29 A 2-kg block rests on a horizontal frictionless table top 1.225 m high. A constant horizontal force F pushes the block from rest at a point 1.5 m from the edge of the table until it leaves the table with a speed of 3 m/sec. Find (*a*) the force F and (*b*) the horizontal distance from the table edge to the point at which the block reaches the floor. *Ans.* (*a*) 6 N; (*b*) 1.5 m

30 A boy tosses a book on a table. It slides off and lands 2.5 ft from the edge of the table, which is 4 ft high. (*a*) Find the horizontal velocity of the book as it left the table. (*b*) If the book had a speed of 8 ft/sec when it was sliding 3 ft from the edge of the table, what was the coefficient of kinetic friction between book and table?

31 What horizontal force is needed to push a 26-kg box up an inclined plane which rises 5 m in a distance of 13 m measured along the plane if the coefficient of friction is 0.25? What hori-

zontal force is required to give this box an acceleration of 2 m/sec² up the plane?

Ans. 157 N; 209 N

32 A block of mass 5 kg resting on a horizontal frictionless surface is connected to a hanging 3-kg block by a cord passing over a frictionless pulley (Fig. 6.9). Find the tension in the cord and the acceleration of the blocks.

33 An Atwood's machine (Fig. 6.8) consists of masses of 2 kg and 1.5 kg suspended from a massless, frictionless pulley. When released from rest, how long is required for the larger mass to descend 2 m? What is the tension in the cord during the descent?

Ans. 1.69 sec; 16.8 N

34 An elevator weighs 2,400 lb. A counterweight of 2,400 lb is attached to it, and the cable between the two passes over a pulley. A person weighing 200 lb steps onto the elevator. If all other forces are neglected, what is the acceleration of the system? The tension in the cable?

35 A brick with a mass of 2.5 kg is given an initial velocity of 6 m/sec up an inclined plane which makes an angle of 37° with the horizontal. If the coefficient of sliding friction is 0.40, find (*a*) the frictional retarding force, (*b*) the deceleration of the brick, (*c*) how far the brick moves up the plane, and (*d*) the acceleration of the brick as it slides back down the plane.

Ans. (*a*) 7.84 N; (*b*) 9.02 m/sec²; (*c*) 2.00 m; (*d*) 2.74 m/sec²

36 A 5-kg block resting on a horizontal surface is connected to a hanging block of 3-kg mass by a cord passing over a light, frictionless pulley. When the system is released, the blocks have an acceleration of 2 m/sec². Find the un-

balanced force on each mass, the tension in the cord, and the coefficient of friction between the 5-kg block and the horizontal surface.

37 Two bodies are suspended by means of a flexible string that passes over a weightless pulley. If one body weighs 13 lb and the other 11 lb, what are the acceleration of the system and the tension in the string? *Ans.* 2.67 ft/sec³; 11.9 lb

38 A mass of 10 kg rests on a horizontal table and is attached to a mass of 4 kg by means of a flexible cord passing over a weightless pulley (Fig. 6.9). If the coefficient of friction between the mass and the table is 0.3, what is the acceleration of the system? The tension in the cord?

39 (*a*) If $m_1 = 2$ kg, $M = 8$ kg, and $m_2 = 4$ kg, find the acceleration of the system and the tension in both cords of the accompanying figure, assuming that there is no friction and the pulleys are massless. (*b*) If the coefficient of friction between the table and mass M is 0.15, find the acceleration and the tension in both cords.

Ans. (*a*) 1.4 m/sec², 22.4 N, 33.6 N; (*b*) 0.56 m/sec², 20.7 N, 37.0 N

40 Find the acceleration and tension in both cords of the accompanying figure if m_1 is 0.2 slug, m_2 is 0.5 slug, M is 0.8 slug, and the coefficient of friction between M and the table is 0.25.

41 Find the acceleration of the blocks and the tension in the cord of the accompanying figure if A has a mass of 4 kg, B has a mass of 5 kg,

Problems 39 and 40

θ is 30°, ϕ is 37°, the planes are frictionless, and the pulley is massless.

> *Ans.* 1.09 m/sec²; 24.0 N

42 The planes of the accompanying figure are frictionless, and the pulley is massless. Find the acceleration of the blocks if A weighs 10 lb, B weighs 5 lb, θ is 30°, and ϕ is 37°.

Problems 41 and 42

43 Body A of mass 1 kg and body B of mass 2 kg are connected by a light string (see accompanying figure). If the coefficients of friction for A and B are, respectively, 0.3 and 0.2, find the acceleration of the system and the tension in the string. *Ans.* 4.05 m/sec²; 0.52 N

44 Body A of the accompanying figure weighs 10 lb, and the coefficient of kinetic friction for it is 0.5. Body B weighs 6 lb, and for it $\mu_k = 0.1$. Find the tension in the connecting cord.

Problems 43 and 44

45 A 20-lb block (A in the accompanying figure) on a 37° inclined plane is connected by a massless string which passes over a frictionless pulley to a 4-lb block B. If the 4-lb block has an upward acceleration of 5 ft/sec², find (*a*) the tension in the cord, (*b*) the frictional force on the 20-lb block, and (*c*) the coefficient of sliding friction.

> *Ans.* (*a*) 4.63 lb; (*b*) 4.25 lb; (*c*) 0.266

46 A 4-kg block A on a rough 30° plane is connected by a massless string which passes over a frictionless pulley to a 3-kg block B as shown in the accompanying figure. If $\mu_k = 0.2$, find the acceleration and the tension in the string.

47 If A has a mass of 6 kg in the accompanying figure, B a mass of 4 kg, θ is 25°, and the acceleration of the system is 0.5 m/sec², find the tension in the cord and the coefficient of sliding friction. Assume the pulley is frictionless.

> *Ans.* 37.2 N; 0.18

48 The angle of the rough plane in the accompanying figure can be varied from 0 to 90°. (*a*) If A has mass m_1, B mass m_2, the pulley is frictionless, and μ_k is the coefficient of kinetic friction between A and the plane, write an equation for the acceleration in terms of m_1, m_2, μ_k, g, and θ. (*b*) If $m_1 = 3$ kg, $m_2 = 2$ kg, and $\mu_k = 0.2$, find the acceleration for $\theta = 0, 20, 40, 60,$ and 80°. (*c*) Over what range of angles does friction either prevent sliding or make the speed decrease?

Problems 45 to 48

49 Show that for the frictionless, massless Atwood's machine of Fig. 6.8, the acceleration of each mass is $g(m_2 - m_1)/(m_2 + m_1)$ and the tension T is $2m_1 m_2 g/(m_1 + m_2)$.

50 A block slides with constant acceleration a down a plane inclined at an angle θ with the horizontal. Find the coefficient of sliding friction μ_k in terms of a, θ, and g.

THUS FAR we have discussed the motion of an object in terms of the forces acting upon it. However, in some cases we may not know what the forces are at every point in the path of a particle, for example, a neutron passing close to a uranium nucleus. In such a case we can still make many predictions about the motion by invoking the principle of conservation of energy, one of the great conservation laws of nature. Before we can apply the conservation of energy in any problem, we must learn what energy is and in what forms it appears. This in turn requires that we define and become familiar with *work* as this word is used in physics.

CHAPTER SEVEN
WORK AND ENERGY

7.1 WORK

The word *work* is a familiar one, used in many everyday senses. In physics, however, it is used in a restricted and carefully defined manner. When a force **F** acts upon a body to produce a displacement **s**, the *work done by the force is defined†ₐ as the product of the displacement and the component of the force in the direction of the displacement.* If the angle between the displacement vector and the force vector is θ (Fig. 7.1), the component of the force in the direction of the displacement is $F \cos \theta$, and the work W is, by definition,

$$W = Fs \cos \theta \qquad (7.1)$$

If the force does not produce a displacement, no work is done in the sense in which work is used in physics. A man holding a 10-lb weight at rest does no work on it. A desk or cement post could hold the weight indefinitely without any difficulty. Although the man holding the weight does *no work on it*, the muscles in his arm do stretch and contract, and thus work is done internally, which may result in fatigue. Similarly, when a man walks at a constant horizontal velocity carrying

† In vector notation $W = \mathbf{F} \cdot \mathbf{s} = Fs \cos \theta$, where $\mathbf{F} \cdot \mathbf{s}$ is called the *scalar* or *dot product* of **F** and **s**.

Figure 7.1
The work W done by a force **F** acting through a displacement **s** is $Fs \cos \theta$, where θ is the angle between **F** and **s**.

a suitcase, he does no work *on the suitcase*, since the force he exerts on the suitcase and the displacement are mutually perpendicular.

7.2 UNITS OF WORK

Since work is measured by the product of force and displacement, its units involve a unit of force multiplied by a unit of length. In the metric system the basic units of force and distance are the newton and the meter. Work is measured in newton-meters; this unit has been named the *joule* (J) in honor of James Prescott Joule, a distinguished British physicist, whose work on the relation between heat and mechanical work was of great importance. *One joule is the work done when a force of one newton acts through a distance of one meter.*

There are many other units in which work may be measured. In the British engineering system force is measured in pounds, and the displacement in feet; the resulting unit of work is the *foot-pound*. *One foot-pound is the work done when a force of one pound acts through a distance of one foot.* Large amounts of work are sometimes reported in ton-miles. Small amounts in the metric system are sometimes measured in ergs; 1 erg is equal to 10^{-7} J.

Example

How many joules of work are done by a force in lifting a mass of 2 kg upward a distance of 3 m?

$$\text{Work in joules} = \text{force in newtons} \times \text{distance in meters}$$
$$F = mg = 2 \text{ kg} \times 9.8 \text{ m/sec}^2 = 19.6 \text{ N}$$
$$W = 19.6 \text{ N} \times 3 \text{ m} = 58.8 \text{ J} \qquad \text{////}$$

Example

A force of 10 lb is used to move a box across a horizontal floor a distance of 5 ft. If the force makes an angle of 30° with the floor, how much work is done?

$$\text{Work} = Fs \cos \theta$$
$$W = 10 \text{ lb} \times 5 \text{ ft} \times 0.866$$
$$= 43.3 \text{ ft-lb} \qquad \text{////}$$

7.3 POWER

In physics, *power is the time rate of doing work;* the average power $\mathcal{P}_{av}$ is equal to work performed divided by the time required:

$$\mathcal{P}_{av} = \frac{\text{work}}{\text{time}} = \frac{W}{t} \qquad (7.2)$$

An engine of high power can do work rapidly.

In the mks system power is measured in joules per second or *watts* (W) (in honor of James Watt, developer of the steam engine). One watt of power is expended when one joule of work is done each second. In the British engineering system power may be expressed in foot-pounds per second, but more often it is given in *horsepower*; 1 hp = 550 ft-lb/sec = 745.7 W.

The origin of the horsepower is of interest. When James Watt tried to sell steam engines for British coal mines, the question arose of how many horses one of these new engines would replace. Watt found that, on the average, the horses were doing about 550 ft-lb of work per second; he called this unit the horsepower. He measured the rate at which his steam engines could work and thus rated them in "horsepower."

Example

A building crane lifts a 1,500-lb steel beam to a height of 44 ft in 10 sec. Find the average power developed.

$$\mathcal{P}_{av} = \frac{work}{t} = \frac{1,500 \text{ lb} \times 44 \text{ ft}}{10 \text{ sec}} = 6,600 \text{ ft-lb/sec}$$

$$= 6,600 \frac{\text{ft-lb}}{\text{sec}} \times \frac{1 \text{ hp}}{550 \text{ ft-lb/sec}} = 12 \text{ hp} \qquad ////$$

Example

An electric motor exerts a force of 400 N on a cable and pulls it a distance of 30 m in 1 min. Find the power supplied by the motor.

$$\mathcal{P}_{av} = \frac{work}{t} = \frac{400 \text{ N} \times 30 \text{ m}}{60 \text{ sec}} = 200 \text{ J/sec}$$

$$= 200 \text{ W} \qquad ////$$

The instantaneous power delivered by a force **F** which acts in the direction of the motion is

$$\mathcal{P} = \lim_{\Delta t \to 0} \frac{F \, \Delta s}{\Delta t} = Fv \qquad (7.3)$$

When the force and velocity are constant, the instantaneous power is equal to the average power. Equation (7.3) is useful in resolving such practical questions as how powerful a motor must be provided to lift an elevator at a specified speed or how many diesel units are required to pull a railroad train at a given speed. If one wishes to make the speed of an airplane or a ship greater, Eq. (7.3) shows that there are basically two things one can do: increase the power and decrease the retarding forces.

Example

A Boeing 747 aircraft has four engines each capable of exerting a thrust of 200,000 N (45,000 lb). When the aircraft is cruising at 1,000 km/hr (641 mi/hr), each engine is delivering 3×10^7 W (40,300 hp). Find the drag. What fraction of its maximum thrust is each engine delivering?

Metric:

$$\mathcal{P} = Fv$$

$$4 \times 3 \times 10^7 \text{ W} = F_{\text{drag}} \times \left(1{,}000 \times \frac{1{,}000}{3{,}600}\right) \text{m/sec}$$

$$F_{\text{drag}} = 4.32 \times 10^5 \text{ N}$$

$$\frac{4.32 \times 10^5}{4 \times 2 \times 10^5} = 0.54 = 54 \text{ percent}$$

British engineering:

$$4 \times 40{,}300 \text{ hp} \times \frac{550 \text{ ft-lb/sec}}{1 \text{ hp}} = F \times 621 \text{ m/hr} \times \frac{1.467 \text{ ft/sec}}{1 \text{ mi/hr}}$$

$$F = 9.76 \times 10^4 \text{ lb}$$

$$\frac{9.76 \times 10^4}{4 \times 45{,}000} = 0.54 \qquad \qquad ////$$

7.4 ENERGY AND ITS CONSERVATION

Energy is defined as the ability or capacity to do work. It occurs in many forms. A swinging hammer can do work by virtue of its motion; energy associated with motion is known as *kinetic energy.* A raised pile driver can do work by virtue of its elevated position; it has what we call *potential energy.* When we buy gasoline and food, we buy chemical energy. We purchase electric energy to obtain work from electric motors as well as for heat and light. The earth's primary source of energy lies in the electromagnetic radiation it receives from the sun. Much of physics involves the relationships between the many forms of energy and the transformations from one form to another.

One of the most important principles of nature and one of the most powerful generalizations of science is the *law of conservation of energy:*

Energy cannot be created or destroyed; it may be transformed from one form into another, but the total amount of energy never changes.

As an example of the transformations which energy may undergo, consider radiation coming to the earth from the sun. Some of this energy falls on plants, where it is transformed into chemical energy through photosynthesis. The energy stored in the plant may be converted eventually into coal or oil, or the plant may be eaten by some animal which utilizes the energy to carry on its existence. Part of the radiant energy from the sun goes into evaporating water from the ocean. The water vapor, lifted high above the earth, eventually returns in the form of rain which may be trapped behind a dam. The water thus stored has potential energy, which may be converted by use of a giant turbine into energy of motion. This kinetic energy may in turn be converted into electric energy in a generator. Electric energy is distributed to houses, where it is converted into heat or light or used to perform work through an electric motor. Throughout all these transformations, the total energy remains constant.

Early in the twentieth century it was found that mass itself could be converted into energy; in this sense mass is one of the forms or manifestations of energy. The sun's mass is decreasing because mass energy is converted into radiant energy in the sun. In nuclear reactors mass is converted into energy. Einstein showed that whenever mass is transformed into another form of energy (or vice versa), the equivalence can be expressed by the equation

$$E = mc^2 \qquad (7.4)$$

where E is the amount of some other form of energy appearing or disappearing, m is the mass disappearing or appearing, and c is the speed of light.

Example

What is the total energy in a gram of matter at rest?

$$m = 10^{-3} \text{ kg}$$

The speed of light c is 3×10^8 m/sec, and $c^2 = 9 \times 10^{16}$ m²/sec².

$$E = mc^2 = 9 \times 10^{13} \text{ kg-m}^2/\text{sec}^2 = 9 \times 10^{13} \text{ J}$$

This energy is sufficient to lift a mass of 1 million kg through a distance of 9,000 km against the action of gravity! ////

Energy is an important commodity in economics and commerce. It is bought and sold in many forms—as coal, petroleum products, food, and electricity. Energy is used to operate automobiles, refrigerators, stoves, washing machines, and television sets. In the past century the amount of energy consumed per person has undergone a remarkable increase. Primitive man used energy only for his food and warmth; modern man has hundreds of appliances which transform energies for his entertainment, comfort, and convenience.

7.5 POTENTIAL AND KINETIC ENERGY

In mechanics we are concerned primarily with two forms of energy—*potential energy* (PE) and *kinetic energy* (KE).

Kinetic energy is the energy a body possesses by virtue of its motion. Any body in motion can set other bodies in motion by colliding with them. The moving head of an axe does work in splitting a log. The bullet leaving the muzzle of a gun has kinetic energy and can do work in penetrating a board.

To find the kinetic energy a body possesses, we consider the work which must be done on the body in order to give it its speed. When the body is stopped, it gives up this amount of energy. By definition, this is its kinetic energy. Consider a mass m, initially at rest, upon which a constant force F is applied through a displacement s in the direction of the force. The work done on the body is Fs. By Newton's second law, $F = ma$; thus the work is equal to mas. Since the acceleration of the body is constant, Eq. (5.7) is applicable and $v^2 = 2as$, since $v_0 = 0$. If we replace as with $v^2/2$ and recall that the work done appears as kinetic energy, the kinetic energy is

$$KE = \tfrac{1}{2}mv^2 \qquad (7.5)$$

In metric units we express the kinetic energy in joules (or the equivalent kg-m²/sec²), while in the British system the kinetic energy is given in foot-pounds (or the equivalent slug-ft²/sec²).

Example

If a 1,000-kg automobile is moving with a speed of 20 m/sec, what is its kinetic energy in joules?

$$KE = \tfrac{1}{2}mv^2 = \tfrac{1}{2}(1,000 \text{ kg})(20 \text{ m/sec})^2$$
$$= 2 \times 10^5 \text{ kg-m}^2/\text{sec}^2 = 2 \times 10^5 \text{ J} \qquad ////$$

Example

What force is required to stop a bullet that has a mass of 15 g and a velocity of 400 m/sec in a distance of 20 cm?

$$\text{Force} \times \text{distance} = \text{change in KE}$$
$$0.20 \text{ m} \times F = \tfrac{1}{2}(0.015 \text{ kg})(400 \text{ m/sec})^2$$
$$= 1,200 \text{ kg-m}^2/\text{sec}^2 \text{ (or joules)}$$
$$F = \frac{1,200 \text{ kg-m}^2/\text{sec}^2}{0.2 \text{ m}}$$
$$= 6,000 \text{ kg-m/sec}^2 = 6,000 \text{ N} \qquad ////$$

The energy that a body (or system of bodies) has by virtue of its position or configuration is called potential energy (Fig. 7.2). When a mass is lifted above the surface of the earth, it has energy because of its position. When a spring is compressed or a bow bent, potential energy is stored up. Other examples of potential energy are found in the mainspring of a watch and in a stretched rubber band. In these cases the material of a body is in a state of strain; because of this strain, the body possesses potential energy.

The measure of the potential energy which a body has because of its elevated position is the work done against gravity in lifting the body from some level chosen as the zero for potential energy. The upward force required is equal to the weight of the body w, and the work done in lifting the body through a height h is given by the product wh; therefore the potential energy is

$$PE = wh = mgh \qquad (7.6)$$

Figure 7.2
Examples of potential energy are the taut bow and the elevated weight.

Example

A block weighing 3 lb is lifted 6 ft against gravity. What potential energy is stored?

$$PE = wh = 3 \text{ lb} \times 6 \text{ ft}$$
$$= 18 \text{ ft-lb} \qquad ////$$

Example

Find the potential energy given to the 50-kg hammer of a pile driver when it is raised 4 m.

$$PE = mgh = 50 \text{ kg} \times 9.8 \text{ m/sec}^2 \times 4 \text{ m}$$
$$= 1{,}960 \text{ J} \qquad ////$$

A second important example of potential energy is that associated with a stretched or compressed spring. Consider the spring of Fig. 7.3. Let its length when there is no load applied be l_0. If we add a load w_1, we find a new length l_1 for the spring. The change in length $l_1 - l_0$ we denote by y_1; in general, we shall represent the extension of the spring by y. If we now double the load on the spring, we find that y is twice as great. The fact that the change in length is proportional to the applied force was first reported by the English physicist Robert Hooke in 1678 in the words "Ut tensio, sic vis" (as the extension, so the force). In general, the force we apply F_a is proportional to the extension y; thus

$$F_a = ky \qquad (7.7)$$

where k is known as the *force constant* of the spring. For many springs Eq. (7.7) is applicable for compression as well as for extension of the spring; y for such a *linear* spring is proportional to, and in the direction of, the applied force.

The potential energy of a stretched (or compressed) ideal spring is the amount of work the spring can do, and this is equal to the work required to distort the spring. To produce a displacement y the applied force F_a is zero at first and increases linearly to ky. The work W done by F_a is the product of the displacement y and the average force $(0 + ky)/2$. The potential energy P stored is equal to this work, and so

$$P = \tfrac{1}{2}ky^2 \qquad (7.8)$$

The potential energy of a stretched or compressed ideal spring is proportional to the square of the displacement y *relative to the configuration in which* $y = 0$. Note that y^2 is positive regardless of whether y is positive or negative. Thus energy is stored either by stretching or by compressing a spring.

It is extremely important to remember that potential energy is always measured relative to some configuration chosen as the zero. For a mass suspended from a spring, the equilibrium position (Fig. 7.4) is the usual choice for this zero configuration. For the gravitational potential energy mgh, the height h is measured relative to some chosen zero level. When we deal with energy of motion, the kinetic energy

Figure 7.3
Elongation of linear spring is proportional to the stretching force.

depends on the reference frame relative to which we measure velocities. A passenger seated in an aircraft traveling 1,000 m/sec has no kinetic energy relative to a set of axes tied to the plane but much kinetic energy relative to axes fixed on the earth.

7.6 ENERGY TRANSFORMATION AND THE WORK-ENERGY THEOREM

The potential energy of a body due to its position depends upon its height h, and this in turn depends on what one wishes to call the zero for height. A book on a table has no potential energy *relative to the tabletop*, but it does have potential energy *relative to the floor* and *relative to mean sea level*. If the book weighs 2 lb and the tabletop is 3 ft above the floor, the potential energy of the book is 6 ft-lb *relative to the floor*. Potential energy is always determined *relative* to some level or configuration which is assigned the value of zero. The choice of this level is at the disposal of the person working the problem.

In many situations we have a simple transformation of potential into kinetic energy with no other forms of energy involved. For example, when a pile driver is lifted into position, it is given potential energy. This potential energy is transformed into kinetic energy before the hammer strikes the pile upon which it is to do work. Similarly, consider water going over a dam (Fig. 7.5). Just above the dam the water has potential energy and (if its velocity is negligible) no significant kinetic energy. As the water falls, its potential energy decreases and its kinetic energy increases. When the water strikes bottom, part of this kinetic energy is transformed into heat.

The swinging of a pendulum illustrates how potential energy may be transformed into kinetic energy and then back to potential energy (Fig. 7.6). When a pendulum is pulled to one side, the mass has no velocity and no kinetic energy, but it does have potential energy. When the mass is released, it swings downward, acquiring kinetic energy and losing potential energy. At the bottom of the swing the pendulum bob has maximum kinetic energy. As it swings upward, this kinetic energy is retransformed into potential energy. A very similar transformation of energy can be traced in the case of a mass on the end of a spring (Fig. 7.4).

Figure 7.4

Relative to the equilibrium position as the configuration of zero potential energy, the system has positive potential energy when the mass on the end of a spring is lifted a distance y. After release, this energy is transformed into kinetic energy and then back to potential energy at the low point of the motion.

Example

A pendulum bob is pulled to one side until its center of gravity has been raised 10 cm above its equilibrium position. Find the speed of the bob as it swings through the equilibrium position.

$$\text{PE at top} = \text{KE at bottom}$$
$$mgh = \tfrac{1}{2}mv^2$$

whence $\quad v^2 = 2gh$
$$= 2 \times 9.8 \text{ m/sec}^2 \times 0.1 \text{ m} = 1.96 \text{ m}^2/\text{sec}^2$$
$$v = 1.4 \text{ m/sec}$$

Consider a boy on a sled at the top of a hill. Let P_i and K_i be the initial potential and kinetic energies, respectively. Suppose that the boy's father pushes on the sled to speed it up, doing an amount of work W_{in}. As the sled goes down the hill, its potential energy decreases, its kinetic energy increases (at least for a while), and the sled does some work W_{out} against friction. The law of conservation of energy requires that the sum of the initial potential energy, the initial kinetic energy, and the *work done on the sled* (the work *input*) be equal at any time to the sum of the potential energy P_f, the kinetic energy K_f, and the *work done by the sled* (the work *output*); thus

$$W_{in} + P_i + K_i = P_f + K_f + W_{out} \qquad (7.9)$$

The law of conservation of energy as expressed in Eq. (7.9) is known as the *work-energy theorem;* it is applicable not only to the sled and to other individual bodies but also to systems of bodies. In the latter case K_i is the sum of the initial kinetic energies, P_i the initial potential energy for the system, W_{in} the *total external work done on the system*, and so forth. In most of our problems one or more of the quantities of Eq. (7.9) will be zero; indeed, we frequently deal with situations in which only one quantity on each side of the equation is nonzero.

7.7 ENERGY IN THE HUMAN BODY

The human body is capable of doing work. The burning of carbohydrates, fats, and proteins provides energy for performing work as well as for maintaining the body temperature and for carrying on its internal vital functions. The pumping of blood through the veins and arteries is an illustration of the internal work performed in the body. This work is done by the heart muscle. For a man at rest the heart normally beats about 70 times per minute and develops a power of about 1.3 W. An average working man uses about 10 million J of energy per day, of which about half goes into maintaining temperature and carrying on internal functions. The remainder is available for the performance of external work.

7.8 SIMPLE MACHINES

To aid us in doing work we frequently make use of simple machines. A machine is a device for overcoming a resisting force at one point by the application of a force at some other point. Consider a specific example: Suppose a man wishes to move a boulder which is too large to shove. He might find a strong stick and use it to move the boulder (Fig. 7.7). If he pulls on one end of the stick with a force F_{in}, a very

Figure 7.5
The potential energy of the water at the top of the dam is converted into kinetic energy at the bottom.

Figure 7.6
Transformation of potential energy into kinetic energy and then back into potential energy.

Figure 7.7
The lever was one of the first simple machines used by man.

F_{in}

Fulcrum

F_{out}

Figure 7.8
The human forearm is a lever with mechanical advantage much less than unity.

F

w

Figure 7.9
In raising the load w, work must be done to lift the lower pulley block and to overcome friction.

much larger force F_{out} is exerted on the boulder. When the stick is in equilibrium, the torque condition gives the relation $F_{in}a = F_{out}b$. If a is 20 times b, F_{out} is 20 times F_{in}. By using this lever, our man can exchange an input force F_{in} for an output force F_{out} 20 times as great.

At first glance it might seem that this violates the principle of conservation of energy, but a little further consideration shows that this is not the case. In order to move the boulder 1 cm, the input force must be exerted for a distance a/b cm. The work done by the man on the stick is equal to the work done on the boulder by the stick. By the use of a simple machine, *it is possible to exchange a small force acting through a large distance for a large force acting through a small distance,* or vice versa.

There are many kinds of simple machines. Among them are levers, inclined planes, pulleys, gear wheels, screws, wheels and axles, the wedge, and the differential pulley. Complicated machines, such as the automobile or the mechanical cotton picker, are composed of large numbers of interconnected simple machines.

7.9 MECHANICAL ADVANTAGE AND EFFICIENCY

As we have seen, a simple machine is a device for providing some output force in return for an input force. *The ratio of the output force F_{out} to the input force F_{in} is the actual mechanical advantage* (AMA) *of the machine:*

$$AMA = \frac{F_{out}}{F_{in}} \tag{7.10}$$

In defining the actual mechanical advantage of the machine, we concern ourselves only with the magnitudes of the forces and not with their directions.

In our example of a man moving a boulder, we used a lever which had a large mechanical advantage. It is by no means true that all useful machines have large mechanical advantages. The simple pulley has a mechanical advantage of essentially unity and serves only to change the direction of the force. However, if we are trying to lift some mortar to the top of a building, it is convenient to be able to stand on the ground and pull downward on the rope to lift the mortar. Some of our most useful machines have mechanical advantages of less than 1. In the human forearm (Fig. 7.8) the output force exerted by the hand is very much smaller than the input force exerted by the muscles, so that the mechanical advantage is much less than unity. In such a machine we have a smaller output force moving through a greater distance than the input force.

In real machines friction opposes the motion, and work must be done to overcome this friction. In addition, it may be necessary to raise some part of the machine which gives no useful work output. For example, in Fig. 7.9 the lower pulley block must be raised when we lift the load w. The work done against friction and in lifting parts of the machine does not appear as useful work output. Therefore, in any real machine the useful work output is less than the work input. The *efficiency* (Eff) *of a machine is the ratio of the useful work output to the work input.*

$$\text{Eff} = \frac{\text{work}_{\text{out}}}{\text{work}_{\text{in}}} = \frac{W_{\text{out}}}{W_{\text{in}}} \qquad (7.11)$$

For an ideal machine, with no friction and massless moving parts, the efficiency would be 100 percent. The mechanical advantage of an ideal machine is called the *ideal mechanical advantage* (IMA). Let d_{out} be the distance moved by the output force and d_{in} the distance moved by the input force. For an ideal machine the efficiency is 100 percent, and $W_{\text{out}} = W_{\text{in}}$; hence, $F_{\text{out}}d_{\text{out}} = F_{\text{in}}d_{\text{in}}$. Therefore, *the mechanical advantage $F_{\text{out}}/F_{\text{in}}$ for an ideal machine is $d_{\text{in}}/d_{\text{out}}$.* This ratio is the *ideal mechanical advantage.*

$$\text{IMA} = \frac{d_{\text{in}}}{d_{\text{out}}} \qquad (7.12)$$

For any real machine the ideal mechanical advantage is greater than the actual mechanical advantage. Indeed, the efficiency can be seen to be equal to the ratio of the actual mechanical advantage to the ideal mechanical advantage by observing that

$$\text{Eff} = \frac{W_{\text{out}}}{W_{\text{in}}} = \frac{F_{\text{out}}}{F_{\text{in}}}\frac{d_{\text{out}}}{d_{\text{in}}} = \frac{\text{AMA}}{\text{IMA}}$$

The concepts of actual mechanical advantage, ideal mechanical advantage, and efficiency may be applied to all types of simple machines.

7.10 ROTATING SYSTEMS

When a pulley is being driven by a belt (Fig. 7.10), the tensions in the straight parts of the belt are not equal. They differ by the friction that is exerted between the belt and the pulley. Let T_2 be the tension in that part of the belt which is moving toward the pulley, T_1 the tension in that part which is moving away from the pulley, and v the velocity with which the belt is moving. The net frictional force between the belt and the pulley is $T_1 - T_2$. The power delivered to the pulley is $(T_1 - T_2)v$, provided the belt does not slip over the pulley.

Example

The tension on one side of a belt is 350 N, and that on the other side is 150 N. The belt is moving at 4 m/sec. Find the power delivered to the pulley in watts.

$$\text{Power} = (350 - 150) \text{ N} \times 4 \text{ m/sec}$$
$$= 800 \text{ N-m/sec} = 800 \text{ W} \qquad \text{////}$$

Engines and motors are rated in terms of the brake horsepower which they develop. This may be measured by using the engine to drive a shaft similar to that of Fig. 7.10 while the belt is held stationary by spring balances which read the tensions in the two sides. Such an arrangement is known as a *Prony brake.* Under these conditions, when the shaft is rotated in the direction shown, T_2 exceeds T_1. The power output, dissipated in heat by friction, is determined by measuring the net frictional force $F = T_2 - T_1$ and the linear speed v of a point on the surface of the shaft and is given by the product Fv.

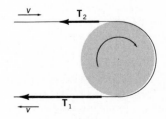

Figure 7.10

A belt-and-pulley system. When the belt drives the pulley as shown here, T_1 exceeds T_2; when the pulley drives the belt, T_2 is larger than T_1.

QUESTIONS

1 Why does a railroad track often wind up a long hill when a road may go directly up and over?

2 An automobile travels with constant velocity along a level road. Is there any resultant force on the car? Is any work being done on the car? Explain.

3 Why does it hurt one's bare hand more to catch a baseball when the hand is held in a fixed position than when the hand moves back as the ball is caught?

4 Give an example of a situation in which a force acts for an extended time but still does no work. Under what condition may a force move and yet perform no work?

5 Why do scissors for cutting paper have long blades and short handles while metal-cutting shears have short blades and long handles?

6 A man is swimming against the current in a river. If he remains stationary relative to the ground, does he do any work? Explain.

7 Why is a high jumper less likely to be injured if he falls in sawdust or sand rather than on hard clay?

8 A stone is thrown from a cliff with an initial speed v_0. How does its speed as it hits a body below depend on the angle at which it is thrown? Explain.

9 What happens to the kinetic energy given to the blood by the pumping action of the heart?

10 One boy tosses a ball to another boy on a moving airplane. Does the kinetic energy of the ball depend on the velocity of the airplane? Explain.

11 Discuss the stable, neutral, and unstable equilibria of a body in terms of potential energy.

12 A barrel is rolled up an inclined plane which makes an angle of 30° with the horizontal. If no work is done against friction, what is the mechanical advantage if the applied force is parallel to the plane and acts through the center (and center of gravity) of the barrel? What is the mechanical advantage if the force acts tangentially to the circumference of the barrel? Explain the difference.

13 If two springs of different stiffnesses but the same length are stretched by the same force, is more work done on the stiffer or the softer spring? Explain. What is the situation if they are stretched through the same distance?

14 An object at rest is dropped on a sidewalk and bounces to a height greater than its original one. How could this occur?

15 A boy walks upward on a downward-moving escalator. If he walks in such a way as to keep his elevation and his potential energy constant, does he do external work on himself? Does he do external work at all? If so, on what?

16 Show that the efficiency of a jackscrew must be less than 50 percent if the jack remains set when no force is applied at the handle.

17 Why is it that for a simple pulley system with two blocks, one fixed and one movable as in Fig. 7.9, the ideal mechanical advantage is the number of strands directly supporting the load?

PROBLEMS

1 A 75-kg man is lifted by an elevator through a distance of 60 m in 30 sec. What is the increase in his potential energy? What average power is expended in raising him?

Ans. 44,100 J; 1.470 kW

2 A man pushes a lawn mower 40 ft in 5 sec by exerting a force of 30 lb at an angle of 53° with the horizon. Find the work done and the average power expended.

3 In how many seconds must a 160-lb man climb 55 ft to the fourth floor of a building if he wishes to expend power at the rate of 1 hp?

Ans. 16 sec

4 If the net retarding force on a ship is 25,000 lb when it is traveling 33 ft/sec (22.5 mi/hr), find the work required to overcome fluid friction in going 1 mi and the power which must be delivered by the screws to overcome this friction.

5 The locomotive of a freight train must exert a force of 40,000 lb on the train as it pulls it along on a level track at a speed of 60 mi/hr (88 ft/sec). Find the power required from the engine and the work done on the train in a distance of 1 mi. *Ans.* 6,400 hp; 2.11×10^8 ft-lb

6 A cylindrical standpipe 30 ft high has an internal diameter of 10 ft. How much work would be required to fill the standpipe with water (*a*) if the water were pumped in at the bottom and (*b*) if it were pumped in at the top? Note that 1 ft³ of water weighs 62.5 lb.

7 How much work is required to hoist an elevator

and its contents with total mass of 3,000 kg to the top of a building 200 m high? What average power is required if this work is done in 50 sec?

Ans. 5.88 × 10⁶ J; 118 kW

8 What is the retarding force on a small plane if 50 hp is required to overcome the drag when the plane is flying level at a constant speed of 120 mi/hr?

9 A 3,200-lb automobile requires 70 hp to overcome retarding forces on a level road when traveling at a speed of 110 ft/sec. Find the net retarding force on the automobile. If the power and retarding force remain the same, find the speed of the car on a 5 percent grade, i.e., one which rises 5 ft in 100 ft of roadway.

Ans. 350 lb; 75.5 ft/sec

10 An escalator is designed to lift 80 people of 75-kg average mass from one floor of a department store to another floor 5 m higher in 1 min. What power is required, assuming that 80 percent of the power goes into lifting people?

11 A force of 40 N stretches a spring 25 cm. What is the force constant of the spring? How much force is required to stretch the spring an additional 15 cm? What is the potential energy of the spring when it is stretched 30 cm?

Ans. 160 N/m; 64 N; 7.2 J

12 A spring balance reads forces in newtons. The scale is 20 cm long and reads from 0 to 50 N. Find the potential energy of the spring (*a*) when it reads 40 N, (*b*) when it is stretched 20 cm, and (*c*) when a mass of 4 kg is suspended from the spring.

13 The spring constant for a spring gun is 50 N/m. If the spring is compressed 10 cm by a 20-g ball in cocking the gun, find the speed of the ball when the trigger is pulled. *Ans.* 5 m/sec

14 A mass of 0.8 kg suspended from a vertical spring produces an elongation of 2 cm. Find the force constant of the spring. What elongation would a force of 20 N produce? What is the potential energy of the spring when the elongation is 2 cm?

15 A pendulum bob on a 1-m cord is pulled over until the cord makes an angle of 37° with the vertical. Find the speed of the bob as it passes through the equilibrium position after being released at rest. *Ans.* 1.98 m/sec

16 Three men using a block and tackle are lifting a safe weighing 3,300 lb to a height of 25 ft. Each man develops ¼ hp. How long will it take to raise the safe if only half the work done by each man is useful?

17 A block falls from a table 80 cm high. It lands on an ideal, massless, vertical spring with force constant 2,100 N/m. The spring is initially 24 cm high, but it is compressed to a minimum height of 10 cm before the block is stopped. Find the mass of the block. *Ans.* 3 kg

18 An 8-lb block is dropped from a height of 2 ft above a tabletop onto an ideal massless spring. The spring is vertical, and its top was initially 1 ft above the table. If the spring constant is 320 lb/ft, what is the length of the spring when the block reaches its lowest point?

19 A block of mass 2 kg starts from rest and slides down a surface which corresponds to a quarter circle of 2.5-m radius (see accompanying figure). (*a*) If the curved surface is smooth, find the speed at the bottom. (*b*) If the speed at the bottom is 5 m/sec, find the energy dissipated by friction in the descent. (*c*) After the block reaches the level region at 5 m/sec, it slides to a stop in 3 m. Find the frictional force.

Ans. (*a*) 7 m/sec; (*b*) 24 J; (*c*) 8.33 N

20 A 4-lb block slides down the quarter circular surface of the accompanying figure. If the radius is 1 ft and the block has a speed of 6 ft/sec at the bottom, find the work done by the block against friction.

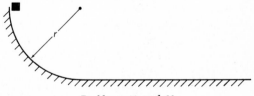

Problems 19 and 20

21 A 20,000-kg aircraft is launched from a carrier by a catapult 50 m long. The speed at the end of the catapult is 60 m/sec. Assume that the acceleration is constant and that the engine thrust just takes care of resistive forces. Find the work done by the catapult and the instantaneous power expended by the catapult just before the plane is launched.

Ans. 3.6 × 10⁷ J; 4.32 × 10⁷ W

22 A ski jumper glides down a 37° slope for 60 ft before taking off from a negligibly short horizontal takeoff. If his speed at takeoff is 45 ft/sec,

find the coefficient of kinetic friction on the slide.

23 The combined mass of a boy and his sled is 40 kg. His father pulls the boy and sled up a slope which rises 14 m in a distance of 50 m measured along the slope. If the frictional force between snow and loaded sled is 48 N, find the work done by the father. If the boy slides back down the slope, find his speed at the bottom.
Ans. 7,890 J; 12.4 m/sec

24 If dissipative resisting forces total 150 lb, find the horsepower necessary to give an automobile weighing 3,200 lb an acceleration of 5 ft/sec² when it has a velocity of 30 mi/hr.

25 A 25,000-kg airplane takes off at a speed of 50 m/sec, and 5 min later it is at an elevation of 3,000 m and has a speed of 100 m/sec. What average power is required during this 5 min if 40 percent of the power is used in overcoming dissipative forces? *Ans.* 4.60 MW

26 A jackscrew (see accompanying figure) with a pitch of 0.125 in. has a handle 2 ft long. A force of 15 lb must be applied when a load of 6,000 lb is being lifted. Calculate the ideal mechanical advantage, the actual mechanical advantage, and the efficiency.

Problem 26

27 In the pulley system shown in Fig. 7.9 a force of 60 lb is required to lift a 100-lb weight. Find the ideal mechanical advantage, the actual mechanical advantage, and the efficiency.
Ans. 2; 1.67; 83 percent

28 The upper end of an inclined ramp 25 ft long is 7 ft above the lower end. A force of 80 lb is required to push a 200-lb box up the incline. Find the coefficient of friction and the work done against friction. If the box were released at the top of the ramp, would it slide down? If so, what would be its speed at the bottom?

29 A boy throws a 0.16-lb stone from the top of a 50-ft cliff with a speed of 40 ft/sec. Find its kinetic energy and speed when it lands in a river below. *Ans.* 12 ft-lb; 69.3 ft/sec

30 A 0.25-lb ball is given an upward velocity of 60 ft/sec. If 10 percent of its energy is lost owing to air resistance during its rise, how high will the ball go?

31 A 48-lb boy on a swing is pulled backward and upward until his center of gravity has been raised 4 ft. What is his potential energy? If the swing is released, what will be his maximum speed?
Ans. 192 ft-lb; 16 ft/sec

32 A 0.5-kg mass is dropped through 1.2 m so that all its potential energy is changed to kinetic energy. Find its energy in joules. Calculate its final velocity by applying the law of conservation of energy. Verify the result by calculating the velocity of a body that falls freely through 1.2 m.

33 A boy, starting from rest, slides down a hill 121 ft high on a sled. If friction is negligible, find the speed of the sled at the bottom. If 19 percent of the energy is dissipated in friction, what will be his speed at the bottom?
Ans. 88 ft/sec; 79.2 ft/sec

34 A pendulum bob has a mass of 0.6 kg. It is suspended by a cord 2 m long which is pulled back through an angle of 30°. Find (*a*) its maximum potential energy relative to its lowest position, and (*b*) its potential energy when the cord makes an angle of 10° with the vertical, (*c*) its maximum speed, and (*d*) its speed when the cord makes an angle of 10° with the vertical, assuming the bob is released at 30°.

35 An inclined plane (see accompanying figure) 41 ft long is used as a simple machine to pull 1,000-lb boxes up to a height of 9 ft. A force of 400 lb up the plane is required to pull the boxes up. If it takes 50 sec to bring a box to the top, find the actual mechanical advantage

of the inclined plane, the ideal mechanical advantage, the efficiency, and the power required.
Ans. 2.5; 4.56; 0.55; 0.596 hp

Problem 35

36 A mass weighing 100 lb rests on an inclined plane that makes an angle of 37° with the horizontal. If the coefficient of friction between the mass and the plane is 0.2, how much work is done in moving the mass up the plane a distance of 10 ft? If the mass slides back down, what will its speed be when it reaches its original position?

37 A water bucket weighing 80 lb is raised by a crank-and-axle arrangement. The axle has a radius of 3 in., and the crank has a radius of 12 in. If a force of 30 lb is required on the crank, find the actual mechanical advantage, the ideal mechanical advantage, and the efficiency.
Ans. 2.67; 4; 0.67

38 A motor drives a hoist which lifts a 5-ton load a distance of 40 ft in 30 sec. The efficiency of the motor is 85 percent, and that of the hoist is 45 percent. What power is supplied to the load? To the hoist? To the motor?

39 A force of 900 N is required on the bar of a capstan to obtain an output force of 8,000 N for lifting an anchor. The axle of the capstan has a radius of 6 cm, and the input force is applied 0.9 m from the axis. Find the work done by the input force in lifting the anchor 10 m in a time of 2 min, the actual mechanical advantage, the ideal mechanical advantage, the efficiency of the capstan, and the power provided by the input force.
Ans. 135,000 J; 8.89; 15; 0.59; 1.13 kW

40 An inclined plane 13 ft long is used to slide a 400-lb box up to a loading platform 5 ft above ground level. A force of 200 lb is required. Find the efficiency of the inclined plane for this job. What force is required to overcome friction? What is the coefficient of friction between the plane and the box?

41 A 3,200-lb car starts from rest and coasts down a uniform grade, at the bottom of which it has a speed of 60 ft/sec. If the car has traveled 2,500 ft and has descended 100 ft, how much

energy was dissipated on the way down? To what average frictional retarding force does this energy dissipation correspond?
Ans. 140,000 ft-lb; 56 lb

42 A 2,400-lb car starts from rest and rolls down a hill 50 ft high and 1,000 ft long as measured along the road. At the bottom of the hill the car's speed is 30 ft/sec. How much energy has been dissipated? To what average frictional retarding force does this energy dissipation correspond?

43 A spring of neutral length 0.4 m and force constant 200 N/m is suspended from a second spring of neutral length 0.3 m and spring constant 500 N/m, which hangs from a hook. If the bottom of the first spring is pulled down until the combined length of the two springs is 1.05 m, find the individual lengths of the two springs and the stretching force required.
Ans. 0.65 m; 0.40 m; 50 N

44 Show that the ideal mechanical advantage of a differential pulley (see accompanying figure) is given by $2R/(R - r)$, where R and r are the radii of the larger and smaller sheaves, respectively. Find the ideal mechanical advantage of a differential pulley in which the radius of the larger pulley is 4.5 in. and that of the smaller pulley is 4 in.

Problem 44

CHAPTER EIGHT
MOMENTUM

IN CHAP. 7 we learned the principle of conservation of energy and saw how useful such a principle can be. Next we shall make use of Newton's second and third laws of motion (Chap. 6) to derive a second great conservation principle—that of the conservation of momentum. With the aid of conservation of momentum we can treat some problems of motion in which we do not have knowledge of the precise forces acting on an object.

8.1 MOMENTUM AND IMPULSE

The *momentum* of a body is a vector defined as the product of its mass m and its velocity $\mathbf{v}$. Consider a constant unbalanced force $\mathbf{F}$ acting on a body of mass m and initial velocity $\mathbf{v}_0$ for a time Δt. By Newton's second law [Eq. (6.6)],

$$\mathbf{F} = \frac{\Delta(m\mathbf{v})}{\Delta t} = \frac{m(\mathbf{v} - \mathbf{v}_0)}{t}$$

from which

$$\mathbf{F}t = m\mathbf{v} - m\mathbf{v}_0 \tag{8.1}$$

The right side of this equation is just the momentum at time t minus the initial momentum, or the change in momentum of the body. The quantity on the left side of the equation, *the product of the force and the time during which the force acts, is called the impulse.* From Eq. (8.1) we see that *the impulse is equal to the change in the momentum.*

In many interactions the net force acting on a body varies during the reaction. For example, when a football is kicked off, the toe of the kicker exerts zero force on the ball until it comes in contact, and the force increases rapidly as the ball is distorted (Fig. 8.1). As the ball returns to its initial shape, the force diminishes. A graph of the force as a function of time might be that of Fig. 8.2. In this case the impulse is given by the area under the force-as-a-function-of-time curve. Alternatively, we can express the impulse as the product of the average force and the time during which the force acts. The average force is indicated by the dashed line of the figure.

In many situations the force varies so rapidly that it is not practical to know what the force is at any particular instant, but the change in the momentum of the body and the total impulse can be measured. If we know how long the impulsive force acts, we can calculate the average value of the force even though we may not know the instantaneous value, i.e., the exact shape of the force curve. We have impulsive forces in play when a pitched ball is hit by a bat or when two billiard balls collide.

Figure 8.1
High-speed x-ray picture of a football being kicked. [*Westinghouse Corporation.*]

Example

93

MOMENTUM

A baseball of mass 0.145 kg is thrown by the pitcher with a speed of 30 m/sec. The bat is in contact with the ball for 0.01 sec and gives it a speed of 40 m/sec in a direction straight toward the pitcher. Find the impulse and the average value of the force.

$$\text{Impulse} = \text{change in momentum}$$
$$= 0.145 \text{ kg } [30 - (-40)] \text{ m/sec}$$
$$= 10.1 \text{ kg-m/sec}$$
$$= 10.1 \text{ N-sec}$$

$$F_{av} = \frac{\text{impulse}}{\text{time}}$$

$$= \frac{10.1 \text{ N-sec}}{0.01 \text{ sec}}$$

$$= 1,010 \text{ N} \qquad \qquad \text{////}$$

Figure 8.2
Force on the football as a function of time.

8.2 THE CONSERVATION OF MOMENTUM

Consider a freely suspended rifle, cocked and ready to fire (Fig. 8.3). When the trigger is pulled, a force $\mathbf{F}_b$ is exerted on the bullet. By Newton's third law, if the gun exerts a force $\mathbf{F}_b$ on the bullet, the bullet exerts an equal and opposite force $\mathbf{F}_g$ on the gun, so $\mathbf{F}_g = -\mathbf{F}_b$. Applying Newton's second law yields

$$\mathbf{F}_b = \frac{\Delta(m_b \mathbf{v}_b)}{\Delta t} = \frac{-\Delta(M_g \mathbf{V}_g)}{\Delta t} = -\mathbf{F}_g$$

Since the forces $\mathbf{F}_b$ and $\mathbf{F}_g$ are not only equal and opposite but also act for exactly the same time, the impulses for the gun and the bullet are equal in magnitude and opposite in direction. Therefore,

$$m_b \mathbf{v}_b = -M_g \mathbf{V}_g \qquad \qquad (8.2)$$

The momentum gained by the bullet is equal in magnitude and opposite in direction to that received by the gun. Since momentum is a vector quantity, the resultant of the momenta is zero, just as it was before the trigger was pulled. The momentum of the system composed of the gun and the bullet has not changed.

It follows directly from Newton's second and third laws of motion that *if two or more bodies interact, the momentum after the interaction is equal to the momentum before the interaction.*

This important principle is known as the *law of conservation of momentum.* It may be stated in the following form: *The total momentum of any system of bodies is unchanged by any interactions between the different members of the system.*

Figure 8.3
The momentum of the system comprising the bullet and the rifle is zero both before and after the bullet is fired. The momentum given the bullet is equal and opposite to that given the rifle.

When a battleship fires a broadside, the shells are given a large momentum in one direction. The battleship recoils with an equal momentum in the opposite direction. When two billiard balls collide, momentum is transferred from one ball to the other, but the resultant momentum after the collision is equal to the momentum before the collision. When two football players collide in midair, the law of conservation of momentum determines which way they fall.

Example

An 80-kg halfback dives over the line of scrimmage at a velocity of 7 m/sec. He is met in midair by a 100-kg linebacker going 5.4 m/sec in the opposite direction. If they fall together, find the horizontal component of their velocity.

$$\text{Momentum before} = \text{momentum after}$$

$$(80 \text{ kg} \times 7 \text{ m/sec}) - (100 \text{ kg} \times 5.4 \text{ m/sec}) = 180 \text{ kg} \times v$$

$$20 \text{ kg-m/sec} = 180 \text{ kg} \times v$$

$$v = \tfrac{20}{180} = 0.11 \text{ m/sec in direction of halfback's initial velocity} \quad ////$$

Consider one aircraft chasing another of about the same maximum speed. If the pursuing craft opens fire, the bullets are given forward momentum, and the plane loses an equal momentum. On the other hand, when the pursued craft opens fire, the bullets have momentum toward the rear, and this aircraft gains speed.

8.3 CENTER OF MASS

When a shell explodes in midair (Fig. 8.4), fragments are ejected in all directions. However, the resultant momentum of all fragments just after the explosion is the same as the momentum of the shell just before the explosion. Indeed, a unique point associated with the system, known as the *center of mass*, continues to move along the same path whether the shell explodes or not. The x coordinate of the center of mass of a system of particles with masses $m_1, m_2, \ldots, m_N$ is defined at any instant by the relation

$$x_c = \frac{m_1 x_1 + m_2 x_2 + \cdots + m_N x_N}{m_1 + m_2 + \cdots + m_N} \tag{8.3}$$

where x_1 is the x coordinate of m_1, etc. Similar relations locate the y

Figure 8.4
The center of mass of the shell fragments moves along the same trajectory before and after the explosion.

Figure 8.5
The constant velocity of a center of
mass (marked by the black cross) is
demonstrated by a timed-sequence
photograph of a thrown wrench.
Intervals are $\frac{1}{30}$ sec. [*From PSSC
"Physics," courtesy of D. C. Heath
and Company.*]

and z coordinates of the center of mass. The sum of the individual masses in the denominator is simply the total mass M of the system of particles.

Consider once again a shell which explodes into N fragments at time t. Just before the explosion let x_c be the x coordinate of the center of mass, m_1 and x_1 the mass and x coordinate of what becomes fragment 1, and so forth. A time Δt later the explosion has occurred and fragment 1 has x coordinate $x_1 + \Delta x_1$, fragment N is at $x_N + \Delta x_N$, and the x coordinate of the center of mass is at $x_c + \Delta x_c$. From

$$\frac{x_c + \Delta x_c}{\Delta t}$$
$$= \frac{m_1(x_1 + \Delta x_1)/\Delta t + m_2(x_2 + \Delta x_2)/\Delta t + \cdots + m_N(x_N + \Delta x_N)/\Delta t}{M}$$

we have for the x component of the velocity of the center of mass

$$v_{xc} = \frac{\Delta x_c}{\Delta t} = \frac{m_1 v_{x1} + m_2 v_{x2} + \cdots + m_N v_{xN}}{M} \tag{8.4}$$

Similar equations give the y and z components of this velocity, and we see that *the velocity of the center of mass of a system of particles is equal to the resultant momentum of the system divided by the mass of the system.*

The forces of explosion for the shell are forces exerted by one part of the system on other parts of the system. Such *internal* forces do not change the resultant momentum of the system. Even though the explosion produces drastic changes in the velocities $\mathbf{v}_1, \mathbf{v}_2, \ldots, \mathbf{v}_N$ of the fragments, it does not affect the velocity of the center of mass. However, the weight of the shell (and of the fragments) is an *external* force which does change the momentum of the system. *The resultant momentum $M\mathbf{v}_c$ of a system remains constant when no external forces act on the system; if external forces act, the center of mass has an acceleration proportional to the resultant external force and inversely proportional to the mass of the system.*

Figure 8.5 is a multiple-flash photograph of a rotating wrench moving freely in the absence of any external force. The center of mass, indicated by the black cross, moves with constant velocity; other mass elements of the wrench have more complicated motions. If an external force were applied to the wrench, it would produce an acceleration of the center of mass. Whether it would change the rotation about the center of mass depends on the line of action of the force. In general, when the resultant force on a body acts through the center of mass, there is no resultant torque to change the rotational velocity about the center of mass; conversely, if the resultant force does not act through the center of mass, there is a resultant torque and the rotational motion is changed.

The acceleration due to gravity is essentially the same at all points occupied by a body of reasonable size. Then the center of mass coincides with the center of gravity (Sec. 4.2).

8.4 ROCKET PROPULSION

Rockets are used for propelling long-range missiles and for placing satellites in orbit. A jet of hot gases is ejected from the combustion chamber of a rocket (Fig. 8.6). Since the rocket exerts a rearward force on the gases, the gases exert an equal and opposite forward force on the rocket. Let **u** be the velocity of the ejected gases relative to the rocket, and Δm the mass of gas ejected at a constant rate in time Δt. Then **u** $\Delta m/\Delta t$ is the time rate of change of momentum of the exhaust gases, which, by Newton's second law of motion, is the force exerted on the gases by the rocket. This, by Newton's third law, is equal and opposite to the reaction force $\mathbf{F}_R$ exerted on the rocket by the gases; $\mathbf{F}_R$, called the *thrust* of the rocket, has the magnitude

$$F_R = u\frac{\Delta m}{\Delta t} \tag{8.5}$$

and a direction opposite to **u**.

Ordinarily the rocket has other forces, such as its weight, acting upon it. If the resultant of these external forces is represented by $\mathbf{F}_E$, the net force acting on the rocket is $\mathbf{F}_E + \mathbf{F}_R$. By Newton's second law,

$$\mathbf{F}_E + \mathbf{F}_R = M\mathbf{a} \tag{8.6}$$

where **a** is the acceleration of the rocket relative to some fixed coordinate system and M is the mass of the rocket. Note that M varies during the flight as fuel is burned or as some portion of the rocket system is jettisoned.

Example

A Thor ballistic missile has a gross weight of roughly 425,000 N at the beginning of flight. It has a rocket engine which produces 600,000 N of thrust. If the exhaust velocity of the gases is 2,060 m/sec, find the rate at which mass is ejected and the acceleration at launch and after 100 sec of operation.

Figure 8.6
Schematic diagram of a rocket engine.

By Eq. (8.5),

$$F_R = u \frac{\Delta m}{\Delta t}$$

$$600,000 \text{ N} = 2,060 \text{ m/sec} \times \frac{\Delta m}{\Delta t}$$

$$\frac{\Delta m}{\Delta t} = 292 \text{ kg/sec}$$

At launch the unbalanced force is the difference between the thrust and the weight. By Newton's second law,

$$600,000 \text{ N} - 425,000 \text{ N} = \frac{425,000}{9.8} \text{kg} \times a$$

$$a = 4.03 \text{ m/sec}^2 = 0.41g$$

After 100 sec of operation 29,200 kg have been ejected, and the mass of the rocket is 14,200 kg. If we assume that air friction is negligible but that the rocket is not so high that g is appreciably different from 9.8 m/sec², the weight of the rocket is 139,000 N, and the unbalanced force is 600,000 N − 139,000 N, or 461,000 N. Then

$$461,000 \text{ N} = 14,200 \text{ kg} \times a$$

$$a = 32.4 \text{ m/sec}^2 = 3.3g \qquad ////$$

8.5 COLLISION PHENOMENA

When two bodies collide, the laws of conservation of momentum and of conservation of energy are both always applicable. However, in some collisions part of the kinetic energy of the bodies is transformed into some nonmechanical form of energy such as heat or sound. In this case the application of the law of conservation of energy to the problem becomes exceedingly difficult, because many kinds of energy may be involved, some of which are difficult to measure. If a steel ball is dropped on a steel plate, the ball and plate are distorted during the action of the impulsive forces, but they return to their original shapes, and the ball springs away from the plate. In this case the bodies are only temporarily deformed, and they regain their original shapes immediately after collision. Bodies which return to their original shapes after collision are said to be *elastic*.

In a perfectly elastic collision, kinetic energy, as well as momentum, is conserved. Collisions between atomic nuclei, atoms, molecules, and electrons are often elastic. Such an ideal situation is never achieved with large-scale objects; we have no perfectly elastic collisions of macroscopic bodies.

Any collision for which the final kinetic energy is less than the initial kinetic energy is said to be *inelastic*. When a bullet is fired into a block of wood, the bullet sticks in the block and the two remain together indefinitely. Such a collision is a *perfectly inelastic* one. The bodies are permanently deformed and never separate. Collisions between billiard balls or between a basketball and floor are not perfectly inelastic since the colliding objects spring apart. Such *semielastic* collisions are very common in nature.

Before collision

At collision

After collision

Figure 8.7
Head-on collision between two moving
spheres.

In every collision momentum is conserved; kinetic energy is conserved only in perfectly elastic collisions. In the next three sections we consider head-on collisions between two bodies of masses m and M (Fig. 8.7) which have initial velocities $\mathbf{v}_i$ and $\mathbf{V}_i$, respectively, before collision, and final velocities $\mathbf{v}_f$ and $\mathbf{V}_f$ after collision. Velocities are vector quantities; therefore, we must consider directions as well as magnitudes. In the equations which follow we adopt the convention that velocities to the right are positive, and velocities to the left negative.

8.6 PERFECTLY INELASTIC COLLISIONS

In a completely inelastic collision the two bodies stick together after they have made contact. As a consequence, both have the same final velocity. Conservation of momentum alone is sufficient to permit us to compute the final velocity if the masses and initial velocities are known:

$$m\mathbf{v}_i + M\mathbf{V}_i = (m + M)\mathbf{V}_f \qquad (8.7)$$

Example

A 2-g bullet is fired into a 2.398-kg block (Fig. 8.8) of wood suspended from a long cord. The bullet is embedded in the block, and the two start off together with a speed of 0.700 m/sec. Find the velocity of the bullet before collision.

$$(0.002 \text{ kg} \times v_i) + (2.398 \text{ kg} \times 0) = 2.400 \text{ kg} \times 0.700 \text{ m/sec}$$

$$v_i = 840 \text{ m/sec} \qquad ////$$

An arrangement such as that of Fig. 8.8 by which the speed of a moving body is measured by capturing it in a massive pendulum bob is known as a *ballistic pendulum*. Since the collision is completely inelastic, Eq. (8.7) is applicable to the collision between body and pendulum bob. The velocity V of the body plus bob immediately after collision may be determined by measuring how much the center of gravity of the system rises in the upward swing of the pendulum. If the center of gravity moves upward a distance h, conservation of energy yields $(m + M)gh = \frac{1}{2}(m + M)V^2$, where m and M are the masses of body and pendulum bob, respectively.

Cords to ceiling

Figure 8.8
The speed of a bullet can be determined by firing it into the bob of a ballistic pendulum and measuring how much the center of gravity of the bob plus bullet is raised at the end of the swing.

8.7 ELASTIC COLLISIONS

In a perfectly elastic collision both momentum and kinetic energy are conserved. If two bodies collide along the line connecting their centers, we may write

$$mv_i + MV_i = mv_f + MV_f \qquad \text{conservation of momentum} \quad (8.8)$$

and

$$\frac{1}{2}mv_i^2 + \frac{1}{2}MV_i^2 = \frac{1}{2}mv_f^2 + \frac{1}{2}MV_f^2 \qquad \text{conservation of energy} \quad (8.9)$$

These equations may be rewritten as

$$m(v_i^2 - v_f^2) = M(V_f^2 - V_i^2) \qquad \text{and} \qquad m(v_i - v_f) = M(V_f - V_i)$$

If we divide the first by the second, we obtain $v_i + v_f = V_f + V_i$ or

$$v_i - V_i = V_f - v_f = -(v_f - V_f) \qquad (8.10)$$

Note that $v_i - V_i$ is the *velocity of approach*, or the velocity of the smaller mass relative to the larger one before the collision, while $v_f - V_f$ is the *velocity of separation*, or the velocity of the smaller mass relative to the larger after the collision. In a perfectly elastic collision the velocity of approach is equal in magnitude to the velocity of separation but opposite in direction.

Example

A 40-g ball traveling east with a speed of 5.0 m/sec has a head-on collision with a 60-g ball traveling 3.0 m/sec west (Fig. 8.7). If the collision is elastic, find the velocities after the collision.

Let us choose east as the positive direction, so velocities to the west are negative. By Eq. (8.8),

$(0.040 \text{ kg} \times 5.0 \text{ m/sec}) + (0.060 \text{ kg} \times -3.0 \text{ m/sec})$
$$= 0.040 \text{ kg} \times v_f + 0.060 \text{ kg} \times V_f$$

$$v_f + 1.5V_f = 0.50 \text{ m/sec} \qquad (A)$$

By Eq. (8.10),

$$5.0 \text{ m/sec} - (-3.0 \text{ m/sec}) = V_f - v_f$$
$$-v_f + V_f = 8.0 \text{ m/sec} \qquad (B)$$

Adding Eqs. (A) and (B) yields

$$2.5V_f = +8.5 \text{ m/sec}$$
$$V_f = +3.4 \text{ m/sec}$$
$$v_f = -4.6 \text{ m/sec}$$

The minus sign means the 40-g ball is moving westward. ////

8.8 SEMIELASTIC COLLISIONS

In a perfectly elastic collision kinetic energy is conserved, and the velocity of approach is equal to the velocity of separation in magnitude but opposite in direction. In a perfectly inelastic collision the velocity of separation is zero, since the two bodies remain together. Most ordinary collisions of macroscopic bodies lie between these two extremes. The bodies spring apart but with kinetic energy reduced. Newton studied collisions between many types of spheres and concluded that in general the velocity of separation is proportional to but less than the velocity of approach:

$$-(v_f - V_f) = e(v_i - V_i) \qquad (8.11)$$

where, for a given pair of materials, e is a constant which lies between 0 and 1. This constant, called the *coefficient of restitution*, is 0 for perfectly inelastic collisions in which the bodies remain together. It is 1 for perfectly elastic collisions. For semielastic collisions it lies between

0 and 1, being near 1 for a golf club striking a golf ball and near 0 for the same club striking a knitted practice ball. If the coefficient of restitution is known, the use of Eqs. (8.8) and (8.11) is sufficient to find the final velocities of the two bodies. Figure 8.9 shows imperfectly elastic collisions of a steel ball and a hardened steel plate. Since the plate is fastened to the earth, M is essentially infinity, and $V_i = V_f = 0$. Hence, $e = -v_f/v_i$.

Example

If the ball of Fig. 8.9 is dropped from a height of 40.0 cm, and if the coefficient of restitution is 0.80, find the height attained on the first bounce.

By conservation of energy, $mgh = \frac{1}{2}mv_i^2$, or $v_i^2 = 2gh$. Therefore, the velocity v_i for the first collision is $\sqrt{2 \times 9.8 \times 0.400} = 2.80$ m/sec. By Eq. (8.11), $-v_f = ev_i$, since V_f and V_i are essentially zero.

$$-v_f = 0.80v_i = 2.24 \text{ m/sec}$$

This initial upward speed will carry the ball to a height h_1 such that $v_f^2 = 2gh_1$ or

$$h_1 = \frac{(2.24)^2 \text{ m}}{2 \times 9.8} = 0.256 \text{ m} = 25.6 \text{ cm} \qquad ////$$

Figure 8.9
Semielastic collisions of a ball. [*From PSSC "Physics," courtesy of D. C. Heath and Company.*]

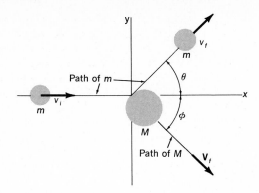

Figure 8.10
Glancing elastic collision of a moving
sphere with another initially at rest.

8.9 COLLISIONS IN SPACE

Thus far we have considered only collisions in one dimension. Typical collisions between billiard balls involve two dimensions, while collisions between air molecules occur in three dimensions. In every case momentum is conserved. Since momentum is a vector quantity, its conservation requires that the components of the resultant momentum be conserved also. Thus, in a three-dimensional collision each component of the resultant momentum remains the same, and we obtain three equations by application of the conservation of momentum.

Consider a body of mass m, moving with a velocity $\mathbf{v}_i$, which has an elastic collision in two dimensions with a body of mass M initially at rest. After the collision the first body moves off at an angle θ to its original path (Fig. 8.10). Let the x axis be determined by $\mathbf{v}_i$. Applying the conservation of momentum yields

x component: $\qquad mv_i + M(0) = mv_f \cos \theta + MV_f \cos \phi$

y component: $\qquad 0 + 0 = mv_f \sin \theta - MV_f \sin \phi$

Conservation of kinetic energy provides the equation

$$\tfrac{1}{2}mv_i{}^2 = \tfrac{1}{2}mv_f{}^2 + \tfrac{1}{2}MV_f{}^2$$

Thus, if m, M, v_i, and θ are given, we have three equations in three unknowns (v_f, V_f, and ϕ).

QUESTIONS

1 Discuss what happens when a novice marksman fires a rifle while holding the butt about a centimeter from his shoulder.

2 Why can a rocket engine accelerate a spacecraft even when there is no air for it to push against?

3 Does the center of mass of a solid body necessarily lie within the body? If not, what are some familiar bodies for which the center of mass is not located within the body?

4 Identical bullets strike a wooden block and a steel block of identical mass. If the bullet sticks in the wood and rebounds from the steel, which block receives the greater impulse? Why?

5 A shell explodes in midair. Is its total momentum greater than before the explosion? Is its kinetic energy changed? Explain.

6 Only a resultant *external* force can change the momentum of the center of mass of a body or system of bodies. If this is true, how can the internal forces of the brakes stop a moving automobile?

7 Discuss qualitatively what happens when a light

sphere collides head on and elastically with a more massive sphere initially at rest. How does this compare with what happens if the more massive sphere is in motion and strikes the light sphere head on? What happens if the two spheres have the same mass?

8 Under what conditions does the center of mass of a body not coincide with its center of gravity?

9 A bee is temporarily imprisoned in a light can which is completely closed. If the container is placed on a sensitive balance, is the weight different when the bee is flying about and when it is at rest? Is the weight affected by anything the bee does?

10 Prove that if a ball makes an elastic collision with a smooth wall, the angle of incidence (between the incident path and a perpendicular to the surface) and the angle of reflection are equal.

11 Can a single body have momentum without having kinetic energy? Without having any energy? Can a body have energy without having momentum? Contrive a simple system of two or three bodies which can have kinetic energy without having momentum.

12 Eighty percent of the weight of a certain guided missile is fuel. Assuming that the thrust exerted by the rocket motor is constant, discuss the factors which influence the acceleration of the rocket as it rises vertically to a height of 300 km.

PROBLEMS

1 A 0.144-kg baseball approaches a batter with a speed of 35 m/sec. The batter lines the ball directly back at the pitcher with a speed of 45 m/sec. Find the change in momentum and the impulse. If bat and ball were in contact for 0.01 sec, find the average force exerted on the ball during this period.
Ans. 11.5 kg-m/sec; 11.5 N-sec; 1,150 N

2 A 160-lb swimmer dives from a 320-lb rowboat initially at rest. If the swimmer leaves the boat with a horizontal speed of 10 ft/sec, find the recoil speed of the boat, neglecting the transfer

of momentum to the water. What was the impulse on the boat?

3 A 2-g bullet is fired from a 2.4-kg rifle with a velocity of 480 m/sec north. Find the momentum of the bullet and the recoil velocity of the rifle, assuming that no other bodies are involved.
Ans. 0.96 kg-m/sec north; 0.40 m/sec south

4 A pitcher exerts an average resultant force of 8 lb for 0.11 sec on a baseball weighing 0.32 lb. Find the average acceleration, the impulse, the speed of the ball, and the magnitude of the momentum.

5 An aircraft with a mass of 12,000 kg catapulted from a carrier has a uniform acceleration of 25 m/sec² over a distance of 40.5 m. Find the resultant force required, the time the force acts, the momentum of the aircraft as it leaves the catapult, and the impulse provided to the aircraft.
Ans. 300,000 N; 1.8 sec; 540,000 kg-m/sec; 540,000 N-sec

6 A bullet of mass 4 g is projected from a gun of mass 7 kg with a speed of 420 m/sec. What is the speed with which the gun recoils?

7 A football of mass 0.42 kg is passed with a velocity of 25 m/sec due south. A defending player lunges at the ball and deflects it so the new velocity is 20 m/sec 37°W of S. Find the magnitude of the impulse. If the player is in contact with the ball for 0.02 sec, what is the magnitude of the average force he exerts?
Ans. 6.3 N-sec; 315 N

8 A system consisting of two masses connected by a massless rod lies along the x axis. A 0.4-kg mass is at $x = 3$ m while a 0.6-kg mass is at $x = 7$ m. Find the x coordinate of the center of mass.

9 Three spheres are located with their centers along a straight line as follows: a 5-kg mass is 0.4 m to the right of a 3-kg mass, and a 6-kg mass is 0.9 m to the right of the 5-kg mass. Find the center of mass of the system.
Ans. 0.3 m to right of 5-kg mass

10 Find the distance from the center of the earth to the center of mass of the earth-moon system if the earth-moon separation is 3.8×10^5 km and the mass of the earth is 81.3 times the mass of the moon. To what fraction of the earth's radius of 6,370 km does this distance correspond?

11 A rigid body consists of a 2-kg mass connected to a 4-kg mass by a massless rod. At $t = 0$ the 2-kg mass is located at $\mathbf{r} = 3\mathbf{i} + 4\mathbf{j}$ m and it has

a velocity of $-4\mathbf{i} + 5\mathbf{j}$ m/sec. The 4-kg mass has $\mathbf{r} = 3\mathbf{i} + \mathbf{j}$ m and $\mathbf{v} = -\mathbf{i} + 5\mathbf{j}$ m/sec. Find the length of the rod, the coordinates of the center of mass at $t = 0$, and the velocity of the center of mass.

Ans. 3 m; $3\mathbf{i} + 2\mathbf{j}$ m; $-2\mathbf{i} + 5\mathbf{j}$ m/sec

12 A freight car weighing 40 tons runs into another freight car of the same weight. One car was stationary, and the other was moving at the rate of 6 mi/hr (8.8 ft/sec). If the cars move off together after collision, with what speed do they move?

13 A 0.005-kg bullet going 600 m/sec strikes a 1.995-kg wooden block which is the bob of a ballistic pendulum. Find the speed at which block and bullet leave the equilibrium position and the height which the center of gravity of the bullet-block system reaches above the initial position of the center of gravity.

Ans. 1.5 m/sec; 0.115 m

14 A 2-g bullet is fired into a 2.998-kg block of wood which is the bob of a ballistic pendulum. If the bob leaves its equilibrium position with a speed of 0.5 m/sec, find the speed of the bullet and the height above the equilibrium position reached by the center of gravity of the block.

15 A 5-g bullet traveling 1,000 m/sec strikes and embeds itself in a 1.995-kg block held on a frictionless table by a spring with constant $k = 200$ N/m. Find the speed of block and bullet immediately after the collision and the distance the spring is compressed.

Ans. 2.5 m/sec; 0.25 m

16 A 3-g bullet is fired horizontally into a 2-kg block of wood suspended from a cord 2 m long. The block swings because of the impact, deflecting the cord to a position 15° from the vertical. Find the initial speed of the bullet.

17 A 4-g bullet is fired horizontally with a speed of 900 m/sec into a 3-kg block of wood at rest on a table. If the coefficient of friction between the block and the table is 0.3, how far will the block slide? What fraction of the bullet's energy is dissipated in the collision itself?

Ans. 24.4 cm; 99.9 percent

18 A 3-g bullet is fired into a 2.997-kg block suspended by a long cord. The bullet remains in the block, which swings until its center of gravity is raised by 1.6 cm. Find the speed of the block and bullet as they leave the equilibrium position of the block. What was the initial speed of the bullet?

19 A ball of 0.4-kg mass and a speed of 3 m/sec in the x direction has a completely elastic collision with a 0.8-kg mass initially at rest. Find the speed of both bodies after the collision.

Ans. -1 and 2 m/sec

20 A proton of mass 1.66×10^{-27} kg collides head on with a helium atom at rest. The helium atom has a mass of 6.64×10^{-27} kg, and it recoils with a speed of 6×10^5 m/sec. If the collision is elastic, find the initial and final speeds of the proton and the fraction of its initial energy transferred to the helium atom.

21 Two perfectly elastic balls, one weighing 2 lb and the other 4 lb, are moving in opposite directions with velocities of 8 and 5 ft/sec, respectively. Find their velocities after head-on impact.

Ans. -9.33 and 3.67 ft/sec

22 Two cars on a linear air track have an elastic collision. They approach each other from opposite directions, each with speed v_0. After the collision the car at the right, which has mass m, is at rest. Find the mass of the car on the left and its speed just after the collision in terms of m and v_0.

23 In the accompanying figure the smaller ball has a mass of 0.2 kg, and the larger one a mass of 0.5 kg. If the smaller one is pulled back and released so that it has a velocity of 4 m/sec just before collision, find the velocities of both balls immediately after they have an elastic collision.

Ans. -1.71 and 2.29 m/sec

24 If the small steel ball in the accompanying figure has a mass of 0.3 kg and the large one a mass

Problems 23 and 24

of 0.5 kg, find the recoil velocity of each ball if the larger one is pulled out and released in such a way that it has an elastic collision when it is moving to the left with a speed of 0.6 m/sec.

25 A mass A of 2 kg moving to the right with a speed of 5 m/sec collides head on with a mass B of 3 kg moving in the opposite direction with a speed of 4 m/sec. After the collision A is moving to the left with a speed of 4 m/sec. Find the velocity of B after collision and the coefficient of restitution. *Ans.* 2 m/sec; 0.667

26 A golf ball is dropped on a sidewalk from a height of 2.5 m and rebounds to a height of 2 m. What was the impulse? Assuming the ball was in contact with the concrete for 0.008 sec, find the average acceleration. What was the coefficient of restitution? The mass of the ball is 45.8 g.

27 A 3-g bullet with a speed of 700 m/sec passes right through a 600-g block suspended on a long cord. The impulse gives the block a speed of 2 m/sec. Find (a) the speed of the bullet after it has passed through the block, (b) the distance which the center of mass rises as the block swings upward after the bullet passes through, (c) the work done by the bullet in passing through the block, and (d) the mechanical energy converted to heat.
Ans. (a) 300 m/sec; (b) 0.204 m; (c) 600 J; (d) 598.8 J

28 A tennis ball bounces down a flight of stairs, striking each step in turn and rebounding to the height of the step above. If the height of each step is 24 cm, find the coefficient of restitution.

29 A 0.4-kg ball traveling 5 m/sec collides head on with a 0.2-kg ball, initially at rest. If the latter proceeds in the original direction with a speed of 6 m/sec, find the speed of the 0.4-kg ball. What was the coefficient of restitution?
Ans. 2 m/sec; 0.80

30 A ball dropped from a height of 4 ft rebounds from a flat surface to a height of 2.25 ft. Find the speed of the ball just before it strikes and just after it leaves the floor. What is the coefficient of restitution?

31 A 0.3-kg block slides down a frictionless hemispherical bowl with a radius of 0.4 m.

At the bottom it collides perfectly inelastically with an 0.6-kg mass initially at rest. What impulse is given the 0.6-kg mass? Find the maximum angle α which the radius vector to the two blocks will make with the vertical after the collision. *Ans.* 0.56 N-sec; 27.3°

32 If the coefficient of restitution for a golf ball dropped on a sidewalk is 0.9, find the height to which the ball returns if it is dropped from a height of 4 ft.

33 A machine gun fires five bullets per second into a target. The mass of each bullet is 3 g, and the velocity 600 m/sec. Find the average force required to hold the gun in position. What is the power delivered to the bullets?
Ans. 9 N; 2.7 kW

34 A machine gun fires 240 bullets each minute with a velocity of 2,000 ft/sec. If the weight of each bullet is 0.016 lb, what horsepower is developed by the gun? What average force is required to hold the gun in position?

35 A ballistic missile has a mass of 1,500 kg on the launch pad. It is fired off vertically, propelled by a rocket engine with a thrust of 16,000 N which ejects gases at the rate of 8 kg/sec. Find (a) the speed of the exhaust gases relative to the rocket engine, (b) the initial acceleration of the missile, and (c) the acceleration 50 sec after launch.
Ans. (a) 2,000 m/sec; (b) 0.867 m/sec²; (c) 4.75 m/sec²

36 A machine gun fires 75 bullets per minute with a velocity of 800 m/sec. If each bullet has a mass of 0.075 kg, what average force is required to hold the gun against the recoil? What power is delivered to the bullets by the gun?

37 A small vernier rocket ejects 1.5 kg of gases with a speed of 2,000 m/sec in order to make a small corrective increase in the speed of a space capsule of 1,800-kg mass. Find (a) the momentum of the exhaust gases relative to the capsule and (b) the change in speed of the space capsule.
Ans. (a) 3,000 kg-m/sec; (b) 1.67 m/sec

38 The Redstone ballistic missile was used by the United States for sending its first astronauts on ballistic trajectories. The Redstone had a rocket engine of 75,000 lb thrust. It weighed 40,000 lb at launch, and its jet exhaust had a speed of 6,000 ft/sec. If the Redstone rose vertically, find its initial acceleration, the rate at which mass was ejected, the acceleration after 20 sec of

operation, and the time at which the acceleration became 5 g's.

39 Find the net thrust of a jet engine traveling 300 m/sec which takes in 32 kg of air each second and burns fuel at the rate of 3 kg/sec, exhausting the 35 kg of combustion products at a speed of 800 m/sec relative to the plane. (To get the net thrust one must take into account the fact that the air burned was not initially moving with the plane.) *Ans.* 18,400 N

40 A rocket has a takeoff mass of 5×10^4 kg, 80 percent of which is fuel. The exhaust speed is 2,000 m/sec. (*a*) Find the minimum burning rate required for lift off the launching pad. (*b*) If the burning rate is 400 kg/sec, find the acceleration at the lift-off and at burnout, assuming the rocket rises vertically and that g remains constant over the flight.

41 A manned space capsule and its propulsion system have a gross weight at launch of 240,000 lb and a total thrust from three engines of 400,000 lb; its engines exhaust 2,000 lb of gases each second. Assuming that g remains constant at 32 ft/sec², that aerodynamic forces are negligible, and that the capsule goes straight up, find (*a*) the velocity of the gases relative to the rocket, (*b*) the equation giving the acceleration of the system as a function of time, and (*c*) the acceleration 40 sec after launch.
Ans. (*a*) 6,400 ft/sec; (*b*) $32(80 + t)/(120 - t)$; (*c*) 48 ft/sec²

42 A body of mass m and speed V has an elastic head-on collision with a mass M at rest. What must M be (in terms of m) if m and M have equal energies after the collision?

43 Sphere A of mass m and velocity v_i has a perfectly elastic collision with an identical sphere B initially at rest. (*a*) Show that for a head-on collision sphere A is left at rest while sphere B is given a velocity equal to v_i. (*b*) If the collision is not head on, show that the angle between the velocities of the spheres after collision is 90°.

44 Show that when a moving particle of mass m collides head on and elastically with a particle of mass M initially at rest, it transfers to the stationary particle the fraction $4mM/(m + M)^2$ of its initial kinetic energy.

45 A spherical mass of 2 kg moving along the x axis with a velocity of 5 m/sec collides elastically with a spherical mass of 6 kg initially at rest. After the collision the 2-kg mass moves off at an angle of 37° with the x axis. Find the speed of both masses after the collision and the angle which the path of the 4-kg sphere makes with the x axis.
Ans. 4.67 m/sec; 1.03 m/sec; 65.8°

46 A ball is dropped from a height h_0 onto a large flat surface (Fig. 8.9). If the coefficient of restitution is e, show that the time T from the first impact until the ball stops bouncing is given by

$$T = \frac{e}{1 - e} \sqrt{\frac{8h_0}{g}}$$

[*Hint:* $1/(1 - e) = 1 + e + e^2 + \cdots.$]

CHAPTER NINE
UNIFORM CIRCULAR MOTION AND GRAVITATION

IN CHAP. 5 we studied trajectories of short-range projectiles, assuming that the acceleration g is constant both in magnitude and direction. Next we discuss another special case of curvilinear motion—that in which a body describes a circular path with uniform speed and an acceleration constant in magnitude but directed always toward the center of the circle. For a space capsule or a planet a typical path is an ellipse with the acceleration changing in magnitude as well as in direction. From the accumulated knowledge of the motions of the planets Newton deduced the law of force which predicts the observed accelerations of the planets and in terms of which we can understand the acceleration g of freely falling bodies. We follow some of the reasoning which led to the formulation of Newton's law of universal gravitation and apply this law to such problems as those involving space satellites.

9.1 UNIFORM CIRCULAR MOTION

When a body moves in a circular path with constant speed, it is said to describe *uniform circular motion*. While the magnitude of the velocity is constant, the direction is always changing. A body describing uniform circular motion has an acceleration $\mathbf{a} = \Delta\mathbf{v}/\Delta t$. In Fig. 9.1 the vector difference $\mathbf{v} = \mathbf{v} - \mathbf{v}_0 = \mathbf{a}\,\Delta t$ is shown for a sizable Δt. If we draw a vector diagram similar to that of Fig. 9.1 for the case of Δt (and hence θ) approaching zero, the acceleration is directed toward the center of the circle. The acceleration is constant in magnitude throughout the circular path, but it varies in direction.

To calculate this acceleration, we proceed as follows. Suppose that the particle passes over the arc $AB = v\,\Delta t$ with constant speed v in time Δt. If θ is small, the chord AB is essentially equal in length to the arc AB. In the velocity vector triangle of Fig. 9.1, the angle between $\mathbf{v}$ and $\mathbf{v}_0$ is also θ (since the velocity of uniform circular motion is always perpendicular to the radius). Therefore, the gray triangles are similar, from which we conclude that $v\,\Delta t/r = a\,\Delta t/v$ and

Figure 9.1

The magnitude of the velocity remains constant in uniform circular motion. The acceleration is always directed toward the center of the circle.

$$a = \frac{v^2}{r} \tag{9.1}$$

The acceleration of a body describing uniform circular motion is directed toward the center of the circle and has the magnitude v^2/r.

9.2 CENTRIPETAL FORCE

If a body has an acceleration toward the center of a circle, there is a net force in that direction. This force is called the *centripetal force*. To find its magnitude, we recall that **F** = m**a** and that $a_c = v^2/r$, whence

$$F_c = ma_c = m\frac{v^2}{r} \tag{9.2}$$

Consider a model airplane constrained to move in a horizontal circle by a control wire. The wire pulls on the airplane toward the center of the circle. By Newton's third law, the airplane must exert an equal and opposite force on the wire. This force, *away* from the center of the circle and *acting on the wire*, is known as the *centrifugal force*.

Note carefully that the force *on the airplane is toward the center* of the circle. If this force were suddenly removed and no unbalanced force acted on the airplane, it would move along a tangent to the circular path. No force acts upon the airplane outward along the radius. The airplane does exert a force outward on the restraint, which in this case is the control wire. This outward force on the restraint is the centrifugal force; it is the reaction to the centripetal force which acts inward on the body describing circular motion.

When an automobile goes around a curve (Fig. 9.2) on a perfectly level road, friction between the tires and roadway is required to provide the necessary centripetal force mv^2/r. If the road is coated with ice, this frictional force may be too small and the car may slide off the road. *There is no force pulling the car outward* along the radius. By Newton's first law, the car would continue to move indefinitely with constant velocity in the absence of any force. If there is a force toward the center of the circle which is inadequate to provide the full centripetal force, the car is accelerated toward the center but not enough to keep it on the road.

If the road is suitably banked, the car may go around the curve without requiring any frictional force. When friction between tires and roadway is absent, the road pushes on the car only in the direction perpendicular to the surface. If the automobile goes around the curve without skidding, the vertical component of this normal force (Fig. 9.3) is exactly the weight of the car and the horizontal component is just mv^2/r, the necessary centripetal force. To satisfy these conditions the angle θ must be such that

$$\tan\theta = \frac{mv^2/r}{mg} = \frac{v^2}{gr} \tag{9.3}$$

Thus the road can be ideally banked for only a single speed. The angle at which the road should be banked depends on the speed of the car but not on its mass.

Figure 9.2
When an automobile goes around a curve, the centripetal force must be provided by the friction of the tires, or by banking the pavement, or by some combination of the two.

Figure 9.3
A curve is perfectly banked for a given speed v when the force exerted perpendicular to its surface by the road has a horizontal component equal to the centripetal force mv^2/r and a vertical component equal to the weight mg of the vehicle.

$$F' = \frac{mv^2}{r} - mg$$

$$F = \frac{mv^2}{r} + mg$$

Figure 9.4

Force exerted by the wings and other surfaces on an airplane at the upper and lower extremities of a circular loop.

9.3 OTHER APPLICATIONS OF CENTRIPETAL FORCE

When an aircraft makes a turn in a horizontal plane, it banks in such a way that the vertical component of the forces on the wings is equal to the weight of the aircraft, while the horizontal component provides the necessary centripetal force. Very-high-speed aircraft cannot make exceedingly sharp turns because it is not feasible to provide the necessary centripetal force.

When an aircraft pulls out of a steep dive, large forces must be exerted by the wings. More than one pilot has pulled the wings off his aircraft when he has tried to pull out of a dive too quickly. Consider an aircraft which is descending at the rate of 280 m/sec and tries to pull out of the dive in a circle of radius 800 m (Fig. 9.4). The centripetal acceleration becomes

$$a_c = \frac{v^2}{r} = \frac{280 \text{ m/sec} \times 280 \text{ m/sec}}{800 \text{ m}} = 98 \text{ m/sec}^2$$

or 10 times the acceleration due to gravity. Thus, the *unbalanced upward force* required on the aircraft is 10 times the weight. At the bottom of the circle the wings must not only provide this centripetal acceleration but overcome the pull of the earth as well. They must provide $10 \times 9.8 \text{ m/sec}^2 = 10$ g's centripetal acceleration plus 1 g to take care of the pull of the earth on the aircraft—a total of 11 g's. The wings of most aircraft are not designed for such heavy loading. A typical fighter aircraft is designed to take accelerations of the order of 8 or 9 g's, while passenger aircraft, bombers, and transports are designed for lower g loadings. At the top of the circular path (Fig. 9.4) the pull of gravity helps to provide the centripetal acceleration, and the net force exerted by the wings is $(mv^2/r - mg)$ downward.

Not only does the aircraft itself have to stand these accelerations, but so do all the objects in the aircraft, including the organs of the pilot's body. Consider a gram of blood in a pilot's brain when the aircraft is performing a 6-g turn. The total force on this gram of blood must be six times the ordinary force exerted if the blood is to remain in place. In the absence of this force the blood may rush out of the pilot's brain, resulting in his "blacking out."

The centrifuge, which is widely used in the separation of liquids of unequal densities, depends for its operation on centripetal force. One type of centrifuge (Fig. 9.5) consists of a wheel that rotates in a horizontal plane. To this wheel are attached buckets that are vertical when the wheel is at rest. However, when the wheel is revolving rapidly, the buckets assume a position such that their axes are almost horizontal. If a mixture of liquids of unequal densities is introduced into the buckets and the wheel is rotated rapidly, the liquids separate, the heavy liquids farther from the axis of rotation, and the lighter liquids nearer to it. This means that the heavier liquids are at the bottom of the buckets when the centrifuge is stopped. The cream separator works on the same principle.

9.4 THE PTOLEMAIC AND COPERNICAN THEORIES

When Newton formulated his laws of motion, he was particularly interested in explaining the movements of the moon, the planets, and other

Figure 9.5
A centrifuge.

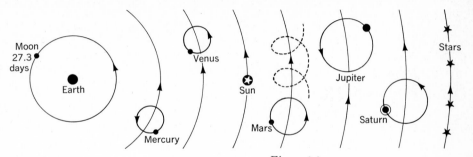

Figure 9.6
In the Ptolemaic system, the earth is at the center of the universe.

heavenly bodies. These motions had fascinated men since the beginning of history. In Greek mythology the daily journey of the sun was attributed to the god Apollo driving a flaming chariot across the sky. In the second century after Christ Ptolemy developed the theory that the stars, the sun, the moon, and the planets all revolve about the earth (Fig. 9.6). He postulated that one might think of the stars as mounted on a great transparent sphere which rotated at a constant rate around the earth at its center. The sun and the moon were mounted on similar, but smaller, spheres which rotated about the earth at rates different from that of the stars. To explain the motions of the planets (*planet* means "wanderer"), which move most of the time in the same direction as the stars but occasionally retrogress, the Ptolemaic theory assumed that each planet moved in a circular path about a center, which in turn followed a much larger circular path about the earth. Small circles superimposed on a large circle were called *epicycles*.

In 1543, Nicholas Copernicus, a Polish monk, published the results of some 35 years of patient study in a celebrated book which showed that the need for epicyclic paths of planets and many of the other complications of the Ptolemaic system disappeared if one took the sun as the center of the universe (Fig. 9.7). The idea of a heliocentric (*helios* means "sun") system as opposed to the Ptolemaic geocentric (*geos* means "earth") system was a great stimulus to the field of astrophysics. Supporting evidence for the Copernican theory grew over the next century.

Galileo constructed a telescope shortly after a Dutch spectacle maker Lippershey discovered in 1608 that he could use two spectacle lenses to form an enlarged image of a distant object. With his telescope Galileo discovered four of Jupiter's moons, mountains on the surface of the moon, Saturn's rings, and dark spots moving across the disk of the sun which gave evidence that the sun was in rotation. Some of these discoveries were readily explained on the basis of the Copernican theory but not by the Ptolemaic theory. Galileo became one of the major supporters of Copernicus, but he soon suffered persecution and imprisonment because his views differed from those held by certain high officials of the church.

The Danish astronomer Tycho Brahe (1546–1609) devoted most of his professional life to making accurate measurements of the positions of stars and planets. His data were analyzed by the mathematical physicist Johan Kepler (1571–1630). Kepler found that the motions of the planets did not always agree exactly with those predicted by the

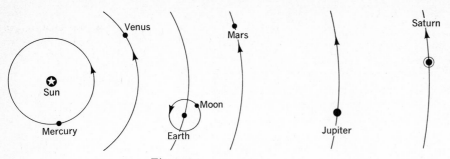

Copernican theory of circular paths. After many years of work he was able to show that the motions of the planets could be predicted by use of three generalizations, which we know as *Kepler's laws.*

1 The orbit of each planet is an ellipse with the sun at one focus.

2 The speed of the planet varies in such a way that the line joining the planet and the sun sweeps out equal areas in equal times (Fig. 9.8).

3 The cubes of the semimajor axes of the elliptical orbits are proportional to the squares of the times for the planets to make a complete revolution about the sun.

Although the general motions of the planets about the sun occur in elliptic orbits, the motions of the earth and of several of the other planets around the sun can be reasonably well described by circular orbits, as can the motion of the moon around the earth. Since the geometrical complexity of circular motion is substantially less than that of general elliptic motion, and since the physical ideas involved in both types are equivalent, we shall focus our attention on circular motion and leave elliptic motion to more advanced treatments. We observe that the circle may be regarded as a limiting case of the ellipse, namely, that in which the two foci merge into one.

9.5 NEWTON'S LAW OF UNIVERSAL GRAVITATION

Let us now apply these ideas and equations to the motion of the earth about the sun. Imagine for the moment that the sun itself is fixed in space. If we assume that the path of the earth is circular, the earth has an acceleration toward the sun (the center of the circle) given by $a_c = v^2/r$. To produce this acceleration there must be an unbalanced centripetal force $F_c = mv^2/r$.

The magnitude of the earth's velocity is $2\pi r/T$, where T is the period (1 year = time for one complete revolution). Therefore, the centripetal force is

$$F_c = m\frac{4\pi^2 r^2}{T^2}\frac{1}{r} = \frac{4\pi^2 mr}{T^2} \tag{9.4}$$

For a circular orbit the semimajor axis is equal to the radius r. By Kepler's third law, T^2 is proportional to r^3, or $T^2 = cr^3$, where c is a constant. Substituting this in Eq. (9.4) yields

$$F_c = \frac{4\pi^2 m}{c}\frac{1}{r^2}$$

so the centripetal force F_c is inversely proportional to r^2. Furthermore, if the proportionality constant for all planets is to be the same, F_c must be proportional to the mass of the body describing the orbit, so that the mass cancels out of Eq. (9.4). Otherwise the relationship between the period and the semimajor axis would involve the mass of the revolving heavenly body. If the mass of one of the bodies is involved in the attractive force between the two, it seems reasonable that the mass of the central body should be involved in a similar way. After extensive reasoning related to the arguments above, Newton proposed his great law of universal gravitation:

Between every two particles in the universe there is a force of gravitational attraction which is proportional to the product of the masses of the two particles and inversely proportional to the square of the distance between them. This force, which acts along the line joining the two particles, is given by

$$F_G = G\frac{m_1 m_2}{r^2} \tag{9.5}$$

where G is the universal gravitational constant. Its value has been measured and is 6.67×10^{-11} N-m²/kg².

When we deal with the gravitational attraction between large objects whose separation is very great compared to the dimensions of either body, we need not calculate the forces between every particle of one body and every particle of the other. If m_1 and m_2 are the masses of the bodies and r the distance between their centers of mass, Eq. (9.5) is directly applicable. Further, Newton was able to show that for any uniform distribution of mass over a spherical shell we may take the entire mass as concentrated at the center and apply Eq. (9.5) directly *for any point outside the sphere*. Thus, even if two uniform spheres touch each other, Eq. (9.5) is valid if r is taken to be the distance between their centers.

The force of attraction between two identical lead spheres of mass 1 kg with centers 1 m apart is only 6.67×10^{-11} N. The gravitational forces between two ordinary man-sized objects are detectable only under the most favorable of conditions. If the interacting bodies are stars or planets, the masses involved are so great that the forces become exceedingly large. Evidence for the validity of Newton's law of gravitation is obtained from astronomical observations. By application of the law, it is possible to predict with great accuracy the motions of the planets and their satellites many years in advance. Such long-term predictions are a severe test of the law.

9.6 DETERMINATION OF THE GRAVITATIONAL CONSTANT AND THE MASS OF THE EARTH

The gravitational constant G was first measured by Cavendish in 1797. He used a torsion balance (Fig. 9.9) in which two spheres of mass m are fixed to the end of a light rod suspended by a long, fine wire to

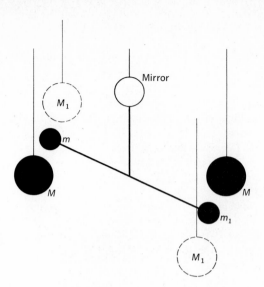

Figure 9.9
Schematic diagram of the apparatus with which Cavendish measured the universal gravitational constant. As the small masses were attracted by the large masses, their displacement was determined from the twisting of the suspending cord. A light beam reflected from the mirror to a scale measured the amount of twisting.

5 kg Hg

0.569 m

5,775 kg lead

Figure 9.10
Jolly's method of measuring the universal gravitational constant.

form a torsion pendulum. When two heavy lead balls are placed at the positions marked M, the rod carrying the two small masses turns to a new position. If the lead balls are placed at M_1, the light rod with the small masses turns in the opposite direction. By measuring the elastic constant of the suspending wire, it is possible to determine the attractive force between the movable masses and the heavy balls.

A method of measuring G which is simple to understand was developed by Jolly. A spherical vessel containing 5 kg of mercury was attached to one arm of a sensitive balance, and the balance put in equilibrium (Fig. 9.10). Then a lead sphere of mass 5,775 kg was rolled beneath the mercury flask with its center about 0.57 m below the center of mass of the mercury. The gravitational attraction between the lead and the mercury was found to be such that a mass of 0.589 mg on the right pan was able to restore equilibrium. All the quantities in Eq. (9.5) were thus known except for G, the value of which was calculated and reported in 1881.

Once the value of the gravitational constant is known, it is possible to calculate the mass of the earth m_E. For the moment, consider the earth to be a perfect sphere of radius $r = 6.4 \times 10^6$ m. The force which the earth exerts on a mass of 1 kg at its surface is 1 kg wt $= 9.8$ N. Therefore, from $F_G = Gmm_E/r^2$,

$$9.8 \text{ N} = 6.67 \times 10^{-11} \frac{\text{N-m}^2}{\text{kg}^2} \frac{1 \text{ kg} \times m_E}{(6.4 \times 10^6 \text{ m})^2}$$

from which $m_E = 6 \times 10^{24}$ kg.

We may also find the mass of the sun M_s. The earth revolves around the sun in an elliptical path which is fairly close to a circle of radius 1.5×10^{11} m (93,000,000 mi). The centripetal force for the earth is provided by the gravitational attraction between the earth and the sun. Thus,

$$F_G = G\frac{M_s m_E}{r^2} = m_E \frac{v_E^2}{r} \qquad (9.5a)$$

and

$$M_s = \frac{v_E{}^2 r}{G} = \frac{(2\pi r)^2 r}{T^2 G}$$

$$= \frac{(2\pi)^2 (1.5 \times 10^{11})^3}{(365.26 \times 86{,}400)^2 \times 6.67 \times 10^{-11}} = 2 \times 10^{30} \text{ kg}$$

Data on the solar system are listed in Table 9.1.

9.7 ARTIFICIAL SATELLITES

Men long dreamed of space stations circling the earth at a height of a few hundred kilometers, and during the 1970s this dream became a reality. Now weather satellites in appropriate orbits transmit to earth periodic "pictures" of the cloud cover over all parts of the globe, while communications satellites relay messages and television programs from one continent to another. Let us find the speed required for a satellite to remain in circular orbit 630 km above the earth's surface (7,000 km from the center of the earth). If m_s is the mass of the satellite,

$$G \frac{m_s m_E}{r^2} = m_s \frac{v^2}{r}$$

or

$$6.67 \times 10^{-11} \frac{\text{N-m}^2}{\text{kg}^2} \frac{6 \times 10^{24} \text{ kg}}{(7 \times 10^6 \text{ m})^2} = \frac{v^2}{7 \times 10^6 \text{ m}}$$

$$v = 7{,}600 \text{ m/sec} \approx 17{,}000 \text{ mi/hr}$$

At this speed the satellite would require a time to encircle the earth of $(2\pi \times 7 \times 10^6 \text{ m})/(7{,}600 \text{ m/sec}) = 5{,}600 \text{ sec} = 1.6 \text{ hr.}$

If the only force on the satellite were gravitational, the satellite would encircle the earth indefinitely. Actually air friction and other small forces gradually reduce the speed, so that eventually it approaches the earth and dissipates its energy in the earth's atmosphere.

Table 9.1
Data on the Solar System

Body	Mass, kg	Average radius,* km	g at surface, m/sec²	Sidereal period, days	Radius of orbit,† km
Moon	7.35×10^{22}	1,738	1.62	27.3	3.8×10^5
Sun	1.97×10^{30}	695,000	274		
Mercury	3.28×10^{23}	2,570	3.9	88	5.8×10^7
Venus	4.82×10^{24}	6,310	8.9	245	1.08×10^8
Earth	5.98×10^{24}	6,370	9.80	365.26	1.50×10^8
Mars	6.37×10^{23}	3,430	3.8	687	2.28×10^8
Jupiter	1.88×10^{27}	71,800	26	4,333	7.78×10^8
Saturn	5.62×10^{26}	60,300	11	1.08×10^4	1.43×10^9
Uranus	8.62×10^{25}	26,700	10	3.07×10^4	2.87×10^9
Neptune	1.0×10^{26}	24,900	14	6.02×10^4	4.5×10^9
Pluto				9.09×10^4	5.9×10^9

* Bodies are not exactly spheres.

† Orbits are actually elliptical; the values quoted are mean distances.

9.8 VARIATION OF WEIGHT WITH POSITION

The weight of an object of mass m is mg, where g is the observed acceleration due to gravity. If the earth were isolated, at rest, and a perfect sphere of radius r, the weight of an object would be the same at all points on its surface. Actually, the earth is somewhat flattened at the poles. Consequently, a body at one of the poles is closer to the center of the earth than an identical body at the equator. This leads to a somewhat greater weight at the poles than at the equator. Another factor which makes the weight of an object less at the equator is the rotation of the earth. At the equator a portion of the gravitational pull is used to supply the centripetal force to keep the body moving with uniform circular motion. Therefore, the force which must be applied to the body to prevent it from falling to the earth is smaller at the equator. The total variation from equator to pole is approximately 0.5 percent.

The gravitational pull is just adequate to produce the required centripetal acceleration for a body in a space capsule in orbit around the earth. The body is "weightless" in the sense that no force is required to hold it in its position in the capsule. An astronaut can place his camera 50 cm in front of him, and it will "float" in that position. Here we have a special case in which the entire gravitational attraction of the earth is required to produce centripetal acceleration, leaving no *net* force on the camera as viewed from the capsule.

The weight of a body varies not only with latitude but also with altitude. The weight is somewhat less atop Pike's Peak than it is in Death Valley. As we go above the earth, the weight of a body decreases gradually as r increases. Figure 9.11 shows how the weight varies with distance from the center of the earth (neglecting effects arising from rotation or revolution). Table 9.2 shows how g differs from place to place at the earth's surface.

In prospecting for oil and for heavy masses of ore below the surface of the earth, careful measurements of g are commonly made at selected points over a fairly large area. Modern gravimeters are sufficiently sensitive for a change in g of a few parts in 100 million to be detectable. The variation in g over a region is plotted on a map, and experts can frequently locate places where drilling for oil or ore is most likely to be successful.

The variations in g at different places on the earth's surface are small enough to be neglected in ordinary activities. However, at the surface of the moon the weight of an astronaut (force with which the moon holds him to its surface) is only one-sixth of that on the earth. Indeed, the gravitational pull on the moon is so small that gas molecules escape readily, and therefore the moon has essentially no atmosphere. On the planet Jupiter a man would be pulled down by a force 2.6 times his weight on earth. Under these circumstances it would be difficult for him to walk about.

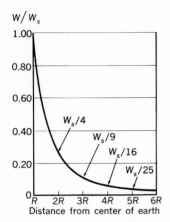

Figure 9.11

The weight of an object decreases as the object moves away from the center of the earth. W_s is the weight at the surface.

Table 9.2
Acceleration Due to Gravity at Sea Level*

Latitude	g, m/sec²
0° (equator)	9.78039
10°	9.78195
20°	9.78641
30°	9.79329
35°	9.79737
40°	9.80171
45°	9.80621
50°	9.81071
55°	9.81507
60°	9.81918
70°	9.82608
90° (pole)	9.83217

*For each meter above sea level g decreases by 3.086×10^{-6} m/sec².

9.9 GRAVITATIONAL POTENTIAL ENERGY

The potential energy of a body of mass m raised a distance h above the surface of the earth is, by Eq. (7.6), mgh. This relation is valid, however, only if the weight mg does not vary significantly over the range of heights involved. Let us now consider the potential energy when the mass is lifted or projected to a height h of hundreds of kilometers above the surface of the earth. Let R be the radius of the earth, M its mass, and r the distance from the center of the earth to a point a distance h above the earth's surface (Fig. 9.12). The potential energy of the mass relative to the earth's surface is the work required to move it from the surface of the earth to $h = r - R$. To calculate this work, we divide the displacement from R to r into many small intervals and calculate the work performed in each of these intervals. From R to r_1 the force varies from GMm/R^2 to GMm/r_1^2. If R and r_1 are close to one another, the average force in the interval from R to r_1 is GMm/Rr_1, and the work done in this interval is, by Eq. (7.1),

$$W_1 = \frac{GMm(r_1 - R)}{Rr_1} = GMm\left(\frac{1}{R} - \frac{1}{r_1}\right)$$

Similarly the work W_2 required to move the mass from r_1 to r_2 is

$$W_2 = GMm\left(\frac{1}{r_1} - \frac{1}{r_2}\right)$$

and so forth. The total work done in carrying m from R to r is

$$W = W_1 + W_2 + \cdots = GMm\left(\frac{1}{R} - \frac{1}{r_1} + \frac{1}{r_1} - \frac{1}{r_2} \cdots + \frac{1}{r_n} - \frac{1}{r}\right)$$

$$= GMm\left(\frac{1}{R} - \frac{1}{r}\right) = \text{PE at } r \text{ relative to 0 at } R \qquad (9.6)$$

In a frictionless atmosphere a projectile fired straight upward with speed v rises until its kinetic energy is transformed into potential energy, or

$$\tfrac{1}{2}mv^2 = GMm\left(\frac{1}{R} - \frac{1}{r_{max}}\right)$$

where r_{max} is the maximum radius attained by the projectile.

Example

A projectile is fired upward with a speed of 5,000 m/sec. How high will it rise above the earth's surface if air friction is ignored?

KE at launch = PE at maximum height

$$\tfrac{1}{2}m \times 2{,}500 \times 10^4 \text{ m}^2/\text{sec}^2 = 6.67 \times 10^{-11} \text{ N-m}^2/\text{kg}^2$$

$$\times m(5.98 \times 10^{24} \text{ kg})\left(\frac{1}{6.37 \times 10^6 \text{ m}} - \frac{1}{r_{max}}\right)$$

from which $r_{max} = 8.0 \times 10^6$ m, corresponding to a height above the earth's surface of about 1.6×10^6 m.   ////

9.10 THE FORMATION OF TIDES

A familiar and important phenomenon closely related to gravitational interactions is that of the tides. The attractive force exerted by the moon raises a tide (Fig. 9.13) on side A nearest to it and simultaneously on side B away from it, in large part because the pull per unit mass is

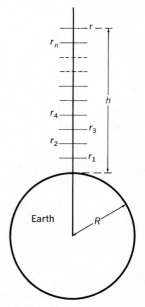

Figure 9.12
The potential energy of a mass at a great height above the earth's surface is determined by dividing the displacement from R to r into many small intervals, calculating the change in potential energy for each interval, and adding the results.

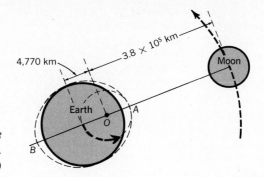

Figure 9.13
How the moon forms the earth's tides.
(Sizes are not to scale.)

greater on the water at A and less at B than the corresponding pull per unit mass nearer the center of the earth. Further, the moon does not revolve about a fixed earth; instead both bodies perform essentially uniform circular motion about the center of mass O of the earth-moon system, which lies almost three-fourths the earth's radius from the center of the earth. Since point B is much farther from O than point A, the centripetal force required for a kilogram of water at B is greater than that for a kilogram at A; consequently g is lowered at B (Sec. 9.8). If lunar attraction were the only factor, tides would have their greatest heights twice in a period of 24.85 hr, as commonly observed in the Atlantic basin. Actually the inertia of the water, the earth's rotation, frictional forces, the interactions with land masses, and other causes complicate the picture. As a consequence, in some places there is only one high tide a day.

Although the sun is much farther from the earth than the moon, the solar mass is so great that there is a solar contribution to the tides which is about four-tenths the lunar contribution. The resulting tide is a combination of the lunar and solar contributions. When the earth, moon, and sun are in line, unusually high tides result.

QUESTIONS

1 Does the centripetal force do any work on a body in uniform circular motion? Why?

2 An airplane makes a complete loop in a vertical plane. Under what condition does the pilot press upward on the seat at the top of the loop?

3 Explain carefully why an automobile may overturn if it goes around a curve too fast. Which wheels leave the ground first?

4 The moon falls about 0.28 ft toward the earth each minute, yet it always stays about the same distance away. Explain how this can be.

5 An astronaut in a stable orbit two earth radii from the center of the earth is weightless in his space capsule. How can this statement be reconciled with Fig. 9.11, which shows a weight one-fourth that at the earth's surface?

6 How would the weight of a radio transmitter vary on a rocket trip to the moon? Would its mass also vary?

7 A ball has a mass of 0.1 kg. If it were taken to the moon, what would its mass be? Would Newton's second law be valid? Would $m = w/g$ be valid? Would the weight of the ball be 0.98 N?

8 When a small continuous chain is set into rapid rotation around the axis of an electric motor and then released, it will roll along a lecture table and bounce over obstacles as though it were a solid hoop. What is the origin of this rigidity? Discuss the behavior of the cowboy's lariat in the same context.

9 Suppose that the moon were to move slowly toward the earth until it came to rest on the earth's surface. Would the center of mass of the moon coincide with its center of gravity? Why not? Which would be closer to the earth?

10 Does a plumb bob at rest necessarily point to the center of the earth? Why not? Why is the vector acceleration due to gravity **g** not ordinarily directed toward the center of the earth? Where on the earth's surface might **g** be expected to point to the center of the earth?

11 The gravitational attraction between the earth and the sun is about 150 times that between earth and moon. Why then does the moon raise larger tides?

12 Earth satellites which revolve with practically the same angular velocity as the earth have been launched as aids to navigation and as relay stations for television and other communication systems. Such a satellite appears stationary to observers on the earth. Could such a satellite remain on the zenith above Chicago for an extended period? Why not? Above what points can such a satellite remain virtually stationary?

13 The earth and the other planets travel about the sun in the same direction and almost in the same plane. Most of the time a more distant planet observed from the earth moves toward the east through the sky, but part of the time it appears to drift westward. Draw a diagram and explain this *retrograde* motion.

14 Recall that Galileo discovered four moons of Jupiter. How could you determine the mass of Jupiter from observations of these moons and their orbits?

15 If a spacecraft in a circular orbit turns on its rocket engines to produce an acceleration in the same direction as its velocity, its average speed in its new orbit is reduced. How could this come about?

PROBLEMS

1 A 1,200-kg car is going around a curve of 150-m radius at a speed of 30 m/sec (67 mi/hr). Find the centripetal acceleration and the centripetal force. *Ans.* 6 m/sec²; 7,200 N

2 A 2,800-lb car begins to skid when traveling 90 ft/sec (61 mi/hr) around a level curve of 400-ft radius. Find the centripetal acceleration, the centripetal force, and the coefficient of friction between the tires and the road.

3 A 160-lb pilot in an aircraft moving at a constant speed of 600 ft/sec pulls out of a vertical dive along an arc of a circle of 4,000-ft radius. Find the centripetal acceleration, the centripetal force, and the total force which the aircraft exerts on the pilot at the bottom of the dive.
Ans. 90 ft/sec²; 450 lb; 610 lb

4 An 8,000-kg aircraft with a constant speed of 150 m/sec (336 mi/hr) pulls out of a vertical dive along an arc of a circle of radius 300 m. Find the centripetal acceleration and the total lift required at the bottom of the dive.

5 A 0.5-kg mass is suspended from a cord 2 m long to form a simple pendulum. If the mass is pulled to one side until it is raised 0.225 m and then released, find the velocity of the mass and the tension in the cord at the bottom of the swing. *Ans.* 2.1 m/sec; 6.01 N

6 People's responses to high accelerations are checked by use of a test chamber at one end of a 7-m horizontal beam which is rotated about a vertical axis at its other end. How many revolutions per minute must the beam make to provide an acceleration of 10 *g*'s (98 m/sec²)?

7 What is the smallest speed an airplane can have making a vertical loop with a radius of 720 m if at the top of the loop objects in the plane are not to start dropping? *Ans.* 84 m/sec

8 What is the highest speed at which an automobile can travel around a curve 100 ft in radius on a level road if the available coefficient of friction between the road and the tires is 0.49?

9 A curve of 200-m radius is banked at 20°. At what speed must an automobile go around this curve if no frictional forces are to be used to keep the automobile on its circular path?
Ans. 26.7 m/sec

10 What is the angle at which a speedway must be banked for cars running at 150 ft/sec if the radius of curvature is 1,000 ft and no frictional force is to be involved?

11 A 0.3-kg ball is whirled in a vertical circle on the end of a string 0.8 m long. At what speed

will the tension in the string be zero at the top of the circle? What will the tension be at the bottom if the tension drops to zero at the very top?   Ans. 2.8 m/sec; 17.6 N

12 A boy weighing 64 lb is swinging in such a way that he describes an arc of 10-ft radius. If his horizontal speed at the lowest point of the swing is 6 ft/sec, what is the total force which the ropes of the swing must sustain at that instant?

13 With what speed must a 1.5-kg pendulum bob swing in the circular path of the accompanying figure if the supporting cord is 1.2 m long and θ is 30°? Find the tension in the cord.

Ans. 1.84 m/sec; 17.0 N

Problems 13 and 14

14 Show that the period T (time for one complete revolution) of a conical pendulum of length l (see accompanying figure) is given by $T = 2\pi\sqrt{(l\cos\theta)/g}$, where θ is the angle between the cord and the vertical.

15 A 0.8-kg pendulum bob hangs from the roof of a moving van. If the van is traveling 30 m/sec around a curve of 200-m radius, find the angle which the cord makes with the vertical when the bob is at rest relative to the van. What is the tension in the cord?   Ans. 24.7°; 8.63 N

16 At what angle should a curve of 150-ft radius be banked if no frictional forces are to be required for a speed of 75 ft/sec? What force would the road exert on a 3,200-lb car?

17 A small cart on the circular loop-the-loop track of the accompanying figure can be approximated as a mass m sliding on a frictionless track. (a) Find the smallest height h in terms of the radius

r from which the mass can be released and still remain in contact with the track throughout the path. (b) If the cart starts from a height of $3r$, what is its acceleration at the top of the loop? With what force does the track press down on the cart at the top? When the cart starts at $h = 3r$, find the horizontal and vertical components of its acceleration at point A on the end of a horizontal diameter.

Ans. (a) 2.5r; (b) 2g; mg; 4g, g

Problems 17 to 20

18 The track in the accompanying figure is frictionless and has a radius of 2 ft. If a 4-lb block is released at A, find its speed at the bottom and the total force exerted on the block by the track at the bottom.

19 If the cart of Prob. 17 starts from a height of $5r$, find its kinetic energy and the force it exerts on the track (a) at the bottom, (b) at the top of the circle, and (c) at point A.
Ans. (a) 5mgr; 11mg; (b) 3mgr; 5mg; (c) 4mgr; 8mg

20 If the track in the accompanying figure has a radius of 0.25 m and is frictionless, find the speed and the force exerted on the track for a 50-g cube released from a height of 1 m (a) at the bottom, (b) at the top, and (c) at point A.

21 Calculate the attractive force of the earth for the moon. How many times greater is the attraction of the earth for the sun?
Ans. 2.03 × 10²⁰ N; 172

22 How many revolutions per day would the earth have to make in order for the weight of a body at the equator to become zero?

23 When a sphere of lead is placed 40 cm from another sphere of lead of mass 10 kg, the attraction of one for the other is found to be 8×10^{-8} N. What is the mass of the first sphere?   Ans. 19.2 kg

24 Find the gravitational attraction between two identical 50-kg lead spheres with centers 0.8 m apart. Compare this force with the earth's gravitational force on one of the spheres.

25 It is proposed to put a space station in a circular orbit at a distance of one-half earth's radius above the earth's surface. What will be the acceleration due to gravity at this station? Find the speed of this station if it is to go around the center of the earth in a circular orbit. How long will it take to make one complete revolution? *Ans.* 4.36 m/sec²; 6,450 m/sec; 9.31 × 10³ sec

26 If a planet had a circular orbit about our sun with a radius four times that of the earth's orbit, what would be its period of revolution and orbital speed?

27 Find the speed of a spacecraft circling the moon in a parking orbit at a radius of 1,800 km (60 km above the lunar surface). What is the period (time for one complete orbit) for this spacecraft?
Ans. 1,650 m/sec; 6,850 sec

28 Find the radius of the orbit of a satellite which revolves about the earth in one sidereal day (86,164 sec). What is its orbital velocity?

29 A satellite with a mass of 1,500 kg is to be placed in a circular orbit of radius 10^7 m (3.63 × 10^6 m above the earth's surface). Find (*a*) the speed of the satellite in orbit and (*b*) the energy which must be provided to the satellite to place it in this orbit.
Ans. (*a*) 6,320 m/sec; (*b*) 6.4 × 10^{10} J

30 Find the centripetal acceleration of an object at the equator due to the rotation of the earth. By how much does the rotation of the earth reduce the weight of an 80-kg man?

31 A rocket is 1,274 km (0.2 earth radius) above the surface of the earth. (*a*) What is *g* at this height? (*b*) Find the weight of a 5-kg mass at this point. (*c*) What is the potential energy of 5 kg at this position relative to the earth's surface as the zero?
Ans. (*a*) 6.81 m/sec²; (*b*) 34 N; (*c*) 5.2 × 10^7 J

32 From the masses and radii given in Table 9.1 estimate the acceleration due to gravity on the surface of (*a*) the moon, (*b*) Venus, (*c*) Jupiter, and (*d*) Mars. What would be the weight of a 2-kg mass at the surface of each?

33 Find the minimum speed with which a bullet must leave the gun of an astronaut on the moon if the bullet is to escape from the moon. If the moon had an atmosphere, what effect would that

have on the required escape velocity? How high would the bullet go if it had one-half the escape velocity? *Ans.* 2,380 m/sec; 5.79 × 10^5 m

34 Imagine a planet of twice the average density of the earth but with half the radius. Calculate the acceleration due to gravity at the surface of this planet.

35 How fast must an aircraft be able to fly due west to keep the sun always on the meridian (*a*) at the equator and (*b*) at 40° latitude? What is the centripetal acceleration in each case?
Ans. (*a*) 1,670 km/hr, 0.034 m/sec²; (*b*) 1,280 km/hr, 0.026 m/sec²

36 Sputnik, the first artificial earth satellite launched by the USSR on Oct. 4, 1957, took 96 min to traverse its almost circular orbit. Calculate its approximate height above the earth's surface and the acceleration due to gravity at that height.

37 Halley's comet has been observed on every passage near the sun at intervals of from 74 to 79 years since 240 B.C. (The variation in the period is due to perturbations produced by the major planets.) It is due about 1986, having appeared in 1910. Assuming that it describes an elliptical path about the sun with a 76-year period, calculate the approximate semimajor axis of its orbit. *Ans.* 2.7 × 10^9 km

38 Show that if a body describes a circular path under the influence of the gravitational inverse-square force law, the square of the period is proportional to the cube of the path radius (Kepler's third law for a circular orbit) and find the proportionality constant in terms of the central mass *M* and universal constants.

39 Show that for any homogeneous spherical planet of radius *R* the period of a satellite orbit with radius *r* only slightly greater than *R* depends only on the density of the planet and is given by $T = 198$ min × $(\rho_w/\rho)^{1/2}$, where ρ_w and ρ are the densities of water and the planet, respectively.

40 (*a*) A planet has a satellite which can be observed by a telescope. If the satellite has a period *T* and a radius (or semimajor axis) *a*, find the mass of the the planet. (*b*) From the

period of the moon and the radius of the moon's orbit, calculate the mass of the earth and compare your value with that quoted in Table 9.1. (*c*) Mars has satellites Phobas and Deimos, which have radii (semimajor axes) of 9.4×10^3 and 2.34×10^4 km and periods of 0.32 and 1.26 days, respectively. Estimate the mass of Mars from these data.

41 A 100-g mass (see accompanying figure) slides down a smooth (frictionless) hemisphere of 0.25-m radius, starting at the top with a very tiny speed. Find the force exerted on the mass by

Problem 41

the hemisphere as a function of the angle θ and the angle at which the mass leaves the surface of the hemisphere.

Ans. $0.98(3 \cos \theta - 2)$ N; 48.2°

42 If a rocket ship moves toward the sun on a path such that it is always on the line connecting earth and sun, show that the new gravitational force will be zero (neglecting the fields of other planets) at a distance d from the earth given by $d = r \sqrt{m_E / M_S}$, where r is the distance from the sun to earth. (*Hint:* $M_S \gg m_E$.)

AN IDEAL RIGID BODY does not change its shape under the influence of forces. Actually, a real solid body undergoes a change in shape or size when subjected to forces. If the forces applied are not too great, the body returns to its original condition when they are removed. The elastic properties of solid bodies such as cables, springs, and beams are of great engineering importance. When an elastic body such as a spring supporting a mass is stretched and then released, internal forces accelerate the mass toward its equilibrium position. When the mass reaches this position, it has kinetic energy and overshoots until restoring forces in the opposite direction bring it to a halt. As a consequence the mass oscillates, performing a harmonic motion.

CHAPTER TEN
ELASTICITY AND HARMONIC MOTION

10.1 ELASTIC DEFORMATION AND HOOKE'S LAW

When a solid object is distorted by external forces, the changes in size or shape depend on the physical properties of the material of which the object is composed. These changes reveal significant and useful information concerning the internal forces between the constituent atoms.

When we hang a weight from a spring, the spring stretches. If we increase the weight, the spring stretches still more. If we remove the weight, the spring returns to its original length. When a diver stands at the end of a diving board, the board is distorted, but when he leaves the board, it returns to its original position. When an archer prepares to shoot an arrow, he bends the bow, which springs back to its original form as the arrow is released. The spring, the diving board, and the bow are examples of elastic objects. *Elasticity is that property of a body by which it experiences a change in shape or volume when a deforming force acts upon it and by which it returns to its original size or shape when the deforming force is removed* (Fig. 10.1).

Materials which do not resume their original shape after being distorted are said to be *inelastic*. Mud, putty, and pastry dough are examples. Lead and solder are relatively inelastic, since it is easy to distort them permanently.

For an ideal spring the extension is directly proportional to the applied force (Sec. 7.5). For other sorts of elastic distortion we find that the magnitude of the deformation is proportional to the force (or

Figure 10.1
Elastic deformation of a golf ball.

torque) producing it. We may generalize Hooke's law and state: *The deformation of an elastic body is directly proportional to the applied force.* Whenever an elastic body is distorted by a force applied by some agent, the body exerts an equal and opposite force on the agent. For example, when we stretch a spring with an applied force F_a, the spring pulls on us with an equal and opposite restoring force F_r. By Hooke's law

$$F_a = ky \quad \text{and} \quad F_r = -ky \tag{10.1}$$

The act of stretching (or compressing) a spring gives it potential energy. It can do work through the action of the restoring force. The potential energy stored for a displacement y is given by Eq. (7.8):

$$P = \tfrac{1}{2}ky^2 \tag{10.2}$$

10.2 VIBRATIONS

When a mass on the end of a spring is pulled down and then released, it moves back and forth with a *vibratory* or *harmonic* motion (Fig. 10.2). Such periodic motions are very common. They occur in all sorts of mechanical structures (such as automobiles, bridges, and buildings), in sound sources, in waves, and in electric oscillations. When a clock pendulum is displaced and then released, a vibratory motion is established. A further illustration of harmonic motion is afforded by clamping one end of a strip of steel (Fig. 10.3) in a vise and displacing the other end. When released, the strip vibrates back and forth with a period which depends on the characteristics of the strip of steel. Other examples of harmonic motion are those of a piston in a steam engine or an automobile and the vibration of a tree limb in a breeze.

Elastic forces lead to these cyclic motions. In the case of a mass on the end of a spring, the restoring force is upward when the mass is displaced downward. The mass is pulled toward its equilibrium position, but when it reaches this position, it has a velocity and continues upward by virtue of Newton's first law. Thus, it overshoots the equilibrium position until there is a displacement in the opposite direction. The mass on the end of the spring moves back and forth until its energy is gradually dissipated in friction of various kinds.

Figure 10.2

Oscillatory displacement of a mass suspended from a spring.

10.3 SIMPLE HARMONIC MOTION

Of the many kinds of vibratory motion, we shall consider quantitatively only the simplest type. *Simple harmonic motion is motion in which the acceleration is proportional to the displacement from the equilibrium position and in the opposite direction.*

In discussing simple harmonic motion, several terms must be defined:

1 The *displacement* is the distance of a body from its equilibrium position.

2 The *amplitude* is the maximum value of the displacement, or the distance from the equilibrium position to the end of the vibrating path.

3 The *period* T is the time necessary for one complete vibration or cycle.

4 The *frequency* ν is the number of complete vibrations made in unit time. The basic unit of one cycle per second has been named the *hertz* (Hz) in honor of Heinrich Hertz, pioneer in radio waves in 1888. Clearly, the frequency is the reciprocal of the period; $\nu = 1/T$.

To develop equations relating the displacement, the amplitude, the frequency, and the acceleration in simple harmonic motion, we make use of the fact that *the shadow of a body which is performing uniform circular motion describes simple harmonic motion.* Consider an object moving around a circle (Fig. 10.4) with uniform speed. Imagine the shadow of this body being projected on the wall at the right by means of a beam of light coming from the left. The shadow moves up and down as the particle moves around the circle with constant speed. Let R be the radius of the circle, which we call the *circle of reference,* and θ the angle between the radius vector to the particle and the line OX. The displacement of the shadow from the "equilibrium position" is

$$y = R \sin \theta \qquad (10.3)$$

Let A represent the acceleration of the particle P toward the center of the circle and a the acceleration of the shadow on the wall. By Eq. (9.1), $A = V^2/R$, where V is the speed of the particle; a is the vertical component of A, which is $A \sin \theta$. Since a is directed downward when y is upward, and vice versa,

$$a = -A \frac{y}{R} = -\frac{V^2}{R^2} y \qquad (10.4)$$

Since neither V nor R changes in magnitude during the motion, V^2/R^2 is a constant. We observe that the *acceleration* of the shadow *is proportional to the displacement and in the opposite direction.* This establishes the fact that *the shadow describes simple harmonic motion.*

10.4 THE PERIOD OF A SIMPLE HARMONIC MOTION

Of all the characteristics of a harmonic motion, the period is one of the easiest to measure; therefore, let us rewrite our expression for the acceleration in terms of this quantity. The period T is the time for the body to make one revolution on the circle of reference; $T = 2\pi R/V$, and $V^2/R^2 = 4\pi^2/T^2$. Therefore,

$$a = -\frac{4\pi^2}{T^2} y \qquad (10.4a)$$

Figure 10.3
Harmonic vibrations of a strip of steel.

Figure 10.4
The shadow of a body moving with uniform circular motion oscillates with simple harmonic motion.

Figure 10.5

The speed v of the shadow is $V \cos \theta$, where V is the speed of the body on the circle of reference.

If ν represents the frequency, we may rewrite Eq. (10.4a) as

$$a = -4\pi^2 \nu^2 y \qquad (10.4b)$$

When Eq. (10.4a) is solved for T, it yields

$$T = 2\pi \sqrt{\frac{-y}{a}} = 2\pi \sqrt{-\frac{\text{displacement}}{\text{acceleration}}} \qquad (10.4c)$$

The quantity under the square-root sign is positive since the displacement and acceleration are of opposite sign. *The period of a simple harmonic motion is independent of the amplitude.* If the amplitude is small, the body does not move far, but it moves slowly, so that the time required to execute one complete vibration is the same as when the amplitude is larger. This interesting fact about simple harmonic motion was discovered by Galileo, who timed the swinging of a great chandelier in the cathedral at Pisa, using his own pulse as a clock.

10.5   VELOCITY IN SIMPLE HARMONIC MOTION

The displacement, the acceleration, and the velocity of a body describing simple harmonic motion are all changing constantly. As we have seen, the shadow of uniform circular motion on a wall (or its projection on a diameter of the circle) describes simple harmonic motion. From Fig. 10.5 we see that the velocity v of a body describing simple harmonic motion is given by

$$v = V \cos \theta = V \frac{\sqrt{R^2 - y^2}}{R} \qquad (10.5)$$

or

$$v = \frac{2\pi}{T} \sqrt{R^2 - y^2} = 2\pi\nu \sqrt{R^2 - y^2} \qquad (10.5a)$$

If we start a clock at the instant the body of Figs. 10.4 and 10.5 passes the line OX on its upward path, the angle θ is given by $\theta = \omega t = 2\pi t / T = 2\pi\nu t$. Hence, by Eqs. (10.3), (10.5), and (10.4b),

$$y = R \sin 2\pi\nu t \qquad (10.6)$$

$$v = V \cos \theta = 2\pi\nu R \cos 2\pi\nu t = 2\pi\nu R \sin\left(2\pi\nu t + \frac{\pi}{2}\right) \qquad (10.7)$$

$$a = -4\pi^2 \nu^2 R \sin 2\pi\nu t = 4\pi^2 \nu^2 R \sin\left(2\pi\nu t + \pi\right) \qquad (10.8)$$

We see (Fig. 10.6) that the velocity reaches its maximum 90° ($\pi/2$ rad) ahead of the displacement. Similarly, it passes through its zeros and reaches its minimum value one-fourth of a period ahead of the displace-

Figure 10.6
How displacement, velocity, and acceleration vary with time in simple harmonic motion.

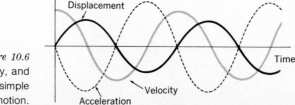

ment. We say that the velocity *leads* the displacement by 90° or $\frac{1}{4}$ period. Similarly, the acceleration leads the velocity by 90° and the displacement by 180°. The displacement, velocity, and acceleration can all be represented by sine (or cosine) waves of the same frequency, but differing in *phase*. Two variables described by sine waves are said to be *in phase* if they reach their maxima, zeros, and minima at the same instant.

Example

A body describes simple harmonic motion with an amplitude of 5 cm and a period of 0.25 sec. Find the acceleration and speed of the body when the displacement is 5, 3, and 0 cm.

By Eq. (10.4a),

$$a = -\frac{4\pi^2}{T^2} y$$

and by Eq. (10.5a),

$$v = \frac{2\pi}{T} \sqrt{R^2 - y^2}$$

The amplitude $R = 5$ cm. When $y = 5$ cm,

$$a = -\frac{4\pi^2 \times 0.05 \text{ m}}{(0.25 \text{ sec})^2} = -31.6 \text{ m/sec}^2$$

$$v = \frac{2\pi}{0.25 \text{ sec}} \sqrt{(0.05 \text{ m})^2 - (0.05 \text{ m})^2} = 0$$

When $y = 3$ cm,

$$a = -\frac{4\pi^2 \times 0.03 \text{ m}}{(0.25 \text{ sec})^2} = -19 \text{ m/sec}^2$$

$$v = \frac{2\pi}{0.25 \text{ sec}} \sqrt{(0.05 \text{ m})^2 - (0.03 \text{ m})^2} = \pm 1.0 \text{ m/sec}$$

When $y = 0$ cm, $a = 0$ and

$$v = \frac{2\pi}{0.25 \text{ sec}}(0.05 \text{ m}) = \pm 1.26 \text{ m/sec} \qquad ////$$

10.6 FORCE AND ENERGY RELATIONS

When a shadow describes simple harmonic motion, of course no force need act on the shadow. However, if a mass such as the bob of a pendulum describes simple harmonic motion, there must be an unbalanced force F_r to produce the acceleration. By Newton's second law,

$$F_r = ma = -4\pi^2 v^2 m y \qquad (10.9)$$

where y is the displacement.

For a mass on the end of a spring, a displacement y brings into play a restoring force $F_r = -ky$, precisely that required to produce simple harmonic motion when the mass is released. If we replace F_r with $-ky$ in Eq. (10.9) and solve for $1/v$, we obtain

$$\frac{1}{v} = T = 2\pi \sqrt{\frac{m}{k}} \qquad (10.10)$$

When the mass has a displacement y, its potential energy is $\frac{1}{2}ky^2$, the work (Eq. 10.2) required to produce this displacement from the equilibrium position. The kinetic energy of the mass is $\frac{1}{2}mv^2$, or $2\pi^2\nu^2(R^2 - y^2)m$ if we make use of the fact that $v = 2\pi\nu\sqrt{R^2 - y^2}$ (Eq. 10.5a). The total energy, the sum of the kinetic and potential energies, remains constant in simple harmonic motion. At the ends of the motion the kinetic energy is zero and the total energy is $\frac{1}{2}kR^2$, while at the equilibrium position the potential is zero and the kinetic energy is $2\pi^2\nu^2R^2m$. Thus,

$$\text{Total energy} = \tfrac{1}{2}kR^2 = 2\pi^2\nu^2R^2m = \tfrac{1}{2}ky^2 + 2\pi^2\nu^2(R^2 - y^2)m \quad (10.11)$$

The total energy of a body performing simple harmonic motion is proportional to the square of the frequency and to the square of the amplitude.

Example

A mass of 0.500 kg stretches a spring 20.0 cm. Find the force constant of the spring. If the mass is pulled down an additional 5.00 cm and then released, find the initial restoring force, the period, and the total energy of the resulting simple harmonic motion.

In this case $F_a = mg = 0.500$ kg $\times$ 9.80 N/kg = 4.90 N, and $y = 0.200$ m. By Eq. (10.1),

$$F_a = ky = -F_r$$
$$4.90 \text{ N} = k \times 0.200 \text{ m}$$
$$k = 24.5 \text{ N/m}$$

When m is displaced 5 cm from its new equilibrium position and released, the initial restoring force is upward and has a magnitude $ky_0 = 24.5 \times 0.05 = 1.23$ N. By Eq. (10.10),

$$T = 2\pi\sqrt{\frac{m}{k}} = 2\pi\sqrt{\frac{0.500 \text{ kg}}{24.5 \text{ N/m}}} = 0.090 \text{ sec}$$

The amplitude $R = 0.0500$ m, since that is the maximum displacement. By Eq. (10.11),

$$\text{Total energy} = \tfrac{1}{2}kR^2$$
$$= \tfrac{1}{2}(24.5 \text{ N/m})(0.0500 \text{ m})^2$$
$$= 0.0306 \text{ J} \qquad \qquad ////$$

If a particle is oscillating in a viscous medium, it does work on the medium and the energy of vibration is correspondingly reduced. For this reason the amplitude diminishes with time (Fig. 10.7). Such vibrations are said to be *damped.*

Figure 10.7
In a damped vibration, the amplitude
decreases with time.

10.7 THE SIMPLE PENDULUM

A simple pendulum affords an illustration of simple harmonic motion. When the pendulum is displaced from its position of equilibrium (Fig. 10.8), the restoring force is $F_r = -mg \sin \theta$. Since $F_r = ma$, the acceleration which this force produces is $a = -g \sin \theta$. The displacement of the pendulum bob from its position of equilibrium N is $S = l\theta$. Hence,

$$\frac{\text{Displacement}}{\text{Acceleration}} = \frac{S}{a} = \frac{l\theta}{-g \sin \theta}$$

For small angles, $\sin \theta = \theta$ when the angle is measured in radians, and for small displacements of the pendulum,

$$\frac{\text{Displacement}}{\text{Acceleration}} = -\frac{l}{g} = \text{constant}$$

Consequently, the pendulum moves with nearly simple harmonic motion for small angles of swing; it deviates from simple harmonic motion for large angular amplitudes. Its period is, by Eq. (10.4c),

$$T = 2\pi \sqrt{\frac{l}{g}} \qquad (10.12)$$

The longer the pendulum, the greater its period; the greater the acceleration of gravity, the shorter the period of the pendulum.

Figure 10.8
A simple pendulum moves with approximately simple harmonic motion.

10.8 STRESS AND STRAIN: YOUNG'S MODULUS

To treat problems in elasticity in a convenient and consistent way, we introduce two new terms, *stress* and *strain*. When we apply a force to distort a body, internal forces within the body resist the distortion. For example, if we try to stretch a light bar (Fig. 10.9), the material at the left of some plane PP' exerts a force **F** to the left on the material at the right of PP'. Since the bar is in equilibrium, the material at the right of PP' exerts an equal and opposite force on the material at the left. The material is said to be under stress. For our bar *the stress is the ratio of the internal force to the area over which the force is distributed*. (When the stress varies from point to point in a material, we must take the limit of this ratio as the area chosen approaches zero

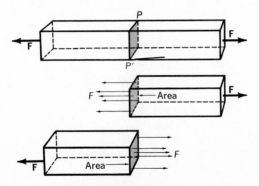

Figure 10.9
The stretching stress in a bar is the ratio of the internal tensile force to the cross-sectional area over which the force is distributed.

to obtain the stress at a point.) We shall consider only situations in which the stress is uniform over the area in question. In this case

$$\text{Stress} = \frac{\text{force}}{\text{area}}$$

As a result of the applied stress, some change in the shape or size of the elastic body is produced. The term *strain* is applied to the relative change occurring in the dimensions or shape of a body when it is subjected to a stress. We shall define three particular types of strain later in this chapter. For all these types of strain Hooke's law states that *the stress applied is directly proportional to the strain produced.*

When a steel cable supports an elevator or a suspension bridge, it is stretched by the load. How much its length changes depends on the load, the cross-sectional area of the cable, its original length, and the material of which it is made. The bigger the load, the larger the stretch expected; the greater the cross-sectional area of the cable, the smaller the stretch. The stress in the cable is the force applied divided by this area. *The stretching strain is the ratio of the change in length to the original length.*

$$\text{Stretching (or longitudinal) strain} = \frac{\text{change in length}}{\text{original length}} = \frac{\Delta l}{l}$$

For an elastic material the stretching stress is directly proportional to the longitudinal strain. *The ratio of the stretching stress to the longitudinal strain is known as Young's modulus* (or sometimes as the *longitudinal* modulus or *stretch* modulus) of elasticity. We denote it by Y.

$$\frac{F}{A} = Y\frac{\Delta l}{l} \tag{10.13}$$

Example

A wire 3.00 m long with a cross section of 0.08 cm² hangs vertically. When a load of 2,000 N is applied to the wire, it stretches 3.9 mm. Find Young's modulus of elasticity.

$$\text{Young's modulus} = \frac{\text{stress}}{\text{strain}} = \frac{F/A}{\Delta l/l} = \frac{2{,}000 \text{ N}/8 \times 10^{-6} \text{ m}^2}{3.9 \times 10^{-3} \text{ m}/3 \text{ m}}$$

$$= 1.9 \times 10^{11} \text{ N/m}^2 \qquad\qquad ////$$

Figure 10.10
The elongation of a wire is approximately proportional to the applied tension until the elastic limit is reached.

10.9 LIMIT OF ELASTICITY

A body that has been deformed and then released returns to its original size and shape provided the stress has not exceeded the *elastic limit*, defined as the maximum stress from which the substance completely recovers its original size and shape. When the elastic limit is exceeded, the body acquires a *permanent set*. The elastic limit differs widely for different materials, being high for steel and low for lead; for any given substance it depends on the temperature. Figure 10.10 shows a typical relation between tension and elongation for a wire. The greatest stress

Table 10.1
Typical Elastic Constants

	Young's modulus		Bulk modulus, N/m²	Shear modulus, N/m²	Elastic limit, N/m²	Ultimate tensile stress, N/m²
Material	lb/in.²	N/m²				
Aluminum	10×10^6	6.9×10^{10}	6.9×10^{10}	2.6×10^{10}	1.3×10^8	1.4×10^8
Brass	14×10^6	9.8×10^{10}	5.8×10^{10}	3.5×10^{10}	3.8×10^8	4.5×10^8
Copper	17×10^6	12×10^{10}	12×10^{10}	4.1×10^{10}	1.5×10^8	3.4×10^8
Medium steel	30×10^6	21×10^{10}	16×10^{10}	8.4×10^{10}	2.5×10^8	4.8×10^8

which can be applied to the wire is the *ultimate tensile stress.* Table 10.1 lists some typical elastic constants.

When a metal rod is stretched beyond its elastic limit and the stress is increased still further, a stage is reached at which the rod begins to stretch rapidly, even though the stress is somewhat decreased. The stress at which this begins is called the *yield point.* Even though the metal is cold, it behaves as if it were in a semifluid state. Once a metal is strained beyond the elastic limit, the strain becomes a function of the time which has elapsed after application of the load. Under such conditions there is a viscous flow which depends both on temperature and on the applied load.

10.10 STIFFNESS AND STRENGTH OF BEAMS

It is important in many engineering applications to be able to calculate the amount of bending that will be produced in a beam by a given load (Fig. 10.11). The deflection depends on the length of the beam, its breadth, and its depth.

$$\text{Deflection} = \frac{\text{load} \times \text{length}^3}{4 \times \text{Young's modulus} \times \text{breadth} \times \text{depth}^3}$$

The stiffer of two beams is not necessarily the stronger. The mere fact that the one beam bends more than the other does not mean that it will hold less. The strength depends on the same dimensions as the

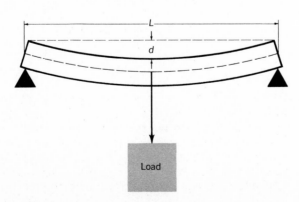

Figure 10.11
Bending of a beam under load.

stiffness, but it depends on them in a different way: the strength is proportional to the breadth and to the square of the depth and inversely proportional to the length.

In the bending of a beam the top layer is shortened since it resists compression, while the lower layer is lengthened and must resist tension. A central layer of the beam remains the same length. Therefore, to make a beam as strong or as stiff as possible for a given amount of material, most of the material is put in the upper and lower layers, and relatively little in the middle. For this reason steel beams are usually manufactured in shapes similar to that of the letter I. They are called I beams and have great strength and stiffness for a given amount of steel.

10.11 VOLUME ELASTICITY

When a solid body such as a piece of iron is immersed in a liquid so that a uniform pressure is applied to its surface, a change in the volume of the solid occurs. By definition,

$$\text{Volume (or bulk) strain} = \frac{\text{change in volume}}{\text{original volume}} = \frac{\Delta V}{V}$$

The corresponding stress (force/area) is taken as $-\Delta p$, with the minus sign appearing because an increase in pressure produces a decrease in volume. The ratio of volume stress to volume strain is called the *bulk modulus* (or *volume modulus*). We denote it by B.

$$B = \frac{\text{volume stress}}{\text{volume strain}} = -\frac{\Delta p}{\Delta V / V} \tag{10.14}$$

Since an increase in pressure produces a decrease in volume, the bulk modulus B is inherently positive.

Example

A sphere of copper having a volume of 100 cm³ is subjected to a pressure of 7×10^6 N/m². Find the change in volume that takes place.

$$-\Delta p = B \frac{\Delta V}{V} \tag{10.14a}$$

$$-7 \times 10^6 \text{ N/m}^2 = 12 \times 10^{10} \text{ N/m}^2 \frac{\Delta V}{100 \times 10^{-6} \text{ m}^3}$$

$$\Delta V = -5.8 \times 10^{-9} \text{ m}^3 = -5.8 \text{ mm}^3 \qquad ////$$

Not only solids, but also liquids and gases exhibit volume elasticity and have bulk moduli defined by Eq. (10.14). The reciprocal of the bulk modulus is called the *compressibility k*. Thus $k = 1/B$. It is customary to list compressibilities rather than bulk moduli for liquids (Table 10.2). The volume modulus of gases is discussed further in Sec. 14.2.

Table 10.2
Compressibility of Liquids

Substance	Temperature, °C	Compressibility per atmosphere
Ethyl alcohol	14	0.0000987
Ethyl ether	0	0.000143
Kerosene	20	0.0000543
Mercury	20	0.0000039
Turpentine	20	0.000075
Water	20	0.000048

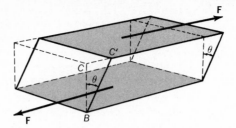

Figure 10.12
The shearing stress is the ratio of the force F to the area of one of the shaded faces; the shearing strain is the angle θ in radians ($\theta = CC'/BC$).

10.12 ELASTICITY OF SHAPE: SHEAR MODULUS

If one cover of a thick book is held firmly on the table, and a force parallel to the top of the table is applied to the other cover, the shape of the book is changed while its thickness and volume remain the same. Such a deformation in shape is called a *shear*. The *shearing stress* is defined as the tangential force per unit area of the surface, and the shearing strain is the angle θ of Fig. 10.12 in radians ($\theta = CC'/BC$). The ratio of the shearing stress to shearing strain is called the *shear modulus S* of the material. (It is also known as the *torsion modulus* or *modulus of rigidity*.)

$$\frac{F}{A} = S\theta \qquad (10.15)$$

The shear modulus is the elastic constant involved when one twists a rod or a tube in an application such as that of the drive shaft of an automobile. Clearly, this elasticity of shape is characteristic of solids only. It is not exhibited by liquids or gases, although liquids do evidence short-range forces between molecules which give rise to interesting surface phenomena.

QUESTIONS

1   If a body such as a bow is distorted until Hooke's law is no longer applicable, does it necessarily follow that the elastic limit has been exceeded? Explain.

2   A spring with force constant k is cut in half. What is the spring constant for one of the halves?

3   How does the period of a simple pendulum with a 200-g steel bob compare with that of a pendulum of the same length but with a 25-g wood bob? What differences in actual behavior would you expect?

4   If one timing device is based on a mass oscillating on the end of a spring and a second one is based on a pendulum, would they keep the same time on the moon as they keep on earth? Explain.

5   Qualitatively, what is the effect on the period of a pendulum of taking it (a) to the North Pole, (b) to the equator, and (c) to the top of Mount Whitney?

6   An unstretched spring has a force constant k. A mass m is hung from it, and equilibrium is achieved. Is the spring constant changed, or is k the same for this new equilibrium situation?

7   Why are precisely simple harmonic motions rare while approximately simple harmonic motions are common?

8   Two bodies of identical mass m are suspended from separate springs with force constants k_1 and k_2, where $k_1 > k_2$. If the bodies are given equal energies, for which spring is the amplitude greater? How do the speeds and equilibrium positions compare?

9   Why is the motion of a simple pendulum not

simple harmonic when the amplitude is large? Is the restoring force too big or too small? Does the period increase or decrease as the amplitude becomes larger?

10 While an elevator has a downward acceleration of $0.2g$, what is the effect on the period of a simple pendulum suspended from the top of the elevator? What is the effect on the period of a mass on the end of a spring?

11 If a hole could be bored through the earth and a body dropped into the hole, it would describe simple harmonic motion with an amplitude equal to the earth's radius. At the center of the earth would the body have any (a) weight, (b) mass, (c) acceleration, or (d) momentum?

12 For a uniform spring one should add one-third the mass of the spring to m in Eq. (10.10) if one wishes a more accurate relation. Why should the mass of the spring be involved in the period?

13 Should steel reinforcing rods in a concrete beam be embedded near the upper or near the lower side of the beam? Explain why.

14 If the collisions between the ball and the plate of Fig. 8.9 were elastic, would the resulting motion be simple harmonic? Would the displacement follow Eq. (10.6) for a half-period?

PROBLEMS

1 A stone is swinging in a horizontal circle 50 cm in diameter, making 20 rev/min. A distant light causes a shadow of the stone to be formed on a nearby wall. What is the amplitude of the motion of the shadow? What is the frequency? What is the period?
    Ans. 25 cm; 0.333 Hz; 3 sec

2 The mass on the end of a spring describes simple harmonic motion with a period of 2.5 sec. Find the acceleration when the displacement is 6 cm.

3 Find the length (in meters) of a pendulum which has a period of 3 sec.   *Ans.* 2.23 m

4 A pendulum executes 25 complete vibrations in 100 sec. Find the length of the pendulum.

5 When a mass of 800 g is suspended from a spring, the spring stretches 12.8 cm. What is the period of oscillation of the mass when it is given a small displacement? What is the frequency?   *Ans.* 0.718 sec; 1.39 Hz

6 A 0.5-kg body performs simple harmonic motion with a frequency of 4 vib/sec and an amplitude of 2 cm. Find the maximum velocity of the body, its maximum acceleration, and the maximum restoring force to which the body is subjected.

7 A ball moves in a circular path of 20-cm diameter with a constant angular speed of 3 rev/sec. Its shadow performs simple harmonic motion on the wall behind it. Find the acceleration and speed of the shadow (a) at the end of the motion, (b) at the equilibrium position, and (c) at a point 8 cm from the equilibrium position.
    Ans. (a) 35.5 m/sec²; 0; (b) 0; 1.88 m/sec; (c) 28.4 m/sec²; 1.13 m/sec

8 A body describing simple harmonic motion has a maximum acceleration of 6π m/sec² and a maximum speed of 2 m/sec. Find T and R.

9 A mass of 0.5 kg is suspended from a spring. A force of 0.5 N pulls the mass downward 0.25 m. If the mass is then released from rest, find (a) the energy associated with the simple harmonic motion, (b) the frequency of the motion, (c) the speed and acceleration of the mass as it passes the equilibrium position, and (d) the speed and acceleration of the mass when its displacement is 0.20 m.
    Ans. (a) 0.0625 J; (b) 0.318 Hz; (c) 0.5 m/sec; 0; (d) 0.3 m/sec; −0.8 m/sec²

10 Identical 0.8-kg masses are suspended from a spring and from a cord. The length of the cord is adjusted until the simple pendulum and the mass on the spring have identical periods of 0.5 sec on the earth, where $g = 9.8$ m/sec². If both were taken to Mars, where $g = 3.9$ m/sec², what would be the period of each?

11 A 0.5-kg mass on the end of a spring has a period of 3 sec and an amplitude of 0.1 m as it describes simple harmonic motion. Find the frequency of the motion, the acceleration of the mass at an end point of its motion, and the maximum velocity attained by the mass. What is the force constant of the spring?
    Ans. 0.333 Hz; 0.439 m/sec²; 0.209 m/sec; 2.19 N/m

12 A 100-g mass on the end of a spring describes simple harmonic motion with an amplitude of 7 cm and a period of 2 sec. Find the maximum

acceleration and maximum velocity of the mass and the force constant of the spring.

13 A simple harmonic motion has a period of 2.5 sec and an amplitude of 10 cm. What is the acceleration at maximum displacement? What is the acceleration 0.5 sec later?

Ans. −0.632 m/sec²; −0.195 m/sec²

14 A mass m suspended from a spring of constant k has a period T. If a mass M is added, the period becomes $2T$. Find M in terms of m.

15 A 0.4-kg mass is suspended from a spring. If the spring is pulled downward with a force of 1 N, it stretches an additional 10 cm. If the spring is released, the mass describes simple harmonic motion with an amplitude of 10 cm. Find the period of the motion and the velocity and acceleration of the mass when the displacement is 8 cm. Find the maximum values of the potential energy and kinetic energy of the vibrating mass, taking the equilibrium position for the zero of potential energy.

Ans. 1.26 sec; 0.3 m/sec; 2 m/sec²; 0.05 J

16 To be acceptable, an 8-kg airplane radio received must pass a shake test in which it is forced to describe simple harmonic oscillations at 100 Hz with a maximum acceleration of 10 g's ($g = 9.8$ m/sec²). Find the maximum displacement, unbalanced force, and velocity.

17 A particle moves with simple harmonic motion along a line between two points that are 16 cm apart. If the period is 2 sec, what is the acceleration of the particle at maximum displacement? What is the speed of the particle as it passes the midpoint? If the displacement is maximum when $t = 0$, write the equations for the displacement and velocity as functions of time. *Ans.* 0.79 m/sec²; 0.251 m/sec; $y = 0.08 \cos \pi t$ m; $v = -0.08 \pi \sin \pi t$ m/sec

18 In a gasoline engine the motion of a piston is approximately simple harmonic. A piston has a mass of 0.8 kg and a stroke (twice the amplitude) of 12 cm. Find the maximum acceleration and the maximum unbalanced force on the piston if it is making 40 complete vibrations each minute.

19 A 1.25-kg mass hangs in equilibrium from a spring of force constant 20 N/m. How far must the spring be stretched to give the mass an acceleration of 4 m/sec² upward when released? What is the period of the motion? Find the speed of the mass 0.1 sec after it passes the equilibrium position.

Ans. 0.25 m; 1.57 sec; 0.92 m/sec

20 A mass of 40 g is moving with simple harmonic motion with 6 cm amplitude and a period of 2 sec. What is the acceleration of the mass one-twelfth of a period after it has passed the midpoint? What is the force acting on the mass? What is the speed?

21 A particle describing simple harmonic motion has a period of 3 sec. If the speed of the particle is 0.6 m/sec when it has a displacement equal to half the amplitude, how long will it take the particle to reach the equilibrium position if it is moving (*a*) toward the equilibrium position and (*b*) away from the equilibrium position? What is the maximum speed of the particle?

Ans. (*a*) 0.25 sec; (*b*) 1.25 sec; 0.69 m/sec

22 A 0.1-kg mass suspended from a spring performs simple harmonic motion with a frequency of 1.5 Hz and an amplitude of 0.05 m. Find the maximum acceleration of the mass, the maximum velocity, the acceleration at the instant the displacement is 0.03 m, and the force needed to stretch the spring 0.05 m.

23 An elevator weighing 18,000 lb is supported by two steel cables, each 1 in.² in cross section. Find the elongation produced in the cables if the elevator is 100 ft from the supporting cylinder.

Ans. 0.36 in.

24 A brass wire with a cross section of 1.2 mm² is 2.4 m long when supporting a load of 2 kg. How much longer will it be when the load is increased to 6 kg?

25 A wire 2 m long with a diameter of 1.6 mm is elongated 0.4 mm when a 6-kg mass is hung on it. What is Young's modulus for the material?

Ans. 1.5 × 10¹¹ N/m²

26 How much will a copper wire 4 m long stretch when a mass of 2 kg is suspended from one end if the cross-sectional area is 0.02 cm²?

27 Find how many cables of 0.8-in.² cross section should be used to support an elevator which weighs 12,800 lb if the elevator will have a maximum acceleration of 10 ft/sec² and the stress is not to exceed one-fifth of the elastic limit of the cable steel, which is 40,000 lb/in.².

Ans. 3

28 A vertical steel column carries a load of 50,000

kg. The area of its cross section is 50 cm², and its length is 8 m. Find the decrease in length produced by this load.

29 A wire made of medium steel is 8 m long and has an area of 0.06 cm². What is the maximum load which this wire can support? What is the greatest load which can be supported without exceeding the elastic limit? If the wire is fastened at its upper end, how far can it be stretched without exceeding the elastic limit?

Ans. 2,880 N; 1,500 N; 9.5 mm

30 The shearing stress on an aluminum rivet is not to exceed one-tenth of the elastic limit of 19,000 lb/in.². How many rivets, each 0.2 in.² in cross-sectional area, are needed to hold two parts of an airplane wing together if each rivet carries the same fraction of the load and the total force on the rivets is 15,000 lb?

31 Find the depth in a lake at which the density of the water is 0.5 percent greater than at the surface. Ignore temperature differences. One atmosphere corresponds to a depth of 10.33 m.

Ans. 1,080 m

32 Find the increase in pressure required to increase the density of medium steel by 0.02 percent.

33 A hydraulic press contains 0.2 m³ of oil of compressibility 2×10^{-5} per atmosphere. Calculate the decrease in volume of the oil if it is subjected to a pressure of 450 atm. Find the bulk modulus of the oil in newtons per square meter.

Ans. 1.8×10^{-3} m³; 5.07×10^9 N/m²

34 Compute the compressibilities of copper and medium steel in atmospheres⁻¹. How do they compare with that of water?

35 Two springs each 15 cm in unstressed length are fastened to pins on the frictionless table 50 cm apart. The force constant k_1 for the spring on the left is 3 N/m, while k_2 for the spring on the right is 6 N/m. The springs are stretched and connected to hooks on opposite ends of a 1-kg mass. It is 5 cm between hooks, so the sum of the extensions of the two springs is 15 cm. (*a*) Find the equilibrium position for the mass. (*b*) What is the effective spring constant for the arrangement? (*c*) Find the period of small oscillations about the equilibrium position.

Ans. (*a*) 5 cm to right of center; (*b*) 9 N/m; (*c*) 2.09 sec

36 Prove that if a particle is simultaneously subjected to simple harmonic motion in both the x and y directions, the resultant motion of the particle is a circle, provided the amplitudes and frequencies of the two harmonic motions are the same and they are 90° out of phase; i.e., the displacement in one direction is maximum when the other displacement passes through zero.

37 A spring of constant k_1 is suspended from a hook. A second spring of constant k_2 hangs from the bottom of the first spring and supports a mass m from its lower extremity. Assume that both springs have negligible mass. Show that when m is displaced from its equilibrium position, it describes simple harmonic motion with a frequency given by

$$\nu = \frac{1}{2\pi} \sqrt{\frac{k_1 k_2}{(k_1 + k_2)m}}$$

38 Show that l m of liquid in a uniform U-tube will, if disturbed, perform simple harmonic motion with a period $T = \pi \sqrt{2l/g}$, provided friction is negligible.

IN CHAP. 4 we found that both the resultant of all forces and the resultant of all torques acting on a body in equilibrium are zero. In the six chapters which followed, we studied the translational motion of a body on which the resultant force is not zero. In this case the body undergoes an acceleration proportional to the force and inversely proportional to its mass. Now we turn to rotational motion, which we develop in terms of several quantities, each of which has an analog in translational motion.

CHAPTER ELEVEN
ROTATIONAL MOTION

11.1 ROTATIONAL VELOCITY

Rotational motion is common in our everyday life. The flywheel of a stationary engine, the armature of an electric motor, and the spinning top all perform rotational motion. The wheel of a moving automobile describes rotational and translational motions simultaneously. In general, the motion of any moving body may be regarded as a translation of the center of mass and a simultaneous rotation about the center of mass.

Consider a flywheel (Fig. 11.1) rotating about the axis O. A straight line at OA on the flywheel rotates from OA to OB in time t. The rate at which the line rotates, called its *angular velocity*, is usually expressed in radians per second, although it might also be measured in revolutions per second, revolutions per minute, or degrees per second. If the rate of rotation is constant, the angular velocity is constant and is equal to the angle turned through divided by the corresponding time. If we represent the angular velocity by ω, the angular displacement by θ, and the time by t, we have

$$\omega = \frac{\theta}{t} \tag{11.1}$$

The angular velocity ω is a vector quantity with its direction along the axis of rotation in the sense indicated by the outstretched thumb of the right hand when the fingers point in the direction of rotation.

If we consider the linear velocity of the point A on the flywheel, we observe that θ, the angle swept through in radians, is equal to the length of arc AB divided by the radius r. The distance s traveled by the point is just the arc length AB, and

$$s = \theta r \tag{11.2}$$

The linear velocity v of the point is given by

$$v = \frac{s}{t} = \frac{\theta r}{t} = \omega r \tag{11.3}$$

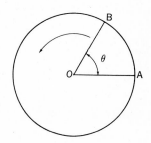

Figure 11.1
A flywheel rotating through an angle θ about an axis through O carries line OA to OB.

For a rigid body the linear velocity of a point increases as the distance from the axis of rotation increases. Note that Eqs. (11.2) and (11.3) are valid *only* if angles are measured in radians.

Example

The spoke of a wheel makes 60 rev/min. It is 50 cm in length. Find the linear velocities of a point on its outer end and of a point halfway between the axis and the outer end.

$$v_e = \omega r_e = 2\pi \text{ rad/sec} \times 50 \text{ cm} = 314 \text{ cm/sec}$$
$$v_m = \omega r_m = 2\pi \times 25 = 157 \text{ cm/sec} \qquad\qquad ////$$

11.2 ANGULAR ACCELERATION

In situations in which rotating bodies are speeding up or slowing down, the angular velocity is no longer constant. The *instantaneous angular velocity* ω is then defined by the relation

$$\omega = \lim_{\Delta t \to 0} \frac{\Delta\theta}{\Delta t} \qquad\qquad (11.1a)$$

The instantaneous angular acceleration α of a rotating object is defined as the rate of change of angular velocity. If the angular acceleration is constant,

$$\alpha = \frac{\Delta\omega}{\Delta t} = \frac{\omega - \omega_0}{t} \qquad\qquad (11.4)$$

where ω and ω_0 are the angular velocities of the body at the times t and 0.

Any point in a rotationally accelerated body has a linear acceleration a_t tangential to its path and given by

$$a_t = \alpha r$$

if we measure α in radians per second per second. This equation gives only the tangential component of the total linear acceleration of the point. The point is moving in a circle and has also a centripetal acceleration a_c equal to v^2/r.

Example

At a certain instant the angular velocity of a wheel is 10 rad/sec. In 20 sec the angular velocity has become 50 rad/sec. What is the angular acceleration?

$$\text{Angular acceleration} = \frac{\text{change in angular velocity}}{\text{time}}$$

$$= \frac{50 \text{ rad/sec} - 10 \text{ rad/sec}}{20 \text{ sec}}$$

$$= 2 \text{ rad/sec}^2 \qquad\qquad ////$$

Example

What are the tangential, centripetal, and total linear accelerations of a particle that is 0.5 m from the axis of a wheel (Fig. 11.2) when the angular acceleration of the wheel is 3 rad/sec² and ω is 2 rad/sec?

$$\text{Tangential linear acceleration} = \alpha r$$

$$a_t = 3 \text{ rad/sec}^2 \times 0.5 \text{ m} = 1.5 \text{ m/sec}^2$$

$$\text{Centripetal linear acceleration} = \frac{v^2}{r} = \omega^2 r$$

$$a_c = (2 \text{ rad/sec})^2 \times 0.5 \text{ m} = 2.0 \text{ m/sec}^2$$

$$\text{Total linear acceleration} = \sqrt{a_t{}^2 + a_c{}^2} = 2.5 \text{ m/sec}^2 \qquad ////$$

$\omega = 2 \text{ rad/sec}; \alpha = 3 \text{ rad/sec}^2$

Figure 11.2
The linear acceleration **a** is the resultant of the tangential acceleration **a**$_t$ and the centripetal acceleration **a**$_c$.

11.3 THE EQUATIONS OF ANGULAR MOTION

The equations of angular motion have the same form as the corresponding equations of translational motion because the angular quantities are defined analogously to the corresponding linear quantities. In the following paragraphs we treat several of the more common special cases.

Case 1 If a body rotates with uniform velocity ω, the angle θ through which it turns in t sec is

$$\theta = \omega t \qquad (11.1b)$$

Example

An electric motor rotates at a constant angular velocity of 1,800 rev/min. Find the angle turned through in 10 sec.

$$\omega = 1,800 \text{ rev/min} \times \frac{1 \text{ min}}{60 \text{ sec}} = 30 \text{ rev/sec}$$

$$= 30 \text{ rev/sec} \times 2\pi \text{ rad/rev} = 60\pi \text{ rad/sec}$$

$$\theta = \omega t = 600\pi = 1,885 \text{ rad} \qquad ////$$

Case 2 If a body begins to rotate from rest with a uniform angular acceleration α, the angular velocity ω at the end of t sec is αt by Eq. (11.4). During the acceleration the average angular velocity is given by

$$\omega_{av} = \frac{\omega_0 + \omega}{2} = \frac{0 + \alpha t}{2} = \frac{1}{2} \alpha t$$

The angle θ swept out in t sec is the average angular velocity times the time, so that

$$\theta = \tfrac{1}{2}\alpha t^2$$

Example

A flywheel starts from rest and has a uniform angular acceleration of 4 rad/sec² for a time of 10 sec. Find the final angular velocity, the average angular velocity during the 10-sec acceleration, and the angle turned through in the 10 sec.

$$\omega = \alpha t = 4 \text{ rad/sec}^2 \times 10 \text{ sec}$$
$$= 40 \text{ rad/sec}$$

$$\omega_{\text{av}} = \frac{\omega_0 + \omega}{2} = \frac{0 + 40}{2} = 20 \text{ rad/sec}$$

$$\theta = \tfrac{1}{2}\alpha t^2 = \tfrac{1}{2} \times 4 \text{ rad/sec}^2 \times 100 \text{ sec}^2$$
$$= 200 \text{ rad} \qquad\qquad\qquad ////$$

Case 3 If a body has an initial angular velocity ω_0 and a uniform angular acceleration α, its angular velocity at the end of t sec is given by

$$\omega = \omega_0 + \alpha t \qquad\qquad (11.4a)$$

If the acceleration is uniform, the average angular velocity is

$$\omega_{\text{av}} = \frac{\omega_0 + \omega}{2} = \omega_0 + \frac{\alpha t}{2}$$

The angle θ swept out is given by

$$\theta = \omega_{\text{av}}t = \omega_0 t + \tfrac{1}{2}\alpha t^2 \qquad\qquad (11.5)$$

If we eliminate t between Eqs. (11.4) and (11.5), we obtain

$$\omega^2 = \omega_0^2 + 2\alpha\theta \qquad\qquad (11.6)$$

11.4 KINETIC ENERGY OF ROTATION

A rotating body has kinetic energy by virtue of the motion of its parts. To find the expression for kinetic energy of rotation, consider Fig. 11.3, which represents a rigid body rotating with an angular velocity ω about an axis through O perpendicular to the plane of the paper. For the particle at A having a mass of m_1 and linear velocity v_1, the kinetic energy is $m_1 v_1^2/2$. Since $v_1 = \omega r_1$, the kinetic energy of the particle at A becomes

$$\frac{m_1 r_1^2 \omega^2}{2} = m_1 r_1^2 \frac{\omega^2}{2}$$

For other particles $m_2, m_3, m_4, \ldots$ moving with velocities $v_2, v_3, v_4, \ldots$ similar expressions are found. The total kinetic energy is

$$\frac{m_1 r_1^2 \omega^2}{2} + \frac{m_2 r_2^2 \omega^2}{2} + \cdots = \frac{\omega^2}{2}(m_1 r_1^2 + m_2 r_2^2 + \cdots)$$

$$= \tfrac{1}{2}\omega^2 \sum_i m_i r_i^2$$

The quantity $\Sigma m_i r_i^2$ is called the *moment of inertia* I of the body about the axis through O. For a rotating body,

$$\text{Kinetic energy} = \tfrac{1}{2}I\omega^2 \qquad\qquad (11.7)$$

This equation is the rotational analog of $\tfrac{1}{2}mv^2$ for the kinetic energy of a mass m moving with a translational velocity v. Note that ω must be expressed in radians per second.

Figure 11.3

The kinetic energy of rotation of a body is the sum of the kinetic energies of all the particles (such as m_1, m_2, . . .) of which the body is composed.

Ring or thin-walled cylinder
$I = MR^2$

Disk or solid cylinder
$I = \frac{1}{2}MR^2$

Annular cylinder or disk
$I = \frac{1}{2}M(R_2{}^2 + R_1{}^2)$

Solid sphere
$I = \frac{2}{5}MR^2$

Slender rod
or meterstick
$I = \frac{1}{12}Ml^2$

Rectangular plate
$I = \frac{1}{12}M(a^2 + b^2)$

Figure 11.4
Moments of inertia of common shapes about the indicated axes through the centers of mass.

11.5 MOMENT OF INERTIA

To obtain the moment of inertia of a rigid body about any axis, we may regard the body as composed of a large number of small mass elements. If we multiply the mass of each element by the square of its distance from the axis and add the contributions of all the elements, we obtain the moment of inertia I:

$$I = m_1 r_1{}^2 + m_2 r_2{}^2 + m_3 r_3{}^2 + \cdots = \sum_i m_i r_i{}^2 \qquad (11.8)$$

Such a summing process for bodies of most shapes involves detailed numerical calculation or the application of integral calculus. However, we can compute the moment of inertia of a simple ring of mass m and radius r about its central axis by direct application of Eq. (11.8). Since all parts of the ring are essentially at the same distance R from the axis, the moment of inertia is $I = MR^2$. In Fig. 11.4 moments of inertia are given for bodies of several simple shapes—in every case for an axis through the center of mass as indicated. The moment of inertia may be expressed in kg-m², slug-ft², or in terms of the product of any mass unit multiplied by the square of a unit of length.

If we know the moment of inertia about an axis through the center of mass, the *parallel-axis* theorem states that the moment of inertia about any axis parallel to the one through the center of mass is given by

$$I = I_0 + Md^2 \qquad (11.9)$$

where M is the mass, I_0 the moment of inertia about an axis through the center of mass, and d the distance from this axis to the *parallel* one about which the moment of inertia is I.

It is sometimes convenient to write the moment of inertia of an extended rigid object in the form $I = Mk^2$, where k is a constant called the *radius of gyration*. Clearly, k corresponds to the radius of a thin circular ring which has the same mass and the same rotational inertia as the rigid body in question.

Example

Find the moment of inertia of a thin rod of mass 2 kg and length 1.6 m about an axis through its center. Also find the moment of inertia about an axis at one end.

For an axis through the center of mass,

$$I_0 = \frac{Ml^2}{12} = \frac{2 \text{ kg} \times (1.6 \text{ m})^2}{12}$$

$$= 0.427 \text{ kg-m}^2$$

For an axis at one end,

$$I = I_0 + Md^2 = 0.427 \text{ kg-m}^2 + 2 \text{ kg} \times (0.8 \text{ m})^2 = 1.71 \text{ kg-m}^2 \quad ////$$

Example

Find the moment of inertia of a 10-kg mass and a 15-kg mass about an axis of rotation which is 0.3 m from the 10-kg mass and 0.4 m from the 15-kg mass. Find the radius of gyration for this system.

$$I = m_1 r_1{}^2 + m_2 r_2{}^2$$
$$= 10 \text{ kg} \times (0.3 \text{ m})^2 + 15 \text{ kg} \times (0.4 \text{ m})^2$$
$$= 3.3 \text{ kg-m}^2$$
$$I = mk^2$$
$$3.3 \text{ kg-m}^2 = 25 \text{ kg} \times k^2$$
$$k = 0.36 \text{ m} \quad ////$$

11.6 COMBINATION OF ENERGY OF TRANSLATION AND ENERGY OF ROTATION

Consider a solid cylinder and a ring which have identical radii R and masses M. When both are placed at the top of an inclined plane (Fig. 11.5), they have identical potential energies Mgh. If the cylinder and

Figure 11.5
A solid cylinder rolls down an inclined plane with a greater speed than a ring, whether or not they have the same mass and outside diameter.

ring are allowed to roll down the plane starting from rest, which will reach the bottom first?

To answer this question, we apply the law of conservation of energy. At the bottom of the inclined plane both the cylinder and the ring have kinetic energy equal to the original potential energy. For each, the kinetic energy is partly associated with the translational velocity of the center of mass and partly with rotation about the center of mass. Let subscript c represent the cylinder, and subscript r the ring. Then

$$Mgh = \tfrac{1}{2}Mv_c^2 + \tfrac{1}{2}I_c\omega_c^2 \quad \text{and} \quad Mgh = \tfrac{1}{2}Mv_r^2 + \tfrac{1}{2}I_r\omega_r^2$$

For a cylinder of mass M and radius R, $I = MR^2/2$, while for the ring $I = MR^2$. Further, $\omega^2 = v^2/R^2$. Therefore, for the cylinder,

$$Mgh = \frac{Mv_c^2}{2} + \frac{Mv_c^2}{4}$$

or

$$v_c^2 = \frac{4gh}{3} \tag{A}$$

For the ring,

$$Mgh = \frac{Mv_r^2}{2} + \frac{Mv_r^2}{2}$$

or

$$v_r^2 = gh \tag{B}$$

The object with the smaller moment of inertia has the greater linear velocity and a smaller fraction of its energy associated with its rotational motion. Further, neither the mass nor the radius appears in Eqs. (A) and (B). Therefore the restrictions that both cylinder and ring have the same radius and mass were quite unnecessary; indeed, Eq. (A) is valid for any homogeneous disk or cylinder, and Eq. (B) for any thin ring.

11.7 NEWTON'S LAWS FOR ROTATIONAL MOTION

Experiments show that the opposition of a body to being set in translation is proportional to the mass of the body and does not depend on the distribution of this mass. We now show for one simple situation that the opposition which a body offers to being set in rotational motion about any axis depends on the moment of inertia about that axis. Consider a small mass m fastened by a light rod to point O, about which the mass m is to be set in rotation (Fig. 11.6). Let us exert on this mass a force F tangential to the circle. The torque τ exerted by force F about the axis through O is Fr, and, since $F = ma$, $\tau = mar$ and

$$\tau = mr^2\alpha = I\alpha$$

In general, to change the angular velocity of a rotating rigid body, we must apply an unbalanced torque, just as we must apply an unbalanced force to change the linear velocity of a body. The angular acceleration is directly proportional to the torque applied:

$$\tau = I\alpha \tag{11.10}$$

where τ is the unbalanced torque acting, α the angular acceleration, and I the *moment of inertia* (or the *rotary inertia*) of the body. The moment of inertia, a measure of the opposition of the body to being

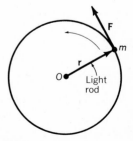

Figure 11.6
The angular acceleration is proportional to the unbalanced torque and inversely proportional to the moment of inertia.

Figure 11.7
Angular velocity or angular acceleration may be represented as a vector along the axis of rotation; its direction can be found by using the right-hand rule.

set into rotation, is analogous to the mass of a body, which is a measure of the opposition of the body to being set into translational motion.

Equation (11.10) expresses Newton's second law for rotary motion, which we have developed from $F = ma$. *If an unbalanced torque acts on a rigid body, an angular acceleration is produced which is proportional to the unbalanced torque and inversely proportional to the moment of inertia of the body.* The angular acceleration is in the direction of the unbalanced torque. As we saw in Chap. 4, the direction of a torque is that in which a right-handed screw moves when acted upon by a torque about its axis. Angular acceleration and angular velocity (but not angular displacement) are vector quantities with directions determined just as for torque (Fig. 11.7).

Example

A flywheel has a moment of inertia of 300 kg-m². Find the torque necessary to produce an angular acceleration of 3 rad/sec².

Torque = moment of inertia $\times$ angular acceleration

$$\tau = 300 \text{ kg-m}^2 \times 3 \text{ rad/sec}^2$$
$$= 900 \text{ N-m} \qquad\qquad ////$$

If we raise the front wheel of a bicycle off the ground and set it spinning, its angular velocity gradually decreases until the wheel comes to rest. Torques due to bearing friction and air friction produce the deceleration. If we reduce the friction, the wheel spins longer; if we could eliminate friction, the wheel would spin indefinitely with constant angular velocity. This is a direct analog of Newton's first law for translational motion and is known as Newton's first law of rotational motion. *A rigid body rotating about a fixed axis continues with constant angular velocity unless acted upon by some unbalanced external torque.* The rotation of the earth on its axis is a reasonable example, although the earth is not quite a rigid body and does not have an exactly fixed axis, and there are some unbalanced frictional torques introduced by tides and other considerations.

Newton's third law of rotational motion states: *For every torque acting on one body there is an equal and opposite torque about the same axis acting upon some other body (or bodies).*

11.8 ANGULAR HARMONIC MOTION

When a heavy disk is suspended by a wire (Fig. 11.8) and rotated through an angle θ from its equilibrium position, there is a restoring torque τ_r which is proportional to the angular displacement:

$$\tau_r = -K\theta \qquad\qquad (11.11)$$

where K is the *torsion constant* of the wire. By Eq. (11.10), $\tau = I\alpha$. Consequently, for this *torsion pendulum,*

$$I\alpha = -K\theta \qquad\qquad (11.12)$$

the angular acceleration is proportional to the angular displacement, and

Figure 11.8
A torsion pendulum describes angular harmonic motion.

angular simple harmonic motion is produced. Equations (10.4a) through (10.11) are applicable to angular harmonic motion if angular displacement, angular amplitude, angular velocity, and angular acceleration are substituted for the corresponding linear quantities and moment of inertia, torque, and torsion constant replace mass, force, and spring constant, respectively. Thus by the analog of Eq. (10.10) for angular harmonic motion

$$\frac{1}{\nu} = T = 2\pi \sqrt{\frac{I}{K}} \qquad (11.13)$$

We have seen that for many of the equations which govern linear motion, there are analogous equations in rotary motion. *These equations are valid when the angles are expressed in radians.* It is unnecessary to memorize many of the equations of angular motion if use is made of the analogy between linear and rotational motions. Table 11.1 lists corresponding quantities in the two kinds of motion. Knowing a relationship between linear quantities, one can often write the appropriate relationship between angular quantities by simply replacing each linear quantity with the corresponding angular one. For example, in linear motion, work = force × displacement. The corresponding expression for angular motion is work = torque × angular displacement. The angular analog of $P = Fv$ is $P = \tau\omega$.

11.9 ANGULAR MOMENTUM

When a mass m such as that of Fig. 11.6 (or a planet) revolves about some axis O, the *angular momentum* **A** of the mass about O is defined as the vector product of the radius vector **r** from O to the mass and the momentum m**v**. The angular momentum has the magnitude

$$A = rmv \sin \theta \qquad (11.14)$$

where θ is the angle between **r** and **v**; the direction of **A** is given by the right-hand rule (Fig. 11.7). Newton's second law for rotational motion may be stated in a form more general than that of Sec. 10.7. *The unbalanced torque acting on a body is equal to the time rate of change of angular momentum.* If there is no unbalanced torque on the body, its angular momentum remains constant.

Not only is angular momentum conserved, but evidence from atomic and molecular physics shows that it is quantized (comes in small, well-defined chunks) as well. Changes in the angular momentum of an

Table 11.1

Linear Quantities and Their Angular Analogs

Linear		*Angular*	
Displacement	s	Displacement	θ
Velocity	v	Angular velocity	ω
Acceleration	a	Angular acceleration	α
Force	F	Torque	τ
Mass	m	Moment of inertia	I

Figure 11.9

Conservation of angular momentum about the vertical axis requires that the angular velocity increase when the moment of inertia is decreased.

atomic system come in units of size $h/2\pi$, where h is Planck's constant and has the magnitude 6.6×10^{-34} kg-m²/sec (or J-sec). This is a minuscule angular momentum, utterly insignificant for macroscopic bodies, but of immense importance in systems of atomic size.

For a rigid body rotating about a fixed axis the angular momentum is the resultant of the angular momenta of all the N atoms of which the body is composed. Since all have the same angular velocity ω,

$$A = \sum_{i=1}^{N} r_i m_i \omega_i r_i = \omega \sum_{i=1}^{N} m_i r_i^2 = I\omega$$

Thus *the angular momentum is the product of the moment of inertia I and the angular velocity ω.*

In a system of nonrigid bodies which interact only with each other, Newton's third law for rotational motion tells us that if body A exerts a torque on body B, body B exerts an equal and opposite torque on body A. Therefore the change in angular momentum of body A is equal to and opposite that of body B. Thus *the angular momentum of the system remains constant in the absence of external torques.* The conservation of angular momentum is one of the key conservation laws of nature, which include conservation of energy and conservation of linear momentum. For example, in Fig. 11.9 a man stands on a platform mounted on ball bearings. In his hands he holds heavy weights. His moment of inertia is greater when his arms are outstretched than when they are folded. If this man is set in rotation with arms outstretched, he rotates with a constant angular velocity as long as he does not change his moment of inertia. If he folds his arms as indicated in the figure, the moment of inertia I of the system (man plus weights) is decreased. Since the angular momentum $I\omega$ remains constant, the angular velocity ω increases. If he stretches his arms again, his angular velocity is reduced. When a diver performs a somersault from the high board, he doubles up to make his moment of inertia small while he is making his turn, and then he decreases his rotational velocity by increasing his moment of inertia through stretching out to his full length.

11.10 GYROSCOPES

There are many kinds of gyroscopes, but all include a body spinning rapidly about an axis. In the gyroscopic compass this gyro element is mounted in such a way that no significant torques are applied. Under these circumstances the vector angular momentum remains constant. The axis of the gyro element always points in the same direction, regardless of the maneuvers of the aircraft or ship in which the gyro is mounted. Thus, it can be made the key element of an automatic pilot.

A gyroscope such as that of Fig. 11.10, which is supported at point O only, does not fall to the ground when the gyro wheel is in rapid rotation. Rather, the entire frame rotates about a vertical axis through O at a constant speed. This motion is called *precession.*

The behavior of the gyroscope can be analyzed in terms of angular momentum. Let the angular momentum of the rotating gyro wheel at $t = 0$ be represented by $\mathbf{A}_0$ (Fig. 11.10). The torque due to the weight

Figure 11.10
Precession of a gyroscope.

of the rotor and frame produces an angular acceleration about the axis OY and, in a brief time Δt, gives the gyroscope a small angular momentum $\Delta\mathbf{A}$ about OY. The resultant angular momentum $\mathbf{A}$ is the vector sum $\mathbf{A}_0 + \Delta\mathbf{A}$. Thus the gyroscope precesses about the axis OZ.

There is an alternative way of looking at the precession of a gyroscope. Suppose that the rotating element of the gyroscope consists of a metal disk in which balls are mounted in slots (Fig. 11.11). When this disk is put into rapid rotation about an axis perpendicular to the plane of the paper, the balls move in the directions indicated in the figure. Now, if a torque is suddenly exerted about a vertical axis, the upper balls continue to move to the left, and the lower ones to the right. Not only does the inertia of each ball keep it going in the same direction, but so also does the inertia of each particle constituting the disk. Hence, the entire wheel tilts, and the axle precesses upward.

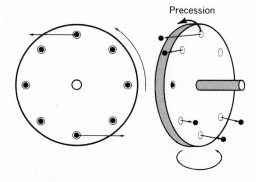

Figure 11.11
Precession occurs when the rotating wheel is turned about a vertical axis; the steel balls fly out of their retaining pockets, and the disk tends to tilt.

An aircraft moving eastward with a single propeller rotating clockwise tends to nose down when it turns south and to nose up when it turns north, since the rotating propeller acts like a gyroscope. If the airplane is traveling eastward and the propeller is rotating clockwise as viewed by the pilot, at any given instant the top blade of the propeller is traveling southward, and the lower blade northward. If at this instant the airplane is turned toward the north, the upper blade continues to move southward, and the lower one northward. As a result, the airplane noses up.

QUESTIONS

1 Discuss the conservation of angular momentum for a diver performing a $2\frac{1}{2}$-turn forward somersault from a high board.

2 A rifle barrel is rifled to impart a spin to the bullet. What is the advantage of this? How does it affect the manner in which the rifle recoils?

3 Why does a typical small helicopter have a small propeller exerting a torque on the craft about a vertical axis? Discuss the torques involved in helicopter flight.

4 If a solid sphere of mass M and radius R raced the ring and cylinder of Fig. 11.5 down the plane, would it reach the bottom first or last? What would be its speed at the bottom?

5 A flywheel rotating with constant angular velocity about a fixed axis is in equilibrium. Is every particle in the flywheel in equilibrium? Explain.

6 Why does an airplane tend to yaw (tilt sideways) when pulling out of a dive? Explain with the aid of a vector diagram.

7 What fraction of the kinetic energy of a rolling billiard ball is rotational? Of a rolling solid cylinder? Of a rolling hoop?

8 Why does an automobile nose downward if it is traveling at a high speed when the brakes are applied? Would this still happen if there were no brakes on the front wheels?

9 The flywheel of an automobile rotates counterclockwise when viewed from the rear. Will the rear wheels skid more readily during a right or left turn? Why?

10 Explain how to distinguish between a hard-boiled egg and a raw one by spinning each on a flat table top.

11 A 150-g meterstick rests on its flatter side, supported by your fingers under each end. A 100-g mass M lies on one end. If both fingers are suddenly removed, M and the stick fall together.

However, if only the finger nearer M is removed, the stick falls away from M even though the latter has acceleration g. Why does this happen?

12 How can a boy in a swing pump himself from a small amplitude to a large one? Discuss in terms of angular momentum and the work done against gravity.

13 Why would you expect the period of a meterstick oscillating as a pendulum about an axis at one end to be longer than the period of a simple pendulum 50 cm long? After all, the center of gravity would be the same distance from the support. Why would you expect the meterstick to have a shorter period than a simple pendulum 1 m long?

14 Tidal friction is calculated to dissipate the earth's rotational energy at the rate of 15×10^{11} W and is primarily responsible for a decrease in the length of the day by 0.0016 sec per century. Eventually both the day and the month should be about 47 of our current days. Why should the month lengthen? Why will the friction due to lunar tides cease when the day and month are of equal duration?

PROBLEMS

1 A small motor starts from rest and attains its rated speed of 1,200 rev/min in 8 sec. Calculate the angular acceleration, assuming it to be uniform. Through how many radians does the motor turn in achieving its rated speed?

Ans. 15.7 rad/sec²; 503 rad

2 A bicycle wheel initially at rest is accelerated uniformly until it attains an angular speed of 15 rad/sec in 5 sec. Find the angular acceleration and the angle turned through in this time.

3 A grinding wheel starts from rest and has a

constant angular acceleration of 10 rad/sec². At $t = 4$ sec find the centripetal and tangential accelerations of a particle 5 cm from the axis, the angular velocity, and the angle turned through.

Ans. 80 m/sec²; 0.50 m/sec²; 40 rad/sec; 80 rad

4 Compute the angular speed, about the axle, of an automobile wheel 14 in. in radius when the car is moving 45 mi/hr (66 ft/sec). What angular acceleration is needed to bring the wheel from rest to this speed in 11 sec?

5 The coefficient of friction between a wooden block and the floor of a merry-go-round is 0.30. If the block is 4 m from the axis, at what angular speed will it slide off?   *Ans.* 0.86 rad/sec

6 An electric motor operates at 1,800 rev/min. Find its angular velocity in radians per second. What is the speed of a point 4 cm from the axis of rotation?

7 One method of determining the speed of a bullet is to fire it through two cardboard disks mounted on a long axle and rotated at high angular speed. If the disks are 1.2 m apart and are rotated at 1,800 rev/min, find the speed of a bullet fired parallel to the axis if it makes holes which are 18° apart.   *Ans.* 720 m/sec

8 If has been proposed that large space stations should be constructed. To provide a pseudo-gravitational field of 0.5 g at a radius of 10 m, what angular velocity would the space station require?

9 A 1,200-kg automobile starts from rest and proceeds around a flat, circular track of 800 m radius with a constant tangential acceleration of 3 m/sec². How long will it take for the centripetal acceleration to reach 4 m/sec²? At that instant what is the resultant acceleration?

Ans. 18.9 sec; 5 m/sec² at 53° with tangent

10 A racing car starts from rest and accelerates uniformly along the track with an acceleration of 8 ft/sec². If the track is circular with a radius of 1,000 ft, at what speed is the centripetal acceleration equal to the tangential acceleration? How long does it take to achieve this speed?

11 The tip of the propeller of an airplane is 1.6 m from the propeller shaft. The moment of inertia of the rotating propeller and associated shaft is 80 kg-m². If the propeller is making 1,200 rev/min and is given an angular acceleration of 3 rad/sec², how long a time is required to increase the angular speed to 1,800 rev/min?

What unbalanced torque acts on the propeller shaft during the acceleration? At what angular velocity will the tip of the propeller have a linear speed equal to the local speed of sound, which is 320 m/sec at the position of the airplane?

Ans. 20.9 sec; 240 N-m; 200 rad/sec

12 The radius of the earth is approximately 6,370 km, and it rotates about its axis once each sidereal day of 86,164 sec. Find the angular velocity of the earth in radians per second, the linear speed due to the earth's rotation of a point near the equator, and the speed of a point on the earth's surface at 40° latitude.

13 The moment of inertia of a body about its axis of rotation is 6 kg-m². A torque of 12 N-m is applied to it. If the body starts from rest, find the angular acceleration. What are the angular velocity and the kinetic energy at the end of 20 sec? *Ans.* 2 rad/sec²; 40 rad/sec; 4,800 J

14 The turbine of a jet engine has a moment of inertia of 120 kg-m². It is to be accelerated uniformly from rest to an angular velocity of 200 rad/sec in 40 sec. Find the angular acceleration, the unbalanced torque required, the angle turned through during the acceleration, and the kinetic energy of the turbine at its final angular velocity.

15 The turbine and associated rotating parts of a jet engine have a total moment of inertia of 40 kg-m². It is accelerated uniformly from rest to an angular velocity of 200 rad/sec in a time of 25 sec. Find (*a*) the angular acceleration, (*b*) the net torque required, (*c*) the angle turned through during the acceleration, (*d*) the work done by the net torque, and (*e*) the kinetic energy of the turbine at its final velocity.

Ans. (*a*) 8 rad/sec²; (*b*) 320 N-m; (*c*) 2,500 rad; (*d*) 8×10^5 J; (*e*) 8×10^5 J

16 Find the torque developed by an airplane engine which rotates a propeller at 1,500 rev/min when supplying 1,200 hp.

17 Find the moment of inertia about an axis through the center of mass and the kinetic energy of a system consisting in two masses of 3 and 4 kg connected by a rod of negligible mass 0.7 m long when the center of mass of the system has a

velocity of 20 m/sec and the system rotates about the center of mass with an angular velocity of 30 rad/sec.   *Ans.* 0.84 kg-m²; 1,780 J

18 If a flywheel has a moment of inertia of 5 kg-m², what angular speed does the wheel attain if 60,000 J of work is done in producing rotational kinetic energy? What torque is required to bring this flywheel to rest in 20 sec?

19 A turbogenerator with a moment of inertia of 150,000 slug-ft² rotates at the rate of 120 rad/sec. The power and load are shut off simultaneously, and friction stops the rotation in 3,000 sec. Find the initial kinetic energy of rotation, the initial angular momentum, and the torque due to friction.
Ans. 1.08×10^9 ft-lb; 1.8×10^7 slug-rad/sec; 6,000 ft-lb

20 A circular hoop has a mass of 0.4 kg and a radius of 50 cm. Find its moment of inertia (*a*) about an axis through the center of the hoop and perpendicular to its plane and (*b*) about an axis through the circumference of the hoop and parallel to the axis through the center.

21 Calculate the moment of inertia of a drive shaft that has a kinetic energy of 12,000 J when it is making 80 rad/sec. Find the radius of gyration if the mass of the wheel is 300 kg.
Ans. 3.75 kg-m²; 0.112 m

22 A body consists of a solid horizontal cylindrical shaft, weighing 150 lb, with a radius of 2 in. and a solid disk having a mass of 8 slugs and a radius of 1.2 ft. The cylinder and the disk are mounted so that they have the same axis. What torque is necessary to give the body an angular speed of 24 rad/sec at the end of 6 sec, starting from rest?

23 A flywheel has a mass of 125 kg, all concentrated in the rim, and a radius of 0.6 m. It is rotating with an angular velocity of 20 rad/sec. How much work was necessary to give the flywheel this angular velocity? If the wheel is stopped in 1 min by a friction brake, what torque is applied?   *Ans.* 9,000 J; 15 N-m

24 A hollow cylinder has a mass of 15 kg and a diameter of 1.2 m. Its mass is concentrated in the rim. It is rolling with a linear speed of 1.6 m/sec. What is its kinetic energy of rotation? Its total kinetic energy?

25 A 30-kg boy stands 2 m from the center of a frictionless playground merry-go-round which has a moment of inertia of 200 kg-m². If the boy begins to run in a circular path with a speed of 0.8 m/sec *relative to the ground*, calculate (*a*) the angular velocity of the merry-go-round and (*b*) the speed of the boy relative to the merry-go-round.   *Ans.* 0.24 rad/sec; 1.28 m/sec

26 A hoop weighing 8 lb and having a radius of 0.5 ft rolls down an inclined plane that makes an angle of 30° with the horizontal. Assuming that there is no slipping, what is the linear acceleration of the hoop? The angular acceleration? The speed of the center of mass after the hoop has rolled a distance of 16 ft?

27 A solid sphere having a diameter of 10 cm rolls down an inclined plane that is 4 m long. What is the angular velocity of the cylinder at the bottom of the plane if it requires 6 sec for it to reach the bottom? What is the angle of the plane?   *Ans.* 26.7 rad/sec; 1.8°

28 What linear velocity is acquired by a solid steel disk that has a radius of 25 cm and a mass of 4 kg when it rolls down an inclined plane which is 2 m long and makes an angle of 10° with the horizontal? What is the acceleration of the disk?

29 A Yo-Yo has a moment of inertia about its center of mass of 4×10^{-5} kg-m² and a mass of 0.1 kg. If the string is wrapped around a center post of 5-mm radius and the Yo-Yo is released from rest, find the angular acceleration and the tension in the string.
Ans. 115 rad/sec²; 0.922 N

30 A 4-lb ring of radius 8 in. rolls down an inclined plane of which one end is 2 ft above the other. If the ring starts from rest at the top, show that its kinetic energy of rotation is equal to its kinetic energy of translation at the bottom. Find the kinetic energy of rotation at the bottom and the speed of the center of mass at the bottom.

31 A pulley wheel has mass M, radius R, and moment of inertia about its center of mass of $0.8MR^2$. It has a cord wrapped around it with one end of the cord fastened to the ceiling. The pulley wheel is released from rest and falls a distance h. (*a*) Consider the motion to be a translation of the center of mass and a rotation about the center of mass. Apply Newton's second law to the translational and rotational motions and calculate the angular acceleration of

the wheel and the tension in the cord. (*b*) Find the speed v of the wheel after it has fallen a distance h.

Ans. (*a*) $0.556g/R$; $0.444Mg$; (*b*) $\sqrt{10gh}/3$

32 A mass of 0.5 kg hangs by a cord wrapped around the axle of a frictionless flywheel. The axle has a radius of 4 cm. The tension in the cord as the mass falls is 4.5 N. Find (*a*) the acceleration of the mass, (*b*) the angular acceleration of the flywheel, and (*c*) the moment of inertia of the flywheel.

33 A meterstick has a small hole bored at the 10-cm mark so that it can be hung from a horizontal nail. If the meterstick is rotated about the nail until it makes an angle of 37° with the vertical and is then released, find its angular velocity as it passes through its equilibrium position, assuming that friction is negligible.

Ans. 2.54 rad/sec

34 What angular acceleration will be imparted to a solid disk that has a radius of 1.5 ft and a weight of 64 lb by a weight of 12 lb hanging from a cord wound around the disk? The axis of the disk is fixed.

35 A 1,200-kg automobile has a 2.4-m wheelbase. Its center of mass is 1.1 m behind the front axle and 0.8 m above the ground. Find the vertical force on the front wheels when the car is at rest and when it is stopping with a deceleration of 4 m/sec². Assume the center of mass does not move relative to the wheels. (*Hint:* Recall that the retarding forces at the road produce a torque which is balanced by a redistribution of the force on the front and rear wheels.)

Ans. 6,370 N; 7,970 N

36 A 3,600-lb automobile has a 9-ft wheelbase, and its center of mass is 4 ft behind the front axle and 2.25 ft above the road. Assuming that the center of mass does not move relative to the wheels, find the vertical force on the front wheels when the car has a uniform deceleration of 16 ft/sec². (*Hint:* See Prob. 35.)

37 A solid uniform disk with an 8-cm radius starts from rest and rolls down a plane 2 m long and inclined at 30° to the horizontal. What is its linear speed at the bottom? Its angular velocity? How long does it take to reach the bottom?

Ans. 3.61 m/sec; 45.2 rad/sec; 1.11 sec

38 Show that if a solid cylinder is to roll down an inclined plane without slipping, the coefficient of static friction between cylinder and plane must be at least $(\tan \theta)/3$, where θ is the angle between the plane and the horizontal.

39 Find the moment of inertia of a thin spherical shell of mass M and the radius R. (*Hint:* Take the difference between the moments of inertia for spheres of radii $R + \Delta R$ and R; then assume that $\Delta R \ll R$.) *Ans.* $\frac{2}{3}MR^2$

40 A bowling ball is released with speed v_0 and no angular velocity just above the alley. It skids until its angular speed reaches a value such that $\omega_f R = v_f$, where R is the radius of the ball and v_f is the speed at which it rolls without skidding. Show that $v_f = \frac{5}{7}v_0$.

CHAPTER TWELVE
FLUIDS AT REST

THUS FAR we have been concerned with the mechanics of particles and rigid bodies. But solid objects represent only a part of our physical universe. Much of the surface of the earth is covered with water, and we live in a great shell of gas which comprises our atmosphere. In this chapter we consider several properties of fluids at rest. By a *fluid* we mean anything which flows; both liquids and gases are fluids. In general, gases are readily compressible, while liquids offer large resistance to efforts to change their volume.

12.1 FLUIDS AND PRESSURE

A liquid occupies a well-defined volume. It has no shape of its own but takes the shape of the vessel containing it. On the other hand, a gas occupies completely any enclosed volume in which it is placed. Liquids yield to a continued application of force that tends to deform them or change their shape. They do, however, manifest wide differences in their readiness to yield to distorting force. Water, alcohol, and ether are very mobile liquids; glycerin is less mobile, and tar still less so.

A fluid exerts forces against the walls of the containing vessel. In order to discuss the interaction between the fluid and the walls, it is convenient to introduce the concept of pressure. *Pressure is defined as the ratio of force to the area on which it acts when the area is perpendicular to the force.*

$$p = \frac{F}{A} \qquad (12.1)$$

More precisely, *the pressure at any point is the ratio of the normal force ΔF exerted on a small area ΔA surrounding the point to this area.* Pressure may be expressed in pounds per square inch, newtons per square meter, etc.—in general, as any unit of force divided by a unit of area.

12.2 PRESSURE IN A LIQUID

If we explore a liquid confined in a vessel with a device for measuring pressure, we find that at a given level in the liquid the magnitude of the force acting on a test area is the same regardless of how the area is oriented. The force is always perpendicular to the area. Further, the deeper we go, the greater the pressure becomes. If p_0 represents the pressure at the surface of the liquid and p the pressure at a depth y, we find that the increase in pressure due to the liquid is directly proportional to the depth y and to the density d of the liquid. Let p_L represent the increase in pressure due to the liquid. Then

$$p_L = p - p_0 = dgy \qquad (12.2)$$

where g is the acceleration due to gravity. That this should be true is easy to understand if we consider the force on an area A at the bottom of a container (Fig. 12.1). This area supports the weight $yAdg$ of the column of fluid above it. The contribution of the liquid to the pressure at A is the weight of this column divided by the area A. Hence $p_L = yAdg/A = dgy$.

In some problems it is convenient to use *weight density* d_w (weight per unit volume) rather than density (mass per unit volume). The weight of a given volume is equal to $d_w V$, and the pressure due to the liquid is yd_w. In the British engineering system we ordinarily use weight density (pounds per cubic foot) rather than mass density (slugs per cubic foot). Hence the relation

$$p_L = p - p_0 = d_w y \qquad (12.2a)$$

is sometimes more convenient than $p_L = dgy$. (Note that, since $w = mg$, $d_w = w/V = mg/V = dg$.)

Although we have derived Eqs. (12.2) and (12.2a) only for the bottom of our container, they are applicable at any point in the liquid at rest. Observe that y is the *vertical* height from the point to the free surface.

Figure 12.1

The increase in pressure due to a liquid at a depth y is proportional to the depth and to the density of the liquid.

Example

The hatch of a submarine is 50 m under the surface of the ocean. If the density of seawater is 1,025 kg/m³, find the pressure at the hatch due to the water and the net force on the hatch if it is a rectangle 0.6 m wide and 1.2 m long. The pressure inside the submarine is the same as that at the surface.

$$\begin{aligned}
p_L = p - p_0 &= dgy \\
&= 1,025 \text{ kg/m}^3 \times 9.8 \text{ N/kg} \times 50 \text{ m} \\
&= 5.02 \times 10^5 \text{ N/m}^2
\end{aligned}$$

The net force on the hatch is the difference between the inward force pA and the outward force p_0A. Therefore,

$$\begin{aligned}
F = pA - p_0A &= (p_L + p_0)A - p_0A = p_LA \\
&= 5.02 \times 10^5 \text{ N/m}^2 \times 0.72 \text{ m}^2 \\
&= 3.61 \times 10^5 \text{ N} \qquad \text{////}
\end{aligned}$$

12.3 PRESSURE OF THE ATMOSPHERE

Just as water is the most widely distributed and most important of liquids, so air is the most important and intimate of gases. It consists for the most part of nitrogen and oxygen. In spite of the fact that there is no chemical union between the two, the composition of the air is extraor-

Figure 12.2

Torricelli showed that the atmosphere would support a column of mercury 76 cm high in a closed tube with vacuum above the mercury. The height of the mercury is independent of the shape or length of the tube.

dinarily constant. Up to a height of 7 mi it contains about 21 parts of oxygen to 78 parts of nitrogen by volume. Besides oxygen and nitrogen, the air contains small amounts of other gases, among which are water vapor, argon, and carbon dioxide. A cubic meter of air weighs over 12.7 N at sea level, although to an ordinary observer the air seems to have no weight and to offer little resistance to bodies moving through it.

Since it has weight, a column of air exerts pressure, just as a column of liquid does. However, Eq. (12.2) is not directly applicable, because the density of a gas, unlike that of a liquid, depends on the pressure (Sec. 14.2). Though a liter of air weighs little, even at ground level, the height of the atmosphere is large, and the weight of all this air pressing on the earth is great.

About 1644 Torricelli, a pupil of Galileo, measured atmospheric pressure in the following way. A glass tube closed at one end was filled with mercury. A finger was then placed over the open end of the tube, the tube inverted in a basin of mercury, and the finger removed from the open end of the tube under the mercury. The mercury in the tube sank until its level was about 76 cm (Fig. 12.2) higher than the level in the basin. The pressure of the atmosphere on the mercury in the basin supported the mercury in the tube. Since the density of mercury is 13,560 kg/m³, the pressure exerted by the atmosphere as given by Eq. (12.2) is $p = dgy = 13,560$ kg/m³ $\times$ 9.80 N/kg $\times$ 0.760 m = 1.013×10^5 N/m² (14.7 lb/in.²). Mercury has 13.6 times the density of water; therefore 1 atmosphere (atm) can support a column of water 13.6×0.760 m = 10.3 m (33.8 ft) high.

Our atmosphere is a great ocean of air which becomes steadily less dense as we go upward. In 1647 Pascal showed that when a Torricelli tube is carried up a mountain, the level of the mercury drops; the greater the height, the greater the decrease in pressure. At 1.6 km elevation the mercury column has fallen 15 cm. At 5.5 km the pressure is half that at sea level. The approximate variation of pressure with altitude is shown in Fig. 12.3. Less than one-quarter of the atmosphere remains above a height of 11 km.

The breathing of animals involves an application of atmospheric pressure. A reduction of pressure is caused by a movement of the diaphragm. The greater pressure of the outside air causes a fresh supply to flow into the lungs. Then air is forced out when the internal pressure is made greater than atmospheric. Sucking and drinking animals take advantage of atmospheric pressure to aid them in these operations. They reduce the pressure in the mouth and allow water to be forced into the mouth by the outside atmospheric pressure.

Many types of pressure gauges read the excess of pressure over atmospheric pressure; this is the so-called *gauge pressure*. When a typical gauge reads 1.65×10^5 N/m² (24 lb/in.²) and atmospheric pressure is 1.01×10^5 N/m² (14.7 lb/in.²), the absolute pressure p is 2.66×10^5 N/m² (38.7 lb/in.²).

Figure 12.3

Variation of atmospheric pressure with height above sea level.

Example

The gauge pressure of water in the water mains is 2.5×10^5 N/m². How much work is required to pump 30,000 m³ of water at atmospheric pressure into the mains?

Work = net force × distance = $\dfrac{\text{force}}{\text{area}}$ × (distance × area)

= net pressure × change in volume

= 2.5×10^5 N/m^2 × 3×10^4 m^3

= 7.5×10^9 J   ////

Figure 12.4
The pressure is the same at all points at a given level in a liquid at rest, regardless of the shape of the containing vessel. In this figure the areas of the openings for A, B, C, and D are the same, but changing one or more of these areas would not affect the conclusion.

12.4 LIQUIDS IN COMMUNICATING VESSELS

Liquids seek their own level in interconnected vessels. When tubes of various sizes and shapes are connected, liquid poured into one of them (Fig. 12.4) comes to the same level in all the tubes (provided surface-tension effects may be neglected). This result is to be expected from the fact that the pressure in the liquid depends on the depth below the free surface. *At all points at the same level within a liquid at rest, the pressure is the same.* If it were not, the liquid would flow from one point to the other until the pressure became equalized.

In Fig. 12.4 the areas of the openings at which vessels A, B, C, and D join the bottom reservoir are the same. The following question may be raised: If the liquid pressure is due to the weight of the fluid above, why is the pressure not greater in vessel D, in which the sides slope outward, than in vessel B, in which the sides slant inward? Certainly vessel D contains a greater weight of water. The answer is that in D the slanting sides exert forces on the liquid which have an upward component. Thus, a portion of the weight of the fluid is actually held up by the sloping sides. On the other hand, in B the force exerted by the slanting sides on the liquid is downward. When the force due to the slanting walls is added to the weight of the fluid, the pressure at the bottom is the same in all four vessels.

Let two liquids that do not react chemically be placed in a bent tube (Fig. 12.5a). When the liquids are at rest, the pressure exerted by the column of lighter liquid is balanced by the pressure due to the column of heavier liquid above the junction level of the liquids. Then, provided surface-tension effects are negligible, $d_1 g y_1 = d_2 g y_2$, from which $y_1/y_2 = d_2/d_1$. *The heights of two liquids above their surface of separation are inversely proportional to their densities.*

If the liquids react chemically, the bent tube may be inverted, and the ends placed in cups containing the liquids (Fig. 12.5b). The air from the upper part of the bent tube is partly removed, and the stopcock closed. The pressure above both liquids is the same, and the atmospheric pressure on the liquids in the open vessels is the same. The

Figure 12.5
The pressure produced by the column of liquid 1 of height y_1 is equal to that due to the column of liquid 2 of height y_2 in both (a) and (b).

Figure 12.6

A pressure change at any point in an enclosed fluid at rest results in an equal pressure change at all other points; this fact is known as Pascal's principle.

difference between the pressure inside the tube and atmospheric pressure is balanced in each case by the rise of the liquid in the tube. These differences in pressure are the same, and again $d_1gy_1 = d_2gy_2$.

Example

If one of the beakers in Fig. 12.5b contains sulfuric acid and the other contains water, and if the height of the column of water is 40 cm when the height of the column of acid is 30 cm, find the density of the sulfuric acid.

$$y_a d_a g = y_w d_w g \qquad d_w = 1,000 \text{ kg/m}^3$$

$$\frac{d_a}{d_w} = \frac{y_w}{y_a} = \frac{0.40 \text{ m}}{0.30 \text{ m}}$$

$$d_a = 1,330 \text{ kg/m}^3 \qquad\qquad ////$$

12.5 PASCAL'S PRINCIPLE

If we increase the pressure at any one point in a fluid which is completely enclosed (Fig. 12.6), the pressure increases by an equal amount at all other points. For example, if we add a mass of 1 kg to piston A of Fig. 12.6, it is found that an equal mass must be added to identical pistons B and C in order to keep them from moving. This is a special case of a general law known as *Pascal's principle. If the pressure at any point in an enclosed fluid at rest is changed, the pressure changes by an equal amount at all points in the fluid.* Note that Pascal's principle does *not* say that the pressure is the same everywhere within the enclosed fluid. The pressure continues to be greater at greater depth. Pascal's principle deals with the change in pressure, not with the absolute pressure.

Consider two cylinders which are connected together and filled with water (Fig. 12.7). If each cylinder is fitted with a piston which moves without friction, and if the two pistons are at the same level, the pressure at each will be the same. If A is the area of the larger piston, and a the area of the smaller one, the force f on the smaller piston is given by pa, while the force F on the larger piston is given by pA. By applying a force f on the small piston, it is possible to produce a force $F = Af/a$ on the larger piston. When the larger piston is raised a distance S, the smaller moves through a greater distance s, where $s = SA/a$.

If the larger and smaller pistons are not at the same level, the pressures on the two pistons are not the same; they differ by dgy, where y is the difference between the heights of the two pistons. If the pressure is increased 10 lb/in.^{2} on the smaller piston, it also increases 10 lb/in.^{2} at the larger piston, but the pressures are not equal. A direct application of these ideas is found in the hydraulic press.

An important application of Pascal's principle occurs in the hydraulic system of an automobile (Fig. 12.8). When the pressure is increased in the master cylinder by pressing on the brake pedal, the pressure increases by an equal amount at every point in the hydraulic system. If all the pistons at the brake shoes have the same cross-sectional area, an equal force is applied at all the brake shoes.

Figure 12.7

Multiplication of force by transmitted pressure in a hydraulic press. The IMA of this simple machine is A/a.

Figure 12.8
Pascal's principle as it is applied in the hydraulic brake system of an automobile.

Pascal's principle applies to all fluids, gases as well as liquids. A typical application of Pascal's principle to gases is the automobile lift pump. Compressed air exerts pressure on the oil in a reservoir. The oil in turn transmits the pressure to a cylinder which lifts the automobile.

Example

Find the minimum gauge pressure which must be supplied to an automobile lift pump with a piston of area 800 cm² to lift an automobile and piston with combined mass of 1,500 kg.

$$\text{Net force} = \text{net gauge pressure} \times \text{area}$$
$$1{,}500 \text{ kg} \times 9.8 \text{ N/kg} = p_g \times 800 \times 10^{-4} \text{ m}^2$$
$$p_g = 1.84 \times 10^5 \text{ N/m}^2 \qquad \text{////}$$

12.6 ARCHIMEDES' PRINCIPLE

It is a matter of common experience that bodies are apparently lighter under water than in air. A fresh egg sinks in pure water but floats in salty water. A piece of iron sinks in water but floats in mercury. If a diver picks up a stone under water and brings it to the surface, he finds that the stone is much heavier above the surface. The principle which explains these observations, discovered by the distinguished Greek mathematician and physicist Archimedes, states that:

A body immersed in a fluid is buoyed up by a force equal to the weight of the fluid displaced.

An experimental verification of Archimedes' principle can be obtained by use of the equipment of Fig. 12.9. A hollow cylindrical cup and a piece of brass turned so that it will just fill the cavity inside the cup are suspended from one arm of a balance, and the weights necessary to restore equilibrium are added to the other pan. When a vessel of water is brought up in such a way that the cylinder C is completely submerged, the side of the balance carrying the cylinder rises, showing

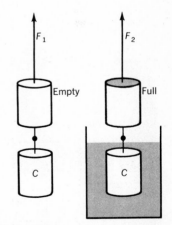

Figure 12.9
Cylinder C is buoyed up by a force equal to the weight of the fluid it displaces; F_2 is equal to F_1 when the container with the volume of C is filled with liquid.

Figure 12.10

The net upward force on the cylinder is equal to the weight of the displaced fluid.

that the water is pushing upward on the cylinder. If water is now poured into the cup until it is full, the original equilibrium of the balance is restored. The cylinder is buoyed up by a force equal to the weight of the water displaced.

Archimedes' principle follows directly from the laws of fluid pressure. If a cylindrical block (Fig. 12.10) is immersed in a vessel filled with liquid, the resultant of the forces on the vertical sides is zero. Upon the upper face of the cylinder there is a downward force $ydgA$ equal to the pressure at the upper surface multiplied by the cross-sectional area A of the cylinder. On the lower face there is an upward force $YdgA$ equal to the pressure at the bottom multiplied by the area. The net upward force is $YdgA - ydgA = (Y - y)Adg = Vdg$, where V is the volume of the cylinder. *The net buoyant force is equal to the weight of the fluid displaced.* The same result holds for a body of any shape in any liquid. Hence, a body immersed in any fluid is lighter by the weight of fluid which it displaces.

Example

An aluminum casting weighs 5.40 lb in air and 3.40 lb when submerged in water. Find the volume of the casting and the weight density of aluminum.

$$\text{Loss of weight} = 5.4 \text{ lb} - 3.4 \text{ lb} = 2.00 \text{ lb}$$

the weight of water displaced. Since 1 ft³ of water weighs 62.5 lb, 2 lb occupies $2/62.5 = 0.032$ ft³.

$$d_w = \frac{w}{V} = \frac{5.40 \text{ lb}}{0.032 \text{ ft}^3} = 169 \text{ lb/ft}^3 \qquad ////$$

Fish are capable of moving toward the surface or into deep water by regulating the quantity of water which they displace and, therefore, the buoyant force. By distending the air bags in their bodies, they can change their volumes and thus change the buoyancy of the water on them. By contracting its air sacs, a fish diminishes its volume and sinks. Similarly, a submarine can submerge by taking water into tanks, thus making the submarine heavier than an equal volume of water. It rises from below the surface by blowing or pumping this water out of the tanks.

Figure 12.11

A sphere floating at the interface between two liquids of different densities. The sphere sinks until the sum of the weight of the first liquid displaced and the weight of the second liquid displaced is equal to the weight of the sphere.

12.7 FLOATING BODIES

When a body floats on a liquid, the buoyant force is equal to its weight. The body sinks until it displaces its own weight (Fig. 12.11) and then remains in equilibrium. This explains why a ship rides lower in the water when loaded than when empty and why it rides higher in salt water than in fresh.

Example

A barge 10 m long and 5 m wide has vertical sides. When two automobiles are driven on board, the barge sinks 5 cm farther into the water. How much do the automobiles weigh?

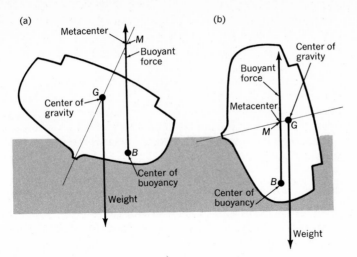

(a)

Metacenter

M

Buoyant force

G

Center of gravity

B

Center of buoyancy

Weight

(b)

Center of gravity

Buoyant force

Metacenter

M

G

B

Center of buoyancy

Weight

Figure 12.12
(*a*) When the center of gravity is below the metacenter M, the boat is stable.
(*b*) When the center of gravity is above the metacenter, the boat is unstable.

Volume of displaced water $= 10 \text{ m} \times 5 \text{ m} \times 0.05 \text{ m}$
$= 2.5 \text{ m}^3$

Weight of water displaced $= 2.5 \text{ m}^3 \times 1{,}000 \text{ kg/m} \times 9.8 \text{ N/kg}$
$= 24{,}500 \text{ N}$

Weight of automobiles $=$ weight of displaced water
$= 24{,}500 \text{ N}$ ////

The buoyant effect of all the displaced fluid may be replaced by a single force acting at the center of mass of the displaced fluid, known as *the center of buoyancy*. For a floating body such as a ship to be in stable equilibrium, there must be a torque which restores the body to its stable position whenever it undergoes an angular displacement. This torque is provided by the combined action of the weight w of the body acting downward at the center of gravity G (Fig. 12.12*a*) and the buoyant force acting upward through the center of buoyancy. The condition for stability will be realized if the metacenter M lies above the center of gravity of the body. The position of the metacenter is determined by the intersection of two lines, one drawn vertically through the center of buoyancy B and the other drawn vertically through the center of gravity G of the body *before displacement*. If the metacenter lies above the center of gravity, there is a restoring torque and the body is in stable equilibrium. If, however, the metacenter lies below the center of gravity, the torque that comes into play when the body is displaced from its normal position increases the displacement further (Fig. 12.12*b*). The body is in unstable equilibrium and turns over.

12.8 DENSITY AND SPECIFIC GRAVITY

Density is the ratio of mass to volume, while weight density is the ratio of weight to volume. Table 12.1 lists the 2 densities of a number of common liquids and solids. When a body has an irregular shape, the volume can be determined by application of Archimedes' principle.

Table 12.1
Densities of Solids and Liquids (20°C)

Substance	Density kg/m³	Density lb$_m$/ft³
Alcohol	789	49.3
Aluminum	2,650	164
Balsa wood	160	10
Brass	8,600	535
Brick	2,100	131
Copper	8,930	555
Cork	240	15
Diamond	3,520	220
Glass (crown)	2,500	156
Glass (flint)	3,700	230
Gasoline	790	49.4
Gold	19,320	1,200
Glycerin	1,260	78.7
Ice (0°C)	917	57.2
Iron (cast)	7,200	450
Iron (wrought)	7,800	480
Kerosene	820	51.2
Lead	11,370	710
Mercury	13,600	840
Oak	800	50
Pine	500	31.2
Silver	10,500	655
Turpentine	870	54.3
Tin	7,290	455
Water (fresh)	1,000	62.5
Water (sea)	1,030	64.4
Zinc	7,150	446.2

When a body is heavier than an equal volume of water (and is insoluble in water), its volume can be determined by finding the loss of weight in water. From the mass and volume the density can be obtained.

Example

A piece of iron has a mass of 0.0780 kg. When submerged in water, it has an apparent weight of 0.666 N. What is the volume of the iron? What is its density?

$$\text{Weight of iron } mg = 0.078 \text{ kg} \times 9.8 \text{ m/sec}^2$$
$$= 0.764 \text{ N}$$

$$\text{Weight of water displaced} = 0.764 \text{ N} - 0.666 \text{ N}$$
$$= 0.098 \text{ N}$$

$$\text{Mass of water displaced } \frac{w}{g} = \frac{0.098 \text{ N}}{9.8 \text{ m/sec}^2}$$
$$= 0.010 \text{ kg} = 10 \text{ g}$$

$$\text{Volume of water displaced } \frac{m}{d} = \frac{0.010 \text{ kg}}{1,000 \text{ kg/m}^3}$$
$$= 10^{-5} \text{ m}^3 = 10 \text{ cm}^3$$

$$\text{Volume of iron} = 10^{-5} \text{ m}^3 = 10 \text{ cm}^3$$

$$\text{Density of iron } \frac{m}{V} = \frac{0.078 \text{ kg}}{10^{-5} \text{ m}^3} = 7,800 \text{ kg/m}^3 \quad \text{////}$$

If the object is lighter than water (and insoluble), its volume can still be determined by this method by using a sinker large enough to pull it below the surface of the water. In this case (Fig. 12.13) the combined weight of the body and the sinker is first determined when the sinker is immersed in water and the body is above the surface of the water. The body is then also submerged, and the combined weight redetermined. The reduction in weight is equal to the weight of water displaced by the body.

Example

A piece of cork has a mass of 0.025 kg. When it is fastened to a sinker and the sinker alone immersed in water, the combined apparent weight of sinker and cork is 1.960 N. When both sinker and cork are immersed, they weigh 0.735 N. What is the density of the cork?

$$\text{Loss of weight due to submerging body} = 1.960 - 0.735 \text{ N}$$
$$= 1.225 \text{ N}$$

$$\text{Mass of water displaced by cork} = \frac{1.225 \text{ N}}{9.8 \text{ m/sec}^2}$$
$$= 0.125 \text{ kg}$$

$$\text{Volume of water displaced} = \frac{0.125 \text{ kg}}{1,000 \text{ kg/m}^3}$$
$$= 1.25 \times 10^{-4} \text{ m}^3$$

Figure 12.13
The buoyant force on the block of wood, which would float if there were no sinker, is the difference between F_1 and F_2.

$$\text{Density of cork} = \frac{\text{mass of cork}}{\text{volume}}$$
$$= \frac{0.025 \text{ kg}}{1.25 \times 10^{-4} \text{ m}^3}$$
$$= 200 \text{ kg/m}^3 \qquad ////$$

The specific gravity of a body is the ratio of its density to the density of water. If the specific gravity of a body is equal to 5, the body weighs 5 times as much as an equal volume of water.

12.9 BUOYANT EFFECT OF THE AIR

Archimedes' principle applies to all fluids. In general, the densities of gases are much smaller than those of liquids, and the buoyant effects are correspondingly reduced. Nevertheless there are many examples of situations in which the buoyant effects of gases—in particular of air—are of considerable importance. The density of air is 1.293 kg/m³ at 0°C and one standard atmosphere pressure.

A simple demonstration of the buoyant effect of air can be made by suspending a lead ball (Fig. 12.14) from one side of a small balance and a large hollow brass sphere from the other side. The hollow sphere is just heavy enough to balance the lead ball when both are in air. If the balance, together with the suspended spheres, is placed under a bell jar and nearly all the air removed by means of a pump, the lead ball no longer balances the hollow sphere. This is because the buoyancy of the air on the hollow brass sphere is greater than on the lead ball; when this lift has been removed, the true weights of the spheres become evident, and the hollow sphere weighs more than the lead sphere.

The gross lifting capacity of a balloon is equal to the weight of the air it displaces. The pressure and density of the air become less at higher elevations. However, as the pressure outside the balloon decreases, the lighter gas inside the balloon expands, thus displacing a larger volume of the less dense air. Ballast may be carried along and thrown overboard when it is desired to make the balloon rise higher. By allowing some of the gas in the balloon to escape, the balloon can be made to descend.

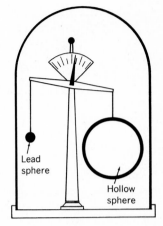

Figure 12.14

A body immersed in air is buoyed up by a force equal to the weight of the air it displaces. The two spheres balance each other at normal air pressure, but when the jar is evacuated, the larger sphere sinks.

Example

A balloon has a volume equal to that of a sphere 15 m in radius. What is the gross mass which it will lift when the density of the air is 1.2 kg/m³?

$$\text{Volume of balloon} = \tfrac{4}{3}\pi(15)^3 = 14,000 \text{ m}^3$$
$$\text{Mass of air displaced by balloon} = \text{volume} \times \text{density}$$
$$= 14,000 \text{ m}^3 \times 1.2 \text{ kg/m}^3$$
$$= 16,800 \text{ kg} \qquad ////$$

12.10 COHESION AND ADHESION

In liquids the molecules move freely with respect to each other but are held together by attractive forces. Not only do the molecules of a liquid cling to each other, but they also cling to the molecules of other sub-

Figure 12.15

A force $2Tl$ is required to hold the movable wire AB in equilibrium against surface tension.

Figure 12.16

How the surface tension of water varies with temperature.

Table 12.2
Surface Tension

Substance	Temperature, °C	Surface tension, N/m
Acetic acid	20	0.0234
Alcohol (ethyl)	20	0.0216
Ether (ethyl)	20	0.0169
Glycerin	18	0.0632
Mercury	18	0.545
Turpentine	20	0.0289
Petroleum	20	0.0259
Water	20	0.0728

stances, as may be seen when a piece of glass is dipped into a vessel of water. The molecules of water adhere to the glass and form a thin film over its surface. *The attraction of like molecules for one another is called cohesion; the attraction of unlike molecules for one another is called adhesion.* It is *cohesive* forces which hold together so firmly the molecules of iron, copper, and other solid substances.

If the molecules of a liquid have less attraction for each other than for the molecules of the solid with which they are in contact, the liquid adheres to the solid and wets it. Here *adhesive forces are greater than cohesive.* When the cohesive forces are greater than the adhesive, the solid is not wet by the liquid. Such is the case when mercury is in contact with glass. If a drop of mercury and a drop of distilled water are placed on a clean glass surface, the water spreads over the glass in a thin layer, while the mercury forms a distorted ball.

If molecules are further apart than a few millionths of a centimeter, the attractive forces are not appreciably great. Intermolecular forces have short range. It takes a moderate force to break a piece of chalk, but after it is broken, the pieces do not adhere if they are pressed back together. Because of the irregular surfaces, not enough molecules are able to exert appreciable attractive forces.

Molecules of water adhere to glass, forming a thin layer over its surface. When we place a little water between two sheets of glass and take care to see that no air is trapped between the plates, it requires a significantly greater force to separate the glass plates than it would if they were dry. The water is acting as an adhesive. Glues are much superior adhesives. Once a glue has set, it may be stronger than the materials it is binding together. If a thin gold foil is placed over a layer of metal and carefully pressed into place, the molecular attraction between the gold and the base metal is strong enough to keep the gold in place without a cement.

12.11 SURFACE ENERGY AND SURFACE TENSION

A greased needle gently placed on the surface of water floats, although the density of the needle is greater than that of the water. Some insects can walk on the surface of a lake or stream, which behaves as though it were covered with a thin elastic film. It requires a force to break this film, the amount of force depending on the molecular attraction characteristic of the liquid.

On the average all molecules of a liquid at rest are in equilibrium under the influence of attractive forces exerted by neighbors. However, if a molecule at a surface begins to leave the liquid, the resultant of these forces pulls it back. A molecule must have considerable energy to escape from the surface.

Consider Fig. 12.15, in which $CEHD$ is a bent wire and AB is a straight wire which may be moved in the direction of the arrow. If we dip this system in a soap solution, we may have the area $AEHB$ covered by a soap film. In order to prevent the soap film from contracting and pulling the side AB toward EH, a force F must be applied. This force is given by $F = 2Tl$, where l is the length of the wire AB and T is the force per unit length. The factor 2 is introduced because the

Figure 12.17
Surface tension creates splash patterns on the surface of milk. [*Courtesy of Harold E. Edgerton.*]

film has two sides. *The surface force per unit length which must be applied to overcome molecular forces is the surface tension.*

If the wire AB is drawn down a distance s, the work done is

$$W = Fs = T(2ls) = T \, \Delta A \qquad (12.3)$$

where $\Delta A = 2ls$, the increase in area counting *both* sides of the film. Thus, T is the ratio of the work required to increase the area against the surface molecular forces to the change in area.

The surface tension of a liquid depends on the temperature (Fig. 12.16). The values of the surface tension for several liquids are listed in Table 12.2.

One way of measuring surface tension is to determine the force required to pull a platinum ring through the surface of a liquid. The total length of liquid surface which must be broken is twice the circumference of the ring. Thus the force required to break the surface is $T \times 4\pi r$, where r is the radius of the ring.

The splashes formed (Fig. 12.17) when drops of milk fall on a surface of milk give an illustration of the effects produced by surface tension. When a drop of oil is placed in a mixture of water and alcohol of the same density, the oil droplet does not rise or fall. Under the action of molecular forces it assumes a spherical form. Lead pellets for a shotgun can be produced by allowing liquid lead to fall from an opening of suitable size. Because of surface forces the freely falling lead drops assume a spherical shape, in which they solidify before they strike anything.

Surface tension plays a vital role in the formation of bubbles. The pressure p inside a bubble in equilibrium is greater than the external pressure p_0 because of surface tension. The forces across the equatorial plane of a stable soap bubble (Fig. 12.18) are balanced; hence, $\pi R^2 (p - p_0) = 4\pi RT$. The excess pressure in the soap bubble is then

$$\Delta p = p - p_0 = \frac{4T}{R} \qquad (12.4)$$

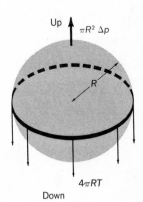

Figure 12.18
Excess pressure Δp inside a soap bubble produces an upward force $\pi R^2 \, \Delta p$ on the top hemisphere of the bubble; this force is balanced by the downward force $2(2\pi R)T$ due to surface tension.

The surface force is $4\pi RT$ for the soap bubble because there are two surfaces to the bubble. A bubble in a carbonated liquid or a bubble of steam in boiling water has only a single surface. In these cases the excess pressure is only half that given by Eq. (12.4), or $2T/R$.

12.12 CAPILLARITY

If a piece of glass tubing of very small bore is thoroughly cleaned and then dipped into water (Fig. 12.19a), the water wets the inside of the tube and rises in it. If the liquid does not wet the tube (Fig. 12.19b) as in the case of mercury in a glass tube, the liquid is depressed. The smaller the bore of the tube, the greater the height to which the liquid rises or the greater the amount which it is depressed. This rise or depression of liquids in tubes of small bore is known as *capillarity*. It is caused by the molecular forces that are responsible for surface energy. If the cohesive forces between molecules of the liquid are greater than the adhesive forces between liquid and wall, the liquid pulls away from the tube and is depressed. If the adhesive forces are greater, the liquid wets the capillary tube and rises.

When a glass tube is thrust into water, the molecules in the surface of the wall just above the water pull up on the molecules of water lying nearest to them and raise them above the level of the water in the vessel. This carries upward a column of water, which is supported by the surface forces. The net upward force available is the vertical component of the surface-tension forces if we are dealing with a vertical cylindrical tube. The angle of contact θ (Fig. 12.19) between the surface of the liquid and the tube depends on the liquid, the gas above, and the kind of tube involved. The upward force per unit length is $T\cos\theta$, and the total upward force is $T\cos\theta$ multiplied by the length of liquid in contact with the tube, which is the circumference. Thus, the net upward force is $2\pi r T\cos\theta$. The liquid rises until the weight of liquid supported, $\pi r^2 h dg$, is equal to this force. Hence, for equilibrium

$$2\pi r T \cos\theta = \pi r^2 h dg \qquad (12.5)$$

or

$$h = \frac{2T\cos\theta}{rdg}$$

For water in clean glass the angle θ is zero, and $\cos\theta$ becomes unity. Whenever a liquid wets a tube thoroughly, θ is zero. When cohesive forces exceed adhesive, θ is greater than 90°, the liquid is depressed in the capillary, and h is negative. For water on paraffin θ is 107°. Water does not wet paraffin. For mercury on glass the angle of contact is approximately 140°. Figure 12.20 shows how mercury and water behave when placed in glass tubes of unequal radii.

Figure 12.19

(*a*) Water wets glass, and the angle of contact θ approaches zero. (*b*) Mercury does not wet glass; the angle of contact θ is 140°.

Figure 12.20

Unequal heights of water and mercury in connecting glass tubes of different radii.

Example

The liquid in a capillary tube rises to the height of 7 cm. The radius of the tube is 0.1 mm, and the density of the liquid 800 kg/m³. If the angle between the liquid and the surface of the tube is zero, find the surface tension of the liquid.

$$2\pi r T \cos \theta = \pi r^2 h d g$$
$$2T \times 1 = 10^{-4} \text{ m} \times 0.07 \text{ m} \times 800 \text{ kg/m}^3 \times 9.8 \text{ N/kg}$$
$$T = 0.0274 \text{ N/m} \qquad ////$$

In recent years the chemical industry has developed a number of *waterproofing* agents which prevent water from wetting a fabric. For these compounds θ is greater than 90°. Similarly, a number of *wetting agents*, or *detergents*, have been developed. These agents change θ to a smaller angle. The addition of a suitable detergent can make water wet paraffin.

There are many illustrations of capillary action in nature. The oil in a lamp rises in the wick by capillary action. Ink spreads in a blotter, and water in a lump of sugar, by this same action. If one end of a towel dips into a bucket of water and the other end hangs over the bucket, the whole towel soon becomes wet.

QUESTIONS

1 Does the force exerted on a dam by an artificial lake depend on how far the water has been backed up? Upon what factors does it depend?

2 An athlete stands on the platform of a sensitive scale. When he takes a deep breath, will his weight as read by the scale change?

3 What physics underlies the act of filling a fountain pen?

4 Why does flame-polishing a piece of glass by putting it in a bunsen flame round off sharp edges?

5 Why do carbonated beverages often foam when a bottle or can is opened?

6 A razor blade can be made to float on water. What forces act on this blade? Is Archimedes' principle applicable?

7 A large soap bubble is formed at one end of a capillary tube, and a small one at the other end. Which will grow at the expense of the other? Why?

8 Why is the radius of the stem of a sensitive hydrometer (Prob. 27) much smaller than the radius of the submerged bulb?

9 An iron ball floats in a beaker half filled with mercury. If water is poured over the mercury until the beaker is filled, will the level at which the ball floats change?

10 What determines the height to which a balloon rises?

11 Why does the pressure inside a soap bubble decrease as it is blown up? How does a rubber balloon behave in this respect? Explain.

12 A beaker partly filled with water rests on the pan of a spring balance. A small boat filled with lead balls floats on the water. If the boat is tipped over so that it and its contents sink to the bottom, does the water level in the beaker change? Does the balance reading change?

13 Water rises to a height of 5 cm in a certain capillary tube. If the tube is broken off 2 cm above the level of the liquid in the main vessel, will water run out the top of the tube? Why not?

14 Explain why a balloon can remain in stable equilibrium at almost any given height in the atmosphere while a propulsionless submarine cannot ordinarily remain in stable equilibrium at a given depth in the ocean. Under what special circumstance might the submarine be in stable equilibrium at some particular depth?

15 Why does a balloon descend when it releases gas, even though its weight is thereby decreased?

PROBLEMS

1 A 60-kg man is skating on ice. Find the pressure exerted by one skate when the other is not touching the ice if the length of the blade in contact with the ice is 0.15 m and the runners are 0.0045 m wide.   *Ans.* 8.7×10^5 N/m^2

2 What is the average pressure due to the weight of a 200-lb person if the effective area of each foot is 18 in.^{2} and the person is standing on both feet? On just one foot?

3 Assuming an average weight density of 995 kg/m³ for the human body, find the volume occupied by an 80-kg swimmer. What pressure due to the water does the swimmer experience at a depth of 2.5 m in fresh water?

$Ans.$ 0.0804 m³; 24,500 N/m²

4 The pressure in an automobile tire is 25 lb/in.² greater than atmospheric pressure. If the wheel on which the tire is mounted supports 800 lb, what area of the tire is in contact with the ground? (Neglect the mechanical strength of the casing.)

5 Find the force on the glass side of an aquarium containing salt water with a density of 1,030 kg/m³ if the glass is 0.75 m wide and the water behind it is 50 cm deep. $Ans.$ 946 N

6 What is the weight of the air in a living room of dimensions 20 by 12 by 8 ft? The weight density of air is 0.081 lb/ft³.

7 A man carries a barometer from the bottom to the top of a building that is 150 m high. At the bottom of the building the barometer reads 76 cm Hg, and at the top it reads 74.61 cm Hg. Find the average density of the air between these heights. $Ans.$ 1.26 kg/m³

8 A lock gate is 20 ft wide and 25 ft high. The height of water on one side is 24 ft and on the other side is 10 ft. What is the net horizontal force on the gate due to water pressure?

9 One end of a closed hydraulic system is 5 m above the other end. The system is filled with oil of density 800 kg/m³. At the higher end, a force of 200 N is applied on an area of 10 cm². What is the net force on a piston with an area of 15 cm² at the lower end? $Ans.$ 359 N

10 What is the reading of a barometer on top of a building that is 120 m high when the barometer at the base of the building reads 74 cm Hg if the average density of air between these heights is 1.25 kg/m³?

11 The apparatus of Fig. 12.5*b* contains kerosene in one arm and distilled water in the other. When the level of the water is at 26.2 cm, the kerosene level is at 32 cm. Find the density of kerosene in both metric and British engineering units.

$Ans.$ 819 kg/m³; 1.59 slugs/ft³

12 A liquid-filled reservoir is connected to two cylinders with closely fitting pistons which are at the same level. One cylinder has a diameter of 0.5 in., while the other has a diameter of 8 in. If a force of 4 lb is exerted on the smaller piston, find the increase in pressure on the larger piston. How large a force is required on the larger piston to keep it at rest if frictional forces are negligible?

13 A submarine is designed to withstand a maximum water pressure of 32,500 lb/ft². If 1 ft³ of seawater weighs 65 lb, find the maximum depth to which the submarine can be taken without exceeding the design pressure. Find the net force on a rectangular hatch of dimensions 3 by 2 ft at this depth. $Ans.$ 500 ft; 195,000 lb

14 The water level in a reservoir standpipe is 250 ft above the lowest part of a town. What is the maximum water pressure in pounds per square inch available due to this head of water?

15 Water and oil are standing in opposite legs of a U-tube open at both ends. Water fills the bottom and stands 28 cm above the oil-water interface. How high does the oil stand above the interface if its density is 800 kg/m³?

$Ans.$ 35.0 cm

16 A submarine is at a depth of 70 ft in seawater which has a density of 2 slugs/ft³. Find the pressure due to the water and the resulting force on a hatch of dimensions 3 by 2 ft.

17 A piece of copper is "weighed" with an equal-arm laboratory balance. When the copper is in air, the balancing mass is 112 g. When it is immersed in water and in kerosene, the balancing masses are 99.5 and 102 g, respectively. Find the densities of copper and kerosene.

$Ans.$ 8,960 kg/m³; 800 kg/m³

18 A barge with vertical sides is used to transport automobiles down the Mississippi River. If 15 cars weighing 3,000 lb apiece are driven onto the barge, which is 50 ft long and 25 ft wide, find the additional volume of water which the barge must displace. How much deeper does the barge ride in the water?

19 The volume of the gas bag of an airship is 140,000 m³. The density of air at 0°C and at atmospheric pressure is 1.29 kg/m³, and that of the hydrogen with which the gas bags are filled is 0.092 kg/m³. If the mass of the airship and its necessary equipment is 80,000 kg, what is the maximum net lift available for a useful load?

$Ans.$ 8.6 × 10⁵ N

20 A truck loaded to a total weight of 5 tons drives onto a ferryboat, causing it to sink $\frac{1}{4}$ in. deeper into the water. What is the area of the horizontal section of the boat at the waterline?

21 A piece of glass with a mass of 341 g is "weighed" when immersed in various liquids. It weighs 247 g in water and 225 g in glycerin. Find the density of the glass and the specific gravity of the glycerin.
Ans. 3,630 kg/m³; 1.23

22 A balloon with a volume of 2.5×10^4 ft³ is filled with helium of weight density 0.011 lb/ft³. The weight density of air is 0.081 lb/ft³. What is the net buoyant force on the balloon?

23 Find the volume and the mass of a specimen of copper that has an apparent weight of 7.5 N when under water.
Ans. 9.65×10^{-5} m³; 0.862 kg

24 A body was "weighed" in water, in oil, and in alcohol. Its loss of weight in water was 1.50 N, in oil 0.98 N, and in alcohol 1.20 N. What is the specific gravity of the oil? Of the alcohol?

25 A piece of wood with a weight of 4.92 N is immersed in water by use of a sinker that weighs 2.55 N in water. The combined weight of the wood and the sinker when both are immersed is 1.68 N. Find the density of the wood.
Ans. 850 kg/m³

26 A piece of cork weighs 1.2 lb. A 4-lb sinker is attached to it. With only the sinker in water, the cork and sinker weigh 4.6 lb. With the sinker and cork both in water, they weigh 0.2 lb. What is the density of the cork? What is its specific gravity? What is the density of the sinker?

27 A hydrometer (see accompanying figure) has a mass of 110 g. The graduated stem BA is 30 cm long and has a cross-sectional area of 0.5 cm², while the bulb below B has a volume of 107.5 cm³. What fraction of the stem BA will be submerged when the hydrometer floats in water? What is the lowest specific gravity the hydrometer can read? What is the specific gravity of a liquid in which the hydrometer floats with the liquid level at B? *Ans.* $\frac{1}{6}$; 0.90; 1.023

28 A thin sheet of metal is bent in the form of a hollow square, and the open square end is dipped into a soap solution whose surface tension is 0.025 N/m. The length of each side of the square is 3.6 cm. With what force does the surface tension pull the metal into the solution?

29 A capillary tube whose inside radius is 0.25 mm is dipped in water. What is the weight of the water raised by capillary action above the normal water level? To what height is the water raised?
Ans. 1.14×10^{-4} N; 5.94 cm

30 Find the work which must be done to increase the outside surface area of a soap bubble by 250 cm² if the surface tension of soap film is 0.030 N/m.

31 A solid glass rod is immersed in water such that its axis is perpendicular to the surface of the water. The rod has a circular cross section and a diameter of 0.64 cm. Find the force exerted on the rod by surface tension.
Ans. 0.00146 N

32 What is the excess pressure inside a soap bubble that is 4 cm in diameter, assuming 0.026 N/m as the surface tension of the soap solution?

33 What is the diameter of a soap bubble when the excess pressure inside is 9.76 N/m² and the surface tension of the soap solution is 0.026 N/m?
Ans. 2.13 cm

34 A capillary tube whose inside diameter is 0.75 mm is dipped in glycerin. The glycerin rises 2.72 cm in the tube. If the density of glycerin

Problem 27

is 1.26 g/cm³, what is its surface tension? Assume the angle of contact is zero.

35 A tube is bent in the form of a U (Fig. 12.20). The diameter of the larger arm is 5 mm, and that of the smaller arm is 0.5 mm. Find the difference in levels when the U-tube contains water.

Ans. 5.35 cm

36 One limb of a U-tube (Fig. 12.20) is 0.5 cm in diameter, and the other is 1 mm in diameter. What will be the difference in surface levels in the two tubes when mercury is poured into them? Take the angle of contact as 140° for mercury.

37 A rectangular block of wood has a cross-sectional area of 120 cm² and a height of 40 cm. It floats vertically in water because to its lower edge is fastened a weight of 600 g of lead. How much of the wood projects out of the water? (Take the density of wood to be 450 kg/m³.)   *Ans.* 17.4 cm

38 A piece of wood with a volume of 900 cm³ floats with eight-tenths of its volume under water. Find the smallest mass of iron of specific gravity 7.8 which will be needed to sink the wood to the bottom.

39 A diver with his suit weighs 300 lb. Blocks of lead with a volume totaling 50 in.³ attached to his shoes just cause him to sink. How many cubic feet of water are displaced by the suit?

Ans. 5.1 ft³

40 Find the volume of cork that must be employed in a life preserver if it is designed to support one-fifth of a man's body out of fresh water, assuming the man weighs 160 lb and has a specific gravity of 1.00 and that all the cork is submerged.

41 A 3-kg mass of wrought iron floats in a beaker of mercury. What fraction of the iron is submerged? If sufficient water is poured into the beaker to cover the iron completely, what fraction of the iron is submerged in the mercury?

Ans. 0.57; 0.54

42 A 0.5-kg plastic sphere has a density of 900 kg/m³. It floats at a water-kerosene interface (Fig. 12.11). What volume V_2 lies beneath the interface?

43 A cylindrical buoy of mass m and radius R floats in water of density d with two-thirds of its height h submerged. If it is depressed a distance $y \ll h$ and released at rest, show that the resulting motion is simple harmonic. Find the frequency.

Ans. $(R/2)\sqrt{gd/\pi m}$

WE HAVE BEEN CONSIDERING the physics of fluids at rest. But fluids also move, and in moving they exert forces on the bodies past which they flow. Fluids also exert forces on bodies which move through them. Indeed, the forces depend only on the relative motion of body and fluid. Hence the forces on an airplane wing at rest in a wind tunnel are identical with those on the same wing moving through air at rest. In this chapter we examine a few of the phenomena associated with moving fluids.

CHAPTER THIRTEEN
FLUIDS IN MOTION

13.1 FLUID FLOW

In hydrodynamics and aerodynamics we are concerned with the motion of bodies through a fluid or the equivalent movement of a fluid past objects. For the latter case imagine many small elements of the fluid which are individually colored so one can see the motions of these "fluid particles" during flow. The trajectory of a single particle is called a *pathline* or a *line of flow*. In general, the velocity of each such particle changes along a line of flow, but let us suppose that every particle passing a given point follows the same pathline and further that the velocity of each particle which passes a given point is the same as that of every other particle as it passes that point. In such a case the flow is described as *steady* or *stationary*. (In nonsteady flow the velocity of particles at a given point is a function of time, and in *turbulent* flow the velocities change erratically from point to point as well as from one time to another.)

In any flow a *streamline* is a curve whose tangent at any point is in the direction of the particle velocity at that point. For steady flow the streamlines coincide with the pathlines. If we consider all the streamlines passing through the boundary of an element of area A (Fig. 13.1), they enclose a region called a *tube of flow*. In steady flow no particles cross the boundaries of a flow tube; every fluid particle which enters at one end leaves at the other.

Imagine a tiny paddle wheel carried along in the tube of flow of Fig. 13.1. In some situations the paddle wheel would travel without rotation; then the flow at the position of the paddle wheel would be *irrotational*. In other flow situations the paddle wheel would rotate, and the flow would be *rotational*. Swirling eddies are an evidence of rotational flow for which angular momentum is an important property. The rotationality is conserved.

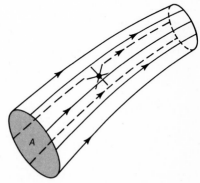

Figure 13.1
A tube of flow bounded by streamlines. The little paddle wheel remains at rest anywhere the flow is irrotational but is subjected to a torque anywhere the flow is rotational.

13.2 PRESSURE IN A MOVING FLUID; BERNOULLI'S THEOREM

As in many branches of physics, it is convenient to begin by considering a highly idealized situation. In this section we discuss the *steady, irrotational* flow of an *incompressible, frictionless* (nonviscous) fluid. Such an analysis was characterized by von Neumann as the study of "dry water," and indeed there is no fluid which behaves exactly like our ideal fluid under all conditions. Nevertheless, the results derived for the ideal fluid apply *in some situations* to many fluids, including air.

Consider an ideal liquid flowing through a pipe of varying cross section (Fig. 13.2). Because the cross section of the pipe is smaller at b than it is at a, the velocity of the liquid must be greater at b than at a in order that the flow of the liquid through each cross section of the pipe may be the same. Consequently, the momentum of the liquid per unit volume increases in going from a to b. According to Newton's second law, a force is necessary to produce a change of momentum. In this case, the force results from a difference in pressure between a and b. The pressure at a is greater than the pressure at b, thereby producing a positive acceleration of the liquid and an increase in its momentum per unit volume. *The pressure in the fluid is smaller where the velocity is greater.*

The change in pressure can be measured by observing the heights of the liquid in transverse manometers at a, b, and c. Since the liquid is *incompressible*, the amount passing through every cross section of the tube per second is the same. Let v_a, v_b, and v_c represent the speeds and A_a, A_b, and A_c represent the cross-sectional areas at a, b, and c, respectively. Then

$$v_a A_a = v_b A_b = v_c A_c \qquad (13.1)$$

For steady flow the speed is inversely proportional to the cross section of the tube.

Example

The cross section of the tube at a is 10 cm², and at b is 2 cm². If the velocity of the stream at a is 1.2 m/sec, what is it at b?

$$\text{Velocity at } b = \text{velocity at } a \times \frac{\text{area at } a}{\text{area at } b}$$

$$= 1.2 \text{ m/sec} \times \frac{10 \text{ cm}^2}{2 \text{ cm}^2} = 6 \text{ m/sec} \qquad ////$$

Figure 13.2

Flow velocity is greater, and pressure less, in regions of reduced cross-sectional area in a horizontal tube of varying cross section.

Figure 13.3
Streamlined flow of a liquid through a tube of variable cross section at different heights.

Consider the case in which all parts of a tube of varying cross section (Fig. 13.3) are not at the same height. Since (1) the liquid is incompressible, (2) the flow is streamlined, and (3) there is no fluid friction (no viscosity), we can use the conservation of energy to derive a relationship between the pressure, velocity, and height at any one point and pressure, velocity, and height at some other point. Let volume V of liquid be forced from a to c. The work done *on* the system in forcing the liquid past a is $F_a S_a = p_a A_a (V/A_a) = p_a V$, where the subscript a represents the value at point a. The work done *by* the system when volume V is forced past point c is $F_c S_c = p_c A_c (V/A_c) = p_c V$. The *net* work done *on* the system is $(p_a - p_c)V$. Since there are no energy losses, we have

Work done on system = energy of volume V at c − energy of V at a

$$(p_a - p_c)V = [\tfrac{1}{2}(dV)v_c^2 + dVgy_c] - [\tfrac{1}{2}(dV)v_a^2 + dVgy_a] \quad \text{(A)}$$

where d is the density of the liquid and dV the mass of the volume V. If we divide both sides of Eq. (A) by V and rearrange the terms, we obtain *Bernoulli's theorem,*

$$p_a + \tfrac{1}{2}dv_a^2 + dgy_a = p_c + \tfrac{1}{2}dv_c^2 + dgy_c = \text{constant} \quad \text{(13.2)}$$

The sum of the pressure, the kinetic energy per unit volume, and the potential energy per unit volume is the same at all points along the same streamline in the case of streamlined flow for a nonviscous, incompressible fluid.

In the flow of a real fluid frictional forces act, and as we move in the direction of the fluid velocity, the sum of the three terms of Eq. (13.2) decreases (Sec. 13.4). Although we have derived Bernoulli's theorem for an incompressible fluid, Eq. (13.2) is approximately true for a gas so long as the speed is well below the speed of sound (about 340 m/sec or 740 mi/hr for air at 20°C).

Example

Water of density 1,000 kg/m³ is passing through the tube of Fig. 13.3. The cross-sectional area at a is twice that at b. At a, $v = 6$ m/sec, $y_a = 10$ m, and $p_a = 140,000$ N/m². If $y_b = 8$ m, find the velocity and pressure at b.

Since the area at a is twice that at b, the speed at b must be twice that at a by Eq. (13.1); hence $v_b = 12$ m/sec. By Eq. (13.2),

$$140,000 \text{ N/m}^2 + \tfrac{1}{2}(1,000 \text{ kg/m}^3)(36 \text{ m}^2/\text{sec}^2)$$
$$+ (1,000 \text{ kg/m}^3 \times 9.8 \text{ m/sec}^2 \times 10 \text{ m})$$
$$= p_b + \tfrac{1}{2}(1,000 \text{ kg/m}^3)(144 \text{ m}^2/\text{sec}^2)$$
$$+ (1,000 \text{ kg/m}^3 \times 9.8 \text{ m/sec}^2 \times 8 \text{ m})$$
$$256,000 \text{ N/m}^2 = p_b + 150,400 \text{ N/m}^2$$
$$p_b = 105,600 \text{ N/m}^2 \qquad \text{////}$$

13.3 APPLICATIONS OF BERNOULLI'S THEOREM

Among the common phenomena involving Bernoulli's theorem are the following.

A Ball in an Air Jet A Ping-Pong ball can be balanced on an air jet with the ball supported by the rate of transfer of momentum from the airstream. When the ball moves slightly to one side, the airspeed is lower and the pressure correspondingly greater on that side than on the other. Consequently there is a net force moving the ball back toward the center of the stream. Similarly, the ball may be held in an inverted funnel (Fig. 13.4) by a downward stream of air. The air speeds up as it passes the constricted region between the ball and the funnel. As a result the pressure there is reduced below atmospheric, and this pressure difference in turn gives rise to a net force holding the ball in the funnel.

The Sprayer The forward stroke of the piston (Fig. 13.5) produces a stream of air past the end of the tube D. The other end of this tube is immersed in the liquid to be sprayed. The stream of air flowing past the open end of the tube reduces the pressure on the liquid in the tube. Atmospheric pressure acting on the surface of the liquid in A raises liquid in the tube D, from which it is carried away by the stream of air. A spray results from this mixture of air with the fine particles of liquid.

Flow of Liquid through an Orifice In Fig. 13.6 liquid is flowing out of an orifice. Both at the top of the liquid and at the orifice the pressure is atmospheric. At the top of the container the liquid has potential energy and no kinetic energy. If we pick the opening as our position of zero height, the escaping liquid has kinetic energy but no potential energy. Applying Bernoulli's theorem to points at the top of the fluid and at the orifice yields

$$p_{\text{atm}} + 0 + dgy = p_{\text{atm}} + \tfrac{1}{2}dv^2 + 0$$

Figure 13.4
A light ball can be supported in an inverted funnel by the Bernoulli pressure reduction.

Figure 13.5
A sprayer in which reduction of pressure at B causes liquid to rise in tube D.

where p_{atm} represents atmospheric pressure, d the density of the fluid, and v the escape velocity.

$$dgy = \tfrac{1}{2}dv^2 \quad \text{or} \quad v = \sqrt{2gy} \tag{13.2a}$$

Figure 13.7 shows streamlines associated with flow through a sharp-edged orifice. In this case the fluid is still being accelerated as it passes the sharp edges at the opening. As the fluid gains speed, the area of flow decreases in accordance with Eq. (13.1). The flow lines converge to give a vena contracta, with the effective area a little less than two-thirds the actual area for a sharp circular orifice. A well-designed nozzle gives a flow which utilizes the actual opening far more effectively.

Two familiar and very important applications of the Bernoulli effect are the lift on the wing of an aircraft and the curving of a rapidly moving, spinning ball. Both depend on the fact that there is some fluid friction, and therefore we introduce the concept of viscosity before we discuss them.

Figure 13.6
Efflux of a liquid from an orifice.

13.4 VISCOSITY

When real fluids are in motion and the shape as well as the position of a given fluid element may be changing, there is an internal friction between adjacent layers moving with different speeds; this internal friction is known as *viscosity*. When two beakers, one containing heavy oil and the other water, are tilted from side to side, the mobility of the water is clearly greater than that of the oil; oil is more viscous than water. Gases have lower viscosity than liquids.

Consider the flow of a fluid over the surface SS' (Fig. 13.8). Whenever a fluid is in contact with a solid surface, there is an infinitesimal layer which does not move relative to the surface regardless of relative motion between fluid and solid. Thus the fluid in contact with SS' is at rest because of adhesion, but each successive layer moves with respect to the layer directly below it with the speed increasing with the distance from SS'. Each horizontal layer is acted upon by the layer above

Figure 13.7
Converging of streamlines to form a vena contracta at a sharp-edged orifice.

Figure 13.8

Displacement of layers of a moving fluid relative to a surface at rest.

with a tangential force in the direction of the motion and by the layer below with a retarding force. These forces arise from momentum transfer between the molecules of the fluid. The dashed figures *abcd* and *efgh* (Fig. 13.8) show the distortion of the fluid as it moves. The upper layer *fg* travels faster than the lower layer *eh*. Consider a tiny imaginary cube represented by *abcd* which becomes *efgh* a short time later. Let F be the tangential force on the plane which cuts the page in *bc* (and later in *fg*). Let A be the area of this surface of the cube, Δl the distance *ab*, and Δv the speed of *bc* relative to *ad*. When Δv is not too great, F/A is proportional to $\Delta v/\Delta l$:

$$\frac{F}{A} = \eta \frac{\Delta v}{\Delta l} \tag{13.3}$$

where η is a constant called the *coefficient of viscosity* (Table 13.1). It is equal to the force per unit area necessary to maintain unit difference of velocity between two parallel planes when the planes are unit distance apart. The coefficient of viscosity of a typical liquid changes with both temperature (Fig. 13.9*a*) and pressure (Fig. 13.9*b*). For gases the viscosity behaves differently; it is proportional to the square root of the temperature and independent of pressure over a broad range. From Eq. (13.3) it is clear that η may have the dimensions newton-seconds per square meter. In many handbooks viscosities are given in poises; 1 poise = 0.1 N-sec/m^2 = 1 dyne-sec/cm^2.

In writing Eq. (13.3) we have assumed that the fluid flows in thin sheets (lamina). Such flow is said to be *laminar*. It is observed only when the velocity is relatively small; at higher velocities the flow becomes turbulent. Since in turbulent flow the velocities of the fluid particles change erratically, the instantaneous velocity at a point is not the same as the time-average velocity. Then, if average velocities are used in Eq. (13.3), the proportionality constant between F/A and $\Delta v/\Delta l$ is much greater than η because small patches of turbulent flow (eddies) also carry transverse momentum from one layer to another.

13.5 LIFT AND DRAG

If a smooth sphere were to move at constant velocity through an ideal nonviscous fluid, theory predicts that there would be no resistance at all! This paradox, attributed to d'Alembert or to Euler, arises because in the absence of viscosity the velocity of the fluid and hence the pressure at every point such as P' on the downstream side would be exactly equal to that at the corresponding point P on the upstream side (Fig. 13.10*a*). The total fluid force on the back side would push the sphere forward just as hard as the fluid force on the front side holds it back. However, in any real fluid viscosity immediately introduces two retarding forces: (1) The fluid molecules in contact with the sphere remain attached, and shear stresses [Eq. (13.3)] cause a tangential decelerating force at the surface; and (2) the pressure at each point P' falls below that at the corresponding point P. The component of the resultant fluid force parallel to the relative velocity between the fluid and any object is called the *drag*.

For real fluids the fluid velocity relative to a solid surface is zero

Table 13.1
Viscosities of Fluids
(All at 20°C except as noted)

	η, N-sec/m^2
Liquids:	
Ethyl alcohol	0.0012
Glycerin	1.48
Mercury	0.0015
Oil:	
SAE 10 at 55°C	0.16–0.22
SAE 20 at 55°C	0.23–0.30
Water	0.001
Gases:	
Air	18×10^{-6}
Argon	22×10^{-6}
Hydrogen	9×10^{-6}
Oxygen	20×10^{-6}

at the boundary. The *velocity gradient* $\Delta v/\Delta l$ and the shear stress have their maximum values at the surface and decrease into the fluid. At the extreme of high velocity and low viscosity the only significant viscous shear is found in a thin zone called the *boundary layer*. Outside this region the velocity changes only slowly with distance, and so the viscous stress is small. For an airplane wing the boundary layer may be less than a millimeter thick.

In Figure 13.10*b* a fast-moving ball is rotating rapidly as it passes through the air. As it spins, the ball drags a thin neighboring layer of air with it. The combination of the spinning motion with a large translational speed results in a concentration of streamlines at side *B*. If we consider the center of the ball as the origin of a coordinate system, the air far from the ball is moving from right to left. The spin of the ball carries air near *A* in the opposite direction, so the resultant airspeed there is less than the speed far from the ball. Near *B* the spin carries air from right to left. Since the speed of the air near *B* is greater than near *A*, the pressure is less at *B* than at *A* by Bernoulli's theorem. Consequently, there is a downward force on the ball which causes it to curve. Such a force component perpendicular to the relative velocity is called *dynamic lift*.

No lift results from symmetrical streamlines such as that of Fig. 13.10*a*. To obtain dynamic lift on an object, an *asymmetric* flow pattern is required, corresponding to higher flow speeds on one side than on the other. An aircraft wing is designed to generate rotational flow when it moves through the air with higher airspeeds above the wing than below (Fig. 13.11*a*). As a consequence, the pressure is less above the wing, and this pressure difference gives rise to a net lift. Although the flow in Fig. 13.11*a* appears to be irrotational everywhere, this is not the case in the boundary layer. If the flow were perfectly irrotational, there would be no lift. In Fig. 13.11*b* the angle between the airfoil and the flow has been increased to the point that a large turbulent wake is produced; this reduces the lift and increases the drag.

(a)

(b)

Figure 13.9

Viscosities of liquids vary with temperature and pressure. (*a*) Viscosity of water decreases as temperature rises. (*b*) Viscosity of a typical oil increases as pressure rises.

(a)

Motion of air relative to ball

(b)

Figure 13.10

(*a*) The motion of the air past a moving ball is symmetric about the velocity vector. If there were no air friction, the pressure at *P'* on the downstream side would be equal to the pressure at the corresponding point *P* on the upstream side. With air friction the pressure is lower at *P'*, and there is a net drag on the ball. (*b*) When the ball is spinning, the spin leads to reduced velocity and increased pressure at *A* and to increased velocity and decreased pressure at *B*.

(a)

(b)

Figure 13.11
(*a*) In flow past a tapered airfoil the flow velocity is increased and the pressure decreased above the airfoil, producing net lift. (*b*) A stalled airfoil in which a large turbulent wake has reduced lift and increased the drag.

13.6 THE REYNOLDS NUMBER AND DYNAMIC SIMILARITY

For speeds considerably less than the speed of sound the nature of the flow about a sphere is determined essentially by the Reynolds number $\mathcal{R}$, a combination of four factors: the density d and the viscosity η of the fluid, the diameter D of the sphere, and the relative velocity v.

$$\mathcal{R} = \frac{dvD}{\eta} \tag{13.4}$$

The Reynolds number is a measure of the ratio of the inertial force to the viscous force on a small element of the fluid; it relates the ratio of the rate of transfer of fluid momentum per unit area to the viscous drag per unit area. When $\mathcal{R}$ is small, viscous forces are dominant and inertial effects subordinate. For large $\mathcal{R}$ inertial reactions outweigh viscous forces, and the flow pattern is quite different. The Reynolds number is dimensionless, having the same value in any consistent system of units.

Not only does the Reynolds number determine the nature of the flow about a sphere, but *flow patterns are the same for any two objects of similar shape at the same Reynolds number* (provided certain conditions are satisfied). This important fact, known as the *principle of dynamic similitude*, permits us to build a model airplane, test it in a small wind tunnel, and make good predictions for a full-sized airplane in the same Reynolds number range. The concept of dynamic similarity is of paramount importance in several branches of engineering.

For low Reynolds number ($\mathcal{R} < 5$) flow is laminar. The streamlines are symmetrical about a sphere or cylinder (Fig. 13.10*a*). Consider the motions of three spheres: a baseball falling in water, a dust particle, and a helium-filled balloon in air. Each is acted upon by (1) an external force mg (its weight) and (2) the net force exerted by the surrounding fluid. This latter force is the resultant of the buoyant force $\mathbf{F}_{\text{buoyancy}}$ associated with Archimedes' principle and the additional force $\mathbf{F}_{\text{fluid dynamic}}$ which comes into play when there is relative motion of sphere and fluid and which always opposes the motion. By Newton's second law, for each sphere

$$\mathbf{F}_{\text{external}} + \mathbf{F}_{\text{buoyancy}} + \mathbf{F}_{\text{fluid dynamic}} = m\mathbf{a} \tag{13.5}$$

[For the helium-filled balloon $\mathbf{F}_{\text{buoyancy}}$ is greater than the weight $\mathbf{F}_{\text{external}}$, so the balloon ''falls'' upward but Eq. (13.5) is still applicable.]

In 1845 Stokes showed that the viscous force on a sphere at low $\Re$ is given by

$$F_{\text{viscous}} = 3\pi\eta vD \qquad \text{for a sphere at low } \Re \qquad (13.6)$$

Since the viscous force is proportional to D while the weight of sphere is proportional to D^3, it is clear why small dust particles have a small terminal velocity, achieved when the resultant force is zero in Eq. (13.5). In the low Reynolds number regime, drag is less on a smooth surface than on a rough one.

13.7 DRAG AT HIGHER REYNOLDS NUMBER

As the relative velocity v increases, there comes a point at which the flow lines no longer follow the surface of a sphere or cylinder all the way back to the center, but instead a pair of small vortices appear just behind the cylinder ($\theta = 180°$). For a cylinder this develops for $5 < \Re < 50$ [D in Eq. (13.4) is now the diameter of the cylinder]. The pressure at the downstream side is less than that at the upstream side; the streamlines begin to separate from the cylinder near $\theta = 180°$. As $\Re$ increases, the separation occurs earlier and earlier (Fig. 13.12), until, when $\Re > 5,000$, the separation begins very near $\theta = 90°$. The entire wake is essentially turbulent.

In these flow regimes the drag can be greatly decreased by streamlining; compare the flow of Fig. 13.12 with that in Fig. 13.11a. It is convenient to write the drag force in the form

$$F_{\text{drag}} = C_D S(\tfrac{1}{2}dv^2) \qquad (13.7)$$

where S is the cross-sectional area of the object normal to the flow, d the density of the fluid, v the relative velocity, and C_D is a factor called the *drag coefficient*. Figure 13.13 shows the drag coefficient for a cylinder as a function of $\Re$. In the low Reynolds number regime C_D decreases with v; recall that for low $\Re$ the drag force is roughly proportional to v, so C_D varies as $1/v$.

In the Reynolds number range $5,000 < \Re < 2 \times 10^5$, C_D remains almost constant at 1.2, but a sharp drop occurs in the vicinity of $\Re = 5 \times 10^5$, where the boundary layer on the upstream side becomes turbulent. As a consequence of the increased kinetic energy and momentum near the boundary, the boundary layer maintains attachment to the cylinder to $\theta \approx 140°$ to $150°$. Increasing the roughness of the

Figure 13.12
Flow around a cylinder at a Reynold's number near 5,000. The streamlines separate from the cylinder near $\theta = 90°$, where θ is zero on the upstream side and 180° on the downstream side of the cylinder.

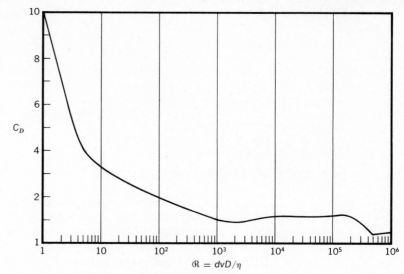

Figure 13.13
Drag coefficient C_D for a circular cylinder with axis normal to the flow.

$$\Re = dvD/\eta$$

cylinder decreases the value of $\Re$ for which the drop in C_D occurs. Over a considerable range of $\Re$ values the drag is less on a blunt body if it is rough than if it is smooth. This is one reason why golf balls are dimpled.

Our discussion pertains to gases as well as liquids as long as the compressibility may be neglected, but that in turn requires that the ratio of v to the local speed of sound be less than 1. This ratio is called the *Mach number*. As v approaches the speed of sound, the drag coefficient increases abruptly. Once the flow is supersonic, C_D decreases again with v. For speeds near or above sound velocity, flows are the same for similar shapes only if both the Mach and Reynolds numbers are the same.

QUESTIONS

1 When an object is dropped from a balloon at high altitude, its velocity may increase for a while and then decrease as it nears the earth. Why does this occur?

2 The surface of a particular river forms a long inclined plane with the speed of the water constant. If a cork is dropped on the surface, will it move down the plane faster than the water? What experiment could you perform to compare the water speed with that of the cork?

3 When there is a steady streamlined flow of liquid through a pipe of varying cross section, does a small volume element of the fluid remain in equilibrium as it moves through the pipe?

4 A tennis ball travels southward, rotating counterclockwise as viewed from above. Which way does it curve? Explain carefully.

5 Discuss the operation of a siphon and a paint sprayer.

6 By applying Bernoulli's theorem to the siphon (Prob. 3) show that the velocity of efflux is given by $v = \sqrt{2gy}$. What limits the height h over which the siphon can operate? If h is less than this limit, will v depend at all on h? If so, why doesn't h appear in the formula for v?

7 Why does the speed of a liquid increase and the pressure decrease when the liquid passes a constriction in the pipe?

8 Over a broad range of pressures the viscosity of gases is independent of the pressure. Why doesn't the viscosity decrease as the number of molecules per unit volume decreases?

PROBLEMS

1 Find the velocity of efflux of water from a hole in the side of a tank if the water surface is 2.5 m above the hole. What mass of water is discharged each second if the area of the hole is 0.25 cm²? (Assume frictionless water and neglect the contraction of the streamlines as they emerge from the hole.) *Ans.* 7 m/sec; 175 g

2 Find the velocity of efflux of water from a hole in the side of a tank when the water in the tank is 9 ft above the hole. If the area of the hole is 0.75 in.², how many cubic feet will be discharged each minute? (Neglect the contraction of the streamlines as they emerge from the hole.)

3 A siphon (see accompanying figure) has a tube with a cross-sectional area of 0.5 cm². Find the velocity of efflux and the volume of water transferred per second if y = 10 cm.

Ans. 1.4 m/sec; 70 cm³

Problems 3 and 4

4 Find the velocity of efflux and the volume of gasoline transferred per minute by the siphon of the accompanying figure if y = 1.44 ft and the area of the tube is 0.03 in.².

5 In a paper machine, paper is formed on a Fourdrinier wire screen, which moves at a speed of 15 ft/sec, by ejecting a mixture of paper fibers and water from a slit orifice onto the screen. What gauge pressure is required at the orifice to match the horizontal speed of efflux to the speed of the wire screen moving just below the slit orifice? Take the density of the mixture (known as *stuff*) to be 2.4 slugs/ft³.

Ans. 270 lb/ft²

6 With what speed will water discharge from the bottom of a tank 25 ft deep through an orifice?

If the orifice has a diameter of 0.5 in., what weight of water will be discharged each second?

7 Water flows at the rate of 8 liters/min from a hole at the bottom of a tank in which the water is 0.9 m deep. Find the effective area of the hole. Find the rate at which the water would escape if an additional pressure of 10⁵ N/m² were applied to the surface of the water.

Ans. 0.32 cm²; 28 liters/min

8 A reservoir is filled with mercury. In its side there is an opening 0.8 mm in diameter. If the opening is 0.4 m below the surface, how many grams of mercury will escape per second?

9 A stream of water projected horizontally from a hose 2.5 m above the ground strikes the ground 6 m away. Find the velocity of efflux and the fluid pressure in the nozzle.

Ans. 8.4 m/sec; 3.53 × 10⁴ N/m²

10 Find the velocity with which water in an enclosed tank will be forced through an orifice in the side of the tank if the orifice is 4 ft below the surface of the water and there is a gauge pressure of 150 lb/ft² on the surface of the water in the tank.

11 Water flowing from an orifice in the side of a tank situated at a height of 6.4 m from the ground strikes the ground 8 m from the foot of the tank. Compute the escape velocity of the water and the gauge pressure in the tank at the level from which the water is squirting.

Ans. 7 m/sec; 2.45 × 10⁴ N/m²

12 The level of the water in the manometer at a (Fig. 13.2) reads 20 cm, and the cross section at a is 15 cm². The cross section of the construction at b is 2 cm². Find the height of the water column at the constriction b, where the velocity is 60 cm/sec.

13 Water in streamline flow is moving with a speed of 5 m/sec through a pipe with a cross-sectional area of 4 cm². If the water gradually descends 10 m and the pipe increases in area to 8 cm², find the pressure and speed of flow at the lower level if the pressure is 1.50 × 10⁵ N/m² at the higher level.

Ans. 2.57 × 10⁵ N/m²; 2.5 m/sec

14 Show that the terminal velocity v_T of a small solid sphere of diameter D and density d falling in a

fluid of density d_f and viscosity η is given by

$$v_T = \frac{D^2 g(d - d_f)}{18\eta}$$

in the low Reynold's number regime where Stokes' law applies.

15 A viscometer consists of two concentric cylinders of large diameter with a thin annular space between them. The cylinders have a mean radius of 5 cm and a length of 12 cm. The thickness of the annular ring is 2 mm. If the outer cylinder is rotated at 50 rev/min while the inner one is held fixed, find the torque on the inner cylinder when the viscometer contains oil of viscosity 0.15 N-sec/m². *Ans.* 0.037 N-m

16 If the viscometer of Prob. 15 is filled with a different kind of oil and a torque of 0.08 N-m is required to rotate the outer cylinder at 40 rev/min when the inner one is fixed, find the viscosity of the oil.

17 Air is streaming past a horizontal airplane wing such that the speed is 400 ft/sec over the upper surface and 300 ft/sec at the lower surface. If the air density is 0.001 slug/ft³, find the difference in pressure between the top and bottom of the wing. If the wing is 30 ft long and has an average width of 6 ft, calculate the gross lift of the wing. *Ans.* 35 lb/ft²; 6,300 lb

18 Water is flowing through the pipe (Fig. 13.2), which has an area of 0.75 ft² at a and an area of 0.5 ft² at b. The pressure at cross section a is 20 lb/in.², and that at b is 15 lb/in.². Find the velocity in each of the cross sections.

19 Water is flowing through a horizontal pipe of varying cross section. The pressure is 8,000 lb/ft² where the cross-sectional area is 4 in.² and 3,000 lb/ft² where the area is 2 in.². What is the rate of flow of water in cubic feet per second? *Ans.* 1.15 ft³/sec

20 A pitot head (see accompanying figure), much used to measure the airspeeds of planes, consists of a *pitot tube* open in the direction of flight and a static tube with holes drilled in the sides to permit its pressure to be the same as that of the surrounding air. When the pitot head moves through the air, the pressure in the pitot line builds up to a value above that of the static line. Show that if Δp is the difference in pressure between the pitot and the static tubes, the speed is given by

$$v = \sqrt{\frac{2\Delta p}{d}}$$

where d is the local density of the air.

21 A plane is flying at a speed of 200 m/sec at an altitude at which the density of the air is 0.8 kg/m³. Find the difference in pressure Δp between the pitot tube and the static tube (see accompanying figure). To how many millimeters of mercury does this correspond? *Ans.* 16,000 N/m²; 120 mm

Pitot tube   Static tube

Problems 20 and 21

22 A constriction in a pipeline similar to that of Fig. 13.2, with the associated pressure gauges, can be used to measure the rate of flow through the line. When so used, it is called a *venturi meter*. Show that the volume Q of fluid flowing through the pipe each second is given by

$$Q = A_a A_b \sqrt{\frac{2(p_a - p_b)}{d(A_a{}^2 - A_b{}^2)}}$$

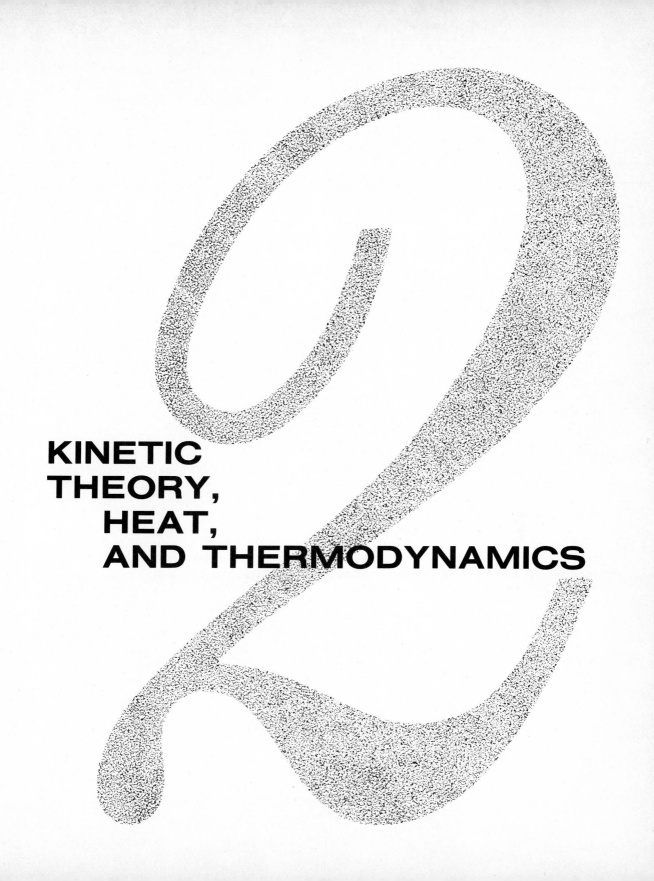

**KINETIC
THEORY,
HEAT,
AND THERMODYNAMICS**

CHAPTER FOURTEEN
KINETIC THEORY AND TEMPERATURE

IN THIS CHAPTER we consider certain facets of the molecular nature of gases, dealing with particles too small to be seen in a typical microscope and of mass too minute to be detected by the most sensitive analytical balance. Development of the hypothesis that a gas consists of a myriad of tiny molecules—widely separated and moving at substantial speeds—leads to an understanding of the gas laws in terms of mechanical interactions and to the deployment of a temperature scale based on the properties of an ideal gas.

14.1 MOLECULAR COMPOSITION OF MATTER

In the fifth century B.C. Greek philosophers, among them Democritus, advanced the theory that all matter was composed of tiny particles called atoms (*atom* means "uncut" or "indivisible"). About 1800, the English chemist Dalton introduced experimental evidence for the existence of atoms and laid the groundwork for modern chemistry. Although we now know that the atom is composed of still smaller particles, we need not consider this fact in discussing the properties of interest here, and we defer discussion of the structure of atoms to later chapters.

One or more atoms may be bound together to form a chemical compound. The smallest unit into which a substance can be divided without chemical decomposition is known as a *molecule*. In gases such as helium and neon the molecule consists of a single atom. In other gases, such as hydrogen and oxygen, two atoms form a molecule, while molecules of carbon dioxide and of water vapor contain three atoms. In the gaseous state the molecules are usually separated by distances large compared with molecular dimensions. At room temperature molecules move with high velocities and have frequent collisions. A gas entirely fills the space of the containing vessel and exerts on it a pressure which results from the change in momentum of the molecules colliding with the walls.

14.2 COMPRESSIBILITY OF GASES: BOYLE'S LAW

When one attempts to decrease the volume of a liquid or a solid, an enormous pressure is required to achieve an appreciable change in volume. The behavior of gases in this respect is quite different. It is relatively easy to compress a gas so that it occupies only one-third of its original volume.

The relationship between the volume of any mass of gas and the pressure was investigated by Robert Boyle in 1660. Boyle's law states that *the product of the pressure and the volume of a given mass of gas is constant if the temperature is not changed.* Thus, if V_0 and p_0 denote the original volume and pressure, while V and p are the final volume and pressure,

$$p_0 V_0 = pV = \text{constant} \qquad (14.1)$$

The mass m of gas, the product of the volume V and the density d, does not change as the pressure is varied, so Eq. (14.1) can be written

$$\frac{p_0 m}{d_0} = \frac{p_1 m}{d_1} \qquad \text{or} \qquad \frac{p_0}{d_0} = \frac{p_1}{d_1} \qquad (14.2)$$

Thus, *when the temperature is constant, the ratio of pressure to density remains constant.* In other words the density is proportional to the pressure. In applying Boyle's law and its corollary Eq. (14.2), it is necessary to use *absolute* pressures; gauge pressures will not do.

Consider a mass of gas initially occupying a volume V at pressure p. Let us keep the temperature constant and change the pressure by a very small amount Δp, giving us a new volume $V + \Delta V$, where ΔV is also very small. Application of Boyle's law yields

$$pV = (p + \Delta p)(V + \Delta V) = pV + p\,\Delta V + V\,\Delta p + \Delta p\,\Delta V$$

Since Δp and ΔV are both small, we may neglect their product in comparison with $p\,\Delta V$ and $V\,\Delta p$. Therefore $p\,\Delta V = -V\,\Delta p$. Comparison with Eq. (10.14) reveals that $p = B$. Thus the *isothermal* (constant temperature) *bulk modulus of a gas is the initial pressure when the change in pressure is small.*

14.3 KINETIC THEORY OF GAS PRESSURE

The molecular nature of matter is well established by chemical and physical evidence. We shall attribute the pressure exerted by a gas to the transfer of momentum to the wall by molecules which are bouncing off. Although we have previously advanced the idea that molecules are in constant motion, we have not discussed experimental evidence for this point of view. Perhaps the simplest and the most direct evidence for the motion of molecules was noted by an English botanist, Robert Brown. With a microscope he observed that very fine particles held in suspension in water were constantly in motion. The smaller the particles, the more freely they moved. The motion is caused by the incessant bombardment of the particles by the molecules of the liquid in which they are suspended. Because of the minute size of the particles, the bombardment on one side is not always equal to the bombardment on the other. Hence, the particles are driven hither and thither.

To explain the physical properties of gases, we make the following tentative assumptions:

1 A gas consists of a very large number of rapidly moving molecules which are so small that the volume of the molecules themselves is negligible compared with the volume of the gas. One justification for this assumption is the fact that the gas can be compressed so that it occupies a small fraction of its normal volume at atmospheric pressure. When 1 cm³ of water at 100°C is converted to steam at 100°C, it occupies a volume of about 1,670 cm³. The distance between molecules in the steam is $\sqrt[3]{1,670}$ (roughly 12) times the distance between molecules in the liquid.

2 When the rapidly moving molecules collide with one another or with the walls of the container, the collisions are elastic. (For polyatomic gases this assumption is too restrictive; it is only necessary to assume that on the average the kinetic energy of translation of the molecules remains constant; see Sec. 15.12.)

3 For an *ideal* gas the attractive forces between molecules are negligible.

4 Under identical conditions of temperature and pressure equal volumes of all gases contain equal numbers of molecules. This hypothesis was advanced in 1811 by Avogadro to explain the results of certain chemical experiments. For any gas at 0°C and one standard atmosphere pressure a mass in grams equal to the molecular weight M occupies a volume of 22.4 liters. This quantity is known as *one mole*. Similarly, the volume occupied under standard conditions by 1 kilomole (M kg) is 22.4 m³. Avogadro's hypothesis implies that M kg of any gas contains the same unique number of molecules, which we call *Avogadro's number*. We designate it by N_A; $N_A = 6.022 \times 10^{26}$ molecules per kilomole.

Let us now calculate the pressure exerted by a gas on the walls of a container as the molecules strike against them. We begin with a single molecule of mass m moving with velocity v_x parallel to the x axis in a cubical box of length L (Fig. 14.1) and compute the average force which this molecule exerts against one of the faces parallel to the YZ plane. As the molecule approaches the wall, the x component of its momentum is mv_x. After the collision the x component of the velocity is reversed, since the collision is elastic. The change of momentum is $2mv_x$. Some time later, the molecule strikes the opposite wall and eventually returns to the first wall. Between successive collisions at this wall the molecule travels a distance $2L$. Therefore, the time between collisions at the same place is $2L/v_x$. The number of collisions per second is $v_x/2L$. Since at each collision the change of momentum is $2mv_x$, the change in momentum per second at one wall for one molecule is

$$2mv_x \frac{v_x}{2L} = \frac{mv_x^{\,2}}{L}$$

Let there be n kmoles of identical molecules in the box. The number of molecules in the box is then $6.022 \times 10^{26}n$ or $N_A n$, since each kilomole contains Avogadro's number N_A molecules. To find the total change in momentum per second at the wall in question, we assume tentatively that all the molecules act independently and add the contributions of all the molecules to obtain

Change in momentum per second $= \dfrac{m}{L}\,(v_{x1}^2 + v_{x2}^2 + \cdots + v_{xNn}^2)$

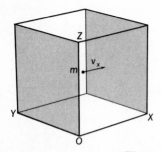

Figure 14.1

A single molecule of mass m moving parallel to the x axis with speed v_x bounces back and forth between the shaded walls.

$$= \frac{nN_A m}{L} \frac{(v_{x1}^2 + v_{x2}^2 + \cdots + v_{xNn}^2)}{N_A n}$$

$$= \frac{nN_A m}{L} \overline{v_x^2}$$

where $\overline{v_x^2}$ represents the average value of the square of v_x for all the molecules in the box.

By Newton's second law of motion the change in momentum per second is equal to the average force. Therefore,

$$F = \frac{nN_A m \overline{v_x^2}}{L} = pL^2$$

since the pressure p is the ratio of the force to the area. Hence,

$$pL^3 = nN_A m \overline{v_x^2}$$

Now L^3 is just the volume V of the box, so

$$pV = nN_A m \overline{v_x^2}$$

For a wall of the box parallel to the XZ plane, $pV = nN_A m \overline{v_y^2}$, and for a wall parallel to the XY plane, $pV = nN_A m \overline{v_z^2}$. We know from experience that the pressure due to the molecular bombardment is the same on all walls (we assume pressure due to the weight of the gas is negligible); thus we expect that $\overline{v_x^2} = \overline{v_y^2} = \overline{v_z^2}$. By the extension of the pythagorean theorem to three dimensions, $\overline{v_x^2} + \overline{v_y^2} + \overline{v_z^2} = \overline{v^2}$, so that $\overline{v_x^2} = \overline{v^2}/3$, and we may write

$$pV = \frac{nN_A m \overline{v^2}}{3} = \frac{2nN_A}{3} \left(\frac{1}{2} m \overline{v^2} \right) \tag{14.3}$$

The product of the pressure and the volume is given by two-thirds of the number of molecules multiplied by the average kinetic energy of translation of the molecules.

Before we relate the average kinetic energy of the molecules to the temperature, it is desirable that we discuss the latter in some detail.

14.4 TEMPERATURE

Our first ideas about temperature come from our senses. By touching a body we may determine whether it is hot or cold. For some purposes our senses give us an adequate description of temperature, but often sensory impressions are unreliable. For example, a room may feel hot to a person who has been outdoors in snow, while it may feel cold to a person entering it from a steam bath. Indeed, sensory impressions of temperature depend greatly on the environment during the recent past.

There is a second situation in which the senses give an unreliable comparison of temperature. If one removes a cardboard container and a metal ice tray from the freezing compartment of a refrigerator, both objects are at the same temperature. Nevertheless, the tray feels much colder to the hand than the cardboard container does, because the metal is a better conductor of heat. In view of the uncertainties associated with out sensations of temperature it is not surprising that scientists have developed objective and reproducible methods for measuring the relative "hotness" of bodies under various conditions.

Temperature is a quantity which cannot be defined in terms of mass, length, and time; it is a fourth fundamental quantity. To measure the temperature of a substance we immerse a *thermometer* in the substance, or at least put the thermometer into intimate contact in such a way that energy can be exchanged freely between the substance and the thermometer. A thermometer is a device with some observable property which changes with temperature; for example, in a mercury thermometer the variable is the length of a mercury column, while in a resistance thermometer the variable is the electric resistance of a conductor. As the thermometer exchanges energy with the substance, the observable property changes until eventually the "reading" becomes constant. Then *thermal equilibrium* has been achieved. By definition, *two systems in thermal equilibrium are at the same temperature.* When the thermometer is in thermal equilibrium with the substance, the substance is at the temperature of the thermometer, and vice versa.

Suppose that a thermometer reads the same when immersed in a beaker of water and when inserted in a hole in a copper sphere. If the sphere is then dropped into the water, it will be found that the temperature of both remains the same. Thus the water and the copper are in thermal equilibrium with each other. *Two systems both in thermal equilibrium with a third system are in thermal equilibrium with each other.* This experimental fact has been called *the zeroth law of thermodynamics.* If the copper had had a higher temperature than the water, experiment would have shown the temperature of the copper to fall and that of the water to rise when the copper was dropped into the water.

14.5 THE FAHRENHEIT AND CELSIUS TEMPERATURE SCALES

The first recorded effort to make an instrument for measuring temperature was that of Galileo about 1593. He took a glass bulb with a long stem and submerged the end of the stem in water (Fig. 14.2). By heating the bulb, he drove some of the air out; as the bulb cooled, water rose in the stem. A change in the temperature of the bulb gave rise to a change in the water level in the stem. Such thermometers were used for many years by physicians and others. The galilean thermometer has several serious handicaps, the most serious of which is that changes in atmospheric pressure also affect the height of the water in the stem.

In the seventeenth century thermometers using water or alcohol sealed in glass tubes were developed. Alcohol is still widely used in inexpensive thermometers. Early in the eighteenth century Fahrenheit introduced thermometers which used mercury as the thermometric substance. These thermometers rapidly won wide acceptance among scientific workers because they were consistent with each other over the whole length of scale and they were convenient, reliable, and reasonably cheap. Fahrenheit elected to call zero on his thermometer "the most intense cold obtained artificially in a mixture of water, of ice, and of sal ammoniac." The temperature of the human armpit he called 96°.

Figure 14.2

Galilean thermometer using air inside the bulb as the working substance.

The choice of these two "fixed points" established the Fahrenheit temperature scale.

The *Celsius*, or *centigrade*, temperature scale, proposed by the Swedish astronomer Celsius about 1742, took as its zero the temperature of a mixture of ice and water under standard pressure. The temperature at which water boils under standard atmospheric pressure was taken to be 100°C.

It was soon found that the fixed points of the Celsius scale could be reproduced easily and with far greater accuracy than the original fixed points of the Fahrenheit scale. The melting point of ice was approximately 32° on the Fahrenheit scale, and the boiling point of water 212°F. Eventually it became standard practice to use the ice point and the boiling point as the fixed points for the Fahrenheit scale.

It is frequently desirable to convert a Celsius temperature to Fahrenheit, and vice versa. This can be done readily if one recalls the fixed points of the two temperature scales. Let F be the temperature on the Fahrenheit scale, and C the temperature on the Celsius scale (Fig. 14.3). The number of Fahrenheit degrees above the freezing point is related to the total temperature difference between the boiling point and the freezing point on the Fahrenheit scale (180°) as the number of Celsius degrees above the freezing point is to 100°C. This proportion may be written

$$\frac{F - 32}{180} = \frac{C}{100} \qquad (14.4)$$

Since 100 Celsius degrees corresponds to 180 Fahrenheit degrees, clearly

$$1 \text{ C degree} = \tfrac{9}{5} \text{ F degree} \qquad (14.5)$$

(Note that 1°C is 33.8°F.) Clearly 1 Celsius degree (C°) is not a temperature, but a temperature interval. Equation (14.4) relates a temperature on the Celsius scale to the corresponding temperature on the Fahrenheit scale, while Eq. (14.5) relates temperature differences. We shall write temperatures as °C or °F, while temperature intervals will be expressed in C° or F°.

14.6 GASES AND THE ABSOLUTE TEMPERATURE SCALE

Gases expand when their temperature is raised and contract when the temperature is lowered. If we keep the pressure on the gas constant, we find that a plot of the volume as a function of temperature gives a straight line (Fig. 14.4) described by

$$V = V_0(1 + \beta_0 t) \qquad (14.6)$$

where V_0 is the volume of the gas at 0°C, V is the volume at some temperature $t\,°C$, and β_0 is a constant called the *coefficient of volume expansion at 0°C*. For any substance and any initial temperature *the coefficient of volume expansion is the change in volume divided by the product of the original volume V_i and the change in temperature*. Thus

$$\beta = \frac{\Delta V}{V_i \, \Delta t} \qquad (14.6a)$$

and $\beta_0 = (V - V_0)/V_0 t = \Delta V/V_0 t$.

Figure 14.3
Comparison of the Celsius (centigrade) and Fahrenheit temperature scales.

Figure 14.4
The volume of a gas as a function of temperature at constant pressure.

The relation expressed by Eq. (14.6) is known as *Charles' law.* By careful experiments, Charles found that the coefficient of volume expansion of all noncondensing gases at constant pressure and 0°C is equal to 0.00367, or $\frac{1}{273}$. *The coefficient of volume expansion is the same at a given temperature for all gases.* Thus, if we extrapolate (Fig. 14.4) a plot of the volume as a function of temperature of any gas to the x axis, the intercept is −273°C, regardless of the original volume of gas taken, the gas chosen, or the pressure. The fact that this intercept of −273°C is the same for different gases, different original volumes, and different pressures suggests this temperature has a very special significance. It was the first indication of *absolute zero* for temperature. Of course, it should not be concluded that if the temperature of the gas were actually lowered to −273°C, it would follow the extrapolated line of Fig. 14.4 and occupy zero volume. Before any such low temperature is reached, the gas liquefies.

If we define a tentative temperature scale by $T = t(°C) + 273$, it turns out to be almost the same as the *absolute* temperature scale devised about 1848 by Lord Kelvin. The Kelvin scale, based on thermodynamics (Chap. 18), is independent of the properties of any particular substance. For more than a century temperatures on the absolute scale were measured in "degrees Kelvin," but the International Union of Pure and Applied Physics now recommends that the word "degrees" be omitted. Thus a temperature of 20°C corresponds to an absolute temperature of 293 K, read "293 kelvins." When Eq. (14.6) is rewritten in terms of absolute temperature, it takes the simpler form

$$V = V_0\left(1 + \frac{t}{273}\right) = V_0\frac{273 + t}{273} = V_0\frac{T}{273} \qquad (14.6b)$$

The volume of a gas at constant pressure is proportional to the absolute temperature T. In general $V_1/V_2 = T_1/T_2$, where V_1 and V_2 are the volumes occupied by the gas at temperatures T_1 and T_2, respectively, when the pressure is kept constant.

If the temperature of a gas is raised and its volume is kept constant, the pressure varies, increasing $\frac{1}{273}$ of its value at 0°C for every rise of 1 Celsius degree. If p_0 represents the pressure at 0°C and p_t the pressure at t°C,

$$p_t = p_0\left(1 + \frac{t}{273}\right) = p_0\frac{T}{273} \qquad (14.7)$$

Figure 14.5 shows the relation between the pressure and temperature of a gas at constant volume. Note that once again the intercept on the temperature axis is −273°C (or more exactly −273.15°C).

14.7 THE GENERAL GAS LAW

From Boyle's law we know that the product of the pressure and the volume of a gas remains constant if the temperature is held fixed. From Charles' law we know that the volume of a gas is proportional to the absolute temperature if the pressure is constant. If the volume is kept constant, the pressure is proportional to the absolute temperature. If we combine these relations, we may write, for any given mass of gas,

Figure 14.5
Pressure as a function of temperature for a gas at constant volume.

$$pV = bT \qquad (14.8)$$

where b is a constant for the particular mass of gas in question. If subscript 1 represents one particular set of conditions for the gas, and subscript 2 another set of conditions, we may write

$$\frac{p_1 V_1}{T_1} = \frac{p_2 V_2}{T_2} \qquad (14.9)$$

In both chemistry and physics it is often convenient to measure quantities of substances in moles. As we saw in Sec. 14.3, *one mole of any gas is that amount of the gas which has a mass in grams equal to the molecular weight*. For example, the molecular weight of gaseous oxygen is 32.0, and hence 480 g of oxygen represent 15.0 moles. A kilomole is 1,000 moles or the amount of the gas which has a mass in kilograms equal to the molecular weight. Thus, since the molecular weight of hydrogen is 2.016, 10 kmoles of hydrogen has a mass of 20.16 kg.

At 0°C and one standard atmosphere (76 cm Hg) pressure, 1 kmole of any gas occupies a volume of 22.414 m³; thus the constant in Eq. (14.8) is the same for 1 kmole of any gas. For n kmoles of any gas,

$$pV = nRT \qquad (14.10)$$

where R is the *universal gas constant*. It has the value

$$R = \frac{1.0133 \times 10^5 \text{ N/m}^2}{273.15 \text{ K}} \times \frac{22.414 \text{ m}^3}{1 \text{ kmole}} = 8{,}314 \text{ J/(kmole)(K)}$$

The relation $pV = nRT$ is the *equation of state* for an ideal gas. It is an excellent approximation for real gases so long as the temperature is far above the boiling point and the pressure is not extremely high.

When we compare Eqs. (14.10) and (14.3), we find that $pV = nRT = \frac{2}{3}nN_A(\frac{1}{2}m\overline{v^2})$, from which

$$\tfrac{1}{2}m\overline{v^2} = \frac{3}{2}\frac{R}{N_A} = \tfrac{3}{2}kT \qquad (14.11)$$

where N_A is Avogadro's number. The ratio $R/N_A = k$ is known as *Boltzmann's constant*. The expression on the left of Eq. (14.11) is the average kinetic energy of translation for the gas molecules, and it is directly proportional to the absolute temperature. Thus, *the absolute temperature is a measure of the average translational kinetic energy of the molecules in a gas*. When we double the absolute temperature, we double this kinetic energy. To double the average velocity of the molecules in a gas sample, we must increase the absolute temperature by a factor of 4.

The average kinetic energy of a molecule of any gas at temperature T is $\frac{3}{2}kT$, where $k = R/N_A = 1.38 \times 10^{-23}$ J/K. At 27°C (300 K), the average kinetic energy of a gas molecule is $\frac{3}{2} \times (1.38 \times 10^{-23}$ J/K$) \times 300$ K $= 6.21 \times 10^{-21}$ J.

Let us now calculate the effective velocity of nitrogen molecules at 27°C. By the "effective velocity" we mean the velocity of a molecule which has the average energy, so $v_{\text{eff}}^2 = \overline{v^2}$. One kilomole of nitrogen has a mass of 28.0 kg, and one molecule a mass of $(28.0/N_A)$ kg $= 28.0$ kg/$6.022 \times 10^{26} = 4.65 \times 10^{-26}$ kg. Hence, by Eq. (14.11),

$$\tfrac{1}{2}m\overline{v^2} = \tfrac{1}{2}mv_{\text{eff}}^2 = \tfrac{3}{2}kT$$

$$\tfrac{1}{2}(4.65 \times 10^{-26} \text{ kg}) \, v_{\text{eff}}^2 = \tfrac{3}{2}(1.38 \times 10^{-23} \text{ J/K})(300 \text{ K})$$

$$v_{\text{eff}}^2 = 26.7 \times 10^4 \text{ m}^2/\text{sec}^2$$

$$v_{\text{eff}} = 517 \text{ m/sec}$$

14.8 DALTON'S LAW OF PARTIAL PRESSURES

We have derived the equations for the pressure exerted by a gas for the case in which the gas contains only molecules of one kind. If the gas contains several different kinds of molecules, each kind rebounds from the walls and contributes to the pressure. The total pressure is the sum of the pressures which each of the various kinds would exert if it occupied the volume alone. We may write

$$p_t = p_1 + p_2 + p_3 + \cdots \tag{14.12}$$

where p_t is the total pressure exerted by all the molecules and $p_1, p_2, p_3, \ldots$ are the pressures which molecules of type 1, 2, 3, ... would exert if each filled the volume alone. This is known as *Dalton's law of partial pressures.*

Air is a mixture of gases, with nitrogen, oxygen, water vapor, carbon dioxide, argon, hydrogen, and many other kinds of molecules present. The atmospheric pressure is the pressure exerted by the nitrogen plus that due to oxygen, plus that due to each of the other atmospheric constituents. Equation (14.11) shows that the average translational kinetic energy is the same for each kind of molecule. If we apply this fact to compare the effective velocities of hydrogen and oxygen molecules at a given temperature, we have

$$\frac{3}{2}\frac{R}{N_A}T = \tfrac{1}{2}m_h v_h{}^2 = \tfrac{1}{2}m_0 v_0{}^2$$

where m_h and v_h are the mass and effective velocity of hydrogen molecules and m_0 and v_0 the corresponding quantities for oxygen molecules. Since $m_0 = 16m_h$, $v_h = 4v_0$. Thus, the effective speed of hydrogen molecules in the atmosphere is four times that of oxygen molecules. The effective speed is also called the *root-mean-square* (rms) *speed,* because it is the square *root* of the *mean* (average) *square* of the speed.

14.9 DISTRIBUTION OF MOLECULAR VELOCITIES IN GASES

In writing the expression for the pressure exerted by the ideal gas, we have assumed that a given molecule travels back and forth striking the walls, without taking into account the effects of collisions with other molecules. Although molecules collide frequently with other molecules (Fig. 14.6), and although the velocity of any individual molecule undergoes sharp changes, the average kinetic energy of the molecules is the same at a given temperature. Some molecules are moving rapidly, and some slowly. Figure 14.7 shows the distribution of speeds for gas molecules.

Figure 14.6
The path of a molecule colliding with other molecules and with the walls of the container.

Figure 14.7
Maxwellian distribution of speeds for gas molecules. Plotted on the ordinate is the relative number of molecules in each 1 m/sec speed range and on the abscissa the ratio of the speed v to the effective (or rms) speed v_{eff}.

The distance a given molecule travels between collisions differs considerably from collision to collision, but its average value is again well defined. The average length of the path over a large number of collisions is called the *mean free path*. For molecules of the air it is of the order of ten-millionths of a centimeter. Table 14.1 lists some of the important molecular quantities for oxygen and hydrogen molecules.

14.10 DEVIATIONS FROM THE GENERAL GAS LAW

In our discussions of gases we have made two important assumptions which are not always justified. We have assumed (1) that the volume of the molecules themselves is negligible in comparison with the total space occupied by the gas and (2) that the attractive forces between gas molecules are negligible. For experiments with gases such as oxygen and hydrogen at room temperature and ordinary pressures, these assumptions are justified. However, if they were true under all circumstances, we would never have molecules sticking together to form liquids or solids. Indeed, substances composed of molecules in which the attractive forces are relatively great are already liquid or solid at room temperature.

Table 14.1
Molecular Quantities at 0°C and 76 cm Hg

Quantity	*Hydrogen*	*Oxygen*
Number of molecules per cubic meter	2.69×10^{25}	2.69×10^{25}
Diameter of each molecule	2.4×10^{-10} m	3.2×10^{-10} m
Mass of each molecule	3.34×10^{-27} kg	5.3×10^{-26} kg
Mean free path	1.8×10^{-7} m	1.0×10^{-7} m
Number of collisions per second	1.00×10^{10}	4.6×10^{9}
Effective speed	1,840 m/sec	461 m/sec
Average speed	1,700 m/sec	425 m/sec
Most probable speed	1,500 m/sec	376 m/sec
Mass per cubic meter	0.0899 kg	1.43 kg
Volume per kilogram	11.1 m³	0.699 m³
Number of molecules in 1 kmole	6.022×10^{26}	6.022×10^{26}

Let us keep constant the temperature of a given mass of a *real* gas and study the behavior of the pressure and volume. If Boyle's law held rigorously, the product pV would be constant for all pressures and volumes so long as the temperature was kept constant. But accurate observations show that the product is not constant when the pressure is varied over a large range. At exceedingly high pressures, the molecules are close together, and the space occupied by the molecules themselves is no longer negligible. Therefore an increase in the pressure of the gas results in too small a decrease in volume, so that pV increases with p instead of remaining constant according to Boyle's law. At exceedingly low pressure the product pV decreases somewhat as p is increased.

Van der Waals showed that the intermolecular attraction and the space occupied by the molecules themselves can be simply taken into account by a modification of the general gas law. According to van der Waals, for 1 kmole of gas,

$$\left(p + \frac{a}{V^2}\right)(V - b) = RT \tag{14.13}$$

where a and b are constants characteristic of a given gas but independent of temperature, pressure, and volume. The constant b is a correction to take account of the fact that the molecules themselves occupy a finite amount of space. A detailed analysis shows that b is about four times the actual volume of the molecules. The term a/V^2 takes account of the attractive forces between the molecules, which have the effect of reducing the volume just as the pressure does.

14.11 DIFFUSION

Evidence supporting the idea that molecules are in constant motion is found in the manner in which the odor penetrates to all parts of a room when a bottle of ammonia is opened. The spread of the odor can be explained on the assumption that the ammonia molecules escape from the bottle and wander about among the air molecules. They are buffeted back and forth, gradually spreading throughout the room. The process by which molecules of one kind penetrate and intermix with molecules of another kind is called *diffusion*.

If a porous jar (Fig. 14.8) is surrounded by another vessel into which hydrogen is introduced, hydrogen diffuses through the porous jar and increases the gas pressure inside it, causing gas to bubble up from the water into which the end of the tube leading from the porous jar is dipped. Molecules diffuse out of the porous walls as well as in, but the heavier air molecules diffuse outward far more slowly than the hydrogen molecules diffuse inward. The reason is the greater speeds of the hydrogen molecules; as we saw in Sec. 14.7, the effective speed of a gas molecule is inversely proportional to the square root of its mass. It is diffusion which keeps the air in such a state of uniform mixture and which accounts for the rapid disappearance of the fumes when a bottle of ether is broken out of doors.

In liquids as well as in gases, the molecules are free to move about. If a little sulfuric acid is placed in a bottle of water, it diffuses

Figure 14.8

The pressure within the porous jar increases because hydrogen diffuses in through its walls faster than the heavier air molecules diffuse out.

to form a uniform concentration. When a lump of sugar is placed in a cup of coffee, the contents of the whole cup are sweetened by the distribution of the sugar molecules throughout the coffee; stirring promotes the rapid mixing of the molecules.

If stirring and convection currents are avoided, diffusion in liquids occurs very slowly. This may be demonstrated by use of a tall glass jar filled with water into which, by means of a thistle tube extending to the bottom of the jar, a solution of copper sulfate is carefully poured until the bottom of the jar is filled to a height of a centimeter or two. If the jar is allowed to stand for some days without being disturbed, the upward diffusion of the copper sulfate can be observed. Gravity tends to keep the copper sulfate on the bottom, but diffusion occurs, and the copper sulfate gradually rises in the jar until finally a uniform distribution is achieved.

When two metals such as copper and nickel are placed in contact, atoms of each gradually diffuse into the other. If the specimens are cut into sections after they have been in contact for some time, the concentrations of the two metals at different distances from the contact plane can be determined. The rate of diffusion depends on the temperature, but in any case it is very slow for solids.

Figure 14.9
Sugar molecules prevent water molecules from leaving through many of the holes in a semipermeable membrane, but they do not prevent water molecules from entering the solution.

14.12   OSMOSIS

When the pores of a membrane are large enough to pass small molecules but not big enough to pass larger ones, the membrane is said to be *semipermeable*. The process of preferential diffusion through a semipermeable membrane is called *osmosis*. For example, if a carrot is hollowed out, and a thick sugar syrup placed in the cavity, there is a net flow of water into the carrot when it is placed in water. If a glass tube is sealed to the cavity, the water may rise to a height of several feet. The carrot serves as a semipermeable membrane (Fig. 14.9) having a great number of small holes through which water molecules may pass either in or out. Sugar molecules, on the other hand, are too large to pass through. The fact that the water level rises in the tube means that more water flows in than flows out. The large sugar molecules act somewhat as valves for many of the openings. A water molecule moving inward may slightly displace the large sugar molecule at *A* in Fig. 14.9 and get into the idealized carrot, but a water molecule coming from the other side can only push the sugar molecule more tightly against the hole. There are many holes through which water molecules pass in either direction, but there are more holes through which water molecules enter than there are by which they leave. As water rises in the tube, the pressure inside increases, eventually becoming great enough so that there is no longer a net inflow. The pressure that just prevents further flow of the solvent is called the *osmotic pressure* of the solution. In dilute solutions in which the molecules are not dissociated, the osmotic pressure is proportional to the concentration of the dissolved substance. In a solution of given concentration, the osmotic pressure increases as the temperature is increased, the rate of increase being the same for all solutions because the increase in average kinetic energy is the same for all kinds of molecules.

When dried fruits such as prunes and raisins are cooked, they swell and burst if the pressure inside becomes sufficiently large. The swelling is due to the fact that the vegetable sacs surrounding the fruit are semipermeable membranes through which there is a net diffusion of water from outside to inside. If marine animals such as oysters are transferred from salt water to fresh water, more water flows into than out of the animal through the membrane that serves as its covering. Dilation and death of the animal result.

QUESTIONS

1 If the speed of every molecule of gas were doubled, what would happen to the temperature? Explain.

2 On the basis of what happens to individual molecules, explain why the temperature of a gas rises when the gas is compressed.

3 What reasons can you advance in support of the molecular hypothesis?

4 What is the temperature (98.6°F) of the normal human body on the Celsius scale?

5 What are the advantages and disadvantages of mercury as a thermometric material?

6 What justification do we have for believing that 50°C should be halfway between 0°C and 100°C on a good mercury thermometer?

7 What types of thermometers do you know about? What kind could be used to measure 4 K? 1400 K?

8 Do more than half the molecules in a gas have a speed less than v_{eff}? Than v_{av}? Than the most probable speed? Which of these speeds corresponds to a molecule with average kinetic energy?

9 Why may there be greater danger associated with the failure of a storage tank for gas at 3,000 N/cm² than with the failure of a water tank in which the pressure is 3,000 N/cm²?

10 Why does the product pV increase slightly as the pressure is raised when pressure and density are very high, even though the temperature remains constant?

11 Why, at very low pressure and density, does pV decrease slightly as p is increased at constant temperature?

PROBLEMS

1 Calculate the effective (rms) speed of molecules of carbon dioxide at standard pressure and 27°C.   *Ans.* 412 m/sec

2 Compute the effective speed of an oxygen molecule under standard pressure at 47°C, making use of the fact that one gram molecular weight (32 g) occupies 22.4 liters under standard conditions (1 atm and 0°C).

3 A Fahrenheit thermometer shaded from the sun on a hot day reads 95°. What is the temperature on the Celsius scale? On the Kelvin scale?
                Ans. 35°C; 308 K

4 The boiling point of pure methanol (methyl alcohol) is 64.65°C. Find the corresponding temperature on the Fahrenheit scale and on the Kelvin scale.

5 The temperature at which mercury boils is 357°C, and that at which it freezes is −40°C. Find the corresponding temperatures on the Fahrenheit scale.   *Ans.* 657°F; −40°F

6 What is the temperature on the Celsius scale when a Fahrenheit thermometer indicates −60°? 5°? 75°?

7 Solid carbon dioxide (Dry Ice) turns to the gaseous phase at −112°F. Find the corresponding temperature on the Celsius scale and the Kelvin scale.   *Ans.* −80°C; 193 K

8 On the Reaumur temperature scale, widely used in France, the ice point is 0°Re and the steam point 80°Re. Extend the relationship

$$\frac{C}{100} = \frac{F - 32}{180}$$

to include a term for the Reaumur scale. Find the Celsius and Fahrenheit temperatures corresponding to 16°Re.

9 Aluminum melts at 659.7°C and titanium at 1800°C. Find the corresponding Fahrenheit temperatures.   *Ans.* 1220°F; 3272°F

10 A thermostat is set to maintain the temperature at 95°C. What is the corresponding temperature on the Fahrenheit scale? On the Kelvin scale?

11 The Scottish engineer Rankine devised an absolute temperature scale based on Fahrenheit degrees. To what Fahrenheit temperature does zero on the Rankine scale correspond? What is room temperature of 70°F on the Rankine scale?   *Ans.* −460°F; 530°R

12 In the early morning, when the temperature is 7°C, the gauge pressure in an automobile tire is 24 lb/in.². In the afternoon, after hard driving, the temperature of the tire is 57°C. If atmospheric pressure has remained constant at 14.5 lb/in.², find the new gauge pressure.

13 A flask with negligible expansion coefficient contains air at a temperature of 27°C and a pressure of 75 cm Hg. Find the pressure in the flask after it is sealed and cooled to −23°C.
                    Ans. 62.5 cm Hg

14 Illuminating gas is stored in a tank designed so that the volume can change but the pressure remains constant. If the tank contains 120,000 ft³ under standard pressure, how much does the volume change when the temperature rises from 7 to 27°C?

15 A sample of gas occupies 500 cm³ at 0°C and 76 cm Hg pressure. What would be its volume if it were heated to 80°C and the pressure increased to 90 cm Hg?   *Ans.* 546 cm³

16 A stratosphere balloon has a partially filled gas bag containing 150,000 ft³ of helium when the barometer reads 75 cm Hg and the temperature is 27°C. Find its volume when the balloon has risen to such a height that the atmospheric pressure is 15 cm Hg and the temperature is −23°C.

17 How many molecules of air remain in each cubic centimeter of a radio tube at 27°C in which the pressure has been reduced to 10^{-9} mm Hg?
                    Ans. 3.2×10^7

18 A McLeod gauge indicates pressure as low as 10^{-5} mm Hg (or 10^{-5} Torr). How many molecules per cubic centimeter would be required to produce this pressure at 27°C?

19 If the effective speed of oxygen molecules at 27°C and 76 cm Hg pressure is 484 m/sec, find (*a*) the effective speed of hydrogen molecules under the same conditions, (*b*) the effective speed of oxygen molecules at 27°C and 50 cm Hg pressure, and (*c*) the temperature at which the effective speed of oxygen molecules is 968 m/sec.
     Ans. (*a*) 1,936 m/sec; (*b*) 484 m/sec; (*c*) 927°C

20 The bag of a partially inflated balloon contains 4,000 ft³ of hydrogen at 76 cm Hg pressure and 27°C. The balloon rises to a height of 5 mi, where the temperature is −23°C and the pressure is 27 cm Hg. Find the volume of the hydrogen under these circumstances.

21 Pure oxygen at 0°C is enclosed in a cylinder having a volume of 7.5 liters. If the pressure in the cylinder is 2,500 cm Hg, what is the mass of the enclosed gas? How much pressure would it exert if the temperature were increased to 750°C?   *Ans.* 0.352 kg; 9,370 cm Hg

22 An air bubble has a volume of 1.5 cm³ at the surface of a lake, where the temperature is 32°C. What was its volume at a depth of 80 m, where the temperature was 7°C? Atmospheric pressure was 10^5 N/m².

23 An air bubble released at the bottom of a pond expands to twice its original volume by the time it reaches the surface. How deep is the pond if atmospheric pressure is 10^5 N/m², the surface is at 27°C, and the bottom of the pond is at 17°C?   *Ans.* 9.52 m

24 A vessel having a capacity of 1.5 liters contains nitrogen under a pressure of 4 atm. It is connected to a vessel of 2 liters capacity containing oxygen at a pressure of 2 atm. If the temperature remains constant, what will be the pressure when the gases have mixed?

25 Find the number of mercury molecules, each with a mass of 3.3×10^{-25} kg and an effective speed of 200 m/sec, that would maintain normal atmospheric pressure against the walls of a containing vessel with a volume of 100 cm³. What is the temperature of the gas?
                 Ans. 2.3×10^{21}; 319 K

26 A helium tank has a volume of 50 liters, is at 27°C, and is filled to a pressure of 1.5×10^7 N/m² (148 atm). A leak develops in the valve, and after a week the pressure is 9×10^6 N/m² when the temperature is 17°C. Find the mass of helium that was in the tank when it was full and the fraction of the original mass which has leaked out.

27 A thoroughly evacuated vessel has a volume of 5.7 liters. It develops a small leak through which 500 million molecules of air enter each second. How long will it take for the pressure of the air in the vessel to reach 1 N/m² if the temperature remains constant at 17°C? Assume the number

of molecules in the vessel at the start is negligible. *Ans.* about 90 years

28 A cylindrical diving bell, open at the bottom, has a volume of 250 ft³ and a cross-sectional area of 20 ft². How high will the water rise in the bell when it is immersed in fresh water to a depth of 50 ft if atmospheric pressure is 14.7 lb/in.² and no air leaves the bell?

29 Find the temperature at which the effective (rms) speed of oxygen molecules is equal to the effective (rms) speed of nitrogen molecules at 0°C. *Ans.* 39°C

30 If a single hydrogen molecule occupies a cubical box 0.05 m on a side and the walls are perfectly smooth, calculate the average force exerted on each wall of the box if the molecule has velocity components $v_x = 1{,}000$ m/sec, $v_y = 800$ m/sec, and $v_z = 1{,}200$ m/sec.

INCREASING THE TEMPERATURE of a gas augments the translational kinetic energy of gas molecules; similarly it increases the random motions of the atoms constituting a solid. For most materials an increase in the internal energy results in an increase in volume and for solids an increase in length; such increases are of great practical importance. We introduce quantitative expressions by which we can calculate the expansion produced by a given change in temperature and by which we can find how much energy must be added to a body to produce such a temperature change.

CHAPTER FIFTEEN
THERMAL EXPANSION AND SPECIFIC HEAT

15.1 LIQUID AND SOLID PHASES

The kinetic theory of matter leads to an understanding of many properties of liquids and solids, as well as of gases. The forces between molecules in an ideal gas are zero unless the molecules are "touching"; in a real gas there are small attractive forces between molecules. As we reduce the temperature of a gas, we reduce the average kinetic energy of translation of the molecules. As the molecules move more and more slowly, attractive forces between molecules, which played a negligible role at high temperatures, become important. Eventually, groups of molecules stick together. The gas begins to liquefy. In the liquid, molecules cling together but are free to move with respect to each other. Liquids do not resist forces tending to change their shape, but they strongly resist forces tending to change their volume. Because the molecules of the liquid can easily be displaced with respect to each other, layers flow relatively free over each other and the liquid assumes the shape of the vessel in which it is placed. In the liquid two or more molecules may stick together and move about as a unit.

When the temperature of the liquid is reduced, the energy of the molecules decreases. Eventually, the translational kinetic energy becomes so small that the molecule is trapped by the attractive forces exerted by its neighbors. Once the molecule is able only to vibrate about some equilibrium position, the material loses its fluid properties and retains its shape. We then speak of it as a *solid*. If each atom of the material is located in a particular place in a regular array, we have a *crystalline solid* (Fig. 15.1*a*) of which diamond, quartz, and calcite are examples. In some cases each atom is confined to a small region about an equilibrium point by the attractive forces of neighboring particles, but no perfectly repetitive structure is formed. This is the case when glass cools and when butter solidifies. The structure is not crystalline but *amorphous* (Fig. 15.1*b*). Solids resist an effort to change their shape

(a)

(b)

Figure 15.1

In a crystalline solid (*a*) the atoms are arranged in a regular pattern which is repeated over and over, while in an amorphous solid (*b*) the atoms are arranged in various ways.

or volume. The interatomic forces between atoms in a solid are reasonably idealized in terms of relatively stiff springs connecting nearest neighbors and weaker springs connecting next nearest neighbors. The forces are such that in a solid it is no longer meaningful to say that a particular atom belongs to a particular molecule. In the solid the molecule is no longer a separate entity; rather, there is an array of atoms, and any one atom is bound to many others.

Under suitable conditions it is possible to change a given material from the solid to the liquid phase and from the liquid to the gaseous phase. Water can be frozen to form ice and evaporated to form steam. The change from one phase to another is discussed in Chap. 16.

15.2 THERMAL EXPANSION OF SOLIDS

Most substances expand when heated (Fig. 15.2). This is not surprising if we remember that heating a solid increases the amplitude of vibration of the atoms and has the effect of moving the atoms somewhat farther apart. If the vibrations were simple harmonic, the average position of each atom would not change, and there would be no expansion. However, the forces which restore an atom to its equilibrium position are greater when the atom approaches too close to a neighbor than when the atom moves an equal distance too far away; the forces are *anharmonic*. As the amplitudes of the vibrations increase, the average distance between atoms becomes slightly larger. The change of length of a typical solid is small—only a few parts in 10,000 for an ordinary daily range of outside temperature—but the forces associated with such expansion can be tremendous and must be allowed for in many structures. The amount of expansion depends on the nature of the substance and on the change in temperature. On railroads small distances are usually left between sections of rail to permit expansion without having the tracks bow. In concrete roads and sidewalks, expansion joints filled with tar are provided. Care is taken in selecting compatible types of glass and metal when one wishes to bring conductors through glass walls into vacuum tubes and light bulbs.

When we heat a solid, the change in length ΔL depends on the material, on the original length L, and on the change in temperature Δt. For most materials we may write the approximate equality

$$\Delta L = \alpha L \, \Delta t \tag{15.1}$$

where α is known as the *coefficient of linear expansion* of the material. From Eq. (15.1) the *coefficient of linear expansion* α *is the change in length per unit length for a one-degree rise in temperature.* Clearly, α is independent of the units in which the length is measured, but it does depend on the temperature unit. A change of 1 Fahrenheit degree corresponds to a change of $\frac{5}{9}$ Celsius degree. Therefore, α per Fahrenheit degree is only $\frac{5}{9}$ of α per Celsius degree. The coefficient of linear expansion varies with the temperature, usually slowly near room temperature but rapidly at very low temperature. The room-temperature coefficients of linear expansion of many solids are listed in Appendix Table B.1. In the construction of a bridge, or indeed in almost any structure, provision must be made for expansion and contraction. One possible arrangement for a small bridge is to mount one end on a roller. As the length of the bridge increases or decreases, this end of the bridge moves forward or back without injuring the piers.

Example

The main cable of a suspension bridge is 1,500 m long at 0°C. If the cable is made of steel, find its length on a hot summer day when the temperature is 35°C.

$$\Delta L = \alpha L \, \Delta t$$
$$= 1.2 \times 10^{-5} \times 1,500 \times 35$$
$$= 0.63 \text{ m}$$

Thus the length is 1,500.63 m at 35°C. ////

15.3 DIFFERENTIAL EXPANSION

Consider a strip of brass and a strip of iron welded together (Fig. 15.3) to form a composite bar. When this bar is heated, it bends because the brass expands more rapidly than the iron; therefore the brass is on the outside of the curve when it is heated. Devices of this kind, called *bimetallic strips*, have found many uses. One of the most familiar is in the thermostat, which can be made to open and close electric circuits as the movable end of the bimetallic strip changes its position (Fig. 15.4). Bimetallic strips are used in a common type of thermometer.

In a watch, the balance wheel (Fig. 15.5) is made of bimetallic strips. As the temperature increases, ends A and B are carried inward, making the moment of inertia of the wheel less; at the same time the expansion of the radius carries the rim of the wheel farther from the center, thereby producing an increase in the moment of inertia. Meanwhile, the elasticity of the hairspring is reduced by the higher temperature. This would make the balance wheel move more slowly if the moment of inertia of the wheel were not reduced to compensate. By proper design the balance wheel can be made to change its moment of inertia just enough so that the period of vibration remains the same over a reasonable temperature range.

15.4 EXPANSION IN AREA

The area of a square of material with sides of length L is L^2. When we raise the temperature by Δt, the length of each side becomes $L + \Delta L$ and the area becomes

$$(L + \Delta L)^2 = L^2 + 2L \, \Delta L + (\Delta L)^2$$

Since ΔL is very small compared with L, we may neglect $(\Delta L)^2$ in comparison with the other terms and write the new area as $L^2 + 2L \, \Delta L$. The change in area ΔA is then given by

$$\Delta A = 2L \, \Delta L = 2\alpha L^2 \, \Delta t = 2\alpha A \, \Delta t \qquad (15.2)$$

From Eq. (15.2) we see that *the coefficient of area expansion is twice the coefficient of linear expansion.*

Figure 15.3
Unequal expansions of metals in a bimetallic strip. The gray material has the lower coefficient of linear expansion.

Figure 15.4
A bimetallic strip AB closes an electric circuit when it bends upward with a temperature change.

Figure 15.5
Balance wheel of a watch.

The area of a hole in a piece of material expands exactly as though it were filled with the material in question, regardless of the shape of the hole or the piece. The pioneers made use of this fact in putting iron tires on the wooden wheels of their wagons. The iron tire was made slightly smaller than the wooden wheel upon which it was to be placed. Then the tire was heated red hot, and, thanks to its expansion, it could be slipped over the wood. As the iron cooled, it shrank around the wood. In modern manufacturing, when one cylinder is to fit inside and be fastened to another cylinder, the two are often "sweated" together. In this process the inner cylinder is made slightly larger than the hole in the outer cylinder. Then the outer cylinder is heated, and the inner one cooled. Because of expansion of the outer cylinder and contraction of the inner one, the outer cylinder may be slipped over the inner. When the two come to the same temperature, they are held together by strong frictional forces.

15.5 VOLUME EXPANSION

Most materials, fluids as well as solids, expand when heated. The change in volume ΔV depends on the substance, the change in temperature Δt, and the original volume V according to the relation

$$\Delta V = \beta V \Delta t \tag{15.3}$$

where β is *the coefficient of volume expansion* of the substance. *The coefficient of volume expansion is the change in volume divided by the product of the original volume and the change in temperature.*

For most substances β varies somewhat with temperature. *For solids the coefficient of volume expansion β is three times the coefficient of linear expansion α*, as the following reasoning indicates. Consider a solid cube of length L on each side. When the temperature is raised an amount Δt, the length of each side becomes $L + \Delta L$. The volume is then $(L + \Delta L)^3 = L^3 + 3L^2 \Delta L + 3L(\Delta L)^2 + (\Delta L)^3$. Since ΔL is very small compared with L, we may neglect the last two terms in comparison with the first two. The initial volume was L^3, and the change in volume is then

$$\Delta V = 3L^2 \Delta L = 3\alpha L^3 \Delta t = 3\alpha V \Delta t \tag{15.3a}$$

Liquids such as alcohol and kerosene expand when heated. Coefficients of volume expansion for several liquids are listed in Table B.2 in Appendix B. However, the most familiar liquid, water, does not behave in so simple a fashion. When warmed from 0 to 4°C, it contracts. As the temperature is raised above 4°, the water expands (Fig. 15.6). Thus a given mass of water has its minimum volume and its maximum density at 4°C under 1 atm pressure. The temperature for maximum density decreases about 0.02 Celsius degree per atmosphere as the pressure is increased. Although the temperature at the bottom of a deep northern lake may be less than 4°C in the winter, it still does not reach the freezing point, which also decreases with increased pressure. As a result the surface may freeze over in winter, but there is still water below. In the summer the upper layers of the lake are warmed, but very deep layers may not increase appreciably in temperature.

Figure 15.6

The volume of 1 kg of water in liters as a function of temperature. The specific volume is minimal at 3.98°C.

15.6 HEAT AND INTERNAL ENERGY

The basic particles of all matter are in constant motion. Atoms in solids vibrate back and forth in complex motions about their equilibrium positions. Molecules in a liquid wander around among the other molecules, having frequent collisions with them and thus exchanging energy. In gases molecules travel about at high speeds and have frequent elastic collisions with their neighbors. *The sum of the kinetic and potential energies associated with the random motion of the atoms of a substance is the internal energy of the substance.* When we heat a material, its internal energy is increased.

The nature of the heating process was speculated upon and argued over for centuries before the answer began to evolve at the end of the eighteenth century. One idea which held wide support for many years was that a fluid called *caloric* entered a body when it was heated and leaked away as the body cooled. When measurements of the mass of a body showed no increase when the body was heated, the proponents of the caloric theory argued that the fluid was massless. As a rival of the caloric theory, there gradually evolved the modern point of view that heat is a form of energy. More specifically, *heat is energy which is transferred between a substance and its surroundings or between one part of the substance and another as a result of temperature differences only.*

One of the decisive experiments which supported the theory that heat is a form of energy was performed by Count Rumford in 1798. He observed that the boring of cannon resulted in a large increase in temperature, although there was no flame or other source of caloric. When the drill was dull, the rise in temperature was exceedingly great and was related to the amount of mechanical work done in the drilling. Thus, adding energy to the cannon by doing work against friction led to the same temperature response as heating it in a furnace.

When we pound a nail with a hammer, the nail becomes hot. As the hammer is stopped, the atoms in the nail are given energy which shows up in the form of increased internal energy. After the hammer has struck, the atoms vibrate with greater amplitudes about their equilibrium positions. When a chisel is ground on an emery wheel, its thermal energy and its temperature are increased. When a moving automobile is stopped, its kinetic energy is transferred to internal energy in the brake drums, tires, and road. In all such cases work is done against friction with a resulting increase in the internal energy of the bodies involved.

We can increase the internal energy of a substance either by adding heat to it or by performing work on it in such a way as to increase the random motion of the atoms. We can raise the temperature of the air in a bicycle pump either by heating the barrel of the pump in a flame or by doing work on the gas by vigorous pumping.

When we put a thermometer in a beaker of water, the random motions of water molecules produce collisions with the molecules of the glass in the thermometer. In turn, the glass molecules exchange energy with the mercury molecules until eventually the "level" of this random motion is the same for the water molecules, the glass molecules, and the mercury molecules. When this occurs, the position of the mercury in the thermometer becomes stationary, and we say the thermometer has come to the same temperature as the water. *Temperature is a measure of the level of internal energy.*

If we put a cool thermometer in hot water, heat flows from the water to the thermometer until the temperatures are equal. In general, whenever two bodies are placed in contact, heat is transferred from the one at higher temperature to the one at lower temperature, just as when two bodies of water are connected, water flows through the connection from the higher surface to the lower one. When two bodies are placed in contact and neither gains heat from the other, the two bodies are at the same temperature by definition, and the levels of internal energy are the same for both. When this condition is satisfied, the bodies are in *thermal equilibrium.*

15.7 ABSOLUTE TEMPERATURE SCALES

If the temperature of a body is a measure of the level of internal energy for the body, there must be a lower limit or an *absolute zero* of temperature. If we remove from the atoms and molecules all their available energy of random motion, we can properly assign them a temperature of absolute zero, since temperature measures the level of the energy of random motion. As we saw in Sec. 14.6, the gas laws had also suggested the existence of such an absolute zero of temperature. As a solid is cooled, the energy associated with random motions of the particles is reduced. At a temperature of 0 K, the motions of all atoms are reduced to their lowest possible value. At one time it was believed that this would correspond to a state of absolutely no motion, but modern quantum mechanics has shown there is necessarily some zero-point energy.

Lord Kelvin established an absolute temperature scale based on thermodynamic reasoning (Chap. 18) which uses degrees of the same size as the Celsius degree. In addition to absolute zero, this scale uses as a fixed point the *triple point of water* (Sec. 16.9), which was taken by international agreement in 1954 to be exactly 273.16 kelvins. The triple point (the temperature at which ice, water, and water vapor are all in equilibrium) is chosen because it is highly reproducible. On the Kelvin absolute thermodynamic scale the ice point is 273.15 K, and the steam point 373.15 K. We use T to indicate absolute temperature, t to represent Celsius or Fahrenheit temperature, and the approximation

$$T(K) = t(°C) + 273 \tag{15.4}$$

The Kelvin scale is essentially equivalent to the ideal-gas scale introduced tentatively in Sec. 14.6.

15.8 HEAT UNITS

Heat is a form of energy and can be measured in the units of mechanics, e.g., joules. Long before this was known, however, heat units based on temperature increases in a known mass of water became well established. In the mks system *the kilocalorie is the quantity of heat required to raise the temperature of one kilogram of water from 14.5 to 15.5°C.* In terms of mechanical-energy units 1 kcal = 4,185.8 J.

Figure 15.7
The specific heat of water as a function
of temperature.

The heat required to raise the temperature of 1 kg of water 1 Celsius degree depends only slightly on the temperature (Fig. 15.7). It takes almost exactly 100 kcal to raise the temperature of 1 kg of water from 0 to 100°C, and in our problems we shall assume that 1 kcal will raise the temperature of 1 kg of water 1 Celsius degree at any temperature in this range.

As used by physical scientists, the *calorie* is the heat required to raise the temperature of one gram of water one Celsius degree. In the biological sciences, the kilocalorie is referred to as the *Calorie* (sometimes not capitalized). Thus, when a book on diet reports that an apple has 70 "calories" of food value and a large cooked egg has 80 "calories," the energies quoted are in the physicists' *kilocalories*.

In the English system heat is measured in British thermal units (Btu). *One British thermal unit is the heat required to raise the temperature of a one-pound mass of water from 58.5 to 59.5°F.* Since 1 lb_m is equal to 0.4536 kg and 1 Fahrenheit degree corresponds to $\frac{5}{9}$ Celsius degree, it follows that 1 Btu is equal to 0.252 kcal. In mechanical units 1 Btu = 778 ft-lb.

The number of units of mechanical energy necessary to produce the same increase in internal energy as one unit of heat energy is called the *mechanical equivalent of heat*. It is represented by J in honor of Joule, who was a pioneer in studying the relationship between work and heat.

$$J = 4{,}186 \text{ J/kcal} = 778 \text{ ft-lb/Btu}$$

One of several experiments devised by Joule to determine J utilizes a well-insulated vessel containing a known mass of water. Within the vessel a series of paddle wheels are rotated to churn the water. The paddle wheels are driven by weights hung over pulleys. When the weights descend, the paddle wheels do work on the water in the vessel (calorimeter), causing the water to increase in temperature. From the weights and the distance through which they descend, the work done on the water can be calculated. From the mass of the water, the mass of the container, and the increase in temperature, the heat corresponding to this amount of mechanical energy can be computed.

15.9 HEAT CAPACITY AND CALORIMETRY

If we add equal amounts of heat to 1 kg of water and to 1 kg of copper, the temperature of the copper goes up far more than the temperature of the water. It takes 0.093 kcal to raise the temperature of 1 kg of copper 1 Celsius degree. *The heat required to change the temperature of a unit mass of a substance one degree is the specific heat of the substance.* Specific heat has the dimensions *energy per unit mass per degree change in temperature;* some examples are kcal/(kg)(C°), Btu/(lb$_m$)(F°), J/(kg)(K). The specific heats of a number of solids and liquids are listed in Appendix Tables B.1 and B.2.

Let Q denote the quantity of heat added to a mass m, and let t_1 and t_2 be the initial and final temperatures, respectively. If c is the specific heat of the material at constant volume,

$$Q = mc(t_2 - t_1) = mc \, \Delta t \qquad (15.5)$$

The quantity $mc \, \Delta t$ of Eq. (15.5) is the increase in the internal energy of the mass.

Example

Find the number of kilocalories required to raise the temperature of 0.100 kg of brass from 25 to 75°C. The specific heat of brass is 0.09 kcal/(kg)(C°).

Heat = mass × sp ht × change in temperature
$= m \times c \times (t_2 - t_1)$
$= 0.100$ kg × 0.09 kcal/(kg)(C°) × (75 − 25)C°
$= 0.45$ kcal   ////

One of the most familiar methods of measuring a quantity of heat is by imparting this heat to a known mass of water and observing the change it produces in the temperature of the water. Experiments of this kind are carried on with the aid of a vessel known as a *calorimeter* (Fig. 15.8) which is carefully insulated so that it neither gains heat from nor loses heat to its surroundings during the experiment. The basic principle underlying all calorimetry experiments is the *law of conservation of energy*, which, for calorimetry, takes the form: *The heat gained by those bodies which gain heat is equal to the heat lost by those bodies which lose heat.*

Example

A 0.450-kg cylinder of lead is heated to 100°C and then dropped into a 50-g copper calorimeter containing 0.100 kg of water at 10°C. The water is stirred until equilibrium is established, at which time the temperature of the whole system is 21.1°C. Find the specific heat of lead.

Heat gained = heat lost
$0.100 \times 1(21.1 - 10) + 0.050 \times 0.093(21.1 - 10)$
$= 0.450c(100 - 21.1)$
$1.16 = 35.5c$
$c = 0.033$ kcal/(kg)(C°)   ////

Figure 15.8
Calorimeter for measuring the heat transferred to the water in the inner container from a heated mass introduced through the opening closed by the plug A. The outer container serves to maintain an almost constant temperature environment for the inner one.

When a body such as a calorimeter is used over and over again in heat experiments, it is often convenient to compute its thermal capacity. *The thermal capacity of a body is defined as the heat required to raise the temperature of the entire body one degree.* This quantity is sometimes known as the *water equivalent.* To find the thermal capacity, it is necessary only to multiply the mass m of the body by the specific heat c.

$$\text{Thermal capacity of body} = mc \qquad (15.6)$$

In the example above, the thermal capacity of the calorimeter is 0.050 kg $\times$ 0.093 kcal/(kg)(C°) = 0.0047 kcal/C°.

15.10 HEAT OF COMBUSTION

The heat of combustion is the heat liberated by burning a unit mass or unit volume of a fuel such as coal or gas. One method of finding it for a solid or liquid fuel is to place a measured quantity in a crucible inside a bell jar which is closed so that the products of combustion can escape only through the openings at the base of the jar. The bell jar is placed inside a calorimeter, the mass of which is known, and this vessel is then filled with a known weight of water. The temperature of the water is determined, and then the fuel is oxidized. A supply of oxygen is admitted through the opening at the top of the bell jar until all the fuel has been burned. The products of combustion bubble up through the water. When the combustion is complete, the temperature of the water is again determined. The heat energy gained by the water and the calorimeter is equal to the chemical energy released in the oxidation.

Example

A 6.8-g sample of coal was burned in a calorimeter like that described above. The mass of the water was 4.54 kg, and its temperature at the beginning of the experiment was 15°C; the final temperature was 26°C. Find the heat of combustion of the coal. The vessel forming the calorimeter has a water equivalent of 260 g (or 0.260 kcal/C°).

$$\text{Energy released} = \text{heat gained}$$

$$\text{Heat of combustion} \times 0.0068 = (4.54 \times 11 \times 1) + (0.26 \times 11 \times 1)$$
$$= 52.8 \text{ kcal}$$

$$\text{Heat of combustion} = \frac{52.8 \text{ kcal}}{0.0068 \text{ kg}} = 7800 \text{ kcal/kg} \qquad ////$$

15.11 HEAT CAPACITIES OF SOLID ELEMENTS

The specific heats of six elements are listed in Table 15.1. When we multiply the specific heat of each element by its atomic weight, we obtain

Table 15.1
Heat Capacities of Solid Elements

Element	Atomic weight	Specific heat, kcal/(kg)(C°)	Specific heat × atomic weight, kcal/(kg at wt)(C°)
Aluminum	27	0.22	5.9
Titanium	47.9	0.14	6.7
Iron	55.8	0.11	6.1
Copper	63.5	0.093	5.9
Tin	118.7	0.054	6.4
Lead	207.2	0.031	6.4

Figure 15.9
Atomic heat capacity, the product of the specific heat and atomic weight, as a function of temperature for carbon (diamond), silicon, and aluminum.

the number in the right column of the table, which is the *heat capacity per kilogram atomic weight*. In 1819 *Dulong and Petit* showed that *for many solid elements the heat capacity per kilogram atomic weight is approximately* 6 kcal/(kg at wt)(C°). This value is very nearly 25 kJ/(kg at wt)(C°) $\approx 3R$, where R is the general gas constant, a fact on which further comment appears in Sec. 15.12. The physical meaning of the *law of Dulong and Petit* is readily seen when we recall that one kilogram atomic weight of any element contains $N_A = 6.02 \times 10^{26}$ atoms; thus, for most solid elements, it takes the same amount of energy to raise the temperature of equal numbers of atoms 1 Celsius degree; on the average we must add about 10^{-26} kcal per atom.

At room temperature there are a few solid elements which are notable exceptions to the law of Dulong and Petit. For example, carbon in the diamond form has a specific heat of only 1.46 kcal/(kg at wt)(C°) at room temperature. However, the value approaches 6 as the temperature is raised (Fig. 15.9). For all elements the specific heat is small at low temperature, approaching zero as the temperature approaches 0 K. Modern quantum theory gives us an understanding of the reasons for the discrepancies between observed specific heats and the law of Dulong and Petit.

15.12 THE SPECIFIC HEATS OF GASES

For gases the specific heat depends on how the heating is carried out. If the volume is kept constant as heat is applied, all the heat goes into increasing the internal energy of the molecules. When the pressure is held constant as heat is added, the internal energy increases as before, but external work is done by the gas as it expands. Consequently it takes more heat to raise the temperature of a unit mass of gas 1 degree at constant pressure than it does at constant volume.

Specific Heat at Constant Volume From Eq. (14.11) we know that $mv^2/2 = 3RT/2N_A$. The total kinetic energy of translation of all the molecules in 1 kmole is $N_A mv^2/2 = 3RT/2$. To increase the temperature by 1 K, we must increase the total kinetic energy of translation to $3R(T + 1)/2$, or by $3R/2$. Hence, for any gas in which all the thermal energy is associated with translational motion, the heat capacity per kilomole is $3R/2 = 2.98$ kcal/(kmole)(C°), since $R = 1.987$ kcal/(kmole)(K). This is just half the value given by the law of Dulong and

Petit for solid elements. Actually the heat which goes to translational kinetic energy is the same for solid elements and gaseous molecules, but the atoms in solids have potential energy equal on the average to the kinetic energy.

For a monatomic gas (or any gas in which all the internal energy of the molecules is associated with translational motion) the specific heat is $3R/2M$, where M $(= N_A m)$ is the mass in kilograms of 1 kmole (numerically M is simply the molecular weight). Thus

$$c_v = \frac{3R}{2M} \tag{15.7}$$

For helium, $M = 4$ kg/kmole, and $c_v = 0.745$ kcal/(kg)(K); for argon, $M = 39.95$ kg/kmole, and $c_v = 0.075$ kcal/(kg)(K).

For diatomic and triatomic gases at room temperature the specific heat is greater than the value $3R/2M$ because the molecule has kinetic energy of rotation as well as of translation. We may think of a diatomic molecule as a dumbbell which is rotating as it moves about (Fig. 15.10). When we raise the temperature, the kinetic energy of rotation increases, as well as the kinetic energy of translation. This is why the specific heat at constant volume for diatomic gases is greater than that predicted by Eq. (15.7). At room temperature the specific heat at constant volume for most diatomic gases is given by $5R/2M$. At higher temperatures it becomes still greater; the additional energy goes into exciting vibrations of the two atoms along the line connecting their centers of mass. For triatomic gases near room temperature, the specific heat is $3R/M$ or more, depending on how much energy is associated with vibration.

Specific Heat at Constant Pressure When a gas is kept at constant pressure, the heat required to increase the internal energy of the molecules is the same as when the gas is kept at constant volume. In addition, however, it is necessary to supply heat energy *to do the external work required to expand the gas*. If a gas expands in a cylinder having at one end a moving piston (Fig. 15.11) by means of which the pressure is held constant, the work W done against the piston in moving it from a to b is $p\,\Delta V$, where p is the pressure and ΔV the change in volume. This may be seen as follows:

$$W = Fs = pAs$$

where F is the force, s the distance through which the piston moves, and A the area of the piston. The product $As = V_2 - V_1$ is the increase in volume ΔV of the gas during expansion. Therefore,

$$W = p\,\Delta V \tag{15.8}$$

This work is represented by the shaded area on the pV diagram of Fig. 15.11.

Consider the work done when 1 kmole of gas is heated 1 K under constant pressure. By the general gas law, $pV = RT$. If we raise the temperature 1° at constant pressure, we have $p(V + \Delta V) = R(T + 1)$. Therefore, $p\,\Delta V = R$. Thus, the external work done by 1 kmole of gas when its temperature is raised 1 K is numerically equal to R. The *heat capacity per kilomole at constant pressure is equal to that at constant volume plus R.*

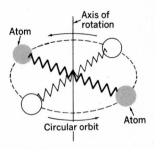

Figure 15.10
A diatomic molecule may have kinetic energy of rotation and vibration as well as kinetic energy of translation.

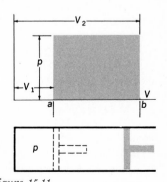

Figure 15.11
The work done by a gas expanding at constant pressure is equal to the product of the pressure and the change in volume.

The energy which must be added per unit mass of gas to do external work is R/M and the specific heat of the gas at constant pressure is R/M greater than the specific heat at constant volume:

$$c_p = c_V + \frac{R}{M} \tag{15.9}$$

The ratio of the specific heat of the gas at constant pressure to the specific heat at constant volume is ordinarily represented by γ. For a monatomic gas c_V is $3R/2M$, and $c_p = 5R/2M$. Therefore,

$$\gamma = \frac{c_p}{c_V} = \frac{5R/2M}{3R/2M} = \frac{5}{3}$$

for a *monatomic* gas.

For typical diatomic gases at room temperature,

$$c_V = \frac{5R}{2M} \quad \text{and} \quad \gamma = \frac{7R/2M}{5R/2M} = 1.4$$

QUESTIONS

1 What is the distinction between temperature and heat? How are they related?
2 Why are specific heats numerically the same in both British and metric units?
3 When a mercury-in-glass thermometer is thrust into boiling water, the mercury first descends and then rises. Why?
4 What is unique about the thermal expansion of water? What are some of the important consequences? Why wouldn't water in glass make a good thermometer?
5 How is the thermal expansion allowed for in (a) concrete pavements, (b) steel bridges, and (c) ordinary light bulbs?
6 An iron rod lies along the diameter of a circular iron hoop and connects opposite sides of the hoop. If the temperature of the whole system is increased, will the hoop remain circular?
7 Does the coefficient of linear expansion depend on what thermometer scale is used? On what units of length are used? Explain.
8 Why is the molar specific heat of a diatomic gas ordinarily greater than the corresponding specific heat of a monatomic gas?
9 How would you expect the specific heat of hydrogen at constant volume to compare with that of oxygen? Why?
10 Nitrogen is sealed in a thin glass container which is placed under a bell jar. If the bell jar is evacuated and the container then broken, what happens to the temperature of the gas? Explain.

11 A stretched rubber hose contracts when its temperature is increased. What does this mean about the coefficient of linear expansion? Can you explain why rubber and some other materials contract with increasing temperature?

PROBLEMS

1 A bar of copper at 25°C is 80 cm long. At what temperature will it be 1 mm longer?
  Ans. 100°C
2 The steel supporting cables of a suspension bridge have a total length of 400 m at 0°C. Find the length at 30°C.
3 The channel span of the steel Ohio River bridge in Cincinnati is 354 m long. Calculate the maximum change in its length if it is subject to extreme temperatures of −20 and 35°C.
  Ans. 0.23 m
4 Steel rails in 20-m lengths are laid in winter at 0°C. How much space must be allowed between consecutive rails to permit expansion in summer at a temperature of 40°C?
5 A steel ring which is 40.00 cm in diameter at room temperature (20°C) is to be shrunk onto a pulley which is 40.08 cm in diameter. Find the temperature to which the ring must be raised so that it will just slip over the pulley.
  Ans. 187°C
6 A steam locomotive had steel driving wheels 70 in. in diameter before the tires were shrunk on. If the diameter of the tire was 69.950 in.

at 20°C, find the temperature to which it had to be heated to make its diameter 70.050 in. so it could be slipped over the wheel.

7 A brass cap is to be "sweated" on a brass cylinder. The inner diameter of the cap is machined to 2.000 cm, while the cylinder has an outer diameter of 2.004 cm. How hot must the cap be before it can be slipped over the cylinder if the cylinder is at 15°C?   *Ans.* 119°C

8 Find the coefficient of areal expansion per Fahrenheit degree for aluminum. An aluminum panel for an aircraft wing has an area of 750.000 in.² at 50°F. What is its area at 110°F?

9 A sheet of brass has an area of 75 cm² at 5°C. What will be its area at a temperature of 85°C?   *Ans.* 75.23 cm²

10 A glass flask holds exactly 500 cm³ of mercury at 15°C. If it is heated to 85°C how many grams of mercury will run out?

11 A glass flask holds exactly 600 cm³. It is filled with ethyl alcohol at 0°C and then heated to 40°C. How much ethyl alcohol runs out?   *Ans.* 25.8 cm³

12 A rod of steel and one of brass have exactly the same length, 1 m, at 0°C. The rods are heated together until they differ in length by 0.6 mm. What is the temperature of the rods?

13 A long U-tube is filled with ethyl alcohol. One arm of the tube is kept at 5°C and the other at 35°C. Find the length of the column in the tube at the higher temperature if the length of the column in the tube at the lower temperature is 70 cm.   *Ans.* 72.31 cm

14 An 80-kg man daily consumes food with a fuel value of about 2800 kcal. If this energy were used to heat 80 kg of water, by how many degrees would the temperature be raised?

15 A soft drink at 0°C has a mass of 400 g. If 1 g of sugar yields 4 Cal (= 4 kcal), what mass of sugar would be required to provide the body with the energy to raise the temperature of the liquid to body temperature of 37°C?   *Ans.* 3.7 g

16 Upper Yosemite Falls is 436 m high. If all the potential energy lost in this fall is converted into heat, how much is the temperature of the water raised?

17 A hiker with his equipment weighs 900 N. If he climbs 500 m up a mountain, and if 20 percent of the energy of the food he consumes goes into potential energy, how many eggs at 80 kcal each would he have to eat to supply himself with food energy?   *Ans.* 7

18 One cubic meter of water falls from the top to the bottom of Niagara Falls, a distance of 161 ft. If all the potential energy lost in the fall is transformed into heat, how much is the temperature of the water raised?

19 A lead bullet that has a velocity of 350 m/sec strikes a target. If 90 percent of the kinetic energy of the bullet is absorbed by the target and the other 10 percent is retained by the bullet as heat, what is the increase in temperature of the bullet?   *Ans.* 48.8 C°

20 A truck and its load together weighs 5 tons. The brakes are used to bring it to rest from a speed of 60 mi/hr. How much heat is developed?

21 How much heat is generated in stopping a flywheel that is making 30 rev/sec about an axis through its center? The flywheel has a moment of inertia of 1.8 kg-m².   *Ans.* 7.64 kcal

22 If 150 g of copper at 240°C is dropped into 280 g of water at 12°C contained in a 100-g copper calorimeter, find the resulting equilibrium temperature.

23 A piece of aluminum with a mass of 40 g is heated in a steam jacket to a temperature of 98.5°C and then plunged into 160 g of water at 12°C, causing the temperature to rise to 16.5°C. What is the specific heat of aluminum, assuming that no heat was lost or gained during the experiment?   *Ans.* 0.219 kcal/(kg)(C°)

24 A piece of lead heated to 100°C is dropped into 150 cm³ of water at 4.4°C. The final temperature is 9.9°C. Compute the mass of the lead.

25 To find the temperature of a furnace, a 200-g piece of platinum was placed inside until temperature equilibrium was reached. The platinum was then dropped into a 200-g copper calorimeter containing 500 g of water at 6°C. If the final temperature of the water was 14°C, find the temperature of the furnace.   *Ans.* 662°C

26 Estimate by use of the law of Dulong and Petit the room-temperature specific heats of selenium, silver, and uranium. (Atomic weights may be found inside the back cover.)

27 Use the law of Dulong and Petit to estimate the specific heats of lanthanum, gold, and nickel. (Atomic weights are listed inside rear cover; the

answers given are experimental values at 18°C
from a handbook.)

Ans. 0.0448, 0.0312, and 0.105 kcal/(kg)(C°)

28 Natural gas from the mains, with a heat combustion of 1,500 Btu/ft³, is used to heat water.
Assuming that half the available heat is wasted,
how much gas will be required to heat 400 lb
of water from 42 to 200°F?

29 The heat of combustion of natural gas is 1,500
Btu/ft³. How many cubic feet of gas are needed
to heat 15 ft³ of water from 40 to 180°F in a
hot-water heater, assuming that 75 percent of
the energy released is absorbed by the water?

Ans. 117 ft³

30 A sample of methyl alcohol of 20-g mass was
burned in a fuel calorimeter containing 9 kg of
water. The water equivalent of the calorimeter
was 800 g. The initial temperature was 11.2°C,
and the final temperature was 22.1°C. Calculate
the heat of combustion of the methyl alcohol.

31 A specimen of coal with a mass of 2.08 g was
burned in a copper calorimeter having a mass
of 1.1 kg. The mass of the water in the calorimeter was 1.6 kg, and the initial temperature was
14.6°C. The final temperature of the water was
23.9°C. Find the heat of combustion.

Ans. 7610 kcal/kg

32 A sample of coal weighing 0.6 oz with a heat
of combustion of 12,000 Btu/lb was burned in
a crucible in a calorimeter. The calorimeter
contained 11 lb of water, and it had an initial
temperature of 48°F. The parts of the calorimeter weighed 6.2 lb and had an average specific
heat of 0.18 Btu/(lb)(F°). To what temperature
was the water raised?

33 Into a 100-g copper calorimeter containing 170 g
of water at 10°C there are dropped simultaneously 60 g of silver at 100°C, 60 g of iron at
80°C, and 15 g of platinum at 90°C. What is
the resulting equilibrium temperature?

Ans. 14.3°C

34 One-tenth of a kilomole of air at 0°C and two
standard atmospheres pressure occupies
1.12 m³. With the pressure kept constant, heat
is added until the temperature reaches 20°C.
Find the volume of the gas at 20°C, the external

work done by the gas, and the heat supplied to
the gas.

35 Find the external work done and the increase
in internal energy if 2 moles (0.056 kg) of nitrogen at 27°C and 1 × 10⁵ N/m² pressure is
expanded at constant pressure until the volume
increases 50 percent. *Ans.* 2,494 J; 6,236 J

36 The specific heat at constant volume of argon,
a monatomic gas, is 0.075 kcal/(kg)(C°). When
2 kg of argon is heated from 27 to 127°C, how
many kilocalories are required if the heating is
done at constant volume? At constant pressure?

37 The specific heats of air at constant pressure
and at constant volume are 0.24 and
0.17 kcal/(kg)(C°), respectively. If 4 kg of air
is heated from 0 to 40°C at constant pressure,
find (*a*) how much external work is done and
(*b*) how much the kinetic energy of the air molecules is increased.

Ans. (*a*) 11.2 kcal; (*b*) 27.2 kcal

38 Argon is a monatomic gas with an atomic weight
of 40. Find the specific heat of argon at constant
pressure and at constant volume. How much
external work is done when 4 kg of argon is
heated from 0 to 5°C at constant pressure? How
much is the internal energy increased?

39 A gram molecular weight (28 g) of nitrogen at
27°C is supplied with 2,000 J of heat. It expands, doing 400 J of work. What is the increase in internal energy? What is the increase
in temperature? *Ans.* 1,600 J; 77 C°

40 Two moles (8 g) of helium at 0°C occupy a
volume of 22.6 × 10⁻³ m³ at a pressure of
2 × 10⁵ N/m². If 400 cal of heat is added to
these moles of helium at constant pressure, find
the final temperature of the gas, the external
work done by the gas, and the increase in the
internal energy of the gas.

41 A cylinder contains 400 cm³ of neon at 0°C and
76 cm Hg pressure. Neon is a monatomic gas
with an atomic weight of 20.2. What is the mass
of this quantity of gas? If the temperature of this
gas is raised to 100°C and the pressure is
reduced to 38 cm Hg, find the new volume of
the gas and the increase in the internal energy
of the gas. *Ans.* 0.36 g; 1,090 cm³; 22.3 J

42 A clock with a steel pendulum beats seconds
correctly when the temperature of the room is
15°C. How many seconds per day will it gain
or lose when the temperature of the room is
25°C?

43 A brass pendulum is adjusted to beat seconds,

i.e., has a period of 2 sec, at 20°C. What will be the gain or loss per day if the temperature of the clock drops to 13°C?

Ans. gains 5.8 sec

44 A steel rod 10 cm long has a cross section of 1.5 cm². What force would be required to extend the bar the same amount as the expansion produced by heating it 25 C°?

45 Compute the hydrostatic pressure required to prevent an aluminum block from expanding when its temperature is increased from 10 to 51°C.

Ans. 2.2×10^8 N/m²

46 A bimetallic strip is made by soldering a strip of brass to a strip of steel. The strips are each 0.2 mm thick and 10 cm long. If one end is held fast, find how far the other end moves if the strip is heated 100 C°. (Assume the strips were initially straight and that after the temperature rise the brass and steel form arcs of concentric circles 0.2 mm apart.)

47 A 40-cm copper rod and a 60-cm aluminum rod

are laid end to end between rigid supports which remain exactly 1 m apart. The rods have identical radii and do not bend. Calculate the stress in each rod and the length of each if the temperature is raised 50°C.

Ans. 9.1×10^7 N/m²; 40.003 and 59.997 cm

48 Show that the stress in a heated rod which is not free to expand (length held constant) is given by $F/A = Y\alpha \Delta t$, where Y is Young's modulus, α the coefficient of linear expansion, and Δt the temperature change.

49 Show that the coefficient of volume expansion of a gas is given by $1/T$, where T is the absolute temperature.

THE SUCCESS of the kinetic theory of gases prompts us to apply the basic ideas to liquids and solids and to changes from one phase to another. Here we can scarcely hope to achieve quantitative understanding because the assumption that interactions between the particles are negligible is no longer valid, and we do not have simple formulas by which to calculate the interatomic forces. Nevertheless, when we apply the ideas qualitatively, we find that they give us considerable insight into the melting of solids and the evaporation of liquids.

CHAPTER SIXTEEN
CHANGE OF PHASE

16.1   MELTING OR FUSION

Removing energy from a gas results in a lowering of its temperature. For many gases liquefaction occurs, and if still more energy is taken away, the liquid solidifies. Such changes in phase are typically reversible, so that adding energy to a solid may lead to its fusion and eventually to its evaporation.

When ice is heated, the temperature rises until it reaches 0°C and then remains stationary until all the ice is melted. Once all the ice is melted, the temperature of the water begins to rise. *The temperature at which the solid changes into liquid is called the melting point.* At the melting point more heat serves to hasten the melting process without any change of temperature.

The temperature at which the liquid changes into the solid state is its *freezing point.* For crystalline substances such as ice or copper, the freezing point and the melting point coincide and are sharply defined. Substances that are not crystalline, such as wax or glass, gradually soften and do not have definite melting points. For certain fats, the melting point is not the same as the freezing point. For example, butter melts between 28 and 32°C and solidifies between 20 and 23°C.

If a pure liquid is carefully protected from mechanical disturbances, it may be cooled below the temperature at which it normally solidifies. This process is known as *supercooling.* Water has been cooled to about −40°C without becoming ice. The liquid at such a temperature is in a state of unstable equilibrium and immediately solidifies if it is disturbed or if a crystal of the solid is dropped into it.

To melt a solid such as ice, it is necessary to supply a given quantity of thermal energy to each unit mass. This is true even though the temperature of the ice at the beginning is the same as the temperature of the water at the end of the process. The ice has a crystalline structure, and the energy is necessary to tear down this structure. Water, in changing back to ice (in general, a liquid in changing back to a solid), gives up the heat that it absorbed in melting. *The heat of fusion of a substance is the heat necessary to convert a unit mass of solid into liquid at the same temperature and pressure.* To change 1 kg of ice to water at 0°C requires 80 kcal; to convert 1 lb_m of ice to water at 32°F requires 144 Btu. Heats of fusion of several solids are listed in Appendix Table B.1.

Example

In an experiment designed to measure the heat of fusion, 25 g of ice at 0°C was dropped into 195 g of water at 30°C contained in a copper calorimeter of mass 100 g. The final temperature was 18°C. Find the heat of fusion of ice.

$$\text{Heat gained} = \text{heat lost}$$

$$0.025L + 0.025 \times 1 \times (18 - 0) = 0.195 \times 1 \times (30 - 18)$$
$$+ 0.100 \times 0.093 \times (30 - 18)$$

where L is the latent heat of fusion and $0.025L$ is the heat needed to melt the ice.

$$0.025L + 0.450 = 2.45$$
$$L = 80 \text{ kcal/kg} \qquad ////$$

16.2 CHANGE OF VOLUME DURING FREEZING

It is a familiar fact that ice floats and that pipes or bottles filled with water burst when frozen. One liter of water makes about 1.09 liters of ice. Cast iron and type metal behave like water in this respect, expanding when they solidify. For this reason they are suitable for making castings in which it is desired to reproduce the detail of a mold. Most metals and other substances contract on solidification. This is one reason why gold and silver coins are stamped rather than cast.

That the forces exerted by freezing water are large may be seen from the fact that they are sufficient to burst strong water pipes on a cold night. These forces are of importance in the formation of soils from rocks. Water penetrates into the crevices in the rocks and freezes. The expansion breaks off fragments, which may eventually become soil. Alternate freezing and thawing of soils tends to pulverize them. Ice forming in the interstices of the soil serves to loosen compact land and give it better tilth.

The fact that water expands when it freezes is responsible for the fact that ice forms on top of a lake or stream rather than the bottom. A layer of ice is an excellent insulation which protects all but shallow water from freezing to the bottom. Fish and other water life can carry on in the water below the ice layer.

Since an increase of pressure causes a body to contract, such an increase favors the liquid phase for any substance which contracts on melting and results in a lower melting point. Ice contracts when it melts; therefore the melting point is lowered by application of pressure. Careful experiments show that this lowering is 0.0075°C for an increase

Figure 16.1

Ice melts at a lower temperature when it is subjected to additional pressure.

Figure 16.2

The freezing point of a salt solution as a function of salt concentration by weight.

Figure 16.3

When a liquid evaporates and the vapor is saturated, the number of molecules escaping the liquid per unit time is equal to the number returning to the liquid per unit time.

of 1 atm of pressure (Fig. 16.1). If a substance expands upon melting, its melting point is raised by the application of pressure.

The effect of pressure on the melting point of ice may be shown by supporting, on a piece of ice, a loop of wire from which a heavy weight is hung. The pressure of the wire lowers the melting point so the ice is in a condition to melt as soon as the necessary heat is supplied. This heat is taken from the water above the wire, causing it to freeze again. This process continues until the wire cuts its way through the ice, leaving the block as solid as it was at the beginning. The process is called *regelation*.

16.3 FREEZING POINT OF A SOLUTION

When any material is dissolved in water, the freezing point of the solution is lower than the freezing point of water. This is also true for other liquids. The freezing point of a solution is lower than that of the pure solvent. This fact is used to prevent the water in the cooling system of an automobile from freezing. For solutions that are not too concentrated, the lowering of the freezing point is proportional to the number of dissolved particles per unit volume. A given number of salt (NaCl) molecules lowers the freezing point more than an equal number of sugar molecules because the salt ionizes, thus producing two particles per dissolved molecule.

When a dilute solution begins to freeze, only solvent freezes out. This makes the remainder of the solution more concentrated and lowers its freezing point still further, until the solution becomes saturated. Then dissolved substance and solvent freeze out in such a way that the concentration of the solution remains unchanged. Figure 16.2 represents the relation between the freezing point and the percentage of salt in a solution. Curve AB shows the relation between temperature and concentration of the salt. The temperature at which the solvent and the dissolved substance crystallize out as a mixture is called the *eutectic temperature*. If the temperature of the solution is higher than the eutectic temperature and the solution is saturated, the dissolved substance crystallizes out as the temperature is lowered (curve BC).

16.4 SATURATED VAPOR

The molecules of a liquid move with a wide range of instantaneous velocities. When the liquid is heated, this range and the average speed increase. Some of the molecules near the surface attain sufficient kinetic energy to escape the forces of attraction which confine the less energetic molecules to the liquid (Fig. 16.3). When the space above the surface of the liquid is enclosed, some vapor molecules return to the surface of the liquid and are captured. As more molecules escape, the number returning to the surface also increases. When the number of molecules returning to the surface is equal to the number escaping, the space above the liquid is said to be *saturated*, and above the liquid we have a *saturated vapor*. The pressure exerted by a saturated vapor depends on the nature of the liquid (Fig. 16.4) and on the temperature. If the temperature is held constant and an attempt is made to increase the pressure on a saturated vapor, part of the vapor condenses. *The saturated vapor pressure of any given liquid depends only on the temperature.*

The saturated vapor pressures of several liquids as functions of temperature are plotted in Fig. 16.5. As the temperature increases, the pressure of the saturated vapor rises. A vapor-pressure curve can be obtained by measuring the difference in level between the surfaces of mercury in the apparatus of Fig. 16.6 as the temperature is increased.

Figure 16.4
When a small volume of liquid is introduced into the Torricelli vacuum above a mercury column, the mercury level is depressed below the vacuum level by the saturated vapor pressure of the liquid.

16.5 EVAPORATION AND BOILING

Even at relatively low temperature the most energetic molecules evaporate from the surface of a liquid. *Evaporation is the escape of molecules from the surface of a liquid.* As the temperature of a liquid is raised, evaporation proceeds more rapidly. Eventually bubbles of vapor form in the liquid and rise to the surface. That temperature at which vapor bubbles form in the volume of the liquid and rise to the surface is the *boiling point* or *boiling temperature.*

We shall now examine the boiling point from a somewhat different point of view. Consider a bubble of vapor within the liquid. This bubble will collapse unless the pressure of the saturated vapor within it is as great as the pressure in the adjacent liquid. The pressure at the bubble is the applied pressure at the surface of the liquid plus the hydrostatic

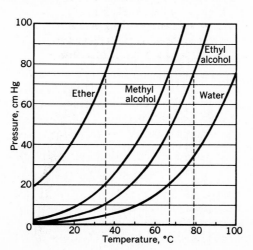

Figure 16.5
Saturated-vapor-pressure curves of four liquids. The dashed lines indicate the boiling points at 1 atm (76 cm Hg) pressure.

Figure 16.6

Apparatus for measuring the saturated vapor pressure of a liquid by recording changes in mercury level with changes of temperature on the thermometer.

pressure due to the liquid above the bubble plus the pressure due to surface tension. Even for a relatively small bubble near the surface, the latter two contributions to the total pressure are normally negligible in comparison with the applied pressure. Thus, *the boiling point is the temperature at which the pressure of the saturated vapor of the liquid is equal to the applied pressure.*

The distinction between *evaporation* and *boiling* should now be clear. Both involve the change of phase from liquid to gas, which is called *vaporization.* Evaporation is vaporization from the surface alone whereas boiling is vaporization within the volume of the liquid. Since the boiling point of a liquid is the temperature at which the vapor pressure is the same as the applied pressure, it follows at once that when the applied pressure is changed, the boiling point also changes. At 1 atm (76 cm Hg) pressure, water boils at 100°C. The curves of Fig. 16.5 show how the boiling points of four familiar liquids vary with pressure. (For water the boiling point is tabulated as a function of pressure in Appendix Table B.3.)

The influence of pressure on the boiling point can be shown by filling a flask half full of water and boiling it vigorously for some time to remove the air from above the water. Remove the flask from the flame and immediately insert a stopper, rendering the flask airtight. Invert the flask, and pour cold water on the bottom. This cold water causes some of the vapor in the flask to condense, and the pressure on the hot water in the flask is reduced sufficiently to allow the water to begin boiling again.

The effect of pressure on the boiling point of water is well illustrated in the action of geysers. If it is 100 ft down from the surface of a narrow column of water to a cavity surrounded by hot rocks, the pressure at these rocks is about 4 atm, three of the four being due to the column of water. At this pressure water boils between 140 and 150°C. Once water in the cavity begins to boil, bubbles rush up the column carrying quantities of water with them, thus reducing the pressure at the cavity and encouraging more rapid boiling until the temperature of the water in the cavity is lowered to the point at which boiling ceases.

In the pressure cooker the increased pressure results in a higher boiling point for water and a higher cooking temperature. On the other hand, when sirups or milk are to be evaporated, they may be placed in "vacuum pans" so that boiling can be carried out at a reduced temperature, thereby eliminating undesirable changes which would occur at the normal boiling temperature.

When a ship is driven by a rapidly rotating propeller, the pressure on one side of the propeller decreases as the angular velocity is increased. When the pressure falls below its saturated vapor pressure, the water flashes into vapor, producing cavities in the liquid. This phenomenon is known as *cavitation.*

16.6 HEAT OF VAPORIZATION

Just as a certain amount of heat is required to convert a unit mass of ice into water without changing its temperature, so a certain amount of heat is required to change a unit mass of water into steam without changing its temperature. When the steam condenses, this heat is liberated.

The heat of vaporization of a substance is the heat necessary to change a unit mass of the liquid to the vapor state without any change in the temperature and pressure. For water at 100°C and 1 atm pressure, it is found to be 540 kcal/kg (970 Btu/lb). It takes more than five times as much heat to change 1 g of water at 100°C into steam at 100°C as it takes to heat that same gram of water from 0 to 100°C. The heats of vaporization of several other liquids are listed in Appendix Table B.2.

The heat of vaporization depends on the temperature (and pressure) at which the change takes place. The higher the temperature, the easier it is to evaporate the liquid; i.e., the higher the temperature, the lower the heat of vaporization (Fig. 16.7).

The following experiment may be used to determine the heat of vaporization of water. Fill a calorimeter with a known mass of water. Pass dry steam from a boiler into the calorimeter. Measure the temperature of the water in the calorimeter at the beginning and after steam has been condensed in it. The mass of condensed steam is found by weighing the calorimeter after the steam has been condensing for a time sufficient to produce the desired rise in temperature. The latent heat of vaporization of water can now be calculated as illustrated below.

Figure 16.7
The heat of vaporization of water decreases as the temperature is raised.

Example

Dry steam at 100°C is condensed in a large calorimeter of water equivalent 0.25 kg, which contains 2.8 kg of water at 5°C. The final temperature of 27°C is reached after 110 g of steam has been condensed. Find the latent heat of vaporization V as given by these data.

$$\text{Heat gained} = \text{heat lost}$$
$$2.8 \times 1 \times (27 - 5) + 0.25 \times (27 - 5) = 0.11V + 0.11 \times 1$$
$$\times (100 - 27)$$
$$61.6 + 5.5 = 0.11V + 8.03$$
$$V = 537 \text{ kcal/kg}$$
$$\text{about 0.5\% low} \quad ////$$

When a liquid is evaporating, heat is taken from the remaining liquid and the surrounding bodies. The withdrawal of this heat may cause their temperatures to decrease. For this reason the evaporation of water sprinkled on the sidewalk cools the air. Evaporation of perspiration from the skin lowers the temperature of the body.

16.7 BOILING POINT OF A SOLUTION

When water has some substance such as sugar dissolved in it, the temperature at which it boils is raised. The amount the boiling point is raised in proportional to the amount of the substance dissolved in the water. In making candies or sirups, the temperature is used as a means of determining the concentration of the solution. By observing

Figure 16.8

The saturated vapor pressure of a solution (dashed line) is lower than that of the solvent (solid line) at any temperature.

the temperature at which the solution is boiling, it is possible to determine when the candy or sirup has boiled long enough. When vegetables, fruits, meats, etc., are boiled in water, some of their contents dissolve in the water. This raises the boiling point slightly, so that the water boils above 100°C. The effect of a dissolved substance on the vapor pressure and the boiling point is shown in Fig. 16.8. The continuous curve gives the relation between the temperature and the vapor pressure of the pure solvent, and the dashed curve gives the same relation for the solution.

When a solid is dissolved in a liquid, the vapor that rises above the liquid does not contain the dissolved substance. To obtain pure water from water containing solids in solution, it is only necessary to evaporate the water and condense the vapor. The solids are left behind.

16.8 SUBLIMATION

If a substance such as solid camphor or solid iodine is left for some time in a closed vessel, vapor which is given off by the solid condenses on the sides of the vessel to form small crystals. The escape of molecules directly from a solid without passing through the liquid phase is called *sublimation*. In like manner, vapor may go directly to the solid state. For most solids, the vapor pressure is very small, but for some our sense of smell tells us that vapor is being given off. The vapor pressure of ice at 0°C is 4.6 mm Hg; as the temperature decreases, the vapor pressure decreases.

Figure 16.9

The saturated vapor pressure of solid carbon dioxide (Dry Ice) increases rapidly with temperature.

The heat of sublimation is the heat necessary to change a unit mass of a substance from the solid of the vapor state without change of temperature. A vapor-pressure curve (Fig. 16.9) showing the vapor pressure when a solid is in equilibrium with its vapor may be plotted for a solid just as for a liquid; it is called a *sublimation curve.*

16.9 TRIPLE POINT

Under certain conditions of temperature and pressure a liquid may be in equilibrium with its vapor. On a temperature-pressure diagram these conditions are represented by the vapor-tension curve OC (Fig. 16.10). Similarly, a solid may be in equilibrium with its vapor along the sublimation curve OA. In like manner the melting curve OB may be drawn showing the relation between the temperature and pressure when the solid is in equilibrium with its liquid. When all these curves are drawn for a substance, they intersect at O, which is called the *triple point.* The triple point indicates the one temperature and pressure at which the solid, vapor, and liquid can exist together without one of them gaining in mass at the expense of the others. At this temperature and pressure, the solid, liquid, and vapor are in equilibrium.

Some substances, such as water, have several solid phases which would be represented by domains to the left of AOB. Water is unusual also in the sense that the melting curve OB leans slightly to the left of vertical, indicating that an increase in pressure lowers the melting point. The triple point for water can be reached by the following experiment. Some water is placed on a watch glass under a bell jar. Evacuation of the bell jar produces boiling, the heat for which is removed from the water, which is cooled. Eventually the temperature falls to the point at which water is simultaneously boiling and freezing. With care one can reach the triple point for water, which is at a temperature of 0.0100°C and a pressure of 4.6 mm Hg. If evacuation is continued and the pressure is reduced below 4.6 mm Hg, only the solid and vapor phases can exist in equilibrium. A wafer of ice remains on the watch glass long after all the liquid has disappeared.

Figure 16.10
Equilibrium between the solid, liquid, and vapor phases of a material occurs at the triple point O.

16.10 THE CRITICAL POINT

If an *unsaturated* vapor is enclosed in a cylinder $HFEG$ (lower part of Fig. 16.11) and piston M is moved in while the *temperature of the vapor remains constant*, the pressure at first increases as the volume is decreased, as represented by the curve AB. Near A, Boyle's law is almost followed, but deviations increase as B is approached. At B the vapor is *saturated*, and liquefaction begins. Along BC the volume decreases while the pressure remains constant (Boyle's law does not apply to saturated vapors). Meanwhile, the quantity of vapor decreases, and that of liquid increases. At C liquefaction is complete. A further application of pressure to the piston causes a small decrease in the volume of the liquid, and the curve CD representing the relation between the pressure and the volume of the liquid becomes very steep. The

Figure 16.11
When a vapor in a cylinder (bottom) is compressed at constant temperature, the vapor begins to liquefy at B and the pressure remains constant as liquefaction occurs from B to C.

Figure 16.12

Isothermal curves for carbon dioxide at six temperatures. Below 32°C the vapor can be liquefied by application of pressure alone.

curve $ABCD$ shows the relation between the volume and the pressure when the *temperature is kept constant;* it is called an *isothermal* curve.

If this process is carried out at higher and higher temperatures, the horizontal part of the curve becomes shorter (Fig. 16.12) until it finally disappears and is replaced by a slight bend in the curve. The point P at which the horizontal part of the isothermal first disappears is known as the *critical point,* and the temperature corresponding to the isotherm through P is called the *critical temperature.* The pressure and specific volume (volume per unit mass) of the vapor at the point P are known as the *critical pressure* and *critical volume,* respectively. At temperatures below the critical temperature, it is possible to liquefy the vapor by the application of pressure *alone;* at temperatures above, it is impossible to liquefy the vapor, however great the pressure. Thus the critical temperature is the highest temperature at which a gas can be liquefied. On Fig. 16.10 the critical temperature and pressure are indicated by point C.

The isothermals of Fig. 16.12 refer to carbon dioxide. Its critical temperature is 31.1°C, and its critical pressure 77 atm. If a heavy-walled glass tube closed at both ends is partly filled with liquid carbon dioxide, the remainder of the tube being filled with the saturated vapor of carbon dioxide, at room temperature there is equilibrium between the liquid and its vapor. As the temperature of the tube is increased, the density of

Table 16.1
Critical Temperatures, Boiling Points, and Freezing Points of Common Gases

Gas	Freezing point, °C	Boiling point, °C	Critical temperature, °C	Critical pressure, atm
Helium	−272	−268.9	−267.9	2.26
Hydrogen	−259	−252.8	−239.9	12.8
Argon	−189	−186.0	−122.0	41.3
Nitrogen	−210	−195.8	−147.1	33.5
Oxygen	−219	−183.0	−118.8	49.7
Water (steam)	0	100.0	374	218

the liquid decreases while the density of the vapor increases. As a result, the meniscus marking the boundary between liquid and vapor becomes less distinct. At the critical temperature the densities of liquid and vapor become equal and the meniscus disappears.

To liquefy a gas, it is necessary to cool it below its critical temperature and to apply sufficient pressure to produce liquefaction (Table 16.1). Most of the so-called "permanent" gases have low critical temperatures and require the application of large pressures. The boiling and freezing points quoted are for a pressure of one standard atmosphere.

QUESTIONS

1 How can pails of water in a small storage room help prevent the freezing of potted plants or fruit during a cold night?

2 How is regelation involved in packing a snow-ball? Why is very cold snow powdery?

3 Can heat be added to a substance without changing its temperature? Under what circumstances?

4 What is the distinction between a vapor and a gas? How does the pressure of a gas vary as a function of volume when the temperature is kept constant? Of a saturated vapor? Of an unsaturated vapor?

5 How does the pressure of air above a water surface affect the saturated vapor pressure of the liquid and the rate of evaporation?

6 If 100 cm³ of saturated pure ether vapor at 25°C exerts a pressure of 53 cm Hg, what pressure is exerted if the volume is reduced to 50 cm³ without changing the temperature? Explain why the pressure is not increased.

7 Why does 100°C steam produce a much worse burn than 100°C water?

8 What role does surface tension play in the violent boiling, called *bumping*, which often occurs when water is heated to boiling in a smooth-walled container? Why is bumping reduced to placing rough, inert crystals in the container?

9 Why do crystalline solids have sharp melting points while amorphous solids usually soften and melt over a broad temperature range?

10 What determines the time between eruptions of a geyser? Explain what conditions must be satisfied for an eruption to occur.

11 Why does the heat of vaporization increase as the temperature is lowered?

PROBLEMS

1 If the specific heat of snow is 0.5 kcal/(kg)(C°), how much heat energy is required to turn 6 kg of snow at −20°C into steam at 100°C?
Ans. 4380 kcal

2 In an experiment designed to measure the heat of fusion of ice, 25 g of ice at −10°C is dropped into 200 g of water at 30°C contained in a copper calorimeter of mass 100 g. The final temperature is 18°C. What value of the heat of fusion do these data yield? Take the specific heat of ice to be 0.5 kcal/kg.

3 A specimen of copper is heated to 150°C and placed on a block of ice. If 46 g of ice is melted, what is the mass of the copper specimen?
Ans. 0.264 kg

4 A 125-g sphere of iron is heated in an oil bath and then placed on a block of ice, causing 25 g of ice to be melted. What was the temperature of the iron?

5 Four ice cubes at 0°C, each of 30-g mass, are dropped in a 50-g glass containing 290 g of water at 20°C. If losses are negligible, how many grams of ice are left when the water and glass have cooled to 0°C? *Ans.* 45 g

6 Ice has a specific gravity of 0.917. What is the weight of 1 ft³ of ice? How much heat is required to melt 3 ft³ of ice at 32°F?

7 How many kilocalories are required to heat 250 g of mercury from 20 to 357°C and to vaporize it at that temperature? *Ans.* 19.8 kcal

8 How much heat is required to change 15 lb of ice at 12°F to steam at 320°F? The specific heat of ice is 0.50 Btu/(lb)(F°) and that of steam is 0.48 Btu/(lb)(F°).

9 A steel sphere with a mass of 80.87 g is placed in an atmosphere of steam at 100°C and, after

reaching equilibrium, is weighed with the water condensed on it. The observed mass is 81.49 g. What was the original temperature of the sphere?
Ans. 62.3°C

10 What power is required to melt 4 kg of ice at 0°C and to raise the temperature of the water to 100°C in 450 sec?

11 Steam at 100°C enters a radiator, and water at 80°C leaves. (*a*) How many grams of steam are required per hour if the radiator delivers 3×10^3 kcal/hr? (*b*) What fraction of the heat is associated with the cooling of the water?
Ans. (*a*) 5,360 g; (*b*) 3.6 percent

12 Two ice cubes, each of mass 30 g and at 0°C, are placed in a 50-g glass which contains 220 g of water. If the glass and the water are initially at 25°C, find the final temperature of the system and the mass of ice remaining, if any.

13 When 50 g of ice at 0°C is dropped into 190.7 g of water at 30°C contained in a 100-g copper calorimeter, the final temperature is 8°C. Find the heat of fusion of ice.   *Ans.* 80 kcal/kg

14 The condenser at an electric generating plant receives steam at 100°C and changes it to water at 50°C. The cooling is done by water, which enters at 10°C and leaves at 40°C. How much cooling water is required for each kilogram of steam condensed?

15 Fifty grams of ice at −20°C is placed in a calorimeter of 25 g water equivalent which initially contains 275 g of water at 40°C. If the specific heat of ice is 0.5 kcal/(kg)(C°), find the heat required to warm the ice to 0°C, the heat required to melt the ice at 0°C without changing its temperature, and the final temperature of the calorimeter and its contents.
Ans. 0.5 kcal; 4 kcal; 21.4°C

16 How many pounds of coal must be burned to produce 800 lb of steam at 300°F if one starts with water at 50°F and if 80 percent of the heat of combustion of the coal goes into the water? Take the heat of combustion of coal to be 11,000 Btu/lb and the specific heat of steam to be 0.5 Btu/(lb)(F°).

17 When 20 g of steam at a temperature of 100°C was passed into an aluminum calorimeter con-

taining an unknown mass of water, the initial temperature of the water was 6°C, its final temperature was 34°C, and the mass of the calorimeter was 300 g. Find the initial mass of water in the calorimeter.   *Ans.* 367 g

18 When 10 g of steam at 100°C is condensed in a 200-g calorimeter containing 150 g of water, the calorimeter and water were initially at 10°C and the final temperature is 40°C. Find the heat lost by the steam, the heat gained by the calorimeter, and the specific heat of the calorimeter.

19 Fifty grams of steam at 100°C was passed into a 400-g copper calorimeter. The calorimeter contained 600 g of water and a certain amount of ice, all at 0°C. The final temperature was 20°C. How much ice was in the calorimeter?
Ans. 183 g

20 A 150-g copper calorimeter contains 350 g of water at 10°C. How many grams of steam at 100°C must be added to raise the temperature to 75°C?

21 A calorimeter contains 250 g of water and 50 g of ice at 0°C. The mass of the calorimeter is 150 g, and its specific heat is 0.12 kcal/(kg)(C°). How many grams of steam at 100°C must be introduced into the water to make the final temperature 40°C?   *Ans.* 27.9 g

22 How much steam at 120°C must enter a calorimeter to turn 150 g of ice at −10°C to water at 35°C? The specific heats of ice and steam are both approximately 0.5 kcal/(kg)(C°). Assume that negligible heat is needed to raise the temperature of the calorimeter.

23 A calorimeter of 30 g water equivalent contains 270 g of water at 10°C. Into this calorimeter 200 g of ice at −50°C is dropped; then 10 g of steam at 120°C is introduced. The specific heats of steam and ice are, respectively, 0.48 and 0.50 kcal/(kg)(C°). Find the final temperature of the system. What remains in the calorimeter?
Ans. 0°C; 336.2 g water and 143.8 g ice

24 If 450 g of ice at 0°C and 50 g of steam at 100°C are mixed with 480 g of water at 30°C in a calorimeter with a water equivalent of 20 g, find the final temperature.

25 It takes 540 kcal, or 2.26×10^6 J, to change 1 kg of water at 100°C to steam at the same temperature at one standard atmosphere pressure. At 100°C, 1 kg of water occupies 0.001 m³, and 1 kg of steam 1.671 m³. Find the external work done in evaporating 1 kg of water

at 100°C and 1 atm pressure. Find the change in the internal energy of the system during the evaporation. *Ans.* 1.69 × 10^5 J; 2.09 × 10^6 J

26 If the specific heat of water vapor at constant pressure is roughly 0.5 kcal/(kg)(C°) and the heat of vaporization at 100°C is 540 kcal/kg, calculate the approximate heat of vaporization of water at 90°C. (*Hint:* By conservation of energy, it should take the same amount of energy to change 1 g of water at 90°C to 1 g of water vapor at 100°C whether the evaporation is done first or last.) Compare your value with that given in Appendix Table B.4. Why is your value only approximate?

CHAPTER SEVENTEEN
HEAT TRANSFER

TO WARM WATER, to melt a solid, or to vaporize a liquid we increase the internal energy of the system by adding heat. In a calorimetry experiment we can find the specific heat of a material by dropping a known mass of the material at high temperature into a known mass of water and determining the equilibrium temperature. In all these cases heat is conveyed from one body to another or from one part of a system to another. In this chapter we study the processes by which these heat transfers are accomplished.

17.1 HEAT-TRANSFER PROCESSES

As we saw in Sec. 15.6, heat is energy in transit from one body to another, or from one part of a substance to another, *by virtue of temperature differences*. There are three natural processes by which heat is transported from one point to another.

1 *Conduction is the process by which heat energy is transferred from particle to particle by collisions or direct interactions.*

When the bowl of a spoon is placed in a hot liquid, heat is conveyed by conduction from the bowl to the handle. Conduction occurs in solids, liquids, and gases. It is the only method by which heat is transferred through opaque solids. The atoms in the hotter part of the material vibrate with greater amplitude than those in the colder part. Atoms with greater thermal energy transfer part of this energy to their neighbors, which in turn transmit energy to their own neighbors. In many materials this exchange occurs through interatomic interactions, but in metals free electrons are the primary agency through which heat energy is transported. These same electrons are responsible for electric conductivity, and it is for this reason that the best thermal conductors are also the best electric conductors.

2 *Convection is heat transfer by the actual movement of the heated material itself.*

Steam produced in a boiler can be transported to radiators throughout a building. The transfer of heat by movement of the hot steam is a convective process. Convection is ordinarily the most important heat-transfer process for liquids and gases. The distribution of heat in practically all houses and buildings is achieved by means of convection currents of heated air, hot water, or steam.

3 *Radiation consists of electromagnetic waves which transmit energy from a source to an absorber.*

The earth's primary source of energy is the radiation from the sun. Part of this energy is in the form of visible light, but much more of it comes in the form of infrared rays which have wavelengths too long for the eye to see. Radiation traverses a vacuum as well as transparent media.

The atoms and molecules of a body at a higher temperature emit energy as radiation, which passes to a cooler body, where it is absorbed and reconverted to thermal energy. Radiation is of primary importance for bodies at high temperatures and for transmission across regions in which there is no material medium. For temperatures only 100 C° or so above or below room temperature, radiation is relatively unimportant compared to conduction and convection for most material mediums.

If a steam radiator provides heat in a room, all three processes play a role in transferring the heat. Conduction brings the heat from the steam to the outside of the radiator. From here the primary heat transfer is by convection, although radiation does play a role in distributing the heat. The first evidence that convection is of principal importance is the fact that a person several feet from the radiator feels equally warm on both sides. It is characteristic of radiation transfer that the energy moves in straight lines from the heat source to the point where it is absorbed. The familiar "roast one side, freeze the other" situation which arises at an outdoor bonfire on a cold night is typical of radiative transfer.

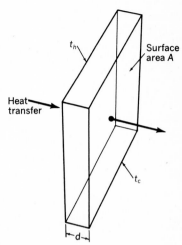

Figure 17.1
The heat transferred per unit time through a slab depends on its thickness, area, material, and on the temperature difference between its surfaces.

17.2 CONDUCTION

If heat is to be conducted through a material, there must be a difference in temperature between two regions. If a constant temperature difference is maintained between the two faces of a slab of material (Fig. 17.1), the heat Q conducted through this slab is proportional to the area A of the slab, the temperature difference $t_h - t_c$ (h for hotter, c for cooler) between the two faces, and the time the temperature difference is maintained. It is inversely proportional to the thickness d of the slab.

$$Q = \frac{kA(t_h - t_c)\ \text{time}}{d} \qquad (17.1)$$

The proportionality constant k for a given material is called the *coefficient of thermal conductivity*. Coefficients for a number of materials are listed in Table 17.1. In the international metric system k *is the quantity of heat energy in joules which passes through one square meter of area in one second when a temperature difference of one Celsius degree is maintained across a thickness of one meter.* Older tables usually list k in cal-cm/(cm²)(sec)(C°). In the British system of units, k is numerically equal to *the number of Btu conducted per hour through an area of one square foot when a temperature difference of one Fahrenheit degree is maintained across a thickness of one inch.*

In problems of thermal conductivity it is important to use the actual temperatures of the slab faces in Eq. (17.1). For example, if one calculates the heat conducted through a glass window when the indoor

Table 17.1
Thermal Conductivities at 20°C
1 J-m/(sec)(m²)(C°) = 1 W/(m)(C°)

Substance	Thermal conductivity k, W/(m)(C°)*
Aluminum	230
Air	0.023
Brass	100
Concrete	1.7
Copper	400
Cork	0.046
Glass	0.63
Ice	1.7
Rock or glass wool	0.038
Sand (white dry)	0.39
Silver	418
Soil (dry)	0.14
Steel	46
Water	0.60
Wood (across grain)	0.13
Zinc	110

* To obtain k in Btu-in./(ft²)(hr)(F°), multiply by 6.9; in cal/(cm)(sec)(C°) divide by 419.

temperature is 22°C and the outdoor temperature is −10°C, one obtains a heat loss which is far too great if one puts 22°C as t_h and −10°C as t_c. Actually, the temperature of the inner surface of the glass under conditions of this sort might well be 12 C° lower than the 22°C of the air in the room, while the outer face of the glass may similarly be some 12 C° or more warmer than the outside air. There are major temperature drops in the thin air layers adjacent to each side of the glass. Equation (17.1) gives acceptable values of Q only when the actual surface temperatures are used.

An apparatus for measuring the coefficient of thermal conductivity of a good conductor such as copper is represented in Fig. 17.2. It consists mainly of a rod, the coefficient of thermal conductivity of which is to be determined. This rod is typically about 30 cm long and has a diameter of about 3 cm. On one end is soldered a copper box through which steam can be passed to make the temperature of that end of the rod about 100°C. Around the other end is soldered a coil of copper tubing through which cold water is circulated. The temperature of this water as it enters is found by means of the thermometer T_3, and its temperature as it leaves is read on the thermometer T_4. The difference in temperature between incoming and outgoing water multiplied by the mass of water gives numerically the quantity of heat that has flowed down the rod. The coil and the rod are carefully insulated with felt to prevent heat losses. Small holes drilled at right angles to the axis of the rod contain thermometers T_1 and T_2 by which the fall of temperature across the thickness d is measured.

Example

In measuring the thermal conductivity of a rod (Fig. 17.2), the following data were obtained. Temperature of incoming water = 20.0°C; temperature of outgoing water = 30.0°C; temperature t_h read by thermometer T_1 = 80.0°C; t_c = 60.0°C; distance between thermometers = 10 cm; area of rod = 20 cm²; mass of water flowing through box = 650 g in a time of 180 sec. Find the coefficient of thermal conductivity.

Figure 17.2
Measuring the thermal conductivity of a
solid rod.

Figure 17.3
The thermal conductivity of sulfur changes when the crystalline form changes.

Heat gained by water = mass of water × temperature change × sp ht
= 0.650 × 10 × 1 = 6.50 kcal = 27,200 J

$$Q = \frac{kA(t_h - t_c)}{d} \times \text{time}$$

$$27,200 = k \times 20 \times 10^{-4} \times \frac{20}{0.10} \times 180 = 72k$$

$$k = \frac{27,200}{72}$$

$$= 380 \text{ J/(m)(sec)(C°)} = 380 \text{ W/(m)(C°)} \qquad ////$$

The ratio of the temperature change Δt to the thickness over which it occurs is called the *temperature gradient*. In the example above it is given by $\Delta t/d = (t_h - t_c)/d = (80 - 60)\text{C°}/0.10 \text{ m} = 200 \text{ C°/m}$.

The thermal conductivity of carbon increases with the temperature. On the other hand, the thermal conductivities of metals ordinarily decrease with an increase of temperature. The thermal conductivity changes abruptly when a substance such as sulfur (Fig. 17.3) passes from one crystalline state to another.

The walls of a house or a refrigerator often consist of two or three layers of different materials. To compute the heat conducted through such a system of layers, we make use of the fact that the only heat which can be conducted through the second layer is that which passes through the first layer, and so forth. When equilibrium is established, the heat conducted per unit area per second through each of the layers is the same.

Example

The wall of a shed in which ice is stored consists of an outer layer of wood 2 cm thick and an inner layer of rock wool 3 cm thick. Find the heat conducted through 50 m² of the wall in 1 hr when the outer wood surface is at 20°C and the inner rock-wool surface is at 5°C. Also find the temperature of the wood-rock-wool interface.

Let t be the temperature of the interface. The heat conducted through both layers is the same, so that

$$Q = \left[\frac{ka\,\Delta t(\text{time})}{d}\right]_{\text{wood}} = \left[\frac{ka\,\Delta t(\text{time})}{d}\right]_{\text{rock wool}}$$

$$= \frac{0.13 \times 50(20 - t) \times 3,600}{0.02} \text{ J}$$

$$= \frac{0.038 \times 50(t - 5) \times 3,600}{0.03} \text{ J}$$

Solving for t yields $t = 17.55°C$, the temperature of the interface. If we put this value into the equation for Q, we obtain

$$Q = \left[\frac{0.13 \times 50(20 - 17.55) \times 3,600}{0.02}\right]_{\text{wood}} = 2.9 \times 10^6 \text{ J}$$

It may be of interest to check this answer by showing that the same heat is conducted through the rock wool.   ////

17.3 CONVECTION

The transfer of heat from one place to another by the motion of the heated substance is known as convection. The convection is said to be *forced* if the heated material is moved by means of fans or pumps, as is the case in a forced-air heating system in a home where fans blow the hot air from the furnace throughout the rooms. *Natural* convection arises from the change in density that takes place when a fluid is heated.

For example, when a gas or a liquid is heated, it expands and becomes lighter than the cold fluid. When water is heated on a stove, the liquid at the bottom is warmed, expands, and rises to the top. Meanwhile the cooler and denser water sinks. The flows thus set up in the liquid are known as *convection currents*. There is no simple formula which permits the calculation of heat transfer by convection in the general case. The heat transferred to or from a surface at one temperature in contact with a fluid at a different temperature depends on many factors, including the geometry and orientation of the surface as well as properties of the fluid.

17.4 RADIATION

Radiation from the sun passes through millions of miles of empty space until eventually it falls upon absorbing matter, where it is transformed into the mechanical energy of random motion. Sunlight falling on an object warms it above the temperature of the surrounding air. In like manner, radiation from a roaring fire warms objects on which it falls.

All bodies, whether cold or hot, radiate. If two bodies are exactly alike in every way except that one is at a higher temperature, the hotter body radiates more energy than the colder one. When a body is at a higher temperature than its surroundings, it radiates more heat than it receives; when it is at a temperature lower than that of its surroundings, it receives more energy than it radiates. The energy radiated by a body increases rapidly as its temperature is raised; it is proportional to the fourth power of the absolute temperature.

The rate at which heat is radiated depends not only on the temperature of the body but also on the character and area of the radiating surfaces. Some surfaces are good radiators, others poor. Generally, polished surfaces are poor radiators; rough, blackened ones are good radiators.

There is an intimate relationship between the rate at which a surface radiates energy and the rate at which the same surface under the same conditions absorbs heat. Good radiators are also good absorbers of heat. That this must be true can be seen easily. Consider a polished reflecting sphere and a rough black sphere in an insulated enclosure. Eventually they reach the same temperature as the enclosure (Fig. 17.4). Radiation from the enclosure falls upon each at the same rate, but the polished surface reflects most of the incident energy, while the blackened sphere absorbs the energy striking its surface. Therefore, the blackened surface is receiving energy at a more rapid rate than the polished one. However, once the bodies and the enclosure are all at the same temperature, they remain at the same temperature. Therefore, the blackened body, which absorbs energy at a greater rate, also radiates at a greater rate. Under equilibrium conditions each body is absorbing energy at the same rate at which it is radiating.

The arguments of the preceding paragraph apply to the situation in which equilibrium has been reached and all bodies are at the same temperature. If the blackened sphere and the polished one are placed in sunlight on a bright day, the blackened sphere attains a substantially higher temperature than the polished one. In this case the bodies and their surroundings (which must include the sun) never attain identical temperatures. However, it is still true that *a good absorber is a good radiator, and a poor absorber a poor radiator.*

Figure 17.4
Two spheres, one black and the other reflecting, remain at the enclosure temperature once thermal equilibrium is reached.

17.5 BLACKBODY RADIATION

The best emitter of radiation is necessarily the best absorber. A perfect absorber of radiation is known as *an ideal blackbody. A blackbody is one which absorbs all the radiation incident upon it.* Although no body with a perfect absorbing surface exists, a surface covered with a thick coating of lampblack absorbs about 99 percent of the heat radiation incident upon it. An almost perfect blackbody can be produced by making a small opening in the wall of a rough cavity (Fig. 17.5). Radiation incident on this small hole enters the cavity, in which it is reflected from wall to wall until completely absorbed. Meanwhile, each point of the surface of the cavity emits radiation which falls on other points of the cavity, where part is reflected and part absorbed. Radiation of all wavelengths is reflected back and forth in this cavity until there is a uniform density of radiation throughout it. If the temperature of the wall is increased, the radiation level is correspondingly increased.

If radiation emerging from the opening of the cavity is examined with suitable instruments, a wide range of wavelengths is found to be present. The energy associated with each wavelength can be measured. The distribution of the energy in this spectrum changes with temperature. The total radiation emitted from such a cavity, called *blackbody radiation*, increases rapidly with temperature. In 1879 Stefan analyzed experimental data obtained by Tyndall and concluded that if R stands for the energy radiated per unit area per second by a blackbody,

$$R = \sigma T^4 \tag{17.2}$$

where σ is a constant equal to 5.67×10^{-8} W/(m²)(K)⁴ and T is the absolute temperature. This is known as *Stefan's law* (or the Stefan-Boltzmann law). If the temperature of a blackbody is doubled, the rate at which energy is radiated is increased by a factor of 16.

Figure 17.5
An ideal blackbody is formed by a cavity of rough internal surface with a very small opening.

Figure 17.6
Blackbody radiation at several temperatures, showing the energy emitted per unit time in a small wavelength range as it varies with wavelength.

If the radiator is not a blackbody, it radiates less than a blackbody of identical size and temperature. If the *absorption factor* α represents the fraction of incident radiation absorbed by the surface of the radiator, the energy radiated per unit area per second is

$$R = \alpha\sigma T^4 \qquad (17.2a)$$

Thus a body which absorbs only half the radiation incident upon it has $\alpha = 0.5$, and it emits only half as rapidly as a blackbody of the same size and shape. For copper α is roughly 0.3; thus a copper radiator radiates only 0.3 as much power per unit area as a blackbody at the same temperature.

As the temperature of a blackbody is increased, the power emitted in every wavelength interval increases, but not in the same proportion. The curves of Fig. 17.6 show the relation between intensity emitted in a constant-wavelength interval and wavelength. As the temperature is increased, the wavelength at which the maximum intensity is radiated moves toward smaller values. If λ_m is the *wavelength at which the radiation is maximum* for the absolute temperature T,

$$\lambda_m T = 2.898 \times 10^{-3} \text{ m-K} \qquad (17.3)$$

This relation is known as *Wien's displacement law*. For temperatures up to 600°C or about 900 K, essentially all the radiation is in wavelengths too long for the human eye to see. However, by the time the temperature reaches 1000 K, the blackbody emits visible radiation (Fig. 17.6) and the body appears dull red. As the temperature is increased, more energy is emitted in the blue and violet regions of the visible spectrum and the color of the body changes from red to orange to yellow and eventually to white.

17.6 THE SURFACE TEMPERATURE OF THE SUN

If we assume that the sun is roughly a blackbody, we can estimate its surface temperature by applying either Stefan's radiation law or the Wien displacement law. Let us consider first the use of Stefan's law. Measurements at the earth reveal that a surface perpendicular to the sun's rays when the sun is directly overhead receives roughly 1,400 W/m². The energy received per unit area per unit time on a surface perpendicular to the sun's rays is called the *solar constant*. The solar constant of 1,400 W/m² is equivalent to about 2 cal/cm²-min. If we assume that the sun's energy is emitted equally in all directions, the total power in watts emitted is given by 1,400 times the area in square meters of a sphere of radius equal to the distance from the sun to the earth. The power radiated by the sun is therefore

$$P = 1,400 \text{ W/m}^2 \times 4\pi \times (1.5 \times 10^{11} \text{ m})^2$$
$$= 4 \times 10^{26} \text{ W}$$

When we divide this by the area of the sun's surface, we obtain, as the energy emitted per square meter per second, the value

$$R = \frac{4 \times 10^{26} \text{ W}}{4\pi \times (7 \times 10^8 \text{ m})^2} = 6.5 \times 10^7 \text{ W/m}^2$$

If we insert this in Stefan's law, assuming the sun is a blackbody, we obtain

$$R = \sigma T^4 \qquad \text{or} \qquad 6.5 \times 10^7 = 5.67 \times 10^{-8} T^4$$
$$T^4 = 1.14 \times 10^{15}$$
$$T = 5800 \text{ K}$$

If the spectral distribution of the radiation from the sun is studied and intensity is plotted as a function of wavelength, it is found that maximum energy is radiated at approximately 0.49 μm. If we put this value in Wien's displacement law, we obtain

$$0.49 \times 10^{-6} \, T = 2.898 \times 10^{-3}$$
$$T = 5900 \text{ K}$$

Both these values have been obtained on the assumption that the sun is a blackbody. Actually, this is not the case. The curves of the sun's radiation fall substantially below the blackbody curves for 5900 K in the short-wavelength region and are substantially above in the far-infrared region.

Our best methods for measuring very high temperatures involve the use of the radiation laws. Instruments called *optical pyrometers* are used to compare the radiation from the body whose temperature is to be measured with the radiation from a calibrated source.

17.7 NEWTON'S LAW OF COOLING

Whenever a body is at a somewhat higher temperature than its surroundings, the rates at which energy is lost due to conduction, convection, and radiation are all roughly proportional to Δt, the temperature difference between the body and its surroundings. Therefore, *the rate at which the body loses temperature is proportional to the difference between its temperature and that of its surroundings.* This is known as *Newton's law of cooling*. Thus, a cup of coffee loses temperature twice as rapidly when it is 80 C° above its surroundings as when it is 40 C° above.

If the body is somewhat cooler than its surroundings, Newton's law is still applicable. In this case both the "rate of cooling" and Δt are negative, since the body gets warmer and its temperature is lower than that of the surroundings.

Example

A pan filled with hot food cools from 92 to 88°C in 2 min when the room is at 20°C. How long will it take to cool from 71 to 69°C?

The average of 92 and 88°C is 90°C, which is 70 C° above room temperature. Under these conditions the pan cools 4 C° in 2 min or 2 C°/min.

$$\frac{\text{Change of temperature}}{\text{Time}} = k\,\Delta t$$

$$\frac{4\text{ C}°}{2\text{ min}} = k(70\text{ C}°) \qquad\qquad \text{(A)}$$

The average of 69 and 71°C is 70°C, which is 50 C° above room temperature. k is the same for this situation as for the original.

$$\frac{2\text{ C}°}{\text{Time}} = k(50\text{ C}°) \qquad\qquad \text{(B)}$$

If we divide (A) by (B), we have

$$\frac{4\text{ C}°/2\text{ min}}{2\text{ C}°/\text{time}} = \frac{k(70\text{ C}°)}{k(50\text{ C}°)}$$

$$\text{Time} = 1.4\text{ min} \qquad\qquad ////$$

17.8 PLANCK'S QUANTA AND THE PHOTON

Near the end of the nineteenth century Max Planck was endeavoring to develop an adequate theory for the wavelength distribution of blackbody radiation (Fig. 17.6). He discovered that he could account quantitatively for blackbody radiation curves if he assumed that energy is radiated in small packets, which he called *quanta* and for which G. N. Lewis later introduced the name *photons*. According to Planck's theory, the energy associated with each *quantum* or *photon* is given by

$$\text{Energy of photon} = W = h\nu \qquad\qquad (17.4)$$

where ν is the frequency of the radiation and h is a universal constant equal to 6.6262×10^{-34} J-sec. In honor of its discoverer, h is called *Planck's constant*.

It was in 1901 that Planck proposed his startling new idea, which now plays a vital role in our understanding of physics in atomic dimensions. Although Planck's theory gave excellent agreement with experimental measurements on blackbody radiation, many people regarded the agreement as fortuitous and considered the proposal that radiant energy comes in chunks or quanta as unacceptable. The objections were based on the misconception that light and other electromagnetic radiation were purely classical wave phenomena. In the early 1800s it was established that light exhibits interference, diffraction, and polarization (Chaps. 27 and 28), which are characteristic wave properties. Virtually a century of experimentation had established that radiation does indeed have wave characteristics. In the 1860s Maxwell had developed the theory of electromagnetic fields and proposed that light consisted of electromagnetic waves, a hypothesis which was strongly supported in 1887 by the experimental work of H. Hertz. Thus it is not surprising that many physicists would regard Planck's proposal that radiation had particle properties as erroneous and retrograde. According to Einstein's extension of Planck's idea, a quantum is emitted by one atom (or oscillator) and absorbed by another as though it were a particle with a well-defined position. On the other hand, a wave emitted from a source spreads out, and by the time it is a centimeter from the source, it is spread over a volume millions of times that of a single atom.

Now we believe that radiation is neither a classical wave phenomenon nor a classical particle phenomenon. It has wave characteristics under some conditions and has particle properties under others. Classical waves and particles are idealizations which are extremely useful, but electromagnetic radiation is neither a pure wave motion nor exclusively particlelike. Instead it exhibits some aspects of each, as we shall see in subsequent chapters.

QUESTIONS

1. A wooden door and a metal doorknob are in thermal equilibrium at 5°C. Why does the metal doorknob feel much colder than the wood?
2. Why are the contents of an automobile radiator less likely to freeze on a cold, clear night if the automobile is under the roof of a carport, even though it is completely exposed to the cold air?
3. When a white plate, a piece of blackened steel, and a piece of reddish copper are heated in a furnace, why do they all glow red as they first begin to be incandescent?
4. A piece of paper tightly wrapped around a copper rod may be held briefly in a bunsen flame without catching fire, while the same paper on a wooden rod would ignite quickly. Why does this happen?
5. Why does an automobile with closed windows get so hot inside when it is parked in the sun?
6. Why wear woolen mittens to keep your hands warm when the thermal conductivity of wool is far greater than that of air?
7. Steam radiators are seldom painted black, although a black radiator should radiate better than one painted a light color. Do you think blackening radiators would increase their effectiveness substantially? Why?
8. If a white china plate with black decorative markings is heated to incandescence, which will be brighter, the markings or the white china? Why?

PROBLEMS

1. If the bottom of an aluminum pan is 2 mm thick and the temperature of the inner surface is 100°C while that of the outer surface is 102°C, find the heat conducted through an area of 100 cm² in 30 sec. *Ans.* 69 kJ
2. The heat exchanger for a reactor is designed to transfer 1.5 megawatts (MW) through a steel conductor 2 mm thick. The temperature on one side of the steel is 140°C, and on the other side 138°C. What area is required?
3. A glass window has an area of 2.5 m² and a thickness of 0.4 cm. The outer surface is at 10.6°C, and the inner surface is at 12.2°C. How much heat flows through the window each hour? *Ans.* 2.27 MJ
4. How many calories will be lost each hour through a glass window 100 cm by 80 cm by 4 mm thick if the inner surface is at 32°F and the outer surface is at 27°F?

5 A brass rod 20 cm long and 1.5 cm² in cross-sectional area is thermally insulated. One end is kept at 100°C in a steam bath, and the other end is cooled by circulating water, as shown in Fig.17.2. In 5 min, 100 g of water enters at 15°C. Two thermometers 10 cm apart along the rod read 80 and 55°C, respectively. Find the heat conducted down the rod each minute and the temperature of the water as it leaves.

Ans. 54 cal/min; 17.7°C

6 A glass window is 0.5 cm thick, 150 cm wide, and 100 cm high. The temperature of the inner surface is 14.1°C, and that of the outer surface is 12.9°C. How much heat is lost per hour through the window?

7 How much water would be evaporated per hour per square meter by the heat which flows through a steel boiler plate 0.3 cm thick when there is a difference in temperature of 1.5 C° between the faces of the plate? Ans. 36.6 kg

8 Find the coefficient of thermal conductivity for asbestos paper if 10 cal flows each second through a slab 2 mm thick and 50 cm² in area when the temperature of one face is maintained at 98°C and that of the other at 36°C.

9 A brass bar has a length of 60 cm and a cross section of 5 cm². One end is kept in a mixture of water and ice at 0°C, consisting initially of 5 g of ice and 25 g of water. The other end is maintained at a uniform higher temperature until all the ice is melted, and then the temperature is raised so as to maintain the same difference in temperature between the two ends of the bar. Neglect losses of heat to other bodies. How many calories are needed to melt the ice? What is the difference in temperature between the two ends of the bar if the ice melts in 15 min? How soon after the ice has melted will the temperature of the water be 80°C?

Ans. 400 cal; 22.3°C; 90 min

10 A brass bar has a length of 20 cm and a cross-sectional area of 2.5 cm². One end is kept in steam at 100°C, and the other in ice at 0°C. Neglecting losses due to radiation, find the number of grams of ice melted in 15 min.

11 Superheated steam is flowing through a steel pipe that has a length of 10 m. The internal diameter of the pipe is 5 cm, and its external diameter is 5.5 cm. If the outside surface of the pipe has a temperature of 106°C and the inner surface a temperature of 110°C, how much heat escapes in 20 min? (Assume the wall of the pipe is approximately a thin slab 0.25 cm thick and of width equal to the average circumference of the pipe.) Ans. 35,000 kcal

12 A large meat-storage refrigerator has a window with an area of 0.3 m². If the window is 1.2 cm thick and the temperature of the inner glass surface is 12°C, how many calories per day are lost through the glass when the outer surface has a temperature of 15°C?

13 A blackbody radiates 20 W when its temperature is 500 K. Find the power radiated if the temperature is raised to 1000 K, and find the wavelength at which the radiation rate is greatest at 1000 K. Ans. 320 W; 2.898 μm

14 A blackbody initially at 27°C is heated to 277°C. The total radiation emitted at the higher temperature is how many times that emitted at the lower temperature? What is the wavelength of the maximum radiation at the higher temperature?

15 If the surface temperature of a star is 5500 K, what is the wavelength of its radiation maximum? What would the surface temperature of a star have to be if it were to have its radiation maximum in the blue at a wavelength of 400 nm?

Ans. 527 nm; 7250 K

16 The surface of a radiation pyrometer receives 0.5 cal/sec from a furnace whose temperature is 627°C. How many calories per second will it receive when the temperature of the furnace is raised to 927°C?

17 If the fireball of a nuclear weapon can be approximated at some instant as a blackbody of 0.5-m radius with a surface temperature of 10^6 K, find the power radiated and the wavelength at which maximum energy is radiated.

Ans. 1.8×10^{17} W; 2.9 nm

18 To what temperature must a blackbody be raised in order to double its total radiation if its original temperature is 927°C?

19 The filament in a tungsten lamp radiates energy at the rate of 150 W when operating at a temperature of 2400 K. A second lamp operating at a temperature of 3000 K radiates five times as much energy. What is the ratio of the area of the second filament to that of the first? What

is the wavelength of the maximum power radiated at each temperature?

Ans. 2.05; 1.20 μm; 0.966 μm

20 Find the heat conducted in 1 min through an area of 2 m² of a layer consisting of 3 cm of cork and 5 cm of wood. The outer surface of the wood is at 25°C, and the opposite cork surface is at 0°C. What is the temperature of the cork-wood interface?

21 A copper-clad, stainless-steel pan has a conducting area of 200 cm² and a 0.1-mm layer of copper on 1 mm of stainless steel. The coefficient of thermal conductivity of stainless steel is 0.05 cal/(cm)(sec)(C°). (*a*) Find the temperature at the copper–stainless steel interface when the outer surface is at 110°C and the inner surface is at 108°C. (*b*) How many calories are conducted through the pan each second?

Ans. (*a*) 109.99; (*b*) 199 cal/sec

22 If it takes 10 min for a glass of water to warm from 39.5 to 40.5°F in a room at 70°F, approximately how long will it take for its temperature to rise from 49 to 51°F?

23 A glass of 20 g water equivalent contains 270 g of water and 10 g of ice. If it takes 30 min for the ice to melt, approximately how much longer will it take for the temperature to rise to 2°C if the surroundings remain at 20°C? Roughly how long will it take for the temperature to rise from 9.5 to 10.5°C? *Ans.* 23.7 min; 22.5 min

CHAPTER EIGHTEEN
THERMODYNAMICS

IN CHAP. 7 we learned that energy is the ability or capacity to do work. A body at a high temperature has internal energy. How can part of this internal energy be used to do useful work? What fraction of the energy is available? The answers to the first of these questions—and there are several—tell us how heat engines operate. The maximum possible efficiency of any particular heat engine is limited by the answer to the second question. In this chapter we consider the operation and efficiencies of various types of heat engines.

18.1 HEAT AND WORK

The heat engines that play such a major role in modern life depend on the transformation of heat energy into work. The hot gases of a jet engine push forward on a moving airplane and thereby do work. An automobile engine drives an automobile only when it is supplied constantly with heat from burning gasoline in the cylinders.

Practically all heat engines depend upon the work done by expanding gases to achieve mechanical work output. The simplest situation to discuss, although a difficult one to use for a heat engine, is that in which a gas expands at *constant pressure* (Sec. 15.12). In this case the work done by the gas is the product of the pressure p and the change in volume $V_2 - V_1$, where V_1 and V_2 represent the volume occupied by the gas before and after the expansion, respectively. If the pressure varies during the expansion, the process may be described by a pV diagram such as Fig. 18.1. The work W done during a small volume increment ΔV is given by Eq. (15.8):

$$W = p\,\Delta V \qquad (18.1)$$

where p is the average pressure appropriate for the small change ΔV. This work is represented by the hatched area in the figure. Clearly, any volume change $V_2 - V_1$ can be regarded as the sum of many small ΔV increments, and the total work done by the gas in producing the volume change $V_2 - V_1$ is represented by the area $ABba$.

The work performed by the expanding gas comes from one or both of two sources: heat supplied to the gas and the internal energy of the gas. In the *thermodynamics* of heat engines we are interested in energy transformations involving the heat supplied to the "working substance" of the engine (usually a gas), the work done by this substance, and changes in its internal energy.

Figure 18.1

The work done by a gas in expanding a small amount ΔV is $p\,\Delta V$, where p is the pressure; when the gas in the cylinder below expands at variable pressure, the work done by the gas is the shaded area under the pV diagram.

When the principle of conservation of energy is used to include heat energy specifically, we may state it in the form: *When heat energy is added to a system, the energy appears either as increased internal energy of the system or as external work done by the system.* This statement of conservation of energy is known as *the first law of thermo-dynamics.*

In the usual case that the working substance is a gas, the first law of thermodynamics requires that

Heat added to gas = increase in internal energy of gas
+ mechanical work done by gas

If ΔQ represents a small amount of heat added to the system,

$$\Delta Q = mc_V \Delta T + p \Delta V \qquad (18.2)$$

where $mc_V \Delta T$ is the increase in internal energy, and $p \Delta V$ is the external mechanical work done by the gas.

18.3 THERMAL EFFICIENCY

In practical heat engines, which achieve mechanical work output through the expansion of hot gases, the variables pressure, volume, and temperature change simultaneously and in a complex manner. To gain quantitative understanding of the transformation of heat into mechanical energy, it is simplest to imagine an engine that uses an ideal gas as working substance and that holds one of these variables constant during each portion of its heat cycle.

As an example of such a cycle, consider a heat engine in which hot gases push a piston back (Fig. 18.2) at a constant high pressure p_1. During this *isobaric process* the gas does the work represented by the area *ABba*. Heat must be supplied to the gas during this expansion. Next let the engine hold the volume of the gas constant and reduce the pressure to p_2 by lowering the temperature of the gas. During this *isovolumic process* no work is done, but heat is removed from the gas. Then let the gas be compressed at constant pressure to its original volume V_1; the work done *on the gas by the piston* is represented by the area *DCba*. During this compression heat must be removed from the gas. Finally, keeping the volume constant at V_1, let the engine heat the gas to its original pressure p_1. This completes a cycle which may be repeated over and over. To find the net useful work done by the gas during one cycle, the work done *on the gases by the piston* must be subtracted from the work done *on the piston by the gases*. In Fig. 18.2 the net work done by the gas is represented by the area *ABCD*. In general, *work done by the gas on the piston* is taken to be *positive*, while *work done on the gas by the piston is negative.* Similarly, heat energy supplied to the gas is taken to be positive, while that removed from the gas is negative.

An important characteristic of any heat engine is its efficiency: *The thermal efficiency of a heat engine is the ratio of the net external work done by the engine to the total heat energy supplied.* In any heat

Figure 18.2
The work done on the piston below *by the gas* in expanding from *A* to *B* is taken as positive, while the work done *on the gas* in compressing it from *C* to *D* is negative.

engine only a fraction of the heat developed by the combustion of fuel is converted into work. In a steam engine much of the heat passes out with the exhaust steam. Gasoline engines are more efficient, but, even so, provision must be made for removing the heat not converted to mechanical work.

Example

If 0.3 kg of coal, which has a heat of combustion of 7600 kcal/kg, is burned in an engine which raises 2,000 kg of water 40 m, what fraction of the heat from the coal has been converted into useful work?

$$\text{Heat supplied} = 0.3 \text{ kg} \times 7600 \text{ kcal/kg} \times 4{,}186 \text{ J/kcal}$$
$$= 9.55 \times 10^6 \text{ J}$$

$$\text{Work done} = \text{weight} \times \text{height} = mgh$$
$$= 2{,}000 \text{ kg} \times 9.8 \text{ N/kg} \times 40 \text{ m} = 7.84 \times 10^5 \text{ J}$$

$$\text{Efficiency} = \frac{\text{mechanical work performed}}{\text{heat supplied}}$$

$$= \frac{7.85 \times 10^5}{9.55 \times 10^6} = 0.082 = 8.2 \text{ percent} \qquad \text{////}$$

18.4 ISOTHERMAL AND ADIABATIC PROCESSES

Consider a gas in a cylinder with walls which are good conductors of heat and with a movable piston at one end. Let the gas expand slowly, maintaining a constant temperature. To do this it is necessary that heat be supplied; otherwise the temperature falls as the gas expands. By the first law of thermodynamics, $\Delta Q = mc_v \Delta T + p \Delta V$. Here we are keeping $\Delta T = 0$, so we must supply heat ΔQ which is equal to the work $p \Delta V$ done by the gas.

A process which is carried out in such a way that the temperature of the working substance remains constant is called an isothermal process. When a gas is compressed isothermally, work is done on the gas, and heat must be removed during the process. The relation between the pressure p and volume V of a gas during isothermal expansion or compression is shown in Fig. 18.3 by the line ST. The equation of this curve is given by Boyle's law: $pV = \text{constant}$.

If a gas is compressed in such a way that no heat is allowed to enter it or to escape from it, the temperature increases. On the other hand, the temperature falls during expansion. An expansion or compression during which no heat enters or escapes may be realized experimentally by enclosing the gas in a cylinder surrounded by nonconducting materials. *A process in which there is no exchange of heat between the substance and its surrounding is called an adiabatic process.*

In terms of Eq. (18.2), an adiabatic process is one for which $\Delta Q = 0$; no heat is added or removed. In this case the external work done by the gas is equal to the decrease in its internal energy: $p \Delta V = -mc_v \Delta T$. The relation between the volume and the pressure of a gas during an adiabatic expansion is shown by the curve LM in Fig. 18.3. The curve for the adiabatic process is steeper than the

Figure 18.3

Curve ST is for an isothermal expansion; curve LM for an adiabatic one. The slope of the adiabatic curve through any point on the pV diagram is greater than the slope of the isothermal curve through the same point.

corresponding curve for the isothermal process. Consider two identical gas samples, one of which undergoes an adiabatic expansion and the other an isothermal expansion. If the pressure, temperature, and volume of the two samples are the same at the beginning, and if they expand to the same final volume, the temperature and pressure of the sample expanded adiabatically are less than the temperature and pressure of the gas expanded isothermally. *When a gas expands adiabatically, the product pV^γ is constant*, where γ is the ratio of the specific heat of the gas at constant pressure to the specific heat at constant volume (Sec. 15.12).

18.5 THE CARNOT CYCLE

To analyze the fundamental principle involved in heat engines Carnot in 1824 discussed an idealized heat engine in which some *working substance* takes heat from a hot *source*, performs some useful work, and then delivers part of this heat to some colder *sink*. At the end of this *cycle* the working substance and all parts of the engine are in the same state as they were in at the beginning.

We shall describe a Carnot cycle using an ideal gas as working substance. Consider a cylinder filled with gas and closed by a movable piston. Suppose that the gas in the cylinder has a volume and pressure represented by point A in Fig. 18.4. If the gas is expanded isothermally in contact with the source which supplies heat to keep the temperature at T_h (h for hot), the volume increases and the pressure decreases until point B is reached. The area under curve AB represents the work done by the gas during this expansion. The cylinder is then insulated so that no heat is added or removed as the gas expands adiabatically until the pressure and volume have the values indicated by point C. During this adiabatic expansion the temperature of the gas falls to T_c (c for cooler), the temperature of the sink. The area under curve BC represents work done *by* the gas during this adiabatic expansion.

Excellent thermal contact with the sink is then established, and the gas is compressed isothermally at temperature T_c until it has the

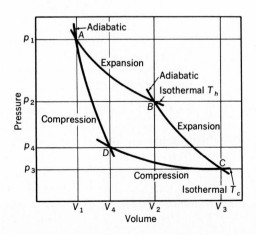

Figure 18.4
The pV diagram for a Carnot cycle is bounded by two isothermals and two adiabatics. The area enclosed in $ABCD$ represents the net external work done by the gas.

pressure and volume indicated by point D. To do this, heat must be removed from the cylinder during the compression. The point D is so chosen that the cycle can be closed by an adiabatic compression along the curve DA. The area under CD represents the *work done on the gas* during the isothermal compression, and that under DA represents work done *on the gas* during the adiabatic compression.

The net work output of the cycle is represented by the area enclosed in $ABCD$. It is the difference between the work done *by the gas* during the isothermal expansion AB and the adiabatic expansion BC and the work done *on the gas* during the isothermal compression CD and the adiabatic compression DA.

During the isothermal expansion along AB at temperature T_h, Q_h units of heat flow into the gas, while during the isothermal compression along CD, Q_c units of heat are rejected by the gas at temperature T_c. During the adiabatic compression along DA and the adiabatic expansion along BC, no heat flows out of the gas or into it. In the full cycle $Q_h - Q_c$ units of heat are transformed into mechanical energy. If W denotes the work done by the engine during this cycle,

$$W = Q_h - Q_c \tag{18.3}$$

18.6 THE EFFICIENCY OF A CARNOT ENGINE

The efficiency of a Carnot engine depends only on the temperatures between which it works. By definition, for a Carnot engine,

$$\text{Eff} = \frac{Q_h - Q_c}{Q_h} \tag{18.4}$$

In establishing his absolute thermodynamic temperature scale, Lord Kelvin showed that

$$\frac{Q_h}{T_h} = \frac{Q_c}{T_c} \tag{18.5}$$

where Q_h and Q_c are the heats transferred in a Carnot cycle at absolute temperatures T_h and T_c, respectively. Hence, the efficiency of a Carnot engine becomes

$$\text{Eff} = \frac{T_h - T_c}{T_h} \tag{18.6}$$

where T_h is the temperature of the hot source providing heat Q_h and T_c the temperature of the cold sink to which the heat Q_c is delivered.

No real engine can have an efficiency greater than that of a Carnot engine working between the same two temperatures. If a real engine could have a greater efficiency than a Carnot engine working between the same temperatures, we could use it to drive a Carnot engine in reverse and transfer net energy from the cold to the hot body with no other change. This would violate the second law of thermodynamics (Sec. 18.8).

Example

Find the efficiency of an ideal engine working between temperatures of 127 and 77°C.

$$\text{Eff} = \frac{T_h - T_c}{T_h} = \frac{400 - 350}{400} = 12.5 \text{ percent} \qquad ////$$

To obtain high efficiencies, we want heat engines to have T_h as high, and T_c as low, as is practical. Modern electric generating plants using steam turbines often employ lakes or rivers to condense the steam, thereby obtaining the lowest practical T_c. When a very large amount of energy Q_c must be rejected at temperature T_c, its delivery to a lake or river may result in *thermal pollution*—an ecologically unacceptable increase in the temperature of the water. In such a case, some other type of sink, such as a cooling tower, is a possible alternative.

18.7 HEAT ENGINES

In the sections above we have discussed the operation of engines of a highly idealized form. It may be instructive to see how some practical heat engines use heat energy to perform mechanical work.

Internal-combustion Engines The four-stroke-cycle *gasoline engine* (Fig. 18.5) is perhaps the most common heat engine. During the *intake stroke* a mixture of gasoline vapor and air enters the partially evacuated cylinder through the intake valve A. When the cylinder is full, the valve closes and the piston moves back, compressing the mixture. When the *compression stroke* is near its end, a spark ignites the mixture. The oxidation of the fuel releases heat energy, resulting in a great increase in the temperature and pressure of the gas in the chamber. As a consequence, there is a large force driving the piston forward during the *power stroke*. In the *exhaust stroke* valve B is open, and the combustion products are forced out. The cycle is then repeated. During each cycle only one of the four strokes is a working stroke during which heat is transformed into work. The working strokes of other cylinders, or a flywheel, provide the energy to carry the piston through its other three strokes. On the compression stroke of a gasoline engine, the temperature of the air and gasoline vapor increases rapidly because external work is done in compressing the gas. If the gas is compressed such that its final volume is less than about one-tenth its original volume, the gasoline may ignite spontaneously because of the high temperature. When such ignition occurs before the cylinder is near the top of its stroke, *knocking* results, with serious loss of efficiency. The ratio of the initial volume of gas to the final volume in the cylinder is called the *compression ratio*. The greater the compression ration, the greater the efficiency of the engine but the greater the difficulty in bringing the explosive mixture safely to the point at which the spark is desired. Modern developments in high-octane fuel have permitted substantial increases in the compression ratios of automobiles, but a serious practical limitation remains.

In the *diesel engine* preignition is avoided by compressing air without fuel vapor. In this way compression ratios of 16 to 1 or more may be obtained; this results in compressed-air temperatures of 500 to 600°C. When the desired compression ratio is reached, fuel is injected into the cylinder, where it ignites spontaneously because of the

(1) Intake

(2) Compression

(3) Explosion

(4) Exhaust

Figure 18.5
Steps in the operation of a four-stroke-cycle gasoline engine.

Figure 18.6
Steam recoiling from the moving blades is redirected by fixed blades against the next set of rotating blades.

high temperature. Thus, no spark is required. The diesel engine can use ordinary fuel oil, or even powdered coal. Because the diesel engine is more efficient than an ordinary gasoline engine, and because it can burn lower-grade fuel, large internal-combustion engines are usually diesels.

Steam Turbines Modern electric generators are often driven by steam turbines in which steam impinges upon a set of rotating blades (Fig. 18.6). The steam is deflected from the rotating blades to a set of fixed blades, where its direction is reversed once more so that it impinges upon a second set of rotating blades fastened to the same shaft as the first set. In a modern steam turbine a number of sets of rotating and fixed blades may be used (Fig. 18.7). As the steam transfers energy to the blades, its temperature drops, the pressure falls, and the volume of the steam increases. To allow for this, the wheels upon which the blades are assembled increase in diameter from the high-pressure end to the exhaust end. The steam may be superheated to above 560°C and supplied at a pressure of 240 atm. Under such circumstances the efficiency may be about 40 percent.

Jet Engines If a firecracker is set off in a tube closed at one end, the tube is driven forward by the increased pressure on the closed end. At the open end the hot gases escape. The momentum transferred to

Figure 18.7
A modern steam turbine. [*General Electric Company.*]

Figure 18.8
Schematic diagram of a turbojet
engine.

the tube is equal to the momentum of the escaping gases in the opposite direction. In jet propulsion it is the reaction from the escaping gases themselves, and not the interaction of the escaping gases with the air, which provides the thrust. This is why jet motors are sometimes called *reaction motors*. For many years jet motors have been used in fireworks for rockets, pinwheels, and other displays. Only in recent years have they been adapted for propelling aircraft and large missiles.

The principles underlying rocket propulsion have been discussed in Sec. 8.4. Rocket engines can operate in a vacuum, since they carry with them all the materials needed for combustion. Other jet engines require oxygen from the air for the burning of fuel. In the *turbojet engine* (Fig. 18.8) fuel is oxidized in a combustion chamber. The resulting hot gases escape through a turbine which drives a compressor to provide the air necessary to burn fuel. Many modern aircraft use turbojet engines. They are particularly suited to aircraft flying at high speeds and high altitudes, where air resistance is low. It is precisely in this region that propellers are not able to exert substantial thrust against the rare air. At low speeds and low altitudes, propeller aircraft are more efficient and have a number of superior features. A combination of the desirable (and undesirable) features of the propeller and jet engine can be achieved by operating a propeller from the shaft of the turbojet engine. Such a combination is called a *turboprop engine*.

18.8 SECOND LAW OF THERMODYNAMICS

In the Carnot engine, the most efficient conceivable engine, a quantity of heat Q_h is taken from the hot reservoir at a temperature T_h. Some of this heat is rejected to a cold reservoir at temperature T_c, and the rest of the energy is converted into work. Only if we have a heat source which is at a temperature T_h greater than that of its surroundings can we get useful work from thermal energy. This idea was formalized by Kelvin in a statement of what he called the second law of thermodynamics: *There is no natural process the only result of which is to cool a heat reservoir and do external work.* This law follows directly from the fact that *heat by itself flows only from bodies at higher temperature to bodies at lower temperature.* Indeed, we may state the second law in the form: *Heat can be made to go from a body at lower temperature to one at higher temperature only if external work is done.* In refrigerators we compel heat to go from a colder to a warmer body, but work must be done to produce this unnatural heat flow in accord with the second law of thermodynamics.

There is a vast store of thermal energy in the ocean. However, we cannot simply take this thermal energy and convert it into work to operate a ship. Only if we have some body at a lower temperature than the ocean to which we can transfer part of the energy can we make an engine operate from the ocean's heat energy. Even then we can convert into work only the *difference* between the heat energy provided at the source and the heat energy rejected to the colder reservoir.

On the other hand, we can always convert mechanical energy into heat energy completely. Thus, the second law of thermodynamics indicates a certain *irreversibility* of natural processes. For example, we can always convert the kinetic energy of an automobile into heat in the brake drums and tires, but we cannot reconvert all this heat energy to kinetic energy.

If we have 50 g of water at 0°C and 50 g at 100°C, we can always mix them to obtain 100 g at 50°C. But, if we have 100 g of water at 50°C, we cannot pour half of it into one vessel and have it at 0°C and the other half into a second vessel and have it at 100°C. This would be in violation of the second law of thermodynamics, which tells us that many natural processes occur only in one direction. The world is filled with irreversible processes; people age, eggs rot, and zinc dissolves in acid.

If the second law of thermodynamics is applicable throughout the universe, and if heat always flows from bodies at higher temperatures to those at lower temperatures, eventually the entire universe will come to the same temperature. Then, although the total energy in the universe remains the same, none of the energy will be available for doing mechanical work. Thus, the second law of thermodynamics tells us that although the total energy is constant, the available energy becomes always less. This implication of the second law of thermodynamics has aroused wide philosophical interest and leads to what philosophers of the nineteenth century called the *Wärmetod* or *heat death*.

18.9 THE REFRIGERATOR AND THE HEAT PUMP

The process of reducing the temperature of a body below that of its surroundings is essentially the reverse of the process employed in a heat engine. One may imagine a Carnot cycle operated in reverse so that the gas is expanded isothermally at a lower temperature T_c and then compressed isothermally at a higher temperature T_h.

In a typical electric refrigerator (Fig. 18.9) liquid Freon is pumped into the cooling coils, where the pressure is reduced. As a consequence, the liquid vaporizes and the gas expands. Both processes remove heat from the surroundings, which are cooled. The gas is pumped out of the cooling chamber, and in an external set of coils it is compressed and liquefied. In both of these processes large amounts of heat are given up. This heat is removed by water cooling in large systems or by air cooling in small systems. The liquid is then ready to repeat the cycle. The work required to transfer heat from lower to higher tempera-

Figure 18.9
Schematic diagram of a refrigerator with refrigerant circulated and compressed by an electric pump.

ture is supplied by the "heat pump." This pump is usually operated by an electric motor, although a gas flame can also be used.

Some modern homes are heated by means of *heat pumps* which remove heat from groundwater or some other heat reservoir and transfer it into the house by means of processes similar to those in the refrigerator. Let us consider a system using water in a well. The heat pump takes heat from the water, which may be at a lower temperature than the house, and transfers or "pumps" it into the house. Just as in the refrigerator, the cycle may involve the vaporization and expansion of a liquid in coils in the water. In the house a compression and liquefaction of the vapor is accomplished with the release of heat there. One of the major advantages of the heat pump is that in the summer it may be used to reduce the temperature inside the house by transferring heat to the water in the well. In this case the vaporization and expansion occur in the house, thereby removing heat, and the compression and liquefaction occur in the well, where the heat is transferred to the water.

QUESTIONS

1 Why does a wire warm up when you bend it back and forth to break it?

2 Two identical samples of nitrogen are compressed until the volume is one-half the original volume, one isothermally and the other adiabatically. Which sample has the higher final pressure? Why is the pressure higher?

3 Can a kitchen be cooled indefinitely by leaving the door of an electric refrigerator open? Explain.

4 Why are the efficiencies of heat engines so low?

5 An electric refrigerator transfers heat from the cold cooling coils to the warm surroundings. Does this violate the second law of thermodynamics? Why not?

6 What is gained by using a high compression ratio in an internal-combustion engine? What are the disadvantages?

7 What are the advantages of jet engines? Why weren't they used more in the past?

8 What are some examples of common irreversible processes in nature?

PROBLEMS

1 A perfect Carnot engine utilizes an ideal gas. The source temperature is 277°C, and the sink temperature is 127°C. Find the efficiency of this engine and find the heat received from the source and the heat released to the sink when 15,000 J of external work is done.
Ans. 0.273; 55,000 and 40,000 J

2 A steam engine develops 2,500 hp. How many Btu of heat must be supplied per minute if the overall efficiency is 10 percent? If the engine operates between 240 and 140°F, what fraction of the efficiency of an ideal Carnot engine does this engine have?

3 Steam is injected into a turbine at 327°C and exhausted at 77°C. If the turbine is 60 percent as efficient as an ideal heat engine, find its overall efficiency. *Ans.* 25 percent

4 A Carnot engine takes 660 kcal from a heat source at 207°C, does some external work, and delivers the balance of the energy to a heat sink at 127°C. Find how much work is done and how much heat is delivered to the sink.

5 A brake is applied to the driving shaft of an engine that develops 5 hp. The brake and the shaft are immersed in a calorimeter so that all the work done against friction is transformed into heat and goes to increase the temperature of the water in the calorimeter. If the weight of the water in the calorimeter is 110 lb, how much will its temperature rise per minute?
Ans. 1.93 F°/min

6 Compute the efficiency of an ideal heat engine operating between 167 and 47°C. If this engine takes 300,000 J from the source, how much work does it perform?

7 A six-cylinder gasoline engine makes 1,200 rev/min. Each piston has an area of

60 cm^2 and a length of stroke of 15 cm. If the average gauge pressure during a power stroke is 6×10^5 N/m^2, find the power developed in watts and in horsepower.

Ans. 32,400 W; 43.4 hp

8 A gasoline engine makes 1,200 rev/min. It has eight cylinders and develops 80 hp. The cylinder is 4 in. in diameter, and the length of the stroke is 5.5 in. Find the average net pressure that is developed during each working stroke.

9 A Carnot refrigerator takes 50 kcal from a freezing chamber at $-23°$C and rejects heat at $27°$C. How much work must be done? How many calories are rejected?

Ans. 41,900 J; 60 kcal

10 At the beginning of the compression stroke of an automobile engine, the gas occupies a volume of 12 in.^{3} at atmospheric pressure. At the end of the compression the pressure is 12.9 atm, and the volume 2.0 in.^{3}. What is the compression ratio? What is the temperature according to the general gas law if the original temperature was $27°$C?

11 A heat pump takes heat from a water reservoir at $7°$C and delivers it to a series of pipes in a house at $27°$C. Assuming that the energy needed to operate the pump is twice that of an ideal Carnot pump, how much mechanical energy is needed to supply the house with 5×10^4 kcal? *Ans.* 2.79×10^7 J

12 If it were possible to make an ideal refrigerator for utilizing a Carnot cycle, how many joules of mechanical energy would be required to remove 5,000 J from the cold compartment at $-13°$C and deliver it to the outside air at $27°$C? How much work would be required to remove this same 5,000 J from a cold compartment at $-63°$C and reject it at $27°$C?

13 How much heat must be added to 20 g of nitrogen gas in a cylinder originally at $0°$C to double the volume if the pressure is kept at 1 atm? How much external work is done? What happens to the difference between the energy supplied and the work done? *Ans.* 1.36 kcal; 1,620 J

14 Air is admitted to the cylinder of a diesel engine at atmospheric pressure and $27°$C. If the gas is compressed adiabatically to one-fifteenth its original volume, find the pressure and temperature after the compression.

15 A gas subject to a constant pressure of 1.5×10^5 N/m^2 is supplied with 30 kcal by an electric heater. If it expands from 0.35 to 0.55 m^3, what is the increase to the internal energy of the gas? What is γ for this gas?

Ans. 9.56×10^4 J; 1.31

16 If 10 liters of helium ($\gamma = \frac{5}{3}$) at $0°$C and 76 cm Hg pressure are compressed adiabatically to a volume of 2.5 liters, find the new pressure and temperature.

17 A quantity of oxygen ($\gamma = 1.4$) is compressed to occupy a volume of 2 liters at 10 atm pressure and $27°$C. If it is allowed to expand adiabatically to 1 atm pressure, find the new volume and temperature. *Ans.* 10.4 liters; 155 K

18 Given that pV^γ is constant for the adiabatic expansion of a given mass of ideal gas from a state characterized by pressure p_1, volume V_1, and temperature T_1 to a state characterized by p_2, V_2, and T_2, show that

$$T_1 V_1{}^{\gamma-1} = T_2 V_2{}^{\gamma-1}$$

and

$$\frac{T_1{}^{\gamma/(\gamma-1)}}{p_1} = \frac{T_2{}^{\gamma/(\gamma-1)}}{p_2}$$

THE WEATHER is determined by the interplay of many factors, most of which we have met in preceding chapters. Radiation from the sun provides the energy to evaporate vast quantities of water from the seas; part of the water vapor condenses to form clouds, rain, and snow. Solar radiation also establishes temperature differences which lead to convection currents and further heat transfer. In this chapter we relate much of what we have been studying to the weather.

CHAPTER NINETEEN
ATMOSPHERIC PHYSICS

19.1 WEATHER AND THE ATMOSPHERE

Nowhere in our daily lives do we meet heat phenomena on a more grandiose scale than in our contacts with the weather. Any understanding of the weather and its patterns has its foundations in the principles of mechanics and thermodynamics. In the solar radiation we have our prime source of energy, part of which establishes the giant convection currents we call *winds* and another part of which goes into evaporating water from the surfaces of lakes and oceans. The water vapor, transported by horizontal and vertical winds to distant places where temperature and other conditions are very different, may condense to form clouds, rain, or snow. The interplay of many energy transformations, including phase transitions, leads to that assemblage of atmospheric conditions which we call the *weather*. The weather at any one point is determined by a myriad of conditions, some general and readily understood, others local and of a very special nature. In the sections which follow we shall try to outline some of the principal considerations. However, before we can discuss weather patterns in any detail, we must become somewhat familiar with the atmosphere of the earth, with how the sun's radiation produces daily and seasonal variations in temperature, and with how the amount of water vapor in the air influences the situation. The science of the atmosphere is *meteorology*, a major area of physical science that is devoted to the study of atmospheric phenomena.

It is only in the lower reaches of the earth's atmosphere, called the *troposphere*, that clouds and storms occur. The troposphere extends to a height of about 11 km in middle latitudes, 18 km in equatorial regions, and 8 km above the poles. The greater height at the equator is due to the higher average temperature there and to the rotation of the earth. Above the troposphere is the *stratosphere*, which extends to a height of approximately 80 km. The stratosphere is particularly well suited for long-distance flying, since the meteorological conditions are more constant there. The stratosphere gets its name from the idea that the region is stratified in layers with roughly constant properties. Above

Figure 19.1

The position of the sun radically affects the energy received each second by any particular area of the earth's surface.

the stratosphere is the *ionosphere*, a region where many molecules are ionized and at which radio waves are strongly reflected.

19.2   THE SUN'S RADIATION

On a clear day a surface placed normal to the sun's rays receives approximately 1,400 W/m² from the sun. If the rays from the sun strike the surface of the earth at an angle other than 90°, the energy in a given bundle of rays is spread out over a larger area (Fig. 19.1). Specifically, the amount of power received is proportional to the intensity of the beam and to the cosine of the angle between the rays and the perpendicular to the surface. Figure 19.2 shows the importance of the angle at which the sun's rays reach the earth; much more energy reaches a unit area in a day during a summer month than during the winter. This is the primary factor which establishes our seasons. In addition to the change in the area over which a given bundle of rays is spread, another factor which helps to diminish the amount of heat from the sun's rays as the sun approaches the horizon is the greater distance the rays must travel in the earth's atmosphere. Consequently, there is somewhat greater absorption.

19.3   GENERAL CIRCULATION OF THE AIR;
THE CORIOLIS ACCELERATION

The inequalities of solar radiation received in different regions of the earth's surface are in part responsible for the general circulation of the

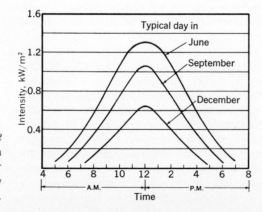

Figure 19.2

How the solar radiation received on a horizontal surface varies in power incident per unit area with time of day at a latitude of approximately 40°N.

atmosphere. If the earth were not in rotation, one might expect a relatively simple circulation pattern in which air in equatorial regions would be heated and rise. The denser air at greater latitudes would then push underneath, and we would have a pattern with air at the surface of the earth flowing toward the equator. The warmed air would rise and flow at high altitudes toward the polar regions, where it would cool and descend. This oversimplified picture is complicated by the rotation of the earth, the effects of which we approach by discussing a far simpler situation.

Consider a large merry-go-round indicated schematically in Fig. 19.3a. When it is at rest, a boy at P_1 can throw a ball to a second boy at P_2 by giving it a velocity v along the line P_1P_2 which establishes the x axis for a coordinate system on the merry-go-round. However, when the merry-go-round is rotating about an axis through O with a constant angular velocity ω, the problem of throwing a ball from P_1 to P_2 becomes more complex. Now if the boy at P_1 gives the ball a velocity v along the line P_1P_2, the ball traces a path such as P_1Q and passes the boy at P_2, as shown in Fig. 19.3b. To an observer on the merry-go-round the ball *appears* to have undergone an acceleration perpendicular to its velocity, although an observer in a stationary reference frame would report that the ball had traveled in a straight line and would correctly attribute the failure of the ball to reach P_2 to the rotation of the merry-go-round. *The apparent acceleration seen by an observer on the rotating merry-go-round is called the Coriolis acceleration* after G. G. de Coriolis (1792–1843), who worked out the equations of motion in the rotating system.

In a qualitative way it is not hard to see why the path of the ball on the merry-go-round is curved. When the merry-go-round is rotating and in the position shown in Fig. 19.3a, the point P_2 is moving north relative to the fixed frame $WSEN$ with a speed ωr_2, while P_1 is moving north with a smaller speed ωr_1. Since there is no net force acting northward on the ball, there is no acceleration in that direction, and when the ball gets a distance r_2 from the axis, it is moving north more slowly than the point on the merry-go-round directly below the ball. Relative to the merry-go-round the ball has experienced an apparent acceleration to its right.

Similarly, a ball thrown from P_2 toward P_1 undergoes a deflection toward its right when the merry-go-round is rotating. It is easy to understand this deflection (Fig. 19.3c) when we realize that when the ball is released at P_2, its true velocity perpendicular to the line P_1P_2 is greater than the velocity of P_1 perpendicular to P_1P_2.

In general, in a rotating coordinate system there is an apparent Coriolis acceleration a_C given by

$$a_C = 2\omega v \sin \phi \qquad (19.1)$$

where ω is the angular velocity of the rotating axes relative to a fixed set of axes, v is the velocity of the body in question measured in the rotating system, and ϕ is the angle between the vectors ω and $\mathbf{v}$. The acceleration is at right angles to both ω and $\mathbf{v}$; it is in the direction in which a right-handed screw would move if it were rotated from $\mathbf{v}$ to ω.

The earth upon which we live and relative to which we observe the motions of bodies forms a rotating coordinate system with an angular

Figure 19.3
The horizontal surface of a large merry-go-round. (*a*) When it is at rest, a boy at P_1 can throw a ball to another boy at P_2 by aiming along the x axis. (*b*) When the merry-go-round is rotating, a ball thrown along the x axis from P_1 follows the path P_1Q, while (*c*) one thrown toward the origin O from P_2 also curves toward the right as viewed by the thrower.

Figure 19.4

The general atmospheric circulation in
the Northern Hemisphere, where winds
deflect to the right; in the Southern
Hemisphere the Coriolis effect leads to
a leftward deflection.

velocity $\omega = 2\pi/84{,}164 = 7.3 \times 10^{-5}$ rad/sec. The Coriolis acceleration in the Northern Hemisphere is to the right for an observer facing
in the direction of the velocity, in the Southern Hemisphere to the left.
It is important to take this apparent acceleration into account in the
accurate calculation of the impact point of a long-range ballistic missile.

The Coriolis acceleration plays a vital role in our understanding
of the circulation of the atmosphere. The earth revolves about its axis
once each day, carrying a point at the equator eastward at a speed
of 470 m/sec and a point at latitude 40° at 360 m/sec. At the poles
there is no eastward velocity. If a mass of air in the Northern Hemisphere
starts moving toward the south, its eastward velocity component is less
than that of points on the earth's surface farther south. As it approaches
these points, it appears to come from the north *and east.* Similarly, a
mass moving northward has a greater eastward velocity component than
points on the earth's surface farther north. When this air reaches these
northern points, it is traveling eastward faster than the ground; therefore,
the wind comes from the south *and west.* The Coriolis acceleration
affects not only the motion of air masses but also the circulation of water.
It is one of the factors involved in explaining the Gulf Stream, the Japan
Current, and other persistent flows in the ocean system.

The combination of unequal heating in equatorial and polar regions
and the rotation of the earth leads to the general atmospheric circulation
indicated in Fig. 19.4. At any given point on the earth's surface there
may be other factors, such as mountains or seas, which profoundly
influence the wind distribution.

19.4 HUMIDITY

The next important factor governing the weather is the humidity, which
is a measure of the amount of water vapor in the air. *The absolute
humidity is the mass of water vapor per unit volume in the atmosphere.*

From the point of view of weather the actual mass of water vapor
per unit volume is considerably less important than how nearly saturated
the air is. At a high temperature the air can hold vastly larger quantities
of water vapor per unit volume than at a low temperature. We can expect
water to condense out of the atmosphere only when the air is saturated
or supersaturated in some region. The ratio of the actual mass of water
vapor per unit volume to the mass per unit volume which the air could
hold at the same temperature if saturated is the relative humidity:

$$Relative\ humidity = \frac{mass\ of\ water\ vapor\ per\ unit\ volume}{mass\ of\ water\ vapor\ per\ unit\ volume\ if\ air\ were\ saturated\ at\ same\ temperature}$$

Since the pressure exerted by the water vapor is proportional to
the mass per unit volume, the relative humidity is also given as the ratio
of the vapor pressure of the water to the saturated vapor pressure at
the same temperature (Appendix Table B.4).

One method of determining the absolute humidity is the chemical
hygrometer. In this instrument a measured volume of air is passed
through a series of U-tubes which are filled with a substance that absorbs

water vapor. These tubes are weighed before and after passing the known volume of air through them. The absolute humidity is obtained by dividing the mass of water absorbed in the tubes by the volume of air passed through.

The relative humidity of the air may be measured with a number of different types of instruments. Some of these depend on the hygroscopic properties of hair or of a thin strip of an appropriate material. When the moisture content of hair changes, its length also changes. Instruments based on this property are convenient but are ordinarily not very accurate. The wet- and dry-bulb hygrometer consists of two thermometers, one of which reads the true temperature of the air. The other thermometer is covered with a wet cloth and reads a temperature that depends on the rate of evaporation of the water into the air. If the air is saturated, the rate of evaporation is no greater than the rate of condensation, and the wet bulb has the same temperature as the dry one. The drier the air, the greater the evaporation from the wet bulb, and the lower the temperature indicated by that thermometer. Charts showing the relative humidity in terms of the readings of the two thermometers have been prepared and may be found in many handbooks.

The moisture of the air is important to our health and comfort. When the relative humidity is high in hot weather, perspiration evaporates slowly from the skin, and we have "sticky" weather. If the relative humidity is low at the same high temperature, water evaporates rapidly from the skin, and the body is cooled because it supplies the heat of vaporization. Extremes of heat and cold are felt less when the humidity is relatively low. The relative humidity has a marked effect on the physical conditions and behaviors of many materials, such as wood and wool. Lack of moisture may cause furniture to shrink and come unglued. An excess of moisture results in swelling. In many forms of manufacturing it is necessary to control the humidity. One of the important functions of any good air-conditioning system is to maintain a desirable humidity.

19.5 CLOUDS, FOG, AND DEW

On a humid summer day water condenses on the outside of a glass filled with ice water. The temperature of the outside glass surface is low, and the air in close proximity is cooled. As the temperature of air falls, the amount of water vapor it can hold is reduced. Eventually air in contact with the glass becomes saturated. If the temperature is lowered still more, water vapor condenses on the cold surface. *The temperature at which the air is saturated is called the dew point.*

Water condenses on any surface which has a temperature lower than the dew point. One way of determining the dew point is to lower the temperature of a polished metal container by evaporating ether from it. When the dew point is reached, a thin film of water is formed on the polished surface.

When the sun goes down and the ground begins to cool, some objects lose heat more rapidly than others. When warm, moist air comes in contact with these objects, some of the moisture is deposited in the form of *dew*. If the temperature is below the freezing point, water vapor is deposited as *frost*. On a typical clear night exposed objects cool faster than the air. As soon as their temperatures fall below the dew point, moisture condenses on them.

If the air cools until it reaches the dew point, moisture condenses out on minute particles in the air. Such a collection of water particles is called a *fog* if the particles are sufficiently small (less than about 30 μm in diameter) so that they do not fall to the earth as rain. Fog often forms in low areas during the night. In the morning, the water particles re-evaporate as the air warms. If fog is formed in a region above us, we speak of it as a *cloud*. Whether we use the term fog or cloud to describe the formation depends on whether we are in it or observing it from some distance. Not all clouds, of course, are formed of water droplets. The high fleecy-white cirrus clouds are composed of tiny ice crystals; they exist where the temperature is below freezing.

A cloud is often found above a tropical island. During the day the sunlight produces greater warming of the island than of the surrounding ocean. As the warm air above the island rises, its temperature falls because the pressure becomes less and the gas expands adiabatically. In a standard atmosphere this expansion results in a drop of temperature of 1 C° for every 110-m increase in altitude. Eventually the air above the island is cooled to the point at which it becomes saturated; here a cloud begins to form. A cloud of this type is characterized by a flat bottom, which suggests that the temperature at that particular height is uniform over a wide area.

19.6 LOCAL TEMPERATURE DIFFERENCES

Along shores and coastal areas movements of air, known as land and sea breezes, are associated with diurnal variations in the temperature. As the sun shines on land adjacent to the sea, the land becomes warmer than the water. Convection currents are set up in the air over the land. As the air rises, the pressure over the land is decreased, and there is a breeze from water to land (Fig. 19.5*a*). At night the land cools more rapidly than the water. Convection currents are again established, but now they are from shore to water (Fig. 19.5*b*).

19.7 THE WEATHER MAP

As a consequence of unequal heating of various parts of the earth's surface and of the rotation of the earth, huge convection currents which we call winds are established. In regions where excessive heating has occurred there are updrafts and reduced pressure, while in cold regions one expects increased density and pressure. Areas with above-average pressure are known as *highs*, and regions of subaverage pressure as *lows*. These regions are shown on typical weather maps by drawing lines known as *isobars*. An isobaric line connects all neighboring places at which the pressure is the same. Such lines are shown in Fig. 19.6. The characteristic pattern of highs and lows in the United States moves from west to east with a speed of roughly 800 km/day.

Barometers measure changes in atmospheric pressure, which may vary from hour to hour and from day to day. These variations are

(a)

(b)

Figure 19.5
Wind directions depend in part on relative temperatures. (*a*) Sea breezes blow from ocean to shore when temperatures are higher on land, and (*b*) land breezes blow from shore to ocean at night when temperatures are lower on land.

Figure 19.6
Weather map showing a well-defined low at Tennessee with associated warm front extending toward the east, cold front to the southwest, and an extensive rain area.

important for weather prediction. When the air contains an unusually large fraction of water vapor, the pressure is ordinarily below normal, since water vapor is lighter than air. Thus a falling barometer is often associated with a storm, while a high atmospheric pressure is characteristic of dry air.

Over fairly extended regions we have large bodies of air which are nearly homogeneous in the horizontal plane with respect to both temperature and pressure. Such a body is called an *air mass*. It is, of course, never homogeneous in the vertical plane; even in the horizontal plane there are gradual variations. An air mass may cover thousands of square miles and extend to heights of many thousand feet. It acquires its characteristics by moving for some time over an area where conditions are reasonably uniform. One which has developed over northern Canada is called a *polar continental* or *polar Canadian* air mass, while one which has originated in the northern regions of the Pacific Ocean is called a *polar Pacific* or *polar maritime* air mass. One which has originated over a tropical land mass is described as *tropical continental*, and one over tropical water as *tropical maritime*.

As an air mass moves, it is modified by heating or cooling, by the addition or removal of moisture, and by mixing with other air. Weather conditions are determined to a great extent by the characteristics of air masses and by the direction of their movements.

Where air masses at different temperatures and pressures meet, the surface of separation is called a *front*. Across a front there is a sharp transition in weather conditions, which may include temperature, pressure, humidity, and wind velocity. Several different types of fronts are recognized (Fig. 19.7).

Figure 19.7
Symbols for fronts.

A *warm front* is a surface on which the direction of motion of the air is such that warm air replaces cold air. On the weather map, warm fronts are represented by black semicircles pointing in the direction toward which the warm air is moving. Figure 19.8 shows a warm front advancing. Because the warm air is less dense, it rides over the colder air, and, in the early stages of the warm front, high clouds are formed by the condensation of moisture in this invading mass of warmer air. As the warm front moves in, lower clouds are formed, and if the warm front involves air of high relative humidity, rain is produced. It ordinarily takes about a day from the first signs of approaching warm air before the warm front passes on the earth's surface.

A *cold front* is a surface along which cold air is replacing warm. It is represented by small black wedges along the front, pointing in the direction in which the cold air is moving. When cold air which moves in under humid, warm air, upward currents of warm air may develop into violent thunderheads. Ordinarily, a cold front moves more rapidly than a warm front, and there is not a great deal of warning before it arrives. As the cold front passes, there is often a hard shower which removes much water vapor from the air and leaves crisp bright weather in its wake. Cold fronts move about 40 km/hr and are steeper and faster than warm fronts. When a cold front overtakes a warm front, the result is an *occluded front*, represented by a combination of cold-front and warm-front symbols pointing in the direction of motion of the front.

A *stationary front* is a surface separating cold and warm air when the surface is not moving.

19.8 CYCLONES AND ANTICYCLONES

An atmospheric condition characterized by a low-pressure region surrounded by closed isobars is called a *cyclone*. As air from the south moves northward (Fig. 19.9), it is deflected to the east in the Northern

Figure 19.8
An advancing warm front.

Hemisphere, while colder air from the north is deflected toward the west. Thus the circulation about a cyclone is counterclockwise in the Northern Hemisphere. Often warmer air moving from the south forms a warm front, and colder air from the north a cold front, as indicated in the figure.

A high-pressure region on the weather map is known as an *anti-cyclone*. About this high-pressure region winds circulate clockwise in the Northern Hemisphere. A high-pressure area is ordinarily characterized by bright, clear weather. The pattern of cyclones and anti-cyclones which migrates across the country is largely responsible for exchange of heat between high and low altitudes.

19.9 TORNADOES

Tornadoes are gigantic whirling funnels of air. The motion of the air is characterized by a cyclonic upward spiral that causes rapid expansion, cooling, and condensation. This condensation forms the dark cloud of the tornado funnel. The width of the storm may vary from 30 m to 2 km, and the forward velocity may vary from 30 to 60 km/hr. The tapering end of the funnel may skip over one area and descend on another. It often has a whirling and serpentine appearance. The whirling velocity of the air causes a decrease in pressure at the center of the funnel. The wind velocity is excessive, both in horizontal and vertical directions. Vertical wind speeds may be as much as 300 km/hr, and horizontal speeds may exceed this figure.

Figure 19.9
Cyclonic circulation about a low in the Northern Hemisphere, showing typical warm and cold fronts.

QUESTIONS

1 How can an air-conditioning system reduce the relative humidity in a building by forcing the air through a spray of water?

2 Why is cyclonic circulation counterclockwise in the Northern Hemisphere and clockwise in the Southern Hemisphere?

3 Why does the barometer fall as moist air moves into a region?

4 Why do ground fogs appear less often on cloudy overcast nights than on clearer nights?

5 Why do the trade winds and the westerlies blow in opposite directions?

6 Why are satellites put into roughly equatorial orbits by firing them eastward rather than westward?

7 Why should the temperature fall less than the ordinary 1 C° for every 110-m increase in altitude if the air is unusually moist?

8 Do raindrops cool down or warm up (*a*) as they fall and (*b*) as they land?

9 Why are the bases of neighboring cumulus clouds often at about the same level? How are such clouds formed?

PROBLEMS

See Appendix Table B.4 for the properties of saturated water vapor.

1 Find the relative humidity and the absolute humidity on a day when the temperature is 30°C and the dew point is 20°C.
   Ans. 55.1 percent; 16.6 g/m³

2 A room has a temperature of 30°C. When a surface in the room is cooled to 10°C, moisture just begins to condense on it. Find the dew point, the approximate absolute humidity, and the relative humidity.

3 A chemical hygrometer gains 0.75 g in mass when 60 liters of air at 23°C is passed through it. Calculate the absolute humidity of the air and the relative humidity if saturated air at 23°C contains 20.6 g/m³.
   Ans. 12.5 g/m³; 61 percent

4 A room is 10 m long, 5 m wide, and 3 m high. The temperature is 20°C. How much water must be evaporated in the room to raise the relative humidity from 20 to 50 percent?

5 How great is the vapor pressure of water in a room at 20°C if the humidity is 55 percent? What mass of water vapor is present in each cubic meter of air?

Ans. 1.27×10^3 N/m²; 9.46 g/m³

6 Assume the density of the atmosphere is 1.29 kg/m³ regardless of the height (of course, the density of the atmosphere does depend on what is above it). If a barometer reads 76 cm Hg at sea level, what is the height of the atmosphere? Convert your answer into miles.

7 The solar constant is 1,400 W/m². How much energy falls each minute on 1 m² of a lake at 40° north latitude on a clear day at high noon in June, when the sun is 23° north of the equatorial plane? In December, when the sun is 23° south of the equatorial plane?

Ans. 80,300 J; 38,100 J

8 Find the amount of water vapor contained in a room which is saturated at 10°C if the room has a volume of 3,000 m³. If the air is heated to 20°C, no water vapor is added, and the pressure remains the same, find the relative humidity.

9 An air-conditioning system delivers 2,900 m³ of air each hour at 17°C and 60 percent relative humidity. To do so, it takes in 3,000 m³ of air at 27°C and 80 percent relative humidity. Find the absolute humidity of the air taken in and the mass of water vapor removed per hour. For saturated air:

Temp., °C	11	13	15	17	19	21	23	25	27
Absolute humidity, g/m³	10	11	12.8	14.5	16.3	18.4	20.6	23.0	25.8

Ans. 20.6 g/m³; 37 kg

10 The relative humidity of an auditorium that has a volume of 5,000 m³ is 30 percent at the beginning of a concert and 70 percent at the end of the concert. If the temperature of the room is assumed to remain at 20°C throughout the concert, how many kilograms of water have been added to the room?

11 Air that is saturated at 30°C rises vertically until its volume is doubled. At the same time, its temperature decreases until it is 10°C. Find the number of grams of water that will condense out of each cubic meter of the air, measured at the original temperature and pressure.

Ans. 11.4 g

12 An air-conditioning system delivers 5,000 m³ of air each hour at 20°C and 40 percent relative humidity. How much water must be removed each hour if the incoming air has a relative humidity of 80 percent at 30°C?

13 One mole of dry air at 27°C and 1 atm pressure expands adiabatically as it rises to 5,400-m altitude, where the pressure is $\frac{1}{2}$ atm. Calculate the temperature at 5,400 m, the decrease in the internal energy of the gas, and the work done by the gas. On the average, how many meters did the air have to rise to be cooled 1 C°? (Compare your result with the value of 1 C° for each 110 m quoted in Sec. 19.5 for a standard atmosphere.)

Ans. 246 K; 1,120 J; 1,120 J; 100 m

14 A body dropped from a height h at the equator will strike the ground a distance $(8h^3/g)^{1/2}\omega/3$ from the spot directly below the point of release if air effects are negligible. In what direction is the impact point displaced? If a steel ball is dropped from a captive balloon 200 m high, what is the distance between the impact point and the spot directly below the release point? Find the Coriolis acceleration just before the ball strikes the ground.

15 An iceberg with a mass of 10^7 kg is moving eastward 15 km/day at latitude 75°N. Find the Coriolis acceleration. What resultant force would be required to produce this acceleration in an inertial frame of reference?

Ans. 2.5×10^{-5} m/sec² south; 250 N

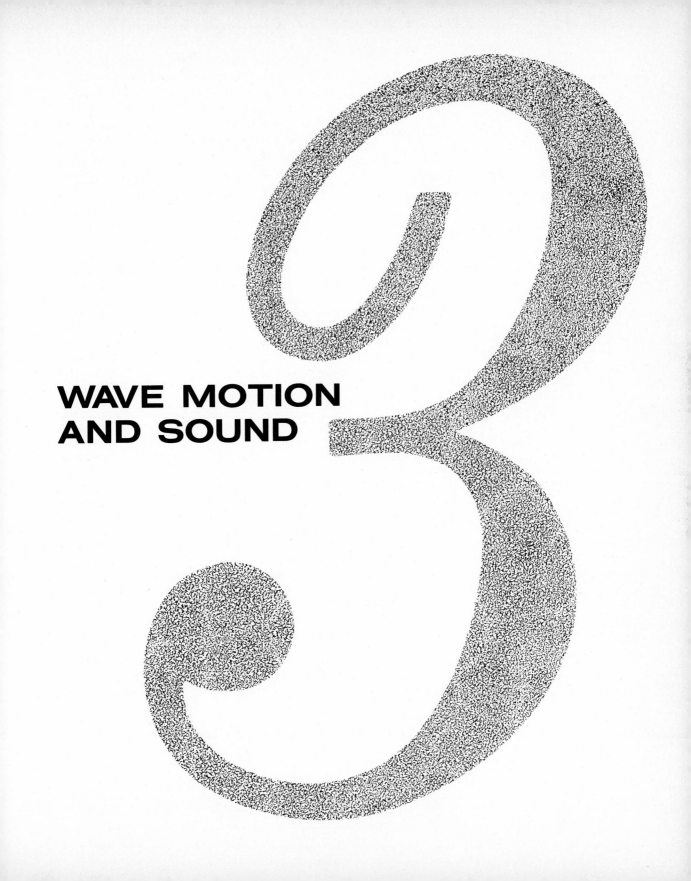

**WAVE MOTION
AND SOUND**

WE HAVE STUDIED several types of motion which can be performed by "particles" of various kinds. Associated with each particle there is an instantaneous position in space as well as mass, kinetic energy, and momentum. A gas molecule or a baseball can be idealized as a point mass in many situations. However, each occupies a volume in space. A fundamental property of classical particles is that no two particles can occupy the same point in space at the same time. Now we turn to a study of classical waves, which have very different properties than classical particles.

CHAPTER TWENTY
WAVE MOTION

20.1 WAVES

An important phenomenon in nature is the transmission of energy from one point to another by wave motion. There are many kinds of wave motions, one of which is the familiar water wave. When a stone is dropped into a pool of still water, the surface is covered with circular wavelets which widen out from the point at which the stone entered the water. The water does not itself move outward from this central point; rather it rises and falls. That such is the case may be seen by observing a floating cork. It moves up and down and back and forth in a roughly elliptical path. The water on which the cork rests goes through this same kind of motion.

In wave motion a disturbance of some kind is propagated. When waves pass through a medium, an individual particle oscillates up and down or back and forth, but it does not progress with the wave. Energy is transferred through the medium, although the particles themselves are not transmitted. In a typical wave motion a vibrating center produces motion in particles of the medium immediately in contact with it; these particles in turn impart motion to their neighbors. In many cases, but by no means always, the particles describe simple harmonic motion about their equilibrium positions, with phases varying along the wave.

20.2 TRANSVERSE AND LONGITUDINAL WAVES

It is convenient to classify waves in terms of how the motion of the individual particles of the medium is related to the movement of the wave itself. In waves produced in a stretched rope (Fig. 20.1) the individual particles move up and down at right angles to the direction in which

Figure 20.1
Wave pulse moving along a stretched string.

the wave itself propagates. A wave of this kind is known as a *transverse wave*, because the particles in the medium move perpendicular (*trans* means "across") to the direction of the wave motion. In the simplest form of transverse wave each particle vibrates in simple harmonic motion with its displacement at right angles to the propagation direction. Light and other forms of electromagnetic radiation show behavior characteristic of transverse waves in many experiments. Waves in stretched strings are an example of a transverse wave motion. *A transverse wave is one in which the vibrations take place at right angles to the direction in which the wave travels.*

If a series of equal masses are joined together by springs (Fig. 20.2) and particle A is moved toward particle B, a wave is propagated down the system. The motion of A toward B compresses the spring between them. B is accelerated toward C, which in turn moves toward D. A compressional wave is produced. Similarly, when mass A is pulled away from B, the spring between them is stretched, and B is accelerated toward A. If A is moved back and forth with simple harmonic motion, a wave motion is propagated along the system with each particle describing simple harmonic motion along the axis of the system. The various masses do not all vibrate in phase. *A wave motion in which the individual particles vibrate back and forth along the direction in which the wave travels is called a compressional or a longitudinal wave.* Sound exhibits the characteristic properties of longitudinal waves.

In some wave motions individual particles do not move back and forth along a single line. For example, in water waves, water molecules move forward and back as well as up and down. When a particle is on a crest A (Fig. 20.3), it moves in the direction in which the wave is moving. In a trough B a particle is moving in the opposite direction. The paths are ellipses or circles. We have in this case a mixed wave, neither longitudinal nor transverse. We can, of course, resolve the motion of the particle into components perpendicular to and parallel to the propagation direction and treat the mixed wave as a combination of a longitudinal and a transverse wave.

There are many other kinds of waves besides those in material mediums in which the disturbances arise out of the displacement of

Figure 20.2
A series of masses joined by springs transmits a longitudinal wave when the particles move along the line connecting the masses.

Figure 20.3
The particles in a water wave describe
circles or ellipses.

particles. If the temperature of one end of a metal rod is first raised gradually, then lowered, raised again, etc., there is set up a succession of changes in the temperature of the rod. These changes travel forward in the rod as a wave of temperature. The daily heating and cooling of the surface of the earth, as it is turned toward and away from the sun, produce waves of temperature that go down into the earth a short distance.

When one end of an ocean cable or telephone wire is suddenly joined to a battery, the change in potential produced in the wire is rapidly transmitted along the conductor. Thus, electric waves may travel through the conductor.

20.3 WAVELENGTH, FREQUENCY, AND VELOCITY

When waves spread out from the center of a disturbance, a surface marking the points which the disturbance has reached is called a *wavefront* (Fig. 20.4). When a drop of water falls on the surface of a pond, the wavefront is a circle that expands continuously. When a small balloon bursts in the air, the wavefront of the sound produced is a sphere with the balloon as the center. The waves of light from a distant star have spherical wavefronts of such large radius that small portions of them may be considered plane.

In a wave the distance from crest to crest (or trough to trough) is called the *wavelength* (Fig. 20.5a). In general, *the wavelength is the distance in the propagation direction between two successive points which are in the same phase of vibration.* Let ν be the frequency of the wave, or the number of complete vibrations made by each particle in unit time. If each wave is of length λ, the total distance traversed by the disturbance in 1 sec is just the number of vibrations per second multiplied by the wavelength, or

$$V = \nu\lambda \qquad (20.1)$$

where V is the velocity of the wave motion. This relation, applicable to all types of wave motion, is one of the most widely used relationships in physics.

Wavefronts
4th sec
3d sec
2d sec
1st sec
S
1
2
3
4

Figure 20.4
Wavefronts from a point source S
generated at intervals of 1 sec.

Example

If the velocity of a disturbance in a steel rod is 5,000 m/sec and the frequency of the vibrations is 2,500 sec^{-1}, (read "2,500 per second") find the wavelength.

$$\lambda = \frac{V}{\nu} = \frac{5{,}000 \text{ m/sec}}{2{,}500 \text{ vib/sec}} = 2.000 \text{ m} \qquad ////$$

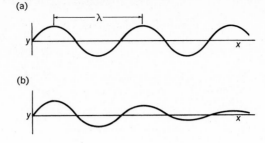

Figure 20.5
Displacement of particles at a particular instant as a function of the distance from the source for (*a*) a plane wave for which the amplitude is constant and (*b*) a spherical wave in which amplitude decreases with distance.

Frequency and wavelength are among the important attributes of a wave; other properties are the velocity of propagation, the amplitude of the disturbance, and the energy and momentum transferred. If the frequency of an ideal wave is to be measured with great precision, it is necessary that the wave have great extension in space. A classically ideal wave has well-defined frequency and amplitude but no specific position; by contrast, a classical particle has no frequency, but its position can be well defined.

20.4 REPRESENTATION OF WAVES

When each particle of a medium through which a wave is traveling describes simple harmonic motion, a common way of representing the wave graphically is to choose two axes (Fig. 20.5) and to plot on the vertical axis the displacement of a particle at a given instant and on the horizontal axis the distance of that particle from the source of the disturbance. Such a curve is an instantaneous "snapshot" of the wave, showing the displacement for all particles of the medium along the direction of propagation. In Fig. 20.5*a* the amplitude is constant, and the intensity of the wave remains the same as we go away from the source. In Fig. 20.5*b* the amplitude and intensity decrease with distance.

For a complex wave with a number of frequencies involved, it is possible to consider the motion of any particle to be the resultant of a number of simple harmonic motions. Thus, a very complex motion may be treated as the sum of a number of simple ones. Although many waves have a far more complicated picture than that of Fig. 20.5*b*, Fourier has shown that all waves can be represented in terms of sine waves similar to that of Fig. 20.5*a*. A Fourier analysis of a complex wave into simple harmonic waves is beyond the scope of this book, but the fact that this can be done means that the arguments and developments which we base on simple sinusoidal curves can be extended to complex waveforms.

When a wave passes through a medium, the individual particles describe simple harmonic motion for which the displacement y of a particle can be represented by

$$y = R \sin 2\pi \nu t \qquad (10.6)$$

where R is the amplitude, ν the frequency, and t the time with its zero chosen as the instant at which the particle passes through the equilibrium

Direction of wave motion ⟶

Figure 20.6
Instantaneous displacement y as a function of distance x at $t = 0$, when the particle at O passes upward through its equilibrium position. The displacement of this particular particle as a function of time is the vertical projection of a point describing uniform circular motion on the circle of reference at the left. After a short time t the displacement at O will be OP, and shortly thereafter a crest will pass O.

position in the positive direction. The displacement of any particle in the medium through which a simple wave is passing depends on both the time and the distance of the particle from the origin. Each successive particle is a little later in phase than the adjacent particle nearer the source. A crest reaches point b, one-fourth wavelength from the origin (Fig. 20.6), one-fourth of a period later than it arrives at O. It reaches point c one-half period later. The crest reaches point d, which is a distance x from point O, a time x/V sec later than it passes O. Let the displacement of O be given by Eq. (10.6). To obtain the displacement at d, we must deduct from t the time x/V. The displacement at a point which is any distance x from point a is given by

$$y = R \sin 2\pi\nu \left(t - \frac{x}{V} \right) \tag{20.2}$$

Although in our plot of displacement as a function of position we obtain a sine curve which suggests a transverse wave motion, this same curve represents a longitudinal wave equally well since the displacement y obeys exactly the same equations. The only difference is that the displacement for longitudinal wave motion is in the direction of propagation, while for a transverse wave motion it is perpendicular.

20.5 THE FLOW OF ENERGY IN A MEDIUM

The energy transmitted in unit time through a unit area perpendicular to the direction of propagation of a wave is called the *intensity* of the wave motion. *The intensity is proportional to the square of the frequency and to the square of the amplitude of the vibrations of the oscillating particles,* as is shown below.

For a particle describing simple harmonic motion, energy is transformed from kinetic energy to potential and back to kinetic. The energy of the particle is $\frac{1}{2}mv_{\max}^2 = 2\pi^2\nu^2R^2m$ by Eq. (10.11). If we consider a wave propagated in the x direction (Fig. 20.7), the wavefront sets in motion every particle which it passes. The intensity of the wave is then given by the total energy supplied to all particles included in a rectangle having a base of unit area and a length equal to the distance traversed by the wave in unit time. The number of particles included in this tube is equal to nV, where n is the number of particles per unit volume and V is the velocity of the wave. Each particle receives energy $2\pi^2\nu^2R^2m$, and the total power transmitted per unit area is the intensity, given by

Figure 20.7
The energy transmitted in 1 sec across the plane zOy, of unit area, occupies a volume of unit cross-sectional area and of length numerically equal to the speed of the wave motion.

$$I = 2\pi^2 \nu^2 R^2 mnV \qquad (20.3)$$

If a disturbance produces waves which are sent out equally in all directions, the wavefronts are spheres with the source as center (Fig. 20.4). If we assume that there is no transformation of the wave energy, the total energy per unit time which passes through any one sphere surrounding the source is exactly equal to the energy per unit time which passes through any other sphere. The energy per unit time passing through a sphere is just the product of the area of the sphere and the intensity I. Therefore, $4\pi R_1^2 \times I_1 = 4\pi R_2^2 \times I_2$, where I_1 and I_2 are the intensities at the spheres of radii R_1 and R_2, respectively, and

$$R_1^2 I_1 = R_2^2 I_2 \qquad (20.4)$$

For a spherical wave the intensity varies inversely as the square of the distance from the source, providing there is no energy loss.

When the energy cannot spread out freely in all directions, the intensity does not vary inversely as the square of the distance. A cheerleader's megaphone is designed to reflect sound waves toward his cheering section. In this case the intensities at equal distances from the source depend strongly on direction. If there is absorption between the source and a receiver, the intensity falls off more rapidly with distance than the inverse-square law suggests, even though the energy may be radiated equally in all directions.

20.6 HUYGENS' PRINCIPLE

When a point source radiates energy equally in all directions, the wavefronts are spherical. If we place a series of obstacles in the medium through which the wave is traveling, these obstacles distort and change the wavefronts. How can we predict where the new wavefront will be a time Δt aften the wavefront has struck an obstacle? The answer was found by Huygens, who observed that in a typical wave motion each particle is set into vibration by a neighboring particle. This led him to postulate that *every point on a wavefront acts as a new source sending out secondary wavelets.*

Consider a spherical wavefront (Fig. 20.8a). Let us, following Huygens, assume that each point on this wavefront is a source of secondary spherical wavelets and ask where the wavefront corresponding to AB will be a time Δt later. About each point 1, 2, 3, 4, ... on the initial wavefront we draw little spheres of radius $V \Delta t$, where V is the velocity of the waves. To find the position of the new wavefront, we find the surface which is tangent to all the secondary wavelets. This surface, CD in Fig. 20.8a, is called the *envelope* of the wavelets. When a plane wave passes through a medium, we can find the new wavefront by applying Huygens' method. The new wavefront is again plane, as shown by Fig. 20.8b.

Huygens' principle may be summarized as follows: *Every point on a wavefront due to any wave motion may be regarded as a secondary source of "Huygens" wavelets which spread out with the velocity of the primary wave. To find the wavefront at any time Δt later, we find*

WAVE MOTION AND SOUND

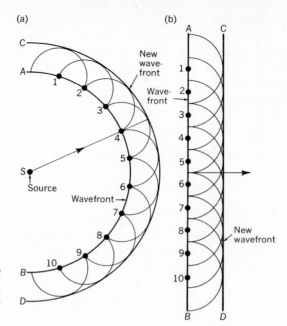

Figure 20.8
Huygens' principle applied to (*a*) a spherical wavefront and (*b*) a plane wavefront.

the forward surface which is tangent to all these secondary wavefronts. This surface gives the new position of the primary wavefront.

At a considerable distance from a point source the waves are almost plane. If these waves fall on a small aperture (Fig. 20.9) in a partition, the aperture acts as a source which sends out almost circular waves in accordance with Huygens' principle. Clearly the waves spread out once more after they leave the aperture, thus bending into the region shadowed by the partition. The bending of waves into the shadow region of obstacles is called *diffraction*. Huygens wavelets traveling to the side from a point source are canceled by destructive interference with wavelets from neighboring points unless the point is at the edge of the

Figure 20.9
A small aperture in a partition serves as a source of almost circular waves.
[*Macalaster Corporation.*]

Figure 20.10
Reflection and diffraction of sound waves can be studied by photographing the wavefronts created by an electric spark. In this case the spark originated at point O on the stage of Royce Hall at the University of California at Los Angeles. Note the complex wavefronts arising from reflection from the proscenium at the left of the stage. [*Courtesy of Vern O. Knudsen and L. P. Delsasso.*]

wavefront, in which case it contributes to the wave spreading into the shadow region.

The question may be asked: Why isn't there also a backward wave if every particle on the wavefront sends out wavelets with the same amplitude in all directions. The absence of this backward wave is explained by the fact that the amplitude of the Huygens wavelets varies according to the *obliquity factor* $(1 + \cos\theta)/2$, being greatest in the forward ($\theta = 0$) direction and diminishing to zero in the backward ($\theta = \pi$) direction.

20.7 THE REFLECTION OF WAVES

An important application of Huygens' principle is to the reflection of waves by an obstacle. When a spherical wave strikes a plane surface, it is reflected, and the curvature of the wavefront is reversed (Fig. 20.10).

Consider a plane wave (Fig. 20.11) falling upon a plane reflecting surface. Let A_1B_1, A_2B_2, and A_3B_3 be consecutive positions of the plane wavefront. Let us apply Huygens' construction to the wavefront A_3B_3 to find the position of the new wavefront. We choose the instant at which the end B_3 of the wavefront reaches the plane surface as the time at which we wish to locate the new wavefront. Let Δt be the interval required for the wave to move from B_3 to the surface. We observe from the figure that the new wavefront is given by RS. Since $V\,\Delta t$ is perpendicular to the wavefront, and since the distance A_3S is the same for the triangles A_3SR and A_3B_3S, the two triangles are similar and equal. Therefore, the angle of reflection r is equal to the angle of incidence i.

In dealing with the reflection of waves, it is often more convenient to follow the path of a ray which moves in the direction of propagation of the wave than to observe the wavefront. For isotropic mediums the ray is perpendicular to the wavefront. If N represents a normal to the

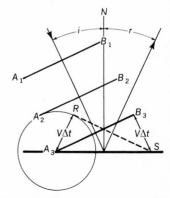

Figure 20.11
Huygens' construction for a plane wave reflected from a plane surface.

surface (Fig. 20.11), the angle of incidence is the angle between the incident ray and N, while the angle of reflection is the angle between the reflected ray and N. The reflected ray is in the plane determined by the incident ray and the normal. This leads us to the two laws of reflection:

1 *The angle of incidence is equal to the angle of reflection.*

2 *The incident ray, the normal, and the reflected ray lie in the same plane.*

20.8 REFRACTION OF WAVES

When waves pass from one medium in which they have a speed V_1 to a second in which their speed is V_2, the directions of the wavefronts and rays change at the boundary between the two media. This phenomenon is known as *refraction*. Figure 20.12 shows the refraction of waves as they pass from one region to another in which their speed is decreased.

Refraction at the interface between two media is readily explained by use of Huygens' principle. In Fig. 20.12 a plane wavefront ABC is just reaching the interface between two media. Let V_1 be the speed of the waves in the medium in which the waves are incident and V_2 their speed in the second medium. While the wavefront goes from BC to EF in the first medium, the wavelets from A travel only as far as D in medium 2. When the wavelets from F reach J, the wavefront is entirely in the second medium, as is represented by GHJ. If Δt is the time required for the wavefront to travel from A to G in medium 2, we have $AG = V_2 \Delta t$ and $CJ = V_1 \Delta t$. Now $\sin i = CJ/AJ$, and $\sin r = AG/AJ$, where i is the angle of incidence, and r the angle of refraction. Therefore $(\sin i)/(\sin r) = (V_1 \Delta t)/(V_2 \Delta t) = V_1/V_2$. The ratio $(\sin i)/(\sin r)$ is called the *index of refraction of medium 2 relative to medium 1*; we represent it by n. Thus

$$n = \frac{\sin i}{\sin r} = \frac{V_1}{V_2} \qquad (20.5)$$

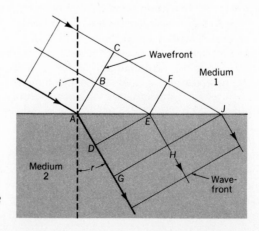

Figure 20.12
Refraction at a plane surface.

For refraction:

1   *The ratio of the sine of the angle of incidence to the sine of the angle of refraction is a constant n, known as the index of refraction.*

2   *The incident ray, the normal, and the refracted ray lie in the same plane.*

Water waves become slower as the water becomes shallower. In this case there is no sharp bending, but the wavefronts bend continuously as the speed changes. On a sloping beach the waves bend so that the wavefronts are almost parallel to the shore regardless of the direction they came from. In general, a ray bends toward the normal to an interface as a wave goes from a region of higher speed to one of lower speed.

20.9 SUPERPOSITION AND INTERFERENCE

It is common to have sound waves from different sources moving through a room at the same time. *When two or more wave motions pass a given point at the same instant, the displacement is the resultant of the displacements which would be produced by each of the waves if it acted separately.* This is a statement of a very important law of nature, sometimes called the *principle of superposition.* It is valid not only for sound but for all other kinds of waves, provided the amplitudes are not too great.

We shall at the moment apply the principle of superposition to the particular case in which two waves are traveling in the same direction at the same time with the same amplitude and the same frequency. If the waves arrive in such a way that crests meet crests and troughs meet troughs (Fig. 20.13), the displacements due to the two waves add and the resultant is a wave of double the original amplitude. These two waves show *constructive interference.* On the other hand, if the two waves arrive at the same point in such a way that crest meets trough and trough meets crest, they cancel one another and we have *destructive interference.*

Two waves of unequal amplitude may still interfere; the resulting amplitude is the resultant of the individual amplitudes. Interference is a fundamental property of wave motions. Indeed, if there is a question

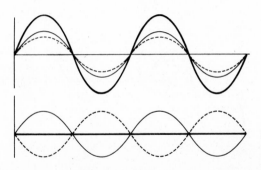

Figure 20.13
Constructive and destructive interference.

Figure 20.14
When a wave in a stretched string is reflected at a wall, a trough is reflected as a crest and a crest is reflected as a trough; there is a 180° phase shift at this reflection.

whether some phenomenon has wave properties, the test which resolves the question in favor of the wave position is one which produces interference between two "rays."

20.10 STANDING WAVES

A particular type of interference of great importance in connection with musical instruments is interference between two waves which are traveling in opposite directions. Consider a long elastic cord which is fixed at one end (Fig. 20.14) while the other end is held in the hand. If the cord is stretched tight and the free end is moved up and down with simple harmonic motion, waves are set up in the cord which travel to the fixed end, where they suffer reflection and travel back to the hand. A crest is reflected as a trough and a trough as a crest at a fixed end.

At any instant two trains of waves are traveling in the cord in opposite directions. If the frequency is chosen properly, the cord ceases to have the appearance of being traversed by trains of waves, but instead vibrates transversely in one or more segments (Fig. 20.15). The behavior of the cord can be explained by consideration of Fig. 20.16. The dotted line represents a wave traveling from right to left, while the broken line represents a similar wave of the same frequency and amplitude traveling from left to right. The resultant disturbance, which arises from the combination of these two trains of waves, is represented by the continuous lines. The resultant at any instant has the form of a wave, but the wave pattern does not move. Each particle vibrates, but the amplitude of vibration varies along the cord. When the frequency is high, all the eye sees is a characteristic blur that appears to remain motionless. This appearance has given rise to the name *standing waves*. At L (Fig. 20.16a), the crests of two component waves are approaching each other. When the two crests coincide, the resulting displacement is maximum. One-quarter of a period later the two components neutralize each other. The crest of one wave meets the trough of the other. At

Figure 20.15
Standing waves in a stretched string.

(a)

———Resultant wave
-- - -▶Incident wave
◀- - - -Reflected wave

(b)

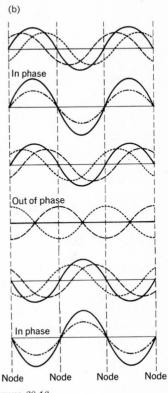

In phase

Out of phase

In phase

Node   Node   Node   Node

Figure 20.16
(*a*) Standing waves result when two
waves of the same amplitude and
frequency travel in opposite directions,
interfering constructively at some points
and destructively at others.
(*b*) Successive positions of incident
and reflected waves producing
standing waves.

this instant the cord is straight. As the waves travel farther in opposite
directions, the portion of the string $N'LN$ is depressed below the horizontal, and after another quarter of a period it has its maximum displacement in the negative direction. At the points N and N' there is never
any displacement. At these points, called *nodes*, the two waves traveling
in opposite directions always interfere destructively. All the particles of
the cord between two adjacent nodes are moving in the same direction
at any given instant, but two adjacent segments of the cord are always
displaced in opposite directions. The length of the segment between
two consecutive nodes is one-half wavelength. The point midway between the nodes at which maximum displacement occurs is called an
antinode (or *loop*).

Standing waves may result from any kind of wave motion. It is
only essential for their production that two waves of the same frequency
and amplitude travel in opposite directions in the medium. Figure 20.17
represents standing waves in a column of air. The short vertical lines
represent layers of air displaced as shown. At each place marked N
there is a node in the displacement; here the air is alternately compressed and rarefied.

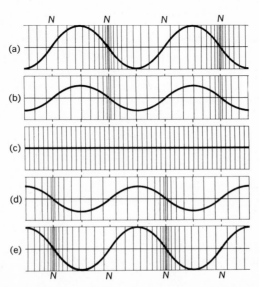

Figure 20.17
Successive stages of standing sound
waves during one-half cycle during
which the nodes (N) representing
rarefactions and compressions in (*a*)
reverse to become compressions and
rarefactions in (*e*).

QUESTIONS

1 What general properties must a body have to be able to transmit longitudinal waves? Transverse waves? Explain why liquids and gases do not transmit transverse vibrations.

2 If the intensity of any kind of wave falls off as $1\ r^2$ [Eq. (20.4)], how does the amplitude vary as a function of r?

3 Two sources produce waves of identical frequency, but one produces waves having three times the amplitude of the others. What is the ratio of the two intensities? If the frequency of a wave motion is doubled while the amplitude is kept constant, how does the intensity of the new wave compare with that of the old?

4 As water waves spread out over the surface of a lake, how does the intensity vary with distance from the source if essentially no energy is dissipated? How does the amplitude vary? Why isn't the inverse-square law followed?

5 What sort of experiments could you perform to prove that energy is associated with a wave motion?

6 What happens to a wave when it comes to an interface between two media?

7 Energy may be transported by particles and by waves. How can one distinguish experimentally between these means of energy transport? What characteristics differ between waves and particles?

8 What points remain stationary in a standing wave? Why don't they move?

9 What conditions must be satisfied to establish a standing-wave pattern?

PROBLEMS

Unless otherwise stated, take the speed of sound in air to be 340 m/sec or 1,100 ft/sec and the speed in water to be 1,450 m/sec.

1 A violin string emits a sound with a frequency of 425 Hz. What is the wavelength of the dis-turbance that passes through the air? What is the period of the vibration?
    Ans. 80 cm; 0.00235 sec

2 A set of circular ripples is produced on the surface of a pond by throwing a stone into the water. At a certain instant, the first crest is 5 m from the point where the stone hit the water, and the sixth crest is 50 cm from the same point. What is the wavelength of the disturbance?

3 When a sound wave is transferred from one medium to another, its frequency does not change. If the wavelength of a disturbance is 34 cm in air, find its wavelength in water, steel, and brass if the speed of sound in steel is 5,000 m/sec and in brass 3,500 m/sec.
    Ans. 1.45 m; 5 m; 3.5 m

4 A tuning fork with a frequency of 750 Hz sends out waves which travel 1,080 ft/sec. How many vibrations does the fork make in the time required for the sound to travel 900 ft?

5 A sounding source with a frequency of 600 Hz sends out waves that travel from air into water. Find the wavelength in each medium.
    Ans. 56.7 cm; 2.42 m

6 Waves travel with a speed of 800 ft/sec through a medium. If 50 waves pass a given point each second, find the wavelength and the period.

7 Water waves are observed passing a certain point at a velocity of 7 m/sec with a distance of 11 m between crests. What is the frequency of the waves?   *Ans.* 0.636 Hz

8 It is 12 m from crest to crest in a system of waves. If 35 waves pass a given point each minute, find the speed of the waves.

9 A swimmer is under water when the starter of a race fires his gun. If the sound ray to the swimmer's ear approaches the water surface at an angle of incidence of 12°, what is the angle of refraction in water?   *Ans.* 62.5°

10 A sound wave from an underwater explosion approaches the surface at an angle of incidence of 60°. What is the angle of refraction in air?

11 What is the speed of waves in a stretched cord if a frequency of 160 Hz produces standing waves with 27.5 cm between adjacent nodes?
    Ans. 88 m/sec

12 Standing waves are produced in a stretched rope. If the distance between successive nodes is 2.5 ft, what is the wavelength? If the waves travel with a speed of 35 ft/sec what is the frequency?

13 Compressional waves with a frequency of 900 Hz are propagated through the air in a tube, the far end of which is closed by a piston. Standing waves are produced. The reflected wave reinforces the source at two successive positions of the piston differing by 19 cm. What is the velocity of the waves in the tube?

Ans. 342 m/sec

14 If a certain particle has a displacement given by $5 \sin 2\pi\nu t$, where $\nu = \frac{1}{12}$ Hz, find the displacement at times 1, 2, 3, . . . , 11, 12 sec. Plot the displacement as a function of time.

15 The displacement of a particle in micrometers is given by $y = 4 \sin 2\pi\nu t$. If $\nu = 10$ Hz, find the displacement and velocity of the particle at times t of 0.02 and 0.06 sec.

Ans. 3.80 μm, 77.7 μm/sec; −2.35 μm, −203 μm/sec

16 A wave is represented by $y = 0.8 \cos 4\pi(t - 0.333x)$, where y and x are in centimeters and t is in seconds. For this wave find the amplitude, period, frequency, and velocity.

17 A transverse wave in a string is represented by $y = 0.02 \sin 4\pi(2x - 15\ t)$, where y and x are in meters and t is in seconds. For this wave find the amplitude, the frequency, the period, the wavelength, the wave speed, and the maximum particle speed.

Ans. 0.02 m; 30 Hz; 0.0333 sec; 0.25 m; 7.5 m/sec; 3.77 m/sec

18 A transverse wave in a stretched wire is described by the equation $y = 0.03 \sin (80\pi t - 0.4\pi x)$, where x and y are in meters and t is in seconds. What are the frequency, amplitude, period, wavelength, and velocity of the wave? Is the wave standing or traveling?

19 If the displacement along a wave is given by $y = 0.05 \sin (20\pi t - 0.1\pi x)$, where distances are in meters and times in seconds, what are the amplitude, frequency, wavelength, and speed of the wave motion?

Ans. 0.05 m; 10 sec⁻¹; 20 m; 200 m/sec

20 If the displacement along a wave is given by the relation $y = 5 \sin 2(t - 6x)$, where x and y are in centimeters and t is in seconds, what are the amplitude, frequency, wavelength, and speed of the wave motion?

21 A wave $y_1 = R \sin (kx - \omega t)$ traveling to the right and a wave of the same amplitude and frequency $y_2 = R \sin (kx + \omega t)$ traveling to the left are superimposed to give a resultant $y = y_1 + y_2$. Show that the resultant is a standing wave $y = 2R \sin kx \cos \omega t$. [*Hint:* Use the formula for $\sin (\theta \pm \phi)$ given in Sec. A.1 of Appendix A.] Show that the wavelength of the resultant wave is given by $2\pi/k$.

CHAPTER TWENTY-ONE
SOUND WAVES

IN CHAP. 20 we discussed many of the general characteristics of wave motions and introduced the vocabulary of waves with words such as wavelength, frequency, amplitude, refraction, reflection, and interference. In the two chapters which follow we treat sound waves and their properties. In this chapter we are concerned with the speed of acoustic waves, the relationships between the physical and the psychophysical properties of the waves, the mechanical aspects of hearing, and the acoustic properties of rooms.

21.1 THE NATURE OF SOUND

Sound has its source in vibrating bodies. Consider the tuning fork of Fig. 21.1. As the tine A swings toward the right, it pushes air molecules and produces a region in which the molecules are crowded together. Such a region is called a *compression* or *condensation*. When the tine swings to the left, a region of reduced pressure called a *rarefaction* is produced. The next swing to the right produces another compression, and so forth. Thus, the sound wave consists of a series of alternate compressions and rarefactions. The molecules move back and forth along the line of propagation, which establishes the *longitudinal* nature of the waves. When the condensations and rarefactions reach the eardrum of a listener, they produce small inward and outward motions of the eardrum, and this starts the physiological processes of hearing.

21.2 THE VELOCITY OF SOUND

The velocity of sound in any medium depends upon its density and its elastic properties. In a wire or rod of a solid material the velocity of sound waves is given by

$$V = \sqrt{\frac{Y}{d}}$$

(21.1)

where Y is Young's modulus and d the density. In fluids the velocity is given by

$$V = \sqrt{\frac{B}{d}}$$

(21.2)

where B is the appropriate bulk modulus. For a gas B is the *adiabatic bulk modulus* given by γp, where p is the pressure and γ is the ratio

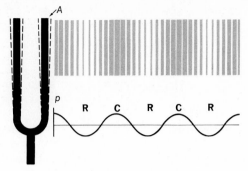

Figure 21.1
Sound waves from a tuning fork consist of alternate compressions and rarefactions. The waves shown below are at some distance from the fork where the wavefronts are almost plane over small arc lengths. Near the fork the wavefronts are more nearly spherical, and the amplitude falls off as $1/r$.

of two specific heats for the gas (Sec. 15.12). It is 1.67 for monatomic gases, 1.40 for diatomic gases, and 1.28 for many triatomic gases. For any gas

$$V = \sqrt{\frac{\gamma p}{d}} \qquad \left(\text{for air, } V = \sqrt{\frac{1.4p}{d}}\right) \qquad (21.3)$$

The speed of sound in air at 0°C is 331.45 m/sec (1,087 ft/sec). In hydrogen, which has a low molecular weight and therefore low density, the speed of sound at 0°C is about 1,270 m/sec.

According to Boyle's law, the ratio of the pressure to the density of a gas is constant if the temperature of the gas remains unchanged. Since the velocity depends on the *ratio* of pressure to density, it is constant for a given gas so long as the temperature is constant. However, the ratio p/d does depend on the temperature. For 1 kmole of gas of molecular weight M, $d = M/V$ while, by the general gas law (Sec. 14.7), $pV = RT$. Therefore, $d = pM/RT$ and

$$V = \sqrt{\frac{\gamma p}{d}} = \sqrt{\frac{\gamma RT}{M}} \qquad (21.3a)$$

Since the density at constant pressure is inversely proportional to the absolute temperature, the speed of sound in a gas increases with temperature according to the equation

$$V_t = V_0 \sqrt{\frac{T}{273}} = V_0 \sqrt{1 + \frac{t}{273}} \qquad (21.4)$$

where t is the Celsius temperature and V_0 is the speed at 0°C.

Example

Find the velocity of sound in dry air at a 100°C.

$$V_{100} = V_0 \sqrt{1 + \frac{t}{273}} = 331 \sqrt{\frac{373}{273}}$$

$$= 386 \text{ m/sec} \qquad \qquad ////$$

The reason the speed of sound in a gas increases with temperature is that the velocity of the molecules increases. In a sound wave the energy is transferred from one molecule to the next; in a gas the speed of the transfer cannot exceed the speed of the molecules. The speed of sound in a diatomic gas is about 74 percent of the average speed of the molecules.

21.3 FREQUENCIES AND WAVELENGTHS OF AUDIBLE SOUNDS

A typical young human ear can hear frequencies lying between 20 and 20,000 Hz. There is substantial difference in range among various individuals. Since the velocity of sound waves is given by the product of the frequency and the wavelength, and since sounds of all different frequencies have the same velocity in air, the range of wavelengths we hear runs from about 17 m to 1.7 cm. That all frequencies travel with essentially the same speed is clearly shown by the fact that the sounds from all instruments of an orchestra reach a listener at the same time. If velocity depended on frequency, the music would sound different indeed to a person far from an orchestra than to one close by.

Sound waves of higher frequency than the human ear can hear, known as *ultrasonic* sounds, are readily produced and are emitted by many animals. Ultrasonic frequencies can be heard by dogs, birds, and many other species; indeed, the bat locates obstacles and finds its way about by emitting ultrasonic frequencies which it detects as they are reflected from various obstacles.

21.4 PITCH, LOUDNESS, AND QUALITY

The word *sound* is used with two related but separate and distinct meanings. Sometimes we use the word to mean the *sensation due to the stimulation of the auditory nerve centers* and at other times to mean the *longitudinal waves transmitted through elastic mediums*. To the psychologist ''sound'' is usually used in connection with hearing, whereas much of the time in physics ''sound'' is a type of wave motion, even though it may not be heard. For example, if we oscillate a wire at a frequency of 30,000 vib/sec, human ears cannot hear the waves produced. In the sense that longitudinal waves exist in the air there is a sound, but from the point of view of hearing there is none.

To a listener sound is ordinarily characterized by the psychophysical properties *pitch*, *loudness*, and *quality*. These subjective characteristics are, of course, related to physical properties of the waves, namely, the frequencies present, their amplitudes, and their phase relations.

In general, the pitch of a sound depends upon the frequency of the *fundamental*, or the lowest frequency present in the sound wave. As the fundamental frequency of a sound source is increased, the pitch is raised. A simple experiment which shows there is a direct relation between pitch and frequency can be performed with the aid of a siren which consists of a disk with a number of holes uniformly spaced on concentric circles. If a jet of air is directed against the holes while the disk is in rotation, a puff goes through each hole as it passes the jet. When the number of puffs per second is increased by speeding up the

Figure 21.2
Sound waves emitted by a trumpet
blown in different ways.

disk, the pitch becomes higher. Although pitch is determined primarily
by the frequency, it depends to some extent on loudness and quality
as well.

Loudness describes the magnitude of the auditory sensation
produced by a sound. In addition to depending on the intensity and
the physical composition of the sound, it also depends on the auditory
acuity and experience of the individual listener. For a given frequency
the loudness of a sound is closely related to the intensity; it increases
roughly as the logarithm of the intensity. For a given intensity the
loudness depends rather sensitively on the frequency.

Sounds which have the same loudness and pitch may have very
different *tone qualities*. For example, if a piano and a trumpet play the
same note at the same loudness, it is easy to distinguish between them.
The reason is that neither sound source produces a single frequency;
each sends out a group of frequencies. As we have seen, the lowest
frequency is called the *fundamental;* it is this frequency which primarily
determines the pitch. In addition to the fundamental there are present
higher frequencies, called *overtones*. When two notes differ in quality,
or *timbre*, the frequencies, phases, and relative intensities of the various
overtones are typically different; further the temporal evolution of the
tone is important in determining its quality. Figure 21.2 shows how the
characteristics of the sound emitted by a trumpet change according to
the way the musician blows it.

21.5   THE EAR

The process of hearing begins when sound waves impinge upon an ear.
The human ear (Fig. 21.3) may be divided into three distinct parts: the
external ear, the middle ear, and the inner ear. The external ear collects
the sound waves and directs them along the auditory canal to the
eardrum, which is caused to vibrate. These vibrations are carried across
the middle ear by three small bones, called the *hammer,* the *anvil,* and
the *stirrup* because of their shapes. Vibrations of the eardrum are
passed from hammer to anvil to stirrup, which in turn is attached to
the "oval window" separating the middle ear from the inner ear. The
principal parts of the inner ear are the *cochlea* and three *semicircular
canals.* The cochlea, shaped somewhat like the shell of a snail, is filled
with fluid which is set into vibration by the movement of the oval window.
Inside the cochlea are about 2.5 turns of the basilar membrane, to which

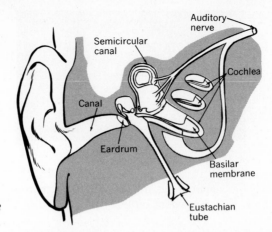

Figure 21.3
Anatomical structure of the human ear.

are connected the roughly 30,000 nerve endings which initiate signals to the brain. Figure 21.4 indicates the location of the nerve endings which show maximum response to a pure tone of 700 Hz and moderate loudness. The semicircular canals are not part of the hearing mechanism but are vital in keeping one's balance.

The human ear is a remarkably sensitive detector of sound waves. At 3,000 Hz the faintest sound the ear can hear has pressure variations of about 2×10^{-5} N/m^2, which corresponds to a displacement amplitude of less than 10^{-11} m, or about one-tenth the diameter of a molecule! Yet it can also hear sounds with pressure variations a million times this great.

The loudness of a sound, as judged by the ear, is approximately proportional to the logarithm of the intensity. This fact has led physicists to measure the intensity levels of sounds on an arbitrary logarithmic scale. The intensity level n of a sound wave *in decibels* (dB) is defined by the equation

$$n = 10 \log \frac{I}{I_0} \qquad (21.5)$$

where I_0 is a reference level, taken as 10^{-12} W/m^2, which is roughly that of the weakest sound which can be heard. For a sound which has an intensity level $n = 60$ dB, the logarithm of I/I_0 is 6, so $I = 10^6 I_0$. In other words a 60-dB sound level corresponds to an intensity 1 million times that of the reference level. For ordinary speech, n ranges from 30 to 70 dB, while for loud music it may reach 100.

The lowest sound level which can produce an audible sound depends on the frequency. The range of intensity levels and the frequencies which can be heard by a typical ear is shown in Fig. 21.5. A sound more intense than that corresponding to the upper curve produces pain rather than hearing.

Figure 21.4
Parts of the cochlea where maximum response occurs for a pure tone of frequency 700 Hz at moderate loudness.

21.6 REFRACTION OF SOUND

When sound waves pass from one medium to another, there is usually a change in velocity. As a consequence, the direction of propagation

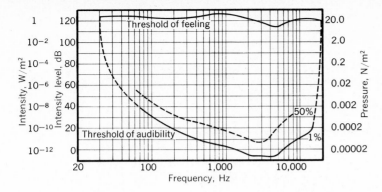

Figure 21.5
Chart of intensity levels showing the threshold of audibility and the threshold of feeling. Only 1 percent of a large number of people tested had the auditory acuity indicated by the lower solid line, while 50 percent could hear intensities represented by the dashed line.

is changed, provided the incident waves meet the surface of separation obliquely.

Since the speed of sound in warm air is greater than it is in cold air, the direction of propagation of the sound changes continuously as it passes from air at one temperature to air at a different temperature. If the air is at rest and the temperature and density are uniform, a wavefront from a point source on the surface of the earth is spherical and the sound travels in straight lines. When the air at the ground is warmer than it is at higher altitudes, the speed of the sound is greater at the surface of the earth and a wavefront is no longer spherical. Since the direction of propagation is perpendicular to the wavefront, the sound is deflected upward (Fig. 21.6) and it cannot be heard for as great distances at it could if this distortion did not take place. When air at the ground is colder than it is at higher altitudes, the sound travels more slowly near the ground. The sound is deflected downward, and the distance at which the sound can be heard is increased. This sometimes happens over a lake at the end of a hot day.

When the wind is blowing, the speed of the sound with respect to the earth is decreased in the direction from which the wind comes and increased in the direction toward which the wind is blowing. Usually near the earth, the higher the altitude, the greater the velocity of the wind and the greater the change in the speed of the sound with respect to the earth. This unequal change in speed causes a distortion in the wavefront. On the windward side of the source the speed is greater at the ground than at points above the ground (Fig. 21.7). This inequality

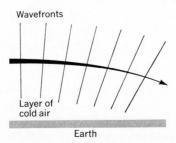

Figure 21.6
Wavefronts are deflected upward when air at the ground is warmer than air higher up and downward when air at the ground is cooler.

Figure 21.7
Wavefronts change shape and
propagation direction in a wind.

causes the wavefront near the ground to be inclined to the vertical and the line of propagation to be directed upward from the earth. On the side of the source toward which the wind is blowing, the speed near the ground is less than at higher altitudes. In this case, the direction of propagation is bent toward the ground, making it possible for the sound to be heard at greater distances.

21.7 REFLECTION OF SOUND

If an observer stands some distance in front of a cliff and produces a sound, the sound is returned to him with its characteristics essentially unchanged. If the observer is 340 m from the cliff, it takes about 2 sec for the sound to return to him. We call the reflected sound an *echo*. The roll of thunder is due to the reflection of the original sound by various bodies including clouds. These echoes reach the observer at different times and produce the rolling continuation of the sound.

When an orchestra is to play out of doors, a large reflecting shell is often provided behind it so that sound waves are reflected toward the audience, thus greatly increasing the loudness.

Sound waves are reflected when they reach the walls of a speaking tube, which thereby prevents them from spreading out. Consequently, the intensity of the sound does not decrease appreciably as the wave advances, and the sound may be heard with only slightly diminished intensity at the other end of the tube many yards from the speaker. In an ear trumpet the waves entering the wide end are gradually diminished in area by reflections in the wall until, at the small end, most of the energy of the incident waves is concentrated over a small area. As a consequence, the intensity is greatly increased.

21.8 ARCHITECTURAL ACOUSTICS

When sound waves are produced in a room, they spread out until they strike the walls, ceiling, or floor. Here they are partly reflected, partly absorbed, and partly transmitted. A hard smooth wall reflects most of the sound; it transmits and absorbs little. On the other hand, a porous, soft material absorbs most of the sound energy and reflects little.

If a steady sound source is maintained for some time in a room, the sound level in the room builds up, as shown in Fig. 21.8. After a short time a state of equilibrium is reached in which the energy lost each second through absorption is equal to the sound power provided by the source. When the source is shut off, the intensity of the sound dies down, as indicated in the second half of the curve in Fig. 21.8.

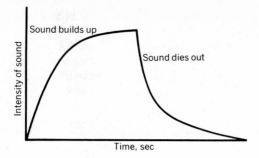

Figure 21.8
Rise and decay of sound in a room.

The echoing and reechoing of sounds in a room because of repeated reflections is known as *reverberation*. *The reverberation time of a room is defined as the time required for the sound level to fall 60 dB after the sound source is shut off.* A drop of 60 dB means that the intensity falls to one-millionth of its original value. If the reverberation time of a room is too long, the sound waves bounce back and forth many times; a speaker's words become blurred because these reflected waves blend with the new sound waves. Such a blending effect is familiar to anyone who has sung in a hard-walled shower room that has a long reverberation time. On the other hand, if the reverberation time is too short, sounds are thin and weak; we say the room is "dead." In general, a somewhat longer reverberation time is desirable for music than for speech. Reverberation times of about 1 sec for a small room and 1.5 to 2 sec for a larger one are desirable for ordinary uses. For making phonograph recordings, a very dead room with virtually no reverberation is used because the desired reverberation properties are supplied by the room in which the listener sits.

The reverberation time of a classroom or auditorium may be estimated by the use of a formula due to Sabine, a pioneer in the field of architectural acoustics. He found that the reverberation time t_R in seconds for a typical room was given by

$$t_R = 0.16 \frac{V}{A} \tag{21.6}$$

where V is the volume of the room in cubic meters and A is the total *absorbing power* of the room and its contents. The total absorbing power of the room is found by adding the contributions of all sound-absorbing surfaces, the contribution of each surface being the product of the exposed area in square meters and the appropriate absorption coefficient. The absorption coefficient, in turn, is the fraction of the incident sound wave energy removed either by absorption or transmission. It is 1.00 for an open window because all the incident energy goes on through and is thus removed from the room. The absorption coefficients of a number of typical materials are listed in Table 21.1.

Example

A shower room has the dimensions 5 by 4 by 3 m. All the walls are of tile, and the door has the same absorption coefficient as tile. Find

Table 21.1
Sound-absorption Coefficients

Open window*	1.00
Brick wall	0.02
Clay tile	0.03
Concrete	0.02
Glass	0.03
Hair felt	0.40
Heavy curtains	0.50
Perforated acoustic ceiling	0.60
Plaster	0.03
Wood	0.05

* Absorbing power of typical person is 0.5 m² of open window.

the approximate reverberation time if one man is showering and singing in this room.

A = (20 + 20 + 12 + 12 + 15 + 15) × 0.03 + the absorbing power of the man, which we shall take to be equivalent to 0.25 m² of open window (at least half the absorption of a person is associated with his clothing).

$$A = 2.82 + 0.25 = 3.07$$

$$t_R = 0.16\,\frac{V}{A} = 0.16 \times \frac{60}{3.07} = 3.1 \text{ sec} \qquad ////$$

Example

A classroom has a volume of 400 m³. There are exposed 600 m² of tile and 500 m² of wood in chairs, doors, and woodwork. There are 30 persons occupying the room. Estimate the reverberation time.

$$A = (600 \times 0.03) + (500 \times 0.05) + (30 \times 0.5 \times 1)$$
$$= 18 + 25 + 15 = 58$$
$$t_R = 0.16 \times \tfrac{400}{58} = 1.1 \text{ sec} \qquad ////$$

21.9 INTERFERENCE OF SOUND WAVES

When two trains of waves pass the same point, the displacement of a particle at this point is the resultant of the displacements which each of the two wavetrains would produce if it acted alone. If the two trains have the same wavelength and direction, the resultant amplitude depends on the amplitude of the two waves and on the phases of the two component disturbances. If the two arrive at the same phase, they reinforce one another, and we have constructive interference; if they arrive one-half wavelength out of phase, the resultant amplitude is the difference between the individual amplitudes. If these amplitudes are equal, the waves cancel one another and we have total destructive interference.

The interference of sound waves can be demonstrated with the apparatus of Fig. 21.9. Sound waves from a source O travel to point D by separate paths ACD and ABD. If the lengths of the two paths are equal, the waves arrive at the ear in phase, and we have constructive interference. If the path ABD is increased in length, the waves travel different distances and the resultant amplitude is less than when the

Figure 21.9
Apparatus for showing interference of
two sound waves which have traveled
paths of different lengths.

interference was constructive. When path ABD is one-half wavelength longer than path ACD, the waves are one-half wavelength out of phase and destructive interference results. As the path ABD is increased still more, the amplitude of the resultant disturbance grows until we have constructive interference once more when path ABD is one wavelength greater than ACD. In general, there is destructive interference when path ABD exceeds ACD by an odd number of half wavelengths and constructive interference when the paths differ by an integral number of wavelengths.

21.10 BEATS

When two steady sound sources of the same frequency are activated together, the intensity of the sound at a given listening point remains constant. If the frequencies are slightly different, however, the intensity of the sound fluctuates. There are bursts of louder sound with comparative silence between them. Each burst occurs when the disturbances from the two sources reinforce one another, while the periods of relative silence occur when the two waves interfere destructively. The fluctuations in intensity when two sound sources of slightly different frequency interfere are called *beats*.

The origin of beats is as follows. Suppose that at a certain instant compressions from both sound sources arrive simultaneously (A of Fig. 21.10). The amplitude of the resultant disturbance is large, and the intensity relatively high. A short time later the more rapidly vibrating source is one-half vibration ahead of the other, and a compression from one source arrives at the same time as a rarefaction from the other (B of Fig. 21.10). The two disturbances interfere destructively, and a minimum intensity results. A little later the more rapidly vibrating source has picked up a full vibration, and once again an intensity maximum is observed.

For two sounds differing in frequency by 1 Hz there is reinforcement once each second, and we observe one beat per second. As the frequency difference is increased, a higher beat rate is produced. In general, the number of beats observed per second is equal to the difference between the frequencies of the two sound sources:

$$\text{Number of beats per second} = N = \nu_1 - \nu_2 \qquad (21.7)$$

where ν_1 and ν_2 are the frequencies of the two sources.

Beats are used in tuning string instruments. As two strings are brought more and more nearly into unison, the number of beats per second becomes less; when no beats are observed, the strings have the same frequency.

Figure 21.10
Beats arise from the superposition of sound waves differing slightly in frequency.

QUESTIONS

1 Why is thunder often so prolonged when the single lightning flash which produced it is almost instantaneous?

2 Why doesn't the inverse-square law apply exactly to the decrease in sound intensity with distance? Is it possible for the intensity to decrease less rapidly than the inverse-square law predicts? Under what circumstances?

3 What evidence is available from everyday life to support the assertion that the speed of audible sounds is independent of frequency?

4 Upon what properties of a transmitting medium does the speed of sound depend? Explain qualitatively why these properties affect the speed.

5 How can the speed of sound in air be independent of the pressure when Eq. (21.3) tells us the speed is given by $\sqrt{1.4p/d}$?

6 Why is the speed of sound in a gas a little less than the average speed of the molecules?

7 How can the position of a gun be determined by using three listening posts and measuring the time of arrival at each post with accuracy?

8 Why is it often true that sounds can be heard for particularly great distances at night or after a rain?

9 Beats are sometimes called *difference tones.* Why is that term used?

10 Draperies and furniture often improve the acoustics of a room. Why? On what factors do the acoustics of the room depend?

11 What is meant by the statement that the sound level in a room is 40 dB? 80 dB?

PROBLEMS

Take the speed of sound in air to be 340 m/sec or 1,100 ft/sec.

1 What is the speed of sound in helium at 0°C and 1 atm pressure if 4 g of helium under standard conditions occupies a volume of 22.4 liters?
Ans. 972 m/sec

2 The vertical walls of a canyon are 2,720 m apart. A man in the canyon fires a gun and hears the echo from the farther wall 3 sec after the echo from the nearer wall. How far is he from the nearer wall?

3 A ship sends signals to a neighboring ship. The sound waves travel by two paths, one in air, and the other in seawater. The signals are heard on the neighboring ship at intervals 6 sec apart. How far is it from one ship to the other if the speed of sound in seawater is 1,450 m/sec?
Ans. 2,660 m

4 At what temperature would the speed of sound in air be 400 m/sec if it is 331.5 m/sec at 0°C?

5 What is the velocity of a compressional sound wave in a steel rod for which Young's modulus of elasticity is 24×10^{10} N/m² and the density is 7,800 kg/m³?
Ans. 5,550 m/sec

6 A timer sets his watch by the report of a gun 100 yd away. What is the error due to the time required for the sound to travel from the gun to his ear?

7 An air-driven turbine breaks up the stream of air into 40 pulses per revolution. What frequency of sound will be heard when the turbine is rotating at the rate of 4,800 rev/min? What is the wavelength of the sound waves generated?
Ans. 3,200 sec⁻¹; 0.106 m

8 Sound travels in water at the rate of 1,450 m/sec. What is the bulk modulus of elasticity of water?

9 A workman strikes the steel rail of a railroad track with a hammer. The sound thus produced reaches an observer through the rail and through the surrounding air. The difference in time is 2 sec. If the speed of sound in steel is 5,000 m/sec, how far is the observer away from the workman?
Ans. 730 m

10 A worker, in pounding a spike, strikes a steel railroad track. A person 0.8 mile along the track hears two reports due to the pounding. One travels through air and one through the rail. What time interval separates the reports if the speed of sound in steel is 16,300 ft/sec?

11 Both loudspeakers of a stereo hi-fi system are sending out, in phase, a pure frequency of 500 Hz. A listener, originally equidistant from the speakers, moves to one side until the note fades to a minimum loudness. If he is then 2.87 m from the closer speaker, how far is he from the farther one?
Ans. 3.21 m

12 Two speakers of a hi-fi set are 3 m apart. When

they are playing a pure single frequency, there is a minimum intensity 4 m immediately in front of each speaker. As one moves along the line 4 m from the line of the speakers, the intensity reaches a single maximum between these two minima. Calculate the frequency.

13 Two strings A and B originally produced the same frequency. The tension of B was released slightly, thereby reducing the frequency, and the strings then produced seven beats per second when sounded together. If the frequency of A is 440 Hz, what is the frequency of B?

Ans. 433 Hz

14 What is the intensity level of a sound of intensity 10^{-8} W/m² relative to the reference level of 10^{-12} W/m²?

15 Acoustic treatment of certain surfaces in a foundry reduced the sound level by 20 dB. How many times greater was the intensity before the treatment than after?　　　　*Ans.* 100

16 The threshold of pain involves an intensity level of roughly 120 dB. What is the corresponding intensity in watts per square meter?

17 If a loudspeaker is regarded as a point source radiating equally in all directions, what acoustic power must it develop to produce a sound level of 60 dB at a distance of 100 m?

Ans. 0.126 W

18 If the intensity due to five violins is five times that due to a single violin, how many decibels is the sound level raised by having five violins playing rather than a single one?

19 A sound wave travels through two branches of a tube which are 1 and 1.3 m long, respectively. List the three lowest frequencies which would suffer destructive interference if the waves were

recombined after traveling through the branches.

Ans. 567, 1,700, and 2,830 Hz

20 An oscillator emits a sound with a wavelength of 0.34 m which is divided into two parts in the apparatus of Fig. 21.9. The two parts travel the same distances, so that constructive interference is experienced. In order to eliminate the sound, one path is increased. What is the shortest distance this path may be lengthened to satisfy this condition? If it is lengthened still further, the intensity increases and then decreases. What total change in path is necessary to obtain the second case of destructive interference? What total change in path is necessary to obtain the third case of destructive interference?

21 An auditorium is essentially a rectangle 40 by 30 by 10 m. It has a perforated acoustic ceiling, concrete floor, and plaster walls. It contains 800 seats, each of which is equivalent to 1 m² of wood. Find the approximate reverberation time of the auditorium when empty and when full.

Ans. 2.3 sec; 1.6 sec

22 A lecture room has a volume of 20,000 m³. If it has 1,200 m² of acoustic ceiling, 2,500 m² of plaster, 1,000 m² of concrete, and 4,000 m² of wood, find the reverberation time when the room is empty and when it holds 500 people.

23 Show that the ratio of the speed of sound in a gas to the effective (rms) speed of the molecules is $\sqrt{\gamma/3}$.

CHAPTER TWENTY-TWO
SOUND SOURCES

WE HAVE BEEN CONSIDERING the propagation and detection of sound waves, but we have not examined precisely how these acoustic waves originate. We turn now to a discussion of how various kinds of sound sources produce the periodic changes in pressure we hear as sound. For some of the simpler sound sources we can predict both the fundamental frequency and the prominent overtones from the length of the source.

22.1 SOUNDING BODIES

Ordinary sound waves are set up by vibrating bodies. A typical source in vibration sends a series of alternate compressions and rarefactions through the air. Most sound sources do not produce a single frequency but a combination of many frequencies. The lowest prominent frequency is the *fundamental;* the higher frequencies are called *overtones*. In vibrating strings and in organ pipes the overtones are usually harmonics of the fundamental. *A harmonic is an overtone which has a frequency that is an integral multiple of the fundamental frequency.* The frequency of the third harmonic is three times that of the fundamental, and so forth.

Most of the overtones of bells, drums, and vibrating plates are not harmonically related to the fundamental. A few sound sources are especially designed to produce as nearly as possible a single frequency, or a *pure tone*. A tuning fork is an example of such a source. It is essentially a rod bent in the form of a U with the central region reinforced to suppress overtones; even so, careful observation often reveals the presence of undesired frequencies.

In most musical sounds, the same frequency is emitted for an appreciable period of time during which the amplitude does not decrease rapidly. In a noise, on the other hand, we usually have a sudden burst of a wide range of frequencies with no regularity and a marked drop in intensity during the emission of a single vibration. Of course, the distinction between noises and musical sounds is not sharp and definite. What is music for one person may be noise to another.

22.2 WAVES IN WIRES AND STRINGS

Many musical instruments make use of stretched strings as sound sources. These strings may be set into vibration by striking, plucking, bowing, or strumming. When the string is displaced at some point and then released, the disturbance is passed from one element of the string to the next and transverse waves proceeding in both directions in the string are produced. These waves are reflected at the fixed ends of

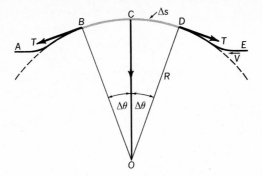

Figure 22.1
Grossly enlarged representation of a transverse wave pulse on a stretched cord.

the string, return in the opposite direction, and go on to the other ends, where they are again reflected. Of the many frequencies into which the initial disturbance may be analyzed, only those which are suitable for establishing standing waves in the string are maintained for any length of time.

The velocity V of a transverse wave along a flexible stretched string depends on the tension T and on the mass per unit length m_l of the string. It is shown below that

$$V = \sqrt{\frac{T}{m_l}} \qquad (22.1)$$

Consider a wave traveling toward the right in the cord AE of Fig. 22.1. Assume that while the pulse, or wave, is moving toward the right with a velocity V, the cord is made to move toward the left with an equal velocity. As a result of these superposed velocities, the wave pulse remains at rest. For simplicity, assume that the small arc $BD = \Delta s$ is circular in form. The component of the tension T along the radius CO is $T \sin \Delta\theta$, so the net force on Δs toward O is $F = 2T \sin \Delta\theta = 2T \Delta\theta$ (approximately). Since $2\Delta\theta = \Delta s/R$, $F = T \Delta s/R$. This force provides the centripetal acceleration required if the mass $m_l \Delta s$ is to traverse the circular path. By Eq. (9.2),

$$m_l \Delta s \frac{V^2}{R} = F = T \frac{\Delta s}{R}$$

from which $V = \sqrt{T/m_l}$.

Of the standing waves established in a stretched string, the one with the longest wavelength is that for which the string vibrates as a single segment (Fig. 22.2). The longest wavelength corresponds to the

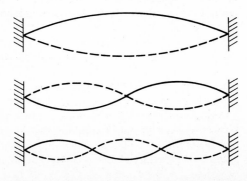

Figure 22.2
Vibrations in a stretched string fixed at both ends, showing the fundamental mode and the first two overtones.

lowest frequency and therefore to the *fundamental* of the string. The standing wave of next longest wavelength is that for which the string vibrates in two segments. In this case the length of the string is equal to the wavelength. The third longest standing wave which can exist in the stretched string occurs when the string vibrates in three segments, the next in four segments, etc.

The allowed modes of vibration for a stretched string are easy to determine if one remembers that both ends of the string are fastened. These ends cannot move and must therefore behave as nodes for any possible standing wave. We are dealing with a situation in which we impose a condition on the ends of the strings, namely, that they remain at rest. Such conditions, known as *boundary conditions*, play a very important role in determining the types of vibrations which are allowed in various kinds of systems. Once we have established the boundary conditions that there be nodes at the ends of the stretched strings, we can readily find the fundamental by finding the longest wave that we can produce which has nodes at both ends. The first overtone corresponds to the next longest possibility, and so on.

If a string is plucked, it vibrates in such a way that both the fundamental and overtones are present. The overtones are all harmonics, and the note from a string consists of the fundamental together with several harmonics.

22.3 HARMONICS OF A STRING

Let L be the length of the string. Then, for the fundamental, the wavelength is $2L$ (Fig. 22.2). Since for any wave motion $V = \nu\lambda$, we have, for the fundamental frequency ν_1,

$$\nu_1 = \frac{1}{2L}\sqrt{\frac{T}{m_l}} \tag{22.2}$$

The frequency of the fundamental depends on length, tension, and mass per unit length. A musical instrument such as the piano affords an illustration of how these factors apply. The strings which play the bass notes are ordinarily heavy, thus having large m_l, and they are also long. Strings for the high notes are short, light, and tightly stretched. In tuning a piano, the tension is varied. In instruments such as the violin in which there are only four strings, the strings are of the same length, but they have different masses per unit length and different tensions. When the violinist wishes to play different frequencies on the same string, he varies the length of the string by placing his finger at the point at which he wishes to create a node.

The first overtone for the vibrating string occurs when $\lambda = L$ (Fig. 22.2), and therefore the frequency of the first overtone is given by

$$\nu_2 = \frac{1}{L}\sqrt{\frac{T}{m_l}} \tag{22.3}$$

The first overtone has a frequency twice that of the fundamental and is therefore the second harmonic. For the second overtone $\lambda = 2L/3$, and the frequency is

$$\nu_3 = \frac{3}{2L}\sqrt{\frac{T}{m_l}} \qquad \text{(second overtone = third harmonic)} \qquad (22.3a)$$

In general, the allowed overtones of stretched strings consist of all harmonics.

Most harmonics blend well with the fundamental to produce pleasing tones. The seventh harmonic is an exception. In order to suppress the seventh harmonic, piano strings are struck at a point where the seventh harmonic would have a node. This assures that there cannot be a perfect node at this point. Violins are often bowed in the vicinity of a node for the seventh harmonic in order to suppress this relatively unpleasant overtone. The superior tonal quality of an expert violinist is associated with his ability to produce not only the desired fundamental but also a combination of overtones which are pleasing. The relative intensities of overtones can be varied over substantial ranges by an expert.

22.4 CLOSED ORGAN PIPES

In an organ pipe the vibrating body is a column of air. At one end of a pipe, the column of air is set into vibration by sending a narrow jet of air toward a thin edge or lip (Fig. 22.3). This end of the pipe is always open so that sound waves can be transmitted from the pipe into the surrounding air. In a closed pipe the other end is blocked off. Thus, a closed organ pipe is closed at one end and open at the other. When air first strikes the lip, a condensation starts down the pipe, is reflected at the closed end, and returns to the lip as a condensation. When it reaches the lip, it pushes the air stream outside the lip, and this starts a rarefaction down the tube. The rarefaction is reflected at the closed end, returns to the lip as a rarefaction, and draws the stream into the pipe again, thereby starting a new compression. Thus the airstream is made to move back and forth across the lip with its period determined by the time required for a compression to travel up and back and a rarefaction to travel up and back. The period is four times the time required for sound to traverse the length of the tube.

An alternative approach to determining the fundamental frequency involves the application of boundary conditions. At the closed end of the pipe an air particle is unable to move forward when a compression comes down the tube. We have a *displacement node at the closed end of an organ pipe*. On the other hand, a particle at the open end is not restrained, and the amplitude of the vibration of a particle is a maximum there. There is always an *antinode at the open end of a pipe;* the displacement there is maximum. The possible standing waves which can be set up in a closed organ pipe are those which have a node

Fundamental tone ($L = \frac{1}{4}\lambda$)

Figure 22.3
Displacement pattern (dashed line) for the fundamental in a closed organ pipe.

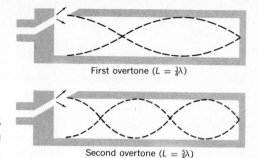

First overtone ($L = \frac{3}{4}\lambda$)

Figure 22.4
First overtone and second overtone in
a closed organ pipe.

Second overtone ($L = \frac{5}{4}\lambda$)

at the closed end and an antinode at the open end. The longest wavelength which can fit into the organ pipe is indicated in Fig. 22.3. In this case the wavelength is four times the length L of the pipe. The frequency ν_1 of the fundamental is given by the ratio of the velocity V of the wave motion, which in this case is the velocity of sound in the gas in the organ pipe, to the wavelength λ. Therefore,

$$\nu_1 = \frac{V}{4L} \tag{22.4}$$

There are other possible modes of vibration for this column of air which have a node at the closed end and an antinode at the open end. The next lowest frequency which can be emitted involves the vibration in which there is one additional node in the pipe (Fig. 22.4). For this case the wavelength is $4L/3$; the *first overtone is the third harmonic,* which has a frequency

$$\nu_3 = \frac{3V}{4L} \tag{22.5}$$

The second overtone is also shown in Fig. 22.4. Its frequency, $\nu_5 = 5V/4L$, is that of the fifth harmonic. The closed pipe can emit any frequency which is an odd integer times the fundamental frequency. Harmonics which have frequencies 2, 4, 6, . . . times the fundamental frequency cannot be sustained in a closed organ pipe. *Closed organ pipes emit only odd harmonics.*

The fundamental depends on the velocity of sound in the gas in the tube. If hydrogen is placed in the tube instead of air, the fundamental frequency is almost quadrupled. For any gas the fundamental depends upon the temperature, since the speed of sound is a function of temperature. As the temperature is raised, the frequencies emitted go up.

22.5 OPEN PIPES

The boundary condition for an open end requires that the oscillation amplitude be maximum. Since both ends of an open pipe are open, there is an antinode at each end. The wavelength of the longest possible standing wave is twice the length of the pipe, as shown in Fig. 22.5. Since $V = \nu\lambda$, the frequency ν_1 of the fundamental is

$$\nu_1 = \frac{V}{2L} \tag{22.6}$$

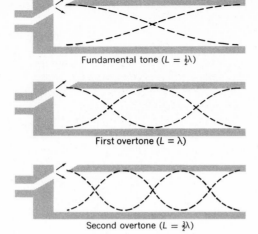

Fundamental tone ($L = \frac{1}{2}\lambda$)

First overtone ($L = \lambda$)

Second overtone ($L = \frac{3}{2}\lambda$)

Figure 22.5
Fundamental, first overtone, and
second overtone of an open organ
pipe.

The fundamental of an open pipe has twice the frequency of the fundamental for a closed pipe of the same length. The next possible mode of vibration (Fig. 22.5) is that in which the pipe has a length equal to one wavelength. This gives rise to the first overtone (second harmonic), with a frequency

$$\nu_2 = \frac{V}{L} \tag{22.7}$$

For the second overtone, $\nu_3 = 3V/2L$, which is the third harmonic. The frequencies of the harmonics of the open pipe are in the ratio of the integers 1, 2, 3, 4, ... For an open pipe it is possible to have all harmonics; only the odd harmonics are emitted by a closed pipe.

Throughout this discussion we have assumed that an antinode occurs exactly at the open end of the pipe. Actually the antinode lies outside this open end by an amount which depends on the diameter of the pipe if round and on its shape and size if it is not round. The length must be corrected slightly for exact calculation of the frequencies.

22.6 MUSICAL SCALES

Some combinations of sounds are pleasing to the ear; others are not. What is felicitous depends in no small measure on the training of the listener. As long ago as 530 B.C. Pythagoras experimented with stretched strings and found that simultaneous notes from strings were harmonious when the ratios of the frequencies were the ratios of two small integers. When one frequency is double another, we say they differ by one *octave;* this ratio of 2 to 1 is a pleasing one. On the other hand, ratios such as 100 to 81 are ordinarily displeasing, at least in part because of the existence of unpleasant beat notes. Modern occidental music is written with scales which consist of notes with frequencies related by fairly simple ratios. We call the ratio of the frequencies of two sounds the *musical interval* of the two notes. The diatonic C major scale is an example of a typical scale which is rather commonly used. The relationships among the notes in this scale are shown in Table 22.1.

A piano can be tuned to play a diatonic scale in any one key, but then the frequencies are not satisfactory for other keys. In order for a piano to be able to play in a number of different keys, it is customary to make certain compromises in which the piano is not tuned perfectly for any key. The compromise usually adopted is known as the *equal-tempered scale*. In one octave of this scale there are 13 notes and 12 intervals; there is a constant ratio between the frequencies of adjacent notes. This ratio is the twelfth root of 2, or 1.05946. The standard frequency of this scale is A = 440 sec^{-1}, so that this note agrees with that on the diatonic C major scale in Table 22.1. No other note in the octave has exactly the same frequency on the two scales.

The physics of music and musical instruments has many fascinating sides. Persons interested in these topics can find a delightful presentation in A. H. Benade, "Horns, Strings, and Harmony" (Anchor Books, Doubleday & Company, Inc., Garden City, N.Y., 1960).

22.7 RESONANCE

If two identical tuning forks are placed a few feet apart and one of them is set into vibration, the second fork also begins to vibrate (Fig. 22.6). This can be shown by stopping the vibrations of the first fork by grasping the tines. Energy has been transferred through the air from the first fork to the second. Because the second fork has the same frequency as the first, the compressions set out by the first fork arrive at just the right time to build up the amplitude of vibration in the second fork. The two forks are said to be in *resonance*. When two bodies are in resonance, a substantial amount of energy can be transferred to one of the bodies through a series of very small, but perfectly timed, impulses from the other. The amplitude is built up just as the amplitude of a child's swing can be built up by small, well-timed pushes.

When marching men cross a bridge, they are often commanded to break step because if a bridge structure happened to be resonant to the frequency of the steps, large and perhaps destructive vibrations could be set up. In automobiles very annoying noises are sometimes produced at certain speeds when some loose object happens to be

Table 22.1
Diatonic C Major Scale
(Based on A = 440 sec^{-1})

Major scale note	C	D	E	F	G	A	B	C′
Name	Do	Re	Mi	Fa	Sol	La	Ti	Do
Frequency	264	297	330	352	396	440	495	528
Interval relative to C	1	$\frac{9}{8}$	$\frac{5}{4}$	$\frac{4}{3}$	$\frac{3}{2}$	$\frac{5}{3}$	$\frac{15}{8}$	2
Musical interval		$\frac{9}{8}$	$\frac{10}{9}$	$\frac{16}{15}$	$\frac{9}{8}$	$\frac{10}{9}$	$\frac{9}{8}$	$\frac{16}{15}$
Musician's interval		Major tone	Minor tone	Semi-tone	Major tone	Minor tone	Major tone	Semi-tone

Figure 22.6
Resonance occurs when two vibrators with the same frequency interact so the vibrations in one induce sympathetic vibrations in the second.

resonant to a small impulse which is received with the proper timing. The designers of aircraft must be careful to avoid flutters and oscillations which may build up to intolerable levels by virtue of resonance.

On the other hand, resonances are sometimes very desirable. If we want a tuning fork to produce a louder sound, we may hold it over a tube which we gradually fill with water until the air column is resonant to the frequency emitted by the tuning fork. This occurs when the air column, closed at one end by the water, satisfies Eq. (22.4).

22.8 THE HUMAN VOICE

When we speak, the lungs, by their bellows action, force a stream of air between the vocal cords, two tightly stretched membranes in the throat, thereby setting them into vibration. The frequency of vibration can be changed by varying the tension in the vocal cords. The vibrations start a train of sound waves through the vocal passages. The tongue, the lips, and the cavities of the chest, nose, and throat interact with this wavetrain. To prove that the quality of the sound is affected by resonant columns in the head, one need only hold one's nose closed while speaking.

The differentiation of speech sounds is nearly all accomplished by the mouth and by positioning the lips. The *voiced sounds* include all the vowel and consonant sounds except p, t, ch, k, f, s, th (thin), and $s\mathring{h}$. The vocal cords do not enter into the production of these latter speech sounds, which arise from vibrations set up in the mouth itself.

Vocal sounds are transmitted through the air by exceedingly complex pressure waves. The amplitudes and frequencies of the various components present in speech sounds vary from one voice to another, but average speech includes frequencies from about 60 to 6,000 Hz. Most of the speech energy is carried by the vowel sounds. The power output of the normal human voice is only about 10^{-5} W.

22.9 THE DOPPLER EFFECT

As an automobile traveling at high speed and sounding its horn passes, the pitch heard by a pedestrian drops sharply. When a source of sound is moving toward an observer or an observer toward a sound source,

Figure 22.7

A listener moving with speed v_L toward a sound source passes in 1 sec all the waves included in the distance numerically equal to $V + v_L$, where V is the speed of the sound.

the pitch heard is higher than the normal pitch. When the source moves away from the observer or the observer from the source, the pitch is lowered.

Consider first the case in which the observer approaches a sound source at rest relative to surrounding air. Then the observer passes more waves each second than he would if he were at rest (Fig. 22.7). Let V represent the speed of sound, and v_L the velocity of the listener toward the sound source. In 1 sec the source sends out ν waves. In this same time the observer passes all the waves included in a distance $V + v_L$. The frequency ν_L heard by the listener is the number of waves he passes each second; thus $\nu_L = (V + v_L)/\lambda$, where λ is the wavelength. Since $\nu = V/\lambda$,

$$\frac{\nu_L}{V + v_L} = \frac{\nu}{V} \tag{22.8}$$

If the observer is moving away from the source, v_L is negative, and the observer frequency ν_L is lower than ν.

If the source is moving (relative to the surrounding air) toward a stationary listener (Fig. 22.8), the wavelength of the sound waves in the air is different in different directions. Let ν be the frequency of the waves passing a point in the air, and ν_S the frequency emitted by the source. In 1 sec the sound waves moving to the right go a distance $V - v_S$ relative to the source, which has emitted ν_S waves in this second. Therefore the wavelength in air is $(V - v_S)/\nu_S = V/\nu$, and

$$\frac{\nu_S}{V - v_S} = \frac{\nu}{V} = \frac{\nu_L}{V} \tag{22.9}$$

since the listener hears the frequency of the waves passing in the air. When the source is moving away from the listener, v_S is negative and the pitch heard is reduced. If both source and observer are moving relative to the air, we may write

$$\lambda_{\text{air}} = \frac{V}{\nu} = \frac{V - v_S}{\nu_S} = \frac{V + v_L}{\nu_L} \tag{22.10}$$

or

$$\frac{\nu_L}{V + v_L} = \frac{\nu_S}{V - v_S} \tag{22.10a}$$

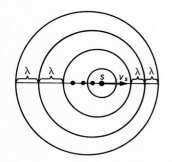

Figure 22.8

When a sound source is moving relative to the air, the wavelength is smaller in the direction of motion and larger in the opposite direction.

Note that both v_S and v_L are measured *relative to the air; both are positive when the source is moving toward the observer and the observer toward the source.*

22.10 SUPERSONIC VELOCITIES AND SHOCK WAVES

When a body travels with a speed greater than that of sound, it is said to be *supersonic*. Figure 22.9 shows a source moving faster than the speed of sound. We observe that the object is beyond the spherical sound waves which it sent out a short time before. However, there exists a surface tangent to all these sound waves which, by Huygens' principle, gives us the position of a compressional wave. This wave accompanying a body traveling at a speed in excess of the speed of sound is called

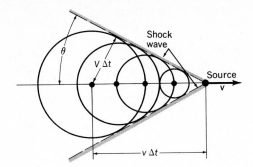

Figure 22.9
Wavefronts produced by a source
moving faster than sound.

a *shock wave*. Figure 22.10 shows a shock wave accompanying a bullet traveling at supersonic speed. Whenever a body travels through a medium at a speed greater than that with which the resulting disturbance is propagated, a similar wave is observed. A common example is the bow wave from a speedboat.

The angle between the direction of motion of the source and the shock wave permits us to compute the velocity of the source, provided the velocity of sound in the medium is known. From Fig. 22.9 we see that $\sin \theta = V/v$, where v is velocity of the source and V the speed of the shock wave.

When a body is moving through the air at subsonic speeds, a compressional wave precedes it, and some of the air particles are moved out of the way. At supersonic speeds the body is traveling faster than the compressional wave. Consequently, there is no preparation in the medium for the oncoming body, and the region of sudden compression which we have called the *shock wave* is produced. A great deal of energy may be associated with this shock wave—energy which comes from the object passing through the air. An aircraft traveling at supersonic speed

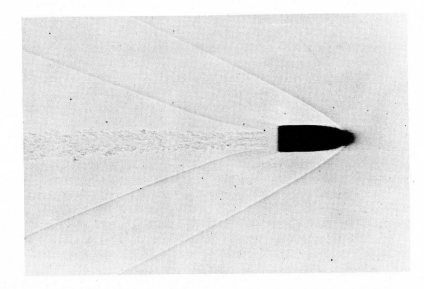

Figure 22.10
Compressional shock wave traveling
with a bullet. [*Olin Mathieson
Chemical Corporation.*]

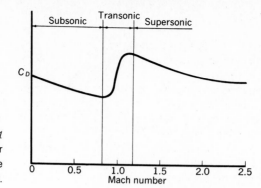

Figure 22.11
The drag coefficient of a particular missile as a function of the missile speed in Mach numbers.

builds up a substantial shock wave. Where the shock wave reaches the earth, it may break windows and cause other damage.

Because of the buildup of a shock wave as an aircraft reaches the speed of sound, the drag on the aircraft increases markedly in this region. In discussing the drag under these conditions, it is customary to compare the speed of the aircraft with the local speed of sound. The latter depends, of course, on the temperature and is substantially lower at high altitudes, where the temperature is low. The ratio of the velocity of the body to the local velocity of sound is called the *Mach number*. Mach 5 means 5 times the velocity of sound. The speeds of missiles and high-velocity aircraft are often quoted in Mach numbers.

If a missile or aircraft is passing through the air, the drag may be written (see Sec. 13.7)

$$F_{\text{drag}} = C_D S(\tfrac{1}{2}dv^2) \qquad (13.7)$$

where S is the cross-sectional area normal to the air stream, d the density of the air, v the velocity of the object, and C_D an empirical factor called the *drag coefficient*. A typical plot of drag coefficient as a function of Mach number for a missile is shown in Fig. 22.11. The drag coefficient depends on the velocity of the moving body and on its size, shape, and smoothness. As one approaches Mach 1, the drag increases rapidly and much faster than v^2. This region of rapid increase in drag coefficient is called the *transonic region*, because it occurs where the speed passes from below to above sonic velocity. The drag coefficient decreases above Mach 1.5, which indicates that once again the drag force is increasing less rapidly than v^2.

When an atomic bomb is exploded in the air, a tremendous shock wave is set up. The shock front represents a moving wall of highly compressed air which may blow down buildings and produce other damage. It has been reported that at a distance of 500 m from the detonation point of a nominal atomic bomb similar to that used at Nagasaki, the pressure of the shock wave is more than four times atmospheric and the speed of the shock wave is about 650 m/sec. The tremendous compressional shock wave is followed by a rarefaction or *suction phase*, which lasts much longer than the pressure phase.

1 When a liquid is being poured into a bottle, why does the pitch of the sound increase as the liquid level rises?

2 Why are the bass strings of a piano wrapped with a close helical winding of other wire?

3 If the wave in an organ pipe is a sound wave, how does the sound get from within the organ pipe to your ear?

4 How may beats be used in tuning a musical instrument such as a piano or a harp? How is such an instrument tuned?

5 What sound sources have overtones which are not harmonics?

6 A violin and an organ are tuned together; what happens if the temperature of the room rises from 20 to 25°C?

7 What is required to observe resonance between two mechanical systems?

8 How are different frequencies produced in a bugle, which is an open air column of constant length?

9 What are the advantages and disadvantages of the equal-tempered scale?

10 Why do marching men break step when crossing a light bridge?

11 If two sounds are an octave apart, how are their frequencies related?

12 If a person takes a deep breath of helium and then speaks, how will the characteristics of his voice be altered? What would be the effect of inhaling carbon dioxide?

13 Can an airplane designed to fly at supersonic speed get its thrust from a propeller? Explain.

14 If two tuning forks have the same frequency, how could you arrange to produce a fluctuating beat frequency from them?

15 Does wind velocity play any role in the doppler effect? Explain quantitatively.

16 If an observer or a source is moving perpendicular to the line joining them, will there be a doppler shift in the frequency of acoustic waves? Explain.

PROBLEMS

Take the speed of sound in air to be 340 m/sec or 1,100 ft/sec.

1 A closed organ pipe is 0.5 m long. Find the frequency of the fundamental and the frequency of the first two overtones.

Ans. 170 Hz; 510 and 850 Hz

2 The whistle of a steamer is in the form of a closed pipe 1 m long. Calculate the frequencies of the fundamental and the first three overtones.

3 An open organ pipe is 85 cm long. What is the wavelength of its fundamental vibration? What is the frequency of its fundamental vibration? What is the frequency of the first overtone? What is the frequency of the third overtone? What is the frequency of the third harmonic?

Ans. 1.70 m; 200 Hz; 400 Hz; 800 Hz; 600 Hz

4 Calculate the lengths in meters of open and closed pipes for a fundamental frequency of 510 Hz.

5 Repeat Prob. 3 for a closed organ pipe which is 85 cm long.

Ans. 3.40 m; 100 Hz; 300 Hz; 700 Hz; 300 Hz

6 Find the frequencies of the fundamental and first two overtones of an open pipe 1.1 ft long.

7 An open organ pipe is 2.2 ft long. Find the frequency of the fundamental and the wavelength of the first overtone. If this pipe is now closed at one end, find the frequency of the fundamental and the wavelength of the first overtone. *Ans.* 250 Hz; 2.2 ft; 125 Hz; 2.93 ft

8 Two open organ pipes of lengths 68 and 70 cm are sounded simultaneously. How many beats per second are produced?

9 Two closed organ pipes of lengths 0.5 and 0.51 m are sounded together and produce three beats per second. What is the velocity of sound in the medium with which the organ pipes are filled? *Ans.* 306 m/sec

10 A sounding tuning fork is held over a vertical glass tube into which water is poured slowly. The remainder of the tube is filled with air. What is the frequency of the tuning fork when the shortest column of air for resonance is 20 cm?

11 Water is poured into a long glass tube closed at one end. What length of air column must be left above the water level if the column is to reinforce and resonate at a frequency of 680 Hz?

Ans. 0.125 m

12 A glass tube open at both ends is so placed that

one end is under water. The tube is adjusted until there is resonance when a sounding tuning fork is held above the open end. If the tuning fork makes 660 vib/sec, what is the shortest length of the tube for resonance?

13 What is the velocity of a transverse wave in a string 150 cm long with a mass of 2.5 g and subject to a tension of 2,400 N? What is the fundamental frequency of this string?

Ans. 1,200 m/sec; 400 Hz

14 A string with a mass per unit length of 0.002 kg/m is stretched by the application of 160 N. What length of string will be required to produce a fundamental frequency of 400 Hz? What tension is required to give this string a fundamental frequency of 500 Hz?

15 A stretched wire 2.5 m long has a fundamental frequency of 300 Hz. What is the velocity of the wave in the wire? What is the wavelength of the fundamental vibration? What is the frequency of the first overtone? What is the frequency of the third overtone? What is the frequency of the third harmonic?

Ans. 1,500 m/sec; 5 m; 600 Hz; 1,200 Hz; 900 Hz

16 A steel piano wire is 1.2 m long and has a fundamental frequency of 250 Hz. If the mass per unit length is 0.01 kg/m, find the tension in the wire. What is the speed of transverse waves in the wire? If the amplitude at the center of the wire is 0.5 mm, find the maximum acceleration of the midpoint.

17 A steel wire has a mass per unit length of 0.005 kg/m and is 1.2 m long. It is stretched by a force of 392 N. Find the speed of transverse waves in this wire. What is the frequency of its fundamental vibration?

Ans. 280 m/sec; 117 Hz

18 A copper wire is stretched between two points 0.6 m apart. It has a fundamental frequency of 200 vib/sec for transverse vibrations. If the wire has an area of 0.2 mm², find (*a*) the stretching stress in the wire, (*b*) the elongation of the wire due to this stress, and (*c*) the frequency of the first overtone.

19 A stretched string made of steel is vibrating at its fundamental frequency of 2,000 Hz. What is the fundamental frequency of a second string made from the same steel but with a diameter twice that of the original and a length twice that of the original if it is stretched by twice the force of the original? *Ans.* 707 Hz

20 Find the frequency heard by a listener at rest when a sound source emitting a frequency of 1,000 Hz goes away at a speed of 85 m/sec.

21 A man on a train which is running at 34 m/sec listens to a siren which has a frequency of 400 Hz. What is the apparent frequency of the sound when the train is approaching the siren? When it is going away from the siren?

Ans. 440 Hz; 360 Hz

22 At what speed in feet per second must a source of sound approach an observer in order for the pitch of each note to be raised by a half tone, i.e., to $\frac{16}{15}$ of the original frequency?

23 Two automobile horns emit the same note frequency 300 Hz. If one of these horns is on a car approaching an observer at 60 mi/hr and the other is on a car moving away from the observer at 60 mi/hr, calculate the frequency heard by the observer in each case.

Ans. 326 Hz; 278 Hz

24 What is the apparent frequency heard by an observer at rest toward whom a police car is approaching at 34 m/sec with its siren emitting a frequency of 450 Hz?

25 A sounding object emits a sound of frequency 550 Hz. An observer heard it at a frequency of 500 Hz. (*a*) If the source is moving, what is its speed? (*b*) If the observer is moving, what is his speed? *Ans.* (*a*) 34 m/sec; (*b*) 30.9 m/sec

26 A passenger standing on the rear platform of a train notes that a warning bell at a grade crossing rings with an apparent frequency of 600 Hz. The observation is made after the train has passed the crossing and while it is moving at 45 mi/hr (66 ft/sec). What is the true frequency of the bell?

27 A siren emitting a frequency of 1,000 Hz moves away from you toward a cliff at a speed of 17 m/sec. Find the frequency of the sound you hear coming directly from the siren and the frequency you hear reflected from the cliff.

Ans. 952 Hz; 1,053 Hz

28 A car moves away from a wall toward an observer with a speed of 20 ft/sec. Find the beat

frequency which the observer hears between a 180-Hz horn on the car and its echo from the building.

29 A train approaching a tunnel in a vertical cliff perpendicular to the tracks sounds its horn, which has a frequency of 250 Hz. If the speed of the train is 110 ft/sec, what is the frequency reflected by the cliff? What frequency does the engineer hear reflected?

Ans. 277.8 Hz; 305.6 Hz

30 Estimate the angle θ in Fig. 22.10 and calculate the approximate speed of the bullet.

31 Find the angle between the shock wave and the path of a bullet which has a speed of 500 m/sec.

Ans. 42.8°

32 Find the notes of a major diatonic scale based on C = 256 Hz.

33 A major diatonic scale is based on F = 360 Hz. Find the frequencies of the other seven notes. Show that for this scale the intervals D/C = 10/9 and E/D = 9/8. How do these intervals compare with those for the C major scale of Table 22.1? In view of the fact that the intervals for the key of C and for the key of F are different, how does one make a piano which can play reasonably well in both keys?

Ans. 405, 450, 480, 540, 600, 675, and 720 Hz

LIGHT

CHAPTER TWENTY-THREE
PROPAGATION AND REFLECTION OF LIGHT

WE TURN NOW to a study of light, one of the first branches of physics to undergo substantial development. In part because of its ancient origins, it will seem at first that there is little connection between light and the parts of physics which we have been studying. However, very soon we find velocity and energy entering the discussion, and a little later we use the physics of wave motion to treat interference, diffraction, and polarization of light. In advanced physics, light becomes a facet of electromagnetism, a topic we introduce in Part 5.

ONE OF OUR PRINCIPAL CONTACTS with the world around us is through light. Not only are we personally dependent on light to convey visual information, but most of what we know about the stars and the solar system is derived from light waves impinging on our eyes and on optical instruments. Similarly, our concept of the structure of atoms comes largely from observing the radiations they emit. We have many instruments which make use of light in their operation. Important among them are the microscope, the telescope, and many kinds of spectrometers.

23.1 THE NATURE OF LIGHT

Much of the early knowledge of light dealt with its general behavior when it strikes materials: how it is reflected and how it is bent when it goes from one medium to another. The ultimate nature of light is not important in studying these phenomena, and the great science which we know as *geometrical optics* was well established before an understanding of the nature of light evolved.

Most objects emit no visible light; they are visible by light which is reflected from them. The sun is our chief source of light and heat, but many other bodies are *luminous* and serve as light sources. Any body heated to a sufficiently high temperature becomes self-luminous through *incandescence*. A gas through which an electric discharge is passed also emits light, the color depending on the gas involved. Entire panels of certain solid materials may be electrically excited to produce light through *electroluminescence*. Other solids emit light when bombarded by electrons or ultraviolet radiation; the television screen is an example of such a *fluorescent* source. A few chemical reactions result in the emission of visible light, even though the chemicals themselves are at a low temperature; the firefly is a *chemiluminescent* source.

During the eighteenth century supporters of Huygens argued that light was a wave motion, while followers of Newton held that light was a stream of particles. Early in the nineteenth century decisive experiments (Chaps. 27 and 28) showed that light exhibits the wave properties of interference, diffraction, and polarization, and for a hundred years it seemed that light was a classical wave phenomenon. Then evidence from blackbody radiation, the photoelectric effect, and x-rays established that light also has properties characteristic of classical particles. Now we hold that light is neither a pure wave phenomenon nor a pure particle phenomenon; rather it is electromagnetic radiation with both wave and particle properties. We also know now that the fundamental constituents of matter (such as electrons), once assumed to be classical particles, similarly display such wave characteristics as interference and diffraction.

The wave properties of visible light are associated with transverse electromagnetic disturbances having wavelengths between roughly 380 and 780 nanometers (1 nm = 10^{-9} m; the nanometer was formerly called the millimicron, abbreviated mμ). In some experiments the wave characteristics of the light dominate; in others the particle properties do. This *wave-particle duality* was a source of great confusion until the development of modern quantum mechanics.

In addition to visible light, there are radiations characterized by longer and shorter wavelengths than the eye can see but with physical properties identical with those of visible light. Sometimes the term *light* is used to denote only radiations which we can see, while at other times we use *light* in a broader sense to mean electromagnetic radiation, regardless of visibility. From the psychophysical point of view, "light" is radiant energy capable of producing a standard response in the human eye. In the broader physical sense, "light" is visible light plus invisible radiations having the same fundamental properties.

The relation between frequency ν, velocity c, and wavelength λ is the same for light as for sound:

$$c = \nu\lambda$$

Yellow light has been found to have a wavelength equal to about 590 nm. Taking the speed of light to be 3×10^8 m/sec, the frequency of yellow light is

$$\nu = \frac{c}{\lambda} = \frac{3 \times 10^8 \text{ m/sec}}{590 \times 10^{-9} \text{ m}} = 5.08 \times 10^{14} \text{ Hz}$$

The wavelength of light is usually expressed either in nanometers or in angstrom units (1 Å = 10^{-8} cm = 10^{-10} m).

Jupiter and
moon

Figure 23.1

By studying the eclipses of Jupiter's
moons Römer was able to estimate the
time required for light to travel the
diameter of the earth's orbit.

Light travels through empty space and through air with a speed of
approximately 3×10^8 m/sec (186,000 mi/sec). In 1 sec light travels
a distance more than seven times around the earth at the equator. It
takes light a little over 8 min to reach us from the sun and 4 years from
the next nearest star. Several methods have been devised for determining the speed of light.

The first evidence that the speed of light is finite came from the
work of the Danish astronomer Römer about 1675. He studied the
periods of Jupiter's moons, four of which are visible in a small telescope.
The innermost of these moons revolves around Jupiter in about 42.5 hr.
Römer found that as the earth moved from B to C (Fig. 23.1) the
measured periods of this moon were all somewhat longer than average;
the periods were shorter than average while the earth moved from G
to H. He concluded correctly that the longer periods were due to the
fact that while the moon revolved around Jupiter, the earth was moving
farther away. The shorter periods occurred when the earth moved closer
to Jupiter during the revolutions of the moon. From his measurements
Römer estimated that 22 min is required for light to travel the diameter
AD of the earth's orbit. Subsequent measurements show that this time
is 1,000 sec.

About this same time the French astronomers Richer and Cassini
measured the diameter AD of the earth's orbit. In 1678 Huygens divided
the Richer-Cassini value of AD by Römer's transit time to obtain the
first value of the speed of light.

One of the most famous determinations of the velocity of light was
that of Michelson. Figure 23.2 is a simplified diagrammatic representation of his experiment. Light from an intense source is focused
by a lens and reflected from an octagonal mirror which, for the moment,
is at rest. This mirror is adjusted so that the beam of light travels to
a stationary mirror and is reflected back to the octagonal mirror and
finally to the eye of an observer. In Michelson's determination a stationary mirror was placed on Mount San Antonio and the octagonal mirror
on Mount Wilson, in California. The distance between mirrors was

Figure 23.2

Simplified representation of Michelson's
apparatus for measuring the speed of
light.

carefully surveyed and found to be 35.4 km. Then the octagonal mirror was put into rapid rotation. Each time one of the eight sides passed the position indicated in the figure, a flash of light was sent to the stationary mirror on Mount San Antonio and returned. If, when it returned, the mirror was in some different position, the light was reflected in some direction other than that of the eye. However, when the speed of rotation of the mirror was made great enough for the mirror to rotate through exactly $\frac{1}{8}$ rev while the light made the trip to Mount San Antonio and returned, the light was reflected to the eye and observed. When this condition was satisfied, the time in which the mirror made $\frac{1}{8}$ rev was exactly equal to the time for the light to travel 70.8 km.

Example

In Michelson's experiment the light path from a rotating mirror to a fixed mirror (Fig. 23.2) is 35.4 km. If an octagonal mirror is rotated at the rate of 31,800 rev/min, the light flashes are observed for the first time. Find the speed of light.

$$\text{Time for 1 rev} = \frac{1}{31,800} \text{ min} = \frac{60}{31,800} \text{ sec}$$

$$\text{Time for } \tfrac{1}{8} \text{ rev} = \frac{60}{31,800 \times 8} \text{ sec} = 2.36 \times 10^{-4} \text{ sec}$$

In this time light travels 70.8 km, so that

$$c = \frac{70.8 \text{ km}}{2.36 \times 10^{-4} \text{ sec}} = 3 \times 10^8 \text{ m/sec (186,000 mi/sec)} \qquad ////$$

Many other methods of determining the speed of light have been devised, several of which are capable of far higher precision than the Michelson measurement. The results of many determinations by many different physicists using a wide variety of methods lead us to believe that the speed of light in free space is 2.997925×10^8 m/sec.

23.3 RECTILINEAR PROPAGATION OF LIGHT

Under ordinary circumstances light travels in straight lines, not bending appreciably around objects. Indeed, we depend on light to tell us when a meterstick is straight. The rectilinear (straight-line) propagation of light is shown by the fact that a small source of light casts sharp shadows. A point source S (Fig. 23.3) illuminates all points on a screen above A and below B. No light arrives at the screen between A and B because

Figure 23.3
An opaque body casts a sharp shadow when illuminated by a point source.

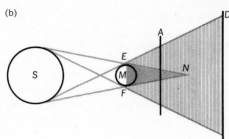

Figure 23.4

(*a*) The shadow of an opaque body illuminated by a source of significant angular dimension consists of an umbra and a penumbra. (*b*) Shadow cast by an object smaller than the source of illumination.

it is stopped by the opaque body *M*. The fact that the boundary of the shadow is sharp shows that light from *S* does not bend appreciably into the shadow of the opaque body.

When the source of light is not small, the boundary of the shadow is no longer sharp (Fig. 23.4*a*). Points on the screen above *A* and below *D* are fully illuminated by *S*. The screen between *B* and *C* receives almost no light. The screen between *A* and *B* and between *C* and *D* is partially illuminated and in a less dense shadow than the part between *B* and *C*. This outer shadow gradually shades off from complete shadow at *B* to complete illumination at *A*, giving the blurred appearance that characterizes most shadows. The region of total shadow is called the *umbra*, while the partially shadowed region is the *penumbra*.

When the source of illumination is larger than the object casting the shadow (Fig. 23.4*b*), the only region of complete shadow is the cone *ENF* having the object *M* as a base. If the screen is placed at *A*, there is a small umbra surrounded by a large penumbra; if at *D*, the umbra is absent.

The best illustration of shadows on a large scale is found in eclipses. When the moon is interposed between the sun and the earth and its shadow falls on the earth, the sun is wholly or partly obscured by the moon and there is a total or partial eclipse of the sun. If the earth is sufficiently near the moon and passes through the complete shadow represented by the cone *FNE* (Fig. 23.4*b*), the eclipse is *total*. If, however, the earth is farther from the moon and passes through only the partial shadow, the eclipse is said to be *annular*. Sometimes the moon passes through the shadow of the earth; there is then an eclipse of the moon.

The pinhole camera illustrates the rectilinear propagation of light. If the pinhole is circular, the light from each point on the object strikes the photographic plate in the form of a small circle (or ellipse). For any other shape of hole each object point illuminates a small area shaped similar to the hole. If the pinhole is large, the areas overlap badly and the image is blurred. The smaller the hole, the sharper the image (unless the hole is so small that diffraction becomes important) but the longer it takes to expose the picture. Note that the image is *inverted*.

23.4 LAWS OF REFLECTION

When a beam of light traveling in a homogeneous medium comes to a second medium, part of the light is reflected. At a silvered surface the fraction of the light reflected is almost 100 percent, while at the surface of clear glass it is only a few percent. In Fig. 23.5, OA is a ray incident on a plane mirror and AR a reflected ray. The angle between OA and AN, the perpendicular or normal to the surface, is called the *angle of incidence;* the angle NAR between normal and reflected ray in the *angle of reflection*. The laws governing reflection (Sec. 20.7) are:

First law of reflection: The incident ray, the reflected ray, and the normal to the surface lie in the same plane.

Second law of reflection: The angle of incidence is equal to the angle of reflection.

When light falls on a rough, opaque surface (Fig. 23.6), the incident light is scattered in all directions. If the surface is so smooth that the distances between successive elevations on the surface are less than about one-quarter the wavelength of the light, there is little random scattering, and the surface is said to be *polished*. Thus, a surface may be polished for radiation of long wavelength but not polished for light of short wavelength. A polished reflector or *mirror* is said to exhibit *regular* (or specular) *reflection*. Rough surfaces produce *diffuse reflection*.

Figure 23.5
When a light ray is reflected from a plane mirror, the angle of incidence i is equal to the angle of reflection r.

Figure 23.6
Irregular or diffuse reflection.

23.5 THE PLANE MIRROR

When a luminous object is placed in front of a plane mirror MM' (Fig. 23.7), a point O on the object sends light in all directions. Several rays which leave this point and strike the mirror are shown in the figure. To an observer in front of the mirror *all* the reflected rays appear to come from the point I behind the mirror. The observer therefore sees a bright spot which appears to be behind the mirror and which we call the *image* of the object. For every point O on the luminous source there is a corresponding point I on the image. The image behind the mirror is a *virtual image,* because the light rays do not actually come from I. However, every ray of light which leaves a point on the object and is reflected from the mirror *appears to come* from the corresponding point on the image.

The ray ON (Fig. 23.7) falls normally on the mirror and is reflected directly back on itself; from the figure triangles ONA and INA are similar and equal. Therefore, the image point is as far behind the mirror as the object point is in front.

Reflection may also be considered in terms of wavefronts and Huygens' principle (Secs. 20.6 and 20.7). Such a treatment leads to

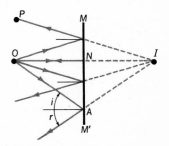

Figure 23.7
The image I of a point source O formed by reflection from a plane mirror is as far behind the mirror as the object O is in front.

the same conclusions so far as the location of the image of an object point is concerned. In terms of this description, spherical waves diverge from each point on the object, have their curvatures reversed (Fig. 20.10) at the reflecting surface, and thereby appear to have originated at the corresponding image point.

Figure 23.8 shows an object in front of a mirror. It is evident from the figure that the image is as far behind the mirror as the object is in front of the mirror and that image and object have the same size. However, *the image is reversed with respect to the reflecting plane.* This is shown more clearly in Fig. 23.9, in which the image $O'X'Y'Z'$ of a set of axes $OXYZ$ is indicated. Note that the axis $O'X'$ points in a direction opposite to that of the axis OX. Such an image is *pre-verted.* Object and image are related in the same way that a right hand is related to a left. If a person salutes with his right hand before a plane mirror, his image appears to salute with the left hand.

Figure 23.8

The image of an extended object formed by a plane mirror is the same size as the object and an equal distance from the reflecting surface.

Figure 23.9

The image formed by a plane mirror is reversed with respect to the reflecting plane.

23.6 THE CONCAVE SPHERICAL MIRROR

A concave spherical mirror is part of a spherical shell with its inner surface polished. The center C of the sphere from which the mirror was taken is called the *center of curvature* of the mirror; the radius of the sphere is called the *radius of curvature* of the mirror. The middle point V of the mirror M (Fig. 23.10) is known as the *vertex,* and the straight line CV through the vertex of the mirror and its center of curvature C is the *principal axis.*

Consider a ray of light AB approaching the mirror parallel to the principal axis. At the mirror the ray is reflected with angle of reflection equal to angle of incidence. Since AB is parallel to the axis, angles i and θ are equal. Triangle FBC is isosceles; therefore BF is equal to FC. If B is not too far from V, BF and VF are nearly equal. Hence, VF is nearly equal to FC; therefore, F is halfway between the mirror and the center of curvature C. This point F is known as the *principal focus* of the mirror. *The distance from the vertex V to the principal focus F is the focal length f of the mirror.* The focal length of a mirror is one-half the radius of curvature R:

$$f = \frac{R}{2} \qquad (23.1)$$

Any ray of light which comes to the mirror parallel to the princi-pal axis is reflected so that it passes through the principal focus F,

Figure 23.10

An incident ray parallel to the principal axis of a concave mirror is reflected so that it crosses the axis at the principal focus.

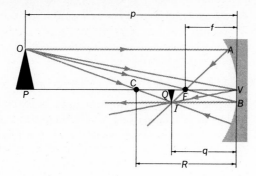

Figure 23.11
A concave mirror forms a real, inverted image of an object outside the principal focus.

provided the angle θ in Fig. 23.10 is small. If a ray of light passes along the line FB, it is reflected back in the direction BA. It is true in general that if a ray takes a certain path through an optical system in one direction, a ray sent backward along the path on which the original ray leaves traverses the same path and comes out along the line on which the original ray entered.

If an object OP is placed before a concave mirror (Fig. 23.11), we can locate an image of point O by drawing a suitable ray diagram. The ray OA parallel to the principal axis is reflected back through F, the principal focus. The ray OF strikes the mirror at point B and is reflected back parallel to the principal axis CV. A ray from O through C comes to the spherical surface along the radius and is reflected directly back on itself. All three reflected rays pass through a common point I. Indeed, if we draw any number of rays which leave point O and are reflected by the mirror, we find that *every ray which leaves point O and strikes the mirror passes through point I.* The point I is the image of point O. In this case it is a *real image*, since the rays of light actually cross at this point. If we put a piece of white paper at I, we see a real image of O. A similar construction can be made for other points on the object; for each object point we can thus locate the corresponding image point. In this case the image is *inverted*.

If an object were placed at IQ, the image would be formed at OP. The points P and Q are called *conjugate points;* an object at one has an image at the other.

Let p, q, and f denote, respectively, the distances of object, image, and principal focus from the mirror. These three distances are related by a simple formula which can be derived with the aid of Fig. 23.11 as follows. The triangles OPV and IQV are similar; hence

$$\frac{OP}{IQ} = \frac{PV}{QV} = \frac{p}{q} \tag{23.2}$$

Also, triangles OPF and BVF are essentially similar, so

$$\frac{OP}{BV} = \frac{PF}{FV} = \frac{p-f}{f} = \frac{p}{f} - 1 \tag{23.2a}$$

Since $BV = IQ$, the left-hand members of Eqs. (23.2) and (23.2a) are equal and

$$\frac{p}{q} = \frac{p}{f} - 1$$

If we divide through by p and rearrange, we obtain

$$\frac{1}{p} + \frac{1}{q} = \frac{1}{f}$$ (23.3)

Example

An object is situated at a distance of 80 cm from a concave mirror of radius of curvature 60 cm. Find the position of the image.

$$p = 80 \text{ cm} \qquad f = \frac{60 \text{ cm}}{2} = 30 \text{ cm}$$

$$\frac{1}{p} + \frac{1}{q} = \frac{1}{f}$$

$$\frac{1}{80} + \frac{1}{q} = \frac{1}{30}$$

$$q = \frac{2,400}{50} = 48 \text{ cm} \qquad \qquad \text{////}$$

When an object is at a great distance, $q = f$, and the image is at the principal focus. As the object is moved toward the mirror, the image moves away. When the object distance is equal to the radius of curvature, the image distance is also the radius of curvature; object and image are identical in size. As the object is moved still closer to the mirror, the image moves away, until it reaches infinity when the object is at the principal focus. From Fig. 23.11 we observe that since the triangles OPV and IQV are similar,

$$\frac{\text{Height of image}}{\text{Height of object}} = \frac{\text{image distance}}{\text{object distance}} = \frac{q}{p}$$ (23.4)

The ratio of the height of the image to the height of the object is known as the (lateral or transverse) *linear magnification.*

Example

Find the size of the image of a body 2.5 cm high when placed 50 cm in front of a concave mirror whose focal length is 20 cm.

$$p = 50 \text{ cm} \qquad f = 20 \text{ cm}$$

$$\frac{1}{p} + \frac{1}{q} = \frac{1}{f}$$

$$\frac{1}{q} = \frac{1}{20} - \frac{1}{50}$$

$$q = 33.3 \text{ cm}$$

$$\frac{\text{Height of image}}{\text{Height of object}} = \frac{33.3}{50} = 0.667$$

$$\text{Height of image} = 0.667 \times 2.5 = 1.67 \text{ cm} \qquad \text{////}$$

If an object lies inside the principal focus of a concave mirror, a diagram like that of Fig. 23.12 indicates that the image lies behind the mirror. It is *upright, virtual,* and *enlarged.* When we solve for q for the case in which p is less than f, we find that it is *negative.*

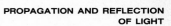
Figure 23.12
A concave mirror forms a virtual, erect, magnified image of an object inside the principal focus.

23.7 THE CONVEX MIRROR

In the case of the convex mirror (Fig. 23.13), a ray of light parallel to the principal axis is reflected away from the principal axis. However, if we extend the reflected ray backward, the line crosses the principal axis at point F, the principal focus for this convex mirror. The principal focus is *virtual*. Rays parallel to the principal axis are reflected away from the axis in such a way that they appear to come from the principal focus. The mirror is a *diverging* mirror, because rays initially parallel diverge after reflection.

  If we place an object in front of a convex mirror (Fig. 23.14), the rays of light which leave a point O on the object and are reflected at the mirror never actually cross, but if we extend these rays behind the mirror, the extensions cross at point I. In a convex mirror a real object forms an *erect, virtual* image, *reduced* in size.

  Equation (23.3) is applicable to the convex mirror, provided that we assign a negative sign to the focal length. Further, for a virtual image formed behind the mirror, the image distance q is negative.

Figure 23.13
A convex mirror has a virtual principal focus.

Example

A bright spot situated 60 cm in front of a convex mirror forms a virtual image 20 cm behind the mirror. Find the focal length of the mirror.

$$\frac{1}{p} + \frac{1}{q} = \frac{1}{f}$$

$$p = 60 \text{ cm} \qquad q = -20 \text{ cm}$$

$$\frac{1}{60} - \frac{1}{20} = \frac{1}{f}$$

$$f = -30 \text{ cm} \qquad \text{////}$$

Figure 23.14
A convex mirror forms a virtual, erect, reduced image of the object at the left.

23.8 THE STANDARD RAYS AND THE SIGN CONVENTION

The same formula can be used for both concave and convex mirrors if we are careful to use the proper signs for various quantities. We establish conventions for choosing these signs by considering a line parallel to the principal axis connecting a point on the object with the mirror.

1 If a ray of light along this line goes from the object to the mirror, the object distance p is positive.

2 The focal length f is positive if this ray parallel to the principal axis strikes the mirror and is reflected so that it passes through the principal focus. If, instead of passing through the principal focus after reflection, it diverges from the principal axis, the focal length is negative.

3 If, after reflection, this ray proceeds toward the image, the image distance q is positive. If, after reflection, the ray goes away from the image, the image distance is negative.

It is of great help in solving optics problems to draw scaled ray diagrams. In such a diagram, the intersection of two rays which leave a point on the object and are reflected from the mirror is sufficient to establish the corresponding point on the image. It is desirable to use three rays and thus guard against errors. We shall place emphasis on three rays, which we call the *standard rays* and which can be drawn with a ruler. These three standard rays and their paths are as follows:

1 The ray which goes from a point on the object along the line parallel to the principal axis leaves the mirror along a line which passes through the principal focus.

2 A ray which leaves a point on the object along a line connecting that point with the principal focus is reflected along a line parallel to the principal axis.

3 A ray from a point on the object through the center of the curvature of the mirror is reflected directly back on itself.

23.9 SPHERICAL ABERRATION AND PARABOLIC MIRRORS

If the width of a mirror (Fig. 23.15) is comparable to its radius of curvature, parallel rays after reflection do not all meet at the single point F, the principal focus. Rays that are reflected from a limited region of the mirror in the neighborhood of its vertex V are brought to a focus at F, but the rays that strike the mirror at points far from the vertex cross the axis at points nearer to the mirror than F. The effect of this *spherical aberration* is to destroy the sharpness of the image formed by the mirror.

If the section of the mirror is a parabola rather than a circle, incident rays parallel to the principal axis are all brought to focus at a single point F, regardless of the size of the mirror. Similarly, all reflected rays which originated at a point source at F leave the parabola parallel to the principal axis. For this reason parabolic mirrors are often used for searchlights. If a light source is placed inside the focus of a parabolic mirror, the rays of light diverge after reflection. If it is outside

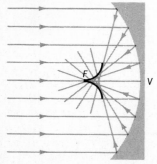

Figure 23.15

Spherical aberration occurs when rays parallel to the principal axis of a spherical mirror of large aperture are reflected so that the outer rays do not cross at the principal focus but over a small region called the *caustic surface*.

the focus, the rays converge after reflection. This fact is used in automobile headlights to give a broad "bright" beam from one filament and a narrow "dim" beam from another.

QUESTIONS

1 Is it possible to photograph a virtual image? Under what circumstances? If you wish to take a photograph of yourself in a plane mirror and you stand 2 m from the mirror, for what distance should the camera be set? Why?

2 What is the shortest plane mirror in which a man 1.8 m tall can see himself completely?

3 If someone asked you why a mirror reverses right and left but not up and down, what would your answer be?

4 How could you locate the center of curvature of a concave mirror with only a sharpened pencil and your eye?

5 An object stands before a spherical mirror which may be either concave or convex. Under what conditions will the image be (a) erect, (b) larger than the object, and (c) real?

6 In Michelson's method of measuring the speed of light, what was the advantage of using eight- and twelve-sided mirrors rather than four- or six-sided ones? What was the disadvantage?

7 A trick mirror makes a man's head look large and his hips small. How is this mirror made?

8 If a ground-glass surface is wetted with water, it becomes more transparent. Why?

9 A burning candle stands between two plane mirrors which touch along a vertical line. How many images of the candle does an observer see if the angle between the mirrors is (a) 90°, (b) 60°, or (c) 30°?

PROBLEMS

Draw a suitable ray diagram for each mirror problem.

1 Standard radio broadcasting stations have an assigned frequency between 535 and 1,605 kHz. The vhf (*very-high-frequency*) television stations (channels 2 to 13) have frequencies between 54 and 216 MHz, while the uhf (*u = ultra*) stations (channels 14 to 83) have fre-

quencies between 470 and 890 MHz. What is the wavelength corresponding to each of the frequencies mentioned?
Ans. 561, 187, 5.56, 1.39, 0.638, and 0.337 m

2 Find the frequencies associated with electromagnetic waves of wavelength (a) 450 nm (visible light), (b) 0.45 Å (x-rays), (c) 45 μm (infrared), and (d) 4.5 km (radio).

3 Light of a certain color has 20,000 waves to the centimeter in air. What is its frequency? Find the number of waves per centimeter when the light is traveling in water if the speed of light in water is three-fourths as great as it is in air and the frequency remains constant. What is the wavelength in each medium?
Ans. 6×10^{14} Hz; 26,700 cm^{-1}; 500 and 375 nm

4 An astronomical unit of distance is the light-year, which is the distance light travels in a year. Compute this distance in kilometers. The nearest star is at a distance of 4 light-years. To how many kilometers does this correspond?

5 If the separation of the mirrors in a Michelson method determination of the speed of light is 30 km and the rotating mirror has eight sides, how rapidly must the mirror be turning when light is first reflected to the eye? *Ans.* 625 rev/sec

6 What minimum speed of rotation is necessary for a 12-sided mirror used in measuring the speed of light with Michelson's arrangement (Fig. 23.2) if the distance from the fixed mirror to the rotating one is 35.4 km?

7 It is desired to use a pinhole camera with film 8 cm high to take a picture of a tree 10 m high. The film is 20 cm from the pinhole. How far should the camera be from the tree to include the full height on the tree? *Ans.* 25 m

8 Fizeau devised a method of measuring the speed of light in which a beam of light passed between two teeth of a toothed wheel, was reflected back by a mirror, and arrived back at the wheel just in time to pass through the next slot between the teeth. Fizeau's wheel has 720 teeth, and the mirror was 8.6 km from the toothed wheel. He observed no light returning through the wheel when it was rotating at 12.6 rev/sec and maxi-

mum brightness of returning light at 25.2 rev/sec. Find Fizeau's value for the speed of light.

9 A plane mirror lies face up, making an angle of 20° with the horizontal. A ray of light shines down vertically on the mirror. What is the angle of incidence? What will be the angle between the reflected ray and the horizontal?

Ans. 20°, 50°

10 Show that a ray of light reflected from a plane mirror rotates through an angle 2θ when the mirror is rotated through an angle θ about an axis perpendicular to both the incident ray and the normal to the surface.

11 An object 1.5 cm high is placed 30 cm from a concave mirror with a radius of curvature of 20 cm. Find the focal length and the location, height, and orientation of the image.

Ans. 10 cm; 15 cm, 0.75 cm, inverted

12 As the position of an object reflected in a spherical-shell mirror of 20-cm focal length is varied, the position of the image varies. Plot the image distance as a function of the object distance, letting the latter change from $-\infty$ to $+\infty$. Where is the image real? Where virtual?

13 An object 2 cm high is located 15 cm in front of a concave mirror which has a focal length of 20 cm. Find the position, size, and character of the image.

Ans. −60 cm; 8 cm; erect and virtual

14 An object 5 cm high is 60 cm from a concave mirror whose radius of curvature is 40 cm. Locate the image. Is it real or virtual? Is it erect or inverted? What is its size?

15 To give him a magnified image of a cavity a dentist holds a small mirror with a focal length of 2 cm a distance of 1.5 cm from a tooth. What is the linear magnification obtained? *Ans.* 4

16 A doctor looks through a small hole at the vertex of a concave mirror to examine a sore throat. If the radius of curvature of the mirror is 40 cm and the light source is 1 m from the mirror, how far from the mirror should the throat be if the light source is imaged on the inflamed area?

17 A man is shaving with his chin 30 cm from a concave magnifying mirror. If the linear magni-

fication is 2.5, find the radius of curvature of the mirror. *Ans.* 100 cm

18 What will be the magnification obtained by using a concave mirror with a focal length 15 in. if the mirror is held 10 in. from the face?

19 Where must an illuminated object be placed with reference to a concave mirror with a radius of curvature of 2 m in order to have its image focused on a screen 5 m from the mirror?

Ans. 1.25 m

20 A magnifying mirror has a radius of curvature of 60 cm. How far must a face be from a mirror if the image is to be erect, virtual, and with linear dimensions three times those of the face?

21 An object 2 cm high is 36 cm from a convex mirror with a radius of curvature of 24 cm. Locate the image. Is it real or virtual? Is it erect or inverted? What is its size?

Ans. −9 cm, virtual, erect, 0.5 cm

22 An object is placed 3 ft from a convex mirror which has a focal length of 1.8 ft. Find the image distance and the ratio of the height of the image to the height of the object.

23 A man's eye is 20 cm from the center of a spherical reflecting Christmas tree ornament which is 8 cm in diameter. Find the image position and the linear magnification.

Ans. −1.78 cm; 0.11

24 A fortune-teller uses a polished sphere of 6-in. radius. If her eye is 10 in. from the sphere, where is the image of the eye?

25 An incandescent lamp is located 2.4 m from a wall. It is desired to throw on the wall an image magnified 2.5 diameters, using a concave mirror. What must be the radius of curvature of the mirror? Where must it be placed?

Ans. 2.29 m; 4 m from wall

26 An object 2 cm high is 30 cm from a concave mirror of 20-cm focal length. Find the image distance and image height. If the object is moved 5 cm farther from the mirror, how far does the image move?

27 An object is placed in front of a concave mirror having a radius of curvature of 25 cm. It is desired to produce first a real image and then a virtual image three times as high as the object. Find the object distance required in each case.

Ans. 16.7 cm; 8.33 cm

28 A short match of length L lies along the axis of a concave spherical mirror of focal length f with its most distant point a distance p from the vertex. Calculate the length L' of the image and

find the *longitudinal magnification L′/L.* Show that if $L \ll p - f$, the longitudinal magnification is equal to the square of the lateral linear magnification [Eq. (23.4)].

29 Show that the path of the reflected ray from O of Fig. 23.7 which passes through point P after reflection is the shortest possible path which reaches the mirror. (This is an example illustrating *Fermat's principle of least time,* which states that the actual path of a light ray between two points is such that it takes less time for the light to traverse this path than it would to traverse any other path which varies slightly from the actual path.)

IN CHAP. 23 we discussed what happens when light is reflected at a surface, and we learned how images are formed by spherical reflectors. Next we examine the bending of light rays as they pass through an interface which separates one medium from another. It is this phenomenon of *refraction* which we must understand if we are to appreciate how lenses form images and how optical instruments such as microscopes and cameras operate.

CHAPTER TWENTY-FOUR
REFRACTION AND DISPERSION

24.1 REFRACTION

When light passes obliquely from one medium to another, there is a change in the direction of propagation. *The bending of light rays as they pass from one medium to another is called refraction.* As a consequence of refraction, a ruler dipped obliquely into water appears to bend sharply at the surface.

Figure 24.1 shows a ray OA incident obliquely on a plane water surface RS. Upon reaching the surface, the ray undergoes a sharp change in direction, proceeding along the line AI. The angle OAN between the incident ray and the normal NN' to the surface is the angle of incidence i, while the angle IAN' is the angle of refraction r. When a ray of light is refracted at a surface, two laws are obeyed:

1 *The incident ray, the refracted ray, and the normal to the surface lie in the same plane.*

2 *The ratio of the sine of the angle of incidence to the sine of the angle of refraction is a constant independent of the angle of incidence.*

The second law of refraction is called *Snell's law* in honor of its discoverer.

The ratio (sin i)/(sin r) depends on the media on the two sides of the interface at which refraction occurs. In order to compare the refraction associated with various media, we refer refraction measurements to a common medium—a vacuum.

The index of refraction of a medium is the ratio of the sine of the angle of incidence measured in vacuum to the sine of the angle of refraction measured in the medium.

When we say the index of refraction of glass is 1.50, we are automatically implying that the light is incident from a vacuum and falls upon the glass. Table 24.1 lists the indices of refraction of a number of common transparent materials. The index of refraction of air differs little from unity; in most situations it makes no significant difference whether light is incident from vacuum or from air.

Figure 24.1
Light ray refracted at the interface RS between air and water.

Table 24.1
**Indices of Refraction
(For light of λ = 589 nm; sodium D radiation)**

Substance	n
Air	1.00029
Carbon dioxide	1.00045
Canada balsam	1.53
Ethyl alcohol	1.36
Carbon disulfide	1.63
Diamond	2.42
Glycerin	1.47
Water	1.333
Crown and plate glass	1.52
Flint glass	1.63
Heavy flint glass	1.66
Quartz	1.54

The correct explanation for refraction was given by Huygens (Sec. 20.8). Whenever a wave motion passes from one medium in which its speed is V_1 to another in which its speed is V_2, the wave pattern changes its direction in such a way that $V_1/V_2 = (\sin i)/(\sin r)$, where i is the angle of incidence, and r the angle of refraction.

In general, for any medium the index of refraction n is given by

$$n = \frac{\sin i \text{ (in vacuum)}}{\sin r \text{ (in medium)}} = \frac{c}{V} \qquad (24.1)$$

where c is the speed of light in vacuum and V the speed in the material.

24.2 RELATIVE INDICES OF REFRACTION

In defining the index of refraction of a material we chose to have light incident from a vacuum. Often an interface separates two transparent media of substantial density, such as glass and water. The laws of refraction are valid in such cases; the ratio $(\sin i)/(\sin r)$ is called the *index of refraction of the second medium relative to the first*. In general, the ratio of the sine of the angle of incidence to the sine of the angle of refraction is equal to the ratio of the speed of light in the first medium V_1 to the speed of light in the second medium V_2:

Index of refraction of medium 2 relative to medium 1

$$= n_{2,1} = \frac{\sin i}{\sin r} = \frac{V_1}{V_2} \quad (24.2)$$

When light goes from one medium to another in which its speed is smaller, it is bent toward the normal; when it goes from one medium to another in which its speed is greater, it is bent away from the normal.

If light moves from medium A into medium B, it is convenient to write the law of refraction in the form

$$n_A \sin \theta_A = n_B \sin \theta_B \qquad (24.3)$$

where n_A and n_B are the indices of refraction and θ_A and θ_B are the angles between the normal and the rays in medium A and medium B, respectively. This relationship applies whether the ray is going from A into B or from B to A.

If the light travels from A to B, the index of refraction of B relative to A is

$$n_{B \text{ relative to } A} = \frac{\sin \theta_A}{\sin \theta_B} = \frac{V_A}{V_B} = \frac{n_B}{n_A} \qquad (24.3a)$$

On the other hand, if the light goes from B to A, the index of refraction of A relative to B is

$$n_{A \text{ relative to } B} = \frac{\sin \theta_B}{\sin \theta_A} = \frac{V_B}{V_A} = \frac{n_A}{n_B} \qquad (24.3b)$$

24.3 REFRACTION THROUGH SLABS WITH PARALLEL FACES

If a ray of light (Fig. 24.2) falls on a plate of glass with parallel faces, it is bent toward the normal as it enters the glass. As it leaves the glass,

Figure 24.2
Refraction and reflection at parallel
plane surfaces.

it is bent away from the normal. Since the faces of the plate are parallel
to each other, the normals to the first and second faces are parallel
to each other, and therefore the angle θ_G between the normal and the
ray in the glass is the same at both surfaces. Consequently, the direction
of the ray emerging from the glass is the same as its direction on
entering; however, the ray is displaced laterally.

Whenever light goes from one transparent medium to another with
different optical properties, there is always a reflected beam as well as
a refracted one. Figure 24.2 shows this schematically. There is reflection at the lower surface of the plate as well as at the upper. Light
which undergoes several reflections is said to be *multiply reflected*.

When a ray of light from air traverses several parallel layers of
transparent substances (Fig. 24.3), it is refracted at each surface. If
it returns to air, its direction is unchanged but it is laterally displaced
from its original path.

Figure 24.3
Path of a light ray through parallel
layers of different materials.

24.4 TOTAL REFLECTION

When a ray of light passes from an optically denser medium such as
water to a rarer medium such as air, it is bent away from the normal
so that the angle of refraction is greater than the angle of incidence
(Fig. 24.4a). If the angle of incidence is made larger, the angle of
refraction increases until it reaches 90°. In this case the refracted ray
travels along the surface of separation between the two media. *The
angle of incidence for which the angle of refraction is 90° is called
the critical angle.*

Example

The index of refraction of water is 1.33. What is the critical angle for
light going from water to air?

Figure 24.4
When light goes from water to air, (*a*)
the refracted ray bends away from the
normal, (*b*) at the critical angle of
incidence the refracted ray is parallel to
the surface, (*c*) the light is totally
reflected when the angle of incidence
exceeds the critical angle.

Application of Eq. (24.3) yields

$$n_{\text{water}} \sin i_c = n_{\text{air}} \sin 90°$$
$$1.33 \sin i_c = 1.00 \times 1.00$$
$$\sin i_c = 0.75$$
$$i_c = 49° \qquad ////$$

When the angle of incidence exceeds the critical angle, light no longer enters the rarer medium and all the incident light is reflected. This type of reflection, known as *total reflection*, takes place at a surface separating an optically rarer from a denser medium when the light comes from the denser to the rarer medium and the angle of incidence exceeds the critical angle. The prism ABC (Fig. 24.5) produces total reflection of the ray OL, since the angle of incidence of 45° exceeds the critical angle for glass.

It would be erroneous to believe that when the critical angle of incidence is reached, there is a sudden transition from all light refracted to all light reflected. At any angle of incidence there is always some reflected light.† At normal incidence the fraction of the incident intensity reflected is given by $[(n_2 - n_1)/(n_2 + n_1)]^2$, where n_1 and n_2 are the indices of refraction of the first and second mediums. As the angle of incidence is increased, a greater fraction of the incident light is reflected and a smaller fraction refracted. It is easy to observe with a pane of glass that as the angle of incidence increases, the fraction of the light reflected becomes larger.

There are many applications of total reflection. Light can be "piped" down a bent tube of glass or Lucite. As the light travels along the tube, it strikes the surface at angles of incidence greater than the critical angle and is totally reflected. Thus the light is transmitted down the tube until it reaches the end, where the angle of incidence is less than the critical angle, and the light emerges. Such "light pipes" may be used to get illumination in surgical operations at places where it is difficult to operate a lamp because of the heat involved. Light can also be piped in streams of water; colored fountains depend on total reflection to keep the light in the water stream until it finally strikes at an angle of incidence smaller than the critical angle and escapes.

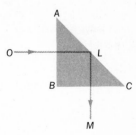

Figure 24.5
Total reflection by an isosceles right-angle prism.

24.5 APPARENT THICKNESS OF A TRANSPARENT BODY

One consequence of refraction is that a body immersed in a transparent medium appears nearer to the surface than it actually is. Let OA and OB represent rays from an object under water (Fig. 24.6); both are bent away from the normal when they reach the surface. An eye at E sees light which appears to come from the image at I; an eye at D would "see" the object at J.

The index of refraction of the water is

$$n = \frac{\sin EAN}{\sin AOV} = \frac{\sin AIV}{\sin AOV} = \frac{AV/IA}{AV/OA} = \frac{OA}{IA}$$

† These remarks apply to ordinary light; the reflection of polarized light is discussed in Sec. 28.4.

Figure 24.6
The apparent depth of an object O in water is less than the actual depth by an amount which depends on the angle of incidence.

Figure 24.7
A mirage may be observed when the air near the ground is substantially hotter than the air higher up.

If the eye receives only rays very near the normal to the surface, we may write OV for OA, and IV for IA. Then

$$n = \frac{OV}{IV} = \frac{\text{actual depth}}{\text{apparent depth}} = \left|\frac{p}{q}\right| \qquad (24.4)$$

The greater the inclination at which the object is viewed, the less its apparent depth.

24.6 ATMOSPHERIC REFRACTION AND THE MIRAGE

Although the index of refraction of air under standard conditions deviates from unity by less than 0.03 percent, there are situations in which atmospheric refraction is far from negligible. One of the most interesting occurs in the *mirage*. On hot sunny days there may be a layer of very hot air in contact with the ground. This air has a lower density and a slightly smaller index of refraction than the air above it. Light from a distant mountain or treetop approaches this layer of hot air at a large angle of incidence and is totally reflected (Fig. 24.7). The light appears to have come from the image just as though it had been reflected from the surface of a lake.

When the air next to the ground is substantially colder than higher air, rays of light are deviated downward (Fig. 24.8). An image of a ship may appear above the ship itself. This phenomenon is called *looming*. It is rather common when an observer is looking over a snowfield or over a body of water which is substantially colder than the air several feet above the surface. Cases have been reported of a lighthouse being seen at a distance of 40 miles when the curvature of the earth would have completely cut it off had there been no looming.

When one looks at an object over the hot burner of a stove or over a hot pavement, one may observe a wavy, shimmering effect. This arises from the bending of the light as it passes from colder to warmer to colder air. The twinkling of stars is in part due to similar phenomena in that the light travels through unstable layers in the atmosphere where the index of refraction is changing significantly.

Figure 24.8
Looming occurs when the air near the surface is substantially colder than the air above.

A ray of light from the sun (or a star) entering the earth's atmosphere obliquely is refracted continuously as it enters denser air. For this reason the sun appears to be at a higher altitude angle than it actually is. When the sun is at the horizon, the correction is about $\frac{1}{2}°$, roughly the angle subtended at the earth by the sun's diameter. As a consequence, the sun is actually below the horizon when we watch it rise or set. When the sun (or moon) is near the horizon, the rays from the lower edge are bent more than those from the upper edge. This produces a shortening of the vertical diameter, so that the sun appears elliptical.

24.7 REFRACTION AT A SPHERICAL SURFACE

Consider a ray OP in medium A which is incident upon a spherical surface of medium B (Fig. 24.9) with center of curvature at C. This ray is refracted at the surface; by Eq. (24.3),

$$n_A \sin \theta_A = n_B \sin \theta_B$$

By definition,

$$\sin \phi = \frac{z}{R} \qquad \tan \alpha = \frac{z}{p+x} \qquad \tan \beta = \frac{z}{q-x}$$

Since $\theta_A + \angle OPC = 180° = \angle OPC + \alpha + \phi$, we have $\theta_A = \alpha + \phi$. Similarly, $\phi = \theta_B + \beta$.

We now confine our derivation to the case in which θ_A, θ_B, α, and ϕ are all small angles. If an angle is small, its sine, its tangent, and the angle itself (in radians) are essentially equal. For this case x is negligible compared with p and q, and we may write

$$\sin \theta_A = \theta_A = \alpha + \phi = \frac{z}{p} + \frac{z}{R}$$

and

$$\sin \theta_B = \theta_B = \phi - \beta = \frac{z}{R} - \frac{z}{q}$$

Then, by Eq. (24.3),

$$\frac{n_A z}{p} + \frac{n_A z}{R} = \frac{n_B z}{R} - \frac{n_B z}{q}$$

or

$$\frac{n_A}{p} + \frac{n_B}{q} = \frac{n_B - n_A}{R} \tag{24.5}$$

The line OC is the principal axis for this optical system. Equation (24.5) is valid only for paraxial rays, i.e., rays which make small

Figure 24.9
Refraction at a spherical surface.

angles with the axis. The radius R is positive when it is measured from the surface to the center of curvature in the direction in which the light is traveling. Thus R is positive for a surface convex to the incident light and negative for a surface concave to the incident light. Similarly, q is positive if the image is on the side of the surface toward which the light is propagating and negative if on the side from which the light approaches.

24.8 REFRACTION BY A PRISM

A wedge-shaped portion of a refracting medium bounded by two plane surfaces is called a *prism*. If the prism is optically denser than the surrounding medium, a ray of light incident on one of the faces is bent toward the normal to that face on entering the prism. On emerging from the opposite face, the ray is bent away from the normal to that face. The angle D (Fig. 24.10) through which the ray is deflected in passing through the prism is called the *angle of deviation*. When the angle at which the ray enters one face is equal to the angle at which it leaves the opposite face, the angle of deviation has its least value and is known as the *angle of minimum deviation*.

By observing the angle of minimum deviation and the angle between the faces of the prism, it is possible to find the index of refraction of the material of which the prism is made. Let A be the angle of the prism, D the angle of minimum deviation, and n the index of refraction. Since the angle between the faces of the prism is the supplement of the angle between the normals HG and NG (Fig. 24.10),

$$A = r + r'$$

The deviation at the first face is $i - r$ and at the second face $i' - r'$. Hence, the total deviation is

$$D = i - r + i' - r' = i + i' - A$$

The angle of deviation is minimal when the angle of incidence is equal to the angle of emergence. In this case, $i = i'$, $r = r'$, $A = 2r$, $D = 2i - A$, and

$$i = \frac{A + D}{2}$$

From the law of refraction,

$$\text{Index of refraction} = n = \frac{\sin i}{\sin r}$$

$$n = \frac{\sin [(A + D)/2]}{\sin (A/2)} \qquad (24.6)$$

Example

For a 60° glass prism, the angle of minimum deviation is found to be 48°. What is the index of refraction of the prism?

$$\text{Index of refraction} = \frac{\sin [(A + D)/2]}{\sin (A/2)}$$

$$n = \frac{\sin 54°}{\sin 30°} = \frac{0.809}{0.500} = 1.62 \qquad ////$$

Figure 24.10
Refraction by a prism.

Figure 24.11
A prism refracts longer wavelengths less than shorter ones, resulting in the dispersion of white light into a spectrum.

24.9 THE DISPERSION OF LIGHT BY A PRISM

When a very narrow beam of white light passes through a prism (Fig. 24.11), it is spread out into a band of colors known as a *spectrum*. Violet light, which has the shortest wavelength of the visible, is bent most, while red, with the longest wavelength, is bent least. *The separation of white light into its component colors is called dispersion.*

Since different colors are deviated different amounts by the prism, it is clear that the index of refraction and the speed of light in the prism must be different for the various colors. In ordinary crown glass, the speed of red light is approximately 1 percent greater than the speed of violet light.

The question may well be raised of why we did not notice dispersion in connection with experiments performed earlier with the deviation of light by a prism. In these experiments a much wider beam of light was used, and the level of illumination in the room was great. Because of the width of the beam, the blue from one part of the beam fell on the red from another, on the orange from still another, and so forth. This led to a white core, but at the edges of the deviated beam there were doubtless evidences of dispersion. However, the colors were so dim compared with the bright central image that they probably were not noticed.

24.10 ACHROMATIC PRISM

When light of more than one color falls on a prism, the emerging beam is not only deviated but spread out into a spectrum. If a second prism is placed just beyond, with its vertex at the base of the first, the second prism tends to gather the different colors together again and combine them into white light. The net dispersion produced by the two prisms is the difference between the dispersions produced by the individual prisms. The deviation produced by the second prism is opposite that produced by the first prism, and the net deviation is the difference between the two. If the two prisms were identical, the net effect would simply be that of passing the light through a parallel-faced plate—no dispersion and no deviation. However, if the prisms are made of different kinds of glass, they may have the same net dispersion but quite different deviations. A flint-glass prism which produces the same dispersion as a crown-glass prism of larger central angle gives a smaller deviation. By placing together crown- and flint-glass prisms which have the same dispersion, it is possible to construct a prism that deviates the light

Figure 24.12
Achromatic prism combination
producing deviation without dispersion.
The red from the incident ray shown
recombines with the yellow, green,
blue, and violet components of
neighboring rays to give white light
once more.

without spreading it into a spectrum (Fig. 24.12). Such a prism is called an *achromatic* prism.

If the angle of the flint-glass prism is increased until the average deviation it produces is equal to that produced by the crown-glass prism, the dispersion of the flint-glass prism is substantially greater. When two such prisms are put together, the combination produces *dispersion without net deviation*. When white light passes through such a combination of prisms, called a *direct-vision spectroscope*, it is dispersed into a spectrum, but its general direction is unchanged.

24.11   THE RAINBOW

No doubt the first evidence of dispersion seen by man was the rainbow. The formation of the rainbow was explained by Descartes about 1637. Let us consider a situation in which the sun is in the west and it is raining in the east. Rays of sunlight enter water droplets (Fig. 24.13), where they are refracted and dispersed. If a ray enters a drop at the proper place, it will be partly reflected at the internal surface of the drop and leave the drop as shown in insert A of the figure. Descartes showed that when the angle θ is 42° red light is strongly reflected, while violet light appears prominently at 40°. The other colors of the spectrum are intensely reflected at angles between 40° and 42°. Each color is strongly reflected at that angle for which it can pass through the droplet with minimum deviation. If an observer stands at O, and if the line SO represents the direction of the incident sunlight, all those raindrops which lie on the surface of a cone of angle 40° about SO send an excess of violet light to the observer, while those which lie on a conical surface of angle 42° about SO return red light.

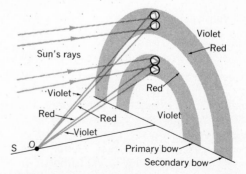

Figure 24.13
Formation of a rainbow.

In addition to the primary bow which we have just discussed, there is sometimes a secondary bow in which the colors are reversed, with violet on the outside and red on the inside. This secondary bow is produced by refraction, dispersion, and two internal reflections inside the raindrop, as shown in insert B. For the secondary bow, the angle between the incident and emergent rays is about 51° for red and 54° for violet.

When exceedingly tiny drops are involved, as may be the case in a garden spray, diffraction effects may become large, and the structure of the bow much more complicated.

QUESTIONS

1 Why is it that when light goes from one medium to another, the wavelength changes but the frequency does not?

2 A fish lies at the bottom of a clear pool. Describe and explain what it sees as it looks in various directions.

3 A swimming pool of clear water is flat where the depth is 1.2 m. If a person stands at the center of this region, why does it appear to him that he is standing at the deepest point?

4 What are the advantages of using a totally reflecting prism (Fig. 24.5) rather than a silvered mirror in an optical instrument such as binoculars or a periscope? What are the disadvantages?

5 What properties should a material have if it is to be particularly useful as a light pipe?

6 How and why does the earth's atmosphere alter the apparent shape of the sun and moon when they are near the horizon?

7 If you were standing waist deep in clear water and wished to spear a fish swimming nearby, would you aim above or below the apparent position of the fish? Why?

8 A clear cylindrical glass filled with water rests on a table. Explain why you cannot see the table through the water surface and the side of the glass but can see your fingers when they are in contact with the glass.

9 Why does the air shimmer over a hot stove? Why do stars twinkle?

10 Why are colors observed in the light from a cut diamond?

11 Very little light is transmitted through chipped ice in a clear glass. When water is added, the system becomes much more transparent even when no significant melting has occurred. Explain why.

12 Why are optical systems using totally reflecting prisms usually designed so the light enters and leaves the surfaces along a normal?

PROBLEMS

1 Find the sine of the critical angle of incidence for an air-water interface and the angle of refraction in water if light is incident from air at an angle of 38°. *Ans.* 48.5°, 27.5°

2 Find the angle of refraction of light incident on a water surface at an angle of 48°. What is the speed of light in water?

3 Calculate the index of refraction of n-propyl alcohol by which light incident from air at an angle of 60° is refracted at an angle of 38.7°.
Ans. 1.39; 2.17×10^8 m/sec

4 Light is passing from air into a liquid and is deviated 22° when the angle of incidence is 65°. Under what conditions will total reflection occur at this interface?

5 A ray of light passes from crown glass to water. What is the critical angle of incidence?
Ans. 61.3°

6 The index of refraction of a glass is 1.60. Find the speed of light in the glass, the angle of refraction in glass if light is incident from water at an angle of 42°, and the critical angle of incidence for a glass-water interface.

7 A glass dish with a plane parallel bottom and refractive index 1.53 is half filled with water. Then carbon disulfide is poured on top of the water. Finally, a glass cover is placed on top of the dish, the cover being flat. A beam of light making an angle of 52° with the vertical is inci-

dent on the horizontal cover. Find the angles which the beam makes with the vertical as it passes through glass, carbon disulfide, water, glass, and air.

Ans. 31°; 28.9°; 36.2°; 31°; 52°

8 The bottom of a crown-glass vessel is a thick plane plate that has an index of refraction of 1.52. The vessel is filled with water. What is the largest angle of incidence in the water for which a ray can pass through the glass bottom and into the air below? How does this compare with the critical angle of incidence for a water-air interface? (See Prob. 1 above.)

9 A fish is 1 m below the surface of a pool. What is its apparent depth when viewed from above?

Ans. 0.75 m

10 A beam of white light is incident on a block of flint glass at an angle of 60°. What is the angular separation in the glass of two rays of light, one of wavelength 486 nm and the other of wavelength 656 nm, if the index of refraction for the former is 1.670 and that for the latter is 1.650?

11 A ray of light is passed through a prism having a refracting angle of 60°. Rotation of the prism causes the ray to be deviated various amounts, the least of which is 45°. Determine the index of refraction of the glass of which the prism is made.

Ans. 1.59

12 What is the index of refraction of a 60° glass prism which produces an angle of minimum deviation of 40°?

13 Given a 55° prism of flint glass, find the angle of minimum deviation. At what angle of incidence must the light strike the prism? Find the deviation if the angle of incidence is 5° less.

Ans. 42.6°; 48.8°; 43.1°

14 A ray of light makes an angle of incidence of 40° with a glass prism of index of refraction 1.5 and refracting angle 60°. Through what angle is the ray deviated by the prism? Does this represent minimum deviation?

15 If light is incident on face AB of the prism of Fig. 24.5 at an angle of 45° (midway between AB and OL), what must be the index of refraction of the prism if the refracted ray is incident on BC at the critical angle of incidence? What happens to the refracted ray at BC if the

index of refraction is greater than this value?

Ans. 1.22; totally reflected

16 A hollow prism is made of plane-parallel plates of glass. The angle between the faces is 60°. What is the angle of minimum deviation when the prism is filled with carbon disulfide and sodium D radiation having a wavelength of 589 nm is passed through it?

17 Show that when a light ray passes through a parallel plate of thickness t (Fig. 24.2), it is displaced sideways by an amount $d = [t \sin (i - r)]/(\cos r)$.

18 Show that Eq. (24.4) follows directly from Eq. (24.5) if the refracting surface is plane, i.e., has a radius of curvature of infinity.

19 Experiment shows that the index of refraction of heavy flint glass is 1.717 for D light (yellow) and 1.742 for F light (blue). Find the angle of dispersion of these two colors produced by a 60° prism if the light strikes the prism with an angle of incidence of 55°.

Ans. 3.4°

20 A fish is swimming in a spherical bowl 50 cm in diameter. If the fish is actually 15 cm from the point on the bowl closest to the observer, what is its apparent distance from that point? (Assume the glass of the bowl is of uniform and negligible thickness.)

21 A glass ashtray has an upper surface which is approximately spherical and concave with a radius of curvature of 20 cm. If the glass has bubbles in it and an index of refraction of 1.60, find the apparent depth of a bubble 2 cm below the surface.

Ans. 1.2 cm

22 Show that if a ray of light is incident on a plane-parallel glass plate of thickness d at a small angle of incidence θ, it emerges from the plate parallel to its original direction but displaced a distance $\theta d(n - 1)/n$, where n is the index of refraction of the glass. (Recall that for small angles $\sin \theta = \theta$ in radians.)

23 A clear plastic ball with a diameter of 10 cm has a flaw which is actually 2 cm under the surface. The index of refraction of the plastic is 1.40. Find the position of the image of the flaw when it is viewed along a diameter passing through the flaw from (a) the near side and (b) the far side.

Ans. (a) −1.47 cm; (b) −10.5 cm

24 Show that the deviation produced by a thin wedge of transparent material of angle A is given approximately by $D = (n - 1)A$ if the angle of incidence is small. (*Hint:* For small angles, $\sin \theta = \theta$ in radians.)

OUR KNOWLEDGE of how light rays are bent when they pass from one material to another can now be applied to lenses, which are the basic elements of most optical instruments. When light rays pass through a lens, they undergo refraction at both surfaces, and, except in very special cases, they emerge traveling in a direction different from that in which they were incident. In this chapter we see how a lens deviates rays to form an image of an object. Much of what we learned about image formation by spherical mirrors is applicable to lenses as well.

CHAPTER TWENTY-FIVE

LENSES

25.1 SIMPLE LENSES

Lenses are bodies of transparent material shaped to converge or diverge a beam of light. Simple lenses are bounded by faces which are small sections of spheres.

Consider a beam of parallel rays falling on the prism in Fig. 25.1a. The rays are deviated by the prism but remain parallel to one another. If we wish to bring these rays together at a point, we must deviate the uppermost ray more than the lowest one. This can be accomplished by grinding the surfaces so that they have the cross section indicated in Fig. 25.1b. If the surfaces of a glass blank are small sections of spheres, they will deviate the upper ray more than any of the others; the higher the ray, the greater the deviation. Of course, this does not guarantee that all incident parallel rays will pass through the same point. However, if we do not use too large a section of the spherical surfaces, the emerging rays all pass close to the same point.

If a lens brings a bundle of parallel rays together at a point focus, the lens is said to be *converging*. If it separates such a bundle of parallel rays, the lens is *diverging*. When glass or plastic lenses are used in air, they are converging if they are thicker in the middle than at the edges, diverging if they are thinner at the middle. Figure 25.2 shows six possible types of spherical lenses with their names.

Consider the lens of Fig. 25.3. The left surface is a section of a spherical surface having point C_1 as its center, while the right surface is a portion of a sphere having point C_2 as its center. *The principal axis of this lens is the line connecting the centers of curvature of the two lens surfaces.* If one of the surfaces is plane, the principal axis is the line through the one center of curvature which is perpendicular to the plane surface.

Rays which approach a converging lens parallel to the principal axis are deviated so that they pass (Fig. 25.4a) through a common point,

(a)

(b)

Figure 25.1
(*a*) Parallel rays deviated by a prism remain parallel. (*b*) A lens with spherical surfaces brings parallel rays to a focus.

Figure 25.2
Lens shapes (from left to right) are
double-convex, plano-convex,
concavo-convex (converging
meniscus), convexo-concave (diverging
meniscus), plano-concave, and
double-concave.

the *principal focus,* on the principal axis. A lens has *two principal foci,* one for light incident from the left, which we denote by F, and the second F' for light incident from the right. For thin lenses the principal foci are equidistant from the optical center of the lens. By a thin lens we mean one whose thickness is negligible compared with the distance to the principal foci and to the objects and images concerned. In this text we confine our attention to thin lenses.

When the rays incident on a lens are parallel, the incident wavefront is a plane perpendicular to the rays. The part of the wavefront which goes through the edges emerges ahead of the central part because it has less distance to go in glass, where the speed is reduced. By Huygens' principle it can be shown that the emerging wavefront is spherical and converges on the principal focus.

In the case of a diverging lens, rays parallel to the principal axis are bent as shown in Fig. 25.4b. These rays never intersect, but they all appear to come from the virtual principal focus F.

25.2 STANDARD RAYS FOR LENSES

In Fig. 25.5 an object OP is a distance p from a converging lens. Every ray of light which leaves point O on the object and goes through the lens passes through point I. In the figure only three rays are shown, but any other ray from point O which traverses the lens proceeds through point I. Therefore, at point I we have an image of point O. For the entire object OP we find an image IQ, which in this case is real and inverted.

Whenever we are concerned with locating the image I of an object point O by ray tracing, we shall use three standard rays. These rays, analogous to those discussed for mirrors, are:

1 A ray from O parallel to the principal axis is deviated so that it passes through the principal focus of a converging lens, or diverges as though it came from the principal focus of a diverging lens.

2 The ray from O which approaches the lens along the line through the principal focus F' for parallel rays coming from the opposite side of the lens is deviated so that it leaves the lens parallel to the principal axis.

3 A ray through the optical center of the lens passes through the

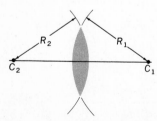

Figure 25.3
The principal axis of a spherical lens is
the line connecting the centers of
curvature of the two lens surfaces.

lens undeviated. For a thin lens we may take the optical center to be the center point of the lens.

In Fig. 25.5 these three standard rays are, respectively, $OAFI$, $OF'I$, and OCI.

25.3 SINGLE LENSES

Let p be the distance from the object to the lens and q that from the image to the lens. The distance from the lens to either principal focus is the *focal length f*. It is easy to show that the equation

$$\frac{1}{p} + \frac{1}{q} = \frac{1}{f} \qquad (25.1)$$

applies as well to lenses as to mirrors.

Equation (25.1) can be derived as follows. The triangles OPC and IQC (Fig. 25.5) are similar, and hence $OP/IQ = CP/CQ = p/q$. Also, triangles ACF and IQF are similar, so that $AC/IQ = FC/FQ = f/(q-f)$. Since $AC = OP$, we have $AC/IQ = OP/IQ = p/q = f/(q-f)$. Therefore, $pq - pf = qf$. If we divide by pqf and rearrange, we obtain $1/p + 1/q = 1/f$.

We note further that $IQ/OP = q/p$. Since IQ is the height of the image, and OP the height of the object, we have

$$\text{Linear magnification} = \frac{\text{height of image}}{\text{height of object}} = \frac{q}{p} \qquad (25.2)$$

For the lens of Fig. 25.5, an object at IQ has its image at OP. Points P and Q are *conjugate points* of the lens—an object at either point has an image at the other.

When an object is inside the principal focus of a converging lens, the image formed is *erect, virtual,* and *enlarged* (Fig. 25.6). In this case, the image distance is negative.

(a)

(b)

Figure 25.4
(*a*) A converging lens deviates rays parallel to the principal axis so that they pass through the principal focus. (*b*) A diverging lens deviates rays incident parallel to the principal axis so they appear to come from a virtual focus.

Figure 25.5
Formation of a real image by a converging lens.

Figure 25.6
A converging lens forms an erect, virtual, enlarged image of an object inside the principal focus.

Figure 25.7
An erect, virtual, reduced image of a real object is formed by a diverging lens.

A real object placed before a diverging lens has an erect, virtual, and reduced image (Fig. 25.7). In applying the general lens formula to the diverging lens, the focal length is taken as *negative*. In Fig. 25.7 the image distance q is also negative.

In calculations involving the lens relation [Eq. (25.1)], one must be careful that the proper signs are taken for object distance, image distance, and focal length. The sign convention used in this text for both lenses and mirrors is as follows:

1   If the direction of an incident ray is from an object point O toward the lens (or mirror), the object distance is positive; if its direction is from the lens toward the object, p is negative.

2   The focal length is positive for a converging lens (or mirror) and negative for a diverging lens (or mirror).

3   The image distance is positive if a ray from O goes from the lens (or mirror) toward the image. It is negative if the ray goes away from the image when it leaves the lens (or mirror).

25.4 COMBINATION OF LENSES

When light passes first through one lens and then through a second, the position of the final image can be calculated by repeated use of the general lens equation. *The object for the second lens is the image formed by the first.* If the second lens is interposed before the rays actually cross (Fig. 25.8), its object is still taken to be the image which the first lens would have formed had the second not been present. In this case the object distance for the second lens is negative, since the light is deviated by this lens before its "object" is formed. If there is a third lens, the image formed by the second lens acts as its object. The final image formed by a very complicated optical system, such as that of a submarine periscope, can be located by successive applications of the relations governing simple lenses and mirrors. In every case the

Figure 25.8
Image formed by a pair of separated lenses.

object for any single component is the image formed by the preceding one. Great care must be taken to assign the proper sign to each distance.

Example

An object 5 cm high is 30 cm from a lens of 10-cm focal length. A diverging lens of focal length −8 cm is placed 9 cm beyond the converging lens (Fig. 25.8). Find the position and height of the final image.

For the first lens,

$$\frac{1}{p} + \frac{1}{q} = \frac{1}{f}$$

$$\frac{1}{30} + \frac{1}{q} = \frac{1}{10} \qquad \frac{\text{height of image}}{\text{height of object}} = \frac{15}{30}$$

$$q = 15 \text{ cm} \qquad \text{height of image} = 2.5 \text{ cm}$$

For the second lens $p = -6$ cm, and

$$\frac{1}{-6} + \frac{1}{q} = \frac{1}{-8} \qquad \frac{\text{height of image}}{\text{height of object}} = \frac{24}{6}$$

$$q = 24 \text{ cm from second lens} \qquad \text{height of final image} = 10 \text{ cm}$$

////

Example

If the diverging lens of the preceding example had been only 4 cm from the first lens, where would the final image have been located?

The image which would be formed by the first lens is still 15 cm from the first lens. The image distance for the second lens is now −11 cm.

$$\frac{1}{p} + \frac{1}{q} = \frac{1}{f}$$

$$\frac{1}{-11} + \frac{1}{q} = -\frac{1}{8}$$

$$q = -29.3 \text{ cm} \qquad ////$$

Example

An object is 25 cm from a diverging lens of 15-cm focal length. A converging lens of focal length 10 cm is 5 cm beyond the diverging lens (Fig. 25.9). Find the position of the final image.

Figure 25.9
Image formation by a combination of a diverging lens and a converging lens.

For the first lens,

$$\frac{1}{p} + \frac{1}{q} = \frac{1}{f}$$

$$\frac{1}{25} + \frac{1}{q} = \frac{1}{-15}$$

$$q = \frac{-75}{8} \text{ cm}$$

For the second lens,

$$p = \tfrac{75}{8} + 5 = \tfrac{115}{8} \text{ cm}$$

$$\frac{1}{p} + \frac{1}{q} = \frac{1}{f}$$

$$\frac{8}{115} + \frac{1}{q} = \frac{1}{10}$$

$$q = 33 \text{ cm} \qquad ////$$

25.5 THE TELEPHOTO LENS

An interesting application of two lenses in series is found in the telephoto lens (Fig. 25.10). In order to produce a large image of a distant object with a single lens, the focal length must be large. This requires a camera of inconvenient length, since the length of the camera must be approximately equal to the focal length of the lens. This difficulty can be avoided by a telephoto lens, which consists of a combination of a converging lens A and a diverging lens B placed at a distance d from each other. The converging lens A would form an image of a distant object just outside its focus F_1, but the lens B causes the image to be formed at F^*. The image formed at F^* is larger than the one that would have been formed at F_1. The magnification of the lens system is increased without increasing the length of the camera too much.

Example

The telephoto lens of Fig. 25.10 is used for taking an action shot at a football game. Lens A has a focal length of 12 cm, and lens B a focal length of -6 cm; they are separated by a distance of 8 cm. If the action is 100 m away, find the position of the final image and the magnification of the system. What focal length would be necessary in a single lens to produce an image of the same size? How long would the camera have to be?

Figure 25.10
A telephoto-lens arrangement.

For lens A,

$$\frac{1}{100} + \frac{1}{q} = \frac{1}{0.12}$$

$$q = 0.1202 \text{ m} = 12.02 \text{ cm}$$

and the size of the image is 0.12/100 that of the object.

For lens B, $p = -4.02$ cm and $f = -6$ cm, and

$$\frac{1}{-4.02} + \frac{1}{q} = \frac{1}{-6}$$

$$q = 12.2 \text{ cm}$$

Height of second image $= \dfrac{12.2}{4.02}$ height of first image

$$= \frac{12.2}{4.02} \times \frac{0.12}{100} \text{ height of original object}$$

Magnification $= 0.00364$

The film is $8 + 12.2 = 20.2$ cm from lens A.

For a single-lens camera to produce the same image size, $p = 100$ m and $q = 0.00364p = 0.364$ m.

$$\frac{1}{p} + \frac{1}{q} = \frac{1}{f}$$

$$\frac{1}{100} + \frac{1}{0.364} = \frac{1}{f}$$

$$f = 0.362 \text{ m}$$

The minimum length of camera is the distance from lens to film, which is 36.4 cm; therefore this camera is almost twice as long as the one using the telephoto lens. ////

A telephoto lens has a fixed effective focal length. For motion pictures and television it is particularly desirable to have a lens system of variable focal length so that the camera can be focused on a scene and then "zoom in" on an item of interest. Such a system, known as a *zoom lens*, requires many lenses as components. The variable focal length is achieved by the axial movement of several lenses simultaneously. They must be moved in such a way that the image is kept in focus and retains the desired brightness. This requires a complicated set of motions of some lenses within the zoom-lens system. A relatively simple zoom lens may have five achromatic doublets (Sec. 25.7) with two of the inner doublets moving together as the effective focal length is varied. A more complex zoom lens may have 18 or more simple lenses, several of which move internally. Modern computers have played an important role in the development of zoom lenses.

25.6 THE LENS-MAKER'S EQUATION

The focal length of a lens depends on the material of which the lens is made and on the radii of curvature of the two surfaces. It is shown below that the focal length is given by

$$\frac{1}{f} = (n - 1)\left(\frac{1}{R_1} - \frac{1}{R_2}\right) \tag{25.3}$$

Figure 25.11
Light rays from object point O are refracted at both surfaces of the lens to produce an image at I.

where n is the index of refraction of the lens material relative to its surroundings, R_1 is the radius of curvature of the surface upon which the light is incident, and R_2 is the radius of curvature of the surface through which the light leaves the lens. *R_1 is positive when the surface is convex* and *negative when the surface is concave to the incident of light*. The same rule applies to R_2. By this convention the radius of curvature of a surface is positive when measured from the surface to the center of curvature in the direction the light is leaving the surface. Thus, for a double-convex lens, the first surface has a positive radius of curvature, and the second a negative one.

Rays from an object point O (Fig. 25.11) are refracted at the first surface of the lens and would form an image at I_1 if everything to the right of the first surface were medium B. By Eq. (24.5),

$$\frac{n_A}{p_1} + \frac{n_B}{q_1} = \frac{n_B - n_A}{R_1} \tag{A}$$

I_1 serves as object for the second surface; since the light goes from the second surface to I_1, the object distance is negative. Thus $p_2 = -(q_1 - t)$, where t is the lens thickness. For a thin lens t is negligible compared to q_1, so $p_2 = -q_1$, and, for the second surface,

$$\frac{n_B}{-q_1} + \frac{n_A}{q_2} = \frac{n_A - n_B}{R_2} \tag{B}$$

Adding Eqs. (A) and (B) yields

$$\frac{n_A}{p_1} + \frac{n_A}{q_2} = (n_B - n_A)\left(\frac{1}{R_1} - \frac{1}{R_2}\right)$$

or

$$\frac{1}{p} + \frac{1}{q} = \left(\frac{n_B}{n_A} - 1\right)\left(\frac{1}{R_1} - \frac{1}{R_2}\right) \tag{25.3a}$$

which is equivalent to Eq. (25.3).

Example

A double-convex lens is made of glass having an index of refraction of 1.5. The front surface has a radius of curvature of 10 cm, and the back surface a radius of curvature of -20 cm (Fig. 25.11). Find its focal length.

$$\frac{1}{f} = (n - 1)\left(\frac{1}{R_1} - \frac{1}{R_2}\right)$$

$$= (1.5 - 1)[\tfrac{1}{10} - (-\tfrac{1}{20})] = \tfrac{1}{2} \times \tfrac{3}{20} = \tfrac{3}{40}$$

$$f = \tfrac{40}{3} = 13.3 \text{ cm} \qquad\qquad ////$$

The reciprocal of the focal length of a lens in meters is the dioptric power of the lens in diopters. Thus, a lens of focal length 50 cm has a power $P = 1/0.5 = 2$ diopters. The powers of lenses used in typical eyeglasses range in magnitude from about 0.2 to 10 diopters.

One advantage of using the powers of lenses is associated with the fact that when two thin lenses are in contact the power of the combination is equal to the sum of the individual powers. If f represents the focal length of the combination of two thin lenses in contact,

$$\frac{1}{f} = \frac{1}{f_1} + \frac{1}{f_2} \longrightarrow \qquad (25.4) \qquad D_c = D_1 + D_2.$$

where f_1 and f_2 are the focal lengths of the two lenses.

25.7 DEFECTS OF LENSES

If we had an ideal lens, every ray of light from any given point on the object would cross at exactly the same point on the image. Further, the image would be similar to the object in every respect. Any circle on the object would be a perfect circle on the image, and any square on the object would lead to a perfect square on the image. If the object were all in one plane, the image would be all in one plane. Actually no such ideal lens exists. For a single lens, the rays from a point on the object ordinarily fall over a small region of the image. If most of the rays from a point on the object cross at the desired image point but some of them cross off on one side, a defect called *coma* occurs. With some lenses the image of a rectangular object (Fig. 25.12) is distorted so that it is *barrel-shaped;* in other cases it may look like a *pincushion*. The detailed explanation of these defects is beyond the scope of this book, but there are three types of image defect which can be readily explained.

Spherical Aberration When rays of light parallel to the principal axis of a spherical lens pass through zones near its edges, they cross the axis nearer the lens than those rays which pass through nearer to the center. Instead of converging to a sharp focal point, they cross the axis at slightly different points and form a blurred image. This defect is known as *spherical aberration.* It can be minimized by using only a small portion of the lens; for a lens of small aperture and long focal length, spherical aberration becomes small. Spherical aberration can be reduced by proper choice of the curvatures of the lens surfaces. When the bending is roughly the same at each lens surface, deviation is minimum, and so is spherical aberration.

Chromatic Aberration When rays of white light parallel to the principal axis pass through an ordinary lens, they are refracted in such a way that different colors are brought to focus at different distances from the lens (Fig. 25.13). Since violet rays are bent more than the red, the focal length for violet light is smaller. A single lens cannot bring all the rays of white light from a point on the object to a single point on the image. As a result the image is not sharp and is likely to have colored rings or markings. This image defect is called *chromatic aberration.*

Object

Pincushion

Barrel

Figure 25.12
The image of an object may suffer from pincushion or barrel distortion.

Crown
glass
Blue
focus
Blue
Red
Red
focus
Red
Blue

Flint
glass
Blue
Red

Red
Blue

Focus

Crown glass   Flint glass

Figure 25.13

Chromatic aberration (top and middle) is corrected by combining two lenses with compensating dispersion but differing deviations to form an achromatic doublet (below).

It may be corrected to a large extent by combining a converging lens of crown glass with a diverging lens of flint glass. These lenses are chosen so that the dispersion of one is equal and opposite to that of the other. In this way no major separation of light into colors occurs. On the other hand, the bending or refractive power of one lens exceeds that of the other; thus there is a resultant bending of the rays in spite of the fact that the lenses deviate the rays in opposite directions. Lenses which are free from dispersion are called *achromatic lenses.* Two lenses, combined to form an *achromatic doublet,* can bring only two colors to exactly the same focus; the other colors are not perfectly corrected. For better correction of chromatic aberration, one may use more lenses.

Astigmatism   Rays of light from an object point far from the principal axis pass through the lens obliquely and do not converge to a common image point. Figure 25.14 shows side and top views of several rays from the object point O. Rays in the vertical plane converge on I_1, while those in the horizontal plane converge on I_2. Thus, the rays from this single point at O focus in two lines I_1 and I_2, called *focal lines.* At a point midway between I_1 and I_2, a roughly circular patch of light corresponds to the closest available approximation to a point image. In quality optical systems astigmatism is corrected by the use of two or more lenses with the proper separation between them.

In a high-quality camera or microscope a combination of several lenses is utilized to perform the function of a single *ideal* lens. The different kinds of glass used in these lenses, their radii of curvature, and their separations are carefully chosen by the optical designer to minimize image defects.

Figure 25.14

Astigmatism and astigmatic focal lines.

1 Why does a lens have two focal points and a mirror only one?

2 What happens to the image formed by a lens if the upper half of the lens is covered? The lower half?

3 What happens to the focal length of a lens when it is immersed in water? Does it matter whether the lens is converging or diverging?

4 If a cylindrical lens is used to form an image of a point source, what will the image look like? Is it really an image in the sense of the discussion in Sec. 25.2?

5 A mercury thermometer is often made in such a way that the mercury column is viewed through a cylindrical lens. Is the lens converging or diverging? How does it function?

6 Under what conditions is a converging spherical lens thicker at the edges than at the center?

7 Can an empty almost spherical fishbowl focus a beam of light? A full one? A full water glass?

8 If a plano-convex lens is used to form an image of a distant object, should the plane or the convex surface be on the side of the object? Why?

9 As a self-luminous object moves closer to a lens, does its image get brighter or dimmer? Why? Explain with equations.

PROBLEMS

Draw a ray diagram whenever appropriate, showing the paths of the three standard rays.

1 A converging lens has a focal length of 18 cm. An object 3 cm high is placed 45 cm in front of the lens. Find the position and height of the image.   *Ans.* 30 cm; 2 cm

2 An object 36 cm from a lens results in a real image 24 cm from the lens. Find the focal length of the lens. If the image is 2 cm high, how high is the object?

3 An object 1.5 cm tall is placed 40 cm from a lens of 60-cm focal length. Find the image distance and the height of the image.   *Ans.* −120 cm; 4.5 cm

4 A converging lens has a focal length of 16 cm. Find the image distance for an object at each of the following distances from the lens: 4, 8, 16, 32, 48, and 80 cm.

5 An object 2.5 cm high stands 50 cm from a diverging lens of −20-cm focal length. Find the position and height of the image.   *Ans.* −14.3 cm; 0.714 cm

6 An illuminated arrow 1.6 cm high is placed 54 cm from a diverging lens of focal length −18 cm. Find the position and the height of the image.

7 The image formed by a converging lens is erect, virtual, and three times as high as the object. If the focal length of the lens is 15 cm, find the object and image distances.   *Ans.* 10 cm; −30 cm

8 A candle is placed at a distance of 0.6 m from a diverging lens with a focal length of −0.4 m. Where is the image located? How high is it relative to the object?

9 An object 1 cm high is 75 cm from a diverging lens with a focal length of 30 cm magnitude. Calculate the image distance and the height of the image.   *Ans.* −21.4 cm; 0.286 cm

10 A plano-convex lens is made of flint glass, which has an index of refraction of 1.69 for violet light and 1.64 for red light. If the radius of curvature of the curved surface of the lens is 20 cm, what is the difference between the focal lengths of the lens for these two colors?

11 A lens made of glass with index of refraction 1.5 has surfaces with radii of curvature of magnitudes 30 and 60 cm. If this is a double-convex lens, what is the focal length? If double-concave, what is the focal length? What is the power in each case?
     Ans. 40 cm; −40 cm; 2.5 and −2.5 diopters

12 An object 2 cm high is located 20 cm in front of a plano-convex lens of 30-cm focal length. The lens is made of glass of index of refraction 1.55. Find the image distance, the height of the image, and the magnitude of the radius of curvature of the convex side of the lens.

13 A double-convex lens is made of glass (index of refraction 1.5) and has radii of curvature of

20 and 30 cm. Find the focal length of this lens in air and in water.   *Ans.* 24 cm; 96 cm

14   A plano-convex lens is made of glass of index of refraction 1.5. The curved surface has a radius of 10 in. An object is placed 50 in. in front of the lens. Where is the image? If the object and lens were both immersed in clear water, where would the image be formed?

15   A glass lens having an index of refraction of 1.52 and radii of curvature of $+40$ and -20 cm is cemented to another glass lens having an index of refraction of 1.66 and radii of curvature of -20 and -60 cm. Find the focal length of the combination and its power.

Ans. 58.8 cm; 1.7 diopters

16   A plano-convex lens has a radius of curvature of 20 cm and an index of refraction of 1.60. A plano-concave lens has the same radius of curvature and an index of refraction of 1.50. What is the focal length of the combination if the two lenses are placed in contact to form a plane slab?

17   An object 2 cm high stands 60 cm in front of a converging lens of 12-cm focal length. Six centimeters beyond the first lens there is a diverging lens with focal length of -15 cm. Find the image distance from the second lens and the height of the final image.

Ans. 22.5 cm; 1.25 cm

18   An object 4 cm high is 120 cm from a converging lens of 40-cm focal length. A diverging lens of focal length -100 cm is placed 20 cm beyond the converging lens. Find the position and height of the final image.

19   Two converging lenses, each of 30-cm focal length, are 20 cm apart. An object 1 cm high is placed 50 cm from one of the lenses. Where is the final image and what is its height?

Ans. 19.4 cm; 0.53 cm

20   A converging lens that has a focal length of 30 cm is in contact with a diverging lens that has a focal length of -40 cm. What are the focal length and power of the combination?

21   An object is 12 cm from a lens of 20-cm focal length, and 10 cm beyond the first lens is a second lens with a focal length of 24 cm. Where is the final image relative to the second lens?

Ans. 60 cm

22   A converging lens and a diverging lens are placed 11 cm from each other. Where will this combination of lenses produce an image of a distant object if the converging lens has a focal length of 20 cm and the diverging lens a focal length of -6 cm?

23   A converging lens with a focal length of 24 cm is placed 40 cm in front of a diverging lens having a focal length of -30 cm. An object is placed 30 cm in front of the converging lens. Find the position and lateral magnification of the image produced by the two lenses.

Ans. -48 cm; 2.4

24   A beam of sunlight falls on a diverging lens of focal length -15 cm; 25 cm beyond this is placed a converging lens of 20-cm focal length. Where should a screen be placed to receive the final image of the sun?

25   Two double-convex lenses have focal lengths of 40 cm. They are separated by 20 cm. An object 5 cm high is placed 100 cm in front of the first lens. Find the position of the image formed by the system of two lenses. How high is the image?   *Ans.* 21.5 cm; 1.54 cm

26   An object is 90 cm from a converging lens of 30-cm focal length. A diverging lens of -60-cm focal length is placed 25 cm beyond the converging lens. Find the position of the final image and the ratio of its height to that of the object.

27   A telephoto lens system (Fig. 25.11) consists of a converging lens of 15-cm focal length and a diverging lens of -8-cm focal length separated by a distance of 10 cm. If the system is used to photograph an object 100 m away, how far must the film be from the converging lens? How far would the film be from the lens in a camera using a single lens if the final image were the same size in both cases?

Ans. 23.5 cm; 40.5 cm

28   Prove that when two thin lenses are in contact, the power of the combination is the sum of the powers of the two lenses. (*Hint:* Assume a distant object and use the fact that the image formed by the first lens acts as object for the second.)

29   Show that if $s = p - f$ (Fig. 25.6) is the distance of an object from the principal focus for parallel rays coming from the opposite side of a lens and

$s' = q - f$ is the distance of the image from the other principal focus, then $ss' = f^2$. (This is known as the *newtonian form* of the lens equation.)

(30) An object and a screen are a distance L apart on an optical bench. A converging lens of focal length f can be at either of two positions to give a sharp image of the object on the screen. Show that the distance of the lens from the object is given by

$$p = \frac{L}{2}\left(1 \pm \sqrt{1 - \frac{4f}{L}}\right)$$

What is the largest value of f for which a sharp image is formed?

CHAPTER TWENTY-SIX
OPTICAL INSTRUMENTS

NOW WE KNOW how lenses and mirrors form images, and, in principle at least, we can trace the path of a ray which falls successively upon any number of reflecting and refracting surfaces. This is the knowledge we require to understand the operation of optical instruments—eyes, cameras, projectors, microscopes, and telescopes—which are the subject of this chapter.

26.1 THE PHOTOGRAPHIC CAMERA

One major application of lenses is in the photographic camera (Fig. 26.1). In a quality camera a combination of lenses acts as a single ideal lens and produces a real image of an external object on a film. The distance between lens and film can be altered to focus the image. In some cameras this adjustment is made by a bellows, and in some by sliding one tube inside another.

The image of a distant object is formed in the focal plane, which contains the principal focus. Therefore, if two cameras of different focal lengths are set up to take the same picture, the lengths of the images are directly proportional to the focal lengths. If both cameras are using the same kind of film, the total amount of light which must reach unit area on each film for proper exposure is the same. The luminous flux entering the camera is directly proportional to the area of the lens opening, which in turn is proportional to the square of the diameter of the circular opening. Consequently, the exposure times of the two cameras are the same if the ratios of focal length to diameter of the opening are the same. *The ratio of focal length to diameter of the opening is called the f number.* When a camera is set at $f/3$, the diameter of the lens opening is one-third the focal length. If the opening is reduced to $f/6$, for the same exposure a time four times as long is required. When a picture must be taken in a very short time, a lens of small f number must be used, and this in turn means a large opening. To produce a lens with a large aperture for which lens defects have

Figure 26.1
A photographic camera.

been satisfactorily corrected is an expensive process. For this reason a camera with an $f/1.5$ lens may cost several times as much as one with an $f/3$ lens.

The smaller the opening through which light is admitted, the greater the depth of focus for a camera. A smaller cone of rays is used to form the image of a point when the opening is small; therefore the circle on which they fall is smaller when a point is not quite in focus. On the other hand, a smaller opening calls for a longer exposure. If one is taking pictures of fast action such as that in a football game, a compromise between short exposure time and great depth of focus must be made.

If the regular lens of a camera is replaced by a wide-angle lens, i.e., one of shorter focal length, a larger area of the object field is focused on the emulsion because the image size of any given object is smaller, as we saw above. On the other hand, a telephoto lens with its longer effective focal length produces a larger image of an object at the cost of a reduced object field.

26.2   THE PROJECTION LANTERN

In the projection lantern the fundamental optical parts are essentially those of a camera. The lantern produces a large image at some distance of a small object close by. All the light which forms the image must come from the object. Since the image has a much greater area than the object, it is necessary that the object be strongly illuminated if a bright image is desired. Basically, the projection lantern (Fig. 26.2) consists of a powerful source of light, large condensing lenses, and a projecting lens. The condensing lenses collect light from the source and send it through the slide, so that this slide is brilliantly illuminated. The objective lens then projects a real image of the slide on the screen. Since the slide is just outside the principal focus of the projecting lens, the image on the screen is enlarged, real, and inverted.

— CONSIST OF
CONDENSING LENSES
PROJECTING LENSES

Example

Find the focal length of the lens which must be used in a lantern to produce the image of a slide 8 cm square upon a screen 3 m square at a distance of 10 m from the lantern. Assume that the image covers the entire screen.

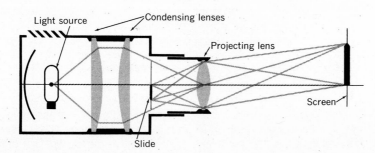

Figure 26.2
A slide projector.

$$\text{Linear magnification} = \frac{q}{p} = \frac{300}{8}$$

Since $q = 1,000$ cm,

$$\frac{q}{p} = \frac{1,000}{p} = \frac{300}{8} \quad \text{and} \quad p = 26.7 \text{ cm}$$

$$\frac{1}{p} + \frac{1}{q} = \frac{1}{f}$$

$$\frac{1}{26.7} + \frac{1}{1,000} = \frac{1}{f}$$

$$\boxed{f = 25.9 \text{ cm}} \qquad ////$$

26.3 THE EYE

Like the camera, the eye can be looked upon as a lighttight enclosure having a lens at one end and a sensitive film of nerve fibers at the other end. The eyeball is roughly a hollow sphere of dense fibrous tissue (Fig. 26.3) about 1.5 cm in diameter. Light enters through a transparent tissue called the *cornea* into a clear fluid, the *aqueous humor*. It then passes through the *lens*, a flexible structure with index of refraction somewhat greater than that of the aqueous humor and of the jellylike *vitreous humor* on the other side. The light is brought to focus on the *retina*, which plays the role of a sensitive screen. The amount of light which enters the eye is regulated by a circular diaphragm, the *iris*, which is responsible for the color of the eye. It is provided with muscles which diminish or dilate the *pupil*, the opening through which the light enters the lens.

The principal bending of light occurs at the cornea, which has a radius of curvature much less than the remainder of the eyeball. The cornea provides roughly two-thirds of the bending required to focus the image of a distant object on the retina, while the lens provides a power of about 20 diopters. When a normal eye is adjusted for very distant objects, it is relaxed, and the lens has its thinnest shape with surfaces of maximum radii of curvature. To bring the image of a near object in focus on the retina, the muscles of the ciliary body which encircles the lens contract, thereby making the radii of curvature smaller, particularly that of the front surface. Thus the focal length of the lens is reduced, and the image focused on the retina.

The remarkable ability of the eye to adjust its focal length is called *accommodation*. The greatest and shortest range at which a given eye can see distinctly are called the *far point* and *near point*, respectively. For a normal eye the far point is at infinity; the near point may be 6 cm for a child and usually increases with age.

26.4 DEFECTS OF VISION

If the image formed by a distant object falls in front of the retina, an eye is *nearsighted* or *myopic* (Fig. 26.4a). This may occur because the lens has too short a focal length, because the eyeball is too long,

Figure 26.3
Human eye.

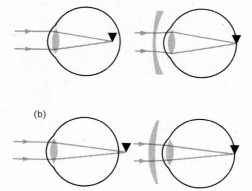

(a)

(b)

Figure 26.4
How eyeglasses correct vision. (*a*) In nearsightedness the image of a distant object falls in front of the retina; a diverging lens enables the eye to form a sharp image on the retina. (*b*) In farsightedness the image of a nearby object falls behind the retina; a converging lens provides correction.

or for some other reason. Such an eye may see a near object distinctly, since the image moves back as the object approaches. To correct for *myopia*, a diverging lens is placed in front of the eye. The image formed by this lens serves as the object for the eye.

Example

A nearsighted eye cannot see objects clearly when they are more than 3 m distant. Find the power of the weakest correcting lens which will just allow this eye to form a sharp image of the moon on the retina.

The image formed by the lens serves as object for the eye. The lens desired must form an image of a distant object 3 m from the eye. Therefore, $p = \infty$ and $q = -3$ m.

$$\frac{1}{p} + \frac{1}{q} = \frac{1}{f}$$

$$\frac{1}{\infty} + \frac{1}{-3} = \frac{1}{f}$$

$$f = -3 \text{ m}$$

$$\text{Power} = \frac{-1}{3 \text{ m}} = -0.33 \text{ diopter} \qquad \text{////}$$

In the *farsighted* or *hyperopic* eye the image of a nearby object is formed behind the retina (Fig. 26.4*b*). This defect may arise because the lens is too flat or the eyeball too short. As the object is moved farther away, the image moves nearer to the retina, so this eye may see distant objects clearly. To correct for farsightedness, a converging lens is placed in front of the eye. This makes the equivalent focal length of the eye shorter.

Example

The near point of an eye is 90 cm. What is the lowest power lens which will permit this eye to see print clearly at a distance of 25 cm?

The image formed by the lens acts as object for the eye. The lens must

(a)   (b)

produce an image of the print at 25 cm at a distance of 90 cm in front
of the eye.

$$p = 25 \text{ cm} \qquad q = -90 \text{ cm}$$

$$\frac{1}{p} + \frac{1}{q} = \frac{1}{f}$$

$$\frac{1}{25} + \frac{1}{-90} = \frac{1}{f}$$

$$f = 34.8 \text{ cm}$$

$$\text{Power} = \frac{1}{0.348} = 2.89 \text{ diopters} \qquad ////$$

Another common defect of the eye is *astigmatism,* which occurs
when the lens or the cornea does not have a truly spherical surface.
The curvature is not the same in different planes containing the axis
of the eye. Then the focal length is not the same in different planes.
When light from an object is received by the normal eye, there is an
attempt by the eye to adjust so that light from all parts of the object
forms sharp images. With an astigmatic eye this is not possible, because
the optical system has different focal lengths for light in different planes.
Consequently, the image formed is indistinct (Fig. 26.5). This defect
may be overcome with lenses that have cylindrical surfaces.

26.5 LUMINOUS FLUX AND THE EYE

The eye has two mechanisms by which it adjusts for changes in the
level of illumination. The first is varying of the diameter of the *pupil*
through the motion of the iris. In bright light the pupil closes down;
in dim light it opens. However, this adjustment covers only a relatively
minor range, and the total amount of light entering the eye can be
changed by the iris by only a factor of about 16. The eye, however,
can see in brilliant sunlight and in dim moonlight, a range of illumination
which covers a factor of millions. The principal mechanism which the
eye uses to adapt for changes in illumination is that of changing the

sensitivity of the retinal surface. To understand this, we must consider the retina in more detail.

Just under the surface of the retina is a mosaic of photosensitive cells, of which there are two kinds, named *rods* and *cones* after their shapes. Both rods and cones have higher indices of refraction than the surrounding material. Light is transmitted down their length by total internal reflection. There are no rods or cones where the optic nerve leaves the eye. The image formed on this region, called the *blind spot,* is not seen. The cones are responsible for color vision; the rods give only achromatic perceptions of white, gray, and black. The rods are grouped in bunches, with each bunch having a connection to the brain, while each cone usually has a private line to the brain. Cones are largely concentrated in the central region of the retina, the *fovea.* As we go away from this central region, the number of cones per unit area decreases. The number of rods per unit area increases to a maximum at 20° from the optic axis and then decreases slowly.

The cones of the foveal area give us acute and color vision under conditions of high to adequate illumination. The solid line of Fig. 26.6 shows how cones respond to equal intensities of various wavelengths. For equal amounts of light energy a yellow-green light of $\lambda \simeq 560$ nm appears brightest to the normal light-adapted eye. One watt of radiant energy at this wavelength corresponds to 680 lumens of luminous flux (Sec. 29.1). Under illuminances lower than about 0.5 lm/m² the cones do not respond well. The rods are sensitive to very small amounts of radiant energy. The dashed curve of Fig. 26.6 shows that the rods have maximum sensitivity at $\lambda \simeq 510$ nm; they are relatively more sensitive to blues and less sensitive to reds than the cones. This is one reason for using blue lights for aisles of theaters and darkened railroad trains.

On the rods the eye builds a photosensitive pigment called *rhodopsin* or *visual purple.* When light strikes a rod, rhodopsin is bleached, and a signal goes to the brain. In bright daylight little rhodopsin is present. In darkness the rhodopsin is built up in the rods, and the eye becomes *dark-adapted.* It takes about 30 min to build up a maximum concentration. When this has been accomplished, the eye is very sensitive to light. This accounts for the blinding effect of bright headlights from an approaching car at night; these same lights would produce no problem at all in daytime.

When the eyes are focused on a distant object, the visual axes are parallel. For viewing a near object there is a movement of the

Figure 26.6
Relative sensitivity of the eye to equal intensities of monochromatic radiation. The solid curve is for cones, the dashed curve for rods.

eyeballs which causes the visual axes to converge. Such a movement is necessary for the image in both eyes to fall on the proper area of the retina. Because of the distance between them, the eyes view an object from slightly different angles. Each retina receives a slightly different picture; this helps us judge distance, solidity, and depth.

26.6 SIMPLE MICROSCOPE

For most distinct vision, an object must be about 25 cm from a "normal" eye. If the object is placed at a greater distance, the image on the retina is smaller, and its details are not seen so distinctly. When the object is placed too near, the image on the retina is blurred unless an additional converging lens is used to aid the eye. A lens used in this way constitutes a *simple microscope* or *magnifying glass*. When the object OP (Fig. 26.7) is inside the principal focus, the lens forms an erect, virtual, magnified image IQ which in turn serves as object for the eye, placed just to the right of the lens. To obtain the greatest advantage, the eye should be as near as possible to the lens. In this way, the field of view is made as large as possible.

When an object is examined with the aid of a simple microscope, the object is brought nearer to the eye than would be possible for distinct vision without the magnifying glass. In Fig. 26.7 the angles subtended at the center of the lens by the object OP and the image IQ are the same. If QE is the distance of most distinct vision, and if the lens were absent, the eye could not see OP distinctly until it were removed to $O'Q$. Hence, by using the lens, the usable visual angle subtended by OP has been increased from α to β. The *angular magnification M* of a microscope is the ratio of the angle subtended at the eye by the image formed by the microscope to the angle subtended by the object when it is at the point of most distinct vision. For the simple microscope of Fig. 26.7 the angular magnification, also known as the *magnifying power*, is given by

$$M = \frac{\beta}{\alpha} = \frac{IQ/QE}{O'Q/QE} = \frac{IQ}{OP} = \frac{QE}{PE}$$

If we take $QE = 25$ cm, the distance of most distinct vision for a normal eye, and observe that PE is the object distance p, we have, by Eq. (25.1), $1/p + 1/(-25 \text{ cm}) = 1/f$, and

$$M = \frac{25 \text{ cm}}{p} = \frac{25 \text{ cm}}{f} + 1$$

for an eye placed as close as possible to the lens.

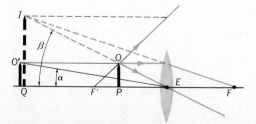

Figure 26.7
A simple microscope produces a virtual, erect, magnified image.

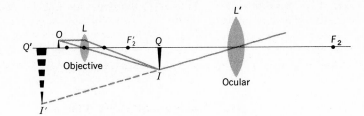

Figure 26.8
Lens scheme for a compound
microscope (not to scale).

It is usually less tiring to use a simple microscope if the final image is at infinity, since then the eye muscles are relaxed. In this case

$$\beta = \frac{OP}{f} \quad \text{and} \quad \alpha = \frac{OP}{25}$$

so that

$$M = \frac{25 \text{ cm}}{f} \implies \text{MAGNIFYING POWER.}$$

With a single converging lens used at the eye, angular magnifications up to about 4 are possible before aberrations become a serious problem. By using eyepieces composed of two lenses it is possible to correct for aberrations and increase the angular magnification to 15 or 20.

26.7 COMPOUND MICROSCOPE

To obtain great magnification, a compound microscope is used. It utilizes a group of lenses acting as an ideal single converging lens L of short focal length (Fig. 26.8). This lens, called the *objective,* produces a real, magnified image IQ of a small object placed just outside its focus. The image IQ is viewed through the eyepiece L', which acts as a simple microscope. This *ocular* produces an enlarged, virtual image $I'Q'$ of the real image IQ.

The magnification of a compound microscope depends on the focal lengths of objective and eyepiece. The objective might have a focal length of 5 mm and form a real image 20 cm from the objective. By Eq. (25.1), $1/p + \frac{1}{200} = \frac{1}{5}$, or $p = \frac{200}{39}$ mm. The linear magnification produced by the lens is then

$$\frac{q}{p} = \frac{200}{\frac{200}{39}} = 39$$

If now the eyepiece has a focal length of 2.5 cm and is adjusted for a final image at infinity, the angular magnification is $25/f = 10$. The overall angular magnification is $39 \times 10 = 390$.

FOR 26.

26.8 ASTRONOMICAL TELESCOPE

The principle of the astronomical telescope (Fig. 26.9) is the same as that of the compound microscope. The objective lens is modified so that the instrument may be used to view distant objects. The objective, a large converging lens of long focal length, forms a real, inverted image IQ of a distant object OP. This real image is formed in the focal plane

Figure 26.9

Lens arrangement of an astronomical telescope (not to scale). Actually IQ is formed at the principal focus of the objective lens, and the final image $I'Q'$ is placed at infinity by adjusting the eyepiece so that IQ is at its principal focus F'_2.

$$\alpha = \frac{O}{P} = \frac{I}{q}$$

of the ocular, by which an enlarged virtual image $I'Q'$ of the real image IQ is produced at infinity.

Because of the great distance of the object from the telescope, the rays from any point are essentially parallel when they reach the objective. Hence, the image formed lies essentially in the focal plane of the objective. The angular magnification of the telescope is determined by the angles that the object subtends and appears to subtend at the eye (Fig. 26.9). The angle α that the object OP subtends at the eye when no lenses are present is $OEP = IEQ$. The angle β that the image $I'Q'$ subtends at the eye is $I'GQ' = IGQ$. For a distant object, with the ocular adjusted to place $I'Q'$ at infinity, F_1, Q, and F'_2 coincide. Then the length of the telescope is $f_0 + f_e$, and

$$\text{Angular magnification} = \frac{\beta}{\alpha} = \frac{QI/f_e}{QI/f_0}$$

$$= \frac{f_0}{f_e} = \frac{\text{focal length of objective}}{\text{focal length of eyepiece}}$$

The image formed by an astronomical telescope is inverted. For terrestrial uses of the telescope it is desirable that the image be erect. This condition is realized by inserting a third convex lens between the objective and the eyepiece in such a way that the image formed by the objective is again inverted before it is viewed by the eyepiece.

Example

The focal length of the objective of a telescope is 150 cm, and the focal length of the eyepiece is 2 cm. Find the angular magnification of the telescope for distant objects.

$$\text{Angular magnification} = \frac{\text{focal length of objective}}{\text{focal length of eyepiece}}$$

$$= \tfrac{150}{2} = 75 \qquad\qquad ////$$

In very large telescopes the objective lens is replaced by a concave mirror. The great mirror in use at the Mount Palomar Observatory has a diameter of 200 in. (5.08 m).

26.9 PRISM BINOCULARS

A terrestrial telescope of reasonable field and magnification becomes inconveniently long. The prism binocular has largely replaced the old

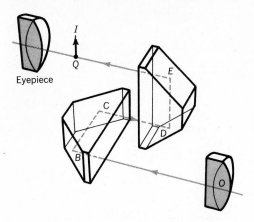

Figure 26.10
Optical scheme of one telescope of
prism binoculars.

"spy glasses." Each side of the binocular achieves a long optical path
by making the light traverse almost three times the length of the binoculars (Fig. 26.10). The beam of light OB from the objective is reflected
internally at B and C by a right-angled prism. In this way its direction
is reversed, and it travels back to a second right-angled prism which
is placed at right angles to the first. Here it is again reflected internally
at D and E and then passes through the eyepiece. The image, after
the reflections by the two right-angled prisms, is restored completely
to the upright position. The eyepiece, serving as a simple microscope,
magnifies this image.

26.10 OPERA GLASS OR GALILEAN TELESCOPE

The opera glass uses an objective which converges the rays from a
distant object toward an image IQ (Fig. 26.11). Before they reach this
image they pass through a diverging lens which deviates rays that were
converging to make them diverge. To an eye on the right-hand side
of the eyepiece, the rays appear to have come from the virtual, erect
image $I'Q'$.

To focus the opera glass, the eyepiece is adjusted so the rays
emerging from it are parallel and the final image is at infinity. Then,

$$\text{Angular magnification} = \frac{\text{focal length of objective}}{\text{focal length of eyepiece}} = \frac{f_0}{f_e}$$

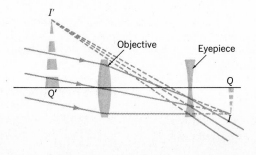

Figure 26.11
Optical scheme of an opera glass. The
final image $I'Q'$ is usually placed at
infinity, in which case Q coincides with
one of the principal foci of the
eyepiece.

The length of the opera glass is $f_0 - |f_e|$; thus the glasses need not be as long as a telescope of comparable magnifying power. A major disadvantage of the opera glass is its small field of view.

QUESTIONS

1 How does the eye accommodate to see objects which are at different distances?

2 How does the eye adjust to permit it to function under conditions of radically different illuminance?

3 There is a blind spot in each eye where the optic nerve leaves the retina. Can you devise a simple experiment to locate the blind spot approximately? Why isn't a person more conscious of the blind spots and bothered by them?

4 When a swimmer is under clear water, why can't he see nearby objects clearly? How do goggles help a swimmer to see more clearly under water?

5 Why is a simple microscope useful when its image gets farther away at the same rate at which the linear dimensions of the image increase?

6 Why do microscopes for children form images with colored edges? How is this avoided in more expensive microscopes?

7 How is the camera similar to the eye? How different?

8 Does a wide-angle lens for a camera have a longer or a shorter focal length than the standard lens for the camera? Why? How does the focal length of a telephoto lens compare with the focal length of the standard lens? What are the advantages and disadvantages of the wide-angle and the telephoto lenses?

9 Why is the depth of focus of a camera at $f/11$ greater than the depth of focus of the same camera at $f/3.5$?

10 Why does the magnification of a telescope increase with the focal length of the objective while the magnification of a compound microscope decreases with increasing focal length?

PROBLEMS

1 A farsighted eye cannot form a sharp image of an object which is closer than 50 cm. Find the focal length and the power of a lens which will permit the eye to form a sharp image of an object 20 cm distant but not of an object closer than this.   *Ans.* 33.3 cm; 3 diopters

2 A farsighted eye has a near point of 45 cm. Find the focal length and the power of the lens which will bring the near point to 15 cm. Where would the near point for this eye be with a lens of power 5 diopters?

3 The near point of an eye is 40 cm. Find the smallest distance at which a slide rule may be held and clearly read if a spectacle lens of $+2.00$-diopters power is worn.   *Ans.* 22.2 cm

4 A nearsighted person cannot see objects clearly when they are more than 50 cm from his eyes. Find the power of the weakest lens which will permit him to see distant objects distinctly. If he puts on a pair of -1.5-diopter glasses, where is his new far point?

5 A myopic (nearsighted) eye has a far point of 60 cm. What is the power of the spectacle lens required for this eye to see distant objects clearly? What would be the far point for this eye with a lens of -1.25 diopters?

          Ans. -1.67 diopters; 2.4 m

6 A thin converging lens of 10-cm focal length serves as a simple microscope. Find graphically by drawing the three standard rays the image of a real object 6 cm from the lens. Check your result by finding the position of the image analytically.

7 A farsighted eye has a far point of infinity and a near point of 80 cm. A plano-convex lens, made of glass whose index of refraction is 1.5, is used to enable the eye to see an object located 25 cm away. What is the largest radius the curved surface may have? Where would the near point and the far point be if this eye were using a 4-diopter lens?

          Ans. 18.2 cm; 19 cm; 25 cm

8 A converging lens used as a reading glass has a focal length of 12.5 cm. What is the angular magnification if the lens is used to produce an image at a distance of 25 cm from the eye?

9 A double-convex lens of 8.33-cm focal length is used as a simple microscope. If it is held close to the eye to form an image 25 cm from the eye,

where should the object be located? What is the angular magnification? What would the angular magnification be if the final image were at infinity?   *Ans.* 6.25 cm; 4; 3

10 The lens of an aerial camera has a focal length of 20 in. What will be the dimensions, on the film plate, of a square of the earth's surface 2 mi on a side, photographed from a height of 5 mi?

11 A camera has a lens of 50-mm focal length. When it is set at $f/3$, what is the diameter of the lens opening? If the setting is changed to $f/9$, how much more exposure time is required? If the camera is originally set to photograph distant objects, how far must the lens be moved to focus on a flower 50 cm away?
                         Ans. 16.7 mm; 9 times; 5.56 mm

12 The lens of a camera has a focal length of 15 cm. It is used to photograph an object that is 3 m from it, and then an object that is 30 m from it. How much must the lens of the camera be moved?

13 A camera lens is set at $f/3$ for a picture which includes a distant point source of light. If the film is actually 0.6 mm too far from the lens for the point source to be in focus, what is the diameter of the circle of light on the film? If the lens had been at $f/11$ what would the diameter have been?   *Ans.* 0.2 mm; 0.055 mm

14 An enlarging camera has a lens with a focal length of 5 in. It is used to enlarge a negative of dimensions 2 by 2.5 in. How far from the lens must the negative be placed in order for the enlargement to be 8 by 10 in.?

15 A metric scale is placed at a distance of 25 cm from the eyes and observed with one eye unaided. The other eye observes a closer similar scale through a converging lens placed close to the eye. A magnified millimeter division appears to be the same size as a 5-mm division seen with the naked eye when the image formed by the lens is also 25 cm from the eyes. Find the focal length of the lens.   *Ans.* 6.25 cm

16 A projection lantern is desired which will throw an image 1.6 m wide of a slide 35 mm wide on a screen 5 m from the lens. What must be the focal length of the projecting lens?

17 The lens of a projection lantern is to be 10 m from a screen on which an image with a height of 1.5 m is desired. The height of the slide is 6 cm. What should be the focal length of the projecting lens?   *Ans.* 38.5 cm

18 The objective of a compound microscope has a focal length of 0.5 cm, and its eyepiece has a focal length of 2 cm. If the distance between the objective and the eyepiece is 22 cm, what is the angular magnification of the microscope when the image is at infinity? Where must the object be placed?

19 A source emits monochromatic light of 560-nm wavelength. If the radiant flux (energy per unit time) through an aperture is 0.3 W, what is the luminous flux through this aperture? If the source had emitted light of 600-nm wavelength, what would the flux have been (Fig. 26.6)?
                         Ans. 204 lm; $\approx$120 lm

20 The eyepiece of a telescope has a focal length of 5 cm. When the final image is viewed at infinity, the objective and the eyepiece are 125 cm apart. Find the angular magnification of the telescope.

21 The world's largest refracting telescope at the Yerkes Observatory has an objective lens 40 in. in diameter with a focal length of 62 ft. If the image formed by this objective is viewed through an ocular (eyepiece) of 3-in. focal length, find the angular magnification for viewing at great distances.   *Ans.* 248

22 What is the diameter of the image of the moon formed at the prime focus by the 200-in. reflector on Mount Palomar if the mirror has a prime focal length of 55 ft and the moon subtends an angle of 0.009 rad at the earth?

23 An 8-mm microscope objective forms an image at a distance of 168 mm. The eyepiece has a focal length of 30 mm. Find the lateral magnification of the objective and the angular magnification of the microscope, assuming the final image is at infinity. If an eye can see two dots as separate when they are 100 μm apart, how close together could two dots be and still be resolved by this microscope?
                         Ans. 20; 167; 0.6 μm

24 A telescope which consists of an objective that has a focal length of 80 cm and an eyepiece with a focal length of 2 cm is used to view an object that is 12 m from the objective. What must be the distance between the eyepiece and the objective for a final image at infinity?

25 The objectives of a pair of binoculars have apertures of 35-mm diameter and focal lengths of 240 mm. The oculars (eyepieces) have focal length of 34 mm. Find the angular magnification when the final image is placed at infinity. Why are these binoculars designated as 7 × 35?

Ans. 7.06; ang. mag. × objective aperture

26 A compound microscope has an objective lens of focal length 5 mm which forms an image 16 cm from the lens. The eyepiece produces a magnification of 8. What is the total magnification?

27 A crude microscope consists of an objective of 0.6-in. focal length and an eyepiece of 2-in. focal length. The lenses are placed 7 in. apart. A person uses the microscope to form an image at infinity. Where must the object be placed? What is the linear magnification of the objective? The angular magnification of the eyepiece? The overall magnification?

Ans. 0.682 in.; 7.33; 5; 37

FROM THE GEOMETRICAL LAWS
of reflection and refraction we have
seen how lenses and optical instruments function. However, there are
limitations on optical instruments due
to the fundamental nature of light.
Further, there are many optical phenomena which cannot be explained in
terms of geometrical laws alone. To
become acquainted with them we turn
now to *physical optics*. We begin by discussing interference of light beams.
This interference shows that light has
wave properties and offers us a
means of measuring the wavelengths
associated with the colors of the visible spectrum which we met in
Chap. 24.

CHAPTER TWENTY-SEVEN
INTERFERENCE AND DIFFRACTION

27.1 DOUBLE-SLIT INTERFERENCE

In 1801 Thomas Young performed a celebrated experiment which led
to the general acceptance of the wave theory of light. Huygens and
others had advanced the wave theory much earlier, but their arguments
were inconclusive until Young showed that light beams produce interference phenomena.

In order to understand Young's experiment, consider a source of
light placed behind a screen containing a narrow slit S (Fig. 27.1).
A second screen having two narrow slits A and B is placed in front
of the first screen in such a way that the openings are equally illuminated
by the light from S. If the illumination on a third screen DF is examined,
it is found that a series of light and dark bands (Fig. 27.2b) result. At
point D, where we might expect the center of a shadow, there is a bright
line. This, of course, means that light from the source S arrives at the
screen at a point which it could not reach if light travels in perfectly
straight lines. Thus, any explanation requires acceptance of the fact
that light bends, to some extent at least, around obstacles. *The bending
of light when it passes an obstacle is called diffraction.*

To explain the pattern of Fig. 27.2b, we accept the tentative hypothesis that light is a wave motion and that Huygens' principle is
applicable. Then we can understand the origin of the bright and dark
lines by the following reasoning. A wavefront from S reaches slits A
and B. Each point on this wavefront acts as a new source of Huygens'
wavelets. Since A and B are equidistant from S, the Huygens' sources
at A and B both send out crests at the same time, and half a cycle
later both send out troughs. The waves from A and B arrive at the
screen DF in such a way that at some points crests meet crests (Fig.

Figure 27.1
When light from a source S passes
through two parallel slits A and B, an
interference pattern is formed on the
screen FD.

27.2a) and troughs meet troughs, thereby giving constructive interference and a bright line. At other points the troughs meet crests and crests meet troughs, resulting in destructive interference and darkness. The bright and dark bands on the screen are called *interference fringes.*

The experiment which we have just described with slits differs from that of Thomas Young in that he used pinholes in the screens rather than slits. In his experiment, interference fringes were also formed, but these fringes were not straight lines.

Let us now see how we can determine the wavelength of light from the interference pattern observed. Note that a bright fringe is produced at D (Fig. 27.1) which is in the heart of the shadow of the slit. This bright line arises from the constructive interference of the light coming through slits A and B. The waves from these two slits arrive in phase, since the light travels exactly the same distance in both cases. For the first bright fringe above D, the light must travel one wavelength farther from slit B than from slit A. For the nth bright fringe above D, the light from B goes n wavelengths farther than that from A. Suppose for the moment that the nth bright fringe is formed at point E. For this particular case the distance BC is n wavelengths if $CE = AE$. On the

(a)

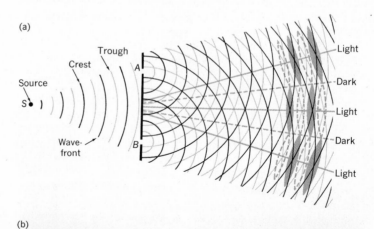

(b)

Figure 27.2
Interference of light from the parallel
slits A and B (a) gives rise to
interference fringes (b).

other hand, for a dark fringe at E the path difference BC must be some odd number of half wavelengths so that the waves interfere destructively.

We can calculate the path difference BC as follows. Let d represent the distance between slit centers, X represent the distance from the double slit to the screen DF, and y represent the distance ED. Then

$$(AE)^2 = X^2 + \left(y - \frac{d}{2}\right)^2 \quad \text{and} \quad (BE)^2 = X^2 + \left(y + \frac{d}{2}\right)^2$$

$$(BE)^2 - (AE)^2 = X^2 + y^2 + yd + \frac{d^2}{4} - X^2 - y^2 + yd - \frac{d^2}{4} = 2yd$$

$$(BE - AE)(BE + AE) = 2yd$$

Then

$$BE - AE = \frac{2yd}{BE + AE} \approx \frac{2yd}{2X}$$

since BE and AE are almost exactly equal to X. Therefore, we have

$$\text{Path difference} = BC \approx \frac{yd}{X} \tag{27.1}$$

The central bright fringe at D occurs for zero path difference. If we count this fringe as zero, the nth bright fringe in either direction comes when the path difference BC is $n\lambda$, which we know from Eq. (27.1) to be yd/X. For the dark fringes, on the other hand, the difference in distance traversed from the two slits must be some odd integral number of half wavelengths. The first dark fringe occurs for $yd/X = \lambda/2$, the second for $yd/X = 3\lambda/2$, and so forth.

We have required that our slit S be illuminated with monochromatic light. If we use white light, the red wavelengths interfere constructively at different places than the blue. As a result, we observe white light at the central point D; then there is a spectrum on each side of D in which the path difference is one wavelength, then another spectrum in which the path difference is two wavelengths, etc. However, in a short distance the spectra overlap seriously, and the illumination appears uniform.

Example

The distance between slits A and B in Young's interference experiment is 0.25 cm. The distance X from the slits to the screen on which the fringes are formed is 100 cm, and the distance from the central bright fringe to the third dark one is 0.059 cm. Find the wavelength of the incident light.

For the third dark fringe the path difference BC must be 2.5λ.

$$2.5\lambda = \frac{yd}{X} = \frac{0.059 \text{ cm} \times 0.25 \text{ cm}}{100 \text{ cm}}$$

$$\lambda = 5.9 \times 10^{-5} \text{ cm} = 590 \text{ nm} = 5900 \text{ Å} \qquad ////$$

The eye responds to waves ranging in wavelength roughly from 3.8×10^{-7} m (violet) to 7.8×10^{-7} m (red). Other detectors respond to other wavelength regions.

27.2 COHERENT SOURCES

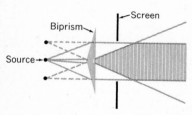

Figure 27.3

Fresnel biprism experiment; light from source S passing through the biprism appears to come from the virtual sources shown, and interference fringes are formed in the gray region.

Shortly after Young's work, Fresnel was able to produce interference fringes similar to those of Fig. 27.2*b* in other ways. One of his methods involves placing a source of light behind a biprism (Fig. 27.3) and finding interference fringes in the crosshatched region where the light which went through the upper prism and the light through the lower prism overlap. Still another method devised by Fresnel involved the use of two mirrors set so that the angle between their surfaces is very near to zero. The Fresnel double-mirror arrangement is shown in Fig. 27.4.

In Young's experiment and in the two interference experiments of Fresnel, light must come from a single source. No interference would be observed in Young's experiment, for example, if one light source were placed behind slit A and another behind slit B. Each source would send out a very large number of waves with completely random phases. At one instant the crest of one wave might meet the trough of another, but an instant later crest might be meeting crest. The light reaching any area would be the sum of the contributions of each of the sources. When two sources send out waves in such a way that there is no regular phase relationship between these waves, the sources are *incoherent*. If there is a regular phase relationship between the waves, the sources are *coherent*.

In the interference experiments which we have discussed, light comes from a single source. A single wavefront is separated into two parts which travel different paths. When the parts are recombined, interference results. In order to produce interference, there must be two "apparent" sources initiating light waves with a well-defined phase relationship. This cannot be achieved with independent light sources. For electromagnetic waves in general it can be done in one of two ways: (1) we may start with a single source, send portions of the wavefronts by different paths, and recombine them in some region or (2) we may drive a source in such a way that it radiates coherently either with a master source or with another source controlled by the same master. In the *laser* excited atoms in a source are led to emit coherent radiation by stimulating emission from the atoms (Sec. 46.10).

27.3 LLOYD'S MIRROR AND PHASE CHANGE
AT REFLECTION

Lloyd was able to obtain interference fringes with a light source and a single mirror in the manner indicated in Fig. 27.5. Light coming directly

Figure 27.4
Fresnel double-mirror method for producing interference.

from the source and light reflected from the plane mirror overlap in the crosshatched region. If a screen is placed at the end of the mirror, one might expect a bright fringe at point P where the screen and mirror come in contact. However, a dark fringe is observed at this point. This fact is explained by a change in phase of one-half wavelength (or 180°) which occurs when light is reflected at the interface between two media if the light approaches the interface through the medium in which it has the higher speed. *When light in a medium is reflected at the surface of a second medium which has a greater index of refraction, there is a half-wavelength phase shift; there is no phase shift when the second medium has a lower index of refraction.*†

It may be helpful to consider an analog to this phase shift at reflection. If a small mass m is set into vibration as a pendulum bob, the x component of the displacement as a function of time is given by the sine wave of Fig. 27.6a. If as m swings through the equilibrium position, it collides with a larger mass M, the laws of conservation of momentum and energy lead to a recoil of m. From the figure we can see that the small mass has missed one-half of an oscillation; one-half wavelength is missing. On the other hand, if a large mass M serves as the bob of the pendulum and it has a collision with a smaller mass m as it swings through the equilibrium position (Fig. 27.6b), the larger mass continues in its original direction with somewhat reduced amplitude. There is no phase shift when the larger mass collides with the smaller one, but a half-wavelength change in phase when the smaller mass collides with the larger.

27.4 INTERFERENCE IN THIN FILMS

Consider the thin air film between two optically plane pieces of glass when the plane surfaces touch at one edge and are slightly separated

---

† This is an oversimplification which leads to correct results in any situation discussed in this text.

Figure 27.5
Lloyd's mirror experiment.

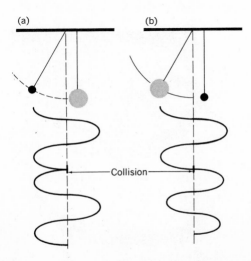

Figure 27.6
Mechanical analog of the 180° phase shift in reflection. The smaller mass reverses direction in collision with a larger mass (a), while the latter continues on after collision with the smaller mass (b).

Figure 27.7
When a thin air wedge is illuminated from above, interference fringes are produced through the superposition of light reflected at the two surfaces of the wedge.

at the other end (Fig. 27.7). When monochromatic light falls normally from above on this air wedge, part of the light is reflected at the upper surface of the wedge and part of the light passes to the bottom, where some of the light is reflected, this time with a change in phase of one-half wavelength. If the air wedge is viewed from above, a series of light and dark bands (Fig. 27.8) is observed.

In the region in which the two plates touch there is a dark band because although the two reflected waves go the same total distance, there is a change in phase for the reflection in which the light goes from air to glass. The first light fringe occurs when the air wedge is one-fourth wavelength thick. The wave reflected from the lower plate goes one-half wavelength farther than the wave reflected at the upper surface, and there is a half-wavelength phase shift at the lower reflection. Therefore, the waves reflected at the two sides of the air wedge meet in phase, and a light band results. When the thickness of the air wedge is one-half wavelength, the wave reflected at the lower plate goes one wavelength farther, but the half-wavelength change in phase on reflection again puts the two waves out of phase by one-half wavelength, and a dark band results. When the air wedge is three-quarters of a

Figure 27.8
Interference by thin air films between an optical flat and gauge blocks. The parallel, equispaced fringes show that the block at the left is plane, while the curves fringes show that the block on the right is worn. [*Do-All Company.*]

wavelength thick, constructive interference occurs once more, and so forth. There is a bright band whenever the thickness of the air wedge is an odd number of quarter wavelengths, and a dark band whenever the thickness of the air wedge is an even number of quarter wavelengths. If n represents the number of a bright fringe, counting the first bright fringe after the point of intersection of the planes as one, the thickness of the wedge at the nth bright fringe is given by $(2n - 1)\lambda/4$.

Vernier and micrometer calipers in precision machine shops can be checked by making certain that they read correctly the width of a working gauge block. To make sure that this block has not been worn down, it is compared with a master block by use of the interference of light waves. The two blocks are placed next to each other on a flat surface, and an optically flat glass plate is placed on top of them (Fig. 27.9). If the two are exactly the same height and the optical flat is in contact with both blocks across their entire upper surfaces, no interference fringes are observed. If the working block is shorter than the master block, parallel fringes are observed, and the amount by which the working block is low can be calculated.

Figure 27.9
Use of interference to test gauge blocks.

Example

When the optical flat of Fig. 27.9 is viewed from above with helium yellow light of $\lambda = 587.6$ nm there is a dark fringe where the flat touches the working block, a second dark fringe one-third the way across the block, a third one two-thirds the way across, and a fourth one just before the master block is reached. How much shorter is the working block than the master?

At the line of contact the separation is a negligible fraction of a wavelength. At the second dark fringe the air wedge is $\lambda/2$ thick, at the third λ, and at the last $3\lambda/2$. Therefore, the working block is $3\lambda/2$ shorter than the master block.

$$\frac{3\lambda}{2} = \frac{3}{2} \times 587.6 = 881 \text{ nm} \qquad \text{////}$$

When a plano-convex lens is placed on an optical flat and illuminated with monochromatic light from above, a series of light and dark rings is observed. This pattern is known as *Newton's rings*. These light and dark rings are analogous to the light and dark straight fringes observed with plane surfaces. In the region where the lens touches the optical flat, there is destructive interference. Where the thickness of the air film is an odd number of quarter wavelengths, a bright circular fringe is seen, while there are dark fringes where the thickness is an integer times $\lambda/2$.

Consider a very thin film of transparent material with surfaces parallel to each other (Fig. 27.10). If one of these surfaces is illuminated by a beam of light of a single wavelength, light is reflected from the upper and lower surfaces. An incident ray such as AB is partly reflected from the upper surface, but most of the light enters the film, and a small part of it is reflected at C. In a similar manner, part of the ray DE is reflected at E, and the remainder of the light enters the film. Some of it is reflected at F, etc. If rays parallel to AB and DE illuminate

Figure 27.10
Interference fringes may be produced when light is reflected from the two surfaces of a thin film and the two beams recombine.

the upper surface XY of the film, parallel rays EK, GH, etc., are reflected from the upper surface. There are also rays which are parallel to EK, GH, etc., which are reflected from the lower surface MN of the film. The rays reflected from the lower surface MN are superposed on those reflected from the upper surface XY. These two sets of reflected beams have nearly the same brightness but differ in phase, because those which are reflected from the lower surface of the film have traveled a distance equal to BCE farther than those reflected at the upper surface, while the latter have undergone a phase shift of one-half wavelength. The two sets of beams may interfere constructively or destructively, depending on the angle of incidence, the wavelength of the light, and the thickness and refractive index of the film. The condition for destructive interference does not occur at the same place for different wavelengths. Hence, when the film is illuminated by white light, certain wavelengths reinforce each other where light of other wavelengths destroy each other. The result is a series of colored fringes. The colors of thin films of oil illustrate this type of interference.

27.5 THE MICHELSON INTERFEROMETER

An interesting example of a device in which part of a wavefront is sent along one path and part along a different one is the Michelson interferometer (Fig. 27.11). Light emerging from a source S falls on a glass plate A, one surface of which is thinly silvered so that it reflects half the light and transmits the remainder. The reflected part goes to mirror C, by which it is reflected, and returns along its original path. The part which went through plate A passes through a second glass plate B, is reflected at mirror D, and returns to mirror A. There it is reflected in such a way that its direction coincides with the direction of the ray which was reflected at mirror C and which subsequently passed through the glass plate A. The wavefront from source S has been split into two wavefronts. Both are received by the eye, one of them after reflection at C, and the other after reflection at D. The plane parallel glass plate B is introduced to compensate for the extra thickness of glass through which the ray reflected at C has passed in reaching the eye.

If the distance from the half-silvered surface at A to mirror C is

(a)

(b)

Figure 27.11
(*a*) Optical paths in a Michelson interferometer and (*b*) a commercial interferometer. [*Atomic Laboratories, Inc.*]

the same as the distance from A to mirror D, the two rays of light travel the same distance, and they are therefore in phase. Under these conditions, the central rays reinforce each other. If, however, the distance AD is greater than the distance AC by one-quarter of a wavelength of the light, the central rays are out of phase and destroy each other. Rays which make small angles with the central ray travel slightly different distances and interfere alternately constructively and destructively (Fig. 27.12). If mirror D is moved along the line AD, the light and dark circles interchange position for each quarter wavelength the mirror is moved.

27.6 DIFFRACTION

When light passes an obstacle, it does not proceed in exactly straight lines, but spreads out somewhat into the geometrical shadow. Ordinarily this effect is small for light waves. However, light does bend slightly

Figure 27.13
When light from a distant,
monochromatic point source passes
through a keyhole aperture in an
opaque screen and the shadow is
observed some distance beyond the
screen, the diffraction pattern below is
found. Above is the shadow predicted
on the assumption of rectilinear
propagation. [*Courtesy of John M.
Stone.*]

around corners in much the same way that water waves bend around obstacles. The effect is prominent when the wavelength is large compared with the size of the obstacle; it is small when the wavelength is short compared with the size of the obstacle. Thus, radio waves and sound waves, which have relatively long wavelengths, bend readily around objects. The wavelengths of visible light are small compared with the sizes of ordinary obstacles; therefore the bending is not ordinarily conspicuous for light waves.

If light always traveled in perfectly straight lines and obeyed only the laws of geometrical optics, a picture like that of the upper half of Fig. 27.13 would be formed by light from a distant monochromatic source passing through a small keyhole aperture some distance from the film. The lower half of Fig. 27.13 shows an actual picture. Not only has there been some spreading of light into the region of the geometrical shadow, but interference between wavelets from various portions of the wavefront passing the aperture has led to a series of light and dark fringes. *The spreading of a wave motion into the geometrical shadow of an object and the other deviations from the predictions of geometrical propagation are called diffraction.* A definition due to Sommerfeld states that diffraction is "any deviation of light rays from rectilinear paths which cannot be interpreted as reflection or refraction."

To account for the fact that light bends around obstacles, we need only invoke Huygens' principle, which tells us that for any kind of wave motion we may regard each point on a wavefront as a new source. Whenever we are dealing with diffraction phenomena, we are treating a case in which various parts of the same wavefront act as coherent sources and the secondary wavelets from these coherent sources interfere. For interference in a Young's double-slit experiment we take two limited sections of a wavefront, send them by different paths, and combine them to get constructive and destructive interference. In diffraction much larger sections of the wavefront may be involved. The general theory calls for finding the net interference effect at a given point for a large number of waves with a continuous variation in phase. The mathematical treatment of such diffraction phenomena is more difficult than that for two-slit interference. For this reason we discuss diffraction in a qualitative way, except for the case of the diffraction grating.

If a series of very fine equidistant parallel slits is ruled on a plate of glass with a fine diamond point, we have a *diffraction grating*. Where the diamond point has made a furrow on the glass, the light cannot pass regularly. Between the furrows, where the surface of the glass is undisturbed, the glass is transparent. The plate of glass is then somewhat like a picket fence. In effect, there are strips through which the light can pass separated by strips through which it cannot pass.

Let AB (Fig. 27.14) represent such a grating on which parallel light is falling so that the direction of the rays is perpendicular to the plane of the grating. Figure 27.14 shows the Huygens wavelets that start out from the slits. These wavelets destroy or reinforce each other, according to whether they are in phase or out of phase at a given point.

Through the slits in the grating come beams of parallel light. If these rays, coming in the direction perpendicular to the plane of the grating, are brought to a focus on the screen PS by the lens, a bright line is formed at K. If the light is viewed in a direction making an angle θ with the normal to the grating, parallel rays of light emerging from the slits in this direction travel unequal distances to reach the screen, and when they are brought to a focus by the lens they may either reinforce or destroy each other. If the angle θ is made such that a ray of light from one slit is one-half wavelength behind the corresponding ray from the adjacent slit, then the rays from one slit are out of phase with the rays from the adjacent slit, and they destroy each other. Hence, there will be darkness. If the angle between the normal to the grating and the ray is increased, the rays from the lower slits will be further and further behind the rays from the upper slits. When the difference in path between corresponding rays from adjacent slits amounts to one wavelength, wavelets from all the slits are again in phase and interfere constructively. If these rays are focused by the lens, a bright image of the slit is formed on the screen at N. Whenever

$$\sin \theta = \frac{n\lambda}{d} \qquad n \text{ an integer} \quad \left[\left(\text{IT CAN'T BE GREATER THAN 1}\right)\right]$$

$$n\lambda = d \sin \theta$$

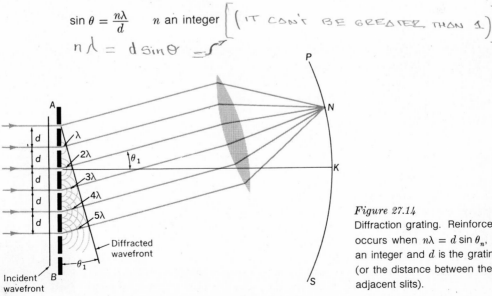

Figure 27.14

Diffraction grating. Reinforcement occurs when $n\lambda = d \sin \theta_n$, where n is an integer and d is the grating space (or the distance between the centers of adjacent slits).

the wavelets from different slits reinforce each other and the screen is illuminated.

The angle θ_1 at which reinforcement is first obtained after leaving the central line K can easily be measured on the divided circle of a spectrometer. For this first reinforcement,

$$\lambda = d \sin \theta_1 \tag{27.2}$$

Since all the quantities in this equation are known except λ, the wavelength of the light can be calculated. This is a very accurate method of measuring wavelengths.

Reinforcement also occurs at other angles θ_n which satisfy the equation

$$n\lambda = d \sin \theta_n \tag{27.3}$$

The integer n gives the *order* of the spectrum. When $n = 2$, the waves from each slit go two wavelengths farther than those from the adjacent slit. In the third order n is 3, and the path difference is 3λ.

Example

For a certain color of yellow light, the angular separation between the central image and the first-order spectrum produced by a plane grating was 17°. The grating has 5,000 rules to the centimeter. What is the wavelength of the light?

Wavelength $= d \sin \theta_1$

$$\lambda = \frac{1 \text{ cm}}{5,000} \times \sin 17°$$

$$= \frac{1}{5,000} \times 0.292$$

$$= 0.0000584 \text{ cm, or } 5840 \text{ Å, or } 584 \text{ nm} \qquad ////$$

27.8 DIFFRACTION BY A STRAIGHT EDGE

Suppose that light is diverging from a narrow slit L (Fig. 27.15a) and that it passes by the straight edge of an opaque screen. If the light were propagated exactly in straight lines, there would be uniform illumi-

Figure 27.15
Light passing a straight edge (*a*) produces a diffraction pattern (*b*) with monotonically decreasing intensity below *O* and a series of fringes in the region *OP*. [*Courtesy of John M. Stone.*]

(a)

Wavefront

S

P

L

Source of light

Opaque screen

S'

O

(b)

nation on the screen above the line LO and complete darkness below it. However, the illumination does not become zero immediately below O but fades away continuously. There is almost complete darkness a small distance below O. Immediately above O the illumination is not uniform but shows a series of bright and dark bands parallel to the edge. The appearance of the fringes thus produced is seen in Fig. 27.15b.

Consider now the case of a fine wire AB (Fig. 27.16a) placed in front of a narrow slit L. The shadow of this wire on the screen MN is found to be bounded on each side by a system of parallel fringes (Fig. 27.16b). Each edge of the wire behaves like a straight edge. Further at the very center of the shadow there is constructive interference of waves diffracted around each side of the wire, so the center is bright relative to the main part of the shadow, though still dark compared with regions CN and DM.

27.9 DIFFRACTION BY A CIRCULAR APERTURE

Another striking illustration of diffraction is observed when light from a luminous point passes through a small circular aperture such as a pinhole. When this aperture is viewed by means of a magnifying glass, there appears a brilliant spot which is surrounded by a series of bright and dark rings, essentially like the circular portions of the diffraction fringes of Fig. 27.13. When images are formed in an optical instrument, even the most perfect optical parts cannot produce a point image of a point source. Instead there is a diffraction pattern consisting of a bright circle surrounded by light and dark rings. Ultimately these diffraction patterns limit the angular separation of two sources which can be distinguished as separate.

Figure 27.16
Light passing a fine wire is diffracted (*a*) producing the diffraction pattern (*b*). [*Courtesy of John M. Stone.*]

QUESTIONS

1 How do the spectra of sunlight formed by diffraction gratings differ from those formed by prisms? Which color is deflected most by prisms? By gratings?

2 An observer sees a green fringe passing through a given point in an oil film. Would other observers looking at the same point necessarily see green there? Explain.

3 Why is it easy to observe interference effects with thin films but difficult with thick ones?

4 How do the colors observed in Newton's rings, oil films, and soap bubbles arise? Explain.

5 Two spectral lines of different colors are observed by means of a grating, and it is seen that the third-order image of one line coincides with the fourth-order image of a second line. What is the ratio of the wavelengths of the two colors?

6 Is the angular separation between a blue line and a red one greater in second order than in first? Why?

7 Would a diffraction grating with 5,000 lines/cm be useful for studying infrared spectra with wavelengths around 3 μm? Explain.

8 Why do coated lenses look purple by reflected light?

9 What happens to the distance between the interference fringes from two slits as the slits are moved closer together?

10 Newton's rings can be observed by transmitted light as well as by reflected light. Would the point of contact be light or dark by transmitted light? Show that the transmitted rings are dark where the reflected ones are bright, and vice versa.

11 Why is the diffraction of sound waves far more frequently observed in our daily experiences than the diffraction of visible light?

PROBLEMS

1 Yellow light of wavelength 5800 Å from a single source is passed through two narrow slits separated by 0.2 mm. On a screen 120 cm away an interference pattern is formed. Calculate the separation of intensity maxima in this pattern.
Ans. 3.48 mm

2 Monochromatic light from a narrow slit illuminates two parallel slits 0.25 mm apart. On a screen 110 cm away interference bands are observed 2.4 mm apart. Find the wavelength of the light.

3 The two parallel slits used for a Young's interference experiment are 0.4 mm apart. The screen on which the fringes are projected is 1.5 m from the slits. What is the distance between the fringes for monochromatic red light of wavelength 600 nm? How far is the third dark fringe from the central bright one?
Ans. 2.25 mm; 5.63 mm

4 A thin air wedge is formed between two optical flats by slipping a piece of paper under one edge of the upper flat. Ten bright fringes are observed per centimeter across the upper flat with light of $\lambda = 500$ nm. What is the thickness of the paper if the flats are 5 cm long?

5 Two pieces of plane glass are placed together with a piece of paper between the two at one edge. When the air film is illuminated from above with cadmium blue light of $\lambda = 4800$ Å and is viewed from above, there is a dark fringe along the line of contact and a series of light and dark fringes. Find the thickness of the air film at the third bright fringe from the line of contact. If there are 10 dark fringes per centimeter of plate, find the angle between the glass surfaces.
Ans. 6000 Å; 2.4 × 10⁻⁴ rad

6 Two glass optical flats are placed together with a piece of paper between the two at one edge. When the air film is illuminated and viewed from above with sodium yellow light ($\lambda = 5900$ Å), there is a dark fringe at the line of contact and a series of light and dark fringes. Find the thickness of the air wedge at the second bright fringe counted from the line of contact. Explain your

solution in terms of path differences and phase shifts.

7 Two plane pieces of optical glass are pressed together at one edge and separated by a fine wire at the other edge. The distance between the wire and the edges of glass that are in contact is 10 cm. When light of wavelength 5000 Å is incident normally on the surface of one of the pieces of glass, interference fringes are observed. If these fringes are 0.15 cm apart, what is the diameter of the wire? What is the thickness of the air wedge at the third bright fringe from the line of contact of the plates?
Ans. 1.67 × 10⁻³ cm; 6.25 × 10⁻⁵ cm

8 Two flat glass plates which are almost parallel produce interference fringes by successive reflections from the surfaces of normally incident light. If the light has a wavelength of 450 nm, what is the difference in thickness that would be indicated in passing from one bright fringe to the next? If there are six bright fringes per centimeter of length along the plates, what is the angular separation between the surfaces?

9 Newton's rings are formed in reflected light by a plano-convex lens balanced upon an optical flat. If the light incident from above has a wavelength of 500 nm, find the thickness of the air film at the second bright ring. In terms of wavelengths how much farther did light reflected from the flat travel than the light reflected from the spherical surface?
Ans. 375 nm; 1.5 wavelengths

10 Newton's rings are observed by reflected light of wavelength 480 nm. The central area is dark and is surrounded by light and dark circles. Find the thickness of the air film at the fourth bright circle. Explain your method.

11 Newton's rings are observed by reflected light of wavelength 589 nm. The central area is dark and is surrounded by light and dark circles. Find the thickness of the air film at the first, third, and fifth bright circles. *Ans.* 147, 736, 1,325 nm

12 When the movable mirror of a Michelson interferometer is moved 0.02 mm, how many fringes pass the reference mark if light of wavelength 5890 Å is used?

13 Light of wavelength 480 nm is incident normally on a wedge-shaped soap film which has index of refraction of $\frac{4}{3}$. By reflected light the top of the film is dark, and below come a series of light and dark lines. What is the thickness of the film at the uppermost light band? Explain how your

answer was obtained, mentioning path differences, phase shifts, etc.   *Ans.* 90 nm

14 The grating in a grating spectroscope is ruled with 500 lines/mm. What is the angle of diffraction of the second-order spectrum of light of wavelength 5000 Å?

15 What is the thinnest film of oil (index of refraction 1.40) floating on water in which green light ($\lambda = 500$ nm in air) is essentially eliminated by destructive interference from a beam of white light incident normally on this film? If the oil film were on glass (index of refraction = 1.50), what thickness would be required?
     Ans. 0, 179 nm; 89.3 nm

16 A lens having an index of refraction of 1.50 has a nonreflecting coating having an index of refraction of 1.20. Find the minimum possible thickness of this coating to give destructive interference in the reflected beam for light of wavelength 5500 Å.

17 A film of oil with an (unrealistically low) index of refraction 1.25 rests on a puddle of water ($n = 1.333$). When yellow light of $\lambda = 500$ nm is incident normally on the film and the reflected light is viewed from above the film, destructive interference is observed. What are the three smallest thicknesses the oil film could have? What are the three smallest thicknesses the film could have for constructive interference?
     Ans. 100, 300, and 500 nm; 0, 200, and 400 nm

18 Oil with index of refraction 1.40 rests on glass (index of refraction 1.50). Green light of $\lambda = 560$ nm is incident normally from above (angle of incidence = 0°). (*a*) Compute the wavelength of the light in the oil and in glass, and (*b*) calculate the thickness of the thinnest layer of oil from which this green light would be essentially eliminated from the reflected beam by destructive interference.

19 When a Michelson interferometer was used to measure the wavelength of monochromatic light, it was found that 300 fringes passed the observing microscope when the movable mirror was displaced 0.066 mm. What was the wavelength of the light?   *Ans.* 440 nm

20 One arm of a Michelson interferometer utilizing light with $\lambda = 500$ nm contains a cell 9 cm long between plane parallel windows. How many fringe shifts would be observed if all the air were

evacuated from this cell? The index of refraction of the air was initially 1.00029.

21 If 4500-Å radiation falls normally on a grating ruled with 5,000 lines/cm, how many orders can be observed on each side of the direct beam? Determine the angles at which the orders are observed.   *Ans.* 4; 13°; 26.7°; 42.5°; 64.2°

22 A glass grating is ruled with 5,400 lines/cm. Light striking the grating normally forms a second-order image diffracted at an angle of 38° from the normal. What is the wavelength of the light?

23 Compute the longest wavelength which can be observed in third order by a transmission diffraction grating with 6,000 lines/cm.
     Ans. 556 nm

24 The fourth-order spectrum contains a certain color diffracted at an angle of 54°. If the grating is ruled with 450 lines/mm, what is the wavelength of the light?

25 A grating is ruled with 6,000 lines/cm. Calculate the angles of diffraction for first-order red and blue light, the wavelengths being, respectively, 7000 and 4000 Å. Determine the angular separation in radians. A lens of 50-cm focal length is placed in the path of light just beyond the grating. Determine the linear separation of the red and blue lines. (This is essentially the basis for a grating spectrometer.)
     Ans. 24.8°; 13.9°; 0.19 rad; 9.5 cm

26 Light is incident normally on a grating which has 250 lines/mm. Find the wavelength of a spectral line for which the deviation in second order is 15°.

27 Show that if AB and DE of Fig. 27.10 are two rays associated with a single wavefront, then destructive interference occurs for reflected light when $2nt \cos r$ is an integer number of wavelengths. Here t is the thickness of the film, n is its index of refraction, and r is the angle of refraction in the film. Show that constructive interference takes place where $2nt \cos r$ is an odd number of half wavelengths.

CHAPTER TWENTY-EIGHT
POLARIZATION AND SPECTRA

THE PHENOMENA of interference and diffraction reveal that light has wave properties. We now show that polarization establishes that these wave characteristics are those typical of transverse waves. The wave aspects of light permit us to disperse white light and other electromagnetic radiation into spectra by use of prisms or diffraction gratings. The study of spectra has been rewarding in many ways, and various branches of spectroscopy continue to employ many physicists in research, in process controls, and in identification of elements and molecules.

28.1 POLARIZATION

When a beam of ordinary light passes through a Polaroid sheet, the unaided eye recognizes only that the intensity has been reduced. If the Polaroid is rotated about an axis normal to its surface, the transmitted beam appears to differ in no way as the rotation proceeds. However, if this transmitted radiation falls on a second Polaroid, a dramatic change in the intensity beyond the second Polaroid occurs as either Polaroid is rotated through 360° (Fig. 28.1). At some angle, and again 180° later, no light is observed beyond the second Polaroid. Ninety degrees from those positions essentially all the light which goes through the first Polaroid passes also through the second. At intermediate angles the intensity is between these extremes. It is evident that light, in passing through the first Polaroid, acquires properties that ordinary light does not possess: the light is *plane-polarized*.

To understand this experiment, consider a stretched string (Fig. 28.2) in which the particles are vibrating perpendicular to the length of the string. If a block of wood with a slot in it is placed over the string, the vibrations are not affected when the slot is parallel to the direction of vibration. However, when the slot is at right angles to this direction, the vibrations do not pass. If the slot makes various angles with the direction of vibration of the string, the component of the vibratory motion parallel to the slot passes through. If a second slot is placed over the string, the vibrations that pass the first slot also pass the second when the two slots are parallel. When the slots are perpendicular to each other, the vibrations that pass the first slot do not pass the second.

Visible light and other electromagnetic radiation have many of the properties of classical transverse waves. The wave model of light involves alternating electric and magnetic fields mutually perpendicular both to each other and to the direction of propagation. In ordinary light

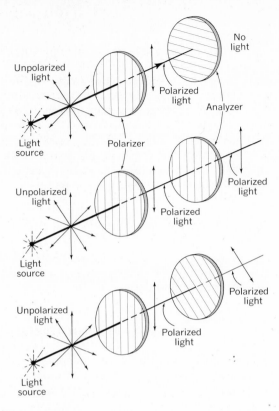

Figure 28.1
Light is plane-polarized by passing it through one Polaroid. All, part, or none of this light passes through a second Polaroid, depending on its orientation relative to the first.

the electromagnetic vibrations occur in all directions in a plane perpendicular to the direction in which the light is traveling (Fig. 28.1); in plane-polarized light the electric vibrations are all in one direction perpendicular to the propagation direction, while the magnetic ones are in the third direction. In discussing polarized light, it is ordinarily desirable to focus attention on the *electric vibrations*, since most common optical phenomena are due to the interaction of the electric vector with the charged particles in matter. The *plane of vibration* is defined as

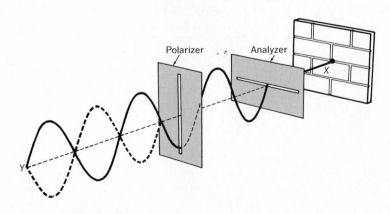

Figure 28.2
Polarization of waves in a stretched string is analogous to the polarization of light.

the plane determined by the electric vibrations and the direction of propagation.

When ordinary light passes through a Polaroid sheet, one component of the vibrations is absorbed, the other component transmitted. Consequently the emerging beam differs from ordinary light in that all the electric vibrations are in one direction and the beam is *plane-polarized*. If it falls on a second Polaroid, that Polaroid transmits only those vibrations which are parallel to its transmission direction. When ordinary light falls on a polarizing agent and plane-polarized light is produced, we call the agent a *polarizer*. To determine whether a beam is polarized or not, we pass the light through a second polarizing agent. When the latter is rotated, there is no change in transmitted intensity if the light is unpolarized, while the intensity goes from maximum to zero if the beam is plane-polarized. A polarizing agent used in this way is called an *analyzer*. If the incident light is partially plane-polarized, the intensity is maximum for one orientation of the analyzer and minimum for another, but there is no zero. When two polarizing agents are arranged with their axes perpendicular so that they cut out all light, they are said to be *crossed*.

28.2 PLANE-POLARIZED LIGHT

When a polarizing agent of large aperture is desired, Polaroid sheets are ordinarily used (Fig. 28.3). A Polaroid consists of two thin sheets of plastic with a thin layer of ultramicroscopic polarizing crystals of

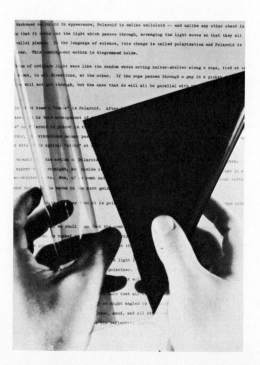

Figure 28.3
Polaroid sheet crossed on itself to show how transmitted light is absorbed in varying degrees.

(a)

(b)

(c)

End-on   Side view

Figure 28.4
Pictorial representation of (*a*) ordinary light, (*b*) plane-polarized light with electric vibrations horizontal, and (*c*) plane-polarized light with the electric vector vertical.

iodosulfate of quinine (also known as *herapathite* after Herapath, who studied the polarizing properties in 1852) between them. The tiny needlelike crystals are aligned with their axes parallel by subjecting them to a strong electric field as the plastic in which they are embedded solidifies. Actually a pair of crossed Polaroids is not perfectly effective for eliminating the far-red and far-violet radiations, so that one can see a little deep purple through crossed Polaroids.

Other materials are also known to produce plane-polarized light; an example of primarily historical interest is a crystal of the mineral *tourmaline*. Tourmaline crystals are generally somewhat colored and are seldom used in modern studies of polarization phenomena. Nicol prisms (Sec. 28.5) are excellent polarizers and analyzers; their disadvantages are high cost and limited aperture.

For discussing polarization we look upon a beam of ordinary light as consisting of a large number of waves, each with its own plane of vibration, with every direction of vibration perpendicular to the rays having equal probability. We represent such a beam, looking head on, by the end-on view of Fig. 28.4*a*. When we pass such a beam through a Polaroid, it comes out plane-polarized. Only an infinitesimal part of an unpolarized light beam contains vibrations in any single direction. However, if we have a polarizer set to transmit only vertical vibrations, approximately half the incident intensity is transmitted. Vibrations which are not exactly vertical also contribute to this beam, since a polarizer transmits all *components* which are parallel to its select direction. We may look on a beam of unpolarized light as being composed of two equal beams plane-polarized at right angles to each other. Any vibration can be resolved into two components, one in each of these directions. Hence, we sometimes represent unpolarized light by showing only two of the many directions of vibration. We use a dot to represent vibration perpendicular to the plane of the paper and arrows to indicate vibrations in this plane (Fig. 28.4).

When a beam of plane-polarized light falls on an analyzer set to transmit a plane of vibrations making an angle θ with that of the incident beam, a portion of the incident beam is transmitted. To calculate the fraction transmitted, we proceed as follows. We resolve (Fig. 28.5) the vibration amplitude A of the incident beam into two components, one parallel to the preferred plane of the analyzer and the other perpen-

Transmitted

A

θ

A_T

Figure 28.5
Resolution of the electric vector into rectangular components.

$$\frac{\sin \theta_i}{\sin \theta_r} = \frac{n_2}{n_1}$$

$$\frac{\sin \theta_i}{\cos \theta_i} = \frac{n}{1}$$

$$\tan \theta_p = n.$$

dicular. The polarizer passes the parallel component and rejects the perpendicular one. The amplitude transmitted A_T is given by

$$A_T = A \cos \theta \tag{28.1}$$

Since the intensity is proportional to the square of the amplitude, the fraction of the incident intensity I transmitted is given by

$$\frac{I_T}{I} = \frac{A_T^2}{A^2} = \cos^2 \theta \tag{28.2}$$

Example

An analyzer is set to transmit vibrations at 37° to those of an incident plane-polarized beam. Find what fraction of the incident intensity is transmitted.

$$A_T = A \cos \theta = A \cos 37° = 0.8A$$

$$\frac{I_T}{I} = \frac{A_T^2}{A^2} = 0.64 \qquad\qquad ////$$

28.3 POLARIZATION BY REFLECTION

When a beam of unpolarized light is reflected at the surface of glass or water, the reflected beam is unpolarized for normal incidence and for grazing incidence. For any other angle of incidence, the reflected ray contains more vibrations parallel to the reflecting surface, while the transmitted beam contains more vibrations in the plane of incidence (Fig. 28.6a). If the angle of incidence is about 45°, an analyzer reveals that some reflected light is transmitted for every orientation of the analyzer, but the intensity is maximum when the analyzer passes electric vibrations parallel to the reflecting surface.

   When the angle of incidence is such that the angle between the reflected and refracted rays is 90° (Fig. 28.6b), the reflected beam is completely plane-polarized. We call the angle of incidence the *polarizing angle* i_p. Since $i_p + r_p + 90° = 180°$, the angle of refraction r_p is the complement of the angle of incidence i_p, so that $\sin r_p = \cos i_p$. If n is the index of refraction of the reflecting material relative to the air, we have

$$n = \frac{\sin i_p}{\sin r_p} = \frac{\sin i_p}{\cos i_p} = \tan i_p \tag{28.3}$$

The relation $n = \tan i_p$ is known as *Brewster's law.*

(a)   (b)

Figure 28.6
Partial polarization by reflection and refraction occurs at most angles (*a*), but at the polarizing angle the reflected beam is completely plane-polarized (*b*).

Although the reflected beam is completely plane-polarized for incidence at the polarizing angle, the refracted beam is only partially plane-polarized. Most of the intensity is associated with the transmitted beam, which includes all vibration components in the plane of incidence and part of those perpendicular to this plane.

Light reflected from metallic and other conducting surfaces is not polarized. A well-silvered mirror may reflect 98 percent of the incident light; both types of polarization are reflected. In order to have plane polarization by reflection, we must have a dielectric medium which is not a good conductor of electricity.

Sunlight reflected from the surface of a lake or from an asphalt pavement is partially polarized. Glasses equipped with polarized lenses oriented so they transmit only vertical vibrations cut out most of the reflected light and only half the unpolarized light. It is for this reason that Polaroid glasses reduce glare which is due primarily to reflected light.

Figure 28.7
In calcite a flash of light at P produces two wave surfaces, a spherical one with electric vibrations perpendicular to the plane containing the optic axis through P and the point of observation and an ellipsoidal one with electric vibrations in this plane.

28.4 DOUBLE REFRACTION

When a crystal of calcite (often called *Iceland spar*) is placed over a page of print, two images are seen in the calcite. If these images are examined through an analyzer, it is found that each of the images is produced by plane-polarized light, with the planes of vibration at 90° to one another. Thus, ordinary light entering the crystal is split into two plane-polarized beams which travel through the crystal in different directions. This phenomenon, known as *double refraction*, is shown by many transparent crystals.

Suppose that a flash of light is created at point P inside the calcite and that we look at the wavefront a very short time Δt later. In glass or water the wavefront is a sphere with radius $V \Delta t$, where V is the speed of light in the medium. In calcite, however, there are two wavefronts (Fig. 28.7), one due to light with its vibration perpendicular to the page and the other due to light with vibrations in the plane of the page. One of these wavefronts is spherical and is called the *ordinary* wavefront, while the other, an ellipsoid of revolution, is the *extraordinary* wavefront. In calcite there is one direction, known as the *optic axis*, in which these two wavefronts travel with the same speed. In other directions the two wavefronts have different speeds, and double refraction results.

For a crystal to show double refraction it must be *anisotropic;* i.e., it must have different properties in different directions. Calcite and quartz are examples of anisotropic crystals. Water and glass are *isotropic*, i.e., have the same properties in all directions. However, a piece of glass can be made anisotropic if it is subjected to stress.

For calcite the optic axis of a typical cleavage crystal is oriented as shown in Fig. 28.7. If we use Huygens' principle to trace light through a calcite crystal, we find that the Huygens wavelets for the extraordinary wave are ellipsoidal rather than spherical (Fig. 28.8). To find the new wavefront for the extraordinary wave, we find the surface which is tangent to these wavelets. This is a plane in the crystal parallel to the ordinary wavefront, but the light which enters the crystal at point A in the figure moves along the line AX which represents the path of the

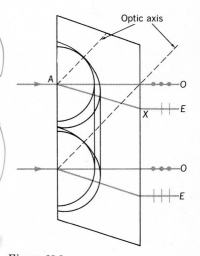

Figure 28.8
Double refraction in calcite showing the Huygens wavelets and the paths of the ordinary and extraordinary rays.

Figure 28.9

Total reflection eliminates the ordinary ray in a Nicol prism, leaving only the plane-polarized extraordinary ray in the emerging beam.

extraordinary ray. The construction for the ordinary ray is just as it would be for an isotropic medium. Thus, an ordinary ray goes straight through the crystal, while the extraordinary ray moves off to one side. If we place a calcite crystal over a dot on a piece of paper and rotate the crystal, the image formed by the ordinary ray stands still, while the image due to the extraordinary ray rotates about it.

28.5 NICOL PRISM

One method of separating the ordinary from the extraordinary ray is by use of a *Nicol prism* (Fig. 28.9). A rhomb of calcite *AMBN* is cut into two parts by a plane *MN*, which makes an angle of about 22° with *MB*. When the cut surfaces have been polished, they are cemented together with Canada balsam, the index of refraction of which is less than that of calcite for the ordinary ray and greater than that of calcite for the extraordinary ray. The refractive indices are Canada balsam 1.55, ordinary ray in calcite 1.658, and extraordinary ray 1.468.

The ray of light *LC* entering the face of the rhomb at *C* is broken into the ordinary and extraordinary rays, which are polarized at right angles to each other as indicated by the dots and crosslines in the figure. When the ordinary ray reaches the surface of separation of the Iceland spar and the Canada balsam at *D* at an angle greater than the critical angle, it is totally reflected and emerges from the rhomb in the direction *OO'*. This surface of the rhomb is ordinarily painted black, and the ray is absorbed in this black coating. The extraordinary ray *CEE'* emerges from the rhomb at the face *BN*. In this way, a beam of plane-polarized light is obtained.

28.6 POLARIZATION BY SELECTIVE ABSORPTION

We have introduced polarization phenomena with the aid of tourmaline crystals and Polaroid sheets. These materials produce plane-polarized light by the phenomenon of *selective absorption*, which is exhibited by a number of minerals and organic compounds. Materials which yield polarized light by selective absorption are not only *anisotropic* and double-refracting, but in addition they strongly absorb one of the polarized beams while transmitting the other.

28.7 POLARIZATION BY SCATTERING

When light rays pass through the air, they interact with air molecules and dust particles. Part of the energy is scattered. Particles with dimensions much smaller than the wavelength of the radiation scatter short wavelengths more strongly than long ones. Indeed, the scattering is proportional to the fourth power of the frequency. Since the violet part of the light from the sun has a frequency approximately double that of red waves, the violet rays are scattered about $2^4 = 16$ times as strongly as the far red.

Light from the sun is scattered in passing through the atmosphere, and some of this scattered light is rescattered to our eyes. Since this scattered light is much richer in blues and violets than in the longer wavelengths, the sky is blue. By the same token the sun, even at midday, looks far more orange than it would if it were viewed from a rocket ship above the earth's atmosphere. Much of the blue and violet has been scattered out before the light reaches the earth's surface. At sunrise and at sunset the sun's rays go through far greater thicknesses of atmosphere; for this reason the sun appears redder at these times.

If the light of the sky is examined through a Polaroid, it is found that this light is partially plane-polarized if the sky is viewed from almost any angle. Light scattered at an angle of 90° with the incident rays is plane-polarized.

28.8 INTERFERENCE WITH POLARIZED LIGHT

If plane-polarized light is incident on a thin sheet of some double-refracting material, the beam is spread into ordinary and extraordinary rays, which travel with different speeds. When they emerge from the double-refracting material, the vibrations are likely to be out of phase. If the light is now passed through an analyzer which transmits only horizontal vibrations, this analyzer selects only the horizontal components of the vibrations from both ordinary and extraordinary rays. These horizontal components may interfere with one another, either constructively or destructively. In particular, if highly convergent light is passed through the double-refracting crystal, the paths of various rays in the crystal differ in length and there is then constructive interference for a given wavelength for some directions and destructive interference for others. Consequently, interference patterns are formed (Fig. 28.10). If the original light contains all the wavelengths in the visible spectrum, the interference pattern is vividly colored. If either the polarizer or analyzer is rotated through 90°, the colors change to the complementary colors, because those for which the interference was initially constructive then have destructive interference, and vice versa.

When a sheet of cellophane is placed under stress, it becomes double-refracting. Interference colors may be observed when such a sheet is placed between two Polaroids. It is possible to determine where the stresses are and how great they are by careful analysis of such an interference pattern. One of the most useful methods of studying the stresses in various structural shapes involves making a plastic model, subjecting it to external stresses, and studying the interference patterns

(a)

(b)

Figure 28.10
Interference patterns from crystals of (*a*) mica and (*b*) quartz placed in converging monochromatic polarized light.

Figure 28.11
Stress distribution in an open-end
wrench made of Photoelastic plastic,
photographed by polarized light.
[*Instrument Division, Budd Company;
Zandman Method.*]

produced (Fig. 28.11). Such studies permit a fairly complete analysis of the stress distributions. Once one knows what stress distribution is most likely to cause difficulty, one can change the shape in such a way as to produce a more desirable distribution.

28.9 ROTATION OF THE PLANE OF POLARIZATION

When a beam of monochromatic plane-polarized light passes through certain substances, the plane of polarization is rotated. Thus, if a plate of quartz cut so that the faces are perpendicular to the axis of the crystal is placed between crossed Nicol prisms, the light is no longer extinguished by the second Nicol prism. If, however, the second Nicol prism is rotated, a new position can be found at which the light is again extinguished. Rotation of the plane of polarization is called *optical activity.*

The angle through which the plane of polarization is rotated depends on the kind of substance interposed between the Nicol prisms, the thickness of the substance, the wavelength of the light, and the temperature. The rotation may be either clockwise or counterclockwise. In this respect there are two kinds of quartz. One kind rotates the plane of polarization clockwise; the other rotates it counterclockwise. Some liquids and gases also cause a rotation of the plane of polarization. Molecules in which a carbon atom is attached by valence bonds to four different atoms or groups of atoms often show optical activity. Optical activity is a matter of importance in organic chemistry and is fundamental to stereochemistry.

28.10 EMISSION SPECTRA

When a beam of white light is passed through a prism, the light is dispersed into a spectrum (Sec. 24.9). Light from an incandescent lamp or a candle contains all the colors of the rainbow (Fig. 28.12). However, if we put salt (NaCl) in the flame of a bunsen burner, the light emitted is a yellow characteristic of sodium. If a lithium salt is put in the flame, a bright red is produced.

The systematic examination of the spectra of various light sources

OPTICAL SPECTRA

E b
5270 5183 5172
F
4861
G
4308

5000

4000

4861   4340   4101

5000   4000

5015   4921   4713   4471   4026

5000   4000

4358   4078   4047

5000   4000

5000   4000

Such diverse and fundamental information on the nature of matter as the composition of distant stars and the structure of atoms and molecules has been obtained by analysis of the light emitted from any incandescent substance.

In the SPECTROSCOPE, such light, passed through a slit and a prism, is broken up into its component wavelengths, which are observed as colored lines, or light of different energies, characteristic of the differences between the various electron energy levels of the atoms. This EMISSION SPECTRUM is CONTINUOUS when the images of the wavelengths are uninterruptedly overlapping; it is a LINE spectrum when only certain specific wavelengths are emitted, as shown here for the elements hydrogen, helium, mercury, and uranium.

On the solar spectrum across the top of this plate appears a series of dark lines — FRAUNHOFER LINES — forming an ABSORPTION SPECTRUM: Some of the light from the intensely hot interior of the sun is absorbed by the cooler gases of its outer layers as the light energies raise the atoms in these layers to higher energy states; bright lines are not, therefore, seen for these changes.

The spectra are calibrated in angstroms (1 A = 10^{-10} m); the letters are arbitrary designations introduced by Fraunhofer for lines important in spectroscopy.

Figure 28.12
White light dispersed into its
component colors by a prism.

was initiated by Kirchhoff about 1859. The spectra of elements gave
us one of the most important clues in the development of modern atomic
theory. There are several ways in which emission spectra may be
classified. We shall group them as follows.

Continuous Spectra When the spectrum from an incandescent solid
or liquid is examined, it shows no regions of darkness (Fig. 28.12). Such
spectra, characteristic of an incandescent lamp and a carbon arc, are
known as *continuous spectra*, since they contain light of every wavelength over a broad region.

Bright-line Spectra When atoms in the gaseous state are excited to
emit radiation, the spectrum consists of a number of *narrow bright lines*
with wavelengths characteristic of the element that emits them. Optical
spectra of hydrogen, helium, mercury, and uranium are shown in the
preceding color insert. Such spectra may be excited by passing an
electric discharge through the gas or by raising the temperature of the
atoms sufficiently. Each element has it own characteristic *bright-line*
spectrum.

Band Spectra When radiation is emitted by excited molecules, the
spectrum often appears in the form of bands of light with regions of
darkness between. Such molecular spectra are often referred to as *band
spectra*. However, when band spectra are studied in spectrometers of
great resolving power, it is found that a band results from a large group
of lines very close together. Overlapping of the lines gives rise to the
band structure.

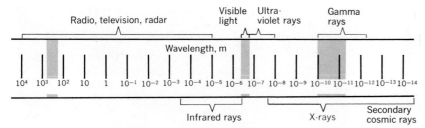

Figure 28.13
The electromagnetic spectrum.

28.11 SPECTROSCOPY

In spectral measurements we are concerned directly with the wave-lengths of the radiation, which range from many kilometers to far less than a billion billionth of a meter. The observation and interpretation of spectra underlies the science of *spectroscopy*, which deals not only with visible light but also with electromagnetic waves of longer and shorter wavelengths including radio, radar, infrared, ultraviolet, x-rays, and gamma rays. The wavelength regions of these radiations are suggested in Fig. 28.13. Actually there is substantial overlapping of regions; the boundaries are not sharp. Except for their inability to excite vision, radiations with wavelengths a little longer than 780 nm (near infrared) or a little shorter than 380 nm (near ultraviolet) have properties very close to those of visible radiation. Infrared and ultraviolet rays are reflected and refracted like visible ones and can expose suitably prepared photographic film. For wavelengths far from those of the optical region, some phenomena occur which are not apparent with visible light.

As an example of how potent a tool spectroscopy is, consider the element *helium;* its spectral lines were observed in light from the sun some 25 years before helium was discovered on the earth. Thanks to study of the spectra of the stars, we know a great deal about their chemical composition, even though no sample has ever been placed in the hands of a chemist. To this day spectroscopy is one of the most active branches of physics and chemistry; both research and industrial applications are legion.

Since each element has a unique spectrum, an examination of

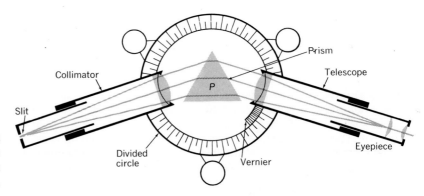

Figure 28.14
Spectrometer, showing how rays made parallel by a collimator are brought to a focus by the telescope.

Figure 28.15
The correspondence of many Fraunhofer lines with the absorption lines of iron vapor is apparent in these emission spectra between 3900 and 4000 Å for iron and the sun (center). [*Courtesy of Mt. Wilson and Mt. Palomar Observatories.*]

the light emitted by a vaporized substance gives direct evidence of its composition. This method is very useful in detecting small quantities of many elements. It is a rapid and sensitive method of analysis. The spectrum of an element can be excited in a variety of ways, among which are passing an electric discharge through a gas containing the element, heating the element in a furnace, and introducing it into an electric arc. The origin of spectral lines is discussed further in Chap. 45.

For the study of optical spectra, a spectroscope or a spectrometer is used. One part of a typical spectrometer (Fig. 28.14) is the collimator, which consists of a tube with a slit at one end and a converging lens at the other. The slit is located at the principal focus of the lens so that the light rays which emerge from the lens are parallel. These rays pass through a grating or a prism which disperses the light to a telescope mounted so that it can rotate about the vertical axis of the instrument. The angle through which the telescope is rotated is read on a divided circle.

28.12 DARK-LINE SPECTRA; FRAUNHOFER LINES

The spectrum of the sun is crossed by a number of fine dark lines known as *Fraunhofer lines* (see color insert following page 372). These lines are produced by absorption in the atmospheres of the sun and the earth. The core of the sun emits white light, which then passes through an atmosphere of less hot vapors and gases. Certain wavelengths corresponding to the spectral lines emitted by various atoms are absorbed and reradiated in random directions in these less hot layers. For example, excited sodium atoms emit strongly two yellow lines, named the D lines by Fraunhofer. Sodium vapor in the solar atmosphere absorbs these wavelengths, which are almost absent when the solar spectrum is examined on the earth. From the Fraunhofer lines it is possible to identify several elements which are relatively abundant in the sun's outer layers. Figure 28.15 compares a portion of the emission spectrum of iron with the corresponding part of the solar spectrum, showing that iron vapor is present in the solar atmosphere.

A dark-line spectrum can be produced in the laboratory by passing an intense beam of continuous radiation through the vapor of some element which absorbs from the continuous spectrum certain wavelengths found in the bright-line spectrum emitted by this element. For example, when light from a carbon arc is sent through sodium vapor, there is strong absorption of the wavelengths of the yellow sodium D lines.

When a sound source is moving toward or away from an observer, the frequency observed is different from the source frequency (Sec. 22.9). A similar effect is observed in light. The doppler effect provides a means of determining the motions of distant stars. If a star is moving away from the earth, the frequency of a spectral line observed at the earth is lower than the emitted frequency so that the observed wavelength is greater than normal; if the star is approaching the earth, the wavelengths are shifted toward smaller values. By measuring the displacements of the lines, the speed of the moving star can be determined.

The doppler effect in light waves differs in several important respects from that in sound waves. In sound waves it is the velocity of the source (or of the observer) *relative to the air* which is of importance. In the case of light waves there is no comparable medium. When source and observer are approaching each other along the line connecting them with relative speed v,

$$\frac{\nu_{observed}}{\nu_{source}} = \frac{\sqrt{c + v}}{\sqrt{c - v}} \qquad (28.4)$$

where c is the speed of light.

The doppler effect is used for measuring speeds of space vehicles and of automobiles. For example, a radar transmitter on a police car sends out very-high-frequency waves, which are reflected by oncoming automobiles. The reflected waves have a different frequency and are "beat" (Sec. 21.10) against the initial frequency, yielding a *beat frequency* equal to the difference between the two. In this case the source is the image of the transmitter reflected from the vehicle, which acts as a plane mirror. Therefore, the speed of the source is twice the speed v of the oncoming vehicle. Let the frequency emitted by the radar be ν, and that received be ν_r. Then Eq. (28.4) yields

$$\frac{\nu_r}{\nu} = \frac{\sqrt{c + 2v}}{\sqrt{c - 2v}} \approx 1 + \frac{2v}{c}$$

since $2v$ is very small compared to c. The beat frequency F is given by

$$F = \nu_r - \nu \approx \frac{2v\nu}{c}$$

Example

A radar transmitter sends out a signal at 100 MHz. An automobile is approaching, and the beat frequency observed is 19 beats/sec. Find the speed of the approaching car.

$$F = \frac{2v\nu}{c}$$

$$F = 19 \text{ beats/sec} \qquad \nu = 10^8 \text{ Hz} \qquad c = 3 \times 10^8 \text{ m/sec}$$

$$19 = \frac{2v \times 10^8}{3 \times 10^8}$$

$$v = 28.5 \text{ m/sec} = 103 \text{ km/hr (64 mi/hr)} \qquad ////$$

QUESTIONS

1   Why is the sun red at sunset? Is this red light polarized or partially polarized? Explain.

2   Why is the sky blue? Is the light from the blue sky polarized or partially polarized? Explain.

3   How should the plane of vibration be oriented in Polaroid sunglasses if they are to be effective in cutting the glare of reflected light from a lake?

4   Why do only anisotropic media exhibit double refraction?

5   How could you make an estimate of the index of refraction of polished black marble by use of a piece of Polaroid and a reflected beam?

6   Proposals have been made to use Polaroid for reducing the dangers of night driving by eliminating the glare of blinding headlights. Explain how this could be done. (Remember that the driver must still see reflected light from his own headlights.) What are the disadvantages of such a system?

7   How does the doppler effect enable us to determine the angular velocity of the sun? To determine the component along the line of sight of the velocity of a distant star?

PROBLEMS

1   Determine the angle of incidence (polarizing angle) for which light reflected from glass of index of refraction 1.52 is plane-polarized.
    Ans. 56.7°

2   Find the angle of incidence for which light reflected from water of refractive index 1.333 is plane-polarized.

3   Determine the critical angle of incidence for the ordinary ray passing from calcite to Canada balsam in a Nicol prism.   *Ans.* 69°

4   Light reflected from a flint-glass surface is plane-polarized when the angle of incidence is 58°. What is the index of refraction of the glass?

5   Ordinary light of intensity I_0 falls on a Polaroid P_1 with its characteristic polarizing direction vertical. An intensity I_1 passes through and falls on a second Polaroid P_2, the polarizing direction of which makes an angle of 37° with the vertical. An intensity I_2 passes through P_2. If the Polaroids pass all components parallel to the polar-

izing direction and none perpendicular, find the ratios I_1/I_0, I_2/I_1, and I_2/I_0.
    Ans. 50, 64, and 32 percent

6   The index of refraction of a plate of flint glass is 1.60. (*a*) Find the polarizing angle of this glass. (*b*) Find the critical angle of incidence for light going from the glass to air.

7   Two Polaroids are crossed. If the analyzer is now rotated through an angle of 40°, what percentage of the plane-polarized light from the polarizer is transmitted, assuming the Polaroid is a perfect transmitter of one polarization and a perfect absorber of the other? What percentage of the ordinary light incident on the polarizer is transmitted through the analyzer?
    Ans. 41.3 percent; 20.7 percent

8   Light of a certain wavelength passes through a Nicol prism and is thus polarized. It then passes through a second Nicol prism whose plane of vibration makes an angle of 65° with the plane of vibration of the first prism. What percentage of the light incident on the second Nicol prism is transmitted?

9   Ordinary light of intensity I_0 falls on a Polaroid P_1 with its characteristic polarizing direction vertical. An intensity I_1 passes through and falls on a second Polaroid P_2, the polarizing direction of which makes an angle of 53° with the vertical. An intensity I_2 passes through P_2 and strikes Polaroid P_3 with polarizing direction horizontal. If the Polaroids pass all components parallel to the polarizing direction and none perpendicular, find the ratios I_1/I_0, I_2/I_1, I_3/I_2, and I_3/I_0. How can any light get through P_3 when it is crossed relative to P_1?   *Ans.* 0.50; 0.36; 0.64; 0.115

10   Two Nicol prisms have their planes of vibration parallel. One of the prisms is then turned so that its plane of vibration makes an angle of 55° with that of the other. What fraction of the amplitude incident on the second Nicol prism is transmitted? What percentage of the light incident on the second Nicol prism is transmitted?

11   A calcite crystal is 0.012 mm thick and cut so its faces are parallel to the optic axis. If the indices of refraction of calcite for $\lambda = 600$ mm

are those given in Sec. 28.6, find the number of ordinary (O) waves and the number of extraordinary (E) waves in the crystal for light incident normally. What is the approximate difference in phase for the O and E waves leaving the crystal?

Ans. 33.16 and 29.36; 7.6π rad or 1.6π rad

12 Two Polaroids are crossed so no light passes the analyzer. If a third Polaroid is placed between these two with its axis making an angle of 30° with that of the polarizer, find the fraction of the plane-polarized light incident on the middle Polaroid which (a) is transmitted through to the analyzer and (b) passes through the analyzer.

IN VISION we are concerned with illumination levels and with the appearance of color, an aspect of light which greatly enriches our seeing. We begin this chapter by introducing the concepts of luminous intensity and illuminance. Then we discuss color and point out some of the mysteries of color vision.

CHAPTER TWENTY-NINE
ILLUMINATION AND COLOR

29.1 STANDARD SOURCES AND LUMINOUS FLUX

To compare light sources and illumination, we make measurements in terms of arbitrary standards. In the early days the standard source was the *British standard candle,* made of spermaceti and burned at the rate of 120 grains per hour. At that time the *lumen* (lm), which is the unit of *light flux,* was the light emerging through an opening made by cutting an area of one square foot out of a spherical surface of radius one foot when a standard candle emitted light at the center of the sphere. These standards are not of current scientific value, but modern standards have been chosen in such a way that we still measure light flux in *lumens* and the luminous intensity of a light source in *candelas* (cd), a unit which evolved from the old candle and is sometimes called the new standard candle.

The *candela* is based on the following arbitrary but highly reproducible operation involving a standard source developed by the National Bureau of Standards and adopted by the International Commission of Weights and Measures in 1948. A schematic diagram of the standard source, basically a glowing blackbody cavity at the temperature of melting platinum, is shown in Fig. 29.1. A cylindrical tube of thorium oxide, surrounded by pure platinum at its melting point and with powdered thorium oxide in the bottom, acts as an ideal blackbody radiator. The tube is closed at the top except for a small hole, the area of which is $\frac{1}{60}$ cm². When this source, operating at the freezing temperature of platinum, is observed from above, its luminous intensity of one candela is essentially that of the old standard candle. *The candela is the luminous intensity in the perpendicular direction of a surface of $\frac{1}{60}$ cm² of a blackbody at the temperature of freezing platinum $(1773.5°C)$ under a pressure of $101,325\ N/m^2$.* Luminous intensities of light sources are measured in candelas.

Imagine an ideal point source S of radiation at the center of a sphere (Fig. 29.2) emitting light equally in all directions and having a luminous intensity of I cd. Actually such an isotropic source is not something we can make, but we can discuss its properties. The total luminous flux F emitted is arbitrarily chosen to be $4\pi I$ lm. Thus an isotropic 1-cd source emits 4π lm of flux, and *one lumen is the luminous flux which falls upon one square meter of the area of a sphere of one-meter radius when an isotropic source of one candela is at the center of the sphere.*

Figure 29.1
International standard light source.

Figure 29.2
Ideal point source S at the center of a sphere of radius R. The solid angle ω subtended by an area A of the spherical surface is A/R^2 steradians.

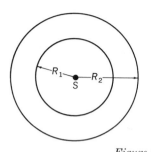

Figure 29.3
The total light flux from point source S which passes through each spherical shell is the same though it falls on different areas. The flux per unit area is inversely proportional to the square of the distance from the source.

The solid angle subtended by 1 m² of the surface of a sphere of 1-m radius is called a *steradian*. The solid angle subtended by an area A of the surface of a sphere of radius R (Fig. 29.2) is given by $\omega = A/R^2$. Thus the entire sphere subtends an angle of 4π steradians. A one-candela source emits one lumen per steradian.

Typical light sources, such as fluorescent and incandescent lamps, emit different luminous fluxes in different directions. Consequently, the apparent source intensity depends on the direction from which it is viewed, just as is the case with the standard source. In Sec. 29.3 we describe a means of measuring the intensity of a practical source.

29.2 ILLUMINANCE

When luminous flux strikes a surface, the surface is illuminated. We define the *illuminance (or illumination)* E of the surface as the *luminous flux incident per unit area*. In the metric system we measure the illuminance in lumens per square meter, or *luxes;* in the British system in lumens per square foot, which are sometimes called *foot-candles*. Consider light proceeding from a point source S of I cd. As the light advances, it is distributed over a larger and larger area. If the medium through which the light travels does not absorb any energy, the total light energy passing any spherical surface concentric with S is constant. The light falling on a unit area of a sphere of radius R is given by the ratio of the luminous flux F emitted to the area $4\pi R^2$ of the sphere:

$$E = \frac{F}{A} = \frac{4\pi I}{4\pi R^2} = \frac{I}{R^2} \tag{29.1}$$

For concentric spheres (Fig. 29.3), the luminous flux F is constant; thus

$$E_1 = \frac{F}{4\pi R_1{}^2} \qquad \text{and} \qquad E_2 = \frac{F}{4\pi R_2{}^2}$$

The amount of light per unit area arriving from a point source varies inversely as the square of the distance from the source.

When the light rays are not incident along a normal to the surface, but make an angle i with the normal (Fig. 29.4), the illuminance is given by

$$E = \frac{I \cos i}{R^2} \tag{29.2}$$

If the eye were equally sensitive to radiant energy of all wavelengths, we could measure the illuminance E of a surface in terms of the amount of energy falling on a unit area of the surface in unit time. However, the eye is more sensitive to some wavelengths than to others (Sec. 26.5). Therefore, if we are interested in light from the point of view of vision, we must compare illuminances on surfaces by use of the eye or an instrument deliberately constructed to respond like the eye. One such instrument is similar to an exposure meter used by photographers. The output of a photoelectric cell is adjusted so that it reads the illuminance directly. The instrument differs from an exposure meter in that suitable filters over the cell give it a response which is proportional to the luminous flux. If the eye and the camera film had

Normal to A

Source

A cos i

Figure 29.4
The illuminance falling on a surface at an angle of incidence i is proportional to cos i.

identical responses to all wavelengths, the same instrument could be used for both purposes.

Table 29.1 lists a few typical illuminances. The figures are approximate; one reason is that different surfaces reflect vastly different fractions of the incident light. A gray surface receiving an illuminance of 50 lm/m² does not appear as bright as a white one under the same illuminance, since more light is reflected back from the white surface. Hence, one may find a lower illuminance acceptable for reading a book printed on good white paper than for reading a newspaper.

The *luminous efficiency* of a light source is the ratio of light flux emitted by the source to the power supplied to the source. Table 29.2 lists some typical luminous efficiencies of modern light sources. For an incandescent lamp, the higher the temperature at which the source operates, the greater the luminous efficiency.

29.3 THE PHOTOMETER

A photometer is a device for comparing the intensities of two sources. To make this comparison, the distances of the sources from a screen are adjusted until they produce the same illuminance. Under these circumstances, if I_1 is the intensity of one source, I_2 that of the other source, and R_1 and R_2 their respective distances from the screen when they produce equal illuminance,

$$\frac{I_1}{R_1{}^2} = \frac{I_2}{R_2{}^2} \qquad (29.3)$$

Example

A standard 32-cd lamp at a distance of 60 cm from a screen gives the same illumination as a lamp of unknown intensity at a distance of 120 cm from the screen. Find the intensity of the unknown lamp.

$$\frac{I_1}{R_1{}^2} = \frac{I_2}{R_2{}^2}$$

$$\frac{32 \text{ cd}}{(0.6 \text{ m})^2} = \frac{I_2}{(1.2 \text{ m})^2}$$

$$I_2 = 128 \text{ cd} \qquad\qquad ////$$

If two sources have different colors, it is impossible to obtain proper balance with a simple photometer. However, if an arrangement is made whereby a screen is illuminated first by one source and then by a second, it is possible to find a frequency of alternation such that the color difference disappears. Then the observer can adjust the source distances until the sources produce equal illuminances. A photometer constructed for this process is called a *flicker photometer*.

29.4 COLOR CLASSIFICATION

There are three psychophysical characteristics in terms of which it is convenient to describe the color of an object. They are *hue, saturation,*

Table 29.1
Typical Illuminances

Situation	Illuminance, lm/m²
Brilliant day, sun overhead	100,000
Overcast day	8,000
Minimum recommended for fine work	500
Night football or baseball	400
Office, classrooms, reading room	300
Street lights	5
Full moonlight	0.2
Starlight	0.0003

Table 29.2
Typical Luminous Efficiencies

Source, W	Luminous flux, lm	Efficiency, lm/W
Tungsten lamp:		
10	80	8
40	470	12
100	1,600	16
1,000	22,000	22
Fluorescent lamp:		
6	240	40
10	450	45
30	1,800	60
40	3,000	75

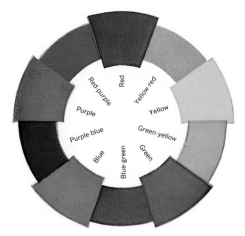

Figure 29.5
Hue circle.

and *lightness* (or, for a self-luminous object such as a light bulb, the *brightness*). The *hue* is specified in terms of the color in a hue circle (Fig. 29.5) which the object most closely resembles. In the system of color notation developed by Munsell, the complete hue circle consists of 100 hues.

The term *saturation* has to do with the extent to which the hue is influenced by the addition of other colors. If blue is the only color present, the saturation approaches unity. On the other hand, if other colors are present, but there is enough blue so that the object has a blue hue, the saturation is low (Fig. 29.6). In mixing paints, for example, a paint of lower saturation can always be made by mixing white with the color. Thus, it is easy to produce a wide range of variations in saturation for the same hue. As we go from zero to 100 percent in

Figure 29.6
Variations in blue hue from a Munsell
color atlas.

Figure 29.7
Color tree vividly shows the three-dimensional relationship between lightness, hue, and saturation.
[*Reproduced, with permission, from the Kodak Data Book "Color as Seen and Photographed." The color tree was produced by Allcolor Company, Inc., New York.*]

saturation, we go from whites and grays to pure colors. Whites and grays are *achromatic* and correspond to zero saturation.

Lightness is connected with the relative amount of light which reaches the eye from the object. If the object is illuminated by white light and reflects all the incident light, it has maximum lightness and appears white; if it reflects none of the light, it is black. It is possible to arrange achromatic grays in a scale of brightness and to match against these grays samples of a given hue which reflect the same fraction of the incident light (Fig. 29.6).

One way of representing colors is in a three-dimensional color cylinder or color tree (Fig. 29.7) in which the lightness (or brightness) is plotted, going from black to grays to brilliant white along an axis. Around this axis can be drawn concentric circles, with the various hues given by the angles, and the saturations by the distance from the central white-black axis.

Ordinary "white" light from the sun or a lamp consists of a mixture of all the visible wavelengths. When it is dispersed by a grating or prism, a continuous spectrum is formed. However, it is not necessary that all visible wavelengths be present to excite the sensation of white light in an observer. Indeed, only two properly chosen wavelengths need be present, and a wide variety of pairs is possible. Two wavelengths (or two colors) which when mixed together give white light are said to be *complementary*. The physical, chemical, and psychological processes associated with vision are highly complex, and what an observer "sees" depends not only on what wavelengths are present but also on many other factors. The seeing process can make distinctions of amazing sublety and may display a color for some object in the visual scene even though no wavelengths ordinarily associated with that color are present. Colors seen in complex images depend not only on the wavelengths incident on the eye but also on intricate feedback mechanisms in the visual chain.

29.5 THE MEASUREMENT OF COLOR

Physicists interested in quantitative measurements of colors have found that it is possible to make reproducible measurements by matching colors projected on two neighboring areas of a reflecting screen. Suppose a projector (Fig. 29.8) throws on area S_1 an arbitrary color X. Let us illuminate the adjacent area S_2 by means of three lanterns, each of which sends out a light of a single narrow wavelength region, A at 425 nm (blue), B at 551 nm (green), and C at 650 nm (red). For almost any color we can find one and only one combination of relative intensities of these three lights which exactly matches the color X. We can then describe color X by three specific numbers representing the amounts of light from A, B, and C.

Although we can match most colors in this way, there are some which cannot be matched by adding everything on S_2. These colors are said to be outside the *gamut* of these *primaries*. However, if we move the proper one of the lamps to the other side and add its light to the unknown color, it is possible to achieve a match. For example, if we try to match a pure green of $\lambda = 500$ nm with the lamps, we find that there is too much red, but if we move lamp C to the other side, we find that we can match $\lambda = 500$ nm plus some red from C with light from lamps A and B. One way of saying this is to say we can match $\lambda = 500$ nm with light from A and B and a *negative* amount of light from C. This, however, introduces the complication of negative numbers for specifying some colors.

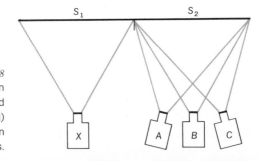

Figure 29.8
Color of almost any hue or saturation thrown on screen S_1 can be matched on screen S_2 by blending (adding) three monochromatic primaries in specific amounts.

Figure 29.9
Spectral-reflectance curves for various objects.

We have chosen for our illuminants blue, green, and red, because with them we can match the widest gamut of colors. In this sense red, green, and blue are *primary* colors. However, if we use any three different wavelengths, we can still match any color if we allow the use of negative contributions and assume that the sources A, B, and C can be varied in intensity over an infinite range.

To avoid the difficulties associated with negative numbers, and for other reasons, the International Commission on Illumination agreed in 1931 on a standard *operational procedure* for determining the color of a surface. (This method can be shown to be equivalent to the type of experiment in which we match the color with three primary sources, except that the sources chosen do not exist in the realm of *real colors*.) The surface is placed in a spectrophotometer. This instrument contains an optical system in which the light from a standard lamp is dispersed into a spectrum by a prism. One narrow range of wavelengths at a time is allowed to fall upon the surface under test, and the amount of light reflected is compared with the amount of the same wavelength range reflected from a standard white surface. A plot is made of the relative reflectance of the surface under test as a function of wavelength. Several such *spectral-reflectance* curves are shown in Fig. 29.9. From the curve it is possible to calculate the coordinates x and y for the color of the surface on the ICI *chromaticity diagram* (Figs. 29.10 and 29.11).

Figure 29.10
Chromaticity diagram shows the spectrum locus and the colors of blackbodies at the temperatures indicated. Point C represents average daylight from an overcast sky and is chosen as the illumination standard.

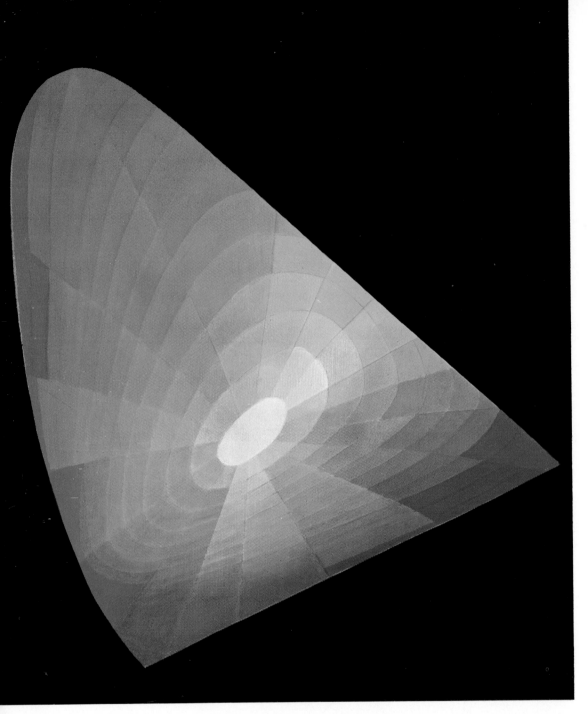

Figure 29.11 Chromaticity diagram has radial boundaries between segments to separate the various hues, with oval boundaries representing lines of constant saturation. The horseshoe-shaped outer boundary corresponds to the positions of pure spectrum colors. [*Courtesy of L. M. Condax and the Eastman Kodak Company.*]

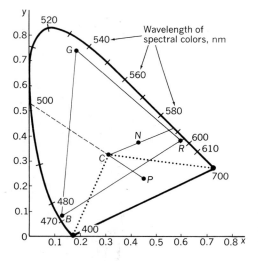

Figure 29.12
Chromaticity diagram showing method of specifying a color in terms of dominant wavelength and purity. The phosphors of color television screens glow in the primaries R, B, and G, permitting mixtures to reproduce any color in the triangle RBG, a close approximation to the full spectrum of visible color.

The horseshoe-shaped boundary (Fig. 29.10) is called the *spectrum locus*, since here we find the pure colors of the continuous spectrum. Point B on the diagram represents a white surface illuminated by sunlight at noon, while point C represents average daylight from an overcast sky. The other points on the inner curve represent the colors of blackbodies at the temperatures indicated on the diagram. Point C is ordinarily chosen as the illumination standard.

Any color above the dotted lines in Fig. 29.12 may be regarded as a mixture of illuminant C and a single spectral frequency. Thus the point N in this figure could be reached by mixing daylight and sodium D radiation. The line CN extended to the spectrum locus gives us 589.3 nm, which is called the *dominant wavelength* for the color specified by N. Since the yellow of point N lies between C and the spectrum color, it is not *saturated yellow*, as is the spectrum color. The distance from C to N divided by the distance from C to the spectrum locus gives the *purity* of the color. For point N the purity is 40 percent.

Any two colors on opposite sides of a line through C on the chromaticity diagram are complementary. The point P below the dotted line is a purple, and not a spectral color. If we extend the line CP, it reaches the spectral curve at 500 nm. Point P is said to have the coordinate 500c, where the c ($=$ complementary) is added to indicate that the point P is on the opposite side of C from the spectrum locus.

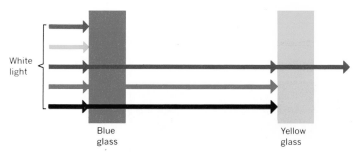

Figure 29.13
Transmission and absorption of white light by yellow and blue glass.

Figure 29.14
Background color affects identical
hues. [*From F. A. Osborne, "Physics
of the Home."*]

We may assign a purity to the color represented by P if we use a straight line connecting the end points of the visible region as the analog of the spectrum locus.

By use of the ICI chromaticity diagram it is possible to specify measurable physical quantities, dominant wavelength and purity, which correspond to the psychological *hue* and *saturation*. From the spectral-reflectance curve for a surface it is possible to compute the *luminous reflectance*, where the word *luminous* means that the value takes into account the visual response of a standard observer and the color char-

Figure 29.15
Spectral-sensitivity curves (above) for
the three cone pigments involved in
color vision, with the colors associated
with the various wavelengths (below).

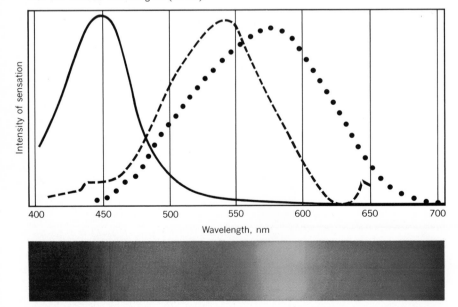

acteristics of the light source. The luminous reflectance is the physical analog of *lightness*.

For color television, phosphors are available which give the primaries R, G, and B of Fig. 29.12 when they are bombarded by an electron beam. The screen is made up of these phosphors. By appropriate variations of intensities, all the colors in the triangle RBG can be obtained. Although this triangle does not include all visible colors, it covers a sufficiently wide range so that rich color communication is possible.

29.6 COLOR OF BODIES

Thus far we have discussed the color of surfaces from which light from a standard illuminant is reflected. Of course, the color of the object depends on the spectral distribution of the incident light. If a white surface is illuminated with red light, it appears red because it reflects only red light to the eye. Since no other colors are present to be reflected, the color of the body is completely determined by this reflected red light. An object that appears red in white light is also red in red light, but it is black in any light which it does not reflect or transmit. Two bodies which have the same color in daylight may not have the same color in lamplight. Light from an ordinary incandescent lamp is richer in red rays than in blue.

A piece of glass which appears blue may actually reflect and transmit other colors such as the neighboring green and violet (Fig. 29.13). Similarly, a yellow glass plate may transmit red and green. If both glasses are placed in a beam of white light, the only color which can pass through both is green. This is an example of color by the *subtractive* process, in which we start with white light and remove various wavelength regions by absorption. The mixing of pigments is another example of a subtractive process. Mixing suitable yellow and blue pigments may give rise to a green paint just as green light passes through the combined filters of Fig. 29.13.

29.7 COLOR VISION

Color vision is known to arise from the stimulation of the cones in the retina. The processes which occur are not completely understood, but current research† in this subject is yielding new insight into the remarkable versatility of the eye in perceiving color. The apparent hue of an object is affected by its surroundings (Fig. 29.14) and, indeed, by the immediate past exposure of the retina. When any area of the retina is subjected to continued stimulation by intense red light and then subjected to white light, the area "sees" blue-green. In general, a retinal area strongly stimulated by any color responds to subsequent white light with a *fatigue image* dominated by the color complementary to the original stimulus.

† Reports of definitive studies in this area may be found in E. H. Land, "Experiments in Color Vision," *Scientific American*, May 1959, p. 84, and in E. F. MacNichol, Jr., "Three-pigment Color Vision," *Scientific American*, December 1964, p. 68.

A reasonable explanation for the ability to match any color with three spots of light (Sec. 29.5), proposed by Young and developed by Maxwell and Helmholtz, is built upon the assumption that there are three types of cone response, each of which is excited to some extent by the whole visual spectrum (Fig. 29.15). Three light-sensitive pigments segregated in three different kinds of receptor cells in the retina have been found: one responsible for sensing blue light has a peak sensitivity at 447 nm (Fig. 29.15), a second sensing green is most sensitive at 540 nm, and the third has maximum sensitivity at 577 nm (yellow) but still senses red well. When all three types of cells are aroused about equally, the sensation of white light is produced. It is not necessary that all wavelengths be present to get the sensation of white light. As we have seen, properly chosen amounts of radiation of only two wavelengths can excite a sensation of perfect white light, since such a pair can excite all three responses roughly equally. Purples arise from excitation of the blue and red sensations without any substantial amount of green. Purple is not a spectral color but a mixture of spectral colors. Yellow is a true spectral color; nevertheless, the simultaneous excitation of the red and green sensations in roughly equal amounts gives the eye the sensation of yellow, even though none of the ''yellow wavelengths'' are present.

The three-color theory is highly successful in dealing with spots of light and is the foundation upon which both color television and color photography have been based. However, matching spots of light is a far cry from seeing colors in complete images under normal conditions. Then the colors of images arise from the interplay of longer and shorter wavelengths over the total visual field. For example, if a photographic transparency of a colored scene taken with red light is projected on a screen with yellow light of $\lambda = 599$ nm, the resulting image is yellow with various areas lighter and darker than others. Now if the image of a second transparency of the same scene taken with green light and projected with yellow light of $\lambda = 579$ nm is superposed, the resulting image includes reds, greens, blues—a broad range of colors, paler and less saturated than the original scene, but still a reasonably faithful reproduction. Thus it appears that color is not determined uniquely by the wavelengths reaching the eye, but rather the various wavelengths bear information by which the eye assigns colors to various objects. Apparently, to see colors the eye needs information concerning long and short wavelengths in a scene, but this information can be brought to the eye by as few as two narrow bands of wavelengths.

29.8 FLUORESCENCE

Although many bodies return only those visible wavelengths which fall upon them, some bodies absorb light of one wavelength and emit light of longer wavelength. This process is called *fluorescence*. Many dyes emit vivid visible radiation under ultraviolet light. Fluorescence is used in a wide variety of applications ranging from the theater to the identification of ores in mining. In fluorescent lamps ultraviolet light from an electric discharge in mercury vapor is absorbed by fluorescent materials, called *phosphors*, which coat the insides of the tubes. By suitable choice of phosphors a wide range of colors of fluorescent light is available.

Some materials continue to emit visible light for some time after exposure to ultraviolet light. Such materials are said to be *phosphorescent*.

QUESTIONS

1 Is the source intensity of an incandescent lamp the same in all directions? What do we mean by source intensity in this case?

2 Do a photometer and a photographic light meter read the same physical quantity? If not, how do they differ?

3 How, in principle, would you calculate the illuminance at a point illuminated by a large, spread-out source such as a fluorescent lamp?

4 The sun is closer to the earth in December than it is in June. Nevertheless, the illuminance on a horizontal surface is greater on a clear summer day than on a clear winter day everywhere outside the equatorial zone. Why?

5 Green color can be seen in (*a*) a green leaf, (*b*) a mercury arc, (*c*) a shirt, and (*d*) a stained-glass window. What is the origin of the green in each case?

6 A "daylight" incandescent bulb has a blue glass envelope. Why?

7 Pieces of cloth appear red, yellow, green, and blue-violet in daylight. How is their appearance modified when seen under candlelight? Under an incandescent light bulb?

8 How can only three colors (and black) be used to print color pictures? Can any color be reproduced in this way?

PROBLEMS

1 It is desired to have an illuminance of 300 lm/m^2 at the working area of a desk. If the light comes from a source 0.9 m away and the angle of incidence is 60°, find the luminous intensity of a source which produces the desired illuminance. *Ans.* 486 cd

2 A small screen is 15 ft from a 90-cd source. The normal to its surface makes an angle of 20° with a line drawn from the source. What is the illuminance?

3 A lens of 5-cm diameter is 50 cm from a lamp which is essentially a point source with a luminous intensity of 160 cd. Find the illuminance at the lens. How many lumens are incident on the lens if the lamp lies on the axis of the lens? *Ans.* 640 lm/m^2; 1.26 lm

4 If a 100-cd incandescent lamp sends out light uniformly in all directions, how many lumens per square foot are received on a screen 6 ft from the lamp when the angle of incidence is 30°?

5 The luminous efficiency of a 200-W light bulb is 20 lm/W. If such a bulb is used as a street-light, what is the illumination on a horizontal street (*a*) 6 m directly below the lamp and (*b*) 8 m from the point directly below the lamp. Assume that there is no reflector and that light is emitted equally in all directions. *Ans.* (*a*) 8.84 lm/m^2 (*b*) 1.91 lm/m^2

6 An arc lamp at a distance of 50 ft from a screen produces the same illumination as a 200-cd lamp at a distance of 8 ft from the screen. What is the luminous intensity of the arc lamp?

7 A photocell actuates a relay which controls the opening and closing of the door of a super-market. At least 0.3 lm must reach the photocell through an opening of 6-cm^2 area if the relay is to operate. What source intensity would be required if a point source is to be used 1.2 m from the relay? (In an actual installation one would use a lamp of much lower candlepower. How would the installation deviate from the setup of the problem?) *Ans.* 720 cd

8 Two lamps of 80 and 20 cd are placed 180 cm apart. At what position between them will the illumination on both sides of an interposed screen be equal?

9 A football field is illuminated at night from six towers, each with twenty-four 2,000-W lamps with luminous efficiencies of 25 lm/W. If one-third of the luminous flux reaches the used area, which is 120 by 75 m, find the average illuminance at the field. *Ans.* 267 lm/m^2

10 A screen is placed 1.5 m from a lamp. When a sheet of smoked glass is placed between the screen and the lamp, the lamp must be moved so that it is at a distance of 0.65 m from the screen in order to produce the same illuminance on the screen. Calculate the percentage of light transmitted by the sheet of glass.

11 A lamp of unknown luminous intensity is placed at a distance of 2.5 m from a 60-cd standard. The proper placement of a photometer screen placed between the sources is found by experiment to be 1.4 m from the standard lamp. What is the luminous intensity of the lamp?

Ans. 37 cd

12 A 40-cd lamp is placed 0.6 m in front of a screen. How far from the screen must a 200-cd lamp be placed for the illuminance on the screen to be three times as great as it was with the first lamp?

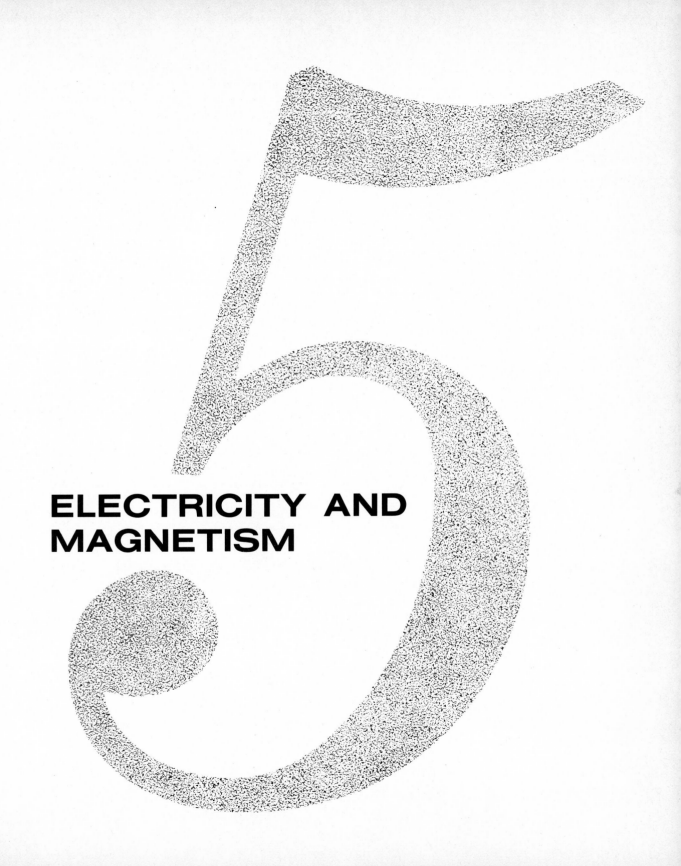

ELECTRICITY AND MAGNETISM

CHAPTER THIRTY
ELECTRIC CHARGES
AT REST

THE TWENTIETH CENTURY may well be called the *age of electricity*. Developments in the nineteenth century made it possible to generate, deliver, and control vast amounts of electric energy at reasonable cost. The results are everywhere apparent. Electric energy is used for lighting, heating, cooking, refrigeration, washing, drying, ironing, and air circulation. Electric energy dominates the world of communication; the telephone, telegraph, radio, and television are direct outgrowths of advances in our understanding of electricity.

LESS OBVIOUS, but no less important, is the role of electricity in our economic development. The automobile and aluminum industries, to mention only two, could not have evolved in their present form until man learned how to generate and control electric energy. So important is electricity in our everyday life that our standard of living would be tremendously lowered without it. Electrical servants have replaced human servants in the average home. In less than a century the science of electricity developed from a curiosity to a dominant position in our technology.

ALTHOUGH most of the common uses of electrical phenomena involve electric charges in motion, the fundamental concepts were first developed in *electrostatics*, the science of stationary electric charges.

30.1 CHARGES BY CONTACT AND SEPARATION

As early as 600 B.C. it was known that amber rubbed with fur has the interesting property of attracting light pieces of straw or paper. A rubber comb run through the hair may exhibit this same property. Such bodies are said to be *electrified*, a term derived from the Greek word for *amber*. Experiments reveal that it is possible to electrify any kind of material

Figure 30.1
Like charges always repel, and unlike
always attract.

by rubbing it with a suitable second material and then separating the two. An electrified body is said to bear an *electric charge*. When a charged body and an uncharged one are brought together, a portion of the charge is likely to be transferred to the uncharged body.

If a small pith ball is charged by contact with an electrified glass rod, the rod and ball repel each other. Similarly, if a pith ball is charged by contact with an electrified rubber rod, they repel each other. However, a ball charged by contact with the electrified *glass* rod is attracted by the electrified *rubber* rod, while the charged *glass* rod attracts a pith ball charged by contact with the electrified *rubber* rod. This simple experiment suggests (see Fig. 30.1) that there are *at least* two different kinds of electric charge. Exhaustive experiments with many substances have led to the conclusion that there are *only* two kinds of charges. The charge on a glass rod rubbed with silk is called *positive*, while that on a rubber rod rubbed with cat's fur is *negative*. From a number of experiments it is possible to establish the qualitative laws of electrostatic reaction: *Like charges repel each other; unlike charges attract each other*.

30.2 THE ELECTRICAL STRUCTURE OF MATTER AND THE FERMI LEVEL

The idea that all matter is composed of tiny particles called *atoms* is well known. Although the word *atom* means "indivisible," atoms have a complex structure. Practically all the mass of an atom is concentrated in a tiny core called the *nucleus*. The nucleus bears a *positive* charge, the magnitude of which depends on the atom in question. All nuclei of a given element bear the same charge. The nucleus with the smallest charge, *1 atomic unit*, is that of the hydrogen atom; the nucleus of ordinary hydrogen has a special name—*proton*. Every other type of atom has a nucleus bearing some integral number of atomic units. All oxygen nuclei have 8 atomic charges, all copper nuclei 29 charges. Indeed, the atomic number of the atom is a measure of how many atomic units of charge its nucleus bears, i.e., of how many protons the nucleus contains.

Outside the nucleus of a neutral atom are *electrons*, the number of electrons being equal to the number of positive charges of the nucleus. Each electron bears one fundamental atomic unit of negative charge, equal in magnitude to the positive charge of the proton. It is sometimes convenient to think of an atom as a submicroscopic solar system, with the nucleus as the sun and the electrons revolving about it much as the planets revolve around the sun (Fig. 30.2). The concept

Figure 30.2
Electron orbits of a copper atom. The nucleus and two innermost electrons are not shown at the center of the atom.

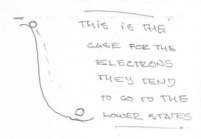

THIS IS THE CASE FOR THE ELECTRONS THEY TEND TO GO TO THE LOWER STATES

of sharply defined orbits must be abandoned in a correct wave-mechanical treatment, but nevertheless the orbit model is a widely used reasoning aid which is capable of yielding plausible values for the average distance of an electron from its nucleus and of the allowed energies of the electrons.

In all atoms more massive than those of helium some of the electrons are closer to the nucleus than others and much more tightly bound. It is the outermost electrons which interact most strongly with neighboring atoms and participate in chemical bonds. Just as water runs downhill, so electrons move to lower energy states. When neutral atoms approach one another, a stable combination may be formed if the energy of the system is reduced by a sharing of outer electrons or of the transfer of an electron from one atom to another.

At 0 K electrons occupy the lowest possible energy states, and even at room temperature all but a minuscule fraction of the electrons in a substance are in the lowest possible energy states. For every material there is a *Fermi energy level* E_F such that the probability that an allowed state of energy E is occupied is given by†

$$f = \frac{1}{e^{(E-E_F)/kT} + 1} \tag{30.1}$$

where e ($= 2.71828$) is the base of the natural logarithms, T is the absolute temperature, and k is the Boltzmann constant (Sec. 14.7). At 0 K, f is 1 for E less than E_F and 0 for E greater than E_F; thus all levels below E_F are filled at 0 K and all levels above E_F are empty.

Consider water filling a swimming pool. The floor and walls play the role of the block of material, while the water is analogous to the electrons and the surface plays the role of Fermi level. When nothing disturbs the surface, all available energy states below this level are filled with water and all higher states are empty. Raising the temperature of a material corresponds to setting up small waves on the water surface. Then at troughs water is missing from levels below the Fermi level and at crests we find water above the Fermi level. When the absolute temperature is nonzero, f is exactly $\frac{1}{2}$ when $E = E_F$. *The Fermi energy E_F is the energy for which the probability of occupancy of an allowed energy state at E_F is $\frac{1}{2}$.*

30.3 EXCHANGE AND CONSERVATION OF CHARGE

ACCORDING TO THE CHARGES RULES GO TO THE ORBITS?

When two electrically uncharged bodies of different composition are brought into intimate contact, it is almost certain that the Fermi level of one was initially higher than that of the other. Then electrons in the first can achieve a lower energy by transferring to the second. Thus electrons move from the substance with higher Fermi level to the substance with lower Fermi level. This transfer lowers the Fermi level in the first and raises it in the second. When full equilibrium is achieved, the Fermi levels coincide. The lower the Fermi level in a material, the greater its affinity for electrons. The transfer of electrons from the first

† Use of this formula involves mathematics beyond the scope of this course. As a consequence there are no problems using it, and it is not expected that the student remember it.

material to the second leaves the latter charged negative and the former positive. The transfer does not go on indefinitely because as the substance with the greater electron affinity (lower Fermi level) accumulates electrons, the excess negative charge repels other electrons, while the substance with the initially higher Fermi level has an excess positive charge and attracts them.

Fermi levels in charge exchange between two substances are directly analogous to water levels in the exchange of water between two interconnected beakers and to temperature levels in the exchange of heat between two interacting bodies. In all three cases there is a transfer of something from the reservoir with higher level to the one with a lower level. Further, when full equilibrium is established, the levels are the same in both reservoirs. Since electrification by contact and separation always involves a transfer of charges from one body to another, the amount of positive charge which appears on one is equal to the negative charge which appears on the other. Whenever a certain positive charge appears on one body, an *equal* negative charge must appear on some other bodies.

Charge is one of the fundamental properties of matter. *The total net charge of any isolated system never changes.* This statement expresses the *law of conservation of charge,* one of the basic laws of nature. This law and the laws of conservation of energy, of momentum, and of angular momentum are of great importance in the interactions of atoms and in the reactions of atomic nuclei.

30.4 CONDUCTORS AND INSULATORS

In some materials, notably the metals, a small fraction of the electrons are not bound to any one nucleus but are free to wander among the atoms. Materials in which charges are free to move about are called *conductors.* In other materials each electron is held by one or two atoms. A material in which charges are not free to move about is called a *nonconductor* or *insulator.* As might be expected, the line between conductors and insulators is not a sharp one. For example, a piece of damp wood is neither a good conductor of electricity nor a good insulator.

In metallic conductors the transfer of charge is by movement of electrons. However, there are many situations in which charges are not conducted by electrons. A solution of sodium chloride in water is a good conductor of electricity, because the molecules break into two parts, a sodium atom with an electron missing and a chlorine atom with one extra electron. These charged particles are called *ions.* The conduction takes place by the movement of both positively charged sodium ions and negatively charged chlorine ions. In liquids and gases, the transfer of charge by ions, both positive and negative, is common.

The early observations of electric charges are made on materials such as amber and glass, which are nonconductors. A piece of glass rubbed with silk obtains a positive charge which does not escape readily because the glass is a good insulator. If a copper rod is rubbed with silk, it also becomes charged, but unless it is carefully insulated the charge is rapidly conducted away. If one holds the copper rod in one's

GLASS → GOOD INSULATOR

RUBBER → GOOD INSULATOR

hand, the charges have no difficulty in traveling along the copper conductor and escaping through the body, since the human body is a fairly good conductor, especially when the skin is moist.

It is possible to charge one end of a hard rubber rod positively and the other end negatively by rubbing one end with a material of great electron affinity and the other end with a material of low electron affinity. Because the rubber is a good insulator, the charges remain on the ends. However, if one tries to do this to a metal rod, the charges do not stay on the ends; rather, there is an immediate flow of electrons. Similarly, if a wire of conducting material is connected between a positively charged metal sphere and a negatively charged sphere, there is a flow of charges. Such a flow is called an *electric current*. In sending electric energy from one point to another, it is standard practice to produce a flow of charges; since this requires some sort of conducting medium, wires of good conductors, usually copper or aluminum, connect homes with power-generating stations.

30.5 COULOMB'S LAW

In 1785 the French physicist Coulomb used a torsion balance (Fig. 30.3) to measure the force between two small charged spheres as a function of the distance between them. He found that *the force between two charges Q_1 and Q_2 is directly proportional to the product of the charges and inversely proportional to the square of the distance r between the charges:*

$$F = k \frac{Q_1 Q_2}{r^2} \tag{30.2}$$

$K = 9. \times 10^9$

where the proportionality constant k depends on the units in which the variables of Eq. (30.2) are measured. Coulomb's law is directly applicable to charged bodies only in situations in which the distance r between charges is large compared with the dimensions of the charged bodies. When two charged objects are very small compared with the distance between them, we may consider the charges as concentrated at points, and we refer to them as *point charges*.

Charge may be measured in a number of units, four of which have been widely used. We shall work with only the first two.

1   The *atomic unit of charge* is the charge on a single proton; the charge on the electron has the same magnitude, but is negative in sign. The atomic unit of charge is exceedingly minute, far too small to be convenient for use in most practical problems in electricity. It is, however, a truly basic unit in that no smaller charge has ever been observed, and all larger charges are believed to be integral numbers of atomic units.

2   The *coulomb* (C) is the practical unit of charge. We shall take it as a new fundamental unit which, with the meter, kilogram, and second, will permit us to formulate the laws of electricity and magnetism in a self-consistent manner. The formal definition of the coulomb in terms of the operations required for a precise measurement of charge involves the use of concepts which are

— Torsion head

— Torsion suspension

Figure 30.3

Coulomb measured the force exerted by the charge of the fixed sphere on the charge of one of the movable spheres of the torsion balance.

developed in subsequent chapters (Sec. 36.8). We can, however, compare the coulomb with the atomic unit of charge:

$$1 \text{ C} = 6.242 \times 10^{18} \text{ atomic units}$$

or   1 atomic unit = 1.60219×10^{-19} C

1 electron

In electrostatics we seldom meet charges as large as 1 C. Typically we shall be dealing with charges of the order of a microcoulomb (1 μC = 10^{-6} C) or even a picocoulomb (1 pC = 10^{-12} C).

3   The *electrostatic unit of charge,* sometimes called the *statcoulomb,* is especially convenient for dealing with electrostatic problems. It is defined in terms of Coulomb's law as follows: The electrostatic unit of charge is that charge which repels an identical charge one centimeter away in vacuum with a force of one dyne (10^{-5} N).

$$1 \text{ C} = 3 \times 10^9 \text{ electrostatic units of charge}$$

4   The *electromagnetic unit of charge,* sometimes called the *abcoulomb,* is defined in terms of the magnetic field produced at the center of a circular coil by a current.

$$1 \text{ C} = 0.1 \text{ electromagnetic unit of charge}$$

Complete systems of electrical units have been built around both the electrostatic and the electromagnetic units of charge. These systems are particularly convenient for handling certain special problems. However, it is difficult to learn three distinct systems of electrical units; therefore we shall use only the practical system in the problems of this text.

When the charges Q_1 and Q_2 are in coulombs, the force in newtons, and the distance r in meters, the constant k of Eq. (30.2) for a vacuum has the value 8.9878×10^9 N-m²/C², which in turn is $10^{-7}c^2$, where c is the speed of light in free space in meters per second. It is customary to write Coulomb's law in the form

$$F = \frac{1}{4\pi\epsilon_0} \frac{Q_1 Q_2}{r^2} \qquad (30.2a)$$

where $\epsilon_0 = 8.8543 \times 10^{-12}$ C²/N-m² is called the *permittivity of free space.* From the point of view of Coulomb's law, replacing k with $1/4\pi\epsilon_0$ appears to offer no advantage; however, we shall see in Chap. 33 that Eq. (30.2a) leads to simpler expressions for capacitances; the advantages become still greater in advanced electromagnetic theory.

As we have seen, in vacuum $1/4\pi\epsilon_0 = k$ is almost 9×10^9 N-m²/C². In air the constant in Coulomb's law is less than in vacuum by the factor 1/1.0006. For our problems we shall use

$$F = (9 \times 10^9 \text{ N-m}^2/\text{C}^2) \frac{Q_1 Q_2}{r^2} \qquad (30.2b)$$

for charges in either air or vacuum.

When several charges are present in one region, the force on any one of the charges is equal to the vector sum of the forces which each of the other charges would exert on the first one if it acted independently.

Figure 30.4
Charges of +25, +20, and −8 μC at
the corners of a 3-4-5 right triangle.

VERTICAL COMPONENT = sin θ x F

Example

Charges A, B, and C of +25, +20, and −8 μC, respectively, are arranged as shown in Fig. 30.4. Find the magnitude of the force on charge A.

$$F_{AB} = 9 \times 10^9 \frac{20 \times 10^{-6} \times 25 \times 10^{-6}}{25} = 0.180 \text{ N}$$

$$F_{AC} = -9 \times 10^9 \frac{25 \times 10^{-6} \times 8 \times 10^{-6}}{9} = -0.200 \text{ N}$$

The vertical component of F_{AB} is $\frac{3}{5} \times 0.180 = 0.108$ N; the horizontal component of F_{AB} is $\frac{4}{5} \times 0.180 = 0.144$ N.

The vertical component of the resultant force on A is $-0.200 + 0.108 = -0.092$ N. The horizontal component of the resultant is 0.144 N. The magnitude of the resultant is then

$$\sqrt{(0.144)^2 + (0.092)^2} = 0.17 \text{ N} \qquad \text{////}$$

30.6 ELECTRIC FIELD INTENSITY

Any region in which electric forces can be detected is called an *electric field*.

The *intensity* (or *strength*) **E** *of an electric field at any point is defined as the ratio of the force* **F** *acting on a small test charge q at that point to the charge q:*

$$\mathbf{E} = \frac{\mathbf{F}}{q} \qquad (30.3)$$

It is desirable that a very small test charge be chosen, since a large test charge would change the charge distribution which creates the field and thus distort the very thing it is being used to measure. Clearly, the electric intensity is a vector quantity. Its direction is the direction of the force on a *positive* charge.

The intensity of the electric field at a distance r from an isolated point charge Q is obtained by performing an imaginary experiment in which a small test charge q is placed at the point where the intensity is desired. By Coulomb's law the force on this test charge is $F =$

$Qq/4\pi\epsilon_0 r^2$. Since $\mathbf{E} = \mathbf{F}/q$, we conclude that the electric intensity at a distance r from a point charge Q is given by

$$E = \frac{Q}{4\pi\epsilon_0 r^2} \qquad (30.3a)$$

If an electric field is due to two or more charges, the electric intensity $\mathbf{E}$ at any point is given by the resultant of the electric intensities due to each charge taken individually.

Example

Find the magnitude of the electric intensity at A of Fig. 30.4 due to the charges at B and C.

We have found the magnitude of the force on a charge of 25 μC at this point to be 0.17 N. By Eq. (30.3),

$$E = \frac{F}{q} = \frac{0.17 \text{ N}}{25 \ \mu\text{C}} = 6{,}800 \text{ N/C} \qquad \text{////}$$

Example

Find the electric field strength at the point P of Fig. 30.5 which is 10 cm from a charge of $+6 \ \mu$C and 40 cm from a charge of $-8 \ \mu$C.

The electric field due to the $+6$-μC charge has a magnitude $9 \times 10^9 \times 6 \times 10^{-6}/(0.1)^2 = 54 \times 10^5$ N/C. The field due to the -8-μC charge has a magnitude $9 \times 10^9 \times 8.0 \times 10^{-6}/(0.4)^2 = 4.5 \times 10^5$ N/C. The fields due to both charges are in the same direction—to the right. Therefore, the resultant has a magnitude of 58.5×10^5 N/C and is directed toward the right. \qquad ////

30.7 LINES OF FORCE

The electric field in the vicinity of one or more charged bodies is frequently represented by drawing *lines of force* as an aid in visualizing the field. *An electric line of force is a line drawn so its tangent at every point has the direction of the electric intensity at that point.* Some of the lines of force associated with a pair of equal and opposite charges are shown in Fig. 30.6, while some due to two identical positive charges are indicated in Fig. 30.7. In electrostatics lines of force always begin on positive charges and terminate on negative charges. The lines of Fig. 30.7 terminate on negative charges which might, for example, be on the walls of the room.

We may draw as many lines of force as we wish in picturing an electric field. A reasonable number of lines can give us a visualization of the direction of the force on a small positive test charge any place

Figure 30.5
Electric field strength $\mathbf{E}$ at point P is the resultant of the fields due to the $+6$- and -8-μC charges.

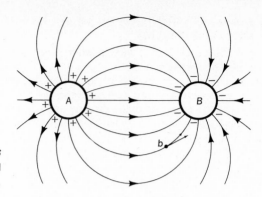

Figure 30.6
Lines of force for a pair of equal and opposite charges.

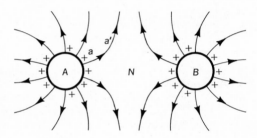

Figure 30.7
Lines of force for two identical positive charges; at the *neutral point* N the field intensity is zero.

in the field. The line of force through a point tells us the direction of the electric intensity at that point but nothing of its magnitude. However, a qualitative estimate of the magnitude of the field can be obtained from a reasonably detailed plot which shows the lines of force approaching one another as the field becomes stronger. (Why does this happen?) Indeed, in an ideal three-dimensional plot, the number of lines of force per unit area perpendicular to the lines would be directly *proportional* to the electric intensity.

30.8 ELECTROSTATIC INDUCTION

If an uncharged conductor B (Fig. 30.8) is brought into the electric field of a positively charged conductor A, the attractive forces due to the excess positive charge on A cause the electrons in B to be pulled toward end C, leaving the farther end D with a deficit of electrons and therefore charged positively. Since B was originally neutral, i.e., contained as much positive as negative electricity, the positive charge on end D is

Figure 30.8
Equal and opposite charges are induced at ends C and D when a positive charge A is brought near the uncharged conductor B.

just equal to the negative charge on end C. Although B bears no net charge, it is attracted by A.

If conductor B is connected to earth by a wire, enough electrons come from the earth to B to neutralize the positive charge at D. Meanwhile the electrons on end C are held fast by the attractive force due to the positive charge on A. If the connection to the earth is broken and B removed from the presence of A, B will have an excess of electrons; it has been negatively charged by electrostatic induction.

A charged sphere in the neighborhood of the surface of the earth (Fig. 30.9) induces a charge on the surface of the earth below it. A charged antenna in the neighborhood of the surface of the earth has a similar effect.

Another illustration involving electrostatic induction is seen in Fig. 30.10. If a metal sphere, charged positively, is introduced into an insulated uncharged hollow sphere, some of the electrons of the hollow sphere are drawn to its inner surface, leaving the outer surface charged positively. If the metal sphere A is placed in contact with the inner surface of the hollow sphere, the electrons from the inner surface of the hollow sphere go over to the sphere A and just compensate the deficit of electrons. This leaves both sphere A and the inner surface of the hollow sphere without a charge, while the outer surface of the hollow sphere is charged positively. If the outer surface is now connected to the earth, it gains a sufficient number of electrons to compensate for its deficit, and it is left uncharged. By experiments of this type Faraday was able to show that the charge induced on the outside of an almost closed hollow conductor is equal to the inducing charge on the inside. As his hollow conductor Faraday used a small ice pail, and the operations described in this paragraph are often referred to as the "Faraday ice-pail experiment."

30.9 THE ELECTROSCOPE

A useful instrument for studying electrostatic phenomena is the electroscope, one form of which consists of two thin leaves of gold foil attached to one end of a metal rod which is terminated at the other end by a metal sphere (Fig. 30.11). When the metal sphere is charged, part of the charge goes to the gold foils, which repel each other and therefore diverge. The greater the charge on the leaves, the greater the divergence.

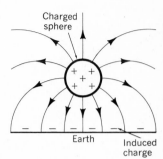

Figure 30.9
A positively charged object induces a negative charge on the earth.

Figure 30.10
Faraday's ice-pail experiment shows that the charge induced on the inside wall of the cavity is equal and opposite to the inducing charge on sphere A, since the external field is not affected by touching A to the inside of the pail and thereby discharging A.

Figure 30.11
A gold-leaf electroscope.

An electroscope can be charged positively or negatively by touching the knob with either a positive glass rod or a negative rubber rod, but there is a substantial danger of transferring so much charge that the gold leaf will be torn off. A safer and easier way to charge an electroscope is by induction. To charge an electroscope positively by induction, a negatively charged rod is brought near the knob, as shown in Fig. 30.12a. Electrons in the knob are repelled by the rod and descend to the leaves. The knob is then grounded by touching it (Fig. 30.12b), and the excess electrons on the leaves run off. The connection to ground is then removed, leaving a net positive charge on the electroscope knob. When the negative rod is removed, these charges rearrange themselves so that both the knob and leaves are positively charged (Fig. 30.12c).

When a negatively charged rod is brought near a positively charged electroscope, some of the electrons on the metal sphere are repelled and move to the leaves, where they reduce the net positive charge and hence the divergence of the leaves (Fig. 30.13). As the negative rod is brought still nearer, more electrons go to the leaves, which eventually lose all their charge and converge. If the negative rod is brought still closer, even more electrons go to the leaves, causing them to diverge once more. In this case, although the total charge on the electroscope is positive, the leaves bear a negative charge.

Figure 30.12
Charging an electroscope by induction.

Figure 30.13
Charge distribution on an electroscope changes but the total charge remains constant as a charged rod comes nearer and nearer.

1 Can two lines of force ever cross? Explain your answer carefully.

2 Why do electrostatic experiments often work poorly in hot, humid rooms?

3 A small charged pith ball is brought near to an uncharged metal sphere, and an attractive force is evident. Why? If the sphere is grounded, the attraction is still greater. Why?

4 When a charged rubber rod is brought near small pieces of cork, some bits of cork may at first cling tightly to the rod, but soon some bits may jump off the rod. What has happened? Why don't all the bits of cork fly off?

5 Does it make any difference which end of conductor B in Fig. 30.8 is grounded by the wire? Justify your answer with an explanation.

6 In what respects are electric and gravitational forces similar? In what respects are they different?

7 How could you charge an electroscope positively with a negatively charged rod? Explain what happens during each step of the process.

8 What does it mean to say that charge is always conserved? After all, you can have a system of several uncharged bodies and make one in which every body is charged.

9 In one form of Cottrell precipitator for capturing smoke particles, a highly negative wire is suspended vertically in a rising column of polluted air. The wire produces a strong inhomogeneous electric field. Why does such an arrangement remove smoke and soot?

10 How can one prove that the force between two charges is proportional to the product of the charges? If someone refused to believe this, what experiment could you devise to convince him?

11 Three point charges are equally spaced along a line. Q_2 and Q_3 are equal in magnitude but opposite in sign. What is the magnitude of Q_1 if the net force on Q_3 is zero? Is the sign of Q_1 the same as that of Q_2 or of Q_3?

PROBLEMS

1 The force on a small test charge is 2.4×10^{-6} N when the charge is placed in an electric field of intensity 6×10^5 N/C. Find the magnitude

of the charge. How many electrons would be required to neutralize this charge?

Ans. 4×10^{-12} C; 2.5×10^7

2 Three charges of $+5$, -3, and $+6$ μC are placed in the same straight line 10 cm apart. What force acts on each charge because of the other two?

3 A point charge of 6×10^{-9} C is located 0.3 m from a point charge of 4×10^{-9} C. Find the force exerted on each charge by the other and the electric field strength at a point 0.2 m from the larger charge on the line connecting the two charges.

Ans. 2.4×10^{-6} N; 2,250 N/C toward larger charge

4 Charges of -6 and $+8$ μC are 2 m apart. Find the force they exert on one another and the electric intensity midway between them.

5 Two identical metal spheres have a mass of 1 kg each. How many electrons must be transferred from one sphere to the other in order to make the electrical attraction between the spheres equal to the gravitational attraction when the spheres are 1 m apart? *Ans.* 5.4×10^8

6 How many electrons must be removed from a small pith ball to give it a charge of 2×10^{-12} C? What is the electric field strength 5 cm from the pith ball?

7 A point charge of 0.015 μC is placed 0.6 m from a point charge of 0.02 μC. What force is exerted on each charge? Find the electric field strength at the point midway between the charges.

Ans. 7.5×10^{-6} N; 500 N/C toward smaller charge

8 In a Bohr model of the hydrogen atom, an electron revolves in a circular orbit of radius approximately 5.28×10^{-11} m about a nucleus which bears a positive charge equal in magnitude to the electronic charge. Find the force on the electron which provides the centripetal acceleration for the uniform circular motion. How large is the centripetal acceleration?

9 The electric intensity in the region between two deflecting plates of an electrostatic deflection television tube is 40,000 N/C. Find the force on an electron passing between these plates.

$W = $ (handwritten) $m \cdot g \cdot z$

$E = $ (handwritten) $\frac{\Delta P}{PB}$ $^{-11} N \cdot \frac{m^2}{kg^2}$

$\frac{E}{m \cdot g}$ (handwritten)

What acceleration does the electron experience if it has a mass of 9.11×10^{-31} kg?

Ans. 6.4×10^{-15} N; 7.03×10^{15} m/sec^2

10 Calculate the ratio of the coulomb electric force to the newtonian gravitational force between two electrons if the mass of an electron is 9.11×10^{-31} kg. What electric intensity is required to balance the gravitational force on an electron?

11 An unknown charge and a charge of -3 μC are 1 m apart. The electric field strength is zero at a point 40 cm from the unknown charge on the line connecting the two charges. Find the unknown charge. *Ans.* -1.33 μC

12 Two point charges repel each other with a force of 4×10^{-4} N when they are 30 cm apart. Find the force if the distance between them is reduced to 15 cm. If one of the charges is -8×10^{-10} C, what is the other?

13 Find the force on a charge of 4×10^{-9} C if it is placed in a uniform field of intensity 55,000 N/C. How much work is done *by the field* if the charge moves 0.3 m in the direction of the field? Perpendicular to the field?

Ans. 2.2×10^{-4} N; 6.6×10^{-5} J, zero

14 A small test charge of 10^{-11} C experiences a force of 2×10^{-6} N when it is 1.5 m from a point charge of unknown magnitude. Find the electric intensity at the test charge due to the unknown charge and the magnitude of the unknown charge.

15 A charge of 0.4 μC is placed in a downward-directed electric field of intensity 20,000 N/C. Find the work required to move the charge (*a*) 0.5 m to the east, (*b*) 0.5 m upward, and (*c*) 2 m upward at an angle of 37° with the vertical.

Ans. (*a*) zero; (*b*) 0.004 J; (*c*) 0.0128 J

16 Charges of -4 and $+6$ μC are placed at two of the vertices of an equilateral triangle with sides 10 cm in length. Find the magnitude of the electric intensity at the third vertex and the force which would act on a charge of 3×10^{-9} C at that vertex.

17 Three charges A, B, and C are located on the same straight line. The distance from A to B is 20 cm, and that from B to C is 40 cm. The charge at A is 0.04 μC, that at B is 0.08 μC, and that at C is 0.06 μC. What force is exerted on A by the charges at B and C if all the charges are positive? What is the electric intensity midway between A and B?

Ans. 7.8×10^{-4} N; 38,160 N/C toward A

18 Charges of $+6$, -4, and $+5$ μC are placed on the corners of a square with sides 20 cm long. Find the magnitude of the electric field strength at the fourth corner, which is diagonally opposite the -4-μC charge.

19 Two small equally charged spheres, each with a mass of 0.24 g, are suspended from the same point by silk fibers 75 cm long. The repulsion between them keeps them 10 cm apart. What is the charge on each sphere?

Ans. 1.32×10^{-8} C

20 Two small pith balls, each weighing 0.002 N, are hung from a common point by nylon threads 15 cm long. When the pith balls are given equal positive charges, they repel each other and stand 18 cm apart, so that each of the supporting threads makes an angle of 37° with the vertical. Find the charge on each pith ball.

21 A charge of 0.2 μC is placed at the origin of a cartesian coordinate system, and a charge of -0.4 μC is placed at $x = 2$ m, $y = 0$. Find the electric intensity at the points (*a*) $x = 4$ m, $y = 0$, and (*b*) $x = 2$ m, $y = 2$ m.

Ans. (*a*) 787.5 N/C toward origin; (*b*) 758 N/C at $-77.9°$ with x axis

22 Charges of 3×10^{-10} C are placed at opposite ends of the hypotenuse of a right triangle with sides 3, 4, and 5 m. Find the magnitude of the electric intensity at the other vertex.

23 Identical 5-μC charges are placed at opposite ends of a diagonal of a rectangle 0.30 by 0.40 m. Find the values of the charges Q_1 and Q_2 which must be placed at the other two corners to make the resultant force on one of the 5-μC charges zero. Will the force on the other 5-μC charge then also be zero? If not, what will its magnitude be?

Ans. -2.56 μC; -1.08 μC; no; 0.85 N

24 An electric dipole consists of a charge $+Q$ separated by a distance d from a charge $-Q$. Find the torque on this dipole when it is placed in a uniform field of intensity **E** (*a*) with its axis perpendicular to **E** and (*b*) with its axis making an angle θ with **E**.

FROM MECHANICS we know that the best approach to many problems is through the concept of energy and its conservation. By using the energy concept we avoid the need of knowing and working with force, which may be changing rapidly in space or in time. In electricity practical problems are conveniently solved by considering changes in the energy of a charge as it moves from place to place, rather than focusing our attention on the force experienced by the charge at each point in its path. In this chapter we use Coulomb's law of force between electric charges to find the potential energy of a small test charge in the field of a point charge. This leads us to the concept of *potential*, one of the key ideas of electricity.

CHAPTER THIRTY-ONE
POTENTIAL

31.1 POTENTIAL ENERGY IN AN ELECTRIC FIELD

When a small test charge $+q$ is moved about in the field of a fixed charge $+Q$ (Fig. 31.1), work must be done to carry the test charge closer to Q. This work goes into increasing the potential energy of the test charge. On the other hand, as q moves away from Q, the electric field does work, and the potential energy of q decreases. In discussing the movement of electric charges, the concept of electric potential energy plays an important role, since many problems in electricity involve the law of conservation of energy, and electric potential energy is one of the forms which must be taken into consideration.

In expressing the electric potential energy of a charge quantitatively, it is necessary to specify some point at which the potential energy is zero, just as it is necessary to specify a zero configuration in dealing with any other form of potential energy. For example, the potential energy of a given mass relative to the earth (Chap. 7) is proportional to the elevation h. In this case it is necessary to decide from what level h is measured. The potential energy of a 1-kg mass 2 m above a table depends on whether one wishes to use the tabletop or some other level to establish the zero for potential energy. In dealing with the particular situation illustrated in Fig. 31.1, it is common to

FIND THE ZERO POINT

Figure 31.1
Work must be done to move a small test charge $+q$ from point A to point B against the electric intensity of charge $+Q$.

consider the potential energy of q to be zero when it is infinitely far from Q. This choice is arbitrary but conventional.

Having chosen the configuration for zero potential energy, we define the potential energy of a charge $+q$ at point A as the work necessary to move the charge from the point at which its potential energy is zero to the point A. If the potential energy of a charge q at a given point is P joules, the potential energy of a charge $3q$ at this same point is $3P$ J, and, in general, for charge nq the potential energy at this point is nP. Since the potential energy of a charge at a given point is directly proportional to the charge itself, it is convenient to introduce the concept of *potential energy per unit charge*. The ratio of the potential energy of a charge at a given point to the charge is called the *potential at the point*. In electrostatics the zero of potential is usually chosen to be at infinity. For this choice the following definition applies: *The potential at a point A is the ratio W/q, where W is the work required to move a small test charge +q from infinity to the point A.*

The practical unit of potential, the *joule per coulomb*, is called the *volt* (V), in honor of Alessandro Volta, whose pioneering work with batteries was a major contribution to the infant science of electricity in 1800. *The potential at a point is one volt when it requires one joule of work to move a positive charge of one coulomb from a point of zero potential to the point in question.*

31.2 THE POTENTIAL DUE TO A POINT CHARGE

The potential at a distance r from a point charge of Q coulombs is the *work per unit charge* required to bring a positive test charge from an infinite distance to the point in question. If q is the test charge being transferred, the repelling force exerted on it by Q is $Qq/4\pi\epsilon_0 r^2$. To move q against this force we must apply a force of this same magnitude toward Q. The element of external work ΔW required to move q a distance Δr is given by $\Delta W = (-Qq/4\pi\epsilon_0 r^2)\,\Delta r$ (Fig. 31.2). If Δr is positive (r increasing), this work is negative: Q provides the required force, and no positive external work need be performed. However, when Δr is negative (r decreasing), external work must be done against the electric field of Q and ΔW is a positive quantity.

Since the repelling force on q varies with r in the same way as the gravitational attractive force on a small mass in the earth's gravitational field (Sec. 9.9), we may use the reasoning of that section to conclude that the work[†] required to move q from infinity to a distance r from charge Q is

$$W = \frac{Qq}{4\pi\epsilon_0}\left(\frac{1}{r} - \frac{1}{\infty}\right) = \frac{1}{4\pi\epsilon_0}\frac{Qq}{r} \qquad (31.1)$$

[†] It is assumed that all operations in this chapter are carried out in air or *in vacuo*.

Figure 31.2
The external work ΔW required to move a small test charge $+q$ a distance Δr in the field of $+Q$ is given by $\Delta W = -(Qq/4\pi\epsilon_0 r^2)\,\Delta r$. When q is moved from A to B, $\Delta \mathbf{r}$ is negative and ΔW is a positive quantity.

This is the potential energy of the test charge relative to zero at $r = \infty$.
The potential at distance r from Q is then

$$V = \frac{W}{q} = \frac{1}{4\pi\epsilon_0}\frac{Q}{r} = (9 \times 10^9 \text{ V-m/C})\frac{Q}{r} \qquad (31.1a)$$

The potential 1 m from a charge of 1 C in vacuum is 9×10^9 V.

Example

Find the potential 2 m from a charge of $+6$ μC.

$$V = \frac{9 \times 10^9 Q}{r} = \frac{9 \times 10^9 \times 6 \times 10^{-6}}{2} = 27{,}000 \text{ V} \qquad ////$$

When there are a number of charges in a region, the potential at any point is the sum of the potentials due to each charge acting alone. Thus if there are three charges Q_1, Q_2, and Q_3 in a region, the potential at a point which is at distance r_1 from Q_1, r_2 from Q_2, and r_3 from Q_3 is simply given by

$$V = \frac{1}{4\pi\epsilon_0}\left(\frac{Q_1}{r_1} + \frac{Q_2}{r_2} + \frac{Q_3}{r_3}\right) \qquad (31.2)$$

The contribution of a negative charge to the potential is negative; the electric field does work in bringing a positive test charge toward a negative charge. A great advantage of working with potential in preference to electric intensities arises from the fact that *potential is a scalar* while the *electric intensity is a vector.*

31.3 THE ISOLATED SPHERE

Consider a metal sphere of radius R, far away from all other bodies and bearing a charge of Q C. The electric field associated with this charged sphere is everywhere radial and, for all points outside the sphere, is indistinguishable from the electric field which would be associated with a point charge Q located at the center of the sphere. Thus, a uniformly charged sphere behaves, *so far as all external points are concerned,* exactly as though all its charge were concentrated at the center. The work necessary to bring a unit charge from infinity to the surface of the sphere is the same as the work which would be required to bring the same unit charge from infinity to the distance R from a point charge Q. The potential of the sphere is therefore given by

$$V = \frac{1}{4\pi\epsilon_0}\frac{Q}{R} = (9 \times 10^9 \text{ V-m/C})\frac{Q}{R} \qquad (31.3)$$

Within the sphere the electric field is zero (see Sec. 31.7). Therefore, no work is required to carry a small test charge from any point on the surface of the sphere to any point within the sphere. *The potential at all points inside the sphere is the same as the potential at the surface.* The electric field and the potential are plotted as a function of r in Fig. 31.3.

[handwritten annotations in right margin:]

$d\gamma = -E\cos\theta\, ds$

$\dfrac{dv}{ds} = -E\cos\theta.$

THE NEGATIVE CHARGES CHANGE THE WHOLE FACE OF V

IT IS AS IN ANY OTHER SURFACE

Figure 31.3

The electric intensity E and potential V at a distance r from the center of a uniformly charged conducting sphere of radius R.

31.4 SURFACE CHARGE DENSITY

When a sphere of radius R bears a charge of Q C, the charge per unit area is given by

$$\sigma = \frac{Q}{4\pi R^2} \tag{31.4}$$

where σ is known as the *surface charge density*.

Consider a large sphere and a small one, both charged to the same potential V and sufficiently far apart for the electric field of either to be negligible at the other. If the charges on the two spheres are then Q and q, and the radii of the two spheres are R and r, respectively, V is given by

$$V = \frac{Q}{4\pi\epsilon_0 R} = \frac{q}{4\pi\epsilon_0 r}$$

Let σ_R and σ_r be the charge densities on the respective spheres. Then

$$V = \frac{4\pi R^2 \sigma_R}{4\pi\epsilon_0 R} = \frac{4\pi r^2 \sigma_r}{4\pi\epsilon_0 r}$$

and

$$R\sigma_R = r\sigma_r \tag{31.5}$$

For isolated spheres at the same potential, the charge densities are inversely proportional to the radii.

The surface of any conductor may be regarded as composed of a large number of spherical segments (Fig. 31.4) of different curvatures. While we may not regard these segments as portions of isolated spheres

(handwritten margin notes) (σ) THIS STRANGE SIGN IS USED TO DETERMINE THE SURFACE CHARGE DENSITY

$\sigma = \frac{q}{4\pi r^2}$

Figure 31.4

As a first approximation, the surface of an arbitrary curved conductor is assumed to be made up of sections of spherical surfaces; charge density becomes greater as the radius of curvature becomes smaller.

Figure 31.5
Flashover on a sealed bushing.
[*General Electric Company.*]

in order to apply Eq. (31.4) to calculate relative charge densities, it remains qualitatively true for convex surfaces such as those of Fig. 31.4 that where the surface has greatest curvature, i.e., smallest radius, the surface charge density is greatest. When a conductor terminates in a sharp point, the surface density at the point is so great that the molecules of air in the neighborhood of the point may become charged with electricity. Since like charges repel, the charged molecules are repelled from the point and the body to which the point is attached is discharged. The ionized molecules may emit light which is readily visible in a darkened room. Such a discharge is called a *corona discharge.* If the field is sufficiently great, flashover occurs (Fig. 31.5).

The fact that charged points allow electricity to escape is used in electrostatic machines, where rows of metallic points conduct electricity from moving to fixed parts of the machine. This fact is also important in the design of lightning rods. The pointed conductors on the rod bring about a silent and gradual discharge from the rod to the clouds. The escape of electricity from these points prevents the accumulation of enough electricity on the building on which the lightning rod is mounted to result in a dangerous disruptive discharge.

31.5 POTENTIAL REFERRED TO THE EARTH

In dealing with potential, it is sometimes convenient to assign zero potential to some conducting body, just as it is convenient to refer the height of a building to the ground level, while the height of a mountain is ordinarily referred to mean sea level. In many situations the potential of the earth is taken as zero. When one speaks of a 110-V lighting circuit or of a 180,000-V transmission line, one ordinarily implies that the earth is taken as zero potential. In a radio or television receiver the chassis is usually regarded as being at zero potential; often it is actually tied to the ground by a direct conductor. In dealing with the electric circuits of an automobile, it is convenient to regard the frame of the car as being at zero potential, although it is ordinarily insulated from the earth by the tires.

In general, one may assign the potential zero to any convenient point. Then the potential at any other point is the ratio W/q, where W is the *work* required to move a small charge q from the point at zero potential to the point in question.

31.6 POTENTIAL DIFFERENCE

An important concept in electrical theory is that of the *potential difference* between two points A and B, which is simply the potential at A minus the potential at B, just as the name *potential difference* implies. Thus, $V_{AB} = V_A - V_B$. Alternatively, *the potential difference between points A and B is the external work per unit charge done against the electrostatic field in moving a small positive test charge from B to A.* The potential difference is independent of the zero of potential. For example, the potential difference between the positive and negative terminals of an ordinary dry cell is 1.5 J/C, regardless of whether the earth or some other point is taken as the zero potential reference. The potential of the positive terminal relative to the negative terminal is +1.5 V, while the potential of the negative terminal relative to the positive is −1.5 V.

Let W be the external work done against an electric field in moving a small positive charge q from point B to point A. The potential difference V_{AB} between these two points is, by the definition above, the ratio of W to q:

$$V = \frac{W}{q} \qquad (31.6)$$

Because of its great importance, we rephrase our definition: *The potential difference between two points is the work per unit positive charge required to move a small test charge from one point to the other.*

Potential difference, as well as potential, is commonly measured in volts (joules per coulomb). *The potential difference between two points is one volt if it requires one joule of external work to move one coulomb of charge from one point to the other.* The potential difference across the terminals of an ordinary automobile battery is 12 V; this means that 12 J of external work is required to transfer 1 C of positive charge from the negative terminal to the positive terminal. This energy is supplied by the chemical reaction within the cells, so that chemical energy is converted into electric energy.

In any electric field the direction of the field is from points at higher potential toward points at lower potential. External work must be supplied to move a positive charge from a point at a lower potential to a point at higher potential. Suppose a small positive test charge is moved from B to A against an electric field and then returned to B. External work is required to transfer the charge from B to A; when the charge returns to B, work is done on the charge by the electric field. Since the force on a negative charge in an electric field is opposite that on

ELECTRIC FIELD AND DIFFERENTIAL POTENCIAL ARE INVERSE, WHILE THE FIRST WORK FROM THE HIGUER TO THE LOWER STATES THE 2nd WORKS FROM THE LOWER TO THE HIGHER STATES

a positive charge, the field moves negative charges from points of lower toward points of higher potential; external work must be done to carry a negative charge from a point of higher to one of lower potential.

31.7 EQUIPOTENTIAL SURFACES

It is often convenient to represent the potential distribution in an electric field by means of equipotential surfaces. An equipotential surface is defined as a surface whose points are all at the same potential. Some equipotentials for a spherical charge distribution are shown in Fig. 31.6. In this particular case the equipotential surfaces are spheres, while the lines of force are radial. The lines of force are always perpendicular at every point to the equipotential surfaces. This follows immediately from the fact that it takes no work to move a small test charge from one point to another on the same equipotential surface. The electric field has no component in the direction of the surface and hence must be perpendicular to it.

In Fig. 31.6, A and B lie on the same equipotential surface, while C lies on another equipotential surface. No work is done in moving a small test charge from A to B along the equipotential surface, since the displacement is at all points perpendicular to the force on the charge. However, if a charge is moved from C to A, work must be done; regardless of what path is chosen, the force due to the field must have a component in the direction of this displacement for at least some part of the movement. The work required to transfer a charge from one point to another is independent of the path. The work required to move a small test charge from C to A is the same whether one goes first along the line of force to the equipotential surface AB and then to A, or from C to A along a straight line, or to B and then over to A.

Lines of force and equipotential surfaces are shown in Fig. 31.7 for the region around two equal point charges of opposite sign.

In a conductor there is a flow of charge if there is a potential difference between any two points. This flow stops as soon as all points in the conductor are at the same potential. In electrostatics, charges are at rest, and therefore there is no potential difference between points in the conductor. The entire conductor is an equipotential surface. Since there can be no electric field within the conductor, there is no net charge within the conductor. *The entire net charge of the conductor*

Figure 31.6
Lines of force (radial) and equipotential surfaces (spheres) for a spherical charge distribution.

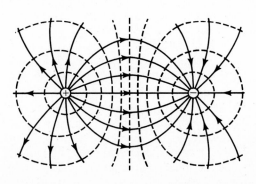

Figure 31.7
Lines of force (solid lines) and equipotential surfaces (dashed lines) near two equal but opposite charges.

Figure 31.8
A 4-MeV Van de Graaff electrostatic
generator. [*High Voltage Engineering
Corporation.*]

resides on the surface. Any excess or deficiency of electrons is at the surface. Since the surface of a conductor is an equipotential surface, and since lines of force are always perpendicular to equipotential surfaces, *lines of force always leave perpendicular to the surface of a conductor* when charges are at rest.

31.8 THE VAN DE GRAAFF ELECTROSTATIC GENERATOR

An important application of the fact that charges go to the outside of a conductor is the Van de Graaff generator (Fig. 31.8), a major tool of nuclear physics. A belt made of an insulating material carries charges at high speed from a charging source into a rounded metal shell (Fig. 31.9), where they are picked off; the charges then go to the outside of the shell. The potential of the high-voltage electrode builds up until charges are lost from the shell and gained at exactly the same rate. Often the entire system is placed in a large tank so that the pressure can be raised to several atmospheres to reduce the loss of charge by sparks and corona.

If the Van de Graaff generator is to be useful for physics experiments, one must utilize the high potential to accelerate some kind of charged particle, or *ion*. A Van de Graaff electrostatic accelerator consists of a Van de Graaff generator, an ion source, and an evacuated tube down which the ions are accelerated. When high-energy electrons are desired, the electron source is usually a hot filament inside the high-voltage electrode, which is made negative. For exciting nuclear reactions by bombarding targets with protons, an ion source involving an electric discharge through hydrogen gas is installed in the high-voltage electrode, which is charged positive. In either case charged particles are accelerated down the evacuated tube to the grounded end, where they are focused on a target.

Figure 31.9
Schematic diagram of a Van de Graaff
electrostatic accelerator.

When a particle of charge q falls through a potential difference V in a vacuum, it gains a kinetic energy Vq. This energy is in joules if V is in volts and q is in coulombs. However, if we express q in atomic units of charge, the energy is given in electron volts (eV). *An electron volt is the kinetic energy gained by a particle bearing one atomic unit of charge in falling through a potential difference of one volt.* The negative charge on the electron and the positive charge on the proton both have the magnitude of one atomic unit of charge. Since the atomic unit of charge corresponds to 1.602×10^{-19} C,

$$1 \text{ eV} = 1.602 \times 10^{-19} \text{ C} \times 1.000 \text{ V}$$
$$= 1.602 \times 10^{-19} \text{ J}$$

Electrostatic generators are highly valuable sources of particles with energies from many thousands of electron volts (keV) to several million electron volts (MeV).

31.9 UNIFORM ELECTRIC FIELD

Consider two large parallel plates, one bearing a charge $+Q$, and the other a charge $-Q$ (Fig. 31.10). The electric field between the two plates (reasonably far from an edge) is constant in both magnitude and direction. Such a field is said to be *uniform*. When a charge q is carried from the lower plate to the upper one, an upward force of magnitude qE is required and the work done against the field is Eqd, where d is the distance between plates. The potential difference V between the plates is $W/q = Ed$. Solving for E yields

$$V = Ed.$$

$$E = \frac{V}{d} \tag{31.7}$$

Electric field strengths may be expressed in units of potential difference divided by distance as well as in terms of the equivalent force divided by charge; they are often quoted in *volts per meter*. (An electric intensity of 1 V/m is exactly the same as one of 1 N/C.) If the test charge q is moved from X to Y (Fig. 31.10) the work done is $W = -(q\mathbf{E}) \cdot \mathbf{s} = -qEs \cos \theta = qEs \cos \varphi$ and the potential difference between Y and X is $V_{YX} = Es \cos \varphi$.

The ratio of the potential difference ΔV between two neighboring points along a line of force to the displacement $\Delta \mathbf{s}$ between the points is a vector called the *potential gradient*. The electric intensity $\mathbf{E}$ is the negative of the potential gradient. The component of the potential gradient in any direction is the negative of the component of the electric intensity in that direction. The negative sign arises from the fact that the potential energy of a positive charge decreases when it is displaced in the direction of the electric vector $\mathbf{E}$.

31.10 THE CHARGE ON THE ELECTRON

The uniform electric field between parallel charged plates was utilized by Millikan and his students to determine the charge on the electron. An extensive series of measurements performed between 1909 and 1917

$$1 \frac{V}{m} = 1 \frac{N}{C}$$

$$\sigma A = q$$

$$D = \sigma = \varepsilon_0 E$$

$$E = \frac{D}{\varepsilon_0} = \frac{\sigma}{\varepsilon_0}$$

$$V = Es = \frac{\sigma}{\varepsilon_0} \cdot s$$

Figure 31.10

The electric field between two parallel charged plates is uniform except at the edges, where fringing occurs.

Figure 31.11

Millikan apparatus for measuring the charge on the electron.

$$\frac{q}{V} = \frac{\sigma\,A}{\frac{\sigma}{\varepsilon_0}\cdot S} = \frac{\varepsilon_0\cdot A}{S}$$

$$C = \frac{q}{V}$$

$$C = \frac{K_e\,\varepsilon_0\,A}{S}$$

produced evidence that every electron bears the same charge. Millikan's apparatus is shown schematically in Fig. 31.11. Two horizontal plates B and C are placed about 1 cm apart. An atomizer A shoots a fine spray of oil into the space above these plates. The drops of oil are so small that they do not settle for a long time. Eventually one or more of the drops finds its way through the opening O in the upper plate. A telescope is focused on this drop so that its movements can be observed over a long period of time. The rate at which the drop falls can be measured by means of micrometer eyepiece in the telescope.

A beam of x-rays is next sent into the air between the horizontal plates. By means of these x-rays, electrons are detached from the atoms of air. One or more of these electrons may be captured by oil drops. When the upper plate B is charged positively and the lower plate C negatively, there is an upward force on a negatively charged oil drop. If the electric field is sufficiently strong, this upward force may equal or exceed the weight of the drop. Under such conditions the drop may be made to hang in the air for a substantial time. If the downward pull of gravity and the upward pull of electrostatic field leave the oil drop essentially at rest, we may write

$$w = Eq = \frac{Vq}{d}$$

where w is the weight of the drop, q the charge on the drop, E the electric field strength, V the potential difference between the plates, and d their separation. By determining w, V, and d, it is possible to calculate the charge q. Frequently, a number of electrons may be attached to the drop, and, as a result, q is several times the charge on a single electron. Careful observations show that whatever charges are present, their magnitudes are always integral multiples of a single elementary charge. This charge e is the smallest known unit of electricity. It is the fundamental unit of electric charge. Every electron bears a charge $-e$, where

$$e = 1.602 \times 10^{-19}\ C$$

The procedure just described is an oversimplification of the method used by Millikan, who actually measured the velocity of the oil drop as it moved upward under the influence of the electric field and the velocity of the same drop as it moved downward due to gravity with the electric field turned off. The calculations are straightforward, but somewhat involved; they lead to the result quoted above.

On a clear day the surface of a level field freely exposed to the sky is negatively charged, so that there is an electric field downward. Between the earth and the ionosphere there is typically a potential difference of roughly 360,000 V. In good weather, the electric field strength at the ground is about 100 V/m. This potential gradient changes continually. In addition to local variations, there are well-defined annual and diurnal variations, which differ at different parts of the surface of the earth. Above the earth's surface, the potential gradient diminishes, and at a height of about 10 km it approaches zero, a fact explained by the rapid increase in the conductivity of the atmosphere at higher altitudes.

The earth's negative charge is believed to come largely from lightning and point discharges under clouds. During thunderstorms the violent, ascending air currents in thunderheads (Fig. 31.12) break large raindrops into small ones. In this process the water droplets become positively charged, and the air molecules negatively charged. Electrification occurs particularly strongly in the area of the thunderhead, where the temperature is a few degrees below zero and tiny ice crystals abound. Water droplets are driven against ice crystals; the ice crystals become strongly negative, while the residue of the water droplets is positive. These tiny droplets are carried to great elevations, where they freeze. The result is that the top of the thunderhead has a large positive charge which eventually spreads out in the upper atmosphere. The region near the 0°C isotherm is strongly negative, while the falling raindrops are positively charged. Most of the lightning flashes to earth are from the lower negative center to ground, and these flashes bring the earth its negative charge. The earth would require 500,000 C of negative charge to produce a potential gradient of 100 V/m over its entire surface. The gradient is not the same all over the earth, and it is actually reversed over substantial areas during lightning storms.

Lightning strokes occur not only between charged clouds and the earth but also between the lower negative charge center and the upper positive region.

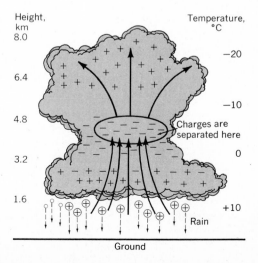

Figure 31.12
Clouds are electrified when ascending air currents in thunderheads produce charges by contact and separation.

QUESTIONS

1 Does an electron released at rest in an electric field move toward regions of higher or of lower potential? Explain.

2 If a body bears a positive charge, is its potential necessarily greater than zero? Devise a situation in which a positively charged body has a negative potential relative to the earth.

3 Why must any net charge on a perfect conductor reside at the surface?

4 Is it possible for the potential at a point to be zero while there is a finite electric intensity at that point? Show a charge configuration which satisfies these conditions.

5 Is it possible for the electric intensity at a point to be zero while the potential at that same point is nonzero? Show a configuration of charge which satisfies these conditions at some point.

6 Can charges be arranged so that both the electric intensity and the potential are zero at some point? Show an arrangement of charges which satisfies these conditions at some point.

7 What is the theory behind the use of lightning rods? Explain how they may save a building from a stroke of lightning.

8 Why must the electric intensity at the surface of an ideal conductor be perpendicular to the surface?

9 A metal cube bears a positive charge. What is the shape of the equipotential surfaces very close to the cube? Very far away from the cube? How does the charge distribute itself over the cube?

10 It is sometimes stated that under static conditions all charges go to the surface of a conductor. What is implied? Do all the electrons and protons of the atoms go to the surface? How would you modify the initial statement to make it less misleading?

PROBLEMS

1 A charge of 16 μC is 40 cm from a charge of 4 μC. Find the potential and the electric field

strength at a point P midway between the two charges. *Ans.* 9×10^5 V; 2.7×10^6 N/C

2 The potential difference between the terminals of an automobile battery is 12 V. How much work is done by the battery in transferring 700 C from one terminal to the other while starting the automobile?

3 A conducting sphere of radius 6 cm bears a charge of 8×10^{-10} C. Find the electric intensity and the potential at the surface of the sphere. *Ans.* 2,000 N/C; 120 V

4 Find the potential at a point 5 m from a charge of 6 μC. How much work is required to bring a 2-μC charge from infinity to this point?

5 A point charge of 8×10^{-10} C is 0.5 m from a charge of -6×10^{-10} C. Find the electric field intensity at the point P midway between the charges, the potential at this same point P, and the work required to bring a charge of 2×10^{-9} C from infinity to point P. *Ans.* 202 N/C; 7.2 V; 1.44×10^{-8} J

6 How much work is required to move a charge of 7 μC from a point where the potential is -5 V to one where the potential is $+150$ V?

7 Two points in the neighborhood of an isolated point charge differ in potential by 75,000 V. How much work is required to carry 6×10^{-10} C from the point of lower to the point of higher potential? If the points are 0.5 and 3 m from the original charge, find the magnitude of this charge. *Ans.* 4.5×10^{-5} J; 5.00 μC

8 Find the potential and electric intensity midway between point charges of -0.06 and 0.12 μC if they are separated by a distance of 3 m.

9 Find the potential difference between two points, one 4 m and the other 1.5 m from a point charge of 2×10^{-9} C. How much work is required to move a charge of 3×10^{-10} C from the 4-m point to the other? *Ans.* 7.5 V; 2.25×10^{-9} J

10 Air at atmospheric pressure breaks down and ionizes when the electric intensity is about 3×10^6 V/m. Find the approximate maximum potential to which a smooth sphere of 0.15 m radius can be charged in air. What charge is required for this potential?

11 A water droplet with a radius of 5×10^{-4} m bears a net charge of 500 electrons. Find its potential. If this droplet coalesces with an identical droplet, also bearing 500 excess electrons, find the potential of the new larger droplet. *Ans.* -0.00144 V; -0.00229 V

12 How much work is required to move a charge

of 2×10^{-9} C from a point 2 m to a point 0.5 m from a point charge of 5 μC? What is the potential difference between these points?

13 In a Millikan oil-drop experiment, the charge on the oil drop is that of four electrons. The oil drop is in the electric field arising from a difference of potential of 3,000 V between two plates that are at a distance of 0.75 cm from each other. Find the force exerted on the oil drop by the electric field. Find the mass of the drop if it is almost in equilibrium in the field.

Ans. 2.56 $\times$ 10^{-13} N; 2.61 $\times$ 10^{-14} kg

14 In a Millikan oil-drop experiment an oil drop of 1.25×10^{-11} g mass is balanced by applying a potential difference of 1,020 V between plates which are 0.8 cm apart. How many elementary charges are on the oil drop?

15 Suppose a solid copper sphere 1.00 cm in diameter had a negative charge arising from one additional electron for each 10^9 copper atoms. If the sphere contained 4.4×10^{22} atoms, what would be the charge on the sphere, the potential of the sphere, and the electric intensity at the surface?
Ans. -7.04 μC; -1.27×10^7 V; -2.53×10^9 N/C

16 The potential of an isolated conducting sphere of radius 6 cm is 3,000 V. Find the charge on the sphere and the potential at a distance of 1 m from the center of the sphere.

17 A 200-eV electron passes through a hole in a screen into a region where there is a uniform electric intensity of 800 N/C. If the velocity vector is in the direction of the field, how far does the electron move before it reverses its direction of motion? What was the initial speed of the electron? *Ans.* 0.25 m; 8.4 $\times$ 10^6 m/sec

18 Find the kinetic energy and the speed of an electron which starts from rest and is accelerated through a potential difference of 5 V.

19 The plates of a cathode-ray tube are parallel and 0.6 cm apart. The potential difference between them is 15 kV. Find the force on an electron passing between the plates. What acceleration is produced if the electron has a mass of 9.11×10^{-31} kg?

Ans. 4 $\times$ 10^{-13} N; 4.4 $\times$ 10^{17} m/sec^2

20 Two large parallel plates are 4 cm apart. If the potential difference between them is 1,200 V, calculate the electric intensity between the plates and the force on a 4×10^{-8} C charge placed anywhere between the plates.

21 A hollow copper sphere has a radius of 8 cm. If it bears a charge of 1.2×10^{-9} C, find the potential and the electric intensity (*a*) inside the sphere, (*b*) at the surface of the sphere, and (*c*) at a point 2 m from the center.
Ans. (*a*) 135 V; zero; (*b*) 135 V; 1,690 N/C; (*c*) 5.4 V; 2.7 N/C

22 How much work is required to carry a charge of 5×10^{-10} C from an infinite distance to a point midway between two identical 8-μC charges 20 cm apart? How much work is required if one of the charges is -8 μC?

23 A metal sphere having a radius of 10 cm carries a charge of 0.12 μC. It is temporarily connected by a conducting wire to a second uncharged metal sphere having a radius of 15 cm. Find the charge remaining on the smaller sphere.
Ans. 0.048 μC

24 Charges of $+6$ μC are placed on diagonally opposite corners of a square with sides 0.5 m long, and a charge of -3 μC is placed at a third corner. (*a*) Find the potential energy of this configuration assuming it was assembled from charges initially at infinity. (*b*) Compute the potential at the fourth corner of the square, assuming $V = 0$ at infinity.

25 Charges of $-2, 3, -4$, and 5 μC are placed at the corners of a square 0.6 m on a side. Find the potential at the center of the square.
Ans. 42,400 V

26 Electric charges of $+6, -6$, and $+2$ μC are placed at the vertices of an equilateral triangle with sides 2 m long. Sketch the configuration and find (*a*) the electric intensity and the potential at the center of the triangle and (*b*) the work required to assemble the configuration if the charges are brought from infinity.

27 Find the potential at the center of a uniformly charged thin ring of radius 9 cm which bears a charge of 2×10^{-9} C. What is the potential on the axis of the ring 12 cm from the center? Find the electric intensity at this point.
Ans. 200 V; 120 V; 640 N/C along axis

28 A rectangle is 4 by 3 m. If charges of $+0.05$ and -0.02 μC are placed at corners separated by 4 m, find the potentials of the other two cor-

ners and the potential difference between them.

29 A spherical conductor of 5-cm radius bears a charge of 2×10^{-8} C. It is surrounded by a concentric sphere of 9-cm radius bearing a charge of -1.5×10^{-8} C. Find (a) the potential difference between the two spheres and (b) the potential and electric intensity at $r = 6$ cm and $r = 10$ cm.

Ans. (a) 1,600 V; (b) 1,500 V; 5×10^4 N/C; 450 V; 4,500 N/C

30 A solid conducting sphere of 0.1-m radius which bears a charge of 12 μC is surrounded by a concentric sphere of 0.2-m radius bearing a charge of -8 μC. (a) Find the electric intensity E at points A, B, and C which are respectively at distances of 0.05, 0.15, and 0.25 m from the center of the spheres. (b) Find the electric potential at these same three points, taking the potential as zero at an infinite distance from the spheres.

31 Find the electric intensity and the potential at the center of a cube with sides of length d (a) if there is a charge $+Q$ at each of the eight corners and (b) if one of the eight charges is changed to $-Q$.

Ans. (a) zero, $4Q/\sqrt{3}\,\pi\epsilon_0 d$; (b) $2Q/3\pi\epsilon_0 d^2$ toward $-Q$, $\sqrt{3}\,Q/\pi\epsilon_0 d$

32 A charge of 8×10^{-10} C is placed on a conducting sphere of radius 5 cm. This sphere is then connected temporarily by a copper wire to an uncharged conducting sphere of radius 3 cm. Find the charge on each sphere.

IN PRECEDING CHAPTERS we have discussed electric charges at rest and the electric fields and potentials associated with the charges. In this chapter we treat the situation in which two neighboring conductors have equal and opposite charges. Such an arrangement, known as a *capacitor*, can store a substantial amount of energy; hence it is a reservoir for both charge and energy. Capacitors are widely used in radio, television, and other electric circuits.

CHAPTER THIRTY-TWO
CAPACITANCE AND DIELECTRICS

32.1 CAPACITORS

A device on which electric charges can be stored is called a *capacitor* or a *condenser*. Capacitors are important components in radio and television circuits, in the ignition systems of automobiles, and in other electrical equipment. The term *condenser* originated in the erroneous idea that electricity was a fluid which could be stored in a suitable container, such as the "Leyden jar," a glass jar with a coating of tinfoil on the inside and another on the outside.

Capacitors have many forms. A typical capacitor consists of two conductors, one charged positively and the other negatively. Charging may be accomplished by transferring the charge from one conductor to the other by means of a battery or other source of potential difference. The charge gained by one conductor is equal to that lost by the other. When we refer to the *charge on a capacitor,* we mean *the magnitude of the charge on either conductor.* Since the two conductors bear equal and opposite charges, the *net* charge is zero.

Let $+Q$ be the charge on the positive conductor, $-Q$ that on the negative conductor, and V the potential difference between the two conductors. *The capacitance C is defined as the ratio of the charge to the potential difference:*

$$C = \frac{Q}{V} \qquad (32.1)$$

From the definition it follows that capacitance may be measured in coulombs per volt. This unit has been named the *farad* (F) in honor of Michael Faraday. *One farad is that capacitance for which a charge of one coulomb will produce a potential difference of one volt.* The farad is a very large unit of capacitance; a capacitance of a few microfarads (μF) is typical, and capacitances of the order of a few picofarads (pF), that is, 10^{-12} F, are common in radio and television circuits.

The capacitance of a pair of conductors is independent of the charge. When the charge is doubled, the potential difference between the conductors is also doubled. The capacitance depends on the size and shape of the conductors, on their relative positions, and on the character of the insulating material between them. Before we develop equations for the capacitances of various arrangements of conductors, it is desirable to consider how the insulating material affects the capacitance.

32.2 THE DIELECTRIC CONSTANT

When two parallel metal plates are insulated from one another and connected to an electroscope (Fig. 32.1), there is a potential difference between the plates if they are charged so that one bears $+Q$ and the other $-Q$ C. The potential difference V may be measured in a variety of ways, e.g., by the divergence of the electroscope leaves. If a sheet of glass is inserted between the two plates, the potential difference between the two plates becomes smaller and the divergence of the electroscope leaves is reduced. Since no charge escaped from the plates while the potential was reduced, the capacitance Q/V of the system with the glass plate in place must be greater than the capacitance without the glass plate.

To understand this phenomenon, consider what happens to the glass plate. The glass is composed of atoms and ions, electrically neutral on the average. When it is placed in the electric field, electrons are attracted by the positively charged conductor, and the nuclei by the negatively charged conductor. The electrons in the glass are not free, but they can undergo slight displacements toward the positive plate, while the nuclei undergo smaller displacements toward the negative plate. The net effect of these minute displacements throughout the glass is to produce a layer of positive charge adjacent to the negative plate and a layer of negative charge next to the positive plate. This negative charge layer reduces the potential of the plate (see Sec. 31.2). Similarly, the positive surface layer near the negative metal plates raises the

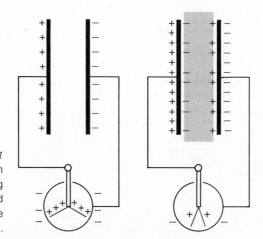

Figure 32.1
The potential difference between capacitor plates is reduced by inserting a dielectric sheet because of the bound surface charges induced on the dielectric.

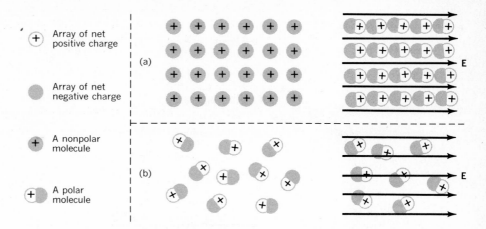

Array of net
positive charge

Array of net
negative charge

A nonpolar
molecule

A polar
molecule

(a)

(b)

E

E

Figure 32.2
Polarization effects in dielectrics. (*a*)
Initially nonpolar molecules, shown
here in a solid, have polarization
induced by an applied field. Silicon,
oxygen, and methane are nonpolar.
(*b*) Polar molecules in a gas have a
random orientation in a field-free
region; application of an electric field
introduces torques which favor
alignment with the field. Thermal
agitation prevents complete alignment.
Water molecules and hydrogen chloride
are polar.

potential of the latter. The potential difference between the metal plates
is decreased. The glass plate increases the capacitance of the capacitor
because it places a layer of negative charge close to the positive conductor and a layer of positive charge close to the negative conductor.
In a very real sense it reduces the *effective* charge on the conductors,
although the actual charge is not changed significantly.

A material is said to be *polarized* when the electric "center of
charge" of the electrons and of the nuclei of a material do not coincide.
This polarization may arise in two distinct ways. If the normal charge
distribution of the molecules is initially symmetrical (Fig. 32.2*a*), the
molecules are initially *nonpolar* and an applied electric field induces a
polarization. If the molecules are already polar (Fig. 32.2*b*) but have
random orientation, the external field produces at least a partial alignment of these polar molecules. In either case, the effect of the polarization is to leave the interior of the dielectric uncharged but to produce
a bound charge on surfaces normal to the field.

In the discussion above, we kept the charge on the two metal
plates constant and found that insertion of a glass sheet resulted in a
reduced potential difference between them. Next consider plates A and
B (Fig. 32.3) connected to a battery which maintains a constant potential
difference between them. If the space between the plates is filled by
a sheet of glass, the induced surface charges on the glass lead to an
increase of the charge $+Q$ on A and $-Q$ on B by a factor of perhaps 5.
Since Q increases when the dielectric is introduced while V is held
constant by the battery, the capacitance of the plates is increased by
the insertion of the dielectric. *The ratio of the capacitance of a capacitor*
with a given material filling the space between conductors to the capaci-
tance of the same capacitor when the space is evacuated is the dielectric
constant† *K of the material.* The dielectric constants of several materials are listed in Table 32.1.

† Some authors prefer the name *specific inductive capacity*, and others use the name *relative*
permittivity.

Figure 32.3
Dielectric between the plates of a
capacitor.

Figure 32.4

A piezoelectric cell in which pressure on the crystal faces produces a potential difference between them.

We have seen that when a slice of dielectric material is placed between two charged plates (or, more generally, in an electric field), there is a polarization of the medium. In some materials, such as quartz and rochelle salt, this displacement of electric charges is accompanied by small changes in the size and shape of the crystal slice, an effect called *electrostriction*. Electrostrictive distortions depend on the orientation of the crystal axes relative to the direction of the electric field.

In view of the fact that the shapes of some crystals change when the internal charges are displaced by an electric field, it is not surprising that changing the shape of a crystal may result in a redistribution of charges. When a thin slice of quartz is compressed (Fig. 32.4), one face becomes positive and the other negative. If the crystal is stretched instead of compressed, the charges on the faces are reversed. Compressing or elongating the crystal results in a potential difference between the faces; this potential difference may be hundreds or even thousands of volts. This phenomenon is known as *piezoelectricity* (*piezo* means "pressure").

If an alternating voltage is applied to a properly sliced quartz crystal, the crystal faces oscillate. By proper choice of the thickness of the slice, the mechanical oscillations can be made to have any desired frequency over a wide range. If a radio-frequency circuit has the same natural frequency as the mechanical oscillations of a quartz crystal, a sharp resonance may be obtained, and the electrical oscillations may be accurately controlled by the mechanical frequency of the quartz crystal. Quartz crystals are often used to control the frequencies of radio and television transmitters.

Table 32.1
Dielectric Constants and Dielectric Strengths

Material	K	Dielectric strength, MV/m
Vacuum	1	
Air (1 atm)	1.006	3
Ammonia (liquid)	22	
Ethyl alcohol (0°C)	28.4	
Transformer oil	2.1	5–15
Water (18°C)	81	
Amber	3	
Barium titanate (25°)	1,200	
Glass	4.8–10	30
Mica	4.5–7.5	200
Paraffined paper	2	40
Polystyrene	2.6	20
Porcelain	6	15
Rubber (hard)	3	21

32.4   THE CAPACITANCE OF AN ISOLATED SPHERE

Consider a single sphere of radius R in vacuum, removed sufficiently far from other bodies so that their influence may be neglected. Let this sphere be charged with Q C, presumably brought to the sphere from an infinite distance. (In this case the second conductor of the capacitor is a sphere of infinite radius, which now bears a charge $-Q$.) According to Eq. (31.3) the potential of the sphere is given by $V = Q/4\pi\epsilon_0 R$. Since $C = Q/V$, it follows immediately that

$$C = 4\pi\epsilon_0 R = \left(\frac{1}{9 \times 10^9} \text{ F/m}\right) R \qquad (32.2)$$

so that C is in farads when R is in meters.

If we imagine all space to be filled with a medium of dielectric constant K, we have

$$C = 4\pi\epsilon_0 KR = \left(\frac{1}{9 \times 10^9} \text{ F/m}\right) KR \qquad (32.3)$$

Obviously, the charge Q has not changed; therefore the increase in C must arise from a decrease in V. The potential of the sphere under these conditions is given by

$$V = \frac{Q}{4\pi\epsilon_0 KR} = (9 \times 10^9 \text{ m-V/C}) \frac{Q}{KR} \qquad (32.4)$$

32.5   CAPACITANCE OF A SPHERICAL CAPACITOR

Figure 32.5 shows a capacitor consisting of two concentric spheres. Let a be the radius of the inner sphere and b the radius of the outer sphere, which is connected to the earth. If the inner sphere bears a charge $+Q$, the outer one must have a charge $-Q$ if its potential is to be zero. (Remember the charge on the inner sphere raises the potential of the outer sphere, so a negative charge is required on the outer sphere to reduce its potential to zero.) If the region between the spheres is filled with a medium of dielectric constant K, the potential V_a of the inner sphere is the sum of the contributions of the charges $+Q$ on the inner sphere and $-Q$ on the outer sphere, so

$$V_a = \frac{Q}{4\pi\epsilon_0 Ka} - \frac{Q}{4\pi\epsilon_0 Kb} = \frac{Q}{4\pi\epsilon_0 K}\left(\frac{1}{a} - \frac{1}{b}\right)$$

Since the potential of the outer sphere is zero, the potential difference V between the spheres is just the potential of the inner sphere V_a, and the capacitance of the two concentric spheres is

$$C = \frac{Q}{V} = \frac{4\pi\epsilon_0 Kab}{b - a} = \frac{Kab}{9 \times 10^9 (b - a)} \qquad (32.5)$$

32.6   CAPACITANCE OF TWO PARALLEL PLATES

If the radii of the two concentric spheres considered above are allowed to increase until they are very large while the difference $b - a$ remains

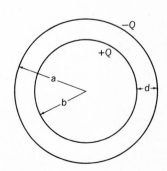

Figure 32.5
Capacitor formed by two concentric spheres.

$$\frac{1}{C_{eq}} = \frac{1}{C_1} + \frac{1}{C_2} + \cdots$$

$$= \frac{C_2 + C_1}{C_1 C_2}$$

$$C_{eq} = \frac{C_1 C_2}{C_2 + C_1}$$

THIS IS FOR CAPACITORS IN SERIES.

constant, the surfaces of the spheres become approximately plane. The product ab becomes almost equal to a^2, since a and b are almost equal when both a and b are very large.

Let S be the area of the sphere of which a is the radius ($S = 4\pi a^2$), and let $d (= b - a)$ be the distance between the spheres (kept constant) as a and b approach infinity. Then

$$C = \frac{4\pi\epsilon_0 Kab}{b - a} = \frac{4\pi\epsilon_0 Ka^2}{d} = \frac{\epsilon_0 KS}{d} = \frac{KS}{9 \times 10^9 \times 4\pi d}$$

If this relation is true for the entire spherical capacitor, it is true for a small portion of it, provided the area of that portion is used instead of the entire area of the sphere. If the spheres are very large, and if a small portion of area A is cut out of the spherical surfaces, the capacitor obtained in this artificial way consists of two parallel plates at a distance d apart, each plate with area A. The capacitance in farads of such a capacitor is

$$C = \frac{KA}{9 \times 10^9 \times 4\pi d} = \frac{\epsilon_0 KA}{d} \tag{32.6}$$

when A is in square meters and d is in meters.

Example

Find the capacitance of a capacitor consisting of two parallel plates that are 0.5 cm apart. Each of the plates has an area of 100 cm². The space between the plates is filled with a medium whose dielectric constant is 3.0.

$$C = \frac{KA}{4\pi d(9 \times 10^9)}$$

$$= \frac{3.0 \times 0.01}{0.005 \times 4\pi \times 9 \times 10^9}$$

$$= 54 \times 10^{-12} \text{ F} = 54 \text{ pF} \qquad ////$$

32.7 PRACTICAL CAPACITORS

Most of the practical capacitors in everyday use represent some modification of the parallel-plate capacitor. An example is the familiar type used for tuning radio and television circuits (Fig. 32.6). The capacitance is varied by changing the effective area of the plates, which is that area close to a plate bearing the opposite charge. Alternate plates are connected together so that there are only two conductors; several plates are used to get a large area in a reasonable space. A further favorable factor in achieving a large area is that both sides of all but the outermost plates are used.

Another common form of capacitor consists of two thin foils of aluminum with a thin sheet of wax-impregnated paper between them. Since both the paper and the foil are flexible, the whole arrangement can be rolled up to form a small compact cylinder. From Eq. (32.6) it follows that the thinner the layer between the conducting plates, the greater the capacitance. However, there is a practical limitation to how

Figure 32.6
Parallel-plate capacitor with variable capacitance.

thin the insulating separator may be, because if it is too thin, a spark may jump through it. The maximum potential difference which an insulating layer will stand can be computed from a knowledge of the thickness and of the *dielectric strength* of the material. *The dielectric strength is the potential gradient at which electrical breakdown occurs.* For common insulators the dielectric strength (Table 32.1) is many millions of volts per meter, so a fraction of a millimeter is enough to support several thousand volts.

Among the materials which are used as insulators in capacitors are impregnated papers, mica, plastics, ceramic materials, glass, and oils. Air is used in the variable capacitor of Fig. 32.6; because of the low dielectric constant, air capacitors have relatively small capacitance for a given area.

In the circuit diagrams of radios and other electric circuits, capacitors are commonly represented by the symbols of Fig. 32.7. An arrow drawn through the capacitor symbol indicates that it is variable; when no arrow is shown, it is implied that the capacitor has a fixed value.

The electrolytic capacitor is a common and inexpensive form. Such a capacitor has one plate of aluminum. The dielectric is a very

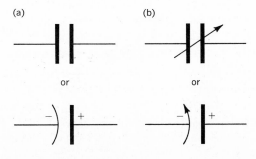

Figure 32.7
Symbols for (*a*) fixed and (*b*) variable capacitors.

thin coating of aluminum oxide on the surface of the plate, and the other conductor is an electrolyte. In such a capacitor d is exceedingly small, and therefore a relatively large capacitance can be provided in a rather small space. However, these capacitors break down at relatively low voltages; furthermore, it is important that the aluminum terminal be made positive, since the aluminum oxide layer conducts when the aluminum is negative.

32.8 ENERGY STORED IN A CAPACITOR

In charging a capacitor, it is necessary to do work to carry the electric charge from one conductor to the other. At the beginning, the two conductors of the capacitor are at the same potential. As charge is transferred from one plate to the other, the difference of potential between the two increases. Suppose that the final potential difference V between the terminals of the capacitor is attained after Q C of electricity has been transferred from one conductor to the other. At the beginning of the charging, the potential difference is zero; at the end, the difference is V. The average potential difference during the charging is $V/2$. The work is equal to the product of the average difference of potential and the quantity of electricity transferred. The energy W stored in the capacitor is given by

$$W = \tfrac{1}{2}QV \tag{32.7}$$

This energy is released when the capacitor is discharged. If the capacitor is allowed to discharge through a wire, the energy is converted to heat in the wire.

Example

A capacitor having a capacitance of 2 μF is charged with 10^{-3} C of electricity. How much energy is stored in it?

$$\text{Energy} = \tfrac{1}{2}QV = \frac{1}{2}\frac{Q^2}{C}$$

$$= \frac{1}{2} \times \frac{10^{-3} \times 10^{-3}}{2 \times 10^{-6}}$$

$$= 0.25 \text{ J} \qquad \qquad ////$$

32.9 CAPACITORS IN PARALLEL

When two or more capacitors are connected in such a way that all the positive conductors are at the same potential V^+ and all the negative conductors at potential V^-, the capacitors are said to be connected *in parallel*. When a number of capacitors (Fig. 32.8) are connected in parallel, the system has a capacitance C equal to the sum of the separate capacitances. This result can be proved as follows.

The capacitors are all charged to the same difference of potential. Let V denote this difference of potential, and let Q_1, Q_2, and Q_3 be the charges on capacitors C_1, C_2, and C_3, respectively. Let Q be the total

Figure 32.8
Capacitors connected in parallel.

charge on all the capacitors. Then

$$Q = Q_1 + Q_2 + Q_3$$

$$Q_1 = C_1 V \qquad Q_2 = C_2 V \qquad Q_3 = C_3 V$$

and

$$Q = CV$$

Substituting yields

$$CV = C_1 V + C_2 V + C_3 V$$

Dividing by V gives

FUNNY THING

$$\boxed{C = C_1 + C_2 + C_3} \qquad (32.8)$$

Hence, the equivalent capacitance of a number of capacitors connected in parallel is the sum of the individual capacitances.

32.10 CAPACITORS IN SERIES

Figure 32.9
Capacitors connected in series.

Two or more capacitors are said to be *in series* when they are connected as shown in Fig. 32.9. When a potential difference V is applied between points A and B, a charge $+Q$ appears on the positive plate of C_1, and a corresponding charge $-Q$ on the negative plate of C_1. The electrons which produce the negative charge on C_1 must come from the plate of C_2 and leave it with a charge $+Q$. The negative plate of capacitor C_2 has a charge $-Q$ which comes from the positive plate of capacitor C_3, which is left with a charge $+Q$. Finally, the negative plate of C_3 bears a charge $-Q$. Thus, *when capacitors are charged in series, the same charge is stored on each capacitor.* The potential difference across the combination of capacitors in series is equal to the sum of the potential differences of the individual capacitors.

Let Q be the charge on each capacitor, C the equivalent capacitance of the capacitors when joined in series, and V_1, V_2, and V_3 the differences of potential between the terminals of C_1, C_2, and C_3, respectively. Since the total difference of potential is equal to the sum of the separate differences of potential,

$$V = V_1 + V_2 + V_3$$

If we divide both sides of this equation by Q and make use of the fact that $Q = Q_1 = Q_2 = Q_3$, we obtain

$$\frac{V}{Q} = \frac{V_1}{Q_1} + \frac{V_2}{Q_2} + \frac{V_3}{Q_3}$$

Since V/Q is the reciprocal of the capacitance,

REMEMBER IT

$$\boxed{\frac{1}{C} = \frac{1}{C_1} + \frac{1}{C_2} + \frac{1}{C_3}} \qquad (32.9)$$

When capacitors are connected in series, the reciprocal of the resultant capacitance is the sum of the reciprocals of the individual capacitances.

Example

A capacitance of 4 μF is connected in series with one of 5 μF. What is the equivalent capacitance of the combination? If 100 V is the potential difference across the combination, find the potential difference across the 4-μF capacitor.

$$\frac{1}{C} = \frac{1}{C_1} + \frac{1}{C_2} = \frac{1}{4} + \frac{1}{5} = \frac{9}{20}$$

$$C = \tfrac{20}{9} = 2.22 \ \mu\text{F}$$

$$Q = CV = 2.22 \times 100 = 222 \ \mu\text{C}$$

$$V_4 = \frac{Q}{C} = \frac{222 \ \mu\text{C}}{4 \ \mu\text{F}} = 55.5 \ \text{V} \qquad\qquad ////$$

32.11 GROUPS OF CAPACITORS

When several capacitors are connected in such a way that some are connected in parallel and others in series (Fig. 32.10), it is possible to replace any parallel group with the equivalent single capacitor of value given by Eq. (32.8). After this is done for the parallel group (or groups), one is left with a problem in series capacitors.

Example

The four capacitors of Fig. 32.10a are 6, 8, 10, and 12 μF, respectively. They are charged to a potential difference V of 120 V. Find the charge and potential difference for each capacitor.

The capacitance of the three parallel capacitors is given by $C_{\|} = 6 + 8 + 10 = 24 \ \mu$F. If we replace these capacitors with their equivalent, our circuit reduces to that of Fig. 32.10b. The capacitance of a 24- and a 12-μF capacitor in series is C_S, where

$$\frac{1}{C_S} = \frac{1}{24} + \frac{1}{12} = \frac{1}{8} \qquad \text{or} \qquad C_S = 8 \ \mu\text{F}$$

The charge on each capacitor of Fig. 32.10b is therefore given by $Q = C_S V = 8 \times 120 = 960 \ \mu$C. $V_{24} = \frac{960}{24} = 40$ V, and $V_{12} = \frac{960}{12} = 80$ V. Note that these potential differences add to 120 V. The 24-μF capacitor, which is equivalent to the 6-, 8-, and 10-μF capacitors in parallel, has a potential difference of 40 V, and so must the three individually. The charges are given as follows: $Q_6 = 6 \times 40 = 240 \ \mu$C, $Q_8 = 8 \times 40 = 320 \ \mu$C, and $Q_{10} = 10 \times 40 = 400 \ \mu$C. Note that these charges add to 960 μC. $\qquad ////$

Figure 32.10
(a) Three capacitors in parellel are connected in series with a single capacitor. (b) Simple series circuit results when the three parallel capacitors in (a) are replaced with one equivalent capacitor C.

(a)

(b)

1 What determines the maximum potential difference which can safely be applied to a capacitor?

2 In how many *apparently* different units has the constant $4\pi\epsilon_0$ been expressed in this chapter? Show that they are all equivalent to $C^2/N\text{-}m^2$, in terms of which the constant was initially introduced with Coulomb's law.

3 Is the condition of a dielectric in a capacitor different when the capacitor is charged from when it is uncharged? In what way?

4 Two parallel metal plates are separated by a 2-mm air gap. The plates are given equal but opposite charges. If a sheet of glass is placed between the plates, what happens to the potential difference between them? Why? What happens to the potential gradient? To the electric field intensity?

5 Why does a parallel-plate capacitor store more charge for a given potential difference when a dielectric (such as mica or glass) is inserted between the plates than when the space is evacuated?

6 Two identical capacitors are connected in parallel and charged to a potential difference of 100 V. If they are now removed from the circuit and the positive conductor of one capacitor is connected to the negative conductor of the second, what is the resulting potential difference? (Some voltage multipliers actually operate in this way.)

7 What is piezoelectricity? How can it be used?

8 Explain why the dielectric constant of the medium outside the two spheres makes no difference in the development of Eq. (32.5).

9 If V is kept constant across the capacitors of Fig. 32.9 as a glass plate is inserted between the plates of C_1, what happens to the charge on each capacitor? To the total stored energy? Is work required to insert the plate? Why?

PROBLEMS

1 A 4-μF capacitor is charged to a potential difference of 110 V. Find the charge on the capacitor plates and the energy stored.

Ans. 440 μC; 0.0242 J

2 A charge of 240 μC is stored in a capacitor at a potential of 800 V. What is the capacitance of the capacitor? What is the energy stored in the capacitor?

3 The potential difference between two clouds is 8 MV. How much electric energy is dissipated if a lightning stroke of 30 C leaps from one cloud to the other? Assume that the potential difference between the clouds decreases to zero and that the system behaves as though the clouds were the plates of a capacitor. *Ans.* 120 MJ

4 What is the capacitance of a sphere of radius 0.2 m? If the dielectric strength for air is 3×10^6 V/m, what is the largest charge which ideally could be placed on this sphere?

5 Find the capacitance of the earth if it is approximately a sphere of 6,400-km radius. If the electric intensity over the entire surface of the earth were 120 V/m directed downward, what charge would the earth have to have? What would the surface charge density be?
Ans. 711 μF; -5.5×10^5 C; -1.06×10^{-9} C/m^2

6 What is the difference of potential between the terminals of a capacitor that has a capacitance of 2.5 μF when the charge on the capacitor is 7.5×10^{-4} C? Find the energy stored in the capacitor.

7 A parallel-plate capacitor has insulation which can withstand an applied potential of 6,000 V. Its capacitance is 0.5 μF. What is the maximum energy the capacitor can store? The maximum charge? If the insulation is paraffined paper, what is its approximate thickness?
- *Ans.* 9 J; 3×10^{-3} C; 0.15 mm

8 Capacitors of 24, 12, and 8 μF are arranged so that they can be connected in series or in parallel. What capacitance is obtained in each case?

9 What is the joint capacitance of three capacitors of 20, 30, and 60 μF when they are connected in series? In parallel? *Ans.* 10 μF; 110 μF

10 A capacitor consists of two parallel plates which are separated by a sheet of mica which is 0.12 mm thick and has a dielectric constant of 6. The area of each plate is 400 cm^2. The capacitor is charged to 400 V. Find the energy stored in the capacitor.

11 A parallel-plate capacitor has a capacitance of

4 pF with air between the plates, and a capacitance of 24 pF with glass between the plates. What is the dielectric constant of the glass in question? If a charge of 6×10^{-10} C is placed on the plates with glass between them, what is the energy stored? If the glass is removed without affecting the charge, what energy is stored? What is the source of the added energy?

Ans. 6; 7.5×10^{-9} J; 4.5×10^{-8} J; work done in pulling out glass

12 A parallel-plate capacitor using mica as the dielectric is to have a capacitance of 30 pF and be able to withstand a potential difference of 6,000 V. What is the minimum thickness of mica required? What is the minimum area the plates of the capacitor may have if the dielectric constant is 6?

13 A capacitor is made up of 200 sheets of tinfoil, each 40 by 30 cm. These sheets are separated by sheets of paraffined paper which are 0.15 mm thick and which have a dielectric constant of 2.5. What is the capacitance of the capacitor when alternate sheets of tinfoil are joined together? *Ans.* 3.52 μF

14 A parallel-plate capacitor consists of two sheets of aluminum, each of area 0.25 m^2, separated by a thin layer of plastic insulation of dielectric constant 3 and thickness of 0.1 mm. Find the capacitance and the charge stored when this capacitor is charged to a potential difference of 500 V.

15 A capacitor is made of two sheets of tinfoil in contact with a plate of glass of dielectric constant 6.5. If the area of each sheet of tinfoil is 120 cm^2 and the thickness of the glass is 0.15 cm, what is the capacitance of the capacitor? *Ans.* 460 pF

16 Three capacitors of capacitances 10, 15, and 30 μF are connected in series across a 90-V source. Find the charge and potential difference associated with each capacitor.

17 Two capacitors, of capacitance 8 and 4 μF, are connected in series across a 60-V battery. Find the charge, the potential difference, and the energy stored for the 4-μF capacitor.

Ans. 160 μC; 40 V; 3.2×10^{-3} J

18 Two capacitors, of capacitances 12 and 6 μF, are connected in series across a potential difference of 150 V. Find the charge, potential difference, and energy stored for the 6-μF capacitor.

19 Three capacitors, of capacitances 5, 7, and 3 μF, are connected in parallel across a 120-V potential difference. Find the total capacitance and the charge on the 7-μF capacitor.

Ans. 15 μF; 840 μC

20 A 15-μF capacitor is connected in parallel with a 10-μF capacitor across a 250-V potential difference. Find the charge on each capacitor and the total energy stored.

21 A 6-μF capacitor is given a charge of 300 μC. (a) Find the potential difference between the conductors and the energy stored in the capacitor. (b) If an uncharged 3-μF capacitor is connected in parallel with the charged 6-μF capacitor, find the new potential difference.

Ans. (a) 50 V; 7.5×10^{-3} J; (b) 33.3 V

22 When a capacitor that has a capacitance of 15 μF and a charge of 600 μC is connected in parallel with an uncharged capacitor, the resulting potential difference is 30 V. Find the capacitance of the second capacitor.

23 Two identical air-dielectric parallel-plate capacitors with $C = 50$ pF are charged in parallel to a potential difference of 150 V and then disconnected from the charging source. A glass plate of dielectric constant 5 is inserted between the plates of one of the capacitors, filling the space completely. Find the new potential difference between the plates and the charge transferred from one capacitor to the other.

Ans. 50 V; 5 nC

24 A 12-μF capacitor is charged to a potential difference of 60 V. It is then disconnected from the source but not discharged. Find the new potential difference if this capacitor is connected in parallel with an uncharged 20-μF capacitor.

25 Three capacitors are connected as shown in the accompanying figure. If the potential difference between A and B is 120 V, and if C_1, C_2, and C_3 are, respectively, 15, 25, and 35 μF, find the capacitance of the combination, the charge on C_2, and the potential difference across C_2.

Ans. 12 μF; 600 μC; 24 V

26 Three capacitors are connected as shown in the accompanying figure to a source of unknown potential difference V between A and B. If C_2 is a 12-μF capacitor charged to a potential difference of 60 V, find V and the charges on

Problems 25 and 26

C_1 and C_3, which have capacitances of 30 and 20 μF, respectively.

27 Find the capacitance of the system of the accompanying figure and the charge on capacitor

$60 u F \leftarrow C_1$

$C_3 \quad \rightarrow 9 uF$

$V = 120V$

$30 uF \leftarrow C_2$

$C_4 \quad \rightarrow 12 uF$

Problem 27

C_2 if C_1, C_2, C_3, and C_4 are, respectively, 60, 20, 9, and 12 μF and the potential difference between A and B is 120 V.

Ans. 8 μF; 600 μC

28 Find the energy required to establish in a parallel-plate air capacitor a uniform electric field of 10^5 V/m over a volume of 1 cm^3.

29 Capacitor $C_1 = 10$ μF is charged to a potential difference of 40 V, capacitor $C_2 = 5$ μF to a potential difference of 60 V, and capacitor $C_3 = 20$ μF to a potential difference of 50 V. They are then connected as shown in the accompanying figure. Find the charge on C_3 after the key is closed. *Ans.* 571 μC

Problem 29

CHAPTER THIRTY-THREE
ELECTRIC CURRENT AND RESISTANCE

THUS FAR IN ELECTRICITY we have confined our attention to electrostatics, the science dealing with electric charges at rest. However, most common applications of electricity involve charges in motion. A flow of charges constitutes an electric current and exists whenever bodies at different potentials are connected by a conducting path. In this chapter we discuss the factors which determine the value of the current which passes through a given conductor.

33.1 CURRENTS AND THEIR EFFECTS

When two metallic spheres, one charged positively and the other negatively, are connected by a copper wire, electrons flow until there is no longer a potential difference between the two spheres. Such a flow of charge is called an electric *current*. The magnitude of the current I is the charge per unit of time that passes any cross section of the wire.

$$I = \frac{Q}{t} \tag{33.1}$$

Current is commonly measured in coulombs per second, or *amperes*, named for the French physicist André Marie Ampère. *The ampere is the current when one coulomb per second passes any cross section of a conductor.*

Electric currents are of great practical importance because of the many ways in which we can use the three principal effects they produce: (1) heating, (2) chemical, and (3) magnetic effects.

Heating by electric currents is utilized in making the filaments of incandescent lamps luminous, in operating electric stoves, and in hundreds of other ways. There are cheaper ways to obtain thermal energy, but none is cleaner or easier to control and handle than electric energy.

Batteries and electroplating are applied electrochemical devices. Aluminum, high-purity copper and silver, and chlorine are among important industrial elements ordinarily obtained by electrolysis.

The magnetic effects of currents are used in giant electromagnets and in tiny electric relays in telephone circuits. The interaction between currents and magnetic fields is fundamental to electric motors and to reproducing the picture on a television receiver.

33.2 THE DIRECTION OF A CURRENT

The charges which are primarily responsible for the current in metallic conductors are negative electrons. However, early in the nineteenth century there was no way to tell whether it was negative or positive charges (or both) which were in motion. In 1747 Benjamin Franklin discussed his experimental results in terms of the flow of an "electric fluid" which we now call "positive charge." Subsequently there came into general use the *convention* that the *direction of the current is the direction in which a positive charge would move under the influence of the electric field.* By definition then, the *conventional* current in a wire flows from a point at higher potential to a point at lower potential, as though the current represented a movement of *positive* charge. Actually, *in metallic conductors* the positive nuclei are not free to move, and the transfer of charge results from a flow of electrons in a direction opposite that of the conventional current. In liquid and gaseous conductors, both positive and negative ions are in motion. In some of the modern high-energy accelerators, such as Van de Graaff generators and cyclotrons, the current may be a movement of positive charges. Obviously no convention could be most convenient for handling every possible situation.

When a constant potential difference is maintained between two points in a conductor, a constant flow of charge results. The current is always in the same direction and is said to be a *direct current.* On the other hand, when the flow of charges is first in one direction and then in the opposite direction, the current is said to be *alternating.* The next five chapters deal primarily with direct currents.

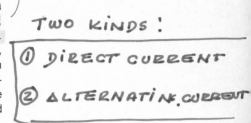

33.3 OHM'S LAW FOR A RESISTOR

To produce a steady current through a conductor, such as the filament of a lamp, an electric field must be established in the conductor. Since the natural flow of charges is always such as to eliminate the electric field, a *steady* current can result *only* if some device such as a battery or generator maintains the field. In this agency, called a source of *electromotive force* (emf), some other form of energy is converted into the electric energy needed to keep the charges flowing. Various sources of emf are discussed in the chapters which follow.

Consider a conductor across which a constant potential difference is maintained by a battery, as indicated in Fig. 33.1. In this figure A represents an ammeter, an instrument for measuring electric current, while V represents a voltmeter, an instrument for measuring potential difference. The common forms of ammeters and voltmeters depend on the magnetic effects of electric currents (Chap. 36) for their operation. In use, the two terminals of the voltmeter are connected to the two points between which one wishes to measure the potential difference, while the ammeter is connected so that all the charges which pass through the device in which the current is to be measured also pass through the ammeter. (In Fig. 33.1 the ammeter passes not only the current through the wire but also the current through the voltmeter; the latter is assumed to be negligible in this case. It is not always negligible.) Let the current through the conductor be I_1 amperes (A) when the potential difference across it is V_1 V. If the battery is replaced by a different one, the potential difference and current may again be read and found to be V_2 and I_2. When V_2 is twice V_1, I_2 is twice I_1. Indeed,

Figure 33.1
A current, read by ammeter A, is produced in resistor R (represented by the sawtooth symbol) by application of a potential difference, read by voltmeter V.

so long as the temperature of the wire is constant, the ratio of the potential difference to the current is constant. The ratio

$$R = \frac{V}{I} = \frac{V_1}{I_1} = \frac{V_2}{I_2} \qquad (33.2)$$

is called the *resistance* of the wire. The resistance is commonly expressed in *ohms*, where *one ohm is that resistance in which a potential difference of one volt produces a current of one ampere.* This unit is named in honor of Georg S. Ohm, who discovered (1825) that the current in a wire is proportional to the potential difference between the ends. Much of our knowledge about resistance is due to the pioneering work of Ohm, who made his own batteries, wires, and meters. The symbol Ω (Greek capital omega) is used as an abbreviation for ohms.

The relation

$$I = \frac{V}{R} \qquad (33.2a)$$

is *Ohm's law for a single resistor*. This law is obeyed within wide limits for metallic and many other conductors. However, for many other nonmetallic conductors the current is *not* proportional to the potential difference.

33.4 JOULE'S LAW OF HEATING

When a potential difference produces a current through a conductor of resistance R, electric energy is converted into thermal energy. The potential difference V represents the energy per unit charge converted from electric to thermal energy in the conductor. The charge Q transported through the conductor in time t is It, by Eq. (33.1). Hence the total energy W dissipated in the conductor in time t is

$$W = VQ = VIt \qquad (33.3)$$

If Ohm's law is obeyed for this conductor, $V = IR$, and

$$W = I^2Rt \qquad (33.3a)$$

The fact that the thermal energy produced by an electric current in a conductor is proportional to the square of the current, to the resistance, and to the time was reported by Joule, and relation [Eq. (33.3a)] is known as *Joule's law of heating*. This formula gives energy in joules when I is in amperes and R in ohms. (To obtain the energy in calories, use the relation 4.186 J = 1 cal.)

Power is defined as the ratio W/t; thus the power is equal to VI. The power is given in watts when the current is in amperes and the potential difference is in volts (volt-amperes = joules per coulomb × coulombs per second = joules per second = watts). With the aid of Ohm's law, we may express the power in the forms

$$P = VI = I^2R = \frac{V^2}{R} \qquad (33.4)$$

It is desirable to have some sort of device to protect electric machines and appliances from excessive currents. One method of furnishing this protection is by means of fuses. A typical fuse consists essentially of a wire that has a low melting point. When an excessive current passes through this fuse wire, the heat generated is sufficient to melt the conductor, thereby opening the circuit. The size of the fuse is so chosen that it melts when the current becomes greater than a preselected amount.

33.5 RESISTIVITY

In his studies of resistance Ohm made wires of various materials, lengths, and areas. He was able to show that for a wire of given material at constant temperature *the resistance is directly proportional to the length and inversely proportional to the cross-sectional area.* This fact may be expressed by the equation

$$R = \rho \frac{l}{A} \tag{33.5}$$

where ρ is called the *resistivity* (or *specific resistance*) of the material. The resistivity depends on the material in question and on its temperature. In the metric system the resistivity of a material is numerically equal to the resistance of a piece of the material one meter in length and one square meter in cross-sectional area. The resistivities of several materials are recorded in Table 33.1.

In the British engineering system the *resistivity* of a material is numerically equal to the resistance of a piece of the material one foot long and one circular mil in area. A circular mil is defined as the area of a circle one one-thousandth of an inch in diameter. The units are usually written as *ohms per (circular) mil-foot*, which is dimensionally incorrect and misleading. The British engineering unit has an advantage of convenience for circular conductors in that the area of a wire d thousandths of an inch in diameter is d^2 cir mils.

Table 33.1
Resistivities and Temperature Coefficients of Resistance
(Approximate values at 20°C)

Material	*Resistivity,* * Ω-m	*Temperature coefficient per Celsius degree*
Aluminum	2.6×10^{-8}	0.0040
Carbon	$3,500 \times 10^{-8}$	-0.0005
Constantan	49×10^{-8}	0.000002
Copper	1.7×10^{-8}	0.00393
Iron	9.7×10^{-8}	0.0058
Manganin	48×10^{-8}	0.0
Silver	1.6×10^{-8}	0.0038
Tungsten	5.5×10^{-8}	0.0047
Glass	$\approx 10^{13}$	
Quartz	$\approx 10^{17}$	

* To obtain ρ in ohms per mil-foot, multiply ρ in ohm-meters by 6×10^8.

33.6 TEMPERATURE COEFFICIENT OF RESISTANCE

The resistance of most conductors increases as the temperature is increased. Figure 33.2 shows how the resistivity of platinum varies with temperature. For many materials, such as carbon, the resistance decreases as the temperature is increased. Very often the change in the resistance of a conductor is roughly proportional to the change in the temperature and to the original resistance R:

$$\Delta R = \alpha R\, \Delta t \qquad (33.6)$$

where ΔR represents the change in resistance and Δt the change in temperature. The proportionality constant α is called the *temperature coefficient of resistance* and is defined as the change in resistance divided by the product of the original resistance and the change in temperature. Approximate values of α for several conductors are listed in Table 33.1.

If R_0 represents the resistance at $0°C$ and α_0 the temperature coefficient at $0°C$, the resistance R_t at temperature t may be written in the form

$$R_t = R_0(1 + \alpha_0 t) \qquad (33.7)$$

since in this case $\Delta R = R_t - R_0$, and $\Delta t = t$. The temperature coefficient of resistance is negative for materials which show a decrease in resistance as the temperature is raised.

For most pure metals the temperature coefficient of resistance is in the neighborhood of 0.004 per Celsius degree for small temperature changes near $0°C$. Resistance is only approximately a linear function of temperature; if we define α as the slope of the curve showing resistivity as a function of temperature, pure metals behave much like platinum (Fig. 33.2), with α roughly proportional to the reciprocal of the absolute temperature. For most alloys α is much smaller than for metals. Indeed, for constantan and manganin, the temperature coefficients of resistance are very near zero. For this reason these two alloys are often used for resistors when it is desirable that the resistance be independent of temperature.

Since the resistance of a wire changes with temperature, it is possible to infer the change in temperature from measurements of re-

$$R_{T2} = R_{T1} + \Delta R$$
$$R_{T1} + \alpha R_{T1}(T_2 - T_1)$$
$$R_{T2} = R_{T1}\left[1 + \alpha \cdot (T_2 - T_1)\right]$$

$$R \propto L$$

FOR A PIECE OF WIRE.

THE R IS PROPORTIONAL TO LENGTH

AND FOR 2 WIRES IN PARALLEL THE RESISTIVITY WILL BE ½ OF THE INITIAL $R \propto \dfrac{L}{A}$

$$R = \rho\,\dfrac{L}{A}$$

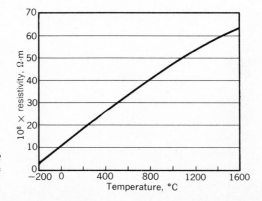

Figure 33.2
The resistivity (or specific resistance) of platinum rises with temperature.

Comparison leads   Resistance coil

Figure 33.3
Resistance thermometer registers
changes in resistance as the
temperature varies.

sistance. A resistance thermometer using platinum wire can be used to determine temperatures over a wide range up to the melting point of platinum. When properly calibrated, such a thermometer gives high precision in the measurement of temperatures. Figure 33.3 shows the essential components of a resistance thermometer.

Example

A resistance thermometer is made of platinum wire. Its resistance in a mixture of ice and water at 0°C was found to be 10 Ω; in a furnace of unknown temperature it was found to be 50 Ω. If the temperature coefficient of resistance of platinum has an average value of 0.0036 per Celsius degree over this range, what was the temperature of the furnace?

$$\text{Temperature of furnace} = t = \frac{R_t - R_0}{\alpha R_0}$$

$$= \frac{50 - 10}{0.0036 \times 10} = \frac{40}{0.036}$$

$$= 1100°C \qquad\qquad ////$$

33.7 RESISTORS IN SERIES

The battery and the resistors of Fig. 33.4 are connected *in series.* Two circuit elements are in series whenever all the charge passing through one of the elements passes through the second. Every electron which passes through the battery of Fig. 33.4 passes through R_1, through R_2, and through R_3 and returns to the battery. The current through each resistor is the same. The current I is the same in all series resistors.

The work necessary to move a coulomb from A to D is the work necessary to move it from A to B plus the work necessary to move it from B to C plus the work from C to D. Thus, the potential difference across the combination of resistors in series is the sum of the potential differences across the individual resistors:

$$V = V_1 + V_2 + V_3$$

The resistance of the combination is, by definition, the ratio of the potential difference V across the combination to the current I:

$$R = \frac{V}{I} = \frac{V_1 + V_2 + V_3}{I} = R_1 + R_2 + R_3 \qquad (33.8)$$

Thus, *the resistance of any combination of resistors connected in series is equal to the sum of the individual resistances.*

Figure 33.4
Three resistors connected in series.

Figure 33.5
Three resistors connected in parallel.

The resistors of Fig. 33.5 are connected *in parallel.* When several conductors are connected between two points so that the current divides between them and then rejoins, they are said to be *in parallel.* An electron in going from B to A of the figure may pass through R_1, R_2, or R_3. The potential difference across each resistor is the same, because the work required to move a charge from A to B is independent of the path chosen. If V is the potential difference between A and B, and V_1, V_2, and V_3 are the potential differences across R_1, R_2, and R_3, respectively, then $V = V_1 = V_2 = V_3$. The current I splits at point A, a part going through each of the parallel resistors. The total charge reaching B each second is equal to that leaving A each second. Therefore,

$$I = I_1 + I_2 + I_3$$

The resistance between A and B is, by definition, the ratio of V to I:

$$R = \frac{V}{I} = \frac{V}{I_1 + I_2 + I_3}$$

or

$$\frac{1}{R} = \frac{I}{V} = \frac{I_1 + I_2 + I_3}{V} = \frac{I_1}{V_1} + \frac{I_2}{V_2} + \frac{I_3}{V_3}$$

and

$$\frac{1}{R} = \frac{1}{R_1} + \frac{1}{R_2} + \frac{1}{R_3} \tag{33.9}$$

When resistors are connected in parallel, the reciprocal of the total resistance is equal to the sum of the reciprocals of the individual resistances.

33.9 QUALITATIVE CONSIDERATIONS IN CONDUCTION

The high conductivity of metals was explained qualitatively by Drude in 1900, shortly after the discovery of the electron. He assumed that when a metal crystallizes in its lattice, one or more electrons are released from each atom. These electrons are free to gain energy from an applied electric field and are responsible for metallic conduction. In 1909 Lorentz extended Drude's theory, considering the electrons to behave like the molecules of a gas with an average kinetic energy of $\frac{3}{2}kT$ [Eq. (14.11)], but his theoretical results were only partially compatible with experimental data. In 1928 Sommerfeld applied wave mechanics and the Fermi distribution function [Eq. (30.1)] to the problem to develop a much better theory. In the Sommerfeld model even at 0 K the average kinetic energy of the "free" electrons is several electron volts, corresponding to an average speed of the order of 10^6 m/sec. When an electric field is applied, these "free" electrons are accelerated and a small drift velocity (≈ 1 mm/sec) is superimposed on the high-speed random motions of the electrons. The current in the metal is thus a small electric "wind" superimposed on a very large random velocity distribution. As the temperature of the metal is increased, the average kinetic energy of the electrons is raised slightly, but by a very small fraction of the electron kinetic energy at 0 K.

Ionized atoms of a pure metal are arranged in a crystalline lattice; because each atom has released one or more electrons, there is an abundant supply of nearly free electrons. When a current passes through the metal, energy is lost by electrons in collisions with the ion cores; this energy shows up as the Joule heat discussed in Sec. 33.4. As the temperature of the metal is increased, movements of the atoms become greater; the chances of an electron colliding with an atom increase, and so does the resistance. The resistance of an ideal crystal lattice would be zero if there were no motion of the atoms, and most of the resistance of pure metals at room temperature is associated with the movements of the ion cores, which are described in terms of *lattice vibrations*. In general, metals have low resistivities and high temperature coefficients of resistance.

Alloys typically have higher resistivities and lower temperature coefficients than pure metals. In alloys the crystal array is less favorable for the free movement of electrons; this leads to a higher resistivity, especially at very low temperatures. However, as the temperature increases, the thermal vibrations of the atoms in the alloy are almost as likely to move a given atom out of the way of a traveling electron as to move it into the way. Consequently, the resistance is considerably less temperature-sensitive than in the case of a pure metal.

Many nonmetallic solids are poor electric conductors; some are excellent insulators. At room temperature insulating materials have practically no free electrons or mobile ions available to carry current, but at sufficiently high temperature the internal energy becomes large enough to free electrons or conducting ions. For ordinary glass, sodium ions, freed by virtue of the increased internal energy, play an important role in conduction at elevated temperatures, but for most insulators it is the internal liberation of electrons which produces conduction at high temperature.

Some materials which are excellent insulators near 0 K require so little energy to provide charge carriers thermally that at room temperature they are fair conductors. Such materials, known as *semiconductors*, typically have $\rho \approx 10^{-4}$ to 10^7 Ω-m, intermediate between metallic conductors ($\rho \approx 10^{-7}$ Ω-m) and good insulators ($\rho \approx 10^{16}$ Ω-m and above). The number of free electrons or other charge carriers increases rapidly with temperature for a typical semiconductor, with a resulting drop in resistance: copper oxide, for example, has a resistance at 70°C which is only one-tenth its resistance at 20°C. In the past 20 years there has been a dramatic expansion of the use of semiconductors in electronic circuits. Among the more common semiconductors are silicon, germanium, selenium, cuprous oxide, and lead sulfide. The physical characteristics of semiconductors, as well as those of metals and insulators, are discussed further in Secs. 48.4 to 48.8.

Carbon is widely used as an electric conductor. It has a negative temperature coefficient of resistance because the number of free electrons available for conduction increases with temperature. The coefficient of thermal conductivity also increases with temperature, and for the same reason; both the electric and thermal conductivities of the best room-temperature conductors are associated with free electrons. It is a well-known fact that good conductors of electricity are also good conductors of heat and that nonconductors of electricity are poor con-

ELECTRICITY AND MAGNETISM

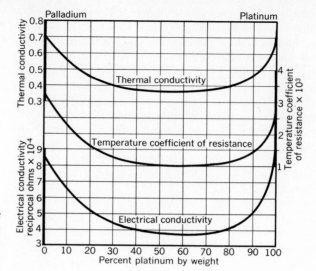

Figure 33.6
Relationship between electric and thermal conductivities for palladium-platinum alloy.

ductors of heat. The Wiedemann-Franz law states that the ratio of the thermal conductivity to the electric conductivity of metals and alloys is proportional to the absolute temperature. That the Wiedemann-Franz law is approximately true for certain alloys is evident from Fig. 33.6.

33.10 SUPERCONDUCTIVITY

At very low temperatures the resistivities of all metals are very much smaller than at room temperature, but the resistivities of some materials drop suddenly to an immeasurably small value. The drop in resistance occurs over an exceedingly small temperature range, as is shown in Fig. 33.7. Metals in which the resistance vanishes at very low tempera-

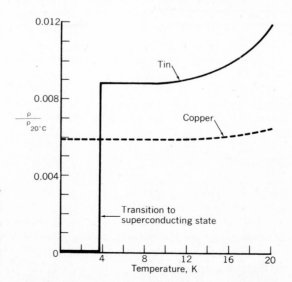

Figure 33.7
Resistivity-ratio curves for tin and copper, showing that tin becomes superconducting, losing all resistance below 3.75 K. Ratios are of resistivity at low temperatures to those at 20°C.

tures are called *superconductors;* when the resistance vanishes, they are said to be in the *superconducting* state. Not all metals become superconducting as the temperature approaches absolute zero. Metals such as gold, platinum, copper, sodium, and iron have resistivities which show no such abrupt changes. Among the elements which become superconductors are niobium, mercury, lead, tin, indium, and aluminum.

QUESTIONS

1 What is the distinction between the *velocity* of an electron in a conductor and the *drift velocity* of the electrons?

2 What do we mean when we say that several circuit elements are in parallel? What is the same for all the elements? Why?

3 What do we mean when we say that two circuit elements are in series? What must be the same for both?

4 A 60-W light bulb and a 100-W bulb are designed to operate on 110 V. Which has the greater resistance? How do you reconcile your answer with the relation $P = I^2R$?

5 In what ways is the current in a conductor analogous to the flow of water through a pipe? At what points does the analogy fail?

6 What are the advantages and disadvantages of strings of Christmas tree lights in series? In parallel?

7 What is meant by a negative temperature coefficient of resistance? Explain why some materials have negative temperature coefficients.

8 The filament of a light bulb is a tungsten wire. If the potential difference across the filament is measured as a function of the current, is Ohm's law obeyed? Why not? Under what conditions would Ohm's law be obeyed by a tungsten filament?

9 Why is the variation of resistance with temperature so different for a pure metal and for a semiconductor?

PROBLEMS

1 A 150-W incandescent lamp operates at 120 V. What current is drawn? What is the resistance of the lamp? How much charge passes through the filament each minute? How many electrons pass through each second?
Ans. 1.25 A; 96 Ω; 75 C; 7.81 $\times$ 10^{18}

2 Find the current drawn by, and the resistance of, an 880-W, 110-V toaster. If electric energy costs 3 cents per kilowatt-hour, how much does it cost to keep the toaster in steady operation for 20 min?

3 An electric heater element is designed to dissipate 1,650 W when connected to a 110-V line. Find the current drawn, the resistance of the heater, the charge which passes through the heater in 40 sec, and the energy transformed to heat in 40 sec.
Ans. 15 A; 7.33 Ω; 600 C; 66 kJ

4 A trouble lamp draws 3 A from a 12-V battery. What is the resistance of the lamp? The power it dissipates? The charge which passes through the lamp in 5 min?

5 A potential difference of 24 V is applied across a resistance of 8 Ω for a period of 3 min. What current is drawn? Calculate the charge which passes through the resistor. How many joules of energy does each coulomb lose in the resistor? What power is dissipated in the resistor?
Ans. 3 A; 540 C; 24; 72 W

6 A transmission line with a resistance of 3 Ω carries a current of 40 A. What is the potential difference between the ends of the line? How much power is dissipated in the line? What charge is transferred each minute?

7 A copper bus bar is 2 cm thick, 3 cm wide, and 75 cm long. Find its resistance and the potential difference between its ends when it bears a current of 2,000 A.
Ans. 2.13 $\times$ 10^{-5} Ω; 0.043 V

8 A silver wire of resistance 3 Ω is drawn through a die so that its length is doubled and its cross section is reduced to one-half of its previous value. Find the new resistance.

9 A ribbon of silver 5 cm long and 1 mm wide is to be made into a resistance of 0.12 Ω. How thick must it be?
Ans. 6.7 μm

10 A piece of wire 8 m long and 0.5 mm in diameter has a resistance of 2 Ω. What length of wire of the same material 0.4 mm in diameter will have a resistance of 2.5 Ω?

ELECTRICITY AND MAGNETISM

11 A 12-Ω resistor has a temperature coefficient of 0.004 per Celsius degree. Initially the resistor is immersed in ice water to keep its temperature constant. (*a*) If a potential difference of 60 V is applied, find the current. (*b*) If the resistor is removed from the ice water and allowed to come to equilibrium with the 60 V still applied, the current falls to 4 A. Calculate the equilibrium temperature. *Ans.* (*a*) 5 A; (*b*) 62.5°C

12 If 1 g of copper and 1 g of aluminum are used to make uniform wires each 5 m long, find the resistances of both wires. What is the ratio of the resistance of the copper wire to that of the aluminum wire? What would be the ratio of the resistance of a copper wire to that of an aluminum wire of the same dimensions?

13 If a lamp filament made of tungsten wire with a cross-sectional area of 5×10^{-9} m^2 is to have a resistance of 12 Ω at 20°C, how long must it be? Find the resistance of this filament at 2020°C. *Ans.* 1.09 m; 125 Ω

14 The field coil of an electric motor has a resistance of 36.9 Ω at 40°C and 31.8 Ω at 0°C. What is its temperature coefficient of resistance?

15 The field coil of a motor draws a current of 2.2 A from a 110-V line when the motor is started and the coil is at 0°C. What is the resistance of the coil at 0°C? If the potential difference across the coil does not change, what current is drawn by the copper field coil at its normal operating temperature of 60°C? *Ans.* 50 Ω; 1.78 A

16 The electric resistance of a tantalum wire is 30 Ω at 20°C, and 40.8 Ω at 120°C. What is its resistance at 200°C if the resistance of tantalum varies linearly with temperature? What is the temperature coefficient of resistance?

17 The resistance of a tungsten lamp filament is 9 Ω at 20°C. If the filament operates at 2220°C, find its resistance. How many times as great is the resistance at operating temperature as the resistance at room temperature?
  Ans. 102 Ω; 11.3

18 A platinum resistance thermometer is used to determine the temperature of an oven. If the average temperature coefficient of resistance of platinum is 0.0037 per Celsius degree and the resistance of the platinum coil is 100 Ω at 0°C, find the temperature when the coil has a resistance of 350 Ω.

19 An oven requires 8 A to heat it to the desired temperature when the applied voltage is 100 V. How much resistance must be inserted in series with the oven in order to keep it at the same temperature if the potential difference is increased to 120 V? How much power is dissipated in this resistor? *Ans.* 2.5 Ω; 160 W

20 An ammeter in series with a battery and a resistance R reads 6 A. When an additional resistance of 5 Ω is inserted in series with the first, the reading of the ammeter is reduced to 4 A. Find the resistance R. The resistances of battery and ammeter are negligible.

21 An airplane deicer operates from a 24-V battery and is capable of melting 0.05 kg of ice per minute. Find the resistance and current when the deicer is in operation. *Ans.* 11.6 A; 2.06 Ω

22 A 6,000-W heating unit in a hot-water heater works 3 hr/day. If electric energy costs 3 cents per kilowatt-hour, how much does hot water cost each 30-day month? How many kilograms of water are heated from 10 to 80°C each day if all the heat dissipated in the heater is effective in warming the water?

23 The belt of a Van de Graaff generator is 0.6 m wide and moves 30 m/sec. If half the charges sprayed on the belt reach the high-potential electrode, how many coulombs must be sprayed onto the belt each second to give a total current to the electrode of 3 mA? Find the charge sprayed on the belt per unit area.
  Ans. 0.006 C/sec; 3.3×10^{-4} C/m^2

24 An electric iron of 1.2-kg mass has an average specific heat of 0.10 kcal/(kg)(C°). The heating unit takes 8 A from a 110-V line. If half the heat is lost by radiation, how long will it take to bring the iron to a temperature of 120°C when it is at 20°C originally?

25 The resistors of the accompanying figure are a portion of a complete circuit, and $R_1 = 8\ \Omega$, $R_2 = 3\ \Omega$, and $R_3 = 6\ \Omega$. The current through R_3 is 0.5 A. Find the current in R_1 and the potential difference between A and B.
  Ans. 1.5 A; 15 V

26 If in the accompanying figure $R_1 = 5\ \Omega$, $R_2 = 12\ \Omega$, and $R_3 = 4\ \Omega$, find the current in R_1 and the potential difference between A and B if the current in R_2 is 0.25 A.

Problems 25 and 26

27 When the light switch of a car is turned on, two headlights, two taillights, and a dash light are connected in parallel with a 12-V battery of negligible internal resistance. If each headlight has a resistance of 3 Ω, each taillight a resistance of 24 Ω, and the dash light a resistance of 48 Ω, find the current through the battery.

Ans. 9.25 A

28 Four 120-V 40-W resistors are connected in parallel. The group is then connected in series with a coil of 90-Ω resistance. How much power is dissipated in each resistor and in the coil of wire when the combination is connected across a 100-V battery?

29 Calculate the resistance between points A and B in the accompanying figure if $R_1 = 24$ Ω, $R_2 = 8$ Ω, $R_3 = 7$ Ω, $R_4 = 12$ Ω, $R_5 = 6$ Ω, and $R_6 = 4$ Ω. *Ans.* 15 Ω

Problems 29 and 30

30 Calculate the resistance between points A and B of the accompanying figure if $R_1 = 30$ Ω, $R_2 = 20$ Ω, $R_3 = 25$ Ω, $R_4 = 24$ Ω, $R_5 = 8$ Ω, and $R_6 = 6$ Ω.

31 Find the resistance between A and B of the accompanying figure if $R_1 = 8$ Ω, $R_2 = 30$ Ω, $R_3 = 4$ Ω, $R_4 = 9$ Ω, and $R_5 = 18$ Ω.

Ans. 15.5 Ω

32 Find the resistance between A and B of the accompanying figure if $R_1 = 9$ Ω, $R_2 = 8$ Ω, $R_3 = 4$ Ω, $R_4 = 5$ Ω, and $R_5 = 20$ Ω.

Problems 31 and 32

33 If a 3-Ω resistor is placed along each of the 12 edges of a cube and connections are made at each corner, find the resistance between two diagonally opposite corners of the cube. (*Hint:* By symmetry each of the remaining six corners has the same potential as two others.)

Ans. 2.5 Ω

CHAPTER THIRTY-FOUR
ELECTRIC CIRCUITS

IN CHAP. 33 we found that the current I through a conductor was given by the ratio of the potential difference V across the conductor to its resistance R. To maintain a steady current in a conductor of fixed resistance, we must provide a constant potential difference. How this can be done is the first topic of this chapter. We clearly need some device which can convert some other kind of energy to electric energy—a device known by the clumsy name "source of electromotive force' or by the abbreviation emf. How can we find the current in each part of a complex circuit involving several resistors and emfs? We learn that this question can be answered by applying two of the great conservation laws of physics—conservation of charge and conservation of energy.

34.1 ELECTROMOTIVE FORCE

To maintain a constant current in a conductor, it is necessary to maintain a steady potential difference across the conductor. This potential difference can be supplied only if some device converts some other form of energy into electric energy. Such a device is called a *source of electromotive force* (abbreviated emf).

There are many kinds of emf. In batteries chemical reactions occur, transforming chemical energy into electric energy. In the giant generators of our electric power plants mechanical energy is converted into electric energy. In a thermocouple it is heat energy, while in the photoelectric cell it is radiant energy which is transformed.

When a charge q receives an energy W in passing through a battery or some other source of electric energy, the emf $\mathcal{E}$ is given by

$$\mathcal{E} = \frac{W}{q} \qquad (34.1)$$

When W is in joules and q in coulombs, the emf is in joules per coulomb, or volts.

When a charge of one coulomb receives one joule of energy upon passing through a source, the source is said to have an electromotive force of one volt. A 12-V battery delivers 12 J of energy to each coulomb which passes through it. Electromotive force and potential difference are measured in the same units. An emf is a particular kind of potential difference, namely, one which arises through the transformation of some other form of energy into electric energy.

In a source of emf not only may some other form of energy be transformed into electric energy, but the reverse process may also occur—electric energy may be converted into another form. For example, in charging a battery, charges are forced through the battery in a direction opposite that in which they go when the battery is discharging; these charges deliver electric energy to the battery, where it is converted into and stored as chemical energy.

34.2 THE CONSERVATION OF ENERGY IN A SIMPLE CIRCUIT

Consider the circuit of Fig. 34.1, in which a battery is connected in series with three resistors. A charge q which passes through the battery gains an amount of energy $\mathcal{E}q$ J. In passing through the resistance R_1, this charge loses energy V_1q J. In passing through resistors R_2 and R_3, this charge loses amounts of energy V_2q and V_3q, respectively. In going once around the complete circuit, the charge loses exactly the same amount of energy as it gains. Therefore,

$$\mathcal{E}q = V_1q + V_2q + V_3q \quad \text{and} \quad \mathcal{E} = V_1 + V_2 + V_3 \quad (34.2)$$

or
$$\mathcal{E} = IR_1 + IR_2 + IR_3 \quad (34.3)$$

From Eq. (34.3) we obtain

$$I = \frac{\mathcal{E}}{R_1 + R_2 + R_3} \quad (34.4)$$

In a simple series circuit the current is equal to the ratio of the emf to the sum of the resistances in the circuit. This statement represents a simplified form of *Ohm's law for a complete circuit.*

[handwritten note: THIS HAPPEN BECAUSE THE ELECTRON COLLISIONS DURING THE PATH.]

[handwritten box: $\mathcal{E} = V = IR.$]

34.3 THE RESISTANCES OF SOURCES OF ELECTROMOTIVE FORCE

Any source of emf, such as a battery or electric generator, has some internal resistance. As a consequence, a current through a battery produces some heating. When the battery is being discharged, the total energy given to a charge q is $\mathcal{E}q$, but a portion of this energy is converted into heat within the battery. If the internal resistance is r, the potential drop in the battery resistance is Ir. The potential difference between the terminals of the battery V_t is the net potential gain.

$$\boxed{V_t = \mathcal{E} - Ir} \quad (34.5)$$

The terminal potential difference is $\mathcal{E}$ *only* when no current is being drawn.

The decrease in the terminal potential difference of a battery when the current drawn is changed is illustrated by the dimming of automobile headlights when the starter is activated. The starter draws a large current from the battery. As a consequence of the increased Ir drop within the battery, the terminal potential is reduced, and the potential

Figure 34.1
A coulomb gains an amount of energy in passing through the battery equal to the energy it loses in the resistors as it goes around the complete circuit.

Figure 34.2

A real battery may be treated as an ideal resistanceless battery and a series resistor r.

difference across the lamps of the car is lower. For an automobile battery of emf about 12 V, the internal resistance may be about 0.005 Ω. The current drawn by the lights is approximately 6 A; thus, when current is being drawn only for the lights, the terminal potential difference of the battery is only a few hundredths of a volt less than the emf. If the current drawn from the battery is increased to 150 A by operating the starter motor, V_t becomes $12 - 0.75$ or 11.25 V. The brightness of the lamps is reduced by such a decrease in terminal potential difference.

For practical purposes we may assume that the real battery is made up of a pure emf and a series resistor, as suggested in Fig. 34.2. Ohm's law for a complete circuit is applicable, but the internal resistance of the battery must be included in the total resistance.

Example

A 45-V battery has an internal resistance of 0.6 Ω. It is connected in a circuit as indicated in Fig. 34.3. Find the current in each resistor.

First, the series equivalent of the two parallel resistors must be found.

$$\frac{1}{R_\parallel} = \frac{1}{12} + \frac{1}{8} = \frac{5}{24}$$

$$R_\parallel = 4.8 \ \Omega$$

Now applying Ohm's law for the complete circuit yields

$$I = \frac{\mathcal{E}}{R} = \frac{45}{0.6 + 20 + 4.8 + 7} = \frac{45}{32.4} = 1.39 \text{ A}$$

This is the current through the battery and the 20- and 7-Ω resistors. The potential difference across the parallel resistors is $1.39 \times 4.8 = 6.67$ V. Therefore, the current in the 8-Ω resistor is $6.67/8 = 0.83$ A, while that through the 12-Ω resistor is $6.67/12 = 0.56$ A. Note that the sum of these currents is 1.39 A. ////

34.4 CHARGING A BATTERY

The terminal potential difference of a battery as it is discharged is given by $\mathcal{E} - Ir$. While the battery is being charged, the terminal potential difference is greater than the emf by an amount Ir, as can be seen from consideration of the energy transformations. In charging the battery, electric energy is converted into chemical energy; the energy per unit charge is $\mathcal{E}$. The charging current produces heat in the battery, and the energy per unit charge required to produce this heating effect is Ir. Thus, the total electric energy which must be delivered to the battery per unit charge is

$$V_t = \mathcal{E} + Ir \tag{34.6}$$

Regardless of the direction of the current through a cell or other source of emf, Joule heat is always produced.

20 Ω

$\mathcal{E} = 45$ V

$r = 0.6$ Ω   12 Ω   8 Ω

7 Ω

Figure 34.3

Circuit composed of a battery and four resistors, two of which are in parallel.

34.5 CELLS IN SERIES

When two or more sources of emf are connected in series, the net emf is the algebraic sum of the individual emfs. If two cells are connected in series in such a way that both would produce a current in the same direction, the emf is the sum of the two emfs; on the other hand, if the two are connected in series in such a way that they would send currents in opposite directions, the net emf is the difference between the two. In the first case the cells are said to be connected in *series aiding,* in the second case in *series opposing.* When a battery is to be charged, it must be connected in series opposing with some other source of emf which supplies electric energy to be transformed into chemical energy.

When several identical cells are connected in series, the total emf is equal to the number of cells multiplied by the emf of a single cell, while the resistance of the battery is equal to the resistance of an individual cell times the number of cells. The type of cell used in ordinary automobile batteries has an emf of approximately 2 V; in order to obtain an emf of 6 V, three cells are connected in series; for 12 V, six cells are required.

[handwritten margin note: TOTAL EMF = NUMBER OF CELLS X EMF OF A SINGLE CELL.]

[handwritten margin note: RESISTANCE = RESISTANCE OF INDIVIDUAL CELL X NUMBER OF CELLS]

34.6 OHM'S LAW FOR A COMPLETE CIRCUIT

In Fig. 34.4 several batteries of negligible resistance are connected in series with a number of resistors. For this circuit one may again apply the fundamental principle that the total energy gained by a charge in going around the complete circuit is equal to the total energy lost.

If a unit charge is carried clockwise around the circuit of Fig. 34.4, the net change in potential energy for the complete circuit is zero, or

$$\mathcal{E}_1 - IR_1 - \mathcal{E}_2 - IR_2 + \mathcal{E}_3 - IR_3 = 0 \qquad (34.7)$$

[handwritten margin note: WHY?]

It is often convenient to mark plus and minus signs at the appropriate ends of all resistors and batteries. The potential change may be considered positive when the test charge is moved from the minus terminal of a circuit element to the positive one and negative when the charge goes from plus to minus. It is not necessary that the charge be carried around the circuit in the direction in which the current flows. Indeed, if the direction of motion is reversed, all signs in Eq. (34.7) are changed, but the sum is still zero. In the circuit of Fig. 34.4, battery 2 is being charged by the other batteries.

The total current, which is the same in all parts of the circuit, may be obtained by solving Eq. (34.7) to obtain

$$I = \frac{\mathcal{E}_1 - \mathcal{E}_2 + \mathcal{E}_3}{R_1 + R_2 + R_3} = \frac{\Sigma \mathcal{E}}{\Sigma R} \qquad (34.8)$$

This equation represents Ohm's law for a complete series circuit, which may be stated as follows:

The current in any series circuit is given by the ratio of the algebraic sum of the emfs to the total series resistance of the circuit.

Ohm's law for a complete circuit is to be distinguished from Ohm's law for a single resistor. The law for a single resistor involves only the potential difference across the resistor, while the law for the complete circuit involves the algebraic sum of the emfs in the circuit and the total resistance of the series circuit. In a series circuit the algebraic sum of the emfs is equal to the sum of the IR drops in the resistors. This follows directly from the law of conservation of energy.

Figure 34.4
Series circuit containing three sources of emf.

When one applies Ohm's law to a complete circuit in which there are parallel resistors, each group of parallel resistors is first replaced with the equivalent series resistor. Then Ohm's law for the circuit is applied directly.

Example

A battery of emf 20 V and internal resistance 1 Ω is connected in series with a 5-Ω resistor, a second battery of emf 8 V and internal resistance 2 Ω which is in series opposing, and a group of three resistors of 12, 6, and 4 Ω resistance in parallel, as shown in Fig. 34.5. Find the current in each resistor and the terminal potential difference in each battery.

We first replace the three parallel resistors with the equivalent single resistor: $R_\| = 2\ \Omega$ from $1/R_\| = \frac{1}{12} + \frac{1}{6} + \frac{1}{4}$. We next apply Ohm's law for the circuit, which gives

$$I = \frac{20 - 8}{1 + 5 + 2 + 2} = \frac{12}{10} = 1.2 \text{ A}$$

for the current in each battery and in the 5-Ω resistor. The potential drop across $R_\|$ is given by $IR_\| = 1.2 \times 2 = 2.4$ V, which remains unchanged if we replace $R_\|$ with the original three resistors. The currents in these resistors are given by V/R as follows: $I_{12} = 2.4/12 = 0.2$ A, $I_6 = 2.4/6 = 0.4$ A, and $I_4 = 2.4/4 = 0.6$ A. Note that the sum is 1.2 A, the current in the main circuit.

The terminal potential difference for the 20-V battery is $\mathcal{E} - Ir = 20 - 1.2 \times 1 = 18.8$ V, while the terminal potential difference for the 8-V battery which is being charged is $\mathcal{E} + Ir = 8 + (1.2 \times 2) = 10.4$ V. ////

34.7   KIRCHHOFF'S LAWS

Circuits ranging from the simplest to very complex networks with many branches and many emfs can be handled by application of two fundamental principles known as *Kirchhoff's laws*. These laws, which apply once a steady state has been reached, have been used in previous

Figure 34.5
Circuit with two opposing batteries connected in series.

discussions, although they have not been specifically named. Kirchhoff's two laws are:

1 *The sum of all currents arriving at any junction in a circuit is equal to the sum of the currents leaving that junction.*

2 *Around any closed loop, the sum of the potential rises is equal to the sum of the potential drops.*

WHEN WE HAVE MANY EQUATIONS, WE CAN SOLVE IT, FINDING THE SIMULTANEOUS SIDES OF THE CIRCUIT.

The first law must apply if we are to avoid an accumulation of charge at any junction in the circuit or a continuing disappearance of charge there. If Kirchhoff's first law were not true and the sum of the currents reaching a junction exceeded the currents leaving it, the charge would build up at this point, and the potential of the junction would change continuously.

Kirchhoff's second law is a special statement of the law of conservation of energy. If a charge gained more energy in going around a closed path than it lost, it would be able to gain more and more energy by repeated traversing of this path. This is obviously not permissible. Once the charge returns to its starting point, its potential energy must be exactly the same as when it started.

When a man takes a hike in the mountains and eventually returns to his starting point, he has climbed up exactly as many meters as he has descended, since he ends at the same altitude at which he started. Kirchhoff's second law is the electrical analog of this mechanical illustration.

In applying Kirchhoff's laws to a problem, it is convenient to carry out the following steps in order:

1 *Assign a direction and a symbol to the current in each independent branch of the circuit.* It is not necessary to worry about which direction to assign the current in a given branch, since if the incorrect assignment is made, the current will turn out to be negative.

DRAW AN ARROW OR SOMETHING LIKE THAT

2 *Place appropriate plus and minus signs at the terminals of every source of emf and every resistor in the circuit.* Remember, that in a resistor the current is from the plus to the minus terminal; thus the choices of current directions in the first step determine the signs of the terminals of all resistors.

POT (+) AND (-) ON THE RIGHT SIDES

3 *Apply Kirchhoff's first law at enough junctions for each current to appear in an equation.* Be sure that each junction equation contains at least one current which has not appeared in earlier equations.

ONE CURRENT → ONE JUNCTION OR MAIN POINT.

4 *Apply Kirchhoff's second law to closed loops until once again every current has been included in at least one equation.* When we go through a circuit element from minus to plus, potential is gained; from plus to minus it is lost. Once again be sure that every new loop equation involves at least one current which has not appeared in a previous loop equation.

5 *Solve the equations for the desired unknowns.*

A familiarity with the use of Kirchhoff's laws may be obtained by studying one or more examples and then by practice on additional problems.

Figure 34.6

Circuit in which there are three branches, each with a different current.

Example

Find the current in all branches of the circuit of Fig. 34.6.

1   Let us designate the currents by I_1, I_2, and I_3 and assume them to be in the directions indicated by the arrows.

2   We place a plus on the higher-potential end of each circuit component, and a minus on the lower-potential end.

3   At point A, $I_1 + I_2 = I_3$, by Kirchhoff's first law.

4   If we start at point A and apply Kirchhoff's second law to the left loop going clockwise, we obtain

$$-15 + 2I_2 + 7I_2 - 3I_1 - I_1 + 40 = 0$$

By going clockwise from A around the right loop, we obtain

$$-8I_3 - 7I_2 - 2I_2 + 15 = 0$$

We now have the equations

$$I_1 + I_2 - I_3 = 0$$
$$4I_1 - 9I_2 = 25$$
$$9I_2 + 8I_3 = 15$$

The solutions are $I_1 = 4$ A, $I_2 = -1$ A, and $I_3 = 3$ A. The fact that I_2 is negative means that the current I_2 is in the direction opposite that assumed. The 15-V battery is being charged.   ////

When several identical sources of emf are connected in parallel, the emf of the combination is the emf of a single source. The net resistance of this combination is equal to the resistance of a single cell divided by the number of cells, since, effectively, all the resistances are in parallel. Cells which are not identical are seldom connected in parallel. Although Kirchhoff's laws are valid in any situation in which currents are constant, it is easier to solve circuits such as that of Fig. 34.7 by the method illustrated below. Note that this method leads to correct results only when the cells are identical, having the same emf, the same internal resistance, and the same lead resistances connecting the sources to the external circuit.

Example

Three identical dry cells with $\mathcal{E} = 1.5$ V and $r = 0.12$ Ω are connected in parallel to the circuit of Fig. 34.7. Find the current in each cell and in each resistor if $R_1 = 5$ Ω, $R_2 = 30$ Ω, and $R_3 = 60$ Ω.

Figure 34.7

Circuit with three identical cells in parallel.

The emf of the three identical cells in parallel is 1.5 V, and the resistance of the combination is 0.04 Ω (three 0.12-Ω resistors in parallel). The resistance of the 30- and 60-Ω parallel combination is 20 Ω. Application of Ohm's circuit law yields

$$I = \frac{1.5}{0.04 + 5 + 20} = \frac{1.5}{25.04} = 0.060 \text{ A}$$

The potential drop across the 30- and 60-Ω resistors is 20 × 0.060, or 1.2 V. Therefore, $I_{30} = 1.2/30 = 0.04$ A, and $I_{60} = 1.2/60 = 0.02$ A. The current through each of the three identical cells is one-third of 0.06 A, or 0.02 A. ////

34.8 THE WHEATSTONE BRIDGE

An accurate and simple method of measuring resistances employs the Wheatstone bridge. The circuit for this bridge is shown in Fig. 34.8. A and B are fixed resistors, the values of which are known. The resistance X whose value is to be determined is connected in the third arm of the bridge, while a known variable resistance R is connected in the fourth arm. The resistance R is varied until there is no current between c and d as indicated by the galvanometer G, an instrument for detecting small currents. When the galvanometer shows no current between c and d, the bridge is said to be *balanced*.

If there is no current in the galvanometer, the potential difference between points c and d must be zero. For this to be true, the potential drop across resistor A must be equal to the potential drop across X, since one end of each of these resistors is at the potential of point a. If I_A is the current in A, and I_X the current in X, the condition for no current in the galvanometer is that $V_{ac} = V_{ad}$ or $I_X X = I_A A$. Similarly, the fall of potential from c to b must be equal to the drop in potential from d to b, which gives $I_R R = I_B B$. The current in R and in X is the same when no current flows through the galvanometer. Similarly, the current in A is the same as that in B. Therefore $I_X X/I_R R = I_A A/I_B B$, or

$$\boxed{\frac{X}{R} = \frac{A}{B}} \qquad (34.9)$$

In the slide-wire Wheatstone bridge the resistors A and B are segments of a wire of uniform cross section and resistivity. The ratio A/B is determined by the position of the movable contact along this wire. In high-quality commercial Wheatstone bridges, the ratio of A to B can be set at one of several values and the final balance made by varying R.

34.9 THE POTENTIOMETER

The potentiometer occupies an important place in electrical measurements because it can be used to measure potential differences with great accuracy and *without drawing any current*. This feature is of great importance in working with sources of emf of high internal resistance and low current capabilities.

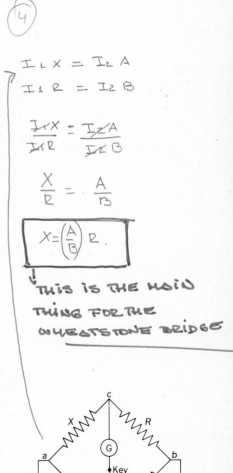

$I_1 X = I_2 A$

$I_1 R = I_2 B$

$\dfrac{I_1 X}{I_1 R} = \dfrac{I_2 A}{I_2 B}$

$\dfrac{X}{R} = \dfrac{A}{B}$

$X = \left(\dfrac{A}{B}\right) R$

THIS IS THE MAIN THING FOR THE WHEATSTONE BRIDGE

Figure 34.8
A Wheatstone bridge circuit.

A schematic diagram of a potentiometer is shown in Fig. 34.9. The wire ab, which is of uniform cross section, carries a current maintained constant by the working battery B. There is a progressive drop in potential along the wire from a to b, directly proportional to the distance from a. To measure an emf whose value $\mathcal{E}_x$ is unknown, this emf is placed in series with a galvanometer and connected as shown in Fig. 34.9, in which point c represents a movable contact. It is important that the emf $\mathcal{E}_x$ oppose the current which the working battery B would produce if $\mathcal{E}_x$ were zero. If the potential difference between a and b is greater than $\mathcal{E}_x$, there is some point c at which the potential difference across ac is equal to $\mathcal{E}_x$. This point can be found by moving the sliding contact until the current through the galvanometer is zero. The potentiometer is then *balanced*.

In the usual application of the potentiometer a standard cell of known emf $\mathcal{E}_s$ is first used in place of $\mathcal{E}_x$, and the balance point is found at some point d. The unknown emf $\mathcal{E}_x$ is substituted for $\mathcal{E}_s$, and the balance point C is found. Then $\mathcal{E}_x$ can be calculated from the relation

$$\frac{\mathcal{E}_x}{\mathcal{E}_s} = \frac{ac}{ad}$$

where ac and ad represent the length of conductor from a to c and from a to d, respectively.

Figure 34.9
A simple potentiometer circuit.

$$E_S = X \, L_S$$
$$E_X = X \, L_X$$
$$\boxed{E_X = E_S \, \frac{L_S}{L_S}}$$

QUESTIONS

1 What is the advantage of connecting identical cells in series?
2 What is the advantage of connecting identical cells in parallel?
3 Why are batteries which are not identical seldom connected in parallel?
4 What is the distinction between the terms *potential difference* and *emf*? Can a battery have an emf and still have no potential difference between its terminals? Under what circumstances?
5 Can the terminal potential difference of a battery exceed its emf? Can the terminal potential difference be negative in the sense that what is normally the + terminal of the battery becomes the − terminal? Explain.
6 Why is it difficult to obtain a good measurement of resistance by use of an ammeter and a voltmeter? Draw circuits to illustrate. Recall that ordinary ammeters and voltmeters have resistance too.
7 Why is a voltmeter rated at 20,000 Ω/V usually more desirable and more expensive than one of the same range rated 5,000 Ω/V?
8 Draw a Wheatstone bridge, and derive the relationship among the resistances when the bridge is balanced.

9 Does interchanging the galvanometer and battery in a Wheatstone bridge affect the balance? Justify your answer with equations.
10 Draw a potentiometer circuit, and explain how it operates. How could the potentiometer be used to calibrate an ammeter with a standard resistor? Draw a suitable circuit.

PROBLEMS

1 Six cells of a storage battery, each with a resistance of 0.015 Ω and an emf of 2.05 V, are connected in series with a coil having a resistance of 6.06 Ω. Find the current and the terminal potential difference of the battery.
Ans. 2 A; 12.1 V
2 A battery of emf 6 V and internal resistance of 0.2 Ω is connected in series with two resistors, one of 4.8 Ω and another of 10 Ω. Find the current, the potential difference across each resistor, and the terminal potential difference of the battery.
3 A 15-Ω resistor bears a current of 0.4 A when it is connected in parallel with two other resistors of 30 and 10 Ω. Find the current in each of these two resistors. *Ans.* 0.2 and 0.6 A

4 A circuit has three parallel branches with resistances of 50, 30, and 80 Ω. When a current of 0.8 A is flowing in the 50-Ω branch, how much current is flowing in each of the other branches?

5 A battery has an emf of 6 V. Determine its internal resistance if a current of 50 A is drawn when an ammeter and leads with a total resistance of 0.1 Ω are connected across the terminals of the battery.   *Ans.* 0.02 Ω

6 In the circuit of the accompanying figure, find the current in R_1 and the charge stored on the capacitor if $\mathcal{E} = 12$ V, $r = 1$ Ω, $R_1 = 11$ Ω, $R_2 = 18$ Ω, and $C = 0.1$ μF.

7 In the circuit of the accompanying figure, find the current in R_1 and the charge stored on the capacitor if $\mathcal{E} = 4$ V, $r = 0.5$ Ω, $R_1 = 4.5$ Ω, $R_2 = 15$ Ω, and $C = 3$ μF.

Ans. 0.2 A; 9 μC

Problems 6 and 7

8 A battery having an internal resistance of 0.2 Ω and an emf of 6 V is used to send a current through a 9-Ω resistance connected in series with two resistances, one of 12 Ω and the other of 4 Ω, in parallel. Find the current in each of the resistances.

9 A 60-V battery with an internal resistance of 1 Ω is to be charged at the rate of 5 A from a 110-V source. What resistance must be connected in series with the battery? What will be the terminal potential difference of the battery during the charging process? Assuming that the emf arises entirely from the conversion of chemical to electric energy, find the rate at which electric energy is converted into chemical energy. At what rate is electric energy converted into heat in the battery? In the series resistor?

Ans. 9 Ω; 65 V; 300 W; 25 W; 225 W

10 A storage battery having an emf of 36 V and an internal resistance of 1.5 Ω is to be charged by connecting it to a 100-V generator. What resistance must be introduced in series with it in order for the charging current to be 8 A? What is the terminal potential difference of the battery?

11 When an external resistance of 75 Ω is connected to the terminals of a battery, the current is found to be 0.5 A. When this resistance is increased to 135 Ω, the current drops to 0.3 A. Find the emf and internal resistance of the battery.   *Ans.* 45 V; 15 Ω

12 Find the current through the batteries and through R_3 of the accompanying figure if $\mathcal{E}_1 = 40$ V, $r_1 = 1$ Ω, $\mathcal{E}_2 = 24$ V, $r_2 = 2$ Ω, $R_1 = 5$ Ω, $R_2 = 9$ Ω, $R_3 = 12$ Ω, and $R_4 = 4$ Ω. Find the terminal potential differences of the batteries and the potential difference between A and B.

13 Find the current through the batteries and through R_4 of the accompanying figure if $\mathcal{E}_1 = 35$ V, $r_1 = 1$ Ω, $\mathcal{E}_2 = 25$ V, $r_2 = 2$ Ω, $R_1 = 9$ Ω, $R_2 = 18$ Ω, $R_3 = 30$ Ω, and $R_4 = 60$ Ω. Find the terminal potential differences of the batteries and the potential difference between A and B.

Ans. 1.2 and 0.4 A; 33.8 and 22.6 V; 1.4 V

14 In the accompanying figure R_4 is 6 Ω and bears a current of 2 A. If $R_3 = 4$ Ω, what current does R_3 carry? If $\mathcal{E}_1 = 24$ V, $r_1 = r_2 = 1$ Ω, $R_1 = 5$ Ω, and $R_2 = 3$ Ω, find $\mathcal{E}_2$.

15 In the accompanying figure $\mathcal{E}_1 = 24$ V and $\mathcal{E}_2 = 6$ V, $r_1 = r_2 = 0.5$ Ω, $R_1 = 2$ Ω, and $R_4 = 6$ Ω. The current in R_3 is 2 A, and that in the main circuit is 3 A. Find R_3 and R_2.

Ans. 3 Ω; 5 Ω

Problems 12 to 15

16 Five storage batteries, each having six cells with an emf of 2 V and an internal resistance of

0.015 Ω per cell, are to be charged in series at the rate of 10 A from a 110-V line. How much resistance must be inserted in series with the batteries?

17 Twelve identical cells, each with an internal resistance of 0.6 Ω, are connected so as to have three parallel groups of four cells in series. The combination sends a 1.5 A through an external resistance of 4 Ω. What is the emf of each cell?

Ans. 1.8 V

18 Six identical cells, each with an internal resistance of 0.24 Ω, are connected in parallel to send a current through an external resistance of 0.76 Ω. How much current will be obtained in the external resistance if each cell has an emf of 1.5 V?

19 In the accompanying figure, $\mathcal{E}_1 = 18$ V, $r_1 = 1\ \Omega$, $\mathcal{E}_2 = 6$ V, $r_2 = 2\ \Omega$, $R_1 = 3\ \Omega$, and $R_2 = 4\ \Omega$. Find the current, the terminal potential difference of each battery, and the potential difference between points A and B.

Ans. 1.2 A; 16.8 and 8.4 V; 13.2 V

Problems 19 and 20

20 In the accompanying figure, $\mathcal{E}_1 = 24$ V, $r_1 = 1\ \Omega$, $\mathcal{E}_2 = 10$ V, $r_2 = 2\ \Omega$, $R_1 = 8\ \Omega$, and $R_2 = 17\ \Omega$. Find the current, the terminal potential difference of each battery, and the potential difference between points A and B.

21 In the accompanying figure, $I_1 = 5$ A, and $I_2 = 3$ A. What is I_3? If $\mathcal{E}_1 = 48$ V, $r_1 = 1\ \Omega$, $\mathcal{E}_2 = 13$ V, $r_2 = 2\ \Omega$, $R_2 = 3\ \Omega$, and $R_4 = 5\ \Omega$, find R_1 and R_3. *Ans.* 2 A; 3 Ω; 9 Ω

22 In the accompanying figure, $I_2 = 2$ A, and

Problems 21 and 22

$I_3 = 5$ A. If $R_1 = 4\ \Omega$, $r_1 = 2\ \Omega$, $R_2 = 5\ \Omega$, $r_2 = 1\ \Omega$, $R_3 = 3\ \Omega$ and $R_4 = 5\ \Omega$, find $\mathcal{E}_1$ and $\mathcal{E}_2$.

23 In the accompanying figure, $\mathcal{E}_1 = 50$ V, $r_1 = 1\ \Omega$, $R_1 = 4\ \Omega$, $\mathcal{E}_2 = 45$ V, $r_2 = 2\ \Omega$, $R_2 = 3\ \Omega$, and $R_3 = 7\ \Omega$. Find the current in each element in the circuit.

Ans. 3 A; 2 A; 5 A

24 In the accompanying figure, $\mathcal{E}_1 = 24$ V, $r_1 = 0.5\ \Omega$, $R_1 = 2\ \Omega$, $\mathcal{E}_2 = 50$ V, $r_2 = 1\ \Omega$, $R_2 = 8\ \Omega$, and $R_3 = 1.4\ \Omega$. Find the current in each element in the circuit.

25 In the accompanying figure $R_3 = 20\ \Omega$ and $I_3 = 2.5$ A. If $\mathcal{E}_1 = 60$ V, $r_1 = 2\ \Omega$, $R_1 = 18\ \Omega$, $r_2 = 1\ \Omega$, and $R_2 = 8\ \Omega$, find I_1 and $\mathcal{E}_2$.

Ans. 0.5 A; 68 V

Problems 23 to 25

26 The current in any element of a network (such as we have been considering) as the result of the simultaneous action of a number of emfs in the network is the sum of the currents which

would exist in this element if each source of emf were considered separately, with all other sources of emf being replaced by their internal resistances. This is a statement of the *superposition theorem*. Find the currents in the circuit of Fig. 34.6 by applying this theorem.

27 Twelve identical resistances of value R are connected to form the edges of a cube. If they are soldered together at the corners, find the effective resistance between opposite corners of one face of the cube.   *Ans.* $3R/4$

28 A battery has an emf of 60 V and an internal resistance of 4 Ω. Find the power dissipated in a variable external series resistor set, successively, at the values 2, 3, 4, 5, and 6 Ω. Plot a rough curve of power dissipated in the load as a function of load resistance to confirm the fact that maximum power is supplied to the load when the load resistance is equal to the internal resistance of the source.

CHAPTER THIRTY-FIVE
CHEMICAL AND THERMAL ELECTROMOTIVE FORCES

IN CHAP. 34 we made use of the fact that batteries convert chemical energy to electric energy, but we made no effort to explain how this energy transformation was accomplished. We now treat some of the basic facts of electrochemistry, in terms of which the functioning of simple batteries can be understood. But batteries are not the only sources of emf in which we are interested. In this chapter we also discuss thermocouples, which convert thermal energy to electric energy.

35.1 ELECTROLYSIS

If two copper plates are inserted into a beaker filled with water and connected to the terminals of a battery (Fig. 35.1), an ammeter in the circuit shows no current. However, if a little copper sulfate ($CuSO_4$) is poured into the beaker, there is a current through the solution. When the copper sulfate dissolves in water, many of the molecules split (or dissociate) into two charged particles called *ions*. One is a copper atom from which two electrons are missing, and the second is a sulfate (SO_4^{--}) ion with two excess electrons. The net charge in atomic units carried by an ion is called the *valence* of the ion. Thus the valence of the Cu^{++} ion is 2, while that of the sulfate ion is -2. In the solution the positively charged copper ions migrate to the negatively charged plate; the negatively charged sulfate ions, to the positive plate.

After a current has passed between the two copper electrodes for some time, the *cathode* (the electrode by which the conventional current leaves) has gained weight and is bright. The *anode* (by which current enters) has lost weight. There has not only been a transfer of electricity through the solution, but copper has been carried from one plate to the other. The Cu^{++} ions which reach the cathode obtain two electrons there and are deposited as neutral copper atoms, while each sulfate ion which goes to the anode gives it the two excess electrons and joins with an atom of copper. The copper sulfate so formed goes into solution, thus keeping the amount of $CuSO_4$ constant. The net effect is a gain of copper by the cathode and a loss of copper by the anode.

The addition of any acid, base, or salt to the pure water in the beaker makes it conducting, provided the solute dissociates into ions. When a current is passed between two platinum electrodes in a dilute solution of sulfuric acid in water (Fig. 35.2), hydrogen is released at the negative electrode and oxygen at the positive electrode. The water is decomposed into its constituents. The volume of hydrogen released is twice that of oxygen. A simplified explanation of the process by which the water is decomposed is as follows. A sulfuric acid molecule in solution splits into one H^+ ion and one HSO_4^- ion. The hydrogen ions

Figure 35.1
In the electrolysis of copper sulfate copper ions migrate to the cathode and sulfate ions to the anode.

are attracted to the cathode, where they receive electrons to form neutral hydrogen atoms. Two atoms promptly form a hydrogen molecule which rises to the top of the collecting tube. An HSO_4^- ion gives up an electron at the positive terminal and then unites with an atom of hydrogen to form sulfuric acid. These atoms of hydrogen are taken from the water, and oxygen is set free. Two atoms of oxygen unite to form a molecule of oxygen gas. The sulfuric acid formed at the anode goes into solution; consequently, the amount of sulfuric acid in the water does not change.

35.2 FARADAY'S LAWS OF ELECTROLYSIS

The quantitative laws of electrolysis may be introduced by considering a number of electrolytic cells connected in series (Fig. 35.3) with a battery so that the same current passes through each cell for the same time. Suppose cell A contains a solution of silver nitrate with silver electrodes, B a solution of sulfuric acid with platinum electrodes, C a solution of copper sulfate with copper electrodes, and D a solution of nickel chloride with nickel electrodes. Passage of a charge through the cells liberates, at the respective cathodes, silver, hydrogen, copper, and nickel. By determining the amount of substance liberated at each cathode as a function of the charge transported, it is possible to confirm two laws of electrolysis, which were discovered by Michael Faraday.

1   *The mass of any substance liberated is proportional to the charge which passes through the cell.* Hence, the mass liberated is proportional to the product of the current and the time.

2   *The masses of different elements liberated by a given charge are proportional to the ratios of atomic weight to valence.*

Suppose that the current exists until 1.008 g of hydrogen is liberated in cell B. Then, in cell A, 107.9 g of silver is deposited; in cell C, 31.77 g of copper; in cell D, 29.35 g of nickel. Whatever current is chosen, and whatever the length of time it exists, the masses deposited at these cathodes always bear the same ratio to each other and are proportional to the quotient obtained by dividing the atomic weight by the valence. The ratio of the atomic weight to the valence of an element is called the *chemical equivalent* or *combining weight*. When an element is monovalent, the chemical equivalent is equal to the atomic weight. If the element is divalent, the chemical equivalent is equal to one-half the atomic weight.

It requires 96,487 C to deposit one gram equivalent wt of any element, and 9.65×10^7 C to deposit one kilogram equivalent weight.

Figure 35.2
The electrolysis of water produces hydrogen at the cathode and oxygen at the anode.

Figure 35.3
Four electrolysis cells connected in series.

A charge of 96,487 C is called one *faraday*. Faraday's two laws of electrolysis can be summarized by the relation

$$m = \frac{Q}{9.65 \times 10^7} \frac{A}{v} \qquad (35.1)$$

where m is the mass of the element liberated in kilograms, Q is the total charge in coulombs, A is the atomic weight of the element, and v is its valence.

The mass deposited is directly proportional to the charge. *The electrochemical equivalent of any substance is defined as the ratio of the mass of the substance deposited to the charge transferred.* If Z represents the electrochemical equivalent, by Eq. (35.1)

$$Z = \frac{m}{Q} = \frac{1}{9.65 \times 10^7} \frac{A}{v} \qquad (35.2)$$

One accurate method for measuring the electric charge which passes through a circuit is to insert in the circuit an electrolytic cell containing a solution of silver nitrate. The mass of silver deposited is a measure of the total quantity of charge passed. Indeed, for many years the ampere was defined as that current which would deposit 0.001118 g/sec from a standard solution of silver nitrate. Since 1948 the ampere has been defined in terms of the interaction between two current-carrying conductors (Sec. 36.8).

35.3 APPLICATIONS OF ELECTROLYSIS

Electrolysis is of great commercial importance. The chromium plating of automobile parts and the silver plating of tableware are examples of the wide variety of commercial plating operations. Practically all our aluminum is produced by the electrolysis of aluminum oxide from a molten mixture. It takes 22 kW-hr of electrical energy to produce a single kilogram of aluminum. Chlorine and many other commercially important elements are obtained by electrolytic processes.

By no means are all electrolytic processes desirable. Electrolysis is an important factor in limiting the life of underground pipes. When a steel pipe is laid near an electrified railroad, the current may find its way into the pipe instead of traveling from the generator to the motors directly through the track. At certain points the water pipe may be eaten away, as is the anode in the electrolysis of copper sulfate. The electrochemical action is complex, but, as part of the process, iron is removed from the pipe in regions where it serves as anode. Electrolytic processes that lead to early disintegration must be carefully guarded against in many types of construction. Electrolytic action is a danger whenever different metals are electrically connected by a conducting solution. Vulnerable areas of some ships are protected by the use of "sacrificial" anodes; in one example an aluminum cable, charged positive relative to the hull, is dragged behind the ship to give electrolytic deposition on the ship in preference to the electrolytic removal of metal which would otherwise occur.

Figure 35.4
Diffusion of ions in a zinc–lead–sulfuric
acid cell.

35.4 CHEMICAL ELECTROMOTIVE FORCES

Electric energy used to send electric current through a solution can produce chemical reactions such as the liberation of hydrogen and oxygen from water. The reverse processes, i.e., chemical reactions providing electric energy, also occur. This fact was discovered about 1800 by the Italian physicist Volta, who built the first batteries.

Most batteries are composed of several cells connected in series. In each cell a chemical reaction converts chemical energy into electric energy. One easily analyzed cell consists of a zinc plate and a lead plate immersed in a dilute solution of sulfuric acid (Fig. 35.4). Zinc ions leave the metal and go into solution. Each zinc ion bears two units of positive charge, having left two electrons with the metal. Thus, the zinc metal becomes negatively charged. As it becomes more negative, it attracts positive zinc ions in the solution. Eventually an equilibrium is reached in which the rate of loss of zinc ions is equal to the rate of return. When this occurs, the potential of the zinc is lower than the potential of the solution. Similarly, lead ions leave the lead plate and go into solution. They bear two unit positive charges and leave the lead plate negative relative to the solution. However, equilibrium for the lead ions is established when the lead is less negative than the zinc. There is now a potential difference between the zinc plate and the lead plate, as shown in Fig. 35.5*a*. If a wire is connected between the zinc and lead plates, electrons flow through the wire from the zinc to the lead. When there is a current through the cell, there is a potential drop in the solution, owing to its resistance (Fig. 35.5*b*). As electrons leave the zinc plate, it becomes less negative; the equilibrium is disturbed, so that more zinc ions go into solution. At the same time the lead plate becomes more negative, and positive ions are attracted back.

Figure 35.5
How the potential varies in a
zinc–lead–sulfuric acid cell (*a*) when
there is no current and (*b*) when a
current is drawn.

Chemical cells can be made from a large number of different materials. The emf depends on the particular materials involved. Generally speaking, the chemically more active metal forms the negative terminal. Table 35.1 shows the electromotive series of metals. It indicates the potentials of various metals relative to a hydrogen electrode formed by bubbling hydrogen gas over a spongy platinum conductor. The values in the table are for standard ion concentrations.

An *ideal* cell utilizing any two of these elements as electrodes has an emf given by the difference between the voltages which are listed in Table 35.1 for the elements in question. For example, the potential of the copper-zinc cell can be predicted by observing that zinc, relative to hydrogen, has a potential of -0.76, while copper has a potential of $+0.34$. A copper-zinc cell has an emf of approximately 1.10 V.

The emf of a cell is determined primarily by the energy released in the chemical reactions per coulomb of charge transferred through the cell. However, it would be erroneous to assume that the only energy transformations occurring in a cell are between chemical and electric energy. When some cells are discharged, part of the chemical energy is transformed into heat; when such cells are charged, this heat is retransformed to chemical energy. Other cells transform heat energy into electric energy during discharge and electric energy into heat energy when they are being charged.

35.5 POLARIZATION OF CELLS

If a zinc strip and a copper strip are inserted into a juicy lemon, the juice serves as electrolyte and a potential difference can be observed between the zinc and the copper. Another simple cell can be made by immersing a zinc plate and a copper plate in a dilute solution of some acid such as H_2SO_4. Neither of these zinc-acid-copper cells performs very satisfactorily as a source of emf because, as current is drawn, hydrogen ions are deposited on the copper. This reduces the emf, since the positive electrode is now essentially hydrogen, rather than copper. Further, hydrogen bubbles on the surface of the plate form an insulating layer and greatly increase the resistance of the cell. This is an example of an effect called *polarization*. (This should not be confused with polarization of a dielectric.) Polarization can occur in many ways in a cell, but its net effect is to reduce the observed emf to a value below that which would be expected on the basis of the electrochemical series. This occurs because one or both of the terminals is coated with some less effective material.

Table 35.1
The Electromotive (or Electrochemical) Series

Element	Potential difference,[*] V
Li	−2.96
Rb	−2.93
K	−2.92
Ca	−2.76
Na	−2.71
Mg	−2.40
Al	−1.70
Zn	−0.76
Fe	−0.44
Cd	−0.40
Ni	−0.23
Sn	−0.14
Pb	−0.12
H	0
Cu	0.34
Ag	0.80
Hg	0.80
Au	1.5

[*] All potential differences are referred to a standard hydrogen electrode taken as zero and are for a temperature of 25°C.

35.6 PRIMARY CELLS

In most cases the chemical reaction in a cell cannot be reversed by changing the direction of the current. A cell in which the chemical reaction is irreversible is known as a *primary* cell, while one which can be charged and discharged repeatedly and reversibly is called a *secondary* cell. The simple zinc-acid-copper cell discussed in Sec. 35.4 is a primary cell. If one attempts to reverse the chemical process by making the zinc electrode positive, zinc ions do not plate out. Instead copper and hydrogen ions are deposited, and the zinc stays in solution. The reaction is not reversible.

The chemical reactions which take place in several kinds of cells of practical or historical importance are discussed below.

The Daniell Cell One of the earliest practical sources of emf was the Daniell cell. It consists of a copper electrode submerged in copper sulfate and a zinc electrode in zinc sulfate. A porous partition separates the zinc sulfate from the copper sulfate such that charges can pass through but the chemicals do not mix readily. The zinc electrode becomes negative and the copper plate positive. When the plates of the Daniell cell are connected through a resistance, electrons flowing through the wire reduce the negative charge on the zinc terminal and permit more zinc to go into solution. The potential of the copper electrode is reduced, and copper ions are deposited from the solution. The net reaction converts zinc and copper sulfate into zinc sulfate plus copper. In this reaction chemical energy is transformed into electric energy.

In the early days of telegraphy the Daniell cell was the standard source of emf. Its output is about 1.1 V, but the terminal potential difference of a cell is somewhat less when current is drawn. The *volt* as the practical unit of potential difference originated with the use of Daniell cells in early telegraph circuits, when the "voltage" was just the number of Daniell cells connected in series.

Dry Cells The Leclanché cell (Fig. 35.6) is a primary cell of interest because a modification, known as the *dry cell*, is widely used. A Leclanché dry cell consists of a zinc can containing a centered graphite, i.e., carbon, rod surrounded by a moist paste of manganese dioxide (MnO_2), zinc chloride ($ZnCl_2$), ammonium chloride (NH_4Cl), and carbon black. Manganese dioxide is a poor conductor, and the carbon black serves to reduce the cell resistance. A thin porous separator prevents direct electric contact between the zinc and the mixture of manganese dioxide and carbon black. The term "dry cell" originates from the fact that the electrolyte is absorbed in the porous materials. As the cell delivers current, zinc is oxidized at the negative terminal and Zn^{++} ions go into solution. At the positive electrode MnO_2 is reduced through complex reactions which seem to vary depending on how much current is drawn from the cell. In the discharge of NH_4^+ ions at the inert carbon electrode, hydrogen is released in the reaction $2NH_4^+ + 2e^- \rightarrow 2NH_3 + H_2$, where e is the abbreviation for electron. The MnO_2 reacts with the hydrogen to produce water and perhaps Mn_2O_3. The NH_3 is immediately absorbed in the water. When too large a current is drawn, hydrogen accumulates rapidly and the cell becomes polarized with the terminal potential difference falling and the cell resistance rising. After a rest period the cell may regain its normal emf of about 1.5 V. Such a cell is well adapted to work in which it is used a short time and then allowed to stand.

The *magnesium dry cell* is constructed in a similar way except that the outer can is magnesium and the electrolyte includes magnesium bromide rather than ammonium chloride. The *mercury dry cell* has a

Figure 35.6
A Leclanché dry cell.

Figure 35.7
A Weston mercury-cadmium standard cell.

zinc negative electrode, a positive electrode of solid mercuric oxide (HgO) mixed with carbon, and an electrolyte of potassium hydroxide (KOH) and zinc oxide. The overall reaction $Zn + HgO \rightarrow ZnO + Hg$ gives an emf of 1.35 V. This cell is widely used in cameras and hearing aids because it provides more energy and has lower internal resistance than a Leclanché cell of the same weight. A further advantage is that it will operate at lower temperature, so it is used in some missile and aviation applications.

Weston Standard Cell   Standard cells offer a means of obtaining definite, known, and constant potential differences. The most widely used of these standard cells is the cadmium or Weston cell (Fig. 35.7). Its emf changes very little with temperature, which is one of the reasons the Weston cell is considered the best standard available. At 20°C its emf is 1.0183 V, and the emf decreases about 4×10^{-5} V for each Celsius degree of temperature increase.

35.7   SECONDARY CELLS

A few of the many possible chemical reactions which may be used for batteries are reversible. The lead storage cell and the Edison cell are the best known among those which can be discharged and recharged repeatedly.

Lead Storage Cell   In the lead storage cell, both the positive and negative plates are made of heavy lead grids full of holes or grooves filled with the active material. The positive plates contain lead peroxide and the negative plates spongy lead. A cell is usually formed of a number of such plates, alternately negative and positive, covered with sulfuric acid. The negative plates are connected together and act as one terminal of the cell, and the positive plates are connected together to form the other terminal. It is customary to have one more negative plate than positive so that every positive plate lies between two negative ones. In this way, both sides of a positive plate are charged or discharged. During the process of recharging, there is a restoration of the peroxide accompanied by an increase in volume, causing a swelling of the plate. Since this swelling takes place equally on both sides of a positive plate, there is little tendency for the plates to warp. If the plates are close together and if, through the use of a large number of plates, the area is large, the cell has a small internal resistance. By increasing the number of plates and making the areas larger, the current capacity of the cell is increased.

The overall chemical reaction taking place in the discharge of the lead storage cell is

$$Pb + PbO_2 + 2H_2SO_4 \rightarrow 2PbSO_4 + 2H_2O$$

When the cell is delivering energy, both the lead peroxide on the positive plates and the spongy lead of the negative electrode are converted to lead sulfate. As this occurs, the emf gradually decreases and the density of the electrolyte falls (Fig. 35.8). Consequently, it is possible to find the state of charge of a battery by measuring the density of the sulfuric acid. When the battery is charged, the reactions are reversed.

Edison Storage Cell A storage battery composed of Edison cells is lighter, more rugged, and longer-lived than a lead storage battery. In the Edison cell the negative plate is a nickel-plated steel grid with a large number of pockets filled with powdered iron and iron oxide (FeO). The positive plate is a nickel-plated steel grid with perforated steel tubes filled with alternate layers of nickel oxides and flaked nickel. The electrolyte is a solution of potassium hydroxide. When the cell is discharging, the nickel oxides are reduced to lower oxides, and the iron is oxidized. When the cell is being charged, the reaction is reversed. The electrolyte enters into intermediate reactions. Its effect in charging and discharging is to transfer oxygen from one plate of the cell to the other. The density changes only slightly during the reactions. The normal emf of an Edison cell is about 1.2 V, varying from 1.4 to 0.9 V as the cell discharges. This variation of emf and the relatively high cost are major disadvantages of Edison cells.

There are two other types of secondary cells which have considerable use. The *nickel-cadmium-alkaline cell* gives a nominal potential difference of 1.34 V from the oxidation of cadmium and the reduction of nickel in $Ni(OH)_4$. The *zinc–silver oxide–alkaline cell* uses a potassium hydroxide electrolyte and has an emf of 1.86 V. In it zinc is oxidized while silver oxide is reduced to silver.

Figure 35.8
As a lead storage cell is discharged, the emf (solid line) and the density (dashed line) of the sulfuric acid both decrease.

35.8 FUEL CELLS

In a battery chemical energy may be converted into electric energy until one or more of the active materials is used up, at which point the battery is ready for discard or, possibly, recharge. In a fuel cell active materials are supplied, and the reaction products removed, continuously; thus the fuel cell can operate as long as the required materials are provided.

In one type of fuel cell hydrogen gas and oxygen are supplied, and energy is released by the oxidation of hydrogen to form water; the process is the reverse of that which occurs in the electrolysis of water. In other kinds of fuel cells chemical energy from the oxidation of various hydrocarbons is converted into electric energy. A major advantage of the fuel cell is its high efficiency—vastly greater than that available when the fuel is burned in a heat engine and then converted to electric energy, chiefly because of the low Carnot efficiency (Sec. 18.6). A hydrogen-oxygen fuel cell at 25°C has a theoretical peak efficiency of 83 percent; other fuel cells have still higher theoretical efficiencies, and practical fuel cells can convert 60 to 70 percent of chemical energy to electric energy.

Although Grove produced the first successful fuel cell in 1839, it is only recently that fuel cells have been developed to the point of economic significance.

35.9 THERMOELECTRICITY: THE SEEBECK EFFECT

When two different materials are in contact, there is ordinarily a potential difference between them. In electrostatics this phenomenon permits us to produce charges by contact and separation. In batteries the potential differences between metals and solutions give rise to the emfs. When two metals are placed in contact, the potential difference between them depends on the metals and on the temperature of the junction.

When two wires of dissimilar metals, e.g., copper and iron, are joined together at the ends to form a closed circuit and one of the junctions thus formed is maintained at a temperature different from that of the other (Fig. 35.9), an electric current is established in the circuit. This effect was discovered in 1821 by Seebeck. The magnitude of the net emf producing the current depends on the two kinds of wire and on the temperatures of the two junctions. If the junctions are at the same temperature, the emfs established are exactly equal and opposite, so that there is no net emf.

Figure 35.9
A thermal emf exists when the temperatures of the two copper-iron junctions are different.

Figure 35.10
The thermal emf of a copper-iron
thermocouple as a function of the
temperature of the hot junction when
the cold junction is at 0°C. The neutral
temperature, the peak emf, and the
inversion temperature depend to some
extent on the purity of the metals.

Consider first a thermocouple made of iron and copper (Fig. 35.9). If one of the junctions is kept at 0°C while the other is heated, a net thermoelectromotive force is produced. When the temperature of the hot junction is raised, the thermal emf first increases at a nearly uniform rate (Fig. 35.10). As the temperature is raised further, the rate of increase becomes less. When the temperature of the hot junction reaches 260°C, a further increase in temperature results in a decrease in the thermal emf. The temperature at which the emf reaches its maximum is the *neutral temperature,* so named because there is neither increase nor decrease of thermal emf with temperature. When the temperature of the hot junction is raised above the neutral temperature, the thermal emf decreases and finally becomes zero at approximately 520°C. If the temperature of the hot junction is still further increased, the current in the circuit reverses direction. The temperature at which the thermal emf passes through zero is called the *inversion temperature.* The curve obtained by plotting thermoelectromotive force as a function of the temperature of one junction (with the second junction kept at constant temperature) is parabolic in shape.

For some materials, such as chromel and alumel, the emf increases continuously as the temperature of the hot junction is raised. For these materials the neutral temperature and the inversion temperature are reached by cooling the variable-temperature junction rather than by warming it.

Thermocouples are convenient for measuring temperatures, since the emf depends on the temperature difference between the junctions in a known way. Ordinarily, one junction is kept at a constant temperature, usually in an ice bath at 0°C. The hot junction need not be anywhere near the cold junction. It may be installed at some inaccessible point, where it would be impossible to place and read a standard mercury thermometer. Thermocouples can be used for measuring temperatures roughly up to the melting points of the materials involved, which may be as high as 2900°F for some couples. Further, they can be made much smaller than glass or metal expansion thermometers, so that they can be used to measure the temperatures of small objects, such as insects or twigs of trees.

It is likely that one would want to make connection to a galvanometer by means of copper wires fastened to the ends of the thermocouple elements. Adding other materials in the circuit has *no effect so long as the junctions of the other metals are all at the same temperature.* Electromotive forces exist at the new junctions, but the algebraic sum of the emfs in the circuit remains the same. The lengths of the wires have no effect on the emf (although they do influence the resistance of the circuit).

The emfs involved in the Seebeck effect are ordinarily small, usually of the order of millivolts. To obtain higher emfs, we may connect a number of thermocouples in series, thus producing a *thermopile*. Thermopiles are often used for measuring radiant energy.

35.10   THE PELTIER AND THOMSON EFFECTS

When two dissimilar metals are placed in contact, electrons diffuse from one metal to the other until the Fermi levels coincide. The resulting potential difference, called the *Peltier emf*, depends on the metals in contact and upon the temperature of the junction.

If the two ends of a wire of any material are at different temperatures, electrons are likely to diffuse more rapidly in one direction than in the other. This unequal diffusion produces a potential difference called the *Thomson emf*. The Seebeck effect, discussed in Sec. 35.9, arises as the result of Thomson emfs between opposite ends of wires of the same material which are at different temperatures and Peltier emfs at the junctions of the metals. Usually the Peltier emfs greatly exceed the Thomson emfs.

When a current is passed through a thermocouple by a battery in the circuit, the Peltier emf at one junction aids the battery, while the Peltier emf at the second junction is in series opposing. At the junction where the Peltier emf is in opposition, heat is evolved not only because of the Joule heating effect but also because of the work done against the emf. At the other junction thermal energy is absorbed and converted into electric energy. The temperature of this junction may be reduced substantially ($\approx$80 C° with some metal-semiconductor junctions). Thus, if a current is sent through a thermocouple, heat is released at one junction and its temperature rises, while heat is absorbed at the other junction with a resultant cooling. This is known as the *Peltier effect*.

QUESTIONS

1   How does the mechanism of conduction in a sodium chloride solution differ from that in a copper wire?

2   What is the source of the electric energy supplied by a chemical cell?

3   A strip of copper and a strip of zinc are inserted in a lemon and connected by a wire. Why do electrons go to the zinc in the lemon while they go away from the zinc in the wire?

4   What factors determine the electromotive force of a battery?

5   How does one get rid of the various metallic impurities in the electrolytic purification of copper?

6   What are the advantages of fuel cells over batteries? The disadvantages?

7   Explain why thermocouples and thermopiles are not more widely used as practical generators of electric energy. Why are they used for space satellites but not for automobiles?

8   A chromel-alumel thermocouple is connected to a galvanometer by copper wires. Under what condition does the presence of the copper wires not change the net emf in the circuit?

1. How long will it take for a current of 4 A to plate 1.7 g of silver on a knife? *Ans.* 380 sec

2. A copper plate of 108.962-g mass is placed in an electroplating bath. A steady current is sent through the bath for 30 min, and the mass of the plate is increased to 110.494 g. What was the current in amperes, assuming copper is divalent?

3. A current of 0.8 A deposits 4.69 g of a monovalent element in 48 min. Find the electrochemical equivalent for the element. What is the element? *Ans.* 2.04 × 10⁻⁶ kg/C; gold

4. Find the electrochemical equivalent for sodium (monovalent), trivalent iron, and divalent cadmium.

5. A current of 1.25 A flows for 48 min through a series of cells containing nickel nitrate, copper sulfate ($CuSO_4$), and silver nitrate. Find the masses of nickel (valence 2), copper (valence 2), and silver (valence 1) deposited.
   Ans. 1.09 g; 1.19 g; 4.02 g

6. An electrolytic cell containing a solution of silver nitrate is connected in series with a cell containing copper sulfate. What mass of Ag^+ ions is deposited on the cathode of the first cell by a charge which deposits 1.5 g of Cu^{++} ions on the cathode of the second cell? What charge has passed through the two cells?

7. Find the volumes of hydrogen and oxygen released under standard conditions by the electrolysis of acidulated water for a period of 1,000 sec with a current of 0.4 A.
   Ans. 46.4 cm³; 23.2 cm³

8. An object that has a surface of 18 cm² is to be plated with silver. What will be the average thickness of the silver if a current of 0.2 A flows for 5 hr?

CHAPTER THIRTY-SIX
MAGNETIC FIELDS OF CURRENTS

ELECTRIC CHARGES at rest exert forces on one another through coulomb interactions. Charges in motion also interact with one another through *magnetic* fields. Currents, of course, are charges in motion, and so magnetic interactions are of great importance whenever we deal with currents. Most of our meters, our electric motors, and the great generators of our electric companies depend on magnetic interactions for their operation. Before we can study the physics underlying these devices, we must gain some knowledge of magnetism and magnetic fields.

36.1 MAGNETS

The mineral lodestone (Fe_3O_4) has the ability to attract other pieces of the same mineral or small pieces of iron. This fact was known in the seventh century B.C. How much earlier it had been discovered we do not know. Lodestones turn with the same side toward the north when they are suspended from cords or floated on corks in water. This property led to the invention of the mariner's compass before the twelfth century.

When a steel knitting needle is stroked from one end to the other with a piece of lodestone, the needle may acquire the property of attracting iron filings and of setting itself along a north-south line when suspended by a string. Such a needle is said to be *magnetized* and is commonly called a *magnet*. Most materials cannot be magnetized in this way; relatively few show attraction for a lodestone or a small magnet. Iron and some of its alloys are by far the best known of magnetic materials; cobalt, nickel, and a number of alloys also exhibit prominent magnetic properties.

One end of a magnetized needle suspended by a cord so that it is free to rotate in any direction normally points in a northerly direction. This end of the needle is commonly called the *north pole* (N pole), an abbreviation for the more fully descriptive *north-seeking pole*. The opposite end of the needle is called the *south pole* (S pole).

If this magnetized needle is dipped into soft-iron filings, the filings cling tenaciously to the ends; relatively few stick to the middle. It is often convenient to think of the magnetic properties of the needle as concentrated in the two ends or poles, although this is an oversimplification.

When two magnetized needles are brought near each other, the two north poles repel each other, as do the two south poles. On the other hand, there is an attractive force between the north pole of one magnet and the south pole of the other. Such observations lead to the conclusion that *like poles repel, unlike poles attract.*

36.2 INTERACTIONS OF CURRENTS AND MAGNETS

In 1819 Oersted discovered that a magnet in the neighborhood of a current-bearing wire undergoes a deflection. A magnetized needle held above a straight wire carrying a current is deflected as shown in Fig. 36.1. If the magnet is held below the wire, the N pole is deflected in the opposite direction. The direction in which a north pole points can always be found by the application of a simple rule, illustrated in Fig. 36.2:

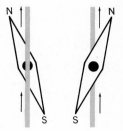

Figure 36.1
Magnetic field of a current deflecting a compass needle.

If the right thumb is pointed in the direction of the conventional current and the fingers are allowed to curl, the direction in which the fingers point is the direction in which the north pole of the needle is deflected by the current.

If current is passed through a cylindrical coil of wire (called a *solenoid*) produced by winding fine wire on a matchstick, the solenoid behaves like a magnet of the same shape. While current is passing through the solenoid, it will pick up small iron filings. It has a north and a south "pole," which exert forces on other poles. The fact that a current-bearing solenoid behaves like a magnet of the same dimensions led Ampère to suggest in 1820 that the forces between iron magnets arise from electric currents in the iron. This point of view has been substantiated by later research and is discussed in Chap. 37.

Magnetic interactions occur among moving charges, whether they are currents in wires, a beam of electrons in a television tube, or the electrons in magnetic materials. The forces so produced are of tremendous practical importance. Electric motors depend on them, as do most of our meters for measuring electric currents and potential differences.

36.3 THE MAGNETIC FIELD

A magnetic field is any region in which forces may be observed to act on small magnets or on small current-bearing elements. *The direction of the magnetic field at any point is the direction of the force experienced by the north-seeking pole of a small test magnet when the test*

Figure 36.2
When the outstretched right thumb is pointed in the direction of the conventional current, the fingers of the right hand curl in the direction of the magnetic field, which is the direction of the force on the north pole of a magnetic needle.

Current into plane of paper

Direction of conventional current

(a)

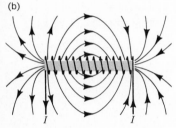

(b)

Figure 36.3

The magnetic fields are similar around (*a*) a bar magnet and (*b*) a current-bearing solenoid of similar shape.

Figure 36.4

A current-bearing conductor lying perpendicular to a uniform magnetic field experiences a force which is perpendicular to both the magnetic field and to its own length. (**B** is directed downward.)

magnet is placed at the point. A small current-bearing solenoid could be substituted for the test magnet.

One may readily explore the magnetic field associated with a large bar magnet by using a small compass needle to find the direction in which the N pole points at various places in the field. A plot of the magnetic field associated with a magnet is shown in Fig. 36.3*a*. In discussing magnetic fields, it is convenient to make use of magnetic lines of force, defined as lines whose tangents give the direction of the magnetic field at every point. It is customary to draw the lines close together where the field is strong, farther apart where the field becomes weaker. The lines of force for a bar magnet are similar to those for a suitably chosen current-bearing solenoid (Fig. 36.3*b*). For the magnetic field associated with a straight current-bearing conductor (Fig. 36.2), the lines of force are circles concentric with the wire.

The magnetic field at any point is characterized not only by a direction, but also by a strength. The vector which describes the field at a point is called the *magnetic induction, magnetic flux density,* or *magnetic intensity.* It is represented by the symbol **B**. To measure **B**, we make use of the fact that there is a force on a current element in a magnetic field.

Consider first the special case of the uniform magnetic field between the pole pieces of the large magnet of Fig. 36.4. If a straight wire bearing a current I is placed perpendicular to the lines of magnetic force, and if l represents the length of the part of the wire in the magnetic field, there is a force **F** on the wire in the direction indicated. We define the magnetic intensity **B** with the equation

$$B = \frac{F}{Il} \tag{36.1}$$

In practical units **F** is measured in newtons, I in amperes, and l in meters. Then **B** has the dimensions *newtons per ampere-meter*, called *teslas* (T) in honor of Nikola Tesla.

In the preceding discussion we stated that the wire must be perpendicular to the direction of the magnetic field. This is very important; indeed, if the wire lies along the lines of force, there is no force at all. If the wire makes an angle θ with the field, the force is given by

$$F = IlB \sin \theta \tag{36.2}$$

Many magnetic fields are not uniform but vary from point to point. The field intensity **B** at any point is the limit of the ratio $\Delta F / I\,\Delta l$ as Δl is made smaller and smaller. Here ΔF represents the small force on a small element of length Δl. The result of using any finite length l of wire is to measure the average intensity over the region covered by l. In actual practice it is often difficult to determine the force on a small element of a current-bearing conductor; therefore other means are ordinarily used to measure **B**.

The force on a current-bearing wire is perpendicular both to the length of the wire and to the magnetic field. For the particular situation of Fig. 36.4 the direction of this force may be found as follows. The lines of flux associated with the current are concentric circles, as indicated. The magnetic field due to the current in the wire exerts a force on the N pole of the magnet in the direction opposite that indicated by **F**. Similarly, the force on the S pole due to the current is out of the paper (opposite to **F**). Since the current in the wire pushes outward on the magnet, the magnet exerts a force inward on the current-bearing wire in accordance with Newton's third law.

The origin of some magnetic fields, such as that of the earth, may not be obvious. In such a case one can find the direction of the force on the wire by observing that on one side of the wire the magnetic field due to the current (Fig. 36.5) reinforces the field already present while on the other side of the wire the two fields are opposite in direction. The force on the wire is directed from the stronger field toward the weaker one, or from the side on which the two fields reinforce to the side on which the two fields oppose.

36.4 MAGNETIC FORCE ON A MOVING CHARGE

The force exerted on a current-bearing conductor in a magnetic field is the resultant of the forces which act on individual moving charges in the conductor. Let us consider a conductor in which there are n charged particles per unit volume, each with charge $+q$ and velocity v perpendicular to a magnetic induction B (Fig. 36.6). The current through the plane PP' is the charge passing per second, which is

$$I = qnAv$$

where A is the cross-sectional area of the wire. The force on a length l of wire is out of the plane of the paper and is given by $IlB = qnAlvB$. Since the number of charges in the length l is nAl, the force on each particle is

$$F = qvB \qquad (36.3)$$

Equation (36.3) gives the force on a single charged particle moving with a velocity v perpendicular to a magnetic field of intensity B. But what is the force if the particle is not moving at right angles to **B**? In this case we resolve the velocity **v** into components $v \sin \theta$ perpendicular to **B** and $v \cos \theta$ parallel to **B**, where θ is the angle between **v** and **B**. There is no force exerted by a magnetic field on a charge moving parallel to the field; only the component of the velocity perpendicular to **B** contributes to the resulting force, which is given by

$$F = qvB \sin \theta \qquad \text{[in vector notation } \mathbf{F} = q(\mathbf{v} \times \mathbf{B})] \quad (36.3a)$$

The force is perpendicular to the plane defined by **B** and **v**. Equation (36.3a) is one of the basic equations of charged-particle physics.

Conductor

Figure 36.5
The force on a current-bearing conductor in a magnetic field is from the stronger toward the weaker resultant field.

Figure 36.6
The force on a current-bearing conductor in a magnetic field results from the forces acting on the individual moving charges.

[Handwritten notes:]
$Fm = q \, V_\perp \, B$

MAGNETIC - LINES FORMS CLOSED PATHS

$B = \dfrac{Fm}{q \cdot V_\perp}$

$B = \dfrac{Fm}{q \cdot V \cdot \sin \theta}$

THE RIGHT HAND RULE IS USED. TO FIND OUT THE FIELD DIRECTION.

The deflection of charged particles by magnetic fields has many uses. The electron beam which sketches the pictures on a television tube is usually directed by magnetic forces. In cyclotrons, betatrons, and synchrotrons, accelerated charged particles are restrained to roughly circular paths by magnetic fields. Electron microscopes and mass spectrographs use magnetic forces to control beams of charged particles. Some of these important instruments are described in later chapters.

Figure 36.7

A rectangular loop of wire bearing a current is subject to a torque in a magnetic field unless its plane is perpendicular to the magnetic lines of force.

36.5 THE MAGNETIC MOMENT OF A COIL

Consider a rectangular loop of wire, bearing a current I, placed in a uniform magnetic field of intensity B with one side of the rectangle perpendicular to the field and another side lying along the field (Fig. 36.7). Let y represent the length of the side perpendicular to the magnetic field and x the length of the side parallel to the magnetic field. Under these circumstances there are no forces on the sides parallel to **B**, but the force on each of the wires perpendicular to the field is given by $F = BIy$. The lever arm about the axis OO' for one of the vertical sides is $x/2$, and the torque τ on this one side is given by $\tau = BIyx/2$. The net torque acting to rotate the loop is twice this, or $BIyx$. The product yx is equal to the area A of the rectangle. The torque τ on the loop is thus

$$\tau = BIA \tag{36.4}$$

Current-bearing coils and magnetized materials in various shapes also experience torques in a magnetic field. The concept of magnetic moment, which we introduce now, is applicable to any object which is subject to a torque when it is placed in a uniform magnetic field. *The magnetic moment M of any object is the ratio of the torque experienced by the object when it is placed in a uniform magnetic field B in such a position that the torque takes on its maximum value to the magnetic induction B*; thus, $M = \tau_{\text{max}}/B$.

For the coil of Fig. 36.7 the torque, maximum when the plane of the coil is parallel to the magnetic field B, is BIA, and hence the magnetic moment is IA. By use of calculus it can be shown that Eq. (36.4) is applicable to a plane loop of any shape.

The magnetic moment M of a current-bearing loop is the product of the current I and the area A of the loop. The magnetic moment

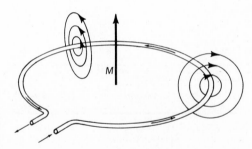

Figure 36.8

The magnetic field and the magnetic moment **M** associated with a current-bearing loop.

is a vector perpendicular to the plane of the coil in the direction in which a right-handed screw would advance if turned in the direction of the current I. It has the dimensions A-m². The magnetic moment associated with a current-bearing loop is shown in Fig. 36.8. The torque τ on a loop of magnetic moment M is given by BM when M is perpendicular to B. When M and B are parallel, the torque is zero. In general, if the angle between M and B is θ, the torque is given by

$$\tau = BM \sin \theta \qquad (36.5)$$

If there is no opposing torque, the loop (Fig. 36.7) rotates under the influence of the magnetic field until the magnetic moment is aligned in the direction of the field.

Thus far we have been discussing single loops. If a coil made by winding N turns in series bears a current I, each turn has a magnetic moment of magnitude IA, and the resultant magnetic moment is N times that of a single loop. Thus, for a coil of N turns, $M = NIA$.

36.6 THE MOVING-COIL GALVANOMETER

If a small current is passed through a coil suspended between the poles of a permanent magnet (Fig. 36.9), the coil rotates until the restoring torque exerted by the suspension is equal to the torque due to the interaction of the current and the magnetic field. This latter torque is proportional to the current. Therefore, the angular deflection of the coil is proportional to the current, provided the restoring torque is proportional to the angular displacement and the magnetic field is perpendicular to **M** for all displacements. This latter condition is reasonably well satisfied for angles up to 30° for the magnet of Fig. 36.9. Thus, a single loop of wire suspended by an elastic suspension in a magnetic field can be used to detect and measure currents. The sensitivity of the instrument may be increased by adding turns to the coil. Such a coil, suspended in a magnetic field, is called a *galvanometer* when it is used to detect and measure small currents. A light pointer or mirror attached to the coil may be used to measure the angular deflection.

36.7 AMMETERS AND VOLTMETERS

When a galvanometer and its scale are so adjusted that the scale readings indicate the current passing through some portion of a circuit in amperes, the meter is called an *ammeter*. A common form of ammeter is shown in Fig. 36.10. It consists of a coil of fine copper wire wound on a light frame which is mounted on jeweled bearings between the poles of a permanent magnet. When a current exists in the coil, it rotates between the poles of the magnet. Two spiral springs, one at the top and the other at the bottom, carry the current into and out of the coil and provide a restoring torque proportional to the angular displacement.

Since a small fraction of an ampere through the coil of an ordinary ammeter produces a full-scale deflection, it is necessary to use a low-resistance *shunt* to carry the remainder of the current. (A shunt is a resistance connected in parallel with a circuit element.) Because the

Figure 36.9

Schematic diagram of a moving-coil galvanometer.

Figure 36.10

The deflection of an ammeter increases as the current through the coil rises.

Figure 36.11
A two-scale voltmeter.

current flowing in the movable coil is always a constant fraction of the full current entering the instrument, the scale can be calibrated so that the pointer indicates the entire current. Any galvanometer can be made to operate as an ammeter by use of a suitable shunt. By choice of some other shunt, the galvanometer can by made into an ammeter of different range.

A galvanometer can also be made into a *voltmeter* by the proper application of an additional resistance (Fig. 36.11). A voltmeter, as its name implies, is an instrument used to measure potential difference. By connecting a high resistance in series with the galvanometer coil, the current passing through the coil can be limited to a value which will not exceed full-scale deflection. Since for a given resistance the current is directly proportional to the potential difference across the instrument, the deflection of the pointer is proportional to the potential difference and the scale may be calibrated to read directly in volts. By suitable choice of the series resistance one can make a voltmeter which gives full-scale deflection for any desired potential difference across its terminals.

Example

A galvanometer requires 0.00015 A to produce a full-scale deflection. The coil has a resistance of 60 Ω. What shunt resistance is needed to convert this galvanometer into an ammeter reading 2 A full scale?

Of the 2 A which enter the ammeter for a full-scale deflection, 0.00015 A must pass through the galvanometer, and the remainder (2 − 0.00015 = 1.99985 A) through the shunt. Since the shunt and galvanometer are in parallel, their potential differences are the same.

$$V_G = V_S \quad \text{and} \quad I_G R_G = I_S R_S$$
$$0.00015 \times 60 = 1.99985 R_S$$
$$R_S = \frac{0.009}{1.99985} = 0.0045 \ \Omega$$

////

Example

What series resistance R is needed to convert the galvanometer of the preceding example into a voltmeter reading 6 V full scale?

When 6 V is impressed across the voltmeter, a current of 0.00015 A must pass through the coil and the series resistance if the deflection is to be full scale. The total resistance of the voltmeter must be

$$R_V = \frac{6}{0.00015} = 40{,}000 \ \Omega$$

But R_V is equal to R plus the resistance of the galvanometer; therefore

$$R = 40{,}000 - 60 = 39{,}940 \ \Omega \qquad ////$$

36.8 THE MAGNETIC FIELD OF A LONG STRAIGHT WIRE

A long straight wire carrying a current has a magnetic field with the lines of force concentric circles about the wire (Sec. 36.2). When the wire lies in air (or some other nonmagnetic material), the magnetic intensity **B** at any point is proportional to the current I and inversely proportional to the distance r from the wire. It can be shown that $B = 2\kappa I/r$, where κ is a constant which has the value 10^{-7} N/A². The magnetic field strength B in teslas (N/A-m) at a distance r m from a long straight wire carrying a current I A is given by

$$B = \frac{2I}{10^7 r} \qquad (36.6)$$

When two long wires parallel to one another bear currents I_1 and I_2 in the same direction, there is an attractive force between them. If the wires are separated by a distance r, the magnetic field at the first wire due to the current in the second is given by $2I_2/10^7 r$. Therefore, the force on a length l of the first wire is given by

$$F = I_1 l B = \frac{2I_1 I_2 l}{10^7 r} \qquad (36.7)$$

The force per unit length is $2I_1 I_2/10^7 r$. That the force is attractive can be seen by studying Fig. 36.12. When the directions of the currents in the two wires are opposite, the force between the wires is repulsive. In this figure we make use of the common practice of representing a current out of the paper with a dot (point of an arrow moving toward you) and a current into the paper with cross feathers (tail of an arrow moving away from you).

Although Eq. (36.7) has been derived for long straight wires, there is an attractive force between any two neighboring conductors which are carrying currents in the same direction. One of the most accurate methods of measuring current involves the determination, in an instrument known as the "current balance," of the attractive force between two precisely made, parallel coils carrying the same current. In the absolute system of units, officially adopted by the United States (*National Bureau of Standards Circular* C 459, 1947), the ampere is defined in terms of measurements made with a current balance. The coulomb is then derived from the ampere as the charge which is carried by a current of one ampere in one second. (For pedagogical reasons we have introduced the coulomb first and then defined the ampere; our units are consistent with those adopted by the National Bureau of Standards.)

When coils carry an alternating current, neighboring turns are attracted to one another by forces which change as the alternating current increases and decreases. The turns may vibrate, moving to-

BETWEEN TWO PARALLEL WIRES.

$$B_1 = \frac{M_0 I_1}{2\pi S}$$

$$F_2 = B_1 L_2 I$$

$$F = \frac{M_0 I}{2\pi S} (1\,meter)\, I_2$$

$$F\,(meter) = M_0 \left(\frac{I_1 I_2}{2\pi S} \right)$$

$$dB = \frac{M_0 dl \sin\theta}{4\pi R^2}$$

Figure 36.12
Two parallel wires attract each other through their magnetic interaction when they bear currents in the same direction and repel each other when the currents are in opposite directions.

gether and then apart. This leads to the familiar and annoying hum associated with many coils carrying alternating currents.

36.9 THE MAGNETIC FIELDS OF OTHER CURRENT CONFIGURATIONS

The magnetic field due to any current configuration is the vector sum of the contributions due to the current in each minute element of length Δl. The contribution $\Delta \mathbf{B}$ to the magnetic field due to the current in a small element of length Δl (Fig. 36.13) has a direction given by Fig. 36.2 and a magnitude

$$\Delta B = \frac{\kappa I \, \Delta l \sin \theta}{r^2} \tag{36.8}$$

where θ is the angle between the element Δl and the vector $\mathbf{r}$ connecting Δl to the point at which $\Delta \mathbf{B}$ is to be measured and κ is 10^{-7} N/A². This equation is an expression of a fundamental rule known as *Ampère's law*. Equation (36.8) is applicable only if the wire is in a nonmagnetic medium. It should be emphasized that the contribution $\Delta \mathbf{B}$ from each element of length Δl is a vector which must be added vectorially to all the other $\Delta \mathbf{B}$'s if one wishes to find the resultant magnetic intensity.

The magnetic field at the center of a circular loop of wire (Fig. 36.8) of radius a is readily computed by use of Eq. (36.8). Since the contributions $\Delta \mathbf{B}$ of the tiny elements of length Δl are all in the same direction and are equal in size, addition of the components gives $B = Il/10^7 a^2$. The length of the wire l is $2\pi a$, so that this equation reduces to

$$B = \frac{2\pi I}{10^7 a} \tag{}$$

If the circular loop is replaced by a circular coil having N turns, the contributions of these turns are all in the same direction. Accordingly, at the center of the circular coil the magnetic intensity $\mathbf{B}$ is given by

$$B = \frac{2\pi NI}{10^7 a} \tag{36.9}$$

One of the common current configurations in magnetic circuits is that of the solenoid, often used for windings on electromagnets. The solenoid provides a uniform field over its entire central section. If it is long compared with its diameter, the magnetic field is almost the same at all points inside the solenoid, except at distances less than one diameter from the ends. At the ends the field falls off. In the central section of the solenoid the magnetic field $\mathbf{B}$ is given by the relation

$$B = \frac{4\pi nI}{10^7} \tag{36.10}$$

where n is the number of turns per meter. It is important to note that n represents the *number of turns per unit length*, while N is used to represent the total number of turns; $n = N/l$, where l is the length of the solenoid. Equation (36.10) is applicable no matter what the cross-sectional shape of the solenoid may be. Solenoids with square or rectangular cross section are common.

Figure 36.13
Ampère's law gives the contribution $\Delta \mathbf{B}$ to the magnetic field at point P due to the current-bearing element of length Δl.

The magnetic fields of a solenoid, a loop, and a single straight wire are shown in Figs. 36.3*b*, 36.8, and 36.2, respectively. In every case *magnetic lines of flux are continuous. They never begin or end;* rather they follow closed curves which encircle the current-bearing conductors. In this respect magnetic lines of force are different from electric lines of force, which begin and end on electric charges.

QUESTIONS

1 Can magnetic lines of force ever cross? Explain.

2 A current-bearing wire has no net static charge. Why may a magnetic field exert a force on the wire in spite of its uncharged condition?

3 A stream of electrons is moving toward the east. If the stream passes through a uniform magnetic field directed upward, in what direction is the electron beam deflected?

4 If there is no deflection of a beam of electrons as they move through a region in space, does this prove that there is no magnetic field in that region? Give two examples of situations in which there could be a field: one with no electric field present and the other with a transverse electric field.

5 How much work is done by a magnetic field on a stream of free electrons moving through it? Justify your answer. Prove that a constant magnetic field does no work on a charged particle passing through it.

6 How could you reverse the direction of motion of a beam of electrons without changing the speed of the electrons at any time?

7 In the equation $\mathbf{F} = q(\mathbf{v} \times \mathbf{B})$, which pairs of the three vectors are always mutually perpendicular? What pair may have any angle between them?

8 How can you convert a galvanometer into an ammeter of specified full-scale reading? Into a voltmeter?

9 In sensitive electronic circuits why are wires carrying equal and opposite currents often twisted together? How does this reduce their resultant magnetic field in neighboring regions?

10 A long helical brass spring hangs vertically with its lower end just making contact with a pool of mercury. If a large current is passed through the mercury and the spring, what will occur? Why? Will the motion of the end of the spring be simple harmonic?

PROBLEMS

1 A long wire bearing a current of 3 A lies perpendicular to a uniform magnetic field between the poles of a large magnet. If 12 cm of the wire lies in the field, and if the force on this length is 0.09 N, find the magnetic intensity B.

Ans. 0.25 T (1 T = 1 N/A-m)

2 Find the force on 6 cm of conductor bearing a current of 5 A if it lies perpendicular to a magnetic field of intensity 1.4 T. What would the force be if the wire made an angle of 60° with the magnetic field?

3 An electron is moving perpendicular to a magnetic field with a speed of 5×10^6 m/sec. If the intensity of the magnetic field is 0.12 T, what force acts on the electron? What is its acceleration?

Ans. 9.61×10^{-14} N; 1.05×10^{17} m/sec²

4 Find the force on and the acceleration of an electron moving 4×10^8 m/sec at right angles to a magnetic field of intensity 0.0009 T in a television picture tube.

5 A coil of 20 turns of wire has an area of 3 cm². It carries a current of 2×10^{-6} A. Find the magnetic moment of the coil. If the magnetic moment is perpendicular to a magnetic field of intensity 1.3 T, what is the torque on the coil?

Ans. 1.2×10^{-8} A-m²; 1.56×10^{-8} N-m

6 A coil of magnetic moment 2.5×10^{-8} A-m² experiences a torque of 7.5×10^{-9} N-m when it is placed in a magnetic field with the magnetic moment normal to the field. Find the magnetic intensity. If the coil has 12 turns and an area of 1.5 cm², what current does it bear?

7 A rectangular galvanometer coil is 1.5 cm high, 1 cm long, and has 30 turns of wire. Find its magnetic moment when it bears a current of 10^{-4} A. What torque is exerted on this coil in a uniform magnetic field of intensity 0.6 T when the plane of the coil (*a*) is parallel to the field,

(b) is perpendicular to the field, and (c) makes an angle of 30° with the field?

Ans. 4.5×10^{-7} A-m^2; (a) 2.7×10^{-7}, (b) 0, and (c) 2.34×10^{-7} N-m

8 A long straight wire bears a current of 4 A. At what distance from the wire is the magnetic field 2×10^{-5} T?

9 Two long straight parallel wires are 10 cm apart. A current of 3 A passes through one wire and a current of 2 A through the other. If the two currents are in the same direction, what is the magnetic intensity at a point midway between the wires? Find the attractive force per meter of length between the wires.

Ans. 4×10^{-6} N/A-m; 1.2×10^{-5} N/m

10 Find the magnetic intensity a distance of 6 cm from a long straight wire bearing a current of 15 A.

11 A solenoid is 30 cm long. It is wound with 6,000 turns of copper wire. What current must be sent through the windings to produce a magnetic intensity of 0.5 T? *Ans.* 19.9 A

12 How many turns of wire must there be in a solenoid 1.1 m long for a current of 4.5 A to produce a magnetic field of 0.006 T at its center?

13 In two concentric solenoids the currents flow in opposite directions. The inner one has 50 turns per centimeter, and the outer one 20 turns per centimeter. What current in the outer coil will be necessary in order to have the field at the center zero when the inner coil is carrying 1.6 A? If no current passes through the outer coil, what is the magnetic intensity at the center?

Ans. 4 A; 0.0101 T

14 A cylindrical solenoid of 3,000 turns, 12 cm in diameter and 1.5 m long, is placed with its axis parallel to the lines of the earth's magnetic field. If the latter has an intensity of 2×10^{-5} T, what current must flow through the coil in order to make the magnetic field at its center zero?

15 Find the magnetic intensity at the center of a circular coil of 25 turns of 3-cm radius when the coil bears a current of 6 A. What is the magnetic moment of the coil?

Ans. 3.14×10^{-3} N/A-m; 0.424 A-m^2

16 A current in a circular loop of wire with a diameter of 8 cm produces a field strength of 3×10^{-5} T at the center of the loop. What is the current? Find the magnetic moment of the loop.

17 Find the force on the current loop of the accompanying figure if $I_1 = 20$ A, $I_2 = 16$ A, $a = 15$ cm, $b = 6$ cm, and $d = 4$ cm.

Ans. 1.44×10^{-4} N to right

18 Find the force on the current loop of the accompanying figure if I_1 and I_2 are 6 and 8 A, respectively, while a, b, and d are 15, 8, and 2 cm, respectively.

Problems 17 and 18

19 A current I of 8 A passes through a long straight wire which makes a semicircular bend of radius r about point C of the accompanying figure. If $r = 4$ cm, find the magnetic intensity B at C and at point P far from the bend.

Ans. 1.03×10^{-4} and 8×10^{-5} N/A-m

Problem 19

20 Find the magnitude and direction of the magnetic induction **B** at the center of curvature of the circular segment of the accompanying figure if the segment is 8 cm long, the radius a is 5 cm, and the current is 10 A.

21 Find the magnitude and direction of the magnetic induction **B** at the center of curvature of the

Problems 20 and 21

circular segment which is bearing a current of 8 A. The radius a is 6 cm, and $\phi = 60°$.

Ans. 1.40×10^{-5} T into paper

22 Find the contribution to the magnetic induction at a, b, and c due to the current I in a short length Δl of conductor (see accompanying figure) if I is 7 A and Δl is 1.5 cm.

23 Find the contribution to the magnetic induction at a, b, and c due to the current I in a short length Δl of conductor (see accompanying figure) if I is 25 A and Δl is 8 mm.

Ans. 8.9×10^{-7}, 3×10^{-7}, and 0 T

Problems 22 and 23

24 A galvanometer has a moving coil with a resistance of 120 Ω and a sensitivity of 1-mm deflection for 4×10^{-7} A. What shunt will be needed to produce a 5-cm deflection for 0.002 A in the main circuit?

25 A long horizontal wire is bent to include a circular loop of $r = 10$ cm (see accompanying figure). If a current of 8 A exists in this wire, what is the magnetic induction B at the center C of the

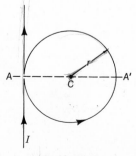

Problem 25

loop? If the loop is twisted about the axis AA' so its plane is vertical, what is the magnitude of B at the center of the loop?

Ans. 3.43×10^{-5} and 5.28×10^{-5} N/A-m

26 A two-scale voltmeter is wired as shown in Fig. 36.11, where the galvanometer has a resistance of 25 Ω and requires a current of 2.0×10^{-3} A for full-scale deflection. Find R_1 and R_2 if the voltmeter is to read full scale for 30 and 100 V.

27 A galvanometer of 10-Ω resistance requires a current of 0.0012 A to produce full-scale deflection. What resistance is required to convert this galvanometer to an ammeter reading 2 A full scale? To a voltmeter reading 30 V full scale?

Ans. $0.012/1.9988$ Ω; $24,990$ Ω

28 A meter movement is a galvanometer with a resistance of 8 Ω, and it requires 0.002 A for full-scale deflection. Find the resistance required to convert this galvanometer to (a) a voltmeter reading 3 V full scale and (b) an ammeter reading 0.4 A full scale.

29 A 20,000-Ω/V voltmeter is connected in series with a 500-Ω/V voltmeter, both of which are calibrated to read 0 to 100 V. What will each read if the voltmeters in series are connected across a 90-V source? *Ans.* 87.8 V; 2.2 V

30 An ohmmeter circuit of a simple multimeter is shown in the accompanying figure. The meter G reads full scale for 0.3 mA and has a resistance of 8 Ω. The battery has an emf of 1.5 V. When there is zero resistance between a and b, the meter reads full scale; when a and b are open, the meter reads zero. What is the value of R_0? What deflection in terms of full scale would result if a 750-Ω resistor were connected between a and b?

Problems 30 and 31

31 The ohmmeter of the accompanying figure has a meter with resistance of 10 Ω, which deflects full scale for 2 mA, and a 1.5-V battery. If there is zero resistance between a and b, the meter reads full scale; when a and b are open, the meter reads zero. Find R_0. What resistance R between a and b would give a current of 1.5 mA (0.75 full-scale reading)? What current would $R = 150 \ \Omega$ give? *Ans.* 740 Ω; 250 Ω; 1.67 mA

32 The trajectory of a charged particle in a uniform magnetic field B_z in the z direction is given by $\mathbf{r}(t) = \mathbf{i}R \cos \omega t + \mathbf{j}R \sin \omega t + \mathbf{k}v \cos \theta \ t$, where $v \cos \theta$ is the z component of the velocity of the ion. This trajectory is a helix. Find R and ω in terms of B_z and $v \sin \theta$, the velocity perpendicular to the z direction.

WE NOW KNOW what a magnetic field is and how magnetic fields are associated with certain current distributions, but we have said little indeed about the magnets of simple compasses and the familiar bar magnets children play with. What are the origin and nature of their magnetic fields? And why does a compass needle point roughly north? These are questions we discuss in this chapter.

CHAPTER THIRTY-SEVEN
MAGNETS

37.1 THE MAGNETIC MOMENT OF A MAGNET

In Chap. 36 we saw that a coil placed in a magnetic field of strength **B** with its magnetic moment **M** perpendicular to the field experiences a torque equal to BM. Similarly, a current-bearing solenoid placed in a magnetic field with its axis perpendicular to the field experiences a torque equal to the resultant of the torques acting on each of the turns of the solenoid. These turns can be regarded as a series of individual coaxial coils. The resulting magnetic moment of the solenoid is NIA, where N is the total number of turns, I the current, and A the cross-sectional area of the solenoid.

$\rightarrow M = NIA$

$\rightarrow M = \dfrac{\tau}{B}$

If we place a small bar magnet (magnetic needle) with its axis perpendicular to a magnetic field (Fig. 37.1), it experiences a torque in a direction such as to align its axis with **B**. The magnetic moment **M** (Sec. 36.5) is the ratio of the torque τ to the field strength **B**. Since the *apparent* magnetic properties of the magnet are associated with the poles at the ends, it is often convenient to think of the torque as arising from the forces acting on the north and south poles, which are separated by a distance equal roughly to the length of the magnet l. Let us define the pole strength p of this magnetic needle as the ratio of the magnetic moment to the length.

$$p = \frac{M}{l} \qquad (37.1)$$

Since the dimensions of the magnetic moment are A-m², the dimensions of pole strength are ampere-meters. The force **F** on a pole p placed in a magnetic field of strength **B** is given by

$$F = Bp \qquad (37.2)$$

A pole has a strength of one ampere-meter if the force acting on it is one newton when the pole is placed in a magnetic field of intensity one tesla (newton per ampere-meter). Observe that the force on the north pole of the magnet is in the direction of **B** while the force on the south pole is opposite to **B**.

If the axis of the magnet makes an angle θ with the magnetic field, the torque is less than if the magnet were perpendicular to the field.

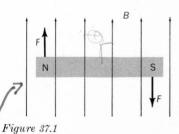

Figure 37.1
A bar magnet with its axis perpendicular to a uniform magnetic field experiences a torque which tends to align the magnet with the field.

$$\tau = Bpl \sin\theta$$
$$M = \frac{\tau}{B} = Pl$$

In general, the torque τ is given by

$$\tau = BM \sin\theta = Bpl \sin\theta \qquad (37.3)$$

$BI A \sin\theta$

as can be seen from Fig. 37.2.

In many problems regarding magnets it is convenient to consider the forces and torques on the magnet as arising from the interaction between magnetic field and point poles. However, it should be remembered that the pole picture is a simplification of the more general point of view that there are forces on current elements in magnetic fields. The torque on the magnet of Fig. 37.2 can be explained in terms of torques acting on elementary current loops. The magnetic moment **M** of the magnet is just that which would be produced by a current $I = M/A$ (where A is the cross-sectional area of the magnet) flowing as a current sheet around the magnet. The point of view of the current loops is rigorous and universally acceptable, but for many problems the force-on-pole concept is simpler and more convenient.

Magnets never have single poles. The magnetic field arises from currents and, as we have seen in Chap. 36, magnetic lines of force are always continuous. The pole of a magnet is a region where a large fraction of lines of force enters or leaves the magnet. Since every line of force which enters a magnet also leaves, it is clear that no magnet can have a single pole.

Lever arm about axis
through center of magnet

Current in

Current out Current loops

Figure 37.2

The torque on a bar magnet whose axis makes an angle θ with a uniform magnetic field is the resultant of all the torques exerted on current loops of infinitesimally small dimensions.

$IA = pl$.

$P = \dfrac{IA}{l}$

37.2 COULOMB'S LAW

When the poles of bar magnets are brought close to each other, they interact. *Like poles repel each other, and unlike poles attract.* Coulomb studied the interaction between magnetic needles with torsion balances and found that the force between two poles varies as the product of the pole strengths and inversely as the square of the distance between the poles:

$$F = \kappa \frac{p_1 p_2}{r^2} \qquad (37.4)$$

When the force is in newtons, pole strength in ampere-meters, and r in meters, κ is 10^{-7} N/A², the same constant which appears in Ampère's law.

Coulomb's law is applicable whenever one can regard the magnetic properties of the needle as concentrated in point poles. This, in turn, requires that the distance between the two magnets be substantially greater than the thickness of either magnet, since otherwise a pole of a magnet cannot be correctly approximated by a point pole. If one wishes to calculate the force which one magnetic needle exerts on another, it is ordinarily necessary to calculate not only the forces between nearest poles but also the forces between the other poles. Each pole of one magnet exerts forces on both the N and S poles of the second magnet. Therefore, to calculate the total force exerted on one magnet by another, it is necessary to compute four forces and add them vectorially.

Example

Two small identical magnets are 8 cm long and have pole strengths of 60 A-m. When they are restrained so they cannot rotate, it is possible to make one "float" 6 cm above the other (Fig. 37.3). Find the force which the lower magnet exerts on the upper one (which is the weight of the upper magnet, of course).

$$F_{NN} = \kappa \frac{p_1 p_2}{r^2} = 10^{-7} \frac{60 \times 60}{(0.06)^2} = 0.100 \text{ N}$$

$$F_{NS} = 10^{-7} \frac{60 \times 60}{(0.1)^2} = 0.036 \text{ N}$$

Similarly $F_{SS} = 0.100$ N, and $F_{SN} = 0.036$ N. The vertical component of F_{NS} is $0.036 \times 0.6 = 0.0216$ N, and the horizontal component is $0.036 \times 0.8 = 0.0288$ N. The horizontal component of F_{SN} exactly balances the horizontal component of F_{NS}. If we take the upward direction as positive, the vertical force on the upper magnet is given by

$$F = 0.100 + 0.100 - 0.0216 - 0.0216 = 0.1568 \text{ N or 16 g wt} \quad ////$$

Figure 37.3
One bar magnet can be floated above another by magnetic repulsion.

Any pole has a magnetic field associated with it. To determine the field intensity **B** at any point, we may imagine that we bring a test pole p_t to the point in question and measure the force **F** on this pole. The intensity **B** is given by F/p_t. To calculate the magnetic intensity due to any group of poles, we find the magnetic intensity which each pole would produce by itself and add all these fields together vectorially. The magnetic field due to one pole p can be calculated readily by Coulomb's law. The force on a test pole p_t at a distance r from the pole p is given by $F = pp_t/10^7 r^2$, and

$$B = \frac{F}{p_t} = \frac{p}{10^7 r^2} \qquad (37.5)$$

Example

A magnet 15 cm long with poles of strength 250 A-m lies on a table. Find the magnitude of the magnetic intensity B at a point P 20 cm directly above the N pole of the magnet (Fig. 37.4).

The magnetic fields are given by

$$B_N = \frac{p_N}{10^7 r^2} = \frac{250}{10^7 (0.2)^2} = 6.25 \times 10^{-4} \text{ T}$$

$$B_S = \frac{p_S}{10^7 r^2} = \frac{250}{10^7 (0.25)^2} = 4.00 \times 10^{-4} \text{ T}$$

The vertical component of B_S is $B_S \cos \theta = 3.2 \times 10^{-4}$ T, and the horizontal component is $B_S \sin \theta = 2.4 \times 10^{-4}$ T. The vertical component of the resultant intensity is $(6.25 - 3.2) \times 10^{-4} = 3.05 \times 10^{-4}$, and the horizontal component is 2.4×10^{-4}. The resultant B is

$$\sqrt{(3.05)^2 + (2.4)^2} \times 10^{-4} = 3.9 \times 10^{-4} \text{ T} \qquad ////$$

Figure 37.4
The magnetic intensity at any point P is the resultant of the fields of the two poles of the magnet.

37.3 THE MAGNETIC FIELD OF THE EARTH

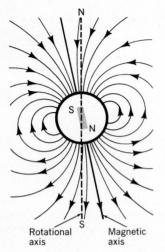

Figure 37.5
The earth's magnetic field.

The usefulness of the compass as a device for determining direction arises from the fact that the earth has a magnetic field which aligns the compass needle. At points high above the earth's surface this field is approximately that which would be produced by a tremendous bar magnet (Fig. 37.5) near, but not at, the center of the earth with its axis making an angle of 11° with the earth's rotational axis, its south-seeking pole at 78.5° north latitude, 69° west longitude, and its north-seeking pole at 78.5° south latitude, 111° east longitude. At the surface of the earth this field is grossly distorted by magnetic materials of the earth's crust, with the result that in 1960 the magnetic poles of the earth as defined by vertical lines of force were located at 76° north latitude, 102° west longitude, and 72° south latitude, 140° east longitude.

In view of the fact that the earth's magnetic axis does not coincide with its axis of rotation, a compass needle points to true north at relatively few regions on the earth's surface. The angle between a free horizontal compass needle and true north is called the *angle of declination* (or sometimes the *variation* or *deviation*). Figure 37.6 shows the declination at various places in the United States. A line on the earth's surface along which a compass needle points to true north is called an *agonic line*. At points east of the agonic line in the United States, the declination is west; at points west of the agonic line, the declination is east. In New York City the declination is about 11.5° west, while in San Francisco it is roughly 18° east.

When a compass needle is free to orient itself along the lines of magnetic force, the needle points downward as well as to the north. The angle between the free magnetic needle and the horizontal plane is called the *inclination* or *angle of dip*. Figure 37.6 also shows lines of constant dip, which are called *isoclinic* lines. The inclination in various parts of the United States varies from about 50 to 80°.

The earth's magnetic field is a relatively weak one. The horizontal component varies approximately from 25×10^{-6} T in the southern part of the United States to 15×10^{-6} T in the northern part. The vertical component varies similarly from about 40×10^{-6} to 55×10^{-6} T as one goes from south to north.

The magnetic isogonic and isoclinic lines, shown in Fig. 37.6 as smooth regular lines, actually have many local irregularities and variations. One would expect the magnetic lines of the earth to be significantly altered by large deposits of iron ores and other magnetic materials, and indeed they are. One method of prospecting involves the careful measurement of the earth's magnetic field over large areas. The vertical component of the earth's field sometimes increases sharply above a body of magnetic ore and is often weaker than normal above an oil deposit.

The earth's magnetic field is not static. Over a period of years the declination at a given point changes substantially. Since 1580 it has varied between 11° east and 24° west at London. To explain the origin of such variations we invoke the "dynamo theory," according to which motions of parts of the earth's fluid core relative to other parts set up currents, which in turn modify the motions. A detailed study of the changes in magnetic field at the surface of the earth suggests that the average speed of the magnetic core material is about 1 m/hr. The

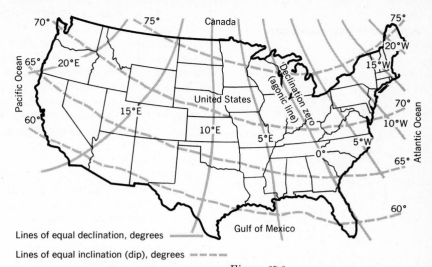

Lines of equal declination, degrees ————

Lines of equal inclination (dip), degrees — — — —

Figure 37.6
Geomagnetic declination (solid lines) and inclination (dashed lines) for the United States.

magnetic field at the earth's surface is a weak clue to the vastly stronger fields in the core.

Roughly 99 percent of the earth's field is due to phenomena in the earth itself; the remainder has its origin in external currents in the ionosphere. In disturbed regions of the solar atmosphere, strong magnetic fields are accompanied by the emission of a tenuous stream of matter, chiefly protons and electrons. When the earth passes through such a stream, the moving charges interact with the earth's magnetic field and its ionosphere, producing a fluctuation in the field at the earth's surface. Magnetic storms of this kind can be partially correlated with sunspots and solar flares, as well as with northern lights (*aurora borealis*) and phenomena of a similar nature on the earth.

37.4 MAGNETIC MATERIALS

Classical studies of the magnetic properties of materials led to an assignment of each substance to one of three classes: *ferromagnetic*, *paramagnetic*, and *diamagnetic*. The assignment was made on the basis of a simple experiment in which a rod or needle of the substance was placed in a strong, nonhomogeneous magnetic field. A few materials interacted strongly with the field and were called *ferromagnetic* (*ferro* for "iron"). Of the common elements only iron, nickel, and cobalt ordinarily reveal prominent magnetic properties. Gadolinium and liquid oxygen are also strongly magnetic; so are a substantial number of compounds and alloys. All these are ferromagnetic in the classical nomenclature.

For the remaining vast majority of materials the magnetic effects were exceedingly small. However, in a strong, nonuniform field each material fell into one of two classes. Some became feebly magnetized in the direction of the field and aligned the axis of the needle with the field. Such substances were called *paramagnetic* (*para-* means "parallel''). Examples are palladium, manganese, and many metallic salts. Other materials became very feebly magnetized in the direction opposite the field. A needle of bismuth aligns itself at right angles to a strong nonuniform magnetic field. Such a material is called *diamagnetic* (*dia-* means ''across''). Most elements and chemical compounds are slightly diamagnetic. Bismuth is the most diamagnetic element. For most practical purposes paramagnetic and diamagnetic materials can be regarded as magnetically inactive. They are not useful for magnets. Ferromagnetic materials are of very great practical importance and utility. They are vital for transformers, electric motors, electric generators, and electromagnets.

Twentieth-century research has led to a microscopic understanding of magnetic phenomena and further classification. We now know that some ''atoms'' have permanent magnetic moments arising from their structures, where ''atom'' implies the atom or ion as it exists in the material. For example, in magnetite (Fe_3O_4 or $FeO \cdot Fe_2O_3$) there are ferric (Fe^{3+}) ions and ferrous (Fe^{++}) ions; the latter contribute about eight-tenths the magnetic moment of the former. In other materials iron ''atoms'' may have electron configurations such that they contribute little or no net magnetic moment. While many compounds and alloys of iron are strongly magnetic, a number, including stainless steel, are not. On the other hand, the Heusler alloys (containing aluminum, copper, and manganese) are ferromagnetic, although none of the component elements are. Why some ''atoms'' are magnetic and others are not is a complex question. However, a qualitative answer can be given in fairly simple terms. Magnetic effects are always associated with moving electric charges. As long ago as 1820, Ampère suggested that within magnetized iron there are circulating currents. It is obvious that these currents cannot be due to a flow of charges around the circumference of the iron core, since such currents would necessarily dissipate heat in the iron as a result of the resistance. Instead, the currents are due to the motions of the atomic electrons. The electrons associated with any given atom are in constant motion, and a moving charge produces a magnetic field.

An electron may establish a magnetic field in two ways. First, we may picture the electron as traveling around the nucleus of the atom in an orbit crudely analogous to the orbit in which the earth revolves about the sun. Second, the electron may be thought of as rotating about its own axis, much as the earth rotates once each day about its axis. A single electron may establish a magnetic field by virtue of its orbital and its spin motions. In almost all atoms the *net* effect of the magnetic fields set up by various electrons is exceedingly small or zero, because the magnetic field set up by one electron counterbalances the fields set up by other electrons. However, in iron, nickel, and cobalt the magnetic fields due to the spinning of the electrons do not cancel for certain of the electrons in an unfilled (M) shell. In these metals, individual ''atoms'' act as tiny magnets.

A magnetic field interacts with electrons in such a way as to induce a magnetic moment which opposes the applied field. This electronic diamagnetism occurs in all substances. If the ''atomic'' magnetic moments of the material are zero or very small, the substance is *diamagnetic*.

When "atoms" of a material have magnetic moments and the interaction between the moments of neighboring "atoms" is small, an applied field may align the moments and *paramagnetism* occurs. If the coupling between neighbors is large enough to order them in the absence of an applied magnetic field, there are three important classifications. When the "atomic" interactions align all the magnetic moments in a small domain, the substance is said to be *ferromagnetic* (Fig. 37.7*a*). However, the coupling may produce two (or possibly more) groups of moments, such that all neighbors of each group are aligned but the two groups are oriented *antiparallel*. If the "atomic" moments of the two groups are equal, the material is *antiferromagnetic* (Fig. 37.7*b*), while if the resultant magnetic moments of the groups is not zero, the substance is *ferrimagnetic* (Fig. 37.7*c*). Many ferrimagnets are poor electric conductors, a quality often desired for devices that operate at high frequencies.

It is unfortunate that the classical and current classifications use the same names but sometimes in different senses. For example, MnO is best described as *ferrimagnetic* in modern usage but is *paramagnetic* in the classical terminology. Similarly, the original magnetic material magnetite ($FeO \cdot Fe_2O_3$) is best described as *ferrimagnetic* in current language but is *ferromagnetic* classically.

(a) Ferromagnetic

(b) Antiferromagnetic

(c) Ferrimagnetic

37.5 MAGNETIZATION BY INDUCTION

In a ferromagnetic material, sometimes individual atoms are the elementary magnets, sometimes combinations or groups of atoms. Although the net magnetic field outside a piece of unmagnetized soft iron is zero, over small volumes of the order of 10^{-6} cm³ most of the elementary magnets are aligned with their N poles in the same direction. There is no field outside the iron because these tiny *domains* are oriented at random. If this iron is placed in a magnetic field, it is magnetized *by induction*. This magnetization occurs through two processes: (1) Domains which are magnetized in the direction of the inducing field grow at the expense of neighboring domains which are magnetized in less favorable directions. (2) The direction of magnetization of an entire domain may be shifted by the simultaneous rotation of the elementary magnets which make up the domain.

Further evidence confirming that the nature of the magnetizing process is the alignment of elementary magnets are the following observations.

1 When a piece of soft iron lies in a weak field, its magnetization is increased by tapping. Tapping the iron gives the elementary magnets a better chance to align themselves with the magnetizing field. On the other hand, if a piece of magnetized material is not in a field, tapping usually disturbs the alignment of some of the domains and thereby reduces the strength of the magnet.

2 Increasing the temperature of the magnet tends to demagnetize it, because adding kinetic energy to the molecules by thermal agitation produces the same effects as tapping or dropping the magnet. Iron heated above 760°C loses its magnetic properties.

Figure 37.7
When the "atoms" of a material have magnetic moments large enough for the interaction with neighboring atoms to produce a repetitive orientation over a domain, a material is (*a*) *ferromagnetic* if all magnetic moments are aligned, (*b*) *antiferromagnetic* if the magnetic moments are arranged in two groups which contribute equal and opposite net magnetization, and (*c*) *ferrimagnetic* if they are arranged in two groups with opposite alignment but one group involves "atoms" with a greater magnetic moment than the other.

Indeed, all magnetic materials have a temperature above which they lose their magnetism. The temperature at which the magnetic properties of a given material disappear is called the *Curie point* of that material.

3 If a long permanent magnet is cut into two equal shorter pieces, the magnetized domains remain aligned. There are now two magnets, each half as long as the original. The pole strength of each new magnet is roughly equal to that of the original magnet. It is easy to understand why magnetic effects seem to be concentrated at the ends or poles when one remembers that throughout the body of the magnet the arrangement of elementary magnets leaves a N pole close to each S pole, so that their effects cancel one another. At one end of the magnet there is a concentration of elementary N poles, and at the other end of S poles. Consequently, the magnetic effects are strong at the ends.

When the N pole of a bar magnet is brought near an unmagnetized needle, the elementary magnets in the needle align themselves in the field. Thus, a S pole is produced near the N pole of the bar magnet. Since the attractive force between the S pole of the needle and the N pole of the bar magnet is greater than the repulsive force between the two N poles, there is a net attractive force. It should be noted that *this attractive force arises only in a nonuniform magnetic field*. In a uniform field the forces on opposite poles are equal and opposite.

While most common magnets have two poles, it is entirely possible for a magnet to have three or more. For example, if the middle of a piece of soft iron is brought up to the strong N pole of an electromagnet, it develops a N pole at each end and a stronger S pole at the middle. When steel gas or water pipes are explored with a small compass needle, several N and S poles are often found along a few feet of the pipe. Such poles are called *consequent* poles.

37.6 PERMEABILITY AND THE MAGNETIZING FIELD

If a long solenoid is bent in the form of a circle until the two ends touch (Fig. 37.8), it becomes a *ring solenoid* or *toroid*. Consider a toroid which has n turns per meter of length and bears a current I. Bending a long solenoid to form a toroid does not change the magnetic field inside; thus the magnetic field in a toroid is given by $B = 4\pi nI/10^7$, according to Eq. (36.10).

Suppose the entire core of the toroid is now filled with iron, and a current I is sent through the windings. How is the value of **B** inside the toroid affected by this iron? A measurement† of **B** shows that it is now several hundred times greater than it was before the iron was added.

Figure 37.8
Magnetic lines of force exist within a toroid or ring solenoid; there is no external field.

† The measurement of **B** in the solid material may be made without disturbing the material in any way by wrapping a few turns of wire around the solenoid as indicated in Fig. 38.3 and observing the induced emf as the field is changed. The operation of such a measuring device, known as a fluxmeter, is described in Chap. 38.

The magnetic field in the iron may be written

$$B = \frac{4\pi K_m nI}{10^7}$$ (37.6)

where K_m is called the *relative permeability*. For any material K_m is the ratio of magnetic intensity in the material to magnetic intensity in vacuum. If we fill the toroid with some other material, such as bismuth, silver, wood, or nickel, and measure the magnetic intensity for a given current in the windings, we can determine its relative permeability (Table 37.1). Those materials with relative permeabilities less than unity are diamagnetic, those with permeabilities slightly greater than unity are paramagnetic, and those with permeabilities much larger than unity are ferromagnetic, according to the classical description.

The magnetic intensity in a toroid depends on two factors: (1) the properties of the medium in the toroid and (2) the influences which are inducing the magnetization of the medium. In this case the inducing field is provided by the current through the turns of the toroid. In other situations it may be provided by magnetic poles or other kinds of current distributions. It is called the *magnetizing field* and is represented by **H**. In the special case of the uniformly filled toroid we have continuous magnetization and no poles; the magnetizing force is the product of the current and the number of turns per unit length. Thus we have

$$H = nI$$ (37.7)

If nI is replaced by H in Eq. (37.6), we obtain

$$B = \frac{4\pi}{10^7} K_m H = \mu H$$ (37.8)

Here μ stands for $4\pi K_m/10^7$ N/A² and is called the *absolute permeability* of the material. Since the relative permeability K_m of a vacuum is unity, the absolute permeability is $4\pi/10^7$ for a vacuum (and for air for practical purposes). It is usually represented by μ_0 and called the *permeability of free space*.

B is different inside the toroid when it is filled with iron because of the magnetization of the iron which occurs when we apply the mag-

THIS IS FROM A CIRCLE.

$$I = H(2\pi R)$$

$$H = \frac{I}{2\pi R}$$

$$B = \frac{\mu_0 I}{2\pi R}$$

$$\mu_0 = \frac{4\pi}{10^7}$$

$$B = \mu_0 H$$

Table 37.1
Relative Permeabilities of Various Substances
(Vacuum = 1.000 by definition)

Material	Relative permeability	Material	Maximum relative permeability
Bismuth	0.99983	Cobalt	250
Silver	0.99998	Nickel	600
Copper	0.99999	Mild steel	2,000
Water	0.99999	Iron (0.2% impurity)	5,000
Air	1.0000004	Silicon iron*	7,000
Aluminum	1.00002	Permalloy	100,000
		Supermalloy	1,000,000

* Used in power transformers.

netizing field. We have seen (Sec. 37.1) that the magnetic moment M of a bar magnet of cross-sectional area A and length l is that which would exist if there were a current M/A flowing around the surface, i.e., a surface current M/lA per unit length. The ratio M/lA is the magnetic moment per unit volume, which we shall indicate with M_v. With magnetic material in the toroid the total magnetic intensity **B** comes from two contributions: μ_0**H** due to the conduction currents in the windings and μ_0**M**$_v$ due to the internal currents in the magnetic material.

$$\mathbf{B} = \mu_0(\mathbf{H} + \mathbf{M}_v) \tag{37.9}$$

In an *ideal* magnetic material **M**$_v$ is proportional to the magnetizing field **H**. From the equation $\mathbf{B} = \mu\mathbf{H} = \mu_0(\mathbf{H} + \mathbf{M}_v)$, we obtain the relation

$$\mu = \mu_0\left(1 + \frac{M_v}{H}\right) = \mu_0 K_m \tag{37.10}$$

Example

A toroid of mean circumference 0.5 m has 500 turns, each bearing a current of 0.15 A. (*a*) Find H and B if the toroid has an air core. (*b*) Find B and M_v if the core is filled with iron of relative permeability 5,000. (*c*) Find the average magnetic moment per iron atom if the density of iron is 7,850 kg/m³.

a By Eqs. (37.7) and (37.8),

$$H = \frac{500 \text{ turns}}{0.5 \text{ m}} \times 0.15 \text{ A} = 150 \text{ A/m}$$

$$B = 4\pi \times 10^{-7} \text{ N/A}^2 \times 150 \text{ A/m}$$
$$= 1.88 \times 10^{-4} \text{ N/A-m (T)}$$

b For iron with $K_m = 5,000$, by Eq. (37.8),

$$B = 4\pi \times 10^{-7} \times 5,000 \times 150 \text{ N/A-m}$$
$$= 0.94 \text{ N/A-m}$$

By Eq. (37.9),

$$0.94 \text{ N/A-m} = 4\pi \times 10^{-7} \text{ N/A}^2 \times (150 + M_v)$$
$$M_v = 7.5 \times 10^5 \text{ A/m (or A-m}^2/\text{m}^3)$$

c One kilogram atomic weight (55.85 kg) of iron has 6.02×10^{26} atoms. Therefore in 1 m³ there are $7,850 \times 6.02 \times 10^{26}/55.85$ atoms. Hence there are 8.48×10^{28} atoms in 1 m³ of iron, which has a magnetic moment M_v of 7.5×10^5 A-m²/m³. The average magnetic moment per iron atom is 8.9×10^{-24} A-m².   ////

For a material such as permalloy, which has a K_m of the order of 100,000, Eq. (37.10) shows that **M**$_v$, the magnetic moment per unit volume due to the "internal currents," is about 99,999 times as great as the magnetizing field due to currents in the external windings.

Only when **M**$_v$ is proportional to **H** is the permeability μ a constant. We shall see in Sec. 37.7 that for ferromagnetic materials μ is not constant over any large range of values of **H**.

If we solve Eq. (37.9) for **H**, we obtain

$$\mathbf{H} = \frac{\mathbf{B}}{\mu_0} - \mathbf{M}_v \qquad (37.11)$$

This is the defining equation† for **H** in the general case.

37.7 THE MAGNETIZATION CURVE

If a toroid is wound around an iron core and a fluxmeter provided to measure changes in magnetic intensity in the core, we can measure **B** as a function of **H** by starting with an unmagnetized core and no current in the windings and increasing the current in the windings step by step. If this is done, we obtain a *magnetization curve* in which the magnetic intensity B is plotted as a function of the magnetizing field $H = nI$. For a typical iron sample, a curve similar to that of Fig. 37.9a is obtained.

The permeability μ is the ratio of B to H, by definition. Figure 37.9b shows how μ varies with H for the particular magnetization curve of Fig. 37.9a. Observe that the permeability is not constant for a ferromagnetic substance, as it is for an ideal magnetic material. Rather it varies over a considerable range. It is not difficult to understand qualitatively why this is true. The iron core is composed of a large number of magnetic domains originally oriented at random. When a tiny current is passed through the windings, the magnetizing field H has a small effect on these domains compared with the influence of neighboring domains; therefore relatively small changes in magnetization occur. As the current is increased, the more favorably aligned domains grow at the expense of the others. The number of elementary magnetic dipoles oriented in the direction of the magnetizing field increases rapidly, as is shown by the steep portion of the magnetization curve. Once most of the elementary magnets are aligned, the region of "hard magnetization" begins. Here the magnetizations of the unaligned domains undergo rotations which bring them into more exact alignment with the applied field. When essentially all the elementary magnets are aligned, the iron is said to be *saturated*. From this point on, increases in H result in small increases in B. When the iron is saturated, the permeability (B/H) falls off rapidly as H is increased.

Observing a portion of the magnetization curve with instruments of great sensitivity will show that the curve is not perfectly smooth but is made up of a number of tiny jumps, called *Barkhausen steps*, owing to the fact that a large number of neighboring elementary magnets along a domain boundary align themselves with the applied field simultaneously, thereby producing sharp little jumps in the magnetization.

Although this picture of magnetism is simplified, it does present the qualitative features in reasonable perspective. During magnetization the sizes of domains change, and domains are reoriented. There may even be an observable change in the length of the magnet. Such a change in length with magnetization is known as *magnetostriction*. Magnetostriction oscillators are used to produce sound waves of very high frequency, particularly in liquid media.

† **B**, **H**, and **M**$_v$ are vectors and do not always have the same direction.

(a)

(b)

Figure 37.9
How (a) the magnetic intensity and (b) the permeability of a sample initially unmagnetized vary with the magnetizing field.

37.8 HYSTERESIS

When an iron core is saturated and the magnetizing field is removed, the magnetization does not fall to zero, because the magnetic domains have been aligned by the magnetization process and help to keep one another aligned. To demagnetize the specimen completely, a magnetizing field in the opposite direction must be applied. If a specimen is magnetized first in one direction and then in another, a plot of the relationship between B and H results in a *hysteresis curve*. Figure 37.10 shows hysteresis loops for three types of ferromagnetic material.

A material which makes a good permanent magnet has a very broad hysteresis loop, since this indicates a high retention of magnetism when the magnetizing field is removed. Such a material would not be good for an electromagnet, which should lose its magnetic properties when the magnetizing field is eliminated. In ac transformers, motors, etc., one ordinarily desires a magnetic material with a hysteresis loop of small area, because it takes energy to reverse the direction of the magnetization. The area of the hysteresis loop is a measure of the energy lost per cycle per unit volume of the material. A transformer made of steel with a broad hysteresis loop would be unsatisfactory because of the large amounts of energy which would be lost in magnetizing and demagnetizing it many times each second.

Soft iron demagnetizes rapidly when the magnetizing field is removed, so it is desirable for electromagnets and transformers. On the other hand, the alloy *alnico*, made of aluminum, nickel, and cobalt steel, is particularly high in retentivity and therefore makes a fine permanent magnet.

37.9 APPLICATIONS OF ELECTROMAGNETS

If a solenoid is wound around a soft-iron core, the iron becomes strongly magnetized when a current is passed through the solenoid and loses most of its magnetism when the current is stopped. There are many uses for such electromagnets. Large ones can pick up rails and operate switches on railroads, as well as lift and move scrap iron and other magnetic materials. Small electromagnets have been developed for extracting particles of iron from the eye and for opening and closing switches in telephone and automobile circuits.

A common use of electromagnetism is in the doorbell. Here a soft-iron vane is attracted by the electromagnet when the switch is

Figure 37.10
Hysteresis curves (*a*) for a piece of soft iron, (*b*) for an alloy which may be made into a permanent magnet, and (*c*) for an alloy suitable for use in a power transformer.

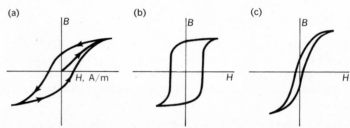

closed. When the iron moves toward the poles of the electromagnet, the clapper strikes the bell, and the electric connection is broken by the movement of the iron vane. The current ceases, and spring action moves the vane back toward its original position. Contact is reestablished, the electromagnet is activated once more, and another cycle begins. The same basic idea is also utilized in the electric horn, except that a diaphragm is set into vibration instead of a clapper. The loudspeakers of many radios, record players, and television receivers (Sec. 47.8) utilize electromagnets as fundamental components.

QUESTIONS

1 If a small permanent magnet is placed in a uniform magnetic field, is there any net force on the magnet? In a nonuniform field? Explain.

2 Why can't one obtain an isolated south pole simply by cutting it off a bar magnet?

3 A steel post is driven into the ground in the United States. Is it likely to be magnetized? Why? At which end is the N pole likely to be? Why?

4 A thin iron rod is deeply notched with a file and then magnetized. After the poles are marked, it is broken in halves. Why was it desirable to notch the rod before magnetizing it? Describe the poles in the two halves if the rod originally had a N pole at one end and a S pole at the other. How could you magnetize the rod so it had a S pole at each end? How would the halves be magnetized after you broke the rod this time?

5 One of two initially identical iron bars is magnetized and retains part of its magnetic moment. How can it be distinguished from the unmagnetized bar in a region in which there is no magnetic field due to the earth or other magnets?

6 What is meant by the "demagnetizing effect of poles"? Explain why a good permanent magnet usually has a "keeper" of soft iron placed between its poles when it is not in use.

7 If a substantial quantity of magnetic ore lay beneath the earth's surface, how would the magnetic field of the earth be affected nearby? Consider both horizontal and vertical components.

PROBLEMS

1 A bar magnet has poles of strength 9 A-m and a magnetic moment of 0.45 A-m². This magnet is placed perpendicular to the earth's magnetic field, which has a horizontal component of 2.5×10^{-5} Wb/m² (or T). (a) Find the torque acting to rotate the magnet about a vertical axis. (b) Find the magnitude of the resultant horizontal magnetic induction at point P on the SN axis 10 cm from the N pole of the magnet.
Ans. (a) 1.125×10^{-5} N-m; (b) 5.59×10^{-5} T

2 What torque is necessary to hold the axis of a magnet at an angle of 40° to the magnetic meridian where the horizontal component of the earth's magnetic field is 2×10^{-5} T? The length of the magnet is 12 cm, and its pole strength is 40 A-m.

3 A magnet has a pole strength of 60 A-m. It is placed at right angles to the earth's magnetic field of 2×10^{-5} T, and a torque of 8×10^{-5} N-m acts on it. Find the magnetic moment of the magnet and the distance between the poles. What would the torque be if the magnet were rotated so its axis made an angle of 30° with the field? *Ans.* 4 A-m²; 6.67 cm; 4×10^{-5} N-m

4 Two magnetic poles have strengths of 40 and 90 A-m, respectively. At what distance in air will the force of attraction between them be 10^{-3} N?

5 A small test magnet has poles of 80 A-m strength separated by 3 cm. Find the force on each pole when this magnet is placed at the center of a circular loop with a radius of 25 cm in which a current of 12.5 A is flowing. (Assume the magnetic intensity is uniform over the volume occupied by the magnet.) What is the resultant force on the magnet? Find the torque on the magnet if its axis is perpendicular to the magnetic field.
Ans. 2.51×10^{-3} N; zero; 7.54×10^{-5} N-m

6 The magnetic declination in central Utah is 15° east. How far from a true-north course would a flier be after traveling 100 mi following the compass without making a correction for declination? Would he be east or west of his course?

7 Find the magnetizing field H and the magnetic flux density B at (a) a point 12 cm from a long straight wire bearing a current of 16 A and (b) the center of a 3,000-turn solenoid which is 25 cm long and bears a current of 1.5 A.
Ans. (a) 21.2 A/m; 2.67×10^{-5} T; (b) 18,000 A/m; 0.0226 T

8 A magnet with poles of 50 A-m strength separated by a distance of 3 cm is placed in the uniform field inside a solenoid with its axis at right angles to the lines of force. If the solenoid is 40 cm long and has 500 turns bearing a current of 2.5 A, what torque does the field exert on the magnet?

9 Find the magnetic induction at a point 12 cm from the center of a magnet on the perpendicular bisector of the line joining the poles if the magnet is 10 cm long and its pole strength is 300 A-m.
Ans. 1.37×10^{-3} T parallel to axis

10 A north pole of 50 A-m strength is placed 20 cm from a south pole 80 A-m strength. How far from the north pole, on a line drawn through the two poles, will the resultant field due to these poles be zero?

11 A bar magnet is 16 cm long, and each pole has a strength of 72 A-m. Find the magnitude of the intensity of the magnetic field at a point 12 cm from the S pole, measured at right angles to the axis of the magnet. *Ans.* 4.18×10^{-4} T

12 A long bar of cobalt steel with a mass of 35 g when placed horizontally over a similar bar, both being equally magnetized, remains suspended at a distance of 0.8 cm above it (Fig. 37.3). What is the pole strength at each end of each bar? (*Hint:* Assume the bars are so long that only the nearest poles interact significantly.)

13 The axes of two magnets are collinear. One has poles of strength 70 A-m separated by 10 cm, and the second has a magnetic moment of 12 A-m² with poles of strength 150 A-m. Find the attractive force between the magnets if the N pole of one is 5 cm from the S pole of the second. *Ans.* 0.331 N

14 Identical magnets 9 cm long with poles of 324 A-m strength lie 12 cm apart with axes parallel (Fig. 37.3). Find the attractive force between the magnets if the N pole of one is opposite the S pole of the other.

15 A bar magnet 15 cm long with a pole strength of 8 A-m is horizontal, at right angles to the earth's magnetic field, with its N pole pointing west. Find the intensity of the magnetic field in a horizontal plane at a point 20 cm west of the N pole. Take the horizontal component of the earth's magnetic field as 2×10^{-5} N/A-m.
Ans. 2.41×10^{-5} N/m 34°W of N

16 The poles of a magnet are 30 cm apart, and each has a strength of 54 A-m. What is the magnitude of the magnetic intensity at a point 24 cm from one pole and 18 cm from the other?

17 An electromagnet has a solenoidal winding 25 cm long with a total of 800 turns. What is the magnetizing field H near the center of the winding and far from any poles if the current is 0.9 A? What is the magnetic induction B at this point if the iron has a relative permeability of 400? *Ans.* 2,880 A/m; 1.45 T

18 An iron ring has a cross section of 0.9 cm² and an average diameter of 25 cm. It is wound with 700 turns of copper wire. If the iron in the core has a relative permeability of 550, what is the magnetic intensity in the iron when a current of 4 A exists in the windings?

19 An iron anchor ring is wound with 600 turns. If the current in this toroidal solenoid is 0.8 A and the relative permeability of the iron core is 300, what is the magnetic induction B, assuming that the toroid has a mean circumference of 40 cm? What is the magnetizing force H? Find the magnetic moment per unit volume of iron and the average contribution to this magnetization per iron atom.
Ans. 0.452 T; 1,200 A/m; 3.59×10^5 A/m; 4.23×10^{-24} A-m²

20 At a place where the horizontal component of the earth's magnetic field is 2×10^{-5} T, a bar magnet 15 cm long with a pole strength of 2.5 A-m is horizontal, at right angles to the earth's field, with its N pole pointing toward the west. Find the direction and intensity of the field in a horizontal plane at a point 10 cm east of the S pole of the magnet.

21 A toroid with 1,200 turns is wound on an iron ring 4 cm² in cross-sectional area, of 80 cm mean circumference and of 1,500 relative permeability. If the windings carry 0.12 A, find (a) the magnetizing field H, (b) the product of H and the length, called the *magnetomotive force*, (c)

the magnetic induction B, (d) the product of B and the cross-sectional area of the ring, called the *flux*, and (e) the ratio of magnetomotive force to flux, called the *reluctance* of the circuit.

Ans. (a) 180 A/m; (b) 144 A; (c) 0.339 T; (d) 1.36×10^{-4} N-m/A (or Wb); (e) 1.06×10^{6} A²/N-m (or A/Wb)

22 A solenoid that is 60 cm long is wound with 1,500 turns of copper wire. An iron rod having a relative permeability of 450 is placed along the axis of the solenoid. What is the magnetic intensity in the rod when a current of 0.7 A flows through the wire? What are the magnetizing field and the magnetic moment per unit volume of the iron? Find the average contribution per iron atom to the magnetization.

CHAPTER THIRTY-EIGHT
INDUCED ELECTROMOTIVE FORCES

IN OUR DISCUSSION of electric circuits thus far, we have used batteries and thermocouples as sources of emf. Electric energy from either of these sources is relatively expensive. Now that we have learned some facts about magnetic phenomena, we are ready to develop the fundamental physics upon which the economic conversion of mechanical energy to electric energy depends. We recall that when an electric current is sent through a conductor, a magnetic field appears near the wire. If a current always produces a magnetic field, is there some way in which a magnetic field can produce a current? There is, and its discovery made feasible the broad use of electric energy in our daily lives.

38.1 THE DISCOVERY OF INDUCED EMFS

The discovery by Oersted in 1819 that a current has an associated magnetic field led a number of physicists to search for some means by which a magnetic field might produce a current. The first observation of such a phenomenon was made in 1830 by Joseph Henry. He used a horseshoe-shaped electromagnet around which he wound a second coil, the terminals of which were connected to a galvanometer (Fig. 38.1). Henry found that when the current through the electromagnet is changed, either increased or decreased, there is a deflection of the galvanometer coil. When the current in the electromagnet is steady, there is no current through the galvanometer. When the current in the magnet is turned on, the deflection of the galvanometer is in one direction; when the current is stopped, the deflection is in the opposite direction.

Several months later Michael Faraday independently discovered the deflection of a galvanometer connected to one coil when the current in an adjacent coil was started or stopped. Faraday published his findings first and is therefore usually credited with the discovery.

Faraday made a thorough study of the phenomenon. He found that the galvanometer in series with the second coil deflected not only when the current in the first coil was started and stopped, but also when the first coil carried a steady current and was moved nearer to or farther from the second. He brought a magnet near the second coil and then withdrew it. The galvanometer needle deflected during the motion of the magnet. When the north pole of a magnet was brought near the

Figure 38.1
Apparatus used by Henry when he discovered electromagnetic induction.

coil, the current flowed in one direction. When this pole was withdrawn, the current flowed in the opposite direction. If a south pole was brought toward the coil, the deflection was in the same direction as when the north pole was withdrawn. Whenever the magnetic lines of force linking the second coil were changed, there was an induced emf which produced a current through the galvanometer. This phenomenon is called *electromagnetic induction*. To develop quantitative relations between change in magnetic field and induced emf, we must become familiar with the meaning of the term *magnetic flux*.

38.2 MAGNETIC FLUX: THE WEBER

Consider an area A (Fig. 38.2a) with its plane perpendicular to a uniform magnetic field of intensity **B**. The magnetic flux Φ through this area is defined as the product of **B** and **A**:

$$\Phi = BA \qquad (38.1)$$

The flux Φ is given in *webers* when B is in teslas (newtons per ampere-meter) and A is in square meters. The weber (Wb) is thus dimensionally equivalent to the newton-meter per ampere. It is named in honor of W. E. Weber (1804–1891), German physicist.

In many situations it is easier to measure the flux through a coil than it is to measure **B** directly. If Φ and the area are known, the average value of **B** is given by Φ/A Wb/m². (The weber per square meter, the tesla, and the newton per ampere-meter are different names for the same unit.)

If **B** is not perpendicular to the plane of the area through which the flux is desired, $\Phi = BA \cos \theta$, where θ is the angle between **B** and the normal to the area (Fig. 38.2b). When **B** is not constant over the area, Eq. (38.1) is applicable if we take the *average* flux density for **B**.

An alternative point of view toward flux was developed by Faraday. We saw in Sec. 36.3 that a magnetic field may be represented by lines of force. If we draw many lines of force in regions where the field is strong and correspondingly fewer where the field is weak, we have a plot which shows not only the direction of the magnetic field at various points, but also its intensity. If we agree to limit the number of lines of force so that we draw one line per square meter where **B** is 1 Wb/m², two lines per square meter where **B** is 2 Wb/m², etc., the number of lines through any large area is just equal to the magnetic flux through that area. Let us call lines drawn according to this convention "lines of flux" to distinguish them from lines of *force*, which give only the direction of the field. In terms of this physical picture, the *flux* through any area is equal to the number of *flux lines* which pass through the area.

(a)

(b)

Figure 38.2
The magnetic flux linking an area is the product of the area and the component of the magnetic intensity perpendicular to the area; $\Phi = AB \cos \theta$.

38.3 FARADAY'S LAW OF ELECTROMAGNETIC INDUCTION

Faraday found experimentally that the magnitude of the induced emf in a single loop was directly proportional to the rate at which the flux linking the loop changed.

Figure 38.3
Iron-core toroid with a fluxmeter that
measures changes in magnetic flux.

$$e = -\frac{\Delta\Phi}{\Delta t} \tag{38.2}$$

where $\Delta\Phi$ is the change in flux occurring in the time Δt. The emf e†
is given in volts when $\Delta\Phi/\Delta t$ is in webers per second. An emf $\Delta\Phi/\Delta t$
is induced in each turn of a coil. If N turns are connected in series,
the total emf in the coil is N times that induced in a single loop. The
significance of the minus sign is discussed in Sec. 38.5.

38.4 THE FLUXMETER

In Sec. 37.6 it was assumed that we can measure **B** inside a toroid
without cutting any holes in the material. We are now in a position to
see how this can be done. Consider the iron-filled toroid shown in Fig.
38.3. Assume the iron is originally unmagnetized and the current in
the winding is zero. If N turns of wire are wound around this toroid
and connected to a suitable galvanometer, we can measure **B** as follows. We pass a small current through the toroid windings. This changes
the magnetic intensity in the iron from zero to some value **B**. If A is
the cross-sectional area of the iron core, the flux in the iron goes from
zero to BA in a time t. According to Faraday's law of induction, there
is, in the fluxmeter windings, as induced emf e given by

$$e = \frac{-N\,\Delta\Phi}{\Delta t} = \frac{-NAB}{t}$$

There is a current in the fluxmeter given by $i = e/R = NAB/Rt$. Now
$it = NAB/R$ is equal to the charge q passing through the fluxmeter.
If we use a suitable galvanometer, we obtain a deflection proportional
to this charge. If we know q, N, and A, we can calculate **B**.

38.5 LENZ'S LAW

The direction of the induced emf is readily predicted by application of
a rule due to Lenz. *The direction of an induced emf is always such*

---

† Note that we have used e rather than $\mathcal{E}$ to indicate the induced emf in this case. In general,
we use $\mathcal{E}$ and I to represent *constant* values, and the lower-case letters e and i to indicate
emfs and currents which *vary in time*.

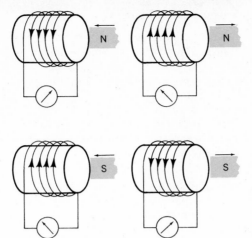

Figure 38.4
The current arising from an induced emf always opposes, through its magnetic effects, the change inducing the emf. Whether the bar magnet is pushed into or pulled out of the coil, the magnetic field of the induced current opposes the motion of the magnet.

that any current it produces opposes, through its magnetic effects, the change inducing the emf.

To illustrate Lenz's law, consider a N pole that is being pushed (Fig. 38.4) toward a coil. The current induced in the coil is in such a direction that its magnetic field opposes the motion of the magnet. In this case, it is directed toward the north pole of the magnet, which requires that the current be counterclockwise. When the N pole is withdrawn, the directions of induced emf and current reverse, since the induced magnetic effect opposes the act which creates it.

Actually, Lenz's law represents one of the many forms in which conservation of energy appears in physics. That this law follows from the conservation of energy may be seen by this reasoning. Changing a magnetic field induces an emf and a corresponding current if there is a conducting path. Thus, electric energy appears as the result of this act. It is the work done against the opposing magnetic field which is transformed into electric energy.

38.6 MOTIONAL ELECTROMOTIVE FORCE

If a wire of length **l** is moved with a velocity **v** perpendicular to a magnetic field of flux density **B** (Fig. 38.5), there is induced in this wire an emf

Figure 38.5
A motional emf is induced by moving a conductor across a magnetic field; in this case **B** is into the page, and the upper end of the wire is positive.

$$\mathcal{E} = vBl \qquad (38.3)$$

Figure 38.6

By moving the wire to the right, an emf is induced which results in a current into the page.

The origin of this emf is in the force exerted on each individual charge moving through the magnetic field. This force, as we saw in Sec. 36.4, is qvB. The electric field strength $E = F/q = vB$. The potential difference across a wire of length l is given by the product of the field strength and the length. Therefore, $\mathcal{E} = El = vBl$. (In the event that **B**, **v**, and **l** are not mutually perpendicular, $\mathcal{E}$ is numerically equal to the volume of a parallelepiped with sides in the directions of **B**, **v**, and **l** and lengths proportional to the magnitudes of the respective quantities.)

In many electric generators the emf is induced by moving wires through a magnetic field at high speed. The direction of the induced emf is such that the force on the resulting current opposes the movement of the wire across the field. From this it follows immediately that the magnetic field of the current produced by the induced emf strengthens the field on the side toward which the wire is moving and weakens it on the opposite side. Once we know the direction of the magnetic field associated with the induced current, we can let the fingers of the right hand curve along the lines of force; the right thumb then points in the direction of the induced emf (see Sec. 36.2). In the situation shown in Fig. 38.6, the induced emf produces a current into the paper when the wire is moved as indicated.

38.7 THE EMF IN A ROTATING LOOP

When a rectangular loop of wire rotates in a uniform magnetic field (Fig. 38.7), the conductors move through the field and a motional emf is induced. The conductors ab and cd cut across the flux lines, while conductors ad and bc do not. In conductor ab there is induced an instantaneous emf

$$e_1 = vBl \sin \theta$$

where θ is the angle between the velocity **v** of the wire and **B**. The direction of the induced emf is from b toward a as ab moves downward. At the same time, in side cd there exists an emf of equal magnitude directed from d to c. The emfs in ab and cd are in series, and the net emf in the loop $badc$ is

$$e = 2vBl \sin \theta = \mathcal{E}_{\max} \sin \theta \qquad (38.4)$$

where $\mathcal{E}_{\max}$ is the maximum emf induced in the loop. For a flat coil consisting of n closely wound loops, both e and $\mathcal{E}_{\max}$ are n times that of a single loop.

The instantaneous emf induced in the coil varies as θ changes. It is zero when the plane of the coil is vertical, since in this position the wires in the coil move parallel to the magnetic intensity **B** and $\sin \theta = 0$. When the plane of the coil is horizontal, the wires move perpendicular to the lines of flux, $\sin \theta = 1$, and the induced emf is maximum. The relation between the position of the coil and the induced emf is evident from Fig. 38.8. As the curve shows, the emf generated rises from zero to a maximum value at 90°, decreases to zero again

Figure 38.7

A varying emf is induced in a rectangular loop rotating in a uniform magnetic field with constant angular velocity.

Figure 38.8
How the emf in the loop of Fig. 38.7
varies with angle as the coil rotates.

at 180°, reverses its direction to reach its largest negative value at 270°, and returns to zero when the rotation is completed. An *alternating emf* is induced in the coil. If such an emf is applied to a circuit, the current alternates with the same frequency as the emf (Chap. 40).

38.8 MUTUAL INDUCTANCE

Consider the neighboring circuits of Fig. 38.9. When the key is pressed, the current in coil A rises and the associated magnetic lines of flux produce a change in the magnetic flux linking coil B. As a consequence, there is an induced emf and a resulting current in coil B. The current in B lasts only as long as the current in A is changing. In such a circuit, coil A is called the *primary* and coil B the *secondary*. When the current in the primary is increasing, the induced emf in the secondary produces a current which, by its magnetic effect, opposes the rise of current in the primary. Note that it is always the *change in current in the primary* which is opposed, not the current itself. Thus, when the current passes its maximum value and is being reduced, the magnetic effect due to the induced current in coil B is such as to keep the primary current at its previous value. The effect which results in an emf being produced in one circuit due to a changing current in another circuit is called *mutual induction*.

The emf e_2 induced in the secondary coil is directly proportional to the rate of change of current in the primary:

$$\mathcal{E}_1 = L \frac{di}{dt}$$

$$e_2 = M \frac{\Delta i_1}{\Delta t} \qquad (38.5)$$

where the constant M is called the *coefficient of mutual inductance*.

The coefficient of mutual inductance between two circuits is the ratio of the emf induced in the second circuit to the rate of change of current with time in the first circuit. When an emf of 1 V is induced in a secondary coil by a current change of 1 A/sec in the primary, the coefficient of mutual inductance is said to be 1 *henry* (H).

Example

The mutual inductance between two circuits is 0.4 H. Find the emf induced in the secondary at an instant when the current is changing at the rate of 90 A/sec in the primary.

$$e_2 = M \frac{\Delta i_1}{\Delta t} = 0.4 \times 90 = 36 \text{ V} \qquad ////$$

Figure 38.9
Mutual inductance between two circuits occurs when magnetic flux from circuit A links circuit B so that current changes in A induce an emf in B.

Figure 38.10

Transformer designed to step up the alternating potential difference across the primary.

If an iron core (Fig. 38.10) is wound with two separate coils, and if an alternating current is maintained in one of them, an alternating emf of the same frequency is induced in the other. Such a device is known as a *transformer*. The first coil is called the *primary*, and the other the *secondary*. Power is transferred from the primary to the secondary by *mutual inductance*.

The change from one ac potential difference to another can be made very efficiently at low frequency by means of a transformer. For example, at 60 Hz the efficiency of a large transformer may be 99 percent, while smaller transformers may be 90 percent efficient. Some power is dissipated in hysteresis and eddy currents induced in the iron by the changing magnetic flux. Also, there is some Joule heating (I^2R) in the windings, but in a well-designed transformer the power delivered to the secondary is almost equal to the power supplied to the primary. If V_p and V_s are the primary and secondary potential differences† and I_p and I_s the corresponding currents,

$$V_p I_p \approx V_s I_s \tag{38.6}$$

Further, for this well-designed transformer the same magnetic flux links both the primary and the secondary windings. Consequently, the ratio of the emf induced in the secondary to that applied to the primary is equal to the ratio of the number of turns in the secondary N_s to the number in the primary N_p:

$$\frac{\mathcal{E}_s}{\mathcal{E}_p} = \frac{N_s}{N_p} \tag{38.7}$$

In an ideal transformer there are no losses, and the induced emfs in the primary and secondary are equal to the corresponding terminal potentials V_p and V_s. From $\mathcal{E}_p = V_p$, $\mathcal{E}_s = V_s$, and Eqs. (38.6) and (38.7), we can write

$$\frac{V_s}{V_p} = \frac{\mathcal{E}_s}{\mathcal{E}_p} = \frac{N_s}{N_p} = \frac{I_p}{I_s} \tag{38.8}$$

Where the emf is large, the current is small; where the emf is small, the current is large. By means of a transformer, a small emf and a large current may be transformed into a large emf and a small current, or vice versa.

In small transformers, where the ratio of transformation is not large, economy of construction and efficiency of operation are obtained by using the same coil for both primary and secondary. Such a transformer is known as an *autotransformer*. The arrangement and connections of the coil are shown in Fig. 38.11. The entire coil AC is the primary of the transformer, and the part between B and C is the secondary. Equation (38.8) is applicable to autotransformers as well as to other types, provided always that heat losses are negligible.

Figure 38.11

Autotransformer designed to step down the potential difference from 550 to 110 V.

† Here we may regard the values of the V's and I's to be either the maximum values attained during the cycle or the "effective" values as defined in Sec. 40.2.

At higher frequencies eddy-current and hysteresis losses in iron become prohibitive, so no iron is used. In such an air-core transformer only a fraction of the magnetic flux from the primary links each turn of the secondary, and the efficiency is low. Under these conditions the transformer equations above are no longer good approximations.

38.10 THE INDUCTION COIL

An illustration of mutual inductance is found in the induction coil, which is constructed as shown in Fig. 38.12. A primary coil made of a few turns of heavy copper wire is wound around an iron core. Insulated from this primary coil is the secondary coil, which is wound on the outside of the primary. The secondary contains a large number of turns of fine, well-insulated wire. By making or breaking the current in the primary at D, an emf is induced in the secondary. In order to make and break the current in the primary, an interrupter similar to that of the doorbell (Sec. 37.10) is connected in the circuit.

When the primary circuit is broken, the current is rapidly reduced to zero and an emf is induced in the secondary. This emf is large, because the number of turns in the secondary is very large and the time in which the primary current is stopped is short. To get the greatest induced emf, the primary current must be stopped as quickly as possible. To effect this, a capacitor C is connected across the gap in the primary. It acts as a storage place into which the charge surges when the circuit is broken. Without this capacitor an arc would be established at the contacts when the circuit is opened. Such an arc pits the contacts and results in a slower stopping of primary current. There is also an emf induced in the secondary when the primary current is rising. This emf is much smaller, because the time required for the current to build up to its maximum value is long compared with the time required to stop the current. The primary current and the induced emf in the secondary are plotted as a function of time in Fig. 38.13.

The spark coil of an automobile is an induction coil similar to the one just described, except that instead of a mechanical vibrator opening and closing the primary circuit a cam on the distributor shaft opens the "points" at the instant a high potential is needed to make a spark between the terminals of a spark plug. The secondary potential is many thousand volts, although only 12 V is available in the primary.

Figure 38.12
An induction coil

Figure 38.13
How the current I_p in the primary of an induction coil and the emf e_s induced in the secondary vary in time.

38.11 SELF-INDUCTANCE

When the current through a circuit such as that in Fig. 38.14 is changing, the magnetic flux linking this circuit is also changing. So long as B is proportional to H, the flux Φ linking the circuit is directly proportional to the instantaneous current i; thus

$$\Phi = Li \tag{38.9}$$

where L is a constant known as the *coefficient of self-inductance*. If, in a time Δt, the current changes by an amount Δi, the flux changes by $\Delta\Phi$ and there is, by Eq. (38.2), an induced emf

$$e = -\frac{\Delta\Phi}{\Delta t} = -L\frac{\Delta i}{\Delta t} \tag{38.10}$$

The negative sign appears in Eq. (38.10) because, by Lenz's law, the direction of the induced emf is always such as to oppose the change in current.

Self-inductance is ordinarily measured in henrys. *The self-inductance of a circuit or component is one henry when there is induced an emf of one volt in that circuit or component at an instant when the current is changing at the rate of one ampere per second.*

THE DIRECTION OF THE $\mathcal{E}$ IS OPPOSING THE CURRENT EITHER FROM GOING DOWN OR TOO HIGH.

$\mathcal{E} = L \dfrac{di}{dt}$

IF THE ARROW GOES DOWN (CURRENT) THEN:

$\mathcal{E} = -L \dfrac{di}{dt}$

Figure 38.14
The flux linking the inductor L changes when the current varies, thereby inducing an emf which opposes the variation in current.

(a)

(b)

Figure 38.15
Inductor circuit (a) for studying the rise and decay in current (b).

The self-inductances of many dc circuits are negligibly small, but a coil with many turns, a large solenoid, or an electromagnet may have a large self-inductance. In ac circuits inductance is of great practical importance, since the current is constantly changing. The emf resulting from self-inductance always opposes the *change of current;* it operates to hold down the current when it is rising and to maintain the current when it is decreasing.

Consider the circuit of Fig. 38.15a. When the key K is closed, the current in the circuit begins to rise. If we apply Kirchhoff's second law to the circuit at any instant, we obtain

IF WE DON'T HAVE ANY RESISTANCE

$$\mathcal{E} - Ri - L\frac{\Delta i}{\Delta t} = 0 \qquad (38.11)$$

$$\boxed{\mathcal{E} - L\frac{di}{dt} = 0}$$

At the instant the switch is closed, i is zero, and $\mathcal{E} = L\,\Delta i/\Delta t$. As time goes on, i increases and eventually the value $I = \mathcal{E}/R$ (Fig. 38.15b) is attained.

If the battery in Fig. 38.15 is suddenly shorted out of the circuit by closing the key S, the emf of self-induction keeps a current passing through the circuit until all the energy stored in the magnetic field of the inductor is dissipated. The current in the circuit is shown as a function of time after closing S in the right half of Fig. 38.15b.

When a current I passes through an inductor L, the energy W stored in the magnetic field is given by

$$W = \tfrac{1}{2}LI^2 \qquad (38.12)$$

as can readily be shown by the application of integral calculus (see derivations in Appendix C). The energy calculated from Eq. (38.12) is expressed in joules when L is in henrys and I in amperes.

QUESTIONS

1   When a switch in a series circuit consisting of a battery, a switch, and an inductor is opened, the induced emf is far larger than when the switch is closed. Why?

2   How does one obtain the thousands of volts needed for the spark in the cylinder of an automobile when one starts with 12 V from a battery?

3   What happens to the work done against the back emf of an inductor in establishing a current?

4   When a sensitive moving-coil galvanometer is to be shipped or carried about, it is customary to short-circuit the coil terminals. Why is this desirable?

5   A dc series circuit includes an inductor and a switch. If 10 A passes through the circuit, is there any limitation on how quickly the current

can be reduced to zero by opening the switch? Why is there often a spark across the terminals of a switch in a dc circuit when the switch is opened?

6 How can a wire-wound resistor be made so it is essentially noninductive?

7 A strongly magnetized rod with its axis vertical is dropped down a long evacuated copper tube, which is vertical and just slightly bigger in diameter than the rod. Why is it that even if it never touches the tube, the rod reaches a constant terminal velocity? What provides the upward force equal to the weight of the rod? What would happen if the rod were shot down the tube with an initial speed greater than the terminal speed reached after dropping?

8 What factors determine the mutual inductance of two coils?

9 A closed conducting plane loop is moved perpendicular to a uniform magnetic field. Is there an induced emf when the plane of the loop is (a) parallel to the motion and (b) perpendicular to the motion? Discuss the proposal that the groundspeed of an airplane be measured by determining the emf induced in a horizontal loop fixed in the plane by the vertical component of the earth's magnetic field.

PROBLEMS

1 A horizontal wire 0.5 m long is falling at a speed of 6 m/sec perpendicular to a uniform magnetic field of 1.1 T (N/A-m) which is directed from east to west. Calculate the magnitude of the induced emf. Is the north or south end of the wire positive? *Ans.* 3.3 V; north

2 An emf of 5 V is obtained by moving a wire 1.2 m long at a rate of 8 m/sec perpendicular to a uniform magnetic field. What is the intensity of the field?

3 A jet aircraft is flying due north at 320 m/sec at a place where the vertical component of the earth's magnetic field is 8×10^{-5} T. Find the potential difference between wing tips if they are

40 m apart. Which tip has the higher potential? *Ans.* 1.024 V; left

4 An axle of a car is 2.4 m long. If the car is moving due south at 25 m/sec at a place where the vertical component of the earth's magnetic field is 9×10^{-5} T, find the potential difference between the two ends of the axle. Which end is positive?

5 A small "search" coil with an area of 1.2 cm² has 50 turns of very fine wire. This coil is placed between the pole pieces of a small magnet and then suddenly jerked out. If the average induced emf is 0.3 V when the coil is pulled to a field-free region in 0.07 sec, what is the magnetic intensity between the poles? *Ans.* 3.5 T

6 A coil of 250 turns with an area of 300 cm² is placed with its plane perpendicular to the earth's field and is rotated in 0.015 sec through a quarter turn, so that its plane is parallel to the earth's field. What is the average emf induced if the earth's field has an intensity of 8×10^{-5} T?

7 The magnetic induction B in the core of a spark coil changes from 1.5 to 0.1 T in 2×10^{-4} sec. If the cross-sectional area of the core is 4 cm², find the original flux through the core. What is the average emf induced in the secondary coil if it has 5,000 turns?
Ans. 6×10^{-4} Wb; 14,000 V

8 The secondary of an induction coil has 15,000 turns. If the flux linking the coil changes from 6.5×10^{-4} to 5×10^{-5} Wb in 1.5×10^{-4} sec, how great is the induced emf?

9 What back emf is induced in a coil of self-inductance 0.008 H when the current in the coil is changing at the rate of 110 A/sec? What energy is stored in the inductor when the current is 6 A? *Ans.* 0.88 V; 0.144 J

10 The current in a circuit changes from 20 A to zero in 0.003 sec. If the average induced emf is 250 V, what is the coefficient of self-inductance of the circuit? How much energy was initially stored in the magnetic field of the inductor?

11 An electromagnet has a self-inductance of 6 H. How much energy is stored in the magnetic field when a current of 5 A exists in the coil? What average emf is induced if the current is reduced to zero in 0.015 sec? *Ans.* 75 J; 2,000 V

12 The coefficient of mutual inductance between two coils is 0.009 H. What emf is induced in the second coil if the current is changing at the rate of 4,000 A/sec in the first coil?

13 Two circuits have a coefficient of mutual inductance of 20 mH. What average emf is introduced in the secondary by a change from 50 to 5 A in 0.006 sec in the primary?   *Ans.* 150 V

14 The secondary of an ideal transformer has 275 times as many turns as the primary. It is used in a 110-V circuit. What is the voltage across the secondary?

15 An ideal transformer has 500 turns on the primary and 25 turns on the secondary. What is the maximum output potential difference if the maximum input voltage is 3,300? If the transformer is assumed to have an efficiency of 100 percent, what maximum primary current is required if a maximum current of 6 A is drawn from the secondary?   *Ans.* 165 V; 0.3 A

16 When the current in the primary of a small transformer is changing at the rate of 500 A/sec, the induced emf in the secondary is 6 V. What is the coefficient of mutual inductance?

17 The circuit of Fig. 38.14 consists of a 40-V battery, a 20-Ω resistor, a 0.05-H inductor, and a key. Find the rate at which current begins to rise when the key is closed, the current at the instant the rate of change of current is 200 A/sec, and the final steady current.

     Ans. 800 A/sec; 1.5 A; 2 A

18 The series circuit of Fig. 38.14 consists of a 30-V battery, a 10-Ω resistor, an 0.08-H inductor, and a switch. Apply Kirchhoff's second law to the circuit to relate the current i and its rate of change $\Delta i/\Delta t$ to the emf, resistance, and inductance of the circuit at any time after the switch is closed. Find the back emf induced in the inductor when the current is 2 A and the energy stored in the inductor at that instant.

19 A toroid with an iron core has an average circumference of 50 cm and a cross-sectional area of 4 cm². When the current in the windings is 0.7 A, the magnetic induction B is 0.9 T. The relative permeability of the coil is 200. A three-turn fluxmeter connected to a galvanometer is used to determine B in the coil. (*a*) Calculate the flux through the solenoid. (*b*) Find the number of turns on the toroidal winding. (*c*) When the key is opened, the current in the winding and the flux drop to zero. Find the coefficient of mutual inductance between the fluxmeter and the toroidal winding.

Ans. (*a*) 3.6×10^{-4} Wb; (*b*) 2,558 turns; (*c*) 0.00154 H

20 A constant potential difference of 45 V is suddenly applied to a coil which has a resistance of 15 Ω and a self-inductance of 0.0075 H. At what rate does the current begin to rise? What is the current at the instant the rate of change of current is 500 A/sec? What is the final current?

21 A ballistic galvanometer with a resistance of 200 Ω gives a full-scale deflection for 7×10^{-4} C of electricity. A coil of 300 turns and 100-Ω resistance is to be constructed to study fields up to 1.4 T by observing deflections produced when the coil is suddenly removed from the field. What is the maximum area allowable for the coil if full-scale deflection is not to be exceeded?   *Ans.* 5 cm²

22 A coil of 100 turns with a radius of 6 mm and a resistance of 50 Ω is placed between the poles of an electromagnet and suddenly removed. A charge of 9×10^{-6} C is sent through a ballistic galvanometer connected to the coil. The resistance of the galvanometer is 160 Ω. What is the intensity of the magnetic field?

23 A toroid of 0.5-m circumference and 5-cm² cross-sectional area has 3,000 turns bearing a current of 0.4 A. It is wound on an iron ring with relative permeability 500. Find the magnetizing field H, the magnetic induction B, the flux Φ, the coefficient of self-inductance L, and the energy stored in the magnetic field.

Ans. 2,400 A/m; 1.51 T; 7.54×10^{-4} Wb; 1.89 mH; 1.51×10^{-4} J

24 Show that for a long narrow toroid of average circumference $2\pi r$, cross-sectional area A, number of turns N, and permeability of core $K_m \mu_0$, the coefficient of self-inductance is given by $L = N^2 \mu_0 K_m A/2\pi r$ and the energy stored per unit volume in the magnetic field of the toroid is $\frac{1}{2}BH$.

CHAPTER THIRTY-NINE
GENERATORS AND MOTORS

MECHANICAL ENERGY can be converted to electric energy through the processes described in Chap. 38. In this chapter we are interested in practical arrangements for generating substantial emfs and transforming large amounts of mechanical energy into electric energy. In some cases the same apparatus can serve as a generator and can also be used to develop mechanical power from electric power. A device which performs this latter function is known as an electric motor.

39.1 INSTANTANEOUS ELECTROMOTIVE FORCE

When a single loop of wire is rotated at constant angular velocity in a uniform magnetic field (Fig. 39.1), the instantaneous emf induced in the loop varies sinusoidally in time (Sec. 38.7). One full cycle is completed each rotation. If ν represents the frequency (or number of cycles completed in unit time), the emf e at any instant is, by Eq. (38.4),

$$e = \mathcal{E}_{\max} \sin \theta = \mathcal{E}_{\max} \sin 2\pi\nu t \qquad (39.1)$$

Both the angle θ and the time t are measured from the instant the plane of the coil is perpendicular to the magnetic intensity B. In this equation and subsequent ones we use lowercase letters to represent the instantaneous values of quantities which vary in time and capital letters to indicate quantities which are constant.

39.2 COLLECTING RINGS

If the loop of Fig. 39.1 is opened at the axis and an outside circuit is connected to the two leads, the induced emf may be applied to the external circuit. To make continuous connection to the outside circuit, the ends of the wire forming the coil are fastened to rings (Fig. 39.2a) mounted on the axis of the rotating loop, and sliding connectors, called *brushes*, complete the circuit. The slip rings rotate with the loop, while the nonrotating brushes are pressed against the rings.

When an alternating emf is applied to a circuit which has resistance R only (no appreciable inductance or capacitance), the current i at any instant is simply the ratio of the applied potential difference to the resistance:

$$i = \frac{e}{R} = \frac{\mathcal{E}_{\max} \sin 2\pi\nu t}{R} = I_{\max} \sin 2\pi\nu t \qquad (39.2)$$

The current and the emf reach their maximum values at the same instant

Figure 39.1
The motional emf induced in the rotating loop is greatest when the effective conductors are moving perpendicular to the magnetic intensity.

and pass through their zeros together. The current is an alternating one and is said to be "in phase" with the emf, since the phase (the argument of the sine function) is the same for both at any time.

39.3 THE COMMUTATOR

The emf induced in a loop rotating in a magnetic field is an alternating one. To obtain a current which is always in the same direction through an external circuit, the terminals of the rotating loop may be joined to a *commutator*, which is a ring divided into two segments, as shown in Fig. 39.2*b*. Against this divided ring press two brushes which are connected to the external circuit and so placed that they slip from one segment of the commutator to the other at the instant the emf of the revolving coil passes through zero.

When the loop of Fig. 39.2*b* is rotating counterclockwise, the wire *ab* which is moving downward has an emf in the direction of the arrow, while the current in the wire *cd* as in the opposite direction. Hence, current leaves the brush *S* and passes through the external circuit in the direction of the arrow from *S* to *T*. After the loop has made one-half revolution, the segment of the commutator which was in contact with brush *T* now makes contact with brush *S* and the other segment makes

Figure 39.2
Connecting rings for generators.
(*a*) Slip, or collection, rings provide continuous connection to an external ac circuit. (*b*) A split-ring commutator delivers unidirectional current to an external circuit.

Figure 39.3
The potential difference between the brushes of the split-ring commutator of Fig. 39.2*b* as a function of the angular position of the coil.

contact with brush T. The wire ab is now moving upward, and current in it is reversed. The wire cd is moving down, and current in it is also reversed. Because the segments of the commutator have reversed their positions, the current still leaves by brush S and passes through the external circuit in the direction in which it flowed originally. In the revolving coil the current alternates; in the external circuit the current and the potential difference are pulsating but *unidirectional* (Fig. 39.3).

39.4 PRACTICAL GENERATORS

In most practical generators several coils are rotated simultaneously in a magnetic field. These coils are connected in such a way as to give the desired emf and current.

Consider a dc generator with two coils whose planes are mutually perpendicular (Fig. 39.4). If the coils are connected in series through a suitable commutator arrangement, the fluctuation in the output emf is much smaller than the fluctuations in output of each coil separately, as shown in Fig. 39.5. By adding more coils, the resultant output can be made constant except for small fluctuations known as the *commutator ripple*.

The wires which rotate past the poles of a generator are ordinarily embedded in slots in an iron core. The rotating system involving the iron core and the wires is known as the *armature*. The purpose of the iron core is to increase the magnetic intensity **B** and thereby to increase the emf. There are emfs induced not only in the copper wires of the armature but in any material which moves in the magnetic field, including the iron of the armature. If the armature were a solid piece of iron, these induced emfs would lead to large currents circulating in the iron and much energy would be dissipated as heat by these current whirlpools which are called *eddy currents*. To reduce the eddy currents, the iron core is constructed from thin sheets or *laminations* separated by thin insulating layers to make the resistance high.

Not only does the practical generator have several coils mounted on the same armature, but these coils often pass several magnetic poles during each revolution. Four- and six-pole generators are common; in

Figure 39.4
Generator with two coils with planes mutually perpendicular.

Figure 39.5
Connecting the commutated outputs of the two coils of Fig. 39.4 in series leads to a steadier output emf than either coil produces alone.

Figure 39.6
Distribution of the magnetic flux in a
magneto.

some applications as many as twenty-four poles are passed in a single
revolution. In a six-pole generator one complete cycle is induced for
each third of a revolution, or three cycles in each revolution.

An emf may equally well be induced in a stationary coil by the
varying of a magnetic field.

39.5 EXCITATION OF THE FIELDS OF GENERATORS

Generators and motors are sometimes classified with regard to the
manner in which their magnetic fields are produced. In the simplest
type of generator the magnetic field may be due to a permanent magnet.
An example of such a generator is the *magneto*, shows in Fig. 39.6.

Most large generators have fields which are produced by electromagnets. Such generators may be classified as self-excited or separately
excited, depending on whether the generator produces its own magnetic
field or uses some other source to excite the electromagnet. Self-excited
generators may be divided into series-, shunt-, and compound-wound
generators.

In the series-wound generator (Fig. 39.7*a*) the current to the
external circuit goes through the field coil, which consists of a few turns
of heavy wire. When the generator is not delivering current, the only
magnetic field is that due to residual magnetism. The greater the current
drawn by the external circuit, the greater the current in the field coils
and therefore the greater the magnetic field in which the armature

Figure 39.7
Generators may be (*a*) series-wound,
(*b*) shunt-wound, or (*c*) compound-
wound.

rotates. When the generator is operated at constant speed, the greater the current, the greater the emf generated. Series-wound generators are relatively uncommon because one does not ordinarily desire an emf which rises rapidly as the current drawn is increased.

In the shunt-wound generator the field coils are in parallel with the external circuit (Fig. 39.7b). When such a generator is operated at constant speed, if the current in the external circuit is increased, the terminal potential difference V_t of the generator decreases for the following reason. The terminal potential is the induced emf $\mathcal{E}$ minus Ir, where r is the resistance of the armature and I the armature current. As I increases, the terminal potential decreases, and so does the current in the field coils (given by V_t divided by the resistance of the field coils). This reduces the magnetic field and the induced emf. Therefore, a shunt-wound generator provides a potential which decreases as the current drawn is increased. Such a current-voltage response is not commonly desired.

When the terminal potential supplied by a generator is to be essentially independent of the current load, a compound-wound generator (Fig. 39.7c) is used. Such a generator has field coils which are composed of a series-wound part and a shunt-wound part. The characteristic of a series-wound field (increased emf with increasing current demand) and the characteristic of a shunt-wound generator (terminal potential decreasing with load current) may be combined to yield a generator with almost any desired variation of terminal potential with load current.

39.6 THE EFFICIENCY OF GENERATORS

The efficiency of a generator is the ratio of the electric energy output of the generator to the energy input, or, alternatively, the ratio of the electric power output to the power input. The power output of the generator is the product of the terminal potential and the current to the external circuit. The power input is greater, of course; it is equal to the power output plus the various losses, such as the I^2R heat loss in the armature and the field coils, hysteresis losses, eddy-current losses, and frictional losses.

39.7 ELECTRIC MOTOR

When a wire bearing an electric current I is placed in a magnetic field **B** so that the length **l** of the wire is perpendicular to the field, the wire experiences a force $F = IlB$ [Eq. (36.2)]. This force is perpendicular to both the magnetic field and the length of the wire. On one side of the wire, the magnetic field due to the current in the wire adds to the field due to the external magnetic circuit, while on the opposite side the two fields oppose. The force on the wire is from the stronger toward the weaker field. If the current is reversed, the force is reversed.

Figure 39.8 shows the cross section of a single current-bearing loop of wire in a uniform magnetic field. Assume that the current enters at a and leaves at b. There results a force downward on wire a and upward on wire b. Thus, there is a torque in a direction such as to

Current out
Force up

-N-

x
a

b

-S-

Current in
Force down

Figure 39.8
The torque on a current-bearing coil arises from forces on the individual conductors.

produce a counterclockwise rotation about the axis of the coil. In an electric motor there are ordinarily many conductors bearing currents perpendicular to a magnetic field. The resulting torque produces rotation of the armature.

There are many kinds of electric motors. That is, there are many ways to convert electric energy into mechanical energy. The underlying principle in most motors is that of a current-bearing conductor experiencing a force in a magnetic field.

39.8 BACK EMF IN A MOTOR

When the conductors in the armature of a motor rotate in a magnetic field, an emf is induced which opposes the current in the conductors. For this reason it is called a *back emf* or a *counter emf*.

If an incandescent lamp is connected in series with a small shunt-wound motor while the armature is held stationary, the lamp glows as if the motor were absent, since the armature resistance is small. When the armature is allowed to revolve, the lamp becomes dim. This is because the emf generated in the armature opposes the impressed emf and thereby reduces the current in the circuit. In a constant magnetic field, the faster the armature revolves, the greater the back emf and the smaller the armature current.

For shunt-wound motors the potential difference across the armature is the potential V_t applied to its terminals. The terminal potential of the armature is equal to the back emf plus the potential drop across the armature resistance r_a due to the armature current I_a:

$$V_t = \mathcal{E}_b + I_a r_a \qquad (39.3)$$

A motor is analogous to a battery under charge, except that electric energy is transformed into mechanical energy rather than chemical energy. If Eq. (39.3) is solved for the armature current, it yields

$$I_a = \frac{V_t - \mathcal{E}_b}{r_a} \qquad (39.3a)$$

The electrical work done against the back emf appears as mechanical energy output of the motor. Indeed, the mechanical power output P of the motor is equal to the product of the back emf $\mathcal{E}_b$ and the armature current:

$$P = \mathcal{E}_b I_a \qquad\qquad (39.4)$$

This follows directly from the fact that $\mathcal{E}_b$ is the electrical work per unit charge performed to produce motion of the conductors, while I_a is the charge per second passing through the conductors.

Since the armature current of a motor depends on the difference between the terminal potential and the back emf, the current is largest when the armature is at rest, for in this case the back emf is zero. In order to keep the current in a large motor from being excessive at the start, it is customary to introduce a series resistor that reduces the current flowing in the circuit. As the speed of the motor increases and the back emf becomes appreciable, this resistance is removed from the circuit. Such a resistance is called a *starting resistance* or a *starting box*.

39.9 SERIES-WOUND AND SHUNT-WOUND MOTORS

When large starting torques are required, motors having field coils which are in series with the armature are used. In such a series-wound motor the back emf is small when the motor is starting and the current drawn is large. Since this same current passes through the series field coils, the magnetic field is also large. The starting torque is proportional to IlB, the force on an individual conductor. As the motor gains speed, the back emf increases and both armature current and field strength decrease. Therefore, the torque is much smaller at high speeds. Series-wound motors are often used for rock crushers and other machines in which particularly large starting torques are desired.

In shunt-wound motors the field coils are in parallel with the armature coils. For a constant applied potential difference, the field strength is constant, and the torque is directly proportional to the armature current, whereas in the series-wound motor the torque is more nearly proportional to the square of the current. Most common small electric motors are shunt-wound.

The efficiency of a motor can be obtained by dividing the mechanical power output of the motor by the electric power input. The difference between the input and output power is dissipated in Joule heat, overcoming air friction, etc.

39.10 DYNAMOS

The fundamental components of many motors are identical with those of certain generators. The single term *dynamo* describes an electric device which can function as either motor or generator. When a dynamo is supplied with a potential difference and energy by an electric source and does mechanical work, it is a motor. When it is driven by a mechanical torque, it develops electric energy and acts as a generator. Not every motor can be operated as a generator, but many can.

There are many applications for a dynamo operating part-time as a motor and the rest of the time as a generator. An example is the dynamo used in diesel-electric locomotives. When the train is set in motion, diesel engines drive generators to supply current to the dynamos on the axles of the driving wheels. These dynamos exert large torques on the wheels and put the locomotive in motion. When it is desired to reduce the speed of the train, the dynamo may be disconnected from its power supply, leaving the armature rotating in a magnetic field as long as the wheels turn. An emf is induced in the armature; if the terminals are connected through a resistor, a current flows. The dynamo is now operating as a generator, converting kinetic energy of the train into electric energy, and thus braking the train. In diesel-electric locomotives the electric energy is dissipated in heat, but in electric locomotives it may be fed back into the power lines.

39.11 WATT-HOUR METER

The Thomson form of recording watt-hour meter (Fig. 39.9) consists of a little shunt motor whose armature turns at a speed proportional to the power supplied to it. The armature is geared to dials which record the energy in kilowatt-hours. The stationary field coils L and M are connected in such a way that the field strength of the motor is proportional to the current to the load. The armature is connected across the line, as a voltmeter would be connected, so that the armature current is proportional to the potential difference. The torque which turns the motor is a constant times the product of the current in the armature and the magnetic field in which the armature turns. Hence, the torque which turns the armature is proportional to the product of the current supplied to the load and the potential at which it is supplied, i.e., to the power received.

Figure 39.9
A watt-hour meter.

To ensure that the motor does not run too fast and to enable it to stop as soon as the current ceases, an electromagnetic brake D is attached to it. This brake consists of an aluminum disk rotating between the poles of a permanent magnet. The eddy currents induced in the disk by its rotation between the poles of the magnet retard its motion and stop it as soon as the current ceases. This type of watt-hour meter can be used with either direct or alternating current.

QUESTIONS

1 How is electric energy used to do external work? Explain how an electric motor may accomplish this task.
2 Why does a dc generator ordinarily have many commutator segments, rather than only two?
3 Does reversing the connections on a dc shunt-wound motor reverse the direction of rotation? Why? If it does not, how can the rotation direction be reversed?
4 Why is iron used in motors and generators? Why not replace the iron by a nonconducting plastic and thereby eliminate the eddy currents in the iron?
5 What is the response of a shunt-wound generator to an increase in load current? Of a series-wound generator? Of a compound-wound generator?
6 Why are series-wound motors used when very high starting torques are desired?
7 How could you make a generator which would have an emf constant in magnitude and direction?

PROBLEMS

1 In a motor a conductor 10 cm long carries a current of 2 A. If it lies perpendicular to a uniform magnetic field of 1.2 T, find the force on the conductor. What torque is produced if the conductor is 5 cm from the axis of the motor?
Ans. 0.24 N; 0.012 N-m
2 Part of the windings of an electric motor is a segment of wire 12 cm long which lies perpendicular to the magnetic field of 1.1 T. Find the force on this segment when it bears a current of 7 A.
3 The armature of a separately excited generator has a resistance of 0.14 Ω. When run at its rated

speed, it yields 132 V on open circuit and 125 V on full load. What is the current at full load? How much power is delivered to the external circuit? What horsepower is needed to drive the generator if its overall efficiency is 85 percent?
Ans. 50 A; 6.25 kW; 9.86 hp
4 A separately excited, constant-speed generator that develops 118 V at no load furnishes only 112 V when a current of 40 A is drawn. What is the resistance of the armature?
5 The armature of a generator has a resistance of 0.16 Ω. When the current through the armature is 5 A, the terminal potential is 224 V. What will be the terminal potential when the current is 45 A, assuming that the field strength and the speed remain unchanged? What is the emf induced?
Ans. 217.6 V; 224.8 V
6 The brush potential of a separately excited generator when it is delivering 6 A is 125 V. When the generator delivers 15 A, the potential difference across the brushes falls to 122 V. What are the induced emf and the resistance of the armature?
7 A rectangular loop of wire 9 by 15 cm bears a current of 8 A. The plane of the coil is parallel to a uniform magnetic field of 1.2 T, with its 15-cm length perpendicular to the lines of flux. Find the force on each side of the loop. Using this force, find the torque on the loop. Find the magnetic moment of the loop and use its value to check the value of the torque.
Ans. 1.44 N, zero; 0.130 N-m; 0.108 A-m²
8 Find the torque on a rectangular loop of wire bearing a current of 12 A when its plane is parallel to the magnetic lines of a field with an intensity of 1.25 T. The dimensions of the loop are 20 cm parallel to the flux and 12 cm at right angles to it.
9 The flux density between the poles of a certain motor is 1.1 T. A conductor carrying 9 A lies in this magnetic field so that it is perpendicular to the field. Find the force on it if 0.1 m of the

conductor is in the field. If the wire moves 2 cm perpendicular to the field in 0.01 sec, and if the field is uniform over this distance, find the work done on the wire. Show that the same work is predicted by finding the induced back emf and multiplying it by the charge transferred in the 0.01 sec.　　　　*Ans.* 0.99 N; 0.0198 J; 0.22 V

10　A shunt-wound motor draws 2.5 A from a 120-V line. The field coils have a resistance of 300 Ω, and the armature resistance is 0.6 Ω. Find the armature current and the back emf. What is the mechanical power output of the motor? How much power is dissipated in I^2R heating in the motor?

11　A shunt motor takes a total current of 8.5 A from 110-V mains. The resistance of the armature is 1.5 Ω, and that of the field coils is 220 Ω. Find the current in the field coils and in the armature, the back emf induced, and the mechanical power output of the motor.
　　　　　Ans. 0.5 and 8 A; 98 V; 784 W

12　A motor running at full load on a 115-V line develops a back emf of 106 V and draws a current of 6 A through the armature. What is the mechanical power output of the motor, disregarding frictional losses? What is the armature resistance?

13　A shunt-wound generator delivers 40 A to an external load at a brush potential of 120 V. The field coils have a resistance of 60 Ω, and the armature has a resistance of 0.15 Ω. If the stray-power loss is 500 W, what is the efficiency of the generator?　　　　*Ans.* 83 percent

14　The armature of a shunt-wound generator has a resistance of 0.16 Ω. The terminal potential difference of the generator is 118 V when the armature current is 50 A. Find the emf of the generator and the power dissipated in heat in the armature.

15　A shunt-wound motor has an armature with a resistance of 0.2 Ω and operates from a 225-V line. What starting resistance is required if the armature current must be limited to 15 A? When the motor is operating normally with the starting resistor out, the armature draws 4 A. Find the back emf and the mechanical power output.
　　　　　Ans. 14.8 Ω; 224.2 V; 897 W

16　A shunt-wound motor with an armature resistance of 0.4 Ω operates on a 120-V circuit. Find the resistance which must be temporarily connected in series with the armature to limit the starting current to 8 A. Find the current drawn

and the mechanical power output when the back emf generated is 118 V.

17　A series-wound motor has an armature with a resistance of 0.5 Ω and field coils with a resistance of 4.5 Ω. The motor draws 10 A at 120 V. Find the power supplied to the motor, the back emf induced in the armature, the power transformed into heat, and the mechanical power output.　　*Ans.* 1.2 kW; 70 V; 500 W; 700 W

18　The armature of a motor connected to a 220-V line draws 35 A at the instant the connection is made, when the motor is at rest. There is a starting resistance of 4 Ω in the circuit. What is the resistance of the armature? What is the counter emf when the current drawn is 15 A (with the starting resistance shorted out)?

19　In the circuit of the accompanying figure, $\mathcal{E}_a =$ 75 V, $r_a = 2$ Ω, $R_2 = 4$ Ω, $\mathcal{E}_b = 30$ V, $r_b =$ 0.5 Ω, $R_3 = 15$ Ω, $R_4 = 10$ Ω, and the current through the generator is 3 A. Find R_1 and the terminal potential differences of the generator and battery.　　　　*Ans.* 2.5 Ω; 69 V; 31.5 V

20　In the circuit of the accompanying figure, $\mathcal{E}_a =$ 80 V, $r_a = 1$ Ω, $R_1 = 6$ Ω, $R_2 = 4$ Ω, $\mathcal{E}_b =$ 35 V, $r_b = 0.5$ Ω, and $R_3 = R_4 = 7$ Ω. Find the current in the main circuit and the terminal potential differences of the generator and battery.

21　In the circuit of the accompanying figure, $\mathcal{E}_a$ is the back emf of a motor. It has a value of 25 V.

Problems 19 to 22

The armature resistance and the resistance of the battery are both 1 Ω, and all other resistances are 6 Ω. Find the emf of the battery if the motor draws 2.5 A. What are the terminal potential and the mechanical power output of the motor? *Ans.* 67.5 V; 27.5 V; 62.5 W

22 In the circuit of the accompanying figure a battery of 48-V emf and 1-Ω internal resistance is providing electric energy to operate a motor which has an armature resistance of 0.5 Ω. If $R_3 = R_4 = 8$ Ω and $R_1 = R_2 = 3$ Ω, what is the back emf of the motor when the armature draws 2 A? What is the mechanical power output of the motor?

WHEN A COIL is rotated in a uniform magnetic field, an alternating emf is generated. Because it is particularly easy to generate alternating emfs, and because transformers make it economically feasible to increase or decrease an alternating potential difference, most of the practical generation and distribution of electric power utilizes alternating currents. Many of the ideas we have treated for the simpler dc circuits can be carried over to ac circuits, but there are also many other factors which must be considered. In this chapter we discuss these factors.

CHAPTER FORTY
ALTERNATING CURRENTS

40.1 THE ALTERNATING EMF

If the armature of a generator is rotated in a uniform horizontal magnetic field, the instantaneous emf induced in a single loop can be represented by a sine curve (Fig. 40.1) in which the angle through which the coil has turned from a vertical plane is plotted on the horizontal axis and the corresponding induced emf on the vertical axis (see Sec. 38.7). The instantaneous emf e is given by

$$e = \mathcal{E}_{max} \sin \theta = \mathcal{E}_{max} \sin 2\pi\nu t \qquad (40.1)$$

where $\mathcal{E}_{max}$ is the maximum emf induced, θ is the angle through which the coil has turned from the position in which the conductors are moving parallel to the magnetic field, ν is the frequency, and t is the time.

It is often convenient to think of the sine wave of Fig. 40.1 as being generated by a vector of magnitude $\mathcal{E}_{max}$ rotating with frequency ν. The instantaneous emf is then represented by the vertical component of this vector, as shown in the figure.

The frequency ordinarily used in electric-power generation and distribution in the United States is 60 Hz. In Europe 50 Hz is a common frequency, while in aircraft and guided missiles 400-Hz generators are

Figure 40.1
An alternating emf varies in time like the vertical component of a vector rotating at the same frequency.

widely used. In telephones and in the audio amplifiers of radio and television sets the frequencies involved are the acoustical frequencies themselves. The frequencies used for radio transmission in the standard broadcast band range from 535 to 1,605 kHz; there are other bands at higher and lower frequencies. Television, radar, and microwave transmission is carried out at much higher frequencies.

40.2 EFFECTIVE VALUE OF ALTERNATING CURRENT

The power dissipated in heat in a resistor by a direct current is I^2R. For a direct current the maximum value, the average value, and the instantaneous value are all the same. For an alternating current, the average value of current over a whole cycle is zero; the current fluctuates between the maximum value in one direction and the same maximum value in the opposite direction. Nevertheless, an alternating current is effective in producing heat in a resistor. If the instantaneous value of the current is i, i^2 is always positive, even though i is negative during one-half of each cycle.

When an instantaneous potential difference $v = V_{max} \sin 2\pi\nu t$ is applied across a pure resistor R, the instantaneous i is $(V_{max} \sin 2\pi\nu t)/R = I_{max} \sin 2\pi\nu t$ (Sec. 39.2). The current is in phase with the applied potential difference, and Ohm's law is directly applicable to instantaneous current and maximum current, provided the appropriate potential difference is used. Thus,

$$i = \frac{v}{R} \qquad \text{and} \qquad I_{max} = \frac{V_{max}}{R}$$

The power dissipated in a resistor R at any instant is vi, the product of the instantaneous values of potential difference and current (Fig. 40.2). Since $v = iR$, the instantaneous power is i^2R, and the average power dissipated over each cycle is $\overline{i^2}R$, where $\overline{i^2}$ is the average value of the square of the current. We call $I = \sqrt{\overline{i^2}}$ the *effective value* of the alternating current. *The effective value of an alternating current is the value of that direct current which would produce heat at the same rate as the alternating current in a given resistor.* An alternating current is said to have an effective value of 2 A when it generates heat in a 1-Ω resistor at the rate of 4 W—the same rate at which a direct current of 2 A produces heat in a 1-Ω resistor. Since the heating effect depends on the square of the current, this is equivalent to saying that the average value of the square of the alternating current is the same as the square of the corresponding direct current. The effective value of an alternating current is found by taking the average of the square of the instantaneous current and then extracting its square root. Sometimes the effective value of the alternating current obtained in this way is called the *root-mean-square* (rms) current.

The effective value of a sine-wave current is given† by

Figure 40.2

Instantaneous values of the potential difference v, current i, and power dissipated in a resistor as functions of time for an alternating potential difference.

$$I_{eff} = \frac{I_{max}}{\sqrt{2}} = 0.707 I_{max} \qquad (40.2)$$

† These relations and several others in this chapter can be obtained readily by use of calculus. The derivations will be found in Appendix C.

where I_{max} is the maximum value of the current. Similarly, the effective value of an alternating potential difference is 0.707 times the maximum; i.e.,

$$V_{eff} = \frac{V_{max}}{\sqrt{2}} = 0.707\,V_{max} \qquad (40.2a)$$

Ordinary ac ammeters and voltmeters are calibrated to read effective values. When we speak of a 110-V ac circuit or of an alternating current of 5 A, we are specifying effective values. Hereafter we use V and I without subscripts to denote effective values.

R	L	C
$\dfrac{V}{I} = R$	$\dfrac{V}{I} = \omega L = X_L$	$\dfrac{V}{I} = \dfrac{1}{\omega c} = X_C$

40.3 ADVANTAGES OF AC TRANSMISSION

One of the major reasons for the widespread use of alternating currents is that it is possible to change from one potential difference to another simply and efficiently by use of transformers (Sec. 38.9). This is highly desirable, since it is most economical to transmit electric power at potentials which are much too high to be used safely in a home or factory.

To illustrate why it is desirable to transmit power at high potentials, consider the following practical problem. A power company must deliver 1,100 kW to a distant city over a transmission line which has a resistance of 2 Ω. How much power is dissipated in the delivery line at 110,000 V and at 1,100 V? From $P = VI$ it is clear that the line must deliver 10 A at 110,000 V or 1,000 A at 1,100 V. The I^2R power loss in this line is 200 W at 110,000 V and 2,000,000 W at 1,100 V. In the latter case more power would be dissipated in the line than is delivered to the ultimate consumer!

If these arguments were the only ones involved, it would be desirable to transmit power at potentials of many million volts. However, as the potential increases, there are serious power losses due to corona, and the difficulties of insulation become severe. For these reasons high-voltage power transmission lines typically operate at potential differences of a few hundred thousand volts, although lines operating at about a million volts have been constructed.

(a)

40.4 CURRENT IN AN INDUCTIVE CIRCUIT

When an alternating potential of frequency ν is applied to a large inductor with negligible resistance (Fig. 40.3a), the current is limited by the inductance. This limitation is due to the fact that, as the current changes, a back emf is induced in the inductance. At every instant this emf opposes the *change in the current*. If we apply Kirchhoff's second law to this situation, we have

$$v = \mathcal{E}_{max}\sin 2\pi\nu t = L\frac{\Delta i}{\Delta t}$$

We note that $\Delta i/\Delta t$, the rate at which the current is changing with time, is greatest when v is maximal, and that when $v = 0$, $\Delta i/\Delta t$ is also zero, which means that the current is either maximal or minimal (Fig. 40.3b).

(b)

Figure 40.3
A circuit (a) containing a source of alternating emf and an inductor of negligible resistance has (b) current i that lags behind the applied potential difference by 90°.

(a)

(b)

Time→

Figure 40.4

A series circuit (*a*) consisting of a source of alternating emf, resistance, and inductance has (*b*) current *i* that lags behind the applied potential difference *v* by an angle θ such that $\tan \theta = 2\pi\nu L/R$.

Thus, when a potential difference is applied across a pure inductance, *the current is not in phase with the potential difference.* Rather it reaches its maximum value when the potential is passing through zero and is zero when the potential difference is at its maximum. The current lags 90° behind the potential difference. With calculus it can readily be shown that

$$i = -\frac{\mathcal{E}_{max}}{2\pi\nu L} \cos 2\pi\nu t = -I_{max} \cos 2\pi\nu t \qquad (40.3)$$

The potential-difference and current curves of Fig. 40.3*b* can be visualized as being generated by two mutually perpendicular vectors rotating together with the same frequency. The maximum current is $\mathcal{E}_{max}$ divided by the quantity $2\pi\nu L$, which is called the *inductive reactance* of the inductor and is indicated by X_L. The effective current is just $I = \mathcal{E}/2\pi\nu L = \mathcal{E}/X_L$. The dimensions of X_L are volts per ampere, or ohms.

When a circuit (Fig. 40.4*a*) contains both resistance and inductance, the current lags behind the potential by some angle less than 90° (Fig. 40.4*b*). Kirchhoff's second law applied to the circuit yields

$$e = \mathcal{E}_{max} \sin 2\pi\nu t = Ri + L\frac{\Delta i}{\Delta t}$$

It is shown in Appendix C that in this case

$$I = \frac{\mathcal{E}}{\sqrt{R^2 + (2\pi\nu L)^2}} \qquad (40.4)$$

and that the current lags behind the potential by an angle θ such that $\tan \theta = 2\pi\nu L/R$. The quantity by which $\mathcal{E}$ must be divided to get I is called the *impedance* of the circuit and is represented by Z. In this case $Z = \sqrt{R^2 + (2\pi\nu L)^2}$. It can be computed as the resultant of R and $2\pi\nu L$, where R and $2\pi\nu L$ are drawn as vectors at right angles to one another as shown in Fig. 40.5*a*. The angle θ by which the current lags behind the emf is the angle between Z and R.

An ac voltmeter connected across the resistor R would read IR. Connected across the inductor L, it would read IX_L. The arithmetical sum $IR + IX_L$ is not equal to the applied effective emf $\mathcal{E}$, since when the potential difference is maximal across the resistor, it is zero across the inductor. Rather, we must add the potentials vectorially, with IX_L 90° ahead of IR as shown in Fig. 40.5*b*, to obtain $\mathcal{E}$. Obviously, the triangle in this figure is similar to the impedance triangle of Fig. 40.5*a*; each side is just I times the corresponding side in the impedance diagram.

Figure 40.5

(*a*) Vector diagram for finding the angle by which the current lags behind the applied emf in an ac series circuit containing inductance and resistance.

(*b*) The vector sum of the potential differences IR and IX_L is equal to the applied emf.

In an ordinary transformer, when no current is being drawn in the secondary, the current in the primary is limited by its inductive reactance. Since the primary of a large transformer ordinarily has a low resistance, the current is controlled almost entirely by the self-inductance when the secondary circuit is open and by the self-inductance and the mutual inductance when there is a current in the secondary.

40.5 THE CAPACITIVE CIRCUIT

When an alternating emf is applied across a capacitor, the current is zero when the applied emf is maximum, because the capacitor is charged to the maximum potential difference. As the potential decreases, the capacitor begins to discharge. The current increases and reaches its maximum value when the emf passes through zero. Thus, the current and the emf are out of phase. In the case of a *pure capacitance* there is a 90° phase difference between the two, with the current leading the emf. The effective current I is given by

$$I = \frac{\mathcal{E}}{1/2\pi\nu C} = \frac{\mathcal{E}}{X_C} \tag{40.5}$$

where C is the capacitance and the quantity $1/2\pi\nu C = X_C$ is called the *capacitive reactance.*

When a circuit (Fig. 40.6a) contains both capacitance and resistance, the current is given by

$$I = \frac{\mathcal{E}}{\sqrt{R^2 + (1/2\pi\nu C)^2}} \tag{40.6}$$

In this case the current leads the emf (Fig. 40.6b) by the angle θ (tan $\theta = X_C/R$). The effective potential difference across the resistance is given by IR, while the effective potential difference across the capacitor is given by IX_C. The vector sum of IR and IX_C gives the applied emf $\mathcal{E}$.

40.6 CIRCUIT CONTAINING INDUCTANCE, CAPACITANCE, AND RESISTANCE

In a series ac circuit containing inductance, capacitance, and resistance (Fig. 40.7a), the current is given by

$$I = \frac{\mathcal{E}}{\sqrt{R^2 + (2\pi\nu L - 1/2\pi\nu C)^2}} = \frac{\mathcal{E}}{Z} \tag{40.7}$$

Note that the impedance may be considered as the *geometric sum* of the resistance, the inductive reactance, and the capacitive reactance with due regard given to the fact that the inductive reactance and the capacitive reactance shift the phase in opposite directions. The impedance can be computed by drawing a vector diagram such as the one in Fig. 40.7b. The angle by which the current lags behind the emf is θ. Clearly, tan $\theta = (X_L - X_C)/R$, sin $\theta = (X_L - X_C)/Z$, cos $\theta = R/Z$. When X_C is larger than X_L, θ is negative, and the current leads the applied emf.

(a)

(b)

Figure 40.6
A series circuit (a) containing capacitance and resistance has (b) current which leads to applied emf by angle θ given by tan $\theta = X_C/R$.

(a)

(b)

Figure 40.7
(a) Series circuit containing resistance, inductance, and capacitance and (b) its impedance, or phasor, diagram.

$V = \omega L I$

$\theta = \tan^{-1} \dfrac{\omega L}{R}$

$I = I\sqrt{R^2 + \omega^2 L^2}$

$I = I.\sqrt{R^2 + X_L^2}$

$V_R = IR.$

$VT = \sqrt{R^2 + \omega^2 L^2} \ \sin(\omega t + \theta)$

$\theta = \tan^{-1} \dfrac{\omega L}{R} = \dfrac{X}{R}$

FOR A COMBINATION LIKE

THE ONE ON THE BACK PAGE.

$V_L = I R_1 + j I \omega L.$

$V_L = I(R_L + j\omega L)$

$V_L = I Z_L.$

$Z_T = Z_L + Z_R + Z_C$

$= (R_L + j\omega L) + (R_C - \dfrac{1}{j\omega C}) + R.$

$(R_L + R_C + R) + j(\omega L - \dfrac{1}{\omega C})$

$z = \sqrt{(R_L + R_C + R)^2 + (\omega L - \dfrac{1}{\omega C})^2}$

Example

In the circuit of Fig. 40.7a, $\mathcal{E} = 24$ V, the total resistance $R = 10 \ \Omega$, $L = 0.05$ H, and $C = 10^{-4}$ F. If the frequency is 60 Hz, find the impedance of the circuit and the current. What would an ac voltmeter connected across R read? Across L? Across C?

$$X_L = 2\pi\nu L = 6.28 \times 60 \times 0.05 = 18.9 \ \Omega$$

$$X_C = \frac{1}{2\pi\nu C} = \frac{1}{377 \times 10^{-4}} = 26.5 \ \Omega$$

$$Z = \sqrt{R^2 + (X_L - X_C)^2}$$
$$= \sqrt{(10)^2 + (18.9 - 26.5)^2}$$
$$= \sqrt{100 + 57.8} = 12.5 \ \Omega$$

$$I = \frac{\mathcal{E}}{Z} = \frac{24 \text{ V}}{12.5 \ \Omega} = 1.9 \text{ A}$$

$$V_R = IR = 19 \text{ V}$$
$$V_L = IX_L = 1.9 \times 18.9 = 36 \text{ V}$$
$$V_C = IX_C = 1.9 \times 26.5 = 51 \text{ V}$$

In this circuit the current leads the emf by an angle given by $\cos \theta = R/Z = 10/12.5 = 0.8$; $\theta = -37°$. (The minus sign means the current leads the emf.)   ////

40.7 RESONANCE IN SERIES CIRCUITS

When $X_C = X_L$, the total reactance is zero and the current is given by $I = \mathcal{E}/R$. In these circumstances the current is in phase with the applied emf, and the circuit is said to be *resonant*. When an emf of constant amplitude but increasing frequency is applied to a series circuit with both inductive and capacitive reactance, the current increases until it reaches a maximum at the resonance frequency. As the frequency rises further, the current becomes smaller. The frequency for which a circuit is resonant can be obtained readily from the requirement that $X_L = X_C$. This gives $2\pi\nu L = 1/2\pi\nu C$, or

$$\nu = \frac{1}{2\pi\sqrt{LC}} \tag{40.8}$$

Physically, in a circuit at resonance, energy is stored during part of the cycle in the magnetic field of the inductor. A quarter cycle later this energy is stored in the electric field of the capacitor. In another quarter cycle it is once again stored in the magnetic field of the inductor. Thus energy is transferred back and forth between inductor and capacitor. The current is in phase with the applied emf because the effects of inductance and capacitance cancel each other. The only net energy supplied to the circuit is that dissipated as heat or radiated.

40.8 POWER IN AC CIRCUITS

The power supplied at any instant in an ac circuit is obtained by multiplying the instantaneous current by the emf at the same instant:

$$p = ei \tag{40.9}$$

where p is the instantaneous power, i the instantaneous current, and e the instantaneous emf. The power varies from instant to instant, since both the current i and the applied emf e change with time.

When the current and the potential difference are in phase, the average power dissipated over one or more cycles is the product of the effective current I and the effective potential difference V (Sec. 40.2). The power is always positive (Fig. 40.2), since the current and the potential difference always have the same sign. When the current and potential difference are out of phase, the power is no longer given by IV. When an emf is applied across a capacitor, the power is negative as much as it is positive, and there is no *net* supply of energy by the generator; the generator supplies energy to charge the capacitor, and then the capacitor returns the energy. For a circuit in which the reactance is primarily inductive, the power curve is similar to that in Fig. 40.8. In this case, the instantaneous power is negative during part of the cycle, which means that some of the energy stored in the inductance is being returned to the power source. In the general case, the power dissipated in a series ac circuit is given by the relation

$$P = \mathcal{E}I \cos \theta \qquad (40.10)$$

where θ is the angle by which the current lags behind the applied emf. The vector diagram from which θ may be found is shown in Fig. 40.7*b*. Cosine θ is called the *power factor* of the circuit.

Power companies are eager to have a power factor as near unity as possible, because in those cases in which the power factor is far different from unity part of the energy supplied is returned each cycle. Since the power company has to provide the I^2R line losses which occur in delivering the power to the customer, this results in a disadvantageous situation. Most power companies charge higher rates for any consumer whose power factor is appreciably different from 1. By the suitable use of additional inductors or capacitors, it is possible to bring the power factors of most loads fairly close to unity.

Figure 40.8
Instantaneous power supplied to a circuit in which applied potential difference v and current i are not in phase.

FOR A CAPACITOR:

$$V = \frac{Q}{C} = \frac{1}{C}\int \overset{\circ}{i}\, dt.$$

$$I = C\frac{dv}{dt}.$$

$$v = V_m \sin \omega t$$

$$\frac{dv}{dt} = \omega V_m \cos \omega t$$

$$\boxed{\overset{\circ}{i} = \omega C V_m \cos \omega t.}$$

40.9 OSCILLATORY DISCHARGE

Consider the circuit of Fig. 40.9*a*, in which we have a capacitor and an inductor. When we close key K_1, the capacitor is charged to a

(a)

(b)

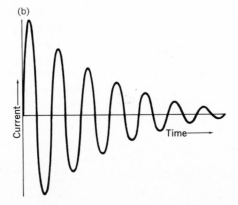

Figure 40.9
(*a*) Circuit for producing the oscillatory discharge of a capacitor through an inductor and (*b*) graph of damped oscillations.

potential difference $\mathcal{E}$ and energy is stored in the capacitor. Let us now open K_1 and close K_2. The capacitor C begins to discharge, and a current is established through the inductor L. Energy is transferred from the electric field of the capacitor to the magnetic field of the inductor. When C is completely discharged, energy is stored in the inductor, which sustains the current until the capacitor has been charged oppositely to its original condition. Now the capacitor discharges by a current in the opposite direction, thereby transferring energy to the inductor, which in turn uses this energy to recharge the capacitor to approximately its original condition.

If there were no resistance in this circuit, the capacitor would be charged to the potential difference with which it started. However, some of the energy is dissipated in heat in the inevitable resistance of the circuit; thus the potential difference does not attain the value $\mathcal{E}$. The capacitor discharges again and again. The current rises and falls, as shown in Fig. 40.9b. After a number of oscillations, all the energy originally stored in the capacitor is dissipated.

The frequency of oscillation is the frequency to which the circuit is resonant. That is the frequency at which the inductive and capacitive reactances are equal, provided the resistance of the circuit is small compared with the reactances. By Eq. (40.8), the frequency of oscillation is given by

$$\nu = \frac{1}{2\pi\sqrt{LC}} \tag{40.11}$$

Example

The oscillating circuit of a radio station has an inductance of 5×10^{-6} H and a capacity of 0.01 μF. Neglecting the effect of resistance, find the natural (resonant) frequency and the wavelength of the electromagnetic waves.

$$\nu = \frac{1}{2\pi\sqrt{LC}} = \frac{1}{2\pi\sqrt{5 \times 10^{-6} \times 1 \times 10^{-8}}}$$

$$= 720{,}000 \text{ Hz}$$

$$\text{Wavelength} = \frac{\text{velocity}}{\text{frequency}} = \frac{3 \times 10^8 \text{ m/sec}}{720{,}000 \text{ Hz}} = 417 \text{ m} \qquad ////$$

QUESTIONS

1 Could one define the ampere of alternating current in terms of chemical effects? Why is the heating effect particularly appropriate for defining the ac ampere?

2 Why is a dc voltmeter unsuitable for measuring an alternating potential difference? How would such a voltmeter respond if the frequency across its terminals were slowly increased from 0.01 to 500 Hz?

3 An ac series circuit contains resistors, an inductor, and a variable capacitor. If the frequency is constant, will increasing the capacitance increase or decrease the current? Explain carefully.

4 Why does the reactance of a capacitor decrease with frequency while that of an inductor increases? Offer physical reasons for this behavior.

5 Why do almost all transformers for 60-Hz alternating current use iron cores, while transformers for frequencies in the megahertz range do not?

6 A good transformer automatically adjusts the

primary current to take care of the load in the secondary circuit. How is this accomplished?

7 Why are ac generators and transformers rated in kilovolt-amperes rather than kilowatts? What is the difference?

8 In what sense is Kirchhoff's second law obeyed for an ac series circuit such as that of Fig. 40.7? In what sense is it not obeyed?

9 Radios are ordinarily tuned to various stations by changing the capacitance of an air capacitor. This is done by varying the effective area of the plates. If one wishes to tune to a station with a higher frequency, must the effective area be increased or decreased?

10 The 12 standard TV channels are located in the 54- to 72-, 76- to 78-, and 174- to 216-MHz bands, while the standard radio band is 535 to 1,605 kHz. With this in mind, give at least one reason why TV channels are switched by changing inductors while radio stations are switched by changing a single variable air capacitor.

11 In Europe electric energy is supplied chiefly at approximately 220 V and at 50 Hz. What advantages and disadvantages does this have over the 60-Hz 110-V circuits common in the United States?

PROBLEMS

1 A pure 40-Ω resistor is connected to a 60-Hz power supply which has a 160-V amplitude. Find the maximum current, the rms (or effective) current, the average current, and the power dissipation. *Ans.* 4 A; 2.83 A; zero; 320 W

2 An alternating current passing through a resistance of 7 Ω produces heat at the rate of 175 W. What is the effective value of the current? Of the potential difference? What are the maximum values?

3 An alternating emf has a frequency of 60 Hz and an effective value of 110 V. What is the maximum value of the emf? What is the instantaneous value of the emf at times $\frac{1}{720}, \frac{1}{360}, \frac{1}{240}$, and $\frac{1}{180}$ sec after it passes the zero value?
Ans. 156 V; 78, 135, 156, and 135 V

4 An alternating emf of frequency 400 Hz has a peak value of 300 V. Find the instantaneous value of the emf at the instants 0.001, 0.002, and 0.004 sec after it passes through its zero value. What is the rms (or effective) value?

5 An ac generating station generates 2 MW and transmits this power through a power line of 2-Ω total resistance. Find the power dissipated in the line if transmission is (a) at 250 kV and (b) at 2.5 kV. *Ans.* (a) 128 W; (b) 1.28 MW

6 At what frequency would a 2-μF capacitor have a reactance of 80 Ω? What inductance would have this same reactance at this frequency?

7 A coil has an inductance of 0.0796 H and a resistance of 40 Ω. What are its reactance and its impedance at 60 Hz? By what angle does the current in this coil lag behind the potential difference? What resistance must be connected in series with this coil to give a total impedance of 100 Ω? *Ans.* 30 Ω; 50 Ω; 37°; 55.4 Ω

8 Find the reactance of a 0.054-H inductor and of a 12-μF capacitor at 60 Hz.

9 An inductor has a resistance of 20 Ω and an inductive reactance of 80 Ω. It is connected in series with an emf of 120 V and a resistor of 40 Ω. Find the current, the potential difference across the inductor (do not forget its resistance), and the potential difference across the resistor.
Ans. 1.2 A; 99 V; 48 V

10 Find the reactance of a 60-nF capacitor at 10 Hz, 10 kHz, 10 MHz, and 10 GHz. Note that a capacitor passes high frequencies readily and offers high reactance to low frequencies.

11 An iron-core coil bearing a 60-Hz current of 3 A dissipates heat at the rate of 180 W. If it has an inductive reactance of 80 Ω, find the inductance, the resistance, and the impedance of the coil and the applied potential difference.
Ans. 0.212 H; 20 Ω; 82.5 Ω; 247 V

12 A pure inductance of 0.3 H and a 50-Ω resistor are connected in series across a 220-V 50-Hz source. Find the current and the angle by which it lags behind the source potential difference.

13 In a 60-Hz series circuit there is a pure capacitor with a reactance of 20 Ω, an inductor with 5-Ω resistance and inductive reactance of 60 Ω, and a 25-Ω resistor. The current is 4 A. Find (a) the impedance of the circuit, (b) the power factor, (c) the potential difference across the inductor, and (d) the inductance L.
Ans. (a) 50 Ω; (b) 0.600; (c) 241 V; (d) 0.159 H

14 A current of 0.4 A is drawn from a 110-V 60-Hz line when a 100-Ω resistor is connected in series with a capacitor. Find the capacitance and the angle by which the current leads the potential difference.

15 A 3.5-μF capacitor and a 35-Ω resistor are connected in series with an 8-V 2,000-Hz emf. Find the current, the angle by which the current leads the emf, and the power dissipated.

Ans. 0.192 A; 33°; 1.29 W

16 Find the capacitive reactance of a 5-μF capacitor at 3,000 Hz. Compute the inductance required to produce series resonance with the capacitor at this frequency.

17 A radio station broadcasts on an assigned frequency of 800 kHz. Find the wavelength of the radio waves sent out. What inductance must be connected in series with a capacitance of 5 pF for resonance at this frequency?

Ans. 375 m; 7.92 mH

18 Signals with a frequency of 1,500 kHz are sent out by a radio station. Find the wavelength. What capacitance is required to produce resonance at this frequency with an inductance of 0.003 H?

19 What capacitance is required with an inductance of 3 μH to form a resonant circuit for a wavelength of 40 cm? *Ans.* 1.5 × 10⁻¹⁴ F

20 A television station operates at a frequency of 200 MHz. What inductance is needed with a capacitance of 0.4 pF to form a circuit resonant to this frequency? What is the wavelength of the radiation?

21 In the ac circuit shown in Fig. 40.7, the effective emf is 50 V, the inductance 0.185 H, the resistance 80 Ω, and the capacitance 8 μF. The frequency is 500/π Hz. (*a*) Find the inductive reactance, the capacitive reactance, and the impedance. (*b*) Find the current and the potential difference across each part of the system. (Note that it is the vector sum of the voltages which adds to 50. The voltage across an inductance or capacitance may exceed the applied voltage!) (*c*) Find the angle by which the current lags the emf. (*d*) Find the power dissipated in the circuit.

Ans. (*a*) 185 Ω; 125 Ω; 100 Ω; (*b*) 0.5 A; V_R = 40 V; V_L = 92.5 V; V_C = 62.5 V; (*c*) 37°; (*d*) 20 W

22 In the ac circuit shown in Fig. 40.7, the effective emf is 130 V, the inductance is 0.32 H, the resistance is 150 Ω, and the capacitance is 1 μF. The frequency is 1,000/π Hz. (*a*) Find the inductive reactance, the capacitive reactance, and the impedance. (*b*) Find the current and the potential difference across each part of the system. (*c*) Find the angle by which the current lags the emf. (*d*) Find the power dissipated in the circuit.

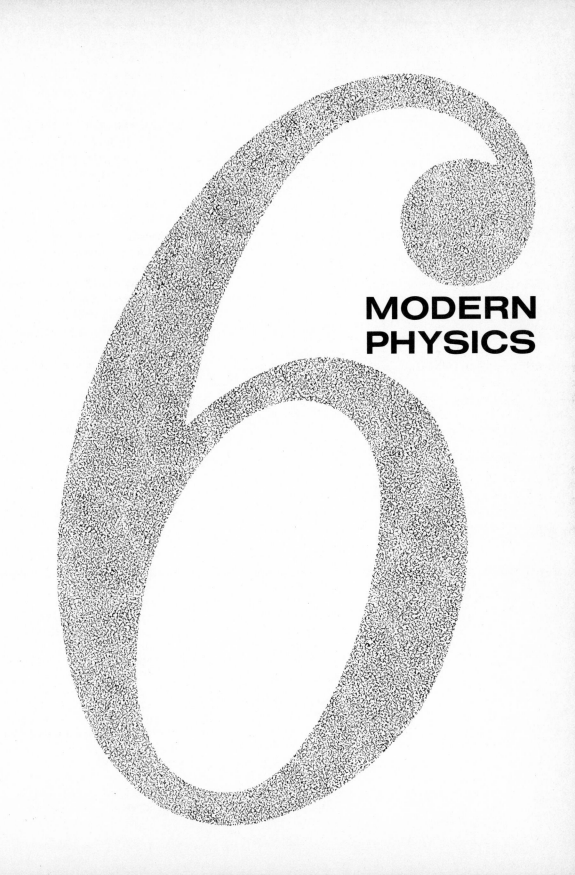

**MODERN
PHYSICS**

CHAPTER FORTY-ONE
RELATIVITY

BY EXPRESSING the laws of electricity in terms of differential equations and introducing some new ideas, Maxwell showed that light was electromagnetic in origin and thus brought unity to optics and electricity. His theory eventually led to new questions about the propagation of light. It was found that while the classical physics we have been studying makes predictions in excellent agreement with observed results for macroscopic bodies and moderate speeds, it fails when one deals with objects of atomic dimensions and with speeds comparable with the speed of light. In 1905 Einstein advanced his celebrated special theory of relativity linking space and time in a new way that resolved the dilemmas regarding the speed of light. The existence of this revolutionary theory has been widely publicized; in this chapter we develop several facets of Einstein's theory and its predictions.

41.1 ELECTROMAGNETIC WAVES

Classically, optics and electricity evolved as independent topics with no apparent interrelationships. However, in 1864 James Clerk Maxwell published a great paper in which he developed what he called a "dynamical theory of the electromagnetic field." His work utilized the electrical researches of Faraday, introduced a missing concept in electric currents, and combined electricity and magnetism to bring forth a new, unified, and fruitful theory. An immediate consequence of Maxwell's differential equations was the prediction that energy could be transmitted by *electromagnetic waves* which should travel with a speed very close to the experimentally known speed of light. Faraday had suggested earlier that light might be an electromagnetic wave, and now Maxwell had a quantitative theory which supported this idea.

In 1887 Heinrich Hertz produced electromagnetic waves, as predicted by Maxwell's theory, with laboratory apparatus. He generated these waves with oscillating circuits, analogous to that of Fig. 40.9, in which spark gaps played the role of switches and an induction coil provided the source of emf. Using a detector which consisted of an incomplete loop of copper wire with a short air gap between the ends, Hertz showed that electromagnetic waves from his oscillator would

Figure 41.1
Electric and magnetic radiation fields a long distance from charges oscillating parallel to the y axis.

produce a spark across the air gap of the detector, provided the dimensions of the latter were such that it was resonant at the frequency of the oscillator. Hertz's work established that electromagnetic disturbances have the wave properties predicted by the Maxwell theory; his experiments were in a sense the beginning of radio.

To understand the origin of the electromagnetic waves sent out by an oscillating current, we must be familiar with a conclusion of electromagnetic theory that an accelerated charge with available energy radiates an electric field parallel to and proportional to the acceleration. Accompanying this electric field is a magnetic field *in phase and perpendicular to it*. Figure 41.1 shows schematically the radiation fields associated with a charge performing simple harmonic oscillations in the y direction. These radiation fields propagate away from the accelerated charge with the speed of light, decreasing in amplitude as $1/r$, where r is the distance from the radiating charge. (The radiation fields exist in addition to the electrostatic field of the charge and the magnetic field due to the velocity of the charge.) Thus the accelerated charge radiates electromagnetic energy, which is propagated at a speed of 3×10^8 m/sec through free space. If the charge oscillates at a frequency in the neighborhood of 6×10^{14} Hz, the radiation is visible to the human eye as light.

41.2 ELECTROMAGNETIC WAVES AND THE ETHER

Radio signals, visible light, and other electromagnetic disturbances have the properties of transverse waves. Late in the nineteenth century physicists were speculating on the question: Through what medium are these waves propagated? Such a question seemed most reasonable; after all, acoustic waves are transmitted through air, transverse waves are transmitted by the strings of a violin, and water waves are propagated at a water-air interface. There was no obvious medium to which one could assign electromagnetic waves, since they were known to pass through a vacuum. However, physicists believed that some medium was necessary for energy to be transmitted through a distance, and this medium was given the name *luminiferous ether*. Does such an ether exist? If it exists, how can one detect it and determine its characteristics? These, too, were questions which haunted physicists during the latter part of the nineteenth century.

Early speculations on the ether revealed that it must have remarkable properties. First of all, to sustain waves with the tremendous speed of light, a material of high rigidity was required. At the same time this rigidity raised problems about how the earth could move through the ether with no apparent retardation—or if one assumed the ether to be transported along with the earth, there was the problem of how the ether-atmospheres of the earth and the sun could be in relative motion without some evidence of shear forces.

41.3 AN ACOUSTIC ANALOG

If a luminiferous ether exists, the speed of light measured in a reference frame at rest with respect to this ether should be different from the speed determined in a reference frame moving with respect to the ether. Thus, the ether should provide an absolute reference frame to which the motions of all other frames and of all bodies in the universe could be referred. It was clear from the beginning that grave experimental difficulties would accompany any effort to resolve the question of how fast the earth moves relative to the ether or, indeed, to make any meaningful measurements whatever regarding the ether.

In order to see one way in which such measurements might be made, let us consider an experiment to determine whether the air through which sound waves are being propagated is moving relative to the earth. This is directly analogous to the problem of determining the ether "wind" associated with the motion of the earth through the ether. However, the acoustic experiment is vastly simpler than one involving light, because the speed of sound is modest, while the speed of light is great. Imagine two observers A and B a distance d apart. Let A send a signal to B when there is a wind blowing from A toward B with a velocity V (Fig. 41.2). The time required for the signal to go from A to B is $d/(c + V)$, where c is the speed of sound relative to air. The time required to transmit a signal from B back to A is $d/(c - V)$, since the propagation speed of the wave relative to B is $c - V$. For example, if $d = 330$ m, $V = 30.0$ m/sec, and the speed of sound in air $c = 330$ m/sec, the time required for a signal to go from A to B is $330/360 = 0.917$ sec, while the time required for the sound to go from B to A is $330/300 = 1.10$ sec. The total time for the traversal A to B and return is 2.017 sec. Clearly, this is different from the 2.000 sec which would be required in still air. While 0.017 sec is not a long time, it is easy with modern techniques to measure much shorter times.

Of course, there are simpler ways of telling whether the air is in motion between A and B, but ideally this type of experiment could

Figure 41.2
Transmission of a sound signal from A to B back in a wind.

establish whether A and B are moving or at rest *relative* to the air. By an experiment similar to this, the speed of the earth in its orbital motion about the sun might be measured relative to the ether, which for the moment we assume to be fixed in the solar system. However, a simple calculation shows that the time difference could not be measured directly by ordinary methods. To see this, we observe that the speed of the earth V in its orbital motion around the sun is 30 km/sec, roughly one ten-thousandth the speed of light c. If the apparatus were at rest relative to the ether, the time T_r for light to go a distance d from A to B and return to A would be $2d/c$. If the earth were moving with a speed V from A toward B through the ether, the time for the round trip T would be $d/(c + V) + d/(c - V)$. The difference between these times $\Delta T = T - T_r$ is given by

$$\Delta T = T - T_r = \frac{d}{c + V} + \frac{d}{c - V} - \frac{2d}{c} = \frac{2d}{c}\frac{V^2}{c^2 - V^2} \quad (41.1)$$

Since $V = 10^{-4}c$, we have $\Delta T = 10^{-8}(2d/c) = 10^{-8}T_r$. Even if d were as great as 1,500 m (about a mile), ΔT would be only 10^{-13} sec—too short a time to measure directly, as Maxwell pointed out in a letter of 1880. However, in that time green light of $\lambda = 500$ nm travels a distance of 60 wavelengths—a distance readily measurable with an interferometer. To detect the presumed motion of the earth through the ether, Michelson devised an ingenious experiment based on an interferometer (Sec. 27.5), which he invented for this purpose.

41.4 THE MICHELSON–MORLEY EXPERIMENT

The ideas underlying Michelson's inspiration involve the following considerations. In the Michelson interferometer a beam of light is split into two parts which are sent by different paths and then recombined. Consider first the transit time of a light beam moving perpendicular to the velocity of the apparatus relative to the ether (Fig. 41.3). If the distance between the half-silvered plate P and the mirror is d, the light pulse travels a distance given by $cT_\perp = 2\sqrt{d^2 + (VT_\perp/2)^2}$, where $T_\perp$

Figure 41.3

An electromagnetic signal would take longer to go from P to M_2 and return than from P to M_1 and return if the Michelson interferometer were moving with velocity V relative to an ether. The position of the interferometer is shown dotted for time $\frac{1}{2}T_\perp$ when the light reaches M_1 and dashed for time $T_\perp$ when the light returns to P.

is the time required for the light to go from P to M_1' and then to P''. Hence,

$$T_\perp = \frac{2d}{c\sqrt{1-(V/c)^2}} \tag{41.2}$$

Let us now compare this time with that required for the portion of the wave which moves parallel to the motion, for which

$$T_\parallel = \frac{d}{c+V} + \frac{d}{c-V} = \frac{2cd}{c^2-V^2} = \frac{2d}{c[1-(V/c)^2]} \tag{41.3}$$

The round trip of the light takes slightly less time moving perpendicular to V than it does moving parallel to V. The difference $\Delta T = T_\parallel - T_\perp$ is small indeed. To a good approximation

$$\Delta T = T_\parallel - T_\perp = \frac{2d}{c}\frac{V^2}{2c^2} \tag{41.4}$$

This is only one-half the time difference of Eq. (41.1), but now the interferometer offers the possibility of a measurement.

In 1881 Michelson and Morley floated a large Michelson interferometer with $d = 11$ m on a pool of mercury. They watched the interference pattern as the interferometer was rotated, thereby interchanging the parallel and perpendicular paths. If the earth were moving through the ether, the interference fringes should have shifted through about 0.37 of a fringe as the interferometer was rotated. No shift was detected!

Michelson and Morley performed this experiment repeatedly over a period of many months. No effect was observed which could be interpreted as showing a motion of the earth through the ether, in spite of the fact that they made measurements throughout the year, when the earth was moving in many different directions relative to the solar system. These results defied satisfactory explanation for 24 years. Michelson first interpreted the experiments as showing that the earth dragged some of the ether along with it, but a subsequent experiment by Lodge showed that a rapidly moving body could not possibly impart to the ether a speed in excess of 0.5 percent of the speed of the body. The most important of the various proposed explanations was that of Lorentz and FitzGerald, who suggested that the arm of the interferometer moving in the direction of the earth's motion shrinks just enough to lead to the null result. As we shall see, this apparently tortured prediction is supported by Einstein's theory of relativity.

41.5 POSTULATES OF THE SPECIAL THEORY OF RELATIVITY

In 1905 Einstein proposed a theory, amazing in its simplicity and genius, which ultimately resolved the Michelson-Morley dilemma. He argued that the reason there is no evidence of relative motion through the ether in any of a number of careful measurements is that any measurement of the speed of light will always yield a value independent of the velocities of both the observer and the source.

Einstein was keenly aware of the successes of classical mechanics, which predicted with great precision the motions of planets, aircraft, and baseballs. It was only when speeds were very great—comparable to that of light—that there was a problem. In classical physics we call a frame of reference in which Newton's laws of motion are valid an *inertial* frame. Any frame of reference which moves with constant velocity relative to an inertial frame is also an inertial frame. Thus, an airplane flying with a constant velocity of 960 km/hr northward relative to an inertial frame provides the framework for another inertial system. Einstein proposed that if two postulates are accepted all the strange results which had been so difficult to interpret could be readily explained. These postulates of the special theory of relativity are:

1 *The laws of physics are the same in all inertial frames.*

2 *The speed of light c is the same for every inertial frame and is independent of any motion of the source.*

An immediate implication of Einstein's postulates is that there exists no preferred frame in the universe; there is no ether relative to which everything can be measured. One inertial frame is equivalent to another so far as the speed of light is concerned. This idea requires, however, that the simple transformations used for centuries in going from one inertial frame to a second must be inadequate, although they work well for slowly moving bodies. The correct transformations must reduce to the classical ones for speeds small compared with the speed of light.

41.6 THE LORENTZ TRANSFORMATIONS

Let us consider first the classical transformations from one inertial frame of reference to a second moving with constant velocity relative to the first. These are known as the *galilean* (or *newtonian*) *transformations*. Suppose we have two inertial frames: $OXYZ$, which we shall regard as at rest, and $O^*X^*Y^*Z^*$, which is moving with a velocity V along the X axis of the rest frame (Fig. 41.4). Let the origins of the two frames coincide at $t = 0$, and assume that time proceeds at the same rate in both frames. If at time t a body is located at a point P specified by (x,y,z) in the rest frame, in the frame $O^*X^*Y^*Z^*$ the body is found at the point (x^*,y^*,z^*), where

$$x^* = x - Vt$$
$$y^* = y \qquad \text{galilean transformations} \qquad (41.5)$$
$$z^* = z$$

and $\qquad t^* = t$

Figure 41.4
Reference frame $O^*X^*Y^*Z^*$ moving with constant velocity V in the X direction of reference frame $OXYZ$.

If a body has a velocity v in the frame $OXYZ$ with components v_x, v_y, v_z, the corresponding components in the $O^*X^*Y^*Z^*$ frame are

$$v_x^* = v_x - V \qquad v_y^* = v_y \qquad v_z^* = v_z \qquad (41.6)$$

These transformations predict that if a beam of light has a velocity c in the X direction of the rest frame, its velocity in the moving frame is $c - V$. But this is not consistent with the results of the Michelson-Morley experiment. Hence we need a new set of transformations which reduce to the galilean transformations for low velocities but permit us to satisfy Einstein's second postulate for very high velocities. It was Einstein's good fortune that such transformations had already been derived by Lorentz. They are

$$x^* = \frac{x - Vt}{\sqrt{1 - V^2/c^2}}$$

$$y^* = y$$
$$z^* = z \qquad \qquad \text{Lorentz transformations} \qquad (41.7)$$

$$t^* = \frac{t - (V/c^2)x}{\sqrt{1 - V^2/c^2}}$$

When V is small compared with c, the Lorentz transformations reduce to the galilean transformations, but as v approaches the speed of light, they become different indeed.

41.7 THE LORENTZ CONTRACTION AND TIME DILATION

Consider a measuring stick of length L^* extending from x_1^* to x_2^* in the moving frame of reference $O^*X^*Y^*Z^*$, so $L^* = x_2^* - x_1^*$ (Fig. 41.5). To an observer in the rest frame the rod has an apparent length $L = x_2 - x_1$, where x_2 and x_1 are measured at the same time t. By Eqs. (41.7), $x_2^* = (x_2 - Vt)/\sqrt{1 - V^2/c^2}$ and $x_1^* = (x_1 - Vt)/\sqrt{1 - V^2/c^2}$. When these are substituted in the expression for the apparent length L in the rest frame, we obtain

$$L^* = x_2^* - x_1^* = \frac{x_2 - Vt}{\sqrt{1 - V^2/c^2}} - \frac{x_1 - Vt}{\sqrt{1 - V^2/c^2}}$$

$$= \frac{x_2 - x_1}{\sqrt{1 - V^2/c^2}} = \frac{L}{\sqrt{1 - V^2/c^2}}$$

or $\qquad L = \sqrt{1 - \frac{V^2}{c^2}}\, L^* \qquad \qquad (41.8)$

Figure 41.5
Frame $O^*X^*Y^*Z^*$, moving with velocity $0.5c$ parallel to the X axis of frame $OXYZ$, has a meterstick lying at rest parallel to the X^* axis; this stick has a length of 0.87 m when measured by an observer at rest in $OXYZ$.

Thus, an object moving relative to an observer appears to be contracted in the direction of motion by the Lorentz factor $\sqrt{1 - V^2/c^2}$.

Example

A meterstick moving parallel to its length at a speed of $0.5c$ passes an observer at rest. What length would the observer find for this meterstick?

By Eq. (41.8),

$$L = \sqrt{1 - \frac{V^2}{c^2}}\, L^* = \sqrt{1 - 0.25} \times 1.000 \text{ m} = 0.866 \text{ m} \qquad ////$$

Next consider the effect of relative motion on a pendulum clock which is fixed at the origin of the rest frame $OXYZ$ and indicates intervals of time $T = t_2 - t_1$. To an observer in the frame $O^*X^*Y^*Z^*$ moving with velocity V past the clock, the period $T^* = t_2^* - t_1^*$. By Eqs. (41.7), since $x = 0$,

$$T^* = \frac{t_2 - t_1}{\sqrt{1 - V^2/c^2}} = \frac{T}{\sqrt{1 - V^2/c^2}} \qquad (41.9)$$

Thus, when an observer is moving relative to a clock (or the clock relative to the observer), the time interval as seen by this observer is longer than that seen by another observer at rest relative to the clock. To an observer at rest a moving clock runs slow.

Example

Muons are unstable particles which have an average lifetime of 2 μsec. If a beam of cosmic-ray muons has a speed of $0.98c$ relative to the earth, find the apparent average life and the distance traversed by the beam during this time.

In the frame in which the muons are at rest, T is 2×10^{-6} sec. Relative to the muons the earth is moving with a speed of $0.98c$. By Eq. (41.9),

$$T^* = \frac{T}{\sqrt{1 - V^2/c^2}} = \frac{2 \times 10^{-6} \text{ sec}}{\sqrt{1 - 0.96}} = 1 \times 10^{-5} \text{ sec}$$

In this time a muon travels

$$s^* = VT^* = 2.94 \times 10^8 \text{ m/sec} \times 1 \times 10^{-5} \text{ sec} = 2{,}940 \text{ m} \qquad ////$$

We now see how observers in fixed and moving frames can find the same speed for light. From the point of view of the observer in the fixed frame, the observer in the moving frame finds the same velocity because metersticks are shortened and clocks run slow in the moving frame.

41.8 VELOCITY TRANSFORMATION

Next consider the values observers in the frames $OXYZ$ and $O^*X^*Y^*Z^*$ would find for the velocity of a body that is moving along the X axis of the rest frame. By the definition of velocity, $v = \Delta x/\Delta t =$

$(x_2 - x_1)/(t_2 - t_1)$. An observer in the moving frame would find $v^* = (x_2^* - x_1^*)/(t_2^* - t_1^*)$, which, by the Lorentz transformations [Eqs. (41.7)], becomes

$$v^* = \frac{(x_2 - Vt_2) - (x_1 - Vt_1)}{(t_2 - Vx_2/c^2) - (t_1 - Vx_1/c^2)} = \frac{(x_2 - x_1)/(t_2 - t_1) - V}{1 - V(x_2 - x_1)/(t_2 - t_1)c^2}$$

or

$$v^* = \frac{v - V}{1 - Vv/c^2} \qquad (41.10)$$

This differs from the galilean velocity transformation by the term vV/c^2, which is negligible for ordinary velocities but becomes highly significant as v approaches c.

Example

A beam of light moving with speed c along the X axis of frame $OXYZ$ (Fig. 41.4) passes an observer in frame $O^*X^*Y^*Z^*$, who determines its speed in his system. Show that he also obtains c, regardless of the value of V, the speed of $O^*X^*Y^*Z^*$ along the X axis of $OXYZ$.

By Eq. (41.10),

$$v^* = \frac{c - V}{1 - Vc/c^2} = c \qquad\qquad \textit{////}$$

Example

A proton moving eastward with a speed of $0.60c$ in a nuclear-physics laboratory passes an electron moving westward with a speed of $0.90c$. Find the speed of the electron relative to a frame of reference riding with the proton.

If we call eastward the positive direction,

$$V = 0.60c \qquad \text{and} \qquad v = -0.90c$$

By Eq. 41.10,

$$v^* = \frac{-0.90c - 0.60c}{1 - (-0.90)(0.60)} = \frac{-1.50c}{1.54} = -0.97c \qquad\qquad \textit{////}$$

An important consequence of the theory of relativity is that according to its postulates no real body can move and no transfer of energy can occur with a speed which exceeds c, the speed of light in free space.

41.9 RELATIVISTIC MASS

If a constant force F acts on a mass m indefinitely, it is clear that Newton's second law $F = ma$ must break down as the speed of the body approaches the speed of light, since it follows directly from Eq. (41.10) that the speed of light represents an upper limit to velocities involving energy transfer. Thus, a constant force does not cause the body to accelerate forever. We may retain $F = ma$ if, as the speed approaches the speed of light, the mass of the body increases, rather than the velocity. Einstein showed by application of the law of conservation of momentum that if m_0 is the mass of a body when it is at rest, its mass m when moving with a velocity v is (Fig. 41.6)

$$m = \frac{m_0}{\sqrt{1 - v^2/c^2}} \qquad (41.11)$$

In terms of this relation, the momentum of a body which has rest mass m_0 becomes

$$\text{Momentum} = mv = \frac{m_0 v}{\sqrt{1 - v^2/c^2}}$$

When a constant force acts on a body as its speed approaches the speed of light, the effect of doing work on the body, and thereby supplying it with energy, is to increase its mass rather than its speed. Einstein showed that this can be reconciled with conservation principles if mass is one of the forms of energy and if the additional energy given to the body is $\Delta W = (m - m_0)c^2$. This idea may be extended so that a body of rest mass m_0 has a rest energy $m_0 c^2$; when its velocity is v and its mass m, its total energy W is mc^2. The kinetic energy (energy associated with motion) is

Figure 41.6
The ratio of the mass of a body in motion to its rest mass as a function of v/c.

$$\text{Kinetic energy} = mc^2 - m_0 c^2 = W - m_0 c^2 \qquad (41.12)$$

$$= m_0 c^2 \left(\frac{1}{\sqrt{1 - v^2/c^2}} - 1 \right) \qquad (41.12a)$$

If we expand this square root assuming that v^2/c^2 is small compared with 1, we obtain

$$\text{Kinetic energy} = m_0 c^2 \left[1 + \frac{1}{2} \frac{v^2}{c^2} + \frac{(-\frac{1}{2})(-\frac{3}{2})}{2!} \frac{v^4}{c^4} + \cdots - 1 \right]$$

$$= \tfrac{1}{2} m_0 v^2 \left(1 + \frac{3}{4} \frac{v^2}{c^2} + \cdots \right)$$

In the limiting case, where v/c is small compared with 1, we obtain the classical formula $\frac{1}{2}mv^2$ for the kinetic energy of a body. As v approaches c, we must turn to the relativistic relation for the kinetic energy as given in Eq. (41.12).

The revolutionary concept that mass is one of the forms of energy has proved to be a cornerstone of modern physics, permitting a new understanding of nuclear physics and leading to the development of nuclear fission as a practical energy source.

41.10 THE MASS OF THE ELECTRON

The first particle for which the relativistic change of mass with velocity was investigated was the electron, the least massive of charged particles. The experimental results are completely in accord with the predictions of the relativity theory. Since high-energy electrons are usually produced by accelerating them through a large potential difference, it is convenient to measure their energies in electron volts. *One electron volt is the energy attained by an electron in falling through a potential difference of one volt* (Sec. 31.8).

Table 41.1 shows how the ratios v/c and m/m_0 vary with the kinetic energy of an electron. In a modern synchrotron, in which electrons are given kinetic energies of over 1 GeV, the mass of the moving electron is greater than the mass of the hydrogen atom at rest!

Example

A Van de Graaff generator accelerates electrons to a kinetic energy of 2 MeV. Find the mass of these electrons. Calculate the speed of the 2-MeV electrons.

$$2 \text{ MeV} = 2 \times 10^6 \text{ eV} \times 1.602 \times 10^{-19} \text{ J/eV}$$
$$= 3.2 \times 10^{-13} \text{ J}$$

$$3.2 \times 10^{-13} \text{ J} = (m - m_0)c^2 = (m - m_0)(9 \times 10^{16})$$

$$m - m_0 = 35.6 \times 10^{-31} \text{ kg}$$

$$m = (35.6 + 9.1) \times 10^{-31} \text{ kg}$$
$$= 44.7 \times 10^{-31} \text{ kg (almost 5 times the rest mass)}$$

$$44.7 \times 10^{-31} = \frac{9.11 \times 10^{-31}}{\sqrt{1 - v^2/c^2}} \qquad \text{by Eq. (41.11)}$$

Squaring both sides and transposing yields

$$2{,}000 \left(1 - \frac{v^2}{c^2}\right) = 83$$

$$\frac{v^2}{c^2} = \frac{1{,}917}{2{,}000} = 0.959 \frac{\text{m}^2}{\text{sec}^2}$$

$$v = 0.98c = 2.94 \times 10^8 \text{ m/sec} \qquad ////$$

41.11 GENERAL RELATIVITY

In our discussion of relativity we have thus far confined our attention to inertial frames of reference. When this restriction applies, we are dealing with what is known as *special relativity*. If we consider frames of reference accelerated relative to an inertial frame, we are led to the *general theory* of relativity, also developed by Einstein. A simple example

Table 41.1

Kinetic energy of electron		$\dfrac{v}{c} = \dfrac{\text{speed of electron}}{\text{speed of light}}$	$\dfrac{m}{m_0}$
eV	J		
0	0	0	1
10^2	1.6×10^{-17}	0.0205	1.0002
10^3	1.6×10^{-16}	0.0625	1.002
10^4	1.6×10^{-15}	0.195	1.02
10^5	1.6×10^{-14}	0.548	1.20
10^6	1.6×10^{-13}	0.941	2.96
10^7	1.6×10^{-12}	0.999	20.6
10^8	1.6×10^{-11}	1.000−	197
10^9	1.6×10^{-10}	1.000−	1,960

of an accelerated frame of reference is that of an elevator which is being accelerated upward. It is a well-known fact that an upward acceleration leads to an increase in the apparent weight of an object in the elevator. However, if an observer were isolated in an elevator on a strange planet, how could he untangle the gravitational contribution to the apparent weight from that associated with the acceleration of the elevator? Einstein concluded that there is no way to distinguish gravitational and acceleration effects in this situation. This conclusion, an important postulate of the general theory of relativity, is known as the *principle of equivalence*.

The special theory leads to most of the relativistic results which are of interest and applicability in modern physics, but the general theory leads to deeper insights and makes predictions which are not derivable from the special theory. We shall mention a few of these predictions.

1 Rays of light passing close to a star should be bent toward it. For example, a ray which just grazes the sun's surface on its way to the earth is deflected through about 1.75 seconds of arc. This observation was one of the first which tested the predictions of general relativity.

2 Physical processes should take place more slowly in regions of low gravitational potential than in regions of high gravitational potential. Thus, atomic vibrations on the sun should appear slowed down relative to their periods on the earth. This leads to the prediction that the spectral lines from atoms of the sun should be shifted very slightly toward the red end of the spectrum when compared with lines from the same elements on the earth.

3 The orbits of the planets should be slightly changed. This effect is important for the planet Mercury, which is closest to the sun. General relativity predicts that, aside from the precession of Mercury's orbit due to the perturbations of other planets, there should be an additional precession of about 43 seconds of arc per century. This precession had been known for many years, but its origin remained a mystery until Einstein developed his general theory. Some questions remain as to the correctness of certain aspects of Einstein's general theory.

QUESTIONS

1 Why don't we observe relativistic effects in our everyday lives?
2 The objection has been raised to the theory of relativity that it violates "common sense." How could you refute this objection?
3 Is it conceivable that an electron beam could move *across* the screen of a TV tube or a cathode-ray tube with a speed greater than the speed of light in free space? Why is this not contradictory to special relativity?
4 What alternatives to the theory of relativity can you propose to explain the results of the Michelson-Morley experiment?
5 For an electron with rest mass m_0 and speed $v = 0.6c$, which is the largest, $\frac{1}{2}mv^2$, $\frac{1}{2}m_0v^2$, or its kinetic energy?
6 By what factor is the density of a body increased when it moves with speed v?

PROBLEMS

1 A river 2 mi wide has a current with a speed of 3 mi/hr. How much longer will it take a boat

with a speed of 5 mi/hr to go upstream 2 mi and return than to go directly across the river and return to the same point? *Ans.* 15 min

2 A river 4 mi wide has a current of 5 mi/hr. How much longer will it take a motorboat with a speed relative to the water of 10 mi/hr to go 4 mi upstream and return than to go straight across the river and return to the same point?

3 How many electron volts of energy must an electron gain to bring its mass to (*a*) $1.1m_0$ and (*b*) $3m_0$? In each case what is the speed of the electron?
Ans. (*a*) 51,100 eV; 1.25×10^8 m/sec; (*b*) 1.022 MeV; 2.83×10^8 m/sec

4 Find the speed at which an electron has a mass 15 times its rest mass.

5 Find the mass of an electron traveling at eight-tenths the speed of light. How many times as great as the rest mass is this value?
Ans. 1.52×10^{-30} kg; 1.67

6 Find the work which must be done on an electron to increase its speed from $0.6c$ to $0.8c$, where c is the speed of light.

7 The rest mass of an electron corresponds to 0.511 MeV. An electron is accelerated to a kinetic energy of 2.555 MeV in a Van de Graaff generator. Find the ratio of the electron's mass to its rest mass and calculate the speed of the electron. *Ans.* 6; 2.96×10^8 m/sec

8 A proton of rest mass 938 MeV (1.67×10^{-27} kg) is given a kinetic energy of 5.628 GeV in a proton-synchrotron. Find the speed and the mass of the proton at this energy.

9 Through what potential difference must protons be accelerated to achieve a speed of $0.600c$? What is the momentum of such a proton? The total energy?
Ans. 235 MeV; 3.76×10^{-19} kg-m/sec; 1.173 GeV

10 Find the kinetic energy, total energy, mass, and momentum of an electron moving at a speed of 0.95 times that of light.

11 An electron is accelerated to a kinetic energy of 51 MeV. Find the ratios m/m_0 and v/c. If an electron at rest can be assumed to be spheri-

cal, find the ratio of the longitudinal to the transverse diameter for the 51-MeV electron.
Ans. 101; 0.99995; 0.0099

12 An arrow passes an observer with a speed 0.8 times that of light. If the rest length of the arrow is 0.7 m, find its apparent length as it passes the observer.

13 The mean life of pions at rest is 2.60×10^{-8} sec. If a beam of pions has a speed of $0.9c$, what will be the observed mean life in the laboratory? *Ans.* 60 nsec

14 The mean life of a muon at rest is about 2 μsec. If muons in cosmic rays have an average speed of $0.99c$, find their mean life and the approximate distance they move before decaying.

15 Imagine that you are at rest in a frame $O^*X^*Y^*Z^*$ which moves horizontally past an inertial frame $OXYZ$ at a speed of $0.8c$. A boy in the latter frame drops a ball which, according to your clock, falls for 0.5 sec. How long would the ball fall as timed by an observer at rest in the $OXYZ$ frame? *Ans.* 0.3 sec

16 Show that Eqs. (41.7) lead to the transformations

$$x = \frac{x^* + Vt^*}{\sqrt{1 - V^2/c^2}}$$

and

$$t = \frac{t^* + Vx^*/c^2}{\sqrt{1 - V^2/c^2}}$$

17 An observer on earth sees two spaceships approaching each other. Each ship is moving with a velocity of 2×10^8 m/sec relative to the earth. How fast will an observer in one of the spaceships find that he is approaching the other?
Ans. 2.77×10^8 m/sec

18 An electron moving to the right with a speed of 2.4×10^8 m/sec passes an electron moving to the left with a speed of 2.8×10^8 m/sec. Find the speed of one electron relative to the other as predicted by (*a*) the newtonian-galilean transformation and (*b*) the relativistic transformation.

19 An alpha particle moving east with a speed of $0.7c$ is passed by an electron moving west with a speed of $0.9c$. Find the speed of the electron relative to the alpha particle.
Ans. 2.94×10^8 m/sec

20 A particle has a rest energy m_0c^2 and a total energy W. Show that the velocity of this particle is $c\sqrt{1 - (m_0c^2/W)^2}$.

21 When an electron leaves the Stanford linear accelerator, its speed is only 10 cm/sec less than the speed of light. Find the mass and the kinetic energy of an electron with that speed. [*Hint*: $c^2 - v^2 = (c + v)(c - v) \approx 2c(c - v)$.]

Ans. 3.53×10^{-26} kg; 19.8 GeV

22 A free particle moving with speed v has rest energy $m_0 c^2$, momentum $p = mv$, kinetic energy K, total energy W (rest energy plus kinetic energy). Given that $m = m_0 / \sqrt{1 - v^2/c^2}$, derive the following relations: (*a*) $p = Wv/c^2$; (*b*) $W^2 = p^2 c^2 + m_0{}^2 c^4$; (*c*) $p^2 c^2 = K^2 + 2m_0 c^2 K$.

CHAPTER FORTY-TWO
ELECTRONS AND POSITIVE IONS

IN CHAPS. 33 TO 40 we discussed dc and ac circuits, considering currents passing through solid or liquid conductors, and in general paying little attention to the motions of the individual charged particles. Now we shall examine the conduction of electricity through an ionized gas, or *plasma*, and the passage of individual ions through an evacuated region. Studies in these areas resulted in the discovery of x-rays, contributed greatly to our understanding of atomic structure, and led to the development of the high-energy accelerators so important in nuclear physics.

42.1 THE EVOLUTION OF MODERN PHYSICS

Before the end of the nineteenth century the electromagnetic nature of light was well established, and the atomic hypothesis of matter was strongly supported by chemical evidence. However, no satisfactory model of an atom and no real understanding of how atoms join to form molecules or solids was yet available. Then, beginning with the discovery of x-rays in 1895, major clues were rapidly uncovered, new ideas were advanced, and the era of *modern physics* was born. Here "modern physics" implies the quantitative description and understanding of phenomena which could not be analyzed in terms of the "classical physics" which had evolved earlier.

Just as the assembling of a jigsaw puzzle usually proceeds by putting together a number of initially separate domains which are subsequently interconnected to give the final picture, so the development of physics has progressed, and continues to progress, by the simultaneous pursuit of apparently unrelated researches, the results of which often lead to a unifying denouement. For example, our current understanding of atomic structure arises from the synthesis of the outcomes of many experiments and calculations. Among the hundreds of important contributions which led to a useful model of the atom, we treat only a selected few which do not require the use of differential equations. In discussing these topics we shall find that experimental results in one area often confirmed and extended ideas which arose in some very different realm. The sequence of presentation follows the chronological order of discovery in some places but departs from it when a more logical presentation seems available. Among the topics which are typically assigned to "modern physics" are the following:

1 The discovery of x-rays in 1895 by Röntgen, together with a treatment of their origin, properties, and uses (see Chap. 43).

2 The discovery of radioactivity in 1896 by Becquerel (Sec. 43.5). Work with natural radioactive sources led Rutherford to propose the nuclear model of the atom in 1911. By the early 1930s accelerators were available for producing nuclear reactions, and nuclear physics had become one of the most active branches of physics.

3 The discovery of the electron as one constituent of the atom. In 1896 Zeeman and Lorentz showed that certain features of the spectra of atoms (Sec. 46.7) indicate that the radiation comes from a negative charge, and they determined the approximate ratio of the charge e on the electron to its mass m. In 1897 J. J. Thomson measured e/m by deflecting a beam of electrons by electric and magnetic fields (Sec. 42.5). About the same time Townsend determined the order of magnitude of the electronic charge e; by 1909 Milliken had developed a method of measuring e capable of good accuracy (Sec. 31.10).

4 The evidence that electromagnetic radiation has particle characteristics as well as wave properties. In 1900 Planck proposed that harmonic oscillators radiate energy only in quanta of energy $h\nu$ (Sec. 17.8). In 1905 Einstein extended Planck's idea to explain the photoelectric effect (Sec. 44.1). In 1913 Bohr combined the nuclear model, the electron, and the particle nature of radiation to provide a model of the hydrogen atom which correctly predicted the spectrum of atomic hydrogen (Secs. 45.3 and 45.4).

5 The idea that electrons and other classical "particles" have wave properties. This hypothesis was advanced by de Broglie in 1924, and the wave equation satisfied by the electrons was proposed by Schrödinger in 1926. Within a few years wave mechanics was used to solve many previously intractable problems and to give some understanding of the structures of atoms and of solids as well as an explanation of the periodic table of the elements.

6 The theory of relativity proposed by Einstein in 1905 (Chap. 41), which revolutionized our thinking about space and time and which introduced the mass-energy relationship.

In the short span of less than 40 years radically new ideas were established in physics, and several of the classical foundation stones had to be abandoned. For example, the law of conservation of mass and the idea that light was *purely* a wave phenomenon lost their fundamental positions and are now recognized to be useful approximations in many situations but not generally valid. Why should the third of a century starting in 1895 have been so fruitful in new discoveries and revolutionary ideas? Surely the presence of intellectual giants was a major factor, but so also was the fact that not until that time were many of the essential tools of scientific investigation available—reliable electric equipment with high-voltage sources, good vacuum pumps, practical photographic emulsions, and so forth. The evolution of the heat engine, the Industrial Revolution, and the growth of higher education brought many minds to bear on the problems of physical science, and an explosive growth ensued.

42.2 GASEOUS IONIZATION

Molecules of air and other gases are normally electrically neutral. Un-ionized air is an excellent insulator, a fact demonstrated by the observation that the leaves of an electroscope may hold their charges many hours. However, an ionized gas is a good conductor. Indeed, if virtually every molecule of the gas has lost one or more electrons, the assemblage of positive ions and free electrons is an excellent conductor. A highly ionized gas is known as a *plasma*—sometimes called the fourth state of matter since it is not solid, liquid, or gas in the usual sense of these words. Actually the major portion of the matter in our universe is in the plasma phase, since this is the state of stars and most interstellar matter. On the earth, conditions of temperature and pressure are such that most matter is solid, liquid, or gaseous, but relatively little of the universe has anything similar to our terrestrial environment.

To ionize a gas molecule, one must provide the few electron volts of energy needed to free an electron. This may be accomplished in a variety of ways, of which collision with an energetic electron or ion is perhaps the most familiar. This is the dominant process in fluorescent lamps, neon tubes, and similar devices in which positive ions are accelerated toward a negative electrode (cathode) and negative ions, chiefly electrons, toward a positive electrode (anode). If the ions have enough energy when they collide with neutral molecules, additional ion pairs are produced.

Charged particles emitted by radioactive nuclei produce ionization in any gas through which they pass. Many devices for detecting and measuring the radiation from fallout or other nuclear disintegrations are fundamentally electroscopes depending on gaseous ionization for their readings.

When gamma rays, x-rays, and ultraviolet radiation pass through gases, electrons are ejected from the atoms, thereby producing positive gas ions and free electrons. Gases drawn from a flame contain many ions. This ionization is greatly enhanced if the flame is fed with a salt such as sodium chloride, which readily breaks into ion pairs. Any gas can be ionized by raising the temperature sufficiently high. The temperature required for a specified fraction of ionized molecules depends on the gas involved. Even neutral molecules can ionize one another if they have enough kinetic energy.

Air at atmospheric pressure "breaks down" under electric fields of about 30,000 V/cm. At reduced pressures the breakdown occurs at lower field strengths. The corona glow from a lightning rod is an example of ionization by a strong electric field. Once such ionization begins, other ions are produced by collisions, and a conducting path between two charged objects is created.

42.3 DISCHARGE IN GASES AT LOW PRESSURES

When an electric discharge is maintained between two electrodes in a long glass tube filled with air (Fig. 42.1), some beautiful and interesting effects are observed as the air is pumped out of the tube. If the potential difference between the electrodes is not much greater than necessary

Anode

Crookes dark space

Negative glow

Cathode

Positive column

Faraday
dark space

Cathode glow

Figure 42.1
Discharge of electricity through a gas
at reduced pressure.

to maintain the discharge, sparks jump between the electrodes at atmospheric pressure. As the pressure in the tube is lowered, a narrow pink streamer appears between the electrodes. As the pressure is reduced further, the streamer expands until it fills almost the entire volume of the tube. By the time the pressure is reduced to 1 mm Hg, the entire gas glows.

As the pressure is reduced still further, one reaches the stage of discharge indicated in Fig. 42.1. A velvety glow, known as the *cathode glow*, covers the surface of the negative electrode. Outside this glow is a region called the *Crookes dark space*. Beyond this is a luminous region known as the *negative glow*, and then a second dark region called the *Faraday dark space*. Then follows another luminous region, known as the *positive column*, which reaches to the positive electrode. The positive column is not perfectly continuous, but may show alternate light and dark layers called *striations* across the path of the discharge. If the distance between the electrodes is increased, the appearance of the cathode region is not much changed; the positive column increases in length to fill the added volume. Practically all the potential difference across the discharge tube occurs between the cathode and the negative edge of the positive column.

The potential difference required to maintain the discharge in a tube depends sensitively upon the pressure. When the air is at atmospheric pressure, a high potential difference is required. As the pressure is reduced, the potential difference required to maintain the discharge decreases steadily (Fig. 42.2). At some relatively low pressure, a minimum in the curve is reached; then, as the pressure is decreased still further, the potential difference required to maintain the discharge increases sharply. Practically all the ionization of the air in the discharge tube occurs as the result of impacts. At high pressure the few residual ions (which may be formed by cosmic rays or by radioactivity) are accelerated by the electric field in the tube, but they travel short distances between collisions and attain relatively low energies. In order to produce more ionization they must have a certain minimum energy. If an ion is to attain this minimum energy between collisions at high pressure, it is necessary that a high electric field be maintained. As the pressure is reduced, the distance an ion travels between collisions is increased and the ion has a greater distance over which to acquire the energy needed to produce ionization by collision. Eventually the pressure becomes so low that ions may go all the way to the collecting electrode without colliding. Then a higher potential difference is required to maintain the discharge; the ions may eject electrons and other ions when they collide with the electrode. These *secondary ions* aid in supporting the discharge.

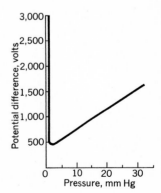

Figure 42.2
Potential difference required to maintain
a discharge in a particular tube as a
function of the pressure.

Figure 42.3
Sharp shadows show that cathode rays
travel in straight lines.

42.4 CATHODE RAYS

As the pressure in a discharge tube is reduced beyond the stage pictured
in Fig. 42.1, the Faraday dark space lengthens and the positive column
shrinks. Eventually a greenish glow spreads over the entire tube. This
is due to fluorescence of the glass as it is bombarded with high-speed
electrons which, Crookes showed, have their origin at the cathode of
the tube. These particles were named *cathode rays* before they were
identified as electrons—indeed, before electrons were "discovered."
Cathode rays have the following properties:

1 *The rays travel in straight lines in the absence of electric and
magnetic fields.* This fact may be demonstrated by inserting some
obstacle in the path of the rays and observing that the geometric
shadow of the obstacle does not show fluorescence (Fig. 42.3).

2 *The rays are deflected by an electric field* (Fig. 42.4).

3 *The rays are deflected by a magnetic field.*

4 *The rays carry negative electric charge.* The direction of the
deflection in both electric and magnetic fields supports this conclusion. Further, by applying a suitable magnetic field in the arrangement of Fig. 42.5, a cathode-ray beam can be deflected so
that it enters the small cylindrical vessel at E which is connected
to a galvanometer G. The galvanometer shows that negative
charge is collected in E when cathode rays enter the cylinder.

5 *The rays emerge normally from the surface of the cathode.* If
the cathode has a concave surface, the cathode rays come to
a focus near its geometric center. If the anode of the discharge
tube is shielded by some obstruction or is off at one side of the

Figure 42.4
The deflection of cathode rays by an
electric field reveals that they bear
negative charge.

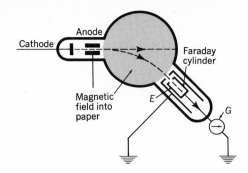

Figure 42.5
Cathode rays are deflected by a magnetic field in such a direction as to confirm that they are charged negative.

tube, the cathode rays strike at points which one would expect to be hit if the rays left perpendicularly from the surface of the cathode.

6 *The rays penetrate small thicknesses of matter.* If a thin window of aluminum is inserted in the tube, the cathode rays pass through the aluminum and make themselves evident by forming luminous streamers in the air beyond the window.

42.5 THE RATIO OF ELECTRONIC CHARGE TO MASS

By using a tube similar to that in Fig. 42.4, J. J. Thomson was able to measure the velocity of cathode rays (electrons). If the plate S of this cathode-ray tube is made positive, the beam is deflected downward. Now if a magnetic field uniform over the area between the plates R and S is applied out of the plane of the paper, it exerts an upward force on the moving electrons. It is possible to choose magnetic and electric fields in such a way that their effects exactly balance. The downward force due to the electric intensity E has a magnitude Ee, where $-e$ is the charge on an electron. The upward force due to the magnetic intensity B has magnitude evB if v is the speed of the electrons. When the beam is undeflected, $Ee = evB$, and

$$v = \frac{E}{B} \tag{42.1}$$

Thomson found that electrons which had been accelerated through a potential difference of 25 kV had a speed of about 8×10^7 m/sec, over one-fourth the speed of light. From the deflection of the electrons by the electric field in the absence of a magnetic field and the speed of the electron, Thomson calculated e/m, the ratio of the charge of an electron to its mass.

Thomson's pioneering experiments in deflecting beams of high-speed electrons by electric and magnetic fields laid the foundation for modern cathode-ray oscilloscopes and television picture tubes.

A more direct way to determine e/m is by measuring the radius of curvature of the path of an electron moving with known velocity perpendicular to a uniform magnetic intensity B. The centripetal force mv^2/r required for the circular path is provided by the force evB exerted on the moving charge by the magnetic field. Therefore,

$$\frac{mv^2}{r} = evB \tag{42.2}$$

or

$$\frac{e}{m} = \frac{v}{Br} \tag{42.2a}$$

The result of many painstaking measurements is that

$$\frac{e}{m} = 1.759 \times 10^{11} \text{ C/kg}$$

When this value is combined with Millikan's much later determination of $e = 1.602 \times 10^{-19}$ C, we find that the mass m of the electron is 9.11×10^{-31} kg.

In Chap. 35 we found that it requires 9.649×10^7 C to deposit electrolytically 1 kg equivalent mass of any element. In the case of hydrogen this corresponds to 1.008 kg. If the cathode rays in a discharge tube bear the same magnitude of charge as a monovalent ion, we can calculate the ratio of the mass M of a hydrogen atom to that of the electron. For the hydrogen atom $e/M = 9.649 \times 10^7$ C/1.008 kg $= 9.575 \times 10^7$ C/kg, from which we have

$$\frac{\text{Mass of hydrogen atom}}{\text{Mass of electron}} = \frac{M}{m} = \frac{1.759 \times 10^{11}}{9.575 \times 10^7} = 1{,}837$$

42.6 POSITIVE IONS

When a potential difference is applied between the terminals of a discharge tube, positive ions are accelerated as well as electrons. If a hole is drilled through the cathode (Fig. 42.6), some of the positive ions stream through the hole, giving a fine line of positive rays. J. J. Thomson used a tube similar to that of Fig. 42.6 to deflect the positive rays by means of electric and magnetic fields, thereby determining the ratio of charge to mass for the various positive ions present. The beam of positive ions passing between the poles of an electromagnet (Fig. 42.7) is deflected horizontally. An electric field parallel to the magnetic flux produces a vertical deflection. As a result, positive ions of different charge-to-mass ratios were deflected to form different parabolic traces on a photographic film. Slow-moving ions of a given kind undergo larger deflections than higher-energy ions of the same kind, primarily because the slower ions spend more time in the deflecting fields.

From the shape and location of such parabolic traces on a photographic plate, the ratio of charge to mass can be determined. In this

Figure 42.6
Positive ions passing through a hole in the cathode are deflected by both electric and magnetic fields.

Figure 42.7
Parabolic trace on a film is formed by positive ions of the same charge-to-mass ratio but different velocities when deflected by parallel electric and magnetic fields. Reversal of the magnetic field produced the other half of the parabolas in the insert.

way Thomson showed in 1907 that neon is composed of atoms of two mass numbers, 20 and 22. This was the first proof that atoms of a given chemical element may have different masses. Atoms of the same element which have different masses are called *isotopes*.

42.7 MASS SPECTROMETERS

Several types of instruments have been developed for the accurate determination of the charge-to-mass ratios of isotopes and complex ions. Such instruments are known as *mass spectrometers* (or *mass spectrographs* if they provide photographic recordings). Figure 42.8 is a schematic diagram of a Bainbridge mass spectrograph. Ions from a source are accelerated through a potential difference and pass through slits S_1 and S_2 into a region in which a uniform electric field E is provided by the plates P_1 and P_2; a uniform magnetic field B perpendicular to the plane of the paper is maintained by a magnet not shown. E and B are adjusted so that only ions having a desired velocity v pass through slit S_3. Ions with too low a speed are deflected toward the negative plate, and those with too high a speed toward the positive plate; only ions with $v = E/B$ are undeflected. An arrangement of crossed electric and magnetic fields of this kind is often called a *velocity selector*.

Once the ions have passed S_3 they are still in the magnetic field, but no longer in an electric field. Consequently the ions move in a circular path with

$$m \frac{v^2}{r} = qvB \qquad (42.2b)$$

From measurements of r, B, and E, a straightforward computation gives q/m for the positive ions. Figure 42.9 shows the circular paths of ion beams of three different masses in a mass spectrometer.

It is not necessary that a mass spectrometer include a velocity selector. If one is satisfied with a somewhat less uniform velocity for

Figure 42.8
A Bainbridge-type mass spectrometer for which there is a uniform magnetic field out of the page over the region enclosed by the dashed rectangle.

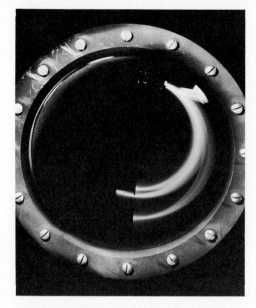

Figure 42.9
Paths of atomic beams of three different masses in a mass spectrometer. [*Westinghouse Corporation.*]

the incident ions, one can simply accelerate the ions through a potential difference V and make them incident directly on S_3. The work Vq done on each ion by the accelerating field is equal to the kinetic energy $\frac{1}{2}mv^2$ gained by the ion, or $v = \sqrt{2Vq/m}$. Simpler mass spectrometers of this kind have been widely used to separate isotopes of various elements.

With the mass spectrometer it is possible to determine directly the ratio of the charge on an ion to its mass. If we know the number of atomic charge units borne by the ion, we can calculate its mass. In obtaining a precise set of mass values for the hundreds of isotopes of the known elements, elaborate comparison measurements are made. This is important because, typically, one can determine the mass difference between two ions of nearly the same q/m with far greater precision than that with which one can determine the mass of either ion by itself.

42.8 AVOGADRO'S NUMBER AND THE MASSES OF ATOMS

The mass spectrometer is by no means the only tool we have for estimating atomic masses. For example, in Chap. 35 we learned that it takes 9.6487×10^7 C to deposit electrolytically 1 kg equivalent mass of any element. Let us now find Avogadro's number, the number of atoms in a kilogram atomic weight. Since a monovalent atomic ion is a neutral atom with one electron missing, the charge on the ion is 1.602×10^{-19} C. Therefore, Avogadro's number N_A is given by

$$N_A = \frac{9.6487 \times 10^7 \text{ C/kg at wt}}{1.602 \times 10^{-19} \text{ C/atom}}$$

$$= 6.022 \times 10^{26} \frac{\text{atoms}}{\text{kg at wt}}$$

The atomic weight of chlorine is 35.5, and the average mass of chlorine atoms is given by $35.5/N_A = 5.89 \times 10^{-26}$ kg. This average mass does not, of course, necessarily represent the mass of every chlorine atom; chlorine has two stable isotopes, and roughly three-fourths of the atoms have a mass of 5.81×10^{-26} kg, while the other one-fourth have a mass of 6.14×10^{-26} kg. The average mass of atoms of any element can be found by dividing its atomic weight by N_A.

QUESTIONS

1 How does the conduction of electricity in a gas differ from conduction in a copper wire? Discuss the similarities and differences.
2 What are the properties of cathode rays?
3 What is "recombination" in electric discharges? Why does it occur?
4 Why does one obtain parabolic traces of ions in the Thomson apparatus (Fig. 42.7) when an ion beam passes through a region in which there are parallel electric and magnetic fields? Why are the faster ions deflected less than the slower ones by the magnetic field when the force on the ion is proportional to v?
5 What is an electron volt? Why is the electron volt a convenient unit for measuring energies in atomic and nuclear physics?
6 How can beams of electrons be focused? What types of electron lenses can you suggest?

PROBLEMS

1 What is the current through a mercury discharge tube operated on a dc potential difference if, each minute, 2×10^{20} electrons and 8×10^{17} singly charged mercury ions pass a given cross section of the tube? What fraction of the current is carried by Hg^+ ions?
  Ans. 0.536 A; 0.4 percent
2 Find the mass of a beryllium atom and a uranium atom if the atomic weight of beryllium is 9.01 and the uranium atom belongs to the isotope corresponding to an atomic weight of 238.
3 If the magnetic induction of the mass spectrograph of Fig. 42.9 is 0.25 T and the electric intensity between the plates is 50,000 V/m, find (a) the speed of ions passed by the velocity selector, (b) the radius of the path of $^{20}Ne^+$ ions (atomic weight = 20) which reach the photo-

graphic plate, and (c) the distance between the line formed by $^{20}Ne^+$ ions and that formed by $^{22}Ne^+$ ions for the same velocity selector.
  Ans. (a) 2×10^5 m/sec; (b) 16.6 cm; (c) 3.32 cm
4 In determining the speed of electrons by Thomson's method it is found that a magnetic field of 6×10^{-3} T is just adequate to compensate for an electric field of 3.5×10^5 N/C. Find the speed.
5 The velocity selector of a mass spectrometer utilizes crossed electric and magnetic fields. The deflection plates are 1.1 cm apart and have a potential difference of 9,900 V between them, and the magnetic induction is 1.2 T. Find the speed of protons which will pass through undeflected. How does this compare with the speed of electrons and of Na^+ ions which would pass through undeflected?
  Ans. 7.5×10^5 m/sec; same
6 An electron with an energy of 180 eV enters a uniform magnetic field of intensity 4×10^{-4} T which is perpendicular to the electron's velocity. Find the radius of the path of the electron and the energy of the electron after it has traveled 3 cm in the magnetic field.
7 Alpha particles (He^{++} ions) of the same energy are passing between two parallel charged plates in a field of electric intensity 6×10^5 N/C. A magnetic field perpendicular to the velocity of the alpha particles and to the electric field is varied until the alpha particles are not deflected. At this point $B = 0.2$ T. Find the force exerted on an alpha particle by the electric field and the speed of the particles.
  Ans. 19.2×10^{-14} N; 3×10^6 m/sec
8 A particle of mass m, charge q, and speed v is moving perpendicular to a magnetic induction B. Show that the momentum of the particle is qBr, where r is the radius of curvature of the path in which the particle moves.
9 An ion of mass m bearing a charge q is accelerated through a potential difference V. It then

enters a magnetic field of flux density B, moving perpendicular to the field. Find the velocity (assumed to be small compared to the speed of light) of the ion and the radius of its path in the magnetic field in terms of m, q, V, and B.

Ans. $\sqrt{2Vq/m}$; $\sqrt{2mV/q}/B$

10 Show that if two ions of the same charge and energy but different mass are passing through a uniform magnetic field, the radii of the paths are proportional to the square roots of the masses.

11 An electron initially at rest is accelerated through a potential difference V. It then passes between two parallel plates, which are oppositely charged to produce a transverse electric intensity of magnitude E. If a magnetic field B is applied in the region between the plates and its intensity and direction are such that the electron is not deflected as it passes between the plates, show that for the electron $e/m = E^2/2VB^2$.

THE DISCOVERY OF X-RAYS represents one of the most important developments in physics and remains
one of the strong arguments for research in pure science. It stimulated
the work which led to the discovery of
radioactivity, which in turn laid the
groundwork for nuclear physics. Both
x-rays and the gamma rays from radioactive nuclei are high-energy electromagnetic radiation similar to visible light but characterized by much
shorter wavelengths. In this chapter
we discuss some of the significant
and useful properties of these rays.

CHAPTER FORTY-THREE
X-RAYS AND GAMMA RAYS

43.1 THE DISCOVERY OF X-RAYS

In 1895 Röntgen was studying the conduction of electricity through
partially evacuated tubes (Sec. 42.3) when he observed a mysterious
radiation which was able to penetrate thin layers of material. He called
this radiation *x-rays*, with the "x" standing for "unknown." Röntgen
promptly performed a broad series of experiments with x-rays and
learned many of the properties of this new kind of radiation. Among
his findings were the following. X-rays produce fluorescence in all kinds
of materials. All substances are somewhat transparent to x-rays, particularly those composed of light elements. Thus paper, wood, and aluminum are almost transparent, while lead and gold are relatively opaque.
Photographic emulsions are sensitive to x-rays, although the eye cannot
see them. The rays are not deflected by either magnetic or electric fields
and hence show no evidence of bearing electric charge. X-rays are
produced whenever a beam of high-energy electrons strikes matter.
Röntgen was unable to concentrate x-rays with lenses or mirrors and
failed to find any evidence of interference or diffraction of x-rays.

We now know that x-rays are electromagnetic radiation similar to
light and ultraviolet radiation, except that the wavelengths are much
shorter. Although the lines of demarcation between the various kinds
of electromagnetic radiation are not sharp, we may think of x-rays as
including radiation of wavelengths shorter than about 10 nm.

Among the earliest x-ray photographs were pictures of the human
hand and forearm. Soon pictures of this kind were utilized in studying
fractures and similar abnormalities. Only three months after Röntgen's
discovery, x-rays were used in connection with a surgical operation.

The news that rays had been discovered which could take pictures
through walls was tremendously exciting. X-rays were front-page news
all over the world in 1896. Grossly exaggerated claims and absurd
statements made some people fear that privacy was lost forever. A
London firm advertised the sale of x-ray-proof underclothing. A bill was

Figure 43.1
Electrons accelerated in a gas-type x-ray tube come from the ionization of the residual gas and from secondary emission of electrons when positive ions strike the cathode.

introduced in one of the state legislatures "to prohibit the insertion of x-rays, or any device for producing the same into, or their use in connection with, opera glasses, or similar aids to vision."

X-rays have found many important applications. They have given the diagnostic methods of physicians, surgeons, and dentists an exactitude that formerly was impossible. X-rays possess properties valuable in the treatment of certain diseases. Malignant cells are somewhat more readily killed by x-rays than normal cells. Under x-ray treatment inflamed glands shrink in size, and various morbid conditions of the blood and skin clear up. By means of x-rays, the metallurgist determines the effect of heat treatment, tempering, rolling, and aging on metals and alloys. Hidden defects in objects and concealed cracks in metals may be revealed by x-rays, which are a major tool in the nondestructive testing of materials.

43.2 THE PRODUCTION OF X-RAYS

Electromagnetic radiation is emitted when electric charges are accelerated (Sec. 41.1). This radiation appears in the form of photons, the energy of which is given by the product of Planck's constant h and the frequency ν. For a typical x-ray photon, λ may by 1 Å, and the corresponding energy 12,400 eV.

The most common means of producing x-rays is to accelerate electrons through high potential differences in vacuum tubes and allow them to strike a target of some heavy element. In Röntgen's original experiments he applied a high potential difference between two elements in a partially evacuated tube. The electrons accelerated came from the ionization of the residual gas molecules and from secondary electrons ejected from the cathode by impact of positive ions. Gas tubes (Fig. 43.1) operating on this principle are still used for producing x-rays.

In 1913 Coolidge developed an x-ray tube which was easier to control than the gas type. In the Coolidge tube (Fig. 43.2) the electrons are emitted from a heated filament F placed in a highly evacuated chamber. By varying the temperature of the filament, the number of electrons emitted can be controlled. These electrons are accelerated and strike a target T from which the x-rays are emitted in every direction, but those which travel into the target are absorbed.

Figure 43.2
A hot filament is the source of electrons in a Coolidge x-ray tube.

Figure 43.3
X-ray diffraction apparatus.

43.3 WAVE PROPERTIES OF X-RAYS

Röntgen's early failure to observe interference and diffraction of x-rays was associated with his apparatus and procedures. The first definite evidence of wave properties for x-rays came out of an experiment devised by Laue in which a crystal played the role of a diffraction grating. When x-rays fall on a body, the atoms scatter the incident radiation. In a crystalline substance the atoms are arranged in a regular way, and definite phase relationships exist between the scattered rays from neighboring atoms. If a narrow pencil of x-rays falls on a crystal C (Fig. 43.3), the scattered rays reinforce one another in certain directions by constructive interference. When a photographic plate is appropriately exposed, a series of spots is formed (Fig. 43.4) where constructive interference occurs. The theory of this interference was worked out by Laue. His colleagues Friedrich and Knipping successfully exposed a photo-

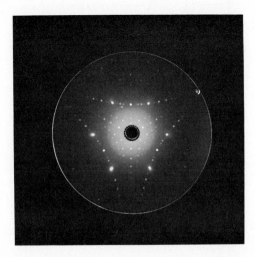

Figure 43.4
Laue photograph of a tungsten crystal.
[*RCA Corporation.*]

Figure 43.5
Constructive interference of the waves
scattered from atoms in adjacent
planes of a crystal occurs when $n\lambda = 2d \sin \theta$.

graphic plate to show the Laue spots in 1912. From the positions of
these spots and the properties of the scattering crystal Laue computed
the range of wavelengths present in their x-ray beam.

Shortly after Laue's work W. L. Bragg proposed a relatively simple
way of looking at the process of diffraction by a crystal grating. He
pointed out that through any crystal a set of equally separated parallel
planes can be drawn which pass through all the atoms of the crystal.
If x-rays are incident on this crystal, each atom scatters the radiation.
By Huygens' principle the condition for constructive interference of the
radiation scattered by every atom lying in a particular plane is that the
angle of incidence be equal to the angle of reflection. To form a Laue
spot, it is necessary not only to have constructive interference of all
scattered wavelets from a given plane of atoms but also to have constructive interference of the radiation scattered from adjacent planes.
For the radiation scattered by the second plane of Fig. 43.5 to interfere
constructively with that scattered by the first plane, it is necessary that
the path difference be an integral number of wavelengths. It can be
seen from the figure that the difference in path is AOB. Let the distance
between adjacent planes be d. Since $AO = OB = d \sin \theta$, it follows

Figure 43.6
Spectrometer for measuring the
wavelengths of x-rays.

(a)

(b)

Figure 43.7

(*a*) Ionization chamber for measuring the intensity of beams of x-rays and other ionizing radiations. (*b*) Current in an ionization chamber as a function of the electric intensity at the collector AC for a particular ionization chamber.

immediately that

$$n\lambda = 2d \sin \theta \qquad (43.1)$$

where n is an integer, λ is the wavelength of the x-rays, and θ is the glancing angle. This important relation is known as *Bragg's law*.

By use of Bragg's law it is possible to measure the wavelengths found in x-ray beams, knowledge of which has contributed greatly to our understanding of atomic structure. Further, by use of x-rays of known wavelength, one can determine the grating space d for crystals of unknown structure. Indeed, by an extension of the ideas of Bragg and Laue, it is possible to infer how atoms are arranged in crystals and in complex molecules, such as proteins.

43.4   THE X-RAY SPECTROMETER

W. H. Bragg, father of W. L. Bragg, developed an x-ray spectrometer (Fig. 43.6) by which the wavelengths of x-rays could be measured. He took an ordinary optical spectrometer, mounted a sodium chloride crystal on the rotating table, and replaced the collimator by an x-ray source and the telescope by an ionization chamber.

An ionization chamber is a device for measuring the number of ion pairs produced in a gas by ionizing radiation or by charged particles. A typical chamber consists of a metal cylinder (Fig. 43.7a) enclosing an insulated metal collector AC which is charged positive relative to the cylinder. When ionizing radiation enters through a "window" at one end, it ionizes the gas in the chamber; electrons are collected by AC while positive ions go to the chamber walls. The current read by the galvanometer G is a measure of the intensity of the ionizing beam. The current through the conducting gas is not proportional to the applied potential difference; it does not follow Ohm's law. As the potential difference between AC and the cylinder is increased, the current through the galvanometer increases (Fig. 43.7b). If the potential gradient is only a few volts per centimeter, many of the ions recombine before they are collected by the plates. However, an electric field of the order of 25 V/cm is ordinarily sufficient to collect virtually all the ions produced,

Wavelength, angstroms

Glancing angle, degrees

Figure 43.8

Bragg's curve showing the current in his ionization chamber as a function of glancing angle θ; the lines a, b, and c are characteristic lines in the L series of platinum which are superimposed on a continuous spectrum.

at which point the saturation current is attained. Once saturation is reached, a further moderate increase in field gives no significant change in current. When the potential gradient is increased far above the value required for saturation, the current increases again, at first slowly, then more rapidly as more and more ion pairs are produced by collisions. At a sufficiently high potential gradient, the gas breaks down and a discharge occurs between the chamber and the collector AC.

Using a beam of x-rays from a platinum target, W. H. Bragg measured the ionization current as a function of glancing angle θ and obtained results shown in Fig. 43.8. When a target of a different element was the source of x-rays, the bumps labeled a, b, and c disappeared and new bumps were found superimposed on the same kind of *continuous* background. Thus the observed x-ray spectrum consisted of bumps or "lines" associated with the particular element of which the target was made and a *continuous spectrum* (Sec. 44.3); the lines which depend on the target element constitute the *characteristic spectrum* (Sec. 45.6).

To calculate the wavelengths of the x-ray lines Bragg needed the grating space d for the NaCl crystals (Fig. 43.9). This can be calculated as follows. If M is the molecular weight of the NaCl, the number of atoms in a mass M is $2N_A$, where N_A is Avogadro's number. The number of atoms per kilogram is $2N_A/M$. To obtain the number of atoms per cubic meter, we multiply by the density ρ to obtain

$$\text{Number of atoms per cubic meter} = \frac{2N_A\rho}{M}$$

The volume per atom is therefore $M/2N_A\rho$. For a cubic crystal we can think of each atom as occupying a cube of length d on a side; thus $d^3 = M/2N_A\rho$, or

$$d = \sqrt[3]{\frac{M}{2N_A\rho}} \qquad \text{for a cubic crystal} \qquad (43.2)$$

Sodium Chlorine Sodium
ion   ion   ion

Figure 43.9
Crystal lattice of rock salt (NaCl).

The molecular weight of NaCl is 58.45, and the density is 2,160 kg/m³. Since $N_A = 6.02 \times 10^{26}$, $d = 2.82 \times 10^{-10}$ m = 2.82 Å.

Bragg measured the wavelengths of many x-ray lines, using sodium chloride crystals. Once the wavelengths of lines are known, it is possible to determine the separations between planes, i.e., the grating spaces, for crystals of more complicated structure.

Example

With an x-ray spectrometer using a rock-salt crystal, the glancing angle for the first reinforcement ($n = 1$) for the Cu $K\alpha$ line is found to be $15.8°$. If the distance between the crystal planes is 2.82 Å, find the wavelength of this line.

$$n\lambda = 2d \sin \theta$$
$$n = 1 \quad \text{and} \quad d = 2.82 \text{ Å}$$
$$\sin \theta = 0.273$$
$$\lambda = 2 \times 2.82 \times 0.273 = 1.54 \text{ Å} = 0.154 \text{ nm} \qquad ////$$

By 1914 it was established that x-rays have wave properties and that they are a form of electromagnetic radiation. Well before this, radioactivity had been discovered, and some radioactive materials were found to emit *gamma* rays, which are of the same fundamental nature as x-rays. Indeed, short-wavelength electromagnetic radiation is called gamma radiation when it comes from a radioactive nucleus and x-radiation when it arises from bombarding a target with high-energy electrons. Since the properties of the rays depend on their wavelengths and not on their origin, we shall discuss radioactivity briefly before we treat the detection and absorption of x-rays and gamma rays.

Electromagnetic radiation has not only wave characteristics but also particle properties, as pointed out in Sec. 17.8. It comes in *photons* with energy $h\nu$ [Eq. (17.4)], where h is Planck's constant and ν is the frequency. Since $\nu = c/\lambda$, the shorter the wavelength the greater the energy associated with a photon. For x-ray and gamma-ray photons the energy ranges from a few hundred to many millions of electron volts. Further evidences of the particle properties of electromagnetic radiation are presented in Chap. 44.

43.5 NATURAL RADIOACTIVITY

Röntgen's discovery of x-rays in 1895 stimulated others to search for new kinds of rays. In 1896 Becquerel found that the compounds of uranium emit radiations that produce a shadow picture on a photographic plate covered with black paper. These radiations have many of the properties of x-rays in that they can produce ionization in gases, affect photographic plates, and pass through thin sheets of substances which are opaque to light.

It was soon found that the property of emitting penetrating radiations is not confined to uranium and its compounds. Minerals containing thorium and several other elements have this same property. Such substances are said to be *radioactive*. A few years after Becquerel's discovery of radioactivity Pierre and Marie Curie were able to isolate the highly radioactive element *radium* from the mineral pitchblende. Many other radioactive elements were discovered later.

Figure 43.10

Alpha and beta particles are deflected in opposite directions by a magnetic field because alpha particles bear positive charges while beta rays are negative; gamma rays are undeflected since they are uncharged. The magnetic field is directed into the page.

When the rays from a radioactive ore are passed through a strong magnetic field, some are bent in one direction, some are deflected in the opposite direction, and some are undeviated (Fig. 43.10). Thus, three different types of radiation are emitted by radioactive substances. These radiations were named *alpha, beta,* and *gamma rays.* The bending of alpha and beta rays in a magnetic field shows that alpha rays carry positive charge and beta rays negative charge while gamma rays are electrically neutral. We know that alpha particles are helium nuclei, beta rays are high-energy electrons, and gamma rays are short-wavelength electromagnetic radiation.

43.6 THE DETECTION AND MEASUREMENT OF HIGH-ENERGY PHOTONS

There are three properties of high-energy photons which are ordinarily used to detect and measure the intensity of x-rays, gamma rays, and photons of the far ultraviolet:

1 *High-energy photons eject electrons from atoms and thereby produce ionization in a gas.* By collecting the ions in an ionization chamber (Fig. 43.7a) one can determine the energy of radiation which has been absorbed in the chamber.

Ionization measurements are used to determine exposures of x-rays and gamma rays. The radiation unit, adapted by the International Congress of Radiology, is called the *roentgen, where the roentgen is that quantity of x- or gamma radiation which produces in dry air a total ionization of 3.33×10^{-10} C per 0.001293 g of dry air.* A roentgen separates 2.08×10^9 ion pairs per cubic centimeter of dry air under standard conditions, which is equivalent to the release of about 88 ergs (8.8×10^{-6} J) of energy per gram of air. Since the roentgen refers to a specific result for x-rays and gamma rays in *air,* it does not imply any specific effect in a *biological system,* nor is it applicable to alpha and beta particles or neutrons. For biological purposes a unit called the *rad* (rd) is defined to be the absorbed dose of any radiation which is accompanied by the release of 100 ergs of energy per gram of absorbing material. For photons in the energy range 0.3 to 3 MeV an exposure dose of 1 R (R is the abbreviation for *roentgen*) gives rise to about 1 rd absorbed dose in tissue.

Although biological effects are produced by all ionizing radiations, the absorbed dose in rads required for a certain effect may be very different for different kinds of radiation. For example, 1 rd of neutrons is far more effective in producing cataracts in the eye than 1 rd of x-rays. This difference in response is taken into account by introducing the *relative biological effectiveness* (RBE), defined for any radiation as the ratio of the absorbed gamma-ray dose in rads to the absorbed dose of the specified radiation which produces the same biological effect. The dose unit for biological effects is called the rem for *roentgen equivalent man,* where

$$\text{Dose in rem} = \text{RBE} \times \text{dose in rads}$$

Exposure of the whole body to a single dose in the range of 100 to 200 rem results in illness but is not likely to be fatal.

For doses between 200 and 600 rem, hospital treatment is likely to save an exposed person, but a dosage of 1,000 rem is usually fatal. Even small doses may cause serious damage; one should never expose oneself needlessly to substantial radiation doses.

2 High-energy photons affect photographic emulsions. Over a fairly wide range the blackening of a photographic plate is proportional to the amount of energy of a given wavelength absorbed in the emulsion.

3 High-energy photons produce fluorescence. In many materials this fluorescence is in the visible spectrum. In medical diagnostics the fluorescent screen is a widely used tool; for example, a fluorescent screen can be used to observe the movement of a mass of bismuth salt as it passes through the digestive tract of a patient. Materials which fluoresce in the ultraviolet when bombarded with high-energy photons are used in scintillation detectors and other devices for studying high-energy photons. When such photons fall on certain crystals, the fluorescence produced is a direct measure of the energy absorbed (Sec. 29.8).

43.7 ABSORPTION OF HIGH-ENERGY PHOTONS

If a sheet of any substance is placed in the path of a beam of high-energy photons such as x-rays, the intensity of the beam is diminished. Let I_0 be the initial intensity of a homogeneous beam and I its intensity after passing through a thickness z of the material. Then

$$I = I_0 e^{-\mu z} \tag{43.3}$$

where μ is the *attenuation coefficient* of the material for the particular wavelength and e is the base of the natural logarithms. The attenuation coefficient varies with the wavelength of the photons, usually increasing as the wavelength increases. It is found experimentally that the absorption depends only on the number and kinds of atoms present in the absorbing layer; it is independent of their physical or chemical state.

A thickness of absorber which is adequate to cut the intensity to one-half is called the *half-value layer*. When a beam of homogeneous x-rays falls on an absorber just thick enough to remove half the beam, a second absorber of the same thickness removes half of what is left, so that one-fourth the initial intensity is transmitted. However, if a non-homogeneous beam is incident, the second absorber removes less than half of what was transmitted by the first. The half-value layer removes more than half of the less penetrating and less than half of the more penetrating wavelengths of a nonhomogeneous beam. Therefore, the radiation which strikes the second absorber is more penetrating on the average than that which struck the first. Thin layers of copper or aluminum are used to remove the softer, i.e., less penetrating, components of x-rays. If the x-rays were being used to treat a deep tumor, these soft x-rays would only produce burns near the skin.

Photons with energies ranging from a few hundred up to a million electron volts are absorbed or removed from a beam primarily by photoelectric absorption (Sec. 46.6) and by scattering. Photons with energies above 1.02 MeV may be absorbed by pair production (Sec. 44.5) as well. We shall invoke the quantum hypothesis to achieve a quantitative understanding of these processes. In one type of scattering we find particularly convincing evidence of the particle properties of electromagnetic radiation; another type of scattering is readily explained in terms of wave properties.

QUESTIONS

1 What evidence exists to show the wave nature of x-rays?

2 X-ray photographs are essentially shadowgraphs. What restrictions must be placed on the target area from which the x-rays originate if the x-ray photograph is to have sharp definition?

3 When x-rays of the digestive track are made, the patient usually is asked to drink a preparation containing heavy atoms (such as barium). What is the purpose and usefulness of this preparation?

4 Why are the very soft x-rays absorbed out of a beam before the x-rays are used to photograph a broken arm? How can the soft x-rays be removed without eliminating the other wavelengths?

5 The gas between two parallel plates is ionized by a constant x-ray beam. If the potential difference between the plates is increased from zero to a very high value, what happens to the current? Explain carefully. Is Ohm's law obeyed?

6 What do we mean when we say that x-rays are not really reflected by a rock-salt crystal? If the process is not reflection, what is it?

7 Why aren't ordinary diffraction gratings with a few thousand lines per centimeter useful for measuring the wavelength of x-rays incident normally in the ordinary way? (They are successfully used at grazing incidence.)

8 What is the difference between 1 roentgen, 1 rad, and 1 rem? Historically the roentgen was defined and used first. Why were the rad and the rem introduced?

PROBLEMS

1 The grating space of calcite is 3.0356 Å. Find the wavelength for which the first-order Bragg "reflection" occurs at 20°, that is, when the incident and emergent radiations are inclined 20° with respect to the surface of the calcite.

What is the angle for second-order "reflection"?
Ans. 2.076 Å; 43°9'

2 Find the angle for first-order Bragg reflection of Cu $K\alpha$ radiation (1.54 Å) from a calcite crystal with a grating space of 3.04 Å. What is the angle for second-order reflection?

3 For Mo $K\alpha$ radiation first-order Bragg reflection from NaCl occurs at an angle of 7.2°. Find the wavelength of this radiation. At what angles are the second- and third-order diffraction maxima?
Ans. 0.707 Å; 14.5°; 22.1°

4 The grating space of mica is 9.963 Å. At what angle would Al $K\alpha$ x-rays with a wavelength of 8.32 Å be reflected? (Such long wavelengths are strongly absorbed in air, so an evacuated spectrometer would be desirable.)

5 Find the grating space of a KCl crystal which has the same crystal structure as NaCl. The density of KCl is 1,990 kg/m³, and the atomic weights of K and Cl are 39.1 and 35.46, respectively.
Ans. 3.14 Å

6 Calculate the grating space of KBr, which has a molecular weight of 119 and a density of 2,750 kg/m³.

7 Potassium iodide (KI) crystallizes as a cubic crystal of density 3,130 kg/m³ and structure similar to that of NaCl. The atomic weights of potassium and iodine are 39.1 and 126.9, respectively. Find the grating space of the KI crystal.
Ans. 3.53 Å

8 Find the grating space of sodium bromide (NaBr) which has a density of 3,203 kg/m³. The atomic weight of Na is 23.0 and that of Br 79.9.

9 Fifty percent of a homogeneous beam of x-rays is absorbed by an aluminum sheet 0.5 cm thick. What percentage of the beam will pass through 1 cm of Al? Through 2 cm? Through 3 cm?
Ans. 25, 6.25, and 1.56%

10 An x-ray beam is studied by means of copper plates. It is found that 0.05 cm of copper passes 50 percent of the beam, 0.1 cm passes 30 percent, and 0.2 cm passes 10 percent. Explain in detail how this result can be reconciled with the results of Prob. 9.

TO THE PHYSICIST OF 1900 it seemed most reasonable to ask: Are x-rays waves or particles? To resolve this kind of question for any beam transferring energy from one place to another, he would perform certain experiments. Evidence of interference or diffraction established the wave nature of the beam. A confirming test for a beam of particles was to show that they could be located in space and that they exhibit rest mass or charge. For a century after Young demonstrated the interference of light, physicists considered light to be a pure wave phenomenon. However, in 1901 Planck found that he had to assume that light was radiated in chunks, or quanta, of energy $h\nu$. Within the next quarter century more evidence was found to show that electromagnetic radiation has particle characteristics as well as wave properties.

IN THIS CHAPTER we begin to examine the evidence for the modern view that "Is a beam composed of waves or particles?" is scarcely a proper question. A beam of electromagnetic radiation exhibits wave properties in many situations, but in others we now find that the photons constituting the beam behave like particles. Particle characteristics are especially prominent for high-energy photons. In Chap. 46 we shall find that classical particles, such as electrons, exhibit wave properties. Neither they nor electromagnetic radiation can be classified as pure wave or pure particle—instead, all show wavelike behavior in some situations and particlelike behavior in others.

CHAPTER FORTY-FOUR
PARTICLE PROPERTIES OF RADIATION

In 1887 Hertz was doing pioneer work in the study of radio waves. He observed that when ultraviolet light fell on the electrodes of a spark gap, a spark jumped at a substantially lower potential difference than usual. A little later it was found that a charged sheet of zinc exposed to ultraviolet radiation loses its charge when negative but retains its charge when positive (Fig. 44.1). We now know that when ultraviolet light falls on a metallic surface, electrons are liberated from the metal. The energy required to eject an electron is provided by the incident electromagnetic radiation. When the metal is negatively charged, these "photoelectrons" are repelled. When the metal is positive, the photoelectrons are attracted back to the metal.

It is not surprising that this *photoelectric effect* was not quickly understood since it was discovered almost a decade before the existence of the electron was established and it was in 1901 that Planck showed (Sec. 17.8) that the blackbody radiation curves could be explained if one assumes that electromagnetic radiation is emitted in photons (or quanta) of energy

$$W = h\nu \qquad (44.1)$$

where h is Planck's constant and ν is the frequency of the radiation. A theory like Planck's, which utilizes some new assumption to explain a phenomenon, is called an ad hoc theory. Such theories are usually viewed with skepticism until the radical assumption is found to have wider application.

The first of much supporting evidence for the validity of Planck's hypothesis came in 1905, when Einstein used it to explain the photoelectric effect, a phenomenon which had been a puzzle to physicists since its discovery. By 1905 it had been established that electrons are emitted from all metals (and many other materials) when ultraviolet light is incident (but not visible light). On the other hand, there are several elements from which visible light ejects photoelectrons. To explain these facts, Einstein adopted the hypothesis that radiant energy comes in quanta of energy $h\nu$. In order to eject a photoelectron from a surface, a photon must have enough energy to free the electron from the atom or material with which it is associated. If it requires an energy W to remove an electron from the metal, and if the photon imparts to it an energy $h\nu$ which is greater than W, the difference in energy appears as kinetic energy of the ejected photoelectron. This explanation of the photoelectric effect in terms of quanta was made by Einstein in 1905

Figure 44.1
When ultraviolet light falls on a zinc plate, photoelectrons are ejected and the electroscope discharges. The glass plate absorbs the ultraviolet, and when it is inserted in the beam, photoemission ceases.

and was one of the early triumphs of the quantum idea. If we let W_{min} represent the smallest energy which can free an electron from the solid, the maximum kinetic energy of ejected electrons is given by Einstein's photoelectric relation:

$$K_{max} = h\nu - W_{min} \qquad (44.2)$$

Many careful experiments have shown that the Einstein relationship is correct. If the frequency of the incident light is increased, the maximum kinetic energy of the ejected photoelectrons increases correspondingly (Fig. 44.2). Similarly, if the frequency of the incident photons is decreased, the energy of ejected photoelectrons decreases until the frequency $\nu_{min} = W_{min}/h$ is reached. For any lower frequency, no photoelectrons are emitted.

If the frequency of the incident light is kept constant while the intensity is changed by increasing the number of photons incident, the number of photoelectrons emitted increases. For a given frequency of radiation the number of photoelectrons ejected per second is directly proportional to the intensity of the incident radiation.

Among the metals from which photoelectrons may be ejected by visible light are barium, cesium, lithium, potassium, sodium, and rubidium. These are elements from which it is particularly easy to remove an electron in chemical reactions. Both the chemical and photoelectric properties of electropositive elements depend on the ease with which an electron can be removed.

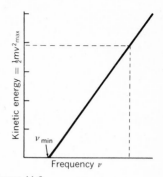

Figure 44.2
The maximum kinetic energy of ejected photoelectrons increases linearly with the frequency of the incident light.

Example

Light of wavelength 700 nm is required to cause the emission of electrons from a potassium surface. What is the energy necessary to remove one of the least firmly bound electrons? Find the kinetic energy and velocity of such an electron if the potassium surface is bombarded with light of wavelength 500 nm.

Energy of 700-nm photon $= h\nu = \dfrac{hc}{\lambda}$

$$= 6.63 \times 10^{-34} \text{ J-sec} \times \frac{3 \times 10^8 \text{ m/sec}}{700 \times 10^{-9} \text{ m}}$$

$$= 2.84 \times 10^{-19} \text{ J}$$

Since this is the lowest-frequency photon which causes photoemission, $W_{min} = 2.84 \times 10^{-19}$ J.

$$\text{Energy of 500-nm photon} = 6.63 \times 10^{-34} \times \frac{3 \times 10^8}{500 \times 10^{-9}}$$

$$= 3.98 \times 10^{-19} \text{ J}$$

$$K_{max} = h\nu - W_{min} = (3.98 - 2.84) \times 10^{-19} \text{ J}$$

$$= 1.14 \times 10^{-19} \text{ J}$$

$$\tfrac{1}{2}(9.1 \times 10^{-31})v_{max}^2 = 1.14 \times 10^{-19}$$

$$v_{max}^2 = 24.8 \times 10^{10} \text{ m}^2/\text{sec}^2$$

$$v_{max} = 5 \times 10^5 \text{ m/sec} \qquad \text{////}$$

44.2 PHOTOELECTRIC CELLS

Figure 44.3

A photoelectric cell.

Devices which utilize the photoelectric effect have many important applications. One is the exposure meter used by the photographer. Here the rate of photoelectron ejection caused by the incident light is a measure of the illumination. Photocells are used to open store doors when people cut off a light beam, to sound burglar alarms, and to turn on street lights whenever the sun fails to provide adequate illumination. Photocells are used in physics, biology, and medicine for measuring the intensity of various kinds of electromagnetic radiation. In a television studio variations in light intensity on a photosensitive surface are used to convert the picture into electric signals in the image orthicon tube.

The sound track of motion-picture films permits varying amounts of light to pass through when the film is run between a light source and a photoelectric cell. The photoelectric cell converts the variations in light intensity into variations in electric current.

There are several different types of photoelectric cells. In some there are a light-sensitive surface which emits electrons and a second electrode, usually maintained at a fairly high positive potential, which collects the electrons. A circuit for such a photocell is shown in Fig. 44.3 with a galvanometer to measure the relative light intensity.

A second type is the *photovoltaic* or *barrier-layer* photocell, in which incident light produces an emf between two terminals. A typical photovoltaic cell (Fig. 44.4) contains a thin metal disk A on which there is a film of light-sensitive material B. In contact with B is a conducting ring or layer C. In one common type A is an iron plate and B is a thin film of selenium over which a thin gold coating is laid to serve as the conductor C. Another type of cell uses a copper plate as A, cuprous oxide as B, and a thin layer of silver as C. When the sensitive surface is illuminated, an emf is generated between the sensitive layer B and the metal disk A. Electrons are driven by the light from the layer B to the metal A across the thin insulating layer. This type of cell is used in exposure meters, since the light produces the emf required to deflect the galvanometer.

Solar cells, widely used to provide energy for satellites and space vehicles, are photovoltaic cells converting radiant energy into electric energy. A familiar form uses a very perfect silicon crystal which is treated with trace impurities to introduce extra positive charges on one side of a thin wafer and extra negative charges on the other side (see Sec. 48.6). Radiant energy disturbs the electrical balance of the charges and thereby induces an emf which continues as long as light strikes

Figure 44.4

An electromotive force is produced when light is incident on a barrier-layer photoelectric cell.

the crystal. Such cells convert roughly 15 percent of the incident radiant energy to electric energy; efficiencies two or more times as great are theoretically possible.

A third type of photoelectric cell utilizes the photoconductivity effect. When photons are absorbed in a material, relatively few electrons leave the material. If a material which is normally an insulator or a poor conductor is irradiated, electrons originally bound to atoms are freed and become available for conduction through the material. The resistance of a piece of germanium, for example, decreases greatly when it is subjected to high illumination. The conductivity of a selenium cell may be 25 or more times greater in bright light than in the dark. The conductivity of a thallium sulfide cell changes by an even larger factor.

44.3   THE CONTINUOUS X-RAY SPECTRUM

Two phenomena from the field of x-rays support the blackbody and the photoelectric evidence for the photon hypothesis. They are the short-wavelength limit of the continuous x-ray spectrum and the Compton effect, which are discussed respectively in this and the following section. As the work of Bragg (Sec. 43.4) and others has shown, the wavelengths of the x-rays emitted when a beam of high-energy electrons strike a target vary over a broad range. There is always a *continuous spectrum* upon which certain lines characteristic of the target element can be superimposed. When the intensity of the continuous radiation in a given wavelength interval is plotted as a function of wavelength for different values of the applied potential across the tube, curves similar to those of Fig. 44.5 are produced. Three features of these curves are immediately apparent:

1   The intensity radiated increases at all wavelengths when the potential difference across the tube is raised.

2   The shortest wavelength emitted at a given potential is sharply defined and decreases as the voltage across the tube increases.

3   As the potential difference across the tube is increased, the wavelength at which the maximum energy is radiated shifts toward shorter wavelengths.

We can understand why the spectrum is continuous and predict quantitatively the short-wavelength limit if we invoke the Planck hypothesis that electromagnetic radiation is emitted in photons of energy $h\nu$. X-rays are produced when the kinetic energy of the incident electrons is transformed into electromagnetic radiation through collisions with atoms of the target. Most of the collisions are glancing ones in which only some moderate fraction of an electron's energy is radiated as a photon. Before being stopped, most electrons have several collisions and produce several photons of widely varying wavelengths. Occasionally an electron has a head-on collision in which it loses all its energy at once. Such collisions are relatively rare and are the ones which produce the x-rays at the *short-wavelength limit*. Clearly, the most energy an electron can lose is all it has. This corresponds to Ve, where V is the potential difference across the tube and e the electronic charge. If we equate this energy to that of the most energetic photon, we obtain

Figure 44.5
Wavelength distribution of the continuous x-ray spectrum from an aluminum target.

$$Ve = h\nu_{max} = \frac{hc}{\lambda_{min}} \qquad (44.3)$$

where c is the speed of light. Solving for λ_{min} yields

$$\lambda_{min} = \frac{hc}{Ve} \qquad (44.3a)$$

This relationship, known as the *Duane-Hunt law* (1915), was used for one of the earliest reliable determinations of Planck's constant h.

In many respects continuous x-ray emission is the inverse of the photoelectric effect. In the production of an x-ray photon the kinetic energy of the incident electron is converted into radiant energy; in the photoelectric effect the radiant energy of a photon is converted, at least in part, into kinetic energy of the electron.

The amount of energy available from each electron is increased as the potential difference across the tube is increased, and we should expect more energy to be radiated per electron. Thus the ordinates in Fig. 44.5 increase as the potential difference across the tube is increased. The area under the curve is proportional to the amount of energy radiated in the form of x-rays. The rest of the energy of the incident electrons appears as heat in the target.

44.4 SCATTERING AND THE COMPTON EFFECT

When a beam of x-rays passes through matter, some energy is scattered out of the beam. According to the classical theory of J. J. Thomson, the atomic electrons are driven to perform simple harmonic motion by the electric intensity E of the incident waves. Since these electrons are accelerated, they radiate a frequency equal to that of the incident waves. Thus the electrons remove energy from the passing wave and reradiate this energy in other directions. Scattering of waves with the frequency of the incident radiation is observed over the entire electromagnetic spectrum; however, it is weak at high frequencies. This unmodified scattering can be treated satisfactorily in terms of classical waves.

In 1923 Compton published the results of careful measurements of the x-ray frequencies scattered by carbon atoms upon which monochromatic x-rays were incident (Fig. 44.6). He found that the scattered beam contained two frequencies, one the same as that of the incident beam, and the second somewhat lower. Figure 44.6 shows the wavelength distribution of the radiation scattered at various angles. At each angle radiation is scattered not only at the wavelength of the incident beam, but also at a longer wavelength whose value depends on the scattering angle θ (the angle between the propagation directions of the incident and scattered radiation).

Compton's measurements showed that the change in wavelength $\Delta\lambda$ is independent of the scattering material but depends on the scattering angle θ according to the relation

$$\Delta\lambda = 0.024(1 - \cos\theta) \quad \text{Å} \qquad (44.4)$$

Compton explained the wavelength shift in the scattering of x-rays

Figure 44.6
The scattering of Mo $K\alpha$ x-rays
($\lambda = 0.707$ Å) at an angle θ produces
two peaks in the scattered radiation,
one at the wavelength of the incident
radiation and the second at a
wavelength greater by
$\Delta\lambda = 0.024(1 - \cos\theta)$ Å.

by assuming that the incident beam of x-rays consists of a stream of photons of energy $h\nu_0$. These photons possess momentum $h\nu_0/c$ as well as energy. Their collisions with electrons (Fig. 44.7) may be described in terms of the laws of conservation of momentum and conservation of energy. Applying these laws to a collision, we obtain, for the relativistic case in which the electron is initially at rest,

Conservation of energy:

$$h\nu_0 = h\nu + (m - m_0)c^2 \qquad (44.5)$$

Figure 44.7
Elastic collision of a photon with a free electron initially at rest. The photon is scattered, and the electron recoils.

Conservation of momentum:

x component:
$$\frac{h\nu_0}{c} = \frac{h\nu}{c}\cos\theta + \frac{m_0 v}{\sqrt{1 - v^2/c^2}}\cos\phi \qquad (44.6)$$

y component:
$$0 = \frac{h\nu}{c}\sin\theta - \frac{m_0 v}{\sqrt{1 - v^2/c^2}}\sin\phi \qquad (44.7)$$

where m and v are the mass and final speed of the electron, m_0 is the rest mass, θ is the angle through which the photon is scattered, and ϕ is the angle between v and the direction in which the photon was incident. With considerable effort these equations, together with Eq. (41.11), can be solved to find the change in wavelength $\Delta\lambda = c/\nu - c/\nu_0$. The result is that

$$\Delta\lambda = \frac{h}{m_0 c}(1 - \cos\theta) = 0.02426(1 - \cos\theta) \qquad \text{Å} \quad (44.4a)$$

Compton's experiments were a clear indication that electromagnetic radiation has particle as well as wave properties. The Compton effect is convincing evidence that radiation in the x-ray region comes in photons or quanta which behave like particles in collisions with electrons.

44.5 PAIR PRODUCTION AND THE POSITRON

Still another phenomenon which requires the photon concept and relativity for explanation was discovered by Anderson in 1932. When photons with energy greater than 1.02 MeV pass through matter, the photon may interact with an atomic nucleus and be transformed into two particles, an electron and a *positron*. A *positron* is a particle identical with an electron except that it bears a positive charge. The higher the energy of a photon, the more probable pair production becomes. It requires a minimum energy of 1.02 MeV to provide the rest mass of the electron and the positron. The positron and electron share whatever energy is left over from the energy of the incident photon. Thus, if pairs are produced by a 4.02-MeV photon, the electron and positron share 3.0 MeV of kinetic energy.

Positrons exist for a very short time. As they move through matter, they lose energy rapidly. As soon as they are stopped, an electron and a positron interact to annihilate each other. Two electron masses vanish, and 1.02 MeV of energy is released. This usually appears in the form of two 0.51-MeV photons moving in opposite directions. The two photons are formed rather than a single 1.02-MeV photon because it is not possible to conserve both energy and momentum for a single photon (unless some additional particle is involved). Both in pair production and in pair annihilation the particle properties of photons are in evidence.

1 If visible light comes in photons, why don't we perceive the discontinuous structure in the light received by the eye?

2 What determines the energy of a photon? Can two photons have the same energy but different wavelengths in air?

3 Ordinary film and x-ray film can be developed in a room illuminated by a red safelight because such films are insensitive to red light. Why are they insensitive to red but not to blue light? Explain in terms of photons.

4 Why can visible light remove photoelectrons from sodium but not from zinc?

5 Why are measurements of the work function W_{min} of a material by determining the energies of photoelectrons very sensitive to the preparation of the surface?

6 When electromagnetic radiation of a given wavelength strikes a surface and ejects photoelectrons, these electrons may have a wide range of energies. Why don't they all have K_{max} of the Einstein equation?

7 How is the short-wavelength limit of the continuous x-ray spectrum related to the photoelectric effect?

8 What factors determine the short-wavelength limit and general shape of the continuous x-ray spectrum as a function of wavelength?

9 Why is the Compton effect readily observed for x-rays but not for visible light? Explain carefully.

10 Is it possible for a Compton recoil electron to have a velocity component opposite that of the incident photon? Is it possible for a Compton scattered photon to have a velocity component opposite that of the incident photon? Explain.

PROBLEMS

1 A photon has an energy of 2.5 eV. Find its frequency and its wavelength in vacuum.

Ans. 6.04×10^{14} sec^{-1}; 496 nm

2 Determine the energy of a photon of mercury green radiation whose wavelength is 5461 Å, both in joules and in electron volts.

3 The eye can detect about 5 photons/cm²-sec of green light ($\lambda = 500$ nm), while the ear can detect about 10^{-13} W/m² under optimum conditions. As a power detector, which is more sen-

sitive and by what factor? *Ans.* The eye; 5

4 It requires 6.5×10^{-19} J to remove one of the least tightly bound electrons from a surface. Find the longest wavelength which will be effective in producing photoelectrons. What is the maximum energy of photoelectrons ejected from this surface by ultraviolet radiation of $\lambda = 150$ nm?

5 The energy W_{min} required to remove one of the least tightly bound electrons from a material is known as the *work function*. The work function of a material is 4 eV. Find the longest wavelength which can produce photoelectric emission from this material and the maximum kinetic energy of photoelectrons emitted when the material is irradiated with photons of 6 eV energy.

Ans. 310 nm; 2 eV

6 The work function W_{min} of a material is 3 eV. Find the longest wavelength which can produce photoelectric emission from this material and the maximum kinetic energy of photoelectrons emitted when the material is irradiated with photons of 5 eV energy.

7 Electrons with a maximum energy of 2.4 eV are emitted from a surface radiated with light of $\lambda = 200$ nm. Find the energy of the incident photons and W_{min}, the smallest energy which can free an electron from the surface.

Ans. 9.93×10^{-19} J; 3.8 eV

8 On a clear day, about 1,400 W of radiant energy from the sun strikes an area of 1 m² if that area is perpendicular to the direction of the sun. If this energy were all in the form of photons of green light of wavelength 5000 Å, how many photons would arrive per square centimeter in 1 sec?

9 Find the maximum kinetic energy in electron volts of the photoelectrons ejected from a potassium photoemitter for wavelengths of 200, 300, 400, and 500 nm if $W_{min} = 2.1$ eV. What is the longest wavelength which will eject photoelectrons? *Ans.* 4.1; 2.0; 1.0; 0.4; 5910 Å

10 A television tube operates at a potential difference of 9,500 V. What is the short-wavelength limit of the x-rays emitted?

11 The shortest wavelength emitted from an x-ray

tube arises when all the kinetic energy of the incident electron is converted into radiant energy. Find the shortest wavelength produced in a tube across which a potential difference of 60,000 V is maintained. *Ans.* 0.207 Å

12 A Coolidge x-ray tube is operating at 150,000 V. With what velocity do the electrons strike the anode? (Relativity correction for the change of mass of electrons with speed must be made.) What is the short-wavelength limit of the x-rays produced?

13 Calculate the short-wavelength limit for the radiation from (*a*) an x-ray tube operated at 20 kV, (*b*) a 30-MeV betatron, (*c*) a 300-MeV synchrotron, and (*d*) a 50-GeV linear accelerator.
Ans. (*a*) 0.62 Å; (*b*) 41.3 fm; (*c*) 4.13 fm; (*d*) 2.48×10^{-17} m

14 X-rays of wavelength 0.092 Å are scattered by carbon. At what angle will the Compton scattered photons have a wavelength of 0.099 Å?

15 Photons of wavelength 0.050 Å undergo Compton scattering by the electrons of some light element. What is the wavelength of a photon Compton scattered at 90° with the incident beam? What is the energy in electron volts of the recoil electrons from these photons?
Ans. 0.0743 Å; 81 keV

16 What is the longest wavelength which a photon can have and still transfer one-half its initial energy to a Compton recoil electron?

17 A 0.511 MeV photon strikes an electron initially at rest head on and is scattered straight backward. Find the kinetic energy of the recoil electron. (*Hint:* 0.511 MeV $= m_0c^2 = h\nu_0 = hc/\lambda_0$.) *Ans.* 0.341 MeV $= \frac{2}{3}m_0c^2$

18 A high-energy photon interacts with a heavy nucleus to produce an electron-positron pair. What was the energy of the photon if $K^+ = 300$ keV and $K^- = 250$ keV, where K^+ stands for the kinetic energy of the positron and K^- the kinetic energy of the electron?

19 A high-energy photon interacts with a lead nucleus to create a positron with a kinetic energy of 0.45 MeV and an electron with a kinetic energy of 0.36 MeV. Find the energy of the incident photon. *Ans.* 1.83 MeV

20 Prove that if a photon is scattered through an angle greater than 60° by a free electron, it cannot produce an electron-positron pair after the scattering event, regardless of the initial energy of the photon.

21 A photon of frequency ν_0 makes a head-on collision with a free electron. Show that the frequency $\nu_{180°}$ of the rebounding photon is given by $\nu_0/(1 + 2h\nu_0/mc^2)$ by applying conservation of momentum and conservation of energy directly to the collision in one dimension.

22 Show that a free electron at rest cannot absorb a photon unless some other particle (such as a nucleus) is involved in the process. (*Hint:* Apply conservation of energy and conservation of momentum.)

23 Show that conservation of energy and conservation of momentum require that pair production occur only when a photon interacts with some other particle; i.e., the photon cannot spontaneously transform to a pair without interacting with some particle.

TO DEVELOP even the simplest atomic model which could predict observed spectral lines required the synthesis of many of the ideas treated in these chapters on modern physics. Needed was the electron, the nucleus, and the concept of the photon—and, indeed, the abandonment of the classical idea that an accelerated charge must emit electromagnetic radiation. With these building blocks Bohr was able to propose a model of the simplest of all atoms, that of hydrogen. In this chapter we consider the optical spectra of hydrogen, Rutherford's proposal of the nuclear atom, and the Bohr model.

CHAPTER FORTY-FIVE
THE HYDROGEN ATOM

45.1 THE BALMER SERIES

Each chemical element emits its own unique spectrum when its electrons are excited in electric discharges or in flames. As we saw in Sec. 28.11, the bright lines characteristic of any particular element offer a means of chemical identification of great simplicity and reliability. It was natural that scientists would endeavor to understand how these lines originate and what determines the particular array of lines associated with a given element.

For an atom which has a spectrum of the complexity of the iron spectrum (Fig. 28.15), the task of explaining quantitatively the origin of all the lines would be forbidding. On the other hand, the visible spectrum of hydrogen (Fig. 45.1) has a challenging regularity and apparent simplicity befitting the simplest and lightest of all atoms. This spectrum consists of an uncountable number of lines which become closer and closer together as the wavelength decreases, until finally at a certain minimum wavelength called the *series limit*, the lines cease. Many physicists had tried in vain to find a magic formula for this series before Balmer in 1885 showed that the wavelengths of these lines are predicted accurately by

$$\lambda = 3.6456 \times 10^{-7} \frac{m^2}{m^2 - 4} \qquad \lambda \text{ in meters}$$

where m is an integer taking the values 3, 4, 5, ... Thereafter this group of lines has been known as the *Balmer series*.

The Balmer formula is successful in giving us the wavelengths of the lines, but what is the physics which underlies this purely empirical relation? It was not until 1913 that this question could be answered in terms of an atom model which we now introduce.

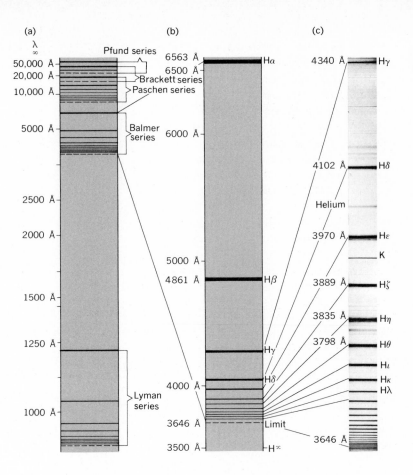

Figure 45.1
(a) The spectrum of atomic hydrogen consists of the Lyman series in the ultraviolet, the Balmer series in the visible region, and several series in the infrared region. (b) The Balmer series in greater detail. (c) A portion of the spectrum of the star Zeta Tauri showing more than 20 lines of the Balmer series. [*Photographed by R. H. Curtiss, University of Michigan.*]

45.2 THE RUTHERFORD MODEL OF THE ATOM

To the early proponents of the atomic theory of matter, atoms were the fundamental building blocks of nature, each type of atom being indivisible and devoid of structure. However, evidence from the discharge of electricity through gases, from the photoelectric effect, and from therm-ionic emission suggests that atoms are not indivisible but rather that the electron is one of the building blocks from which atoms are assembled. The fact that an atom is electrically neutral requires that it have equal amounts of positive and negative charge. Toward the end of the nineteenth century Thomson suggested that the positive electricity is distributed uniformly over a sphere which has the same radius as the atom (about 10^{-10} m). Inside this sphere he imagined a number of electrons which would correspond roughly to plums in a pudding. Photons, collisions with other atoms, and thermal collisions could eject electrons from such an atom. However, this model was not successful in explaining the origin of spectral lines; further, it soon became evident that atoms must be constructed very differently from Thomson's model.

In 1911 Rutherford fired alpha particles (He^{++} ions) from radioac-tive materials through thin metal foils and observed how they were

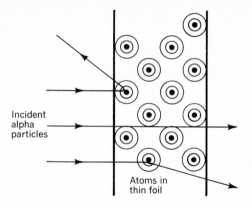

Incident
alpha
particles

Atoms in
thin foil

Figure 45.2
Alpha-particle scattering; very large
repulsive forces are required to scatter
particles backward.

deflected (Fig. 45.2). From the Thomson model one would expect only
small deflections, because the positive and negative electricity in the
atom, distributed over a relatively large volume, should exert no great
forces to deflect the alpha particles. Rutherford's experiments showed
that a surprisingly large number of alpha particles were scattered backward by the foil. The alpha particles which went through were given
much larger changes in direction than could be explained on the basis
of the Thomson model. Rutherford showed that such significant scattering at large angles was to be expected if all the positive electricity of
the atom were concentrated at almost a point. He calculated how the
number of alpha particles scattered should depend on angle for various
kinds of foils and for alpha particles of different energies. His predictions
were in excellent agreement with measurements made by Geiger and
Marsden.

In the Rutherford model all the positive electricity is concentrated
in a very small volume called the *nucleus*. Surrounding this core are
the electrons, held to the nucleus by electrostatic attraction. The radius
of even the largest nucleus is only about 10^{-14} m, while the radius of
a typical atom is nearer to 10^{-10} m. Thus all the positive charge and
almost all the mass of the atom are concentrated in about 10^{-12} the
volume of the atom.

45.3 THE BOHR ATOM

In 1913 Niels Bohr made a great forward stride in explaining the structure
of atoms. Bohr imagined that a hydrogen atom was composed of a
nucleus called a *proton* and a single electron which revolves about the
proton in a circular orbit (Fig. 45.3). The centripetal force required to
keep the electron in this orbit is provided by the coulomb electrostatic
attraction between the proton and electron:

$$\frac{mv^2}{r} = \frac{1}{4\pi\epsilon_0} \frac{e^2}{r^2} \tag{45.1}$$

where m, v, and e are the mass, velocity, and charge of the electron
and r is the radius of the orbit. Bohr further postulated that the only

Orbit of electron

+ Proton

Electron spinning
about axis

values which the angular momentum of the electron could have were
integral multiples of $h/2\pi$, where h is Planck's constant.

$$mvr = n\frac{h}{2\pi} \tag{45.2}$$

where n is an integer. This condition Bohr adopted from the radiation
laws which Planck had developed several years earlier.

Solving Eqs. (45.1) and (45.2) simultaneously for r yields

$$r = 4\pi\epsilon_0 \frac{h^2 n^2}{4\pi^2 m e^2} = \frac{\epsilon_0 h^2}{\pi m e^2} n^2 \tag{45.3}$$

The smallest orbit in which the electron can exist is that given
by $n = 1$. If $n = 1$ with the other constants given their proper values,
$r_1 = 5.29 \times 10^{-11}$ m. This agrees well with measured values of the
radius of the hydrogen atom. The next smallest allowed radius is four
times as great, the third possible radius is $9r_1$, and so forth. Normally,
the hydrogen electron is in the orbit corresponding to $n = 1$, but on
occasions it may be "excited" to one of the larger orbits.

The total energy of the hydrogen electron is the sum of the kinetic
and potential energies. The kinetic energy $mv^2/2$ is, by Eq. (45.1),

$$\text{Kinetic energy} = \tfrac{1}{2}mv^2 = \frac{e^2}{8\pi\epsilon_0 r}$$

while the potential energy, assumed to be zero when proton and electron
are an infinite distance apart, is given by Eq. (31.1):

$$\text{Potential energy} = Ve = -\frac{e^2}{4\pi\epsilon_0 r}$$

from which

$$\text{Total energy} = \text{kinetic} + \text{potential energy} = -\frac{e^2}{8\pi\epsilon_0 r}$$

$$= -\left(\frac{1}{4\pi\epsilon_0}\right)^2 \frac{2\pi^2 m e^4}{h^2 n^2} = -\frac{m e^4}{8\epsilon_0^2 h^2 n^2} \tag{45.4}$$

The minus sign means the electron is bound to the nucleus and cannot
escape unless it is given enough additional energy.

The hydrogen electron has its lowest possible energy when it is
in the orbit characterized by $n = 1$. If the electron is in any other orbit,
it may be expected to jump to an orbit of lower energy (Fig. 45.4).
When it makes such a jump, a photon of radiation is emitted. The energy

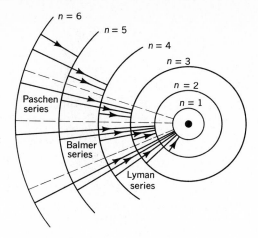

Figure 45.4
Bohr orbits and transitions for the
excited hydrogen atom (not to scale).

of the emitted photon is $h\nu$, where ν is the frequency. It is equal to the energy lost by the electron as it goes from a level A characterized by the integer n_A to another level B characterized by n_B.

$$\text{Energy radiated} = h\nu = \frac{me^4}{8\epsilon_0^2 h^2}\left(\frac{1}{n_B^2} - \frac{1}{n_A^2}\right) \qquad (45.5)$$

Combining Eq. (45.5) with the relation $c = \nu\lambda$, we obtain

$$\frac{1}{\lambda} = \frac{\nu}{c} = \frac{me^4}{8\epsilon_0^2 h^3 c}\left(\frac{1}{n_B^2} - \frac{1}{n_A^2}\right) = R\left(\frac{1}{n_B^2} - \frac{1}{n_A^2}\right) \qquad (45.6)$$

where R is called the Rydberg constant. It has the value 1.09737×10^7 m^{-1}.

45.4 ENERGY LEVELS OF HYDROGEN

According to Bohr's theory, an electron bound to a proton to form a hydrogen atom can exist only in certain orbits characterized by a total energy given by Eq. (45.4). Modern quantum mechanics does not support the assumption of well-defined electron orbits, but it does predict that the electron must be in one of the energy states (or levels) given by Eq. (45.4). In an electric discharge through hydrogen, electrons are torn free from hydrogen atoms. Eventually the electrons and protons recombine; in this recombination the electron may be captured in one of the many possible energy levels (Fig. 45.5). Atoms with electrons in one of the higher states are said to be *excited*. The electron of an excited hydrogen atom normally makes successive transitions to lower states until it finally reaches the stable level corresponding to $n = 1$. When it jumps from one state to another, radiation is emitted.

The Balmer series arises when electrons from higher levels jump to the level with $n = 2$, as is shown schematically in Figs. 45.4 and 45.5. According to Bohr's formula [Eq. (45.6)], the wavelengths of the series of lines emitted when electrons jump from higher states to that charac-

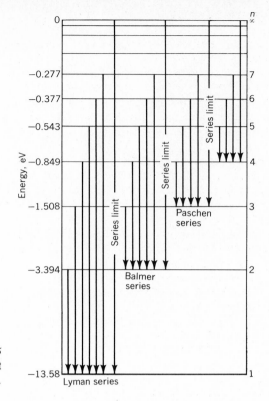

Figure 45.5
Energy-level diagram for hydrogen (not to scale).

terized by $n_B = 2$ are given by

$$\frac{1}{\lambda} = R\left(\frac{1}{4} - \frac{1}{n_A{}^2}\right)$$

or

$$\lambda = \frac{1}{R}\frac{4n_A{}^2}{n_A{}^2 - 4} = 3.65 \times 10^{-7}\frac{n_A{}^2}{n_A{}^2 - 4}$$

This agrees with Balmer's formula if we let $n_A = m$.

We expect electrons eventually to make transitions to the state with $n = 1$, since this is the normal state for the electron. For such a transition Eq. (45.5) shows that the emitted frequencies should be in the ultraviolet region. Lyman studied the ultraviolet spectrum of hydrogen in 1906 and found, just as Bohr theory predicts, a series of lines corresponding to electron jumps from excited levels to the level $n = 1$. This series we call the *Lyman* series in honor of its discoverer.

What should happen if electrons in higher excited states make transitions to levels $n = 3$? Equation (45.5) predicts that the frequencies would be in the infrared region of the spectrum. In 1908 this series was discovered by Paschen. Later another series in the far infrared was discovered by Brackett. The wavelengths of the lines in these series are accurately given by Eq. (45.6) when $n_B = 3$ for the Paschen series and $n_B = 4$ for the Brackett series, while n_A takes on all larger integral values.

Consider an ionized helium atom. Here we have a nucleus with a charge of two fundamental units and a single electron. If we apply the Bohr theory to this system, the relations developed for hydrogen are applicable, provided we use $2e$ for the nuclear charge. The energies become four times as large in magnitude, and *the predicted frequency of the line emitted for any given transition is four times that for hydrogen.* Experiments reveal that this is indeed the case, except for very small shifts which can be explained.

Neutral lithium has three electrons. If two are removed, the remaining *doubly ionized lithium* is found to have a spectrum like that of hydrogen except that all frequencies are greater by a factor of 9. Similarly *triply ionized beryllium* and *quadruply ionized boron* have spectral series like those of hydrogen with frequencies respectively 16 and 25 times greater. Spectral lines for more than a dozen one-electron systems have been observed and for nuclear charge Ze the frequency is Z^2 times that of the corresponding hydrogen line. In general, if we have a nucleus with positive charge Ze and a single electron, the force between electron and nucleus is $Ze^2/4\pi\epsilon_0 r^2$, and the radius of an allowed orbit is (by the method of Sec. 45.3)

$$r = \frac{\epsilon_0 h^2}{\pi m e^2}\frac{n^2}{Z} \qquad (45.3a)$$

while the energy is

$$\text{Total energy} = -\frac{m e^4}{8\epsilon_0^2 h^2}\frac{Z^2}{n^2} \qquad (45.7)$$

Thus the radius is $1/Z$ times and the energy Z^2 times the corresponding value for hydrogen.

45.6   CHARACTERISTIC X-RAYS AND MOSELEY'S LAW

An early triumph of Bohr's theory was its successful application in explaining the wavelengths of certain characteristic x-ray lines. As we saw in Sec. 43.4, W. H. Bragg found that the x-ray spectrum emitted by targets consists of continuous radiation upon which lines characteristic of the target element are superimposed. The *characteristic lines* of each element are numerous, and their wavelengths depend on the atomic number of the target element in a fairly regular way.

Shortly after Bragg reported his work, the English physicist Moseley measured the wavelengths of characteristic lines from a number of elements (Fig. 45.6) and found that the strongest line observed for each of these elements had a frequency given by the empirical relation

$$\nu = 2.52 \times 10^{15}(Z - 1.13)^2 \qquad \text{Hz} \qquad (45.8)$$

This strong line observed by Moseley is called $K\alpha$, and we now discuss its origin in terms of the Bohr model.

Suppose that a single electron circling a nucleus of charge Ze in the orbit characterized by $n_A = 2$ were to make a transition to the $n_B = 1$ orbit. Bohr theory predicts for such a transition that a photon

Atomic
numbers

Figure 45.6

Drawing of Moseley photographs of K x-ray spectra. Brass is an alloy of copper and zinc ($Z = 30$).

Figure 45.7

Moseley plots of the square roots of
the frequencies (arbitrary units) of K
lines as functions of the atomic number
of the emitting element.

of energy W should be emitted, where, by Eqs. (45.5) and (45.7),

$$W = h\nu = \frac{me^4 Z^2}{8\epsilon_0{}^2 h^2}\left(\frac{1}{n_B{}^2} - \frac{1}{n_A{}^2}\right) = \frac{me^4 Z^2}{8\epsilon_0{}^2 h^2}\left(\frac{1}{1^2} - \frac{1}{2^2}\right) \quad (45.9)$$

If this equation is solved for ν with the known values of h, m, and e
inserted, we obtain

$$\nu = 2.46 \times 10^{15} Z^2 \qquad \text{Hz} \qquad\qquad (45.9a)$$

Subsequent work has shown that this line is indeed associated with the
transition of an electron from a level with $n = 2$ to a level with $n = 1$.
A plot of the square root of the frequency as a function of atomic number
for this line is shown in Fig. 45.7.

Moseley also observed a line called $K\beta$ associated with the transition $n_A = 3$ to $n_B = 1$. This line has a frequency which corresponds
well with what one would expect if one placed $n_A = 3$ in Eq. (45.9).
The square root of the frequency of $K\beta$ as a function of atomic number
is also shown in Fig. 45.7.

Both the slight difference between the constants of Eqs. (45.8)
and (45.9a) and the correction to the atomic number in Eq. (45.8) can
be explained quantitatively as well as qualitatively in terms of effects
due to other electrons. A state characterized by $n = 1$ is called a K
state and the associated electron a K electron. A K electron moves
in a force field which is not very different from that of the nuclear charge
Ze alone. The energy, however, is affected by the second K electron
in particular and, to a smaller extent, by the other electrons. The net
effect of all the other electrons is to make the effective nuclear charge
appropriate to Eq. (45.9) about $Z - 1$ rather than Z.

Characteristic x-ray lines involving an electronic transition from
a state of higher n to the $n = 1$ level are called K-*series* lines. Other
families of x-ray lines have also been observed. Those corresponding
to transitions from higher n to the $n = 2$ level are called L lines (Fig.
45.8), and electrons with $n = 2$ are known as L electrons. Lines corresponding to transitions to levels characterized by $n = 3$ are called M
lines, etc. Figure 45.9 shows the wavelengths at which the principal
K, L, and M lines are found for various elements.

The x-ray spectra of elements permit us to make an unambiguous
assignment of atomic number to every element. At one time there was
uncertainty whether potassium or argon had the higher atomic number.

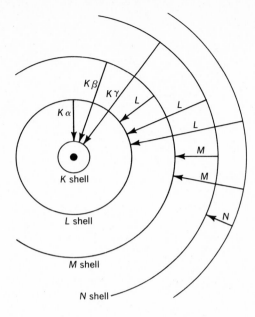

Figure 45.8
Transitions involved in the emission of characteristic x-ray lines in terms of the Bohr model, in which the fine structure of the shells is ignored.

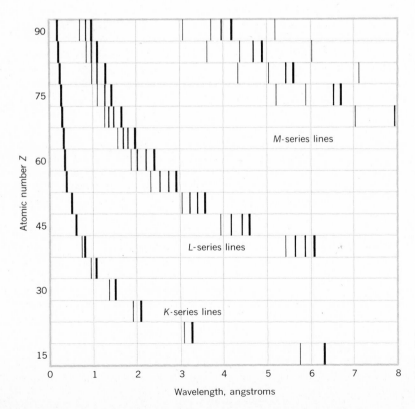

Figure 45.9
Wavelengths of some of the more prominent characteristic x-ray lines of the elements; only a small fraction of the L and M lines is shown.

Argon has a greater mass, but the chemical properties suggest that potassium has the higher atomic number. X-rays have shown clearly that argon has atomic number 18 while potassium has 19. X-rays resolved a similar paradox in connection with cobalt and nickel.

45.7 ELECTRON SPIN

The $K\alpha$ x-ray lines observed by Moseley actually show up as double lines when examined under higher resolution. In 1925 Uhlenbeck and Goudsmit showed that these and many other features of atomic spectra could be explained by assuming that an electron not only revolves around a nucleus, but has a property, called *spin*, which for our purposes may be visualized as a rotation about its own axis (see Fig. 45.3). A determination of the spin angular momentum about any axis always results in one of two values: either $+h/4\pi$ or $-h/4\pi$, where h is Planck's constant. For a single electron traversing an orbit, the total angular momentum is the vector resultant of the orbital angular momentum and the spin angular momentum. Essentially the spin either adds to or subtracts from the orbital angular momentum an amount $h/4\pi$. The energy is slightly different in the two cases for the following reason: The spinning charge produces a magnetic moment, so the spinning electron resembles a tiny magnet. From the point of view of the electron, the nucleus revolving about the electron produces a magnetic field; the energy of the electron-magnet depends on whether its magnetic moment is aligned with this field or opposed to it.

In the optical spectra of sodium and the other alkalies, the energy associated with a given orbit has two values, depending on whether the spin angular momentum of the excited electron is parallel or antiparallel to the orbital angular momentum. Such a splitting of energy levels may give rise to series of double lines. The sodium D lines are examples of such a doublet; they have wavelengths of 5890 and 5896 Å.

QUESTIONS

1 What is a spectral series? What is a series limit?

2 In the Bohr theory what is the significance of the fact that the allowed sharp levels all have negative energy?

3 Is the gravitational attraction between the proton and the electron of a hydrogen atom significant compared with the coulomb attraction? Find the ratio of coulomb to gravitational attraction for an electron in the first Bohr orbit. Does this ratio change from orbit to orbit?

4 Suppose for the moment that all electrons do not have the same charge but that there are three different charges. What effect would this produce in the hydrogen spectrum? Which do you regard as the more convincing evidence that all electrons bear the same charge, the Milliken oil-drop data or the sharpness of lines in the spectra of elements such as hydrogen? What do the spectral results suggest so far as the masses of electrons are concerned?

5 What is the experimental evidence that all the positive charge and most of the mass of an atom are concentrated in a nucleus?

6 What is the maximum number of photons that can be emitted when a hydrogen atom excited to an $n = 5$ level goes to the ground state? What is the minimum number? (Assume there are no nonradiative transitions.)

7 Can a photon with an energy greater than the binding energy of the hydrogen electron (13.6 eV) be absorbed by a hydrogen atom? Explain quantitatively.

PROBLEMS

1 The ionization potential of a hydrogen atom is 13.58 eV. Use this figure to find the energies of the hydrogen levels with $n = 1$, $n = 2$, and $n = 3$ and to determine the wavelength of the lowest frequency line in the visible (Balmer) spectrum of hydrogen.

   Ans. -13.58, -3.40, and -1.51 eV; 656.3 nm

2 Use the Balmer formula to calculate the wavelengths of the first four lines of the Balmer series of hydrogen.

3 Compute the longest and shortest wavelengths in the Lyman series of hydrogen. What are the energies of photons with these energies?

   Ans. 121.6 nm; 91.2 nm; 10.2 eV; 13.6 eV

4 Calculate the wavelengths of the first two lines and of the series limit of the Balmer series for doubly ionized lithium.

5 Calculate the wavelengths of the first two lines and of the series limit for the Lyman series of ionized helium. *Ans.* 30.4, 25.7, and 22.8 nm

6 Find the speed of an electron in the first Bohr orbit. How many revolutions does it make each second? Show that the ratio of the speed of an electron in the first Bohr orbit to the speed of light is $\alpha = e^2/2\epsilon_0 hc = \frac{1}{137}$. This ratio is known as the *fine-structure constant*.

7 The electron of a Bohr model hydrogen atom revolves around the proton with a speed of 2.19×10^6 m/sec at a radius of 5.29×10^{-11} m. To what current does this correspond? Find the magnetic induction at the proton due to the motion of the electron. Compute the magnetic moment due to this orbital motion. (This magnetic moment is known as a *Bohr magneton*.)

   Ans. 0.00105 A; 12.5 T; 9.3×10^{-24} A-m²

8 The orbital electron of a singly ionized helium atom is attracted by the nucleus, which has a positive charge twice that of the electron. Find the radius of the first Bohr orbit of ionized helium and the speed of an electron in this orbit.

9 Calculate the momentum of the photon emitted when a hydrogen atom makes a transition from a state characterized by $n = 2$ to that for which $n = 1$. Find the recoil speed of the hydrogen atom.

   Ans. 5.44×10^{-27} kg-m/sec; 3.26 m/sec

10 Find the energies of the photons and the wavelengths of the first three lines of the Paschen series for hydrogen.

11 Calculate the angular velocity of a solid sphere of radius 5×10^{-15} m and the mass of an electron if its angular momentum were $h/4\pi$. What would be the tangential velocity at the equator of such a sphere?

   Ans. 5.8×10^{24} rad/sec; 2.9×10^{10} m/sec ($\approx 100c$)!

12 (*a*) Show that the frequency of revolution of an electron in a circular Bohr orbit of hydrogen is given by $\nu = me^4/4\epsilon_0^2 n^3 h^3$. (*b*) Show that for large n the frequency of revolution is approximately equal to the frequency emitted according to Eq. (45.5) when the electrons make a transition from state n to state $n - 1$. (This problem is an example of *Bohr's correspondence principle*, which implies that when n is large, quantum physics makes the same predictions as classical physics.)

13 Show that, on the basis of the approximations which led to Eq. (45.9), the slope of the Moseley plot (Fig. 45.7) for $K\beta$ is expected to be $\sqrt{\frac{32}{27}}$ times that for $K\alpha$.

14 Show that if a neutral hydrogen atom at rest is struck by a moving hydrogen atom, the incident atom must have at least 20.4 eV of energy to raise the atom initially at rest to its first excited state. Why does it take 1.5 times the ionization energy just to excite the atom? (*Hint:* Momentum must be conserved.)

CHAPTER FORTY-SIX

WAVE MECHANICS AND ATOMIC STRUCTURE

EFFORTS TO UNDERSTAND the structure of more complex atoms and the properties of solids by extending the Bohr postulates met with limited success. Great advances in these areas required first the recognition that the electron has wave properties as well as particle properties and then the development of a "wave mechanics" capable of predicting the "behavior" of an electron. Although use of this new mechanics involves mathematical skills well beyond those expected here, many of the important ideas can be discussed without carrying through difficult calculations. We now introduce some experimental evidence of wave properties for electrons and explore some of their consequences.

46.1 DE BROGLIE'S HYPOTHESIS

The brilliant success of the Bohr theory in explaining the spectra of hydrogen and the K-series x-ray lines makes it one of the great triumphs of modern physics. In spite of its success, the Bohr theory is not entirely correct. Indeed, one of its key postulates—that the angular momentum mvr of the electron in its orbit is $nh/2\pi$—is not true. The Bohr theory is a magnificent contribution to physics, not because it is correct in detail, but because it opened the way for modern quantum mechanics, in which the electron can no longer be regarded as a tiny particle describing sharp orbits about a nucleus. Instead the electron is found to exhibit wave characteristics which are extremely important in treating its behavior in atoms and in solids.

The first man to suggest that electrons, and indeed other classical particles as well, should have wave properties was Prince Louis Victor de Broglie. In 1924 he proposed that since the electromagnetic "waves" of classical physics possessed particle attributes as well as wave attributes, atomic particles might be expected to exhibit wave behavior under proper conditions. His principal argument was based on the fact that nature reveals many symmetries. One possible symmetry was that both electromagnetic radiation and classical particles have both particle and wave properties. If particles have wave properties, how can one predict the wavelengths? Here de Broglie argued that the Compton effect showed that photons had momentum $h\nu/c$, or simply h/λ. Hence one should assign a wavelength $\lambda = h/(\text{momentum})$ to photons, and perhaps to particles as well. For a particle of mass m and velocity v, the resulting *de Broglie wavelength* is

$$\lambda = \frac{h}{mv} \qquad (46.1)$$

whether the particle is an airplane or an electron. For macroscopic bodies this wavelength is so tiny that we have no practical means of measuring it. However, for electrons and other atomic particles, wavelengths of measurable magnitude are available. Measurements on such particles confirm the de Broglie hypothesis.

Example

What is the de Broglie wavelength of an electron traveling at a speed of 10^7 m/sec? Through what potential difference must an electron fall to achieve this speed?

$$\lambda = \frac{h}{mv} = \frac{6.63 \times 10^{-34} \text{ J-sec}}{9.1 \times 10^{-31} \text{ kg} \times 10^7 \text{ m/sec}}$$

$$= 7 \times 10^{-11} \text{ m} = 0.7 \text{ Å}$$

$$Ve = \tfrac{1}{2}mv^2$$

$$V \times 1.60 \times 10^{-19} \text{ C} = \tfrac{1}{2} \times 9.1 \times 10^{-31} \times 10^{14} \text{ J}$$

$$V = 280 \text{ J/C} = 280 \text{ V} \qquad \text{////}$$

(a)

46.2 WAVE PROPERTIES OF ELECTRONS

The first experimental confirmation of de Broglie's hypothesis was made by Davisson and Germer in 1927. They bombarded a single crystal of nickel with a narrow pencil of low-voltage electrons incident perpendicular to the crystal face and determined the intensity of the "reflected" electrons as a function of the angle at which they bounced off the crystal. There were striking maxima and mimima in this angular distribution which could be quantitatively explained if the electrons had the wavelengths predicted by Eq. (46.1).

(b)

An entirely different experiment by G. P. Thomson also supported the de Broglie hypothesis and emphasized the similarity between electrons and x-rays of the same wavelength. If a pencil of monochromatic x-rays passes through a *powdered* crystalline substance placed at C in Fig. 43.3, the x-rays produce a diffraction pattern on the photographic plate. This pattern consists of a series of concentric circular fringes produced by interference of the scattered x-rays (Fig. 46.1*a*). If a stream of electrons, all of the same velocity, passes through a thin film of metal, a similar set of rings is produced (Fig. 46.1*b*). In both cases one can compute the positions of the observed fringes on the assumption that the incident beams consist of waves of appropriate wavelength—for the electrons, the de Broglie wavelength $\lambda = h/mv$.

The resolving power of an optical microscope is limited by the wavelength of the light used. For two neighboring particles to be separable, light of very short wavelength must be used if the particles are very close together. Unfortunately we do not have lenses which can focus x-rays and thereby permit us to build an x-ray microscope. How-

Figure 46.1
(*a*) X-ray diffraction pattern of aluminum. [*Courtesy of Mrs. M. H. Read, Bell Telephone Laboratories, Inc.*] (*b*) Electron-diffraction pattern of a thin film of cesium iodide. [*Courtesy of L. H. Germer, Bell Telephone Laboratories, Inc.*]

ever, when electrons are accelerated through a potential difference in the neighborhood of 10^5 V, their wavelengths are only a few hundredths of an angstrom. With such wavelengths, particles that are as close as 10^{-9} m can be separated and photographed by an electron microscope. The electrons are controlled and focused by electric and magnetic fields.

Electrons are by no means the only classical "particles" for which wave properties have been observed. Other subatomic particles, and even some molecules, have been found to have wavelengths given by Eq. (46.1). Neutrons, which are neutral particles of mass slightly greater than that of the proton, have been studied by use of Bragg diffraction; slow "thermalized" neutrons have wavelengths in the same range as x-rays, and this property makes possible the use of neutron spectrometers and permits physicists to obtain beams of neutrons all of which have virtually the same speed.

46.3 WAVES MECHANICS AND THE UNCERTAINTY PRINCIPLE

In 1926 Schrödinger discovered his famous wave equation which underlies modern wave mechanics, a great body of theory which takes account of both the wave and particle attributes of electrons. Although the formulation of wave mechanics involves differential equations and relatively advanced mathematical techniques, some of the ideas can be discussed here. When Schrödinger applied his wave equation to the hydrogen problem, he obtained the same values for the energy of the allowed states as Bohr had found [Eq. (45.4)], but his values for the orbital angular momentum did not agree with the values postulated by Bohr. Schrödinger found that for a given n the orbital angular momentum $mvr = lh/2\pi$, where l is an integer, less than n or zero, while Bohr had taken the incorrect larger value $nh/2\pi$. Further, for a hydrogen electron in a given n state, the wave mechanics predicts that the electron, far from being confined to a circular orbit, might be much closer to or much farther from the nucleus than the appropriate Bohr radius, although the latter is a good measure of the average separation. In wave mechanics the concept of sharp orbits is completely abandoned. Modern wave mechanics, or quantum mechanics, is a fascinating topic. It has led to tremendous advances in physics and in chemistry as well as being a great stimulus to modern philosophy.

One of the startling principles of quantum mechanics, advanced by Heisenberg in 1927, is that there is a basic uncertainty in nature which represents an ultimate limit to the precision with which processes can be described. One form of the Heisenberg *uncertainty principle* can be stated as follows.

In any experiment in which we endeavor to measure both the position and the momentum of a particle, there is an inherent indefiniteness in the results such that if Δx is the limit of uncertainty in the x coordinate of the position, and $\Delta(mv_x)$ is the uncertainty in the corresponding component of the momentum, the product of these uncertainties must be at least of the order of magnitude of Planck's constant h divided by 4π.

$$\Delta x\,\Delta(mv_x) \gtrsim \frac{h}{4\pi} \qquad (46.2)$$

A similar indefiniteness accompanies any measurement of an energy. If Δt is the time available for the energy measurement, there is an inherent uncertainty ΔW in the energy such that

$$\Delta W\,\Delta t \gtrsim \frac{h}{4\pi} \qquad (46.3)$$

The constant h is so small that for macroscopic bodies these quantum indefinitenesses do not trouble us. However, when we deal with atomic sizes, the uncertainty principle is extremely important.

The uncertainty principle is a fundamental part of wave mechanics. It tells us that we cannot simultaneously know both the precise position and the exact velocity of a particle such as an electron. As a consequence, an atom model based on sharp orbits for electrons at precisely defined distances from a nucleus can only be a rough approximation. Actually there is a finite possibility of finding the electrons over a wide range of positions. For a hydrogen electron in its lowest energy state, wave mechanics predicts that the most probable distance of the electron from the nucleus is the Bohr radius, but there is a modest probability that the electron is as much as twice that distance away.

46.4 THE PAULI EXCLUSION PRINCIPLE

When the Schrödinger equation is applied to multielectron atoms, the resulting differential equation is awesome and only approximate solutions are available. However, for some atoms the approximations are extremely good. It is found that one can describe complex atoms reasonably well if one thinks of each electron as existing in a hydrogenlike state and moving in the electric field due to the nucleus and to the average contributions of the other electrons. A still more accurate description includes the effects associated with the magnetic fields due to the motions of the electrons and to their intrinsic magnetic moments. In this very rough model we imagine each electron with its own orbital and spin angular momenta moving in a fuzzy average hydrogenlike "orbit."

When we treat complex atoms from the point of view of hydrogenlike "orbits," we find that we may specify the state of each electron in terms of four numbers,† called *quantum numbers*, each of which has a physical significance. These quantum numbers are:

1 The principal quantum number n, which appears in Eq. (45.7) and is a rough measure of the average distance of the electron from the nucleus. The energy of a state depends strongly on n, which takes on integral values 1, 2, 3, 4, ...

2 The azimuthal quantum number l, which is a measure of the orbital angular momentum in units of $h/2\pi$. It takes on integral values $n-1$, $n-2$, . . . , 0. Thus, when $n=3$, l can be 2, 1, or 0.

---

† There are several possible ways of choosing these four numbers; our choice here is based on convenience for describing the inner-electron configurations. A different choice is preferable for discussing the visible spectra of light elements.

3   The quantum number j, which measures the total angular momentum of the electron, spin plus orbital, in units of $h/2\pi$. Its values are limited to the positive half-integers $l - \frac{1}{2}$ and $l + \frac{1}{2}$.

4   The quantum number m_j, which specifies how the total angular momentum is oriented in space relative to some direction prescribed by a magnetic field or some other kind of field. It takes on values $j, j - 1, j - 2, \ldots, -j$.

In 1925 W. Pauli proposed an important rule, known as the *Pauli exclusion principle*, which may be stated for our purposes in the form: *No two electrons in any one atom may have the same four quantum numbers*. Pauli's great generalization has many applications and forms one of the foundation stones for our understanding of atomic structure.

A neutral atom in its lowest energy state has each electron in the lowest energy level accessible to it. We have seen in Sec. 45.3 that for the ground state of hydrogen the single electron is in the state characterized by $n = 1$. For helium with its two electrons the ground-state configuration has both electrons with $n = 1$. This requires that $l = 0$ and $j = \frac{1}{2}$ for each electron, so one electron has $m_j = \frac{1}{2}$ and the other must have $m_j = -\frac{1}{2}$. It is not possible to have more than two electrons in any one atom with $n = 1$, since that would require that two of the three have the same four quantum numbers—a clear violation of the exclusion principle, to which no exceptions are known. Two electrons with $n = 1$ form a closed *shell*, known in x-ray terminology as the *K shell*.

46.5   THE PERIODIC TABLE

The arrangement of atoms in the periodic table (inside back cover) can be explained in terms of the Pauli exclusion principle. In the preceding section we have seen that hydrogen and helium constitute the first period (row) in this chart. The next element is *lithium*, atoms of which have three electrons. The third of these electrons cannot have $n = 1$ but goes into the lowest available state for which $n = 2, l = 0$. After lithium comes beryllium ($Z = 4$) with two electrons with $n = 1, l = 0$ and two with $n = 2, l = 0$. For boron ($Z = 5$) the fifth electron has $n = 2, l = 1$. Altogether there can be six electrons in an atom with $n = 2, l = 1$; two with $j = \frac{1}{2}$ and $m_j = \frac{1}{2}$ or $-\frac{1}{2}$ and four with $j = \frac{3}{2}$ and $m_j = \frac{3}{2}, \frac{1}{2}, -\frac{1}{2},$ or $-\frac{3}{2}$. As we fill the $n = 2, l = 1$ subshell we have the elements boron, carbon ($Z = 6$), nitrogen (7), oxygen (8), fluorine (9) and finally neon (10). At this point the $n = 2$ (or L) shell is full.

Sodium has 11 electrons, of which 10 form a neon core; the eleventh is in the state characterized by $n = 3, l = 0$. As we proceed through the table of the elements (see inside back cover), each successive element has one more electron than the preceding and this electron takes always the lowest energy state which is not yet occupied. After all states with $n = 3, l = 0,$ and $l = 1$ are filled, the next elements are potassium ($Z = 19$) and calcium ($Z = 20$), for which the last electrons have $n = 4, l = 0$, even though levels with $n = 3, l = 2$ remain empty. However, as we proceed up the periodic table, the $n = 3$ shell is quickly filled, and this gives rise to the first long period in the periodic table.

(The period is the principal quantum number n of the outermost electrons.) A detailed explanation of why shells of higher n begin to fill before all the possible levels of lower n are filled is beyond the scope of this text. It is found that (with the exception of a few irregularities) the levels continue to fill in the order: all levels with $(n = 4, l = 1)$, followed by $(n = 5, l = 0)$, $(n = 4, l = 2)$, $(n = 5, l = 1)$, $(n = 6, l = 0)$, $(n = 4, l = 3)$, $(n = 5, l = 2)$, $(n = 6, l = 1)$, $(n = 7, l = 0)$, $(n = 5, l = 3)$. Quantum mechanics, including the Pauli exclusion principle, has given us an understanding of the periodic table, and of atomic structure in general, which has led to great advances in physics and chemistry.

The periodic table had its origins in the work of Mendeleev in Russia and (independently) Meyer in Germany, about 1868 to 1870. The arrangement was based on the observation that various chemical and physical properties recur periodically throughout the sequence of the elements in order of their atomic weights. Elements of similar properties were placed in the same column, with the lighter ones above the heavier. We now note a few correlations between chemical and physical properties and the assignment of numbers associated with the Pauli exclusion principle.

The elements lithium, sodium, potassium, rubidium, cesium, and francium all have a single electron alone in a new shell. They are all electropositive elements with similar chemical properties; they react vigorously with water to release hydrogen. Physically they are all metals with shiny luster and high electric and thermal conductivities. Immediately preceding these *alkali metals* in the periodic table are the *noble gases*, helium, neon, argon, krypton, xenon, and radon, all of which are relatively inert chemically. Except for helium, all have outer electron configurations with eight electrons in their outer shell corresponding to the filling of all levels of this shell with $l = 0$ and $l = 1$. The elements which directly precede the inert gases—fluorine, chlorine, bromine, iodine, and astatine—are called the *halogens*; all are nonmetals, and all are strongly electronegative, forming salts when they combine with the alkali metals. In general, those elements which have similar electron configurations in their outer shells have similar chemical and physical properties.

One physical characteristic of the elements which can be correlated with the way in which electrons are added to fill the various shells is the *ionization potential*, or the energy required to remove one of the least tightly bound electrons from an atom, shown graphically in Fig. 46.2. Note that for potassium, sodium, lithium, etc., the ionization

Figure 46.2
Ionization potentials of the elements.

potential is exceptionally low compared with that of neighboring atoms. This arises from the fact that in each case there is a single electron in a new outer shell. On the other hand, if there are eight electrons in any given outer shell, a peculiarly stable electronic configuration occurs in which the element is chemically inert—the case for the noble gases. Many other properties show a similar periodicity, among them atomic diameter, compressibility, melting point, and coefficient of thermal expansion. A particularly important example of this periodicity is found in the visible spectra of the elements, which arise for any element when one of the outer electrons of an atom is excited from its normal state to an excited one of higher energy. Atoms with similar electron configurations exhibit similar spectra.

46.6 LOW-LYING ATOMIC ENERGY LEVELS

In the preceding section we have seen how the Pauli exclusion principle can be used to give us insight into the periodic table and the chemical and physical properties associated with the outer electrons configurations of atoms. But can we find any evidence that the four quantum numbers n, l, j, and m_j are useful for describing the inner structure of atoms? Actually the characteristic x-ray spectra (briefly treated in Sec. 45.6) are emitted when an inner electron makes a transition to a lower energy state made vacant by knocking an electron out of this state. A detailed explanation of the many strong characteristic x-ray lines can be made by appropriate use of the quantum numbers n, l, and j on which the energy depends; in the absence of an external field the energy does not depend on m_j. Simpler and more direct evidence comes from the photoelectric absorption of x-rays.

With the aid of an x-ray spectrometer and the continuous spectrum from an x-ray tube, a monoergic beam of x-ray photons can be produced. If a thin foil of some element is placed in the beam, the intensity is reduced from I_0 to I. If t is the thickness of the foil, the absorption coefficient is given by $[\ln (I_0/I)]/t$. Figure 46.3 is a plot of absorption coefficient for a lead foil as a function of wavelength. At 0.14 Å there is a sharp drop, called the K absorption edge, due to the fact that photons of wavelength greater than 0.14 Å do not have enough energy to eject one of the K electrons. A K electron is characterized by the quantum numbers $n = 1$, $l = 0$, $j = \frac{1}{2}$.

As the wavelength is increased, the absorption increases, until at 0.78 Å the photons no longer have enough energy to eject electrons with $n = 2$, $l = 0$, $j = \frac{1}{2}$, known in x-ray terminology as L_I electrons. At 0.81 Å comes the L_II absorption discontinuity associated with two electrons having $n = 2$, $l = 1$, $j = \frac{1}{2}$ ($m_j = \pm\frac{1}{2}$) while the L_III edge at 0.95 Å is due to four electrons with $n = 2$, $l = 1$, $j = \frac{3}{2}$ ($m_j = \frac{3}{2}, \frac{1}{2}, -\frac{1}{2}, -\frac{3}{2}$). At longer wavelengths the absorption increases steadily until one comes to five edges between 3 and 5 Å associated with the $n = 3$ (M) electrons. Thus in a sufficiently heavy atom one can find an absorption discontinuity associated with each possible value of n, l, and j and *no others*. Table 46.1 lists the energy required to eject an electron from each of the states characterized by all possible values of l and j for $n = 1$, 2, and 3.

Figure 46.3
The photoelectric absorption coefficient
of lead as a function of x-ray
wavelength.

Further confirmation of these low-lying energy states comes from measuring the kinetic energies of photoelectrons ejected from a thin film of some element. If the film is bombarded with photons of energy $h\nu$, the kinetic energy (KE) of a photoelectron ejected from the K shell is given by

$$KE = h\nu - W_K \qquad (46.4)$$

where W_K is the binding energy of the electron in the K shell. If the electron comes from the L_{III} shell, its kinetic energy is $h\nu - W_{L_{III}}$ and

Table 46.1
Energies Required to Eject an Electron from X-ray Levels of Seven Elements, in keV*

n	l	j	X-ray	Al ($Z = 13$)	Cu ($Z = 29$)	Mo ($Z = 42$)	Ag ($Z = 47$)	W ($Z = 74$)	Pb ($Z = 82$)	U ($Z = 92$)
1	0	$\frac{1}{2}$	K	1.562	8.996	20.036	25.556	69.637	88.163	115.80
2	0	$\frac{1}{2}$	L_I	0.1154	1.104	2.872	3.811	12.115	15.892	21.795
2	1	$\frac{1}{2}$	L_{II}	0.00730	0.955	2.633	3.529	11.559	15.231	20.974
2	1	$\frac{3}{2}$	L_{III}	0.00727	0.935	2.528	3.356	10.219	13.061	17.193
3	0	$\frac{1}{2}$	M_I		0.122	0.509	0.718	2.821	3.860	5.556
3	1	$\frac{1}{2}$	M_{II}		0.078	0.413	0.603	2.574	3.566	5.187
3	1	$\frac{3}{2}$	M_{III}		0.076	0.396	0.572	2.279	3.076	4.306
3	2	$\frac{3}{2}$	M_{IV}	0.003	0.234	0.373	1.870	2.591	3.725
3	2	$\frac{5}{2}$	M_V		0.231	0.367	1.807	2.482	3.556

* λ in angstrom units is equal to $12.398/(E$ in keV).

so forth. Thus, a knowledge of the energy $h\nu$ of the incident photon and the kinetic energy of an ejected photoelectron permits the determination of the energy required to remove the electron from its bound state.

Example

X-ray photons of 20.000 keV energy fall on a lead foil and eject photoelectrons, one group of which has a kinetic energy of 6.939 keV. (*a*) What is the binding energy associated with the level from which these electrons are ejected? (*b*) What kinetic energy would a photoelectron from the L_I level have?

a
$$\text{KE} = h\nu - W$$
$$6.939 \text{ keV} = 20.000 \text{ keV} - W$$
$$W = 13.061 \text{ keV}$$

which corresponds to the L_{III} level.

b
$$\text{KE} = h\nu - W_{L_I} \quad \text{(see Table 46.1)}$$
$$= 20.000 \text{ keV} - 15.892 \text{ keV}$$
$$= 4.108 \text{ keV} \qquad \text{////}$$

The characteristic x-ray lines of an element arise from electron transitions from one of the lower levels to a vacancy in a level still lower. From the energy of these photons and the known energy levels of the atom one can determine the initial and final states of the electron making the transition. Data from absorption edges, the kinetic energies of photoelectrons, and the wavelengths of characteristic spectral lines are all mutually compatible with each other and with the requirements of the Pauli exclusion principle.

46.7 THE ZEEMAN AND STARK EFFECTS

The energies of the low-lying energy states are not affected by the application of ordinary electric and magnetic fields, but this is not the case with the energy states of the outer electrons. For example, when a source of light is placed between the poles of a powerful electromagnet, a single spectral line may break up into several components (Fig. 46.4). This separation of a spectral line into components by the action of a magnetic field is known as the *Zeeman effect*. The number of components depends on the particular spectral line and is not the same when the light is viewed in the direction of the magnetic field as when the light is viewed at right angles to this direction. Because of the interaction between the external applied magnetic field and the magnetic fields due to the electrons in the atom, the energy levels in the atom are split up into a number of additional levels. Transitions between these new levels give rise to additional spectral lines. Studies of the Zeeman effect have proved invaluable in resolving problems in spectroscopy and in the development of a detailed understanding of many facets of atomic structure.

Figure 46.4
A magnetic field splits a single spectral line (above) into several components (below). These lines were photographed after the light had passed through a polarizer, which reduced the relative intensities of the outer lines. [*Courtesy of G. H. Dieke.*]

An electric field can also split spectral lines. In a sufficiently large electric field, a single spectral line is replaced by a number of components. The greater the electric field, the greater the separation of these components. This separation of spectral line into components by an electric field is known as the *Stark effect* (Fig. 46.5) after its discoverer. The splitting of the spectral line in the figure varies from zero at the top of the figure, where there is no electric field, to its greatest value at the bottom, where the electric field is largest. Data on the amount of separation for different spectral lines give valuable information about the atoms emitting the lines.

Figure 46.5
Stark effect for a line of the helium spectrum, showing the splitting due to an electric field. [*After Foster.*]

46.8 GASEOUS MOLECULES

That atoms attract one another is clear from the fact that (1) two or more atoms may join together to form a stable gaseous molecule and (2) larger aggregates of atoms are bound together to form liquids and solids. The force between a pair of atoms depends on the outer electron arrangements, and it becomes stronger as their electronic charge clouds begin to overlap. Much progress has been made in understanding interatomic forces since the advent of wave mechanics. We must limit our discussion to molecules consisting of only two atoms. What forces hold simple diatomic molecules together? There is no single answer to this question; there are several kinds of binding, of which we consider two.

Ionic Binding Some molecules are composed of a positive ion bound to a negative ion by coulomb electrostatic attraction. An example is potassium chloride (KCl), in which a potassium atom surrenders its outer (or valence) electron to a chlorine atom. The K^+ ion and the Cl^- ion are held together by *ionic* binding.

Covalent Binding Many gas molecules are composed of two identical atoms; examples are the hydrogen, oxygen, nitrogen, and chlorine molecules. Here we can scarcely invoke the concept of ionic binding, since there is no reason for one of two identical atoms to become positive and the other negative. The search for an alternative mechanism ended with modern wave mechanics. The development of the theory is beyond the scope of this book, but in a qualitative way it is possible to discuss some of the results. In a hydrogen molecule (H_2) the binding comes about from the sharing of two electrons by two hydrogen atoms. This gives rise to an attractive force associated with the electron *exchange*. The occurrence of exchange forces is a strictly wave-mechanical phenomenon for which there is no classical analog. The pair of electrons involved in the exchange must have oppositely directed spins (Sec. 45.7). If the spins of the two electrons are aligned, the atoms repel each other and no molecule is formed. In a similar fashion the binding of oxygen and nitrogen diatomic molecules rises from the sharing of pairs of electrons between the atoms. Binding arising from this type of electron sharing is called *covalent* or *homopolar*.

The extremes of ionic and cavalent binding are reasonably well illustrated by potassium chloride and hydrogen molecules, respectively.

Figure 46.6
Energy of a KCl molecule as it varies with the separation of the K⁺ and Cl⁻ ions.

However, there are many molecules for which the binding is intermediate between covalent and ionic binding. In these molecules the shared electrons are closer to one of the atoms much more of the time than to the other atom; examples are carbon monoxide and water molecules.

An important characteristic of any diatomic molecule is its *dissociation energy*, the energy required to break the molecule into its parts. This, in turn, is equal to the binding energy, the energy released when the two atoms are brought together to form the molecule. Typical binding energies of molecules are of the order of a few electron volts: for potassium chloride 4.4 eV, for sodium chloride 4.24 eV, and for hydrogen 4.48 eV.

For any pair of atoms there exists an equilibrium interatomic distance at which the energy is minimum (Fig. 46.6). If the two atoms are slightly farther apart, a net restoring force draws them closer. On the other hand, if the atoms approach more closely than the equilibrium distance R_0, strong forces drive them apart.

If the only forces between two atoms were attractive, the atoms would eventually coalesce. Actually, as atoms approach one another, strong repulsive forces arise. Classically, one imagines the atoms to be solid spheres which cannot penetrate one another. However, the atom of modern physics has a tiny nucleus with an electron cloud about it. This picture is used to explain the fact that the atoms occupy a well-defined volume in terms of the Pauli exclusion principle. If we try to bring a second atom so close that its electrons begin to penetrate the occupied shells of the first atom, the Pauli exclusion principle requires that some of the electrons go to higher energy states. This requires energy and gives rise to a repulsive force.

46.9 MOLECULAR SPECTRA

In Chap. 45 we considered the spectra emitted by electron transition in individual atoms. Molecules, composed of two or more atoms, also have spectra. Typically, molecular spectra are considerably more complex than atomic spectra, with radiation emitted in three spectral ranges corresponding to different kinds of transitions between molecular energy states. Starting with the longest-wavelength region, we explain these spectra as follows.

Rotational Spectra A simple diatomic molecule can be visualized as a dumbbell (Sec. 15.12). If such a molecule is put into rotation, classically we would expect it to emit radiation of the same frequency as the rotation. However, quantum mechanics limits the rotational energy levels (Fig. 46.7) to those given by the relation

$$E_J = \frac{h^2}{8\pi^2 I} J(J + 1) \qquad (46.5)$$

where I is the moment of inertia of the molecule about its center of mass, h is Planck's constant, and J is the rotational quantum number, which takes on values 0, 1, 2, 3, ... Radiative transitions occur only when J increases or decreases by 1 ($\Delta J = \pm 1$). The rotational energy levels are spaced in such a way that the rotational lines consist of a fundamental frequency and integral multiples of that frequency, all in the far infrared region.

Vibration-Rotation Spectra A diatomic molecule is less analogous to a rigid dumbbell than it is to two masses connected by a strong spring. Such masses can perform vibrations relative to one another under the elastic forces which bind them. The frequencies associated with such vibrations are typically much higher than those associated with pure rotation. Consequently, the vibrational spectra appear in the nearer infrared region. Any time its vibrational spectrum is excited, a molecule has much more than enough energy for the rotational spectra; thus we ordinarily observe rotational motion superimposed on the vibrational.

Electronic Spectra At still higher frequencies we find evidence of transitions of electrons, associated now with the molecule, between allowed quantum states. Since the energies involved are much greater than those associated with either vibration or rotation, electron excitation is generally accompanied by vibrational and rotational excitation. For any single electron state there are many vibrational and rotational states in which the molecule may exist. Consequently, each type of transition between electron energy states produces a large number of closely spaced lines. Such spectra, called *molecular bands* (Fig. 46.8), are often found in the visible and ultraviolet regions.

46.10 LASERS AND MASERS

As we have seen, both atoms and molecules have sharply defined allowed energy states. When one of them is in an excited state, it undergoes a series of transitions which bring it eventually to its lowest, or *ground*, state. Many of these transitions, but not necessarily all, involve the emission of a photon. Typically, an isolated excited atom or molecule emits a photon spontaneously and quite independently of its neighbors. The resulting radiation from many such emitters is *incoherent*—a disorganized jumble of waves for which the directions of emission and the phase relationship between the photons are entirely random, even though the frequency may be the same. However, if a photon of a given frequency interacts with an excited atom about to emit this same frequency, the atom may be *stimulated* to radiate energy in the same phase and in the same direction as that of the incident photon. The stimulated radiation from many excited atoms may be *coherent*, with the waves in phase and traveling in the same direction.

Figure 46.7
Lowest rotational levels of a diatomic molecule with moment of inertia I.

Figure 46.8
Bands in the spectrum of nitrogen. [*Courtesy of G. H. Dieke.*]

Thus, if enough identical radiating systems can be excited in some region, a weak beam of the proper frequency can grow to a strong one by stimulating the excited systems to radiate their energy in the same phase and direction, thereby giving a beam of high temporal and spatial coherence.

A device which produces a beam of coherent light by stimulating emission from a large population of excited radiators is called a *laser*, an acronym for *l*ight *a*mplification by *s*timulated *e*mission of *r*adiation. The first stimulated emission amplifiers operated in the microwave region and received the name *masers* with the *m* coming from *microwave*. However, *maser* is now a general term for stimulated-emission devices which operate over a broad spectral range, including microwave, infrared, visible, and ultraviolet frequencies. For example, lasers are often called optical masers.

To operate any maser it is necessary (1) to produce an unnaturally large number of excited systems ready to emit the desired radiation and (2) to stimulate the emission in such a way as to obtain a coherent beam in the desired direction. Many techniques have been devised for achieving these objectives. As an example, we shall consider the ruby laser, which was the first to operate (1960). Ruby is aluminum oxide in which a few aluminum atoms are replaced by chromium atoms; the latter form Cr^{3+} ions in the crystal and are responsible for the ruby color. In the first laser a cigarette-sized ruby rod was placed on the axis of a powerful xenon flash lamp which provided several thousand joules of light each flash. Part of this light was absorbed by the Cr^{3+} ions, raising many to the bands of levels B_1 and B_2 (Fig. 46.9), from which they quickly decay to the upper laser level E_2 by nonradiative transitions in which the energy goes into heating the ruby. An excited ion remains in this state for a few milliseconds before undergoing a spontaneous transition to the ground state by the emission of a 694.3-nm photon. However, when a beam of this frequency is passing through the crystal, the excited ions are stimulated to emit in phase with the beam. By polishing both ends of the ruby rod, silvering one end, and half-silvering the other, it is possible to produce particularly large amplification for radiation in the axial direction, some of which escapes through the partially silvered end to form the laser beam.

The process of providing the required number of excited atoms in a laser is known as *pumping*. In the ruby laser there must be enough optical pumping to put more ions in level E_2 than remain in E_1 if laser operation is to be achieved. For 1,000 J of electric energy supplied to the flash lamp there is perhaps 10 J of laser output in a period of 100 μsec—about 100 kW of highly directional, monochromatic radiation. In other lasers peak powers over 1,000 MW have been obtained and power densities in excess of 10^{19} W/m². Such almost incredible optical beams make it possible to detect light reflected from the moon and to burn through a 1-cm steel plate with a single shot. In gas lasers pumping comes from an electric discharge through the gas, and in semiconductor lasers passing a current through a p-n junction provides the pumping action. Hundreds of different laser and maser systems have been developed to provide intense beams of coherent radiation over a broad spectral range. These devices have found many uses and opened the door to a whole new world of research and application.

Figure 46.9

Simplified energy-level diagram for the Cr^{3+} ion in ruby. Ions in the ground state are pumped by light from a xenon flash bulb to the bands of levels B_1 and B_2, from which they quickly decay by nonradiative transitions to the metastable level E_2. Passing radiation stimulates the transition to the ground level E_1, and coherent radiation is emitted at a wavelength of 694.3 nm.

1 Why do lithium, sodium, potassium, and rubidium have similar chemical properties and similar spectra?

2 What is the Pauli exclusion principle? How does it help us understand the periodic table?

3 What features of the Bohr model of hydrogen are retained in wave mechanics? What features are abandoned?

4 Why don't we observe more evidences of the wave properties of particles in everyday life?

5 The atomic weight of argon is greater than that of potassium. Why do we believe potassium has the greater atomic number?

6 What properties of elements show the periodicity of the periodic table?

7 Physicists sometimes use the rough approximation that the energy needed to remove a K electron from an atom of atomic number Z is $100Z^2/8$ eV. Why is this a better approximation for heavy elements than $13.6Z^2$ eV, which corresponds to Eq. (45.7)?

8 How can the energies of the low-lying atomic levels be determined? What sorts of checks on these levels are available?

9 In terms of the discussion of Sec. 46.6, explain why there are three absorption edges for the L shell and five for the M shell. [An additional fact you need to know is that the energy of a state depends on n, l, and j but not on m_j (unless a magnetic or electric field is applied).]

10 What is the distinction between coherent and incoherent radiation?

11 What are some of the advantages of lasers and masers over previous sources? What possible uses and applications can you envision for lasers and masers?

PROBLEMS

1 Find the velocity and the de Broglie wavelength of an electron which has a kinetic energy of 200 eV. *Ans.* 8.39×10^6 m/sec; 0.867 Å

2 When ultraviolet photons of a single energy fall on a surface of a metal which has a work function of 4.5 eV, the most energetic photoelectrons have an energy of 2 eV. Find the speed and the de Broglie wavelength of the ejected electrons. What is the wavelength of the incident photons?

3 Find the de Broglie wavelength of an 8-MeV proton. *Ans.* 10.1 fm

4 Find the momentum and de Broglie wavelength of an electron in the second Bohr orbit of hydrogen. How does the de Broglie wavelength compare with the circumference of the orbit?

5 Find the de Broglie wavelength associated with electrons of energies of 4 eV and 4 keV. With neutrons of energies 0.04 and 40 eV. *Ans.* 6.13 Å; 0.194 Å; 1.43 Å; 0.0452 Å

6 Find the speed and energy of a proton which has a de Broglie wavelength of 1.5 Å.

7 Find the energy in electron volts required to strip a scandium atom ($Z = 21$) of its last electron, assuming the other 20 have been removed. How does this compare with the energy required to excite the K x-ray lines of scandium (about 4.5 keV)? Why the difference? *Ans.* 6.0 keV; other electrons change potential

8 If the K, L, and M energy levels of platinum lie at roughly 78, 12, and 3 keV, respectively, compute the approximate wavelengths of the $K\alpha$ and $K\beta$ lines. What minimum potential difference across an x-ray tube is required to excite these lines? At approximately what wavelength is the K absorption edge? The three L edges?

9 Use the data of Table 46.1 to calculate the wavelength of the $K\alpha_1$ ($K–L_{III}$; electron makes transition to K vacancy from L_{III} level), $K\beta_1$ ($K–M_{III}$), and $L\alpha_1$ ($L_{III}–M_V$) lines of silver (47). *Ans.* 0.558 Å; 0.496 Å; 4.15 Å

10 If Mo $K\alpha_1$ photons (0.707 Å) fall on a copper foil, find the kinetic energy of photoelectrons ejected from the K and L_{III} shells of copper.

11 A muonic atom consists of a nucleus of charge Z, a captured negative muon, and $Z - 1$ electrons. The smaller Bohr orbits of the muon are much closer to the nucleus than any of the Bohr orbits of the electrons. The muon has the same charge as an electron but 207 times the mass. Assuming the electrons do not affect the muon significantly, find the radius of the first Bohr orbit for a muon captured by an aluminum nucleus ($Z = 13$). Find the energy of the muon in orbits characterized by $n = 1$ and $n = 2$ and the wavelength of the photon emitted when the muon makes the transition from the higher of these states of the lower.

Ans. 2×10^{-14} m; 4.76×10^5 eV; 1.19×10^5 eV; 3.5×10^{-12} m

12 The moment of inertia of an HCl molecule is about 2.72×10^{-47} kg-m². Find the energies of the first two excited states relative to the ground state and the wavelengths of the first (longest λ) two lines of the rotational spectrum.

13 The moment of inertia of a H_2 molecule in its ground state is about 4.55×10^{-48} kg-m². Find the lowest two excited rotational levels for the H_2 molecule and the wavelengths of the first (longest λ) two lines of the rotational spectrum.

Ans. 2.44 and 7.33×10^{-21} J; 81 and 40.5 μm

14 Prove that the de Broglie wavelength of an electron cannot exceed the short-wavelength limit of the x-ray photons which could be produced by this electron. [*Hint:* $Ve = (m - m_0)c^2$.]

15 The moment of inertia of a potassium chloride molecule is 2.38×10^{-45} kg-m². Find the energy of the three lowest excited rotational levels for KCl molecules. What wavelengths are emitted when transitions occur between these levels?

Ans. 4.67, 14.02, and 28.04×10^{-24} J; 0.0213 and 0.0142 m

RESEARCH in the passage of electrons through evacuated regions and their deflection by electric and magnetic fields gave us a knowledge of the mass and charge of the electron. But it did far more than this; it laid the foundations for radio and television and the vast electronics industry. In this chapter we consider the physics of simple vacuum tubes and their use for producing, amplifying, and detecting electromagnetic waves of radio and television frequencies.

CHAPTER FORTY-SEVEN
VACUUM-TUBE ELECTRONICS

47.1 THERMIONIC EMISSION

In 1883, when Thomas Edison was developing the incandescent lamp, he made an important discovery. He observed that if he had a third electrode (Fig. 47.1) in one of his lamps, there was a current to this electrode when it was *positive* relative to the incandescent filament, but not when it was *negative*. *When materials are heated to a high temperature, electrons are emitted.* In Edison's experiments these electrons were attracted to the positive electrode, and a current was registered. When the third electrode was negative, the electrons were repelled. The emission of electrons from heated surfaces is *thermionic emission;* it is sometimes called the *Edison effect.*

The qualitative explanation of thermionic emission is relatively simple. In every material there are some electrons which are less tightly bound than others. In metals, for instance, there are electrons which are free to move about, although they cannot leave the surface of the metal unless they are given additional energy. One way of providing this energy is by raising the temperature. As the temperature goes up, an occasional electron gets enough energy to escape. At higher temperature more electrons leave, some of which have not only enough energy to escape, but enough to have relatively high speeds after they break free of the material. Figure 47.2 shows schematically that both

Figure 47.1
A hot wire emits electrons, which go to electrode P when it is positive relative to the emitter.

Figure 47.2
As the temperature of a filament rises, both the number of electrons emitted and their average speed increase.

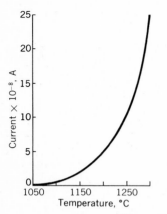

Figure 47.3
Thermionic emission increases rapidly with a rise in temperature.

Figure 47.4
The space charge of electrons near an emitter increases as the temperature is raised.

the number and the average speed of escaping electrons increase as the temperature is raised.

Thermionic emission is analogous in many ways to evaporation from the surface of a liquid. In both cases, particles which happen to obtain an unusually large amount of energy when they are at the surface may escape. The theoretical work of Richardson showed that the current density J (current I emitted per unit area A) from a heated surface is given by the relation

$$J = \frac{I}{A} = aT^2 e^{-b/T} \qquad (47.1)$$

where a and b are constants, T is the absolute temperature, and e is the base of the natural system of logarithms. The constant a is the same for all pure metals; b varies from metal to metal and has particularly low values for those elements which are good thermionic emitters.

The number of electrons emitted per unit area per second increases rapidly as the temperature is increased (Fig. 47.3). If we are to be able to measure the total emitted current, the potential difference between the emitting surface and the positive collecting plate must be fairly high. If it is not, there is a cloud of electrons around the emitting surface which repels additional electrons and drives many of them back to the filament. This electron cloud is called a *space charge* (Fig. 47.4). If the collecting plate is made only slightly positive, the current is limited by this space charge and bears no simple relation to the total number of electrons emitted from the filament. Figure 47.5 shows how the current collected by the plate depends on the potential for three temperatures of the filament. In each case the maximum current is determined by the rate at which electrons are emitted. When every electron emitted is collected, the current has reached its *saturation* value. As the temperature of the emitting surface is raised, the saturation current increases. At low plate potentials the current is *space-charge limited* and does not depend on the temperature so long as the temperature is sufficiently high. Clearly, the current to the collecting plate does not obey Ohm's law.

The oxides of barium and strontium are copious electron emitters

Figure 47.5
The electron current to a collecting plate as a function of the plate potential relative to the filament for three filament temperatures.

at relatively low temperatures. Many radio tubes and other electronic devices which operate at low plate potentials use cathodes coated with such oxides. Where high potentials are required, the bombardment of the negative cathode by positive ions may destroy the oxide surface. If the potentials are not particularly high, a thoriated tungsten filament is commonly used as the emitter. It is produced by adding a small amount of thorium oxide to tungsten. When this material is heated, a layer of thorium is formed on the surface of the tungsten and provides an excellent emitting surface. Where the filament is subject to bombardment by high-energy positive ions, thoriated tungsten fails and pure tungsten emitters are used.

(a)

47.2   THE DIODE

A two-element vacuum tube containing an electron-emitting cathode and a plate is called a *diode*. Such a tube conducts electricity freely when the plate is positive but not when the plate is negative. A diode is useful for changing an alternating current to a direct one. When it is used in this way, it is called a *rectifier*.

Suppose we wish to charge a battery but we have available only an alternating potential difference. The direct application of an alternating potential difference to the battery results in no useful charging, since the electric energy stored during one half cycle is returned during the second half cycle. However, if we provide a circuit similar to that of Fig. 47.6*a*, there is a current through the battery only when the plate of the rectifier tube is positive. Thus there is a current in the proper direction to charge the battery. Figure 47.6*b* shows the alternating potential difference provided for the circuit and the current. In this circuit current exists only during half a cycle. This is therefore called *half-wave rectification*.

The circuit shown in Fig. 47.7 provides *full-wave rectification*. When the upper terminal of the ac transformer is positive, electrons go from the cathode to the upper plate of the tube and then through the load. Conversely, when the lower end of the transformer is positive,

(b)

Figure 47.6
(*a*) Circuit of a diode rectifier used to give a unidirectional current through a load and (*b*) graph of the resulting current when an alternating potential difference is applied.

Figure 47.7
Full-wave rectification circuit used to provide the current shown below to the load.

Figure 47.8

A filtered rectifier produces an almost constant current (solid line below) through the load, in contrast to the current variation in the absence of the filter (dashed curve).

electrons go to the lower plate. The current through the load is always in the same direction.

To make the current steady, one can introduce a filter consisting of one or more capacitors in parallel with the load and one or more inductors (called *chokes* when used in this way) in series with the load (Fig. 47.8). By suitable choice of capacitors and inductors it is possible to provide almost any desired degree of constancy in the current. The general action of the capacitors is to store energy in the electrostatic field when the output from the tube is at its maximum and to release this energy to maintain the current as the output from the tube falls to zero. An inductor connected in series with the load stores energy in its magnetic field when the current is maximum and releases this energy to maintain the current as the current decreases.

47.3 THE TRIODE

In 1907 De Forest had the brilliant idea of introducing a *grid* between the cathode and the plate of a diode to control the current to the plate. This grid consists of an open mesh or helix around the filament, as shown in Fig. 47.9. By keeping the grid at a potential slightly negative relative

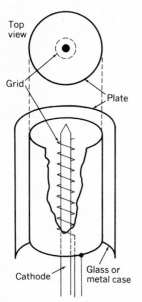

Figure 47.9
A simple triode.

Figure 47.10
Circuit for measuring the static characteristics of a triode.

Figure 47.11
Relation between plate current and grid potential in a triode when the plate potential is held constant.

to the cathode, it is possible to prevent the grid from collecting electrons (Fig. 47.10). Because the grid is close to the electron-emitting cathode, a small change in grid potential makes a significant change in the number of electrons which pass the grid and reach the plate.

If the plate potential of a triode is kept constant and the grid potential is varied, the plate current varies as a function of grid potential as shown in Fig. 47.11. (Note that all potential differences are measured relative to the filament.) If the grid is made sufficiently negative, the current to the plate may be cut off entirely.

Figure 47.12 shows the plate current of a typical triode as a function of plate potential for three different values of the grid potential. Note that at points x and y in this figure the plate current is exactly the same. From x to y the grid potential decreases from -2 to -5 V, while the plate potential increases from 90 to 140 V. Thus, an increase in the plate potential of 50 V has compensated for a decrease in grid potential of 3 V. To put it another way, a reduction in the grid potential of 3 V requires an increase of 50 V in plate potential to keep the plate current constant. Under these conditions of operation, the grid is $\frac{50}{3}$

Figure 47.12
Plate current in a triode as a function of plate potential for three grid potentials.

Figure 47.13
Circuit using a triode as an amplifier.

or 16.7 times as effective in controlling the plate current as is the plate. We say that the *amplification factor* is 16.7.

If when the grid potential of a triode is changed by a small amount $-\Delta V_g$, it requires a change in plate potential ΔV_p to keep the plate current constant, the amplification factor μ is defined as

$$\mu = \frac{\Delta V_p}{-\Delta V_g} \qquad I_p \text{ constant} \tag{47.2}$$

47.4   THE TRIODE AS AN AMPLIFIER

The fact that small changes in the grid potential can produce large changes in the plate current in a triode permits us to obtain large potential variations in the plate circuit when the grid potential is changed by small amounts. Consider the circuit shown in Fig. 47.13. If we apply a small alternating signal to the grid, the plate current changes as indicated in Fig. 47.14. As a result of this change in plate current, the

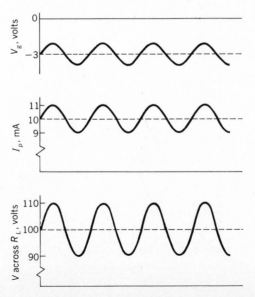

Figure 47.14
Variation of grid potential, plate current, and potential difference across the load resistor R_L.

potential difference across the plate load resistor R_L varies. For such a circuit we define the *voltage gain* as

$$\text{Voltage gain} = \frac{V_{\text{out}}}{V_{\text{in}}} = \frac{R_L \, \Delta i_p}{\Delta v_g} \qquad (47.3)$$

where R_L is the resistance of the plate resistor, and Δi_p and Δv_g are the variations in plate current and grid potential, respectively.

It is possible to apply the changes in potential across the plate (or load) resistor to the grid of a second amplifier tube. In this way signals can be amplified by a factor of many hundreds.

47.5   THE TRIODE OSCILLATOR

In Sec. 40.9 we learned that when a capacitor is discharged through an inductor of low resistance, damped oscillations occur with a frequency ν given by Eq. (40.8): $\nu = 1/(2\pi\sqrt{LC})$. The oscillations are damped because energy is dissipated in heat and sent out as radiation, while no energy is supplied to the circuit. Steady oscillations of constant amplitude can be maintained in such a circuit provided energy is supplied to compensate for the losses. It is possible to use a triode to produce such stable oscillations by providing energy to the oscillating circuit from the plate-potential supply of the triode. One of the ways in which this can be done is indicated in Fig. 47.15. Here the condenser C_1 and the inductor L_1 form the oscillating circuit. The oscillations produce variations in the potential of the grid. This, in turn, results in variations of the plate current. The varying plate current passes through the inductor L_2 and induces, by mutual induction, an emf in the coil L_1, thereby supplying energy to the oscillating grid circuit. The energy is fed back into the grid circuit in such a way as to provide for the losses in this circuit. In this way oscillations of constant amplitude and of frequency determined by the grid-circuit elements are maintained.

If one wishes the oscillation frequency to remain very constant, one uses for C_1 a capacitor which has as dielectric between its plates a carefully cut crystal of quartz. In an oscillating electric field, such a crystal undergoes mechanical vibrations (see Sec. 32.3), the frequency of which depends on the thickness of the quartz. If the oscillating circuit and the crystal are resonant to essentially the same frequency, the oscillator assumes the frequency of the mechanical vibrations of the quartz. In such a case the oscillator is *crystal-controlled*.

Figure 47.15
Simple oscillator circuit using a triode.

47.6 THE RADIO TRANSMITTER

In preceding sections we have seen how simple diode and triode vacuum tubes operate. We now consider how vacuum tubes may be used in simple radio circuits. Since the detailed operation of a radio transmitter or receiver involves many complexities, we emphasize only the basic physics of their operation.

All radio and television stations send out electromagnetic waves. The frequencies involved vary from many thousand to many million hertz. In principle, a simple oscillator such as that of Fig. 47.15 could produce these oscillations; in practice, more sophisticated circuits are used. A primitive type of radio transmitter is shown in Fig. 47.16. It consists of an oscillator, an antenna, and a *modulator*. The oscillator frequency is determined by the resonant grid circuit. For the standard radio band the frequencies range from 535 to 1,605 kHz. The oscillations in the grid circuit produce, through mutual induction, an oscillating current in the inductor L_3 which is part of the antenna circuit. The oscillator thus "drives" the antenna, setting up in it an alternating current of frequency characteristic of the oscillator. The current surging back and forth in the antenna produces the rapidly changing electromagnetic field which we know as radio waves and which can be detected at great distances from the transmitter. In Sec. 41.1 we discussed the electromagnetic waves sent out by an oscillating current (Fig. 41.1) such as the one established in the antenna.

If a simple alternating current from an oscillator is used to drive an antenna, the transmitter sends out a constant-amplitude signal at a high frequency, called the *carrier frequency*. If we wish to convey intelligence by such a signal, we must interrupt it or modify it in some way. The most primitive method of doing this is to turn the oscillator on and off, thus sending a series of dots and dashes. However, it is

Figure 47.16
Simple radio transmitter.

Figure 47.17
(*a*) Unmodulated carrier wave, (*b*) audiofrequency wave, (*c*) modulated carrier wave, and (*d*) rectified wave.

also possible to send information which leads to accurate reproduction of sound waves. If we speak into the microphone of Fig. 47.16, there are set up in the grid circuit additional small voltage fluctuations which vary the amplitude of the oscillations. When we wish to send a 1,000-Hz note over the transmitter, we "modulate," i.e., vary, the amplitude of the carrier wave at a frequency of 1,000 times a second. Since the carrier frequency is many thousand times as high as the highest sound frequency we wish to transmit, the carrier wave makes many thousands of oscillations in each period of the sound wave. Figure 47.17 shows the general appearance of the unmodulated carrier wave, of the sound frequency with which we modulate this wave, and of the modulated wave which is sent out by the transmitter. This particular method of transmitting intelligence is called *amplitude modulation*. There are, of course, other ways in which the carrier wave can be modified to convey intelligence. One of the possibilities is to introduce fluctuations in the frequency, keeping the amplitude constant. This is *frequency modulation*.

47.7 THE RADIO RECEIVER

The electromagnetic waves sent out by a radio transmitter fall upon the antenna of the receiver shown schematically in Fig. 47.18. These electromagnetic waves set up oscillations in the antenna which induce small potential variations in the grid circuit of the receiver. There are many transmitters from which signals may be received. It is in the grid circuit L_2C_2 that the selection of the signal to be heard is made. By varying the capacitance C_2, the grid circuit L_2C_2 can be made resonant to any desired carrier frequency. Then oscillations in the antenna corresponding to that particular frequency establish changes in potential in the grid circuit, while oscillations of all other frequencies produce negli-

Figure 47.18
Simple radio receiver.

gible effects because the impedance of the circuit is small for other frequencies.

The oscillations of potential across C_2 in the resonance circuit are applied to the grid of the triode in Fig. 47.18. If this grid is made sufficiently negative, the plate current increases substantially when the grid potential changes in the positive direction but decreases relatively little (or not at all) when the changes are in the negative direction. In this case the output of the plate circuit is similar to Fig. 47.17*d*. The variations in plate current at radio frequency are such that the human ear cannot detect them; however, the average plate current over many carrier-wave oscillations fluctuates with the modulation frequency. Thus the current through the earphones oscillates with the modulation frequency, and sound is heard.

To drive a loudspeaker, much more energy must be provided than can be expected from the output of a single triode of the type indicated. In a modern radio receiver several more tubes may be used to provide this additional power. The signal coming into the initial resonant circuit is amplified before the detection process is performed. Then the audio frequency may be amplified several times before it is finally supplied to a *power amplifier* which drives the loudspeaker.

47.8 THE LOUDSPEAKER

Figure 47.19
Dynamic loudspeaker.

A loudspeaker is a device which transforms audio-frequency variations in electric current into sound waves. Figure 47.19 shows schematically the principal parts of a dynamic loudspeaker. A light thin coil of wire is attached to the apex of a light cone. The coil is placed in a magnetic field. Current from the power amplifier is fed to this coil; the fluctuating current produces fluctuating forces upon the coil, causing it to vibrate. The cone vibrates with the coil and produces waves in the surrounding air.

47.9 THE KENNELLY–HEAVISIDE LAYERS

Radio reception from distant stations is usually superior at night. Sometimes reception is excellent at distances of several thousand miles when a receiving station only a few hundred miles away may not be able to detect the signal. In order to explain these phenomena it was suggested by Kennelly and by Heaviside that there are layers of ionized air in the upper atmosphere which act as refractors and reflectors of radio waves. Ions are produced in the outer layers by a variety of processes, chiefly by ultraviolet light from the sun. The height of the ionized layers changes from time to time; as a result, the transmission of electric waves varies with time. In general, the ionizing layers form at different altitudes at different times of day, being higher and more uniform at night.

47.10 THE CATHODE-RAY OSCILLOSCOPE

One of the most useful of all types of vacuum tubes is the so-called *cathode-ray tube*. A heated cathode (Fig. 47.20) emits electrons. These electrons are accelerated to a cylindrical anode, which has a hole in its center through which the cathode rays pass. The electron beam may be controlled by the grid. Beyond the anode is a beam of rapidly moving electrons of well-defined energy.

This beam passes between a pair of flat plates P_1 whose plane is horizontal. If a potential difference is applied between these plates, an electric field is created which may move the beam of electrons up or down. Next the beam passes between a pair of plates P_2 whose plane is vertical. If a potential difference is applied between these plates, the beam may be moved to the right or left. The beam of electrons strikes a screen coated with a fluorescent material that emits light where the electrons strike. The position of the point on the screen at which electrons strike can be controlled by the potentials applied to the deflecting plates.

In one common application of the oscilloscope the horizontal deflectors P_2 have applied to them a sawtooth potential variation which sweeps the beam at a constant rate from left to right and then suddenly switches the beam back to the left edge. If a sinusoidal potential difference is applied to the deflecting plates P_1, the beam traces a sine curve on the face of the oscilloscope. If some other type of potential variation is applied to the plates P_1, the picture on the oscilloscope screen reproduces the fluctuations in potential across the deflecting plates.

When a cathode-ray tube is combined with electronic circuits for

Figure 47.20
Cathode-ray tube.

amplifying signals and applying potentials to the horizontal and vertical deflecting plates, the unit is called a *cathode-ray oscilloscope*. The cathode-ray oscilloscope is one of the most convenient, versatile, and powerful tools of modern physics.

47.11 TELEVISION

In the television receiver a cathode-ray tube in which horizontal and vertical deflections are usually produced by magnetic rather than electric fields is used. The magnetic deflection is obtained through a pair of coils, one on each side of the picture tube, that produce a horizontal magnetic field, and another pair, one above and one below the neck of the tube, that produce a vertical field. By sending high-frequency alternating currents through these coils it is possible to make the cathode-ray beam move back and forth across the tube, producing the familiar lines of the television picture.

If one moved the cathode-ray beam across the face of the oscilloscope, then down a little and across the screen again, etc., only a number of white lines would be drawn on the face of the screen. However, by controlling the intensity of the beam by means of the grid in the tube, it is possible to draw lines which vary in their lightness, depending on the potential applied to the grid at the particular instant the spot is drawing a given segment of the line. Thus the cathode-ray beam draws lines of varying brightness. There are 525 horizontal lines drawn to make a complete picture. In order to transmit pictures of moving objects, it is necessary that the picture be "redrawn" many times a second. In current television practice the even-numbered lines are drawn in $\frac{1}{60}$ sec and the odd-numbered lines in the next $\frac{1}{60}$ sec. Thus, 30 complete pictures or "frames" are drawn each second.

At the sending station the information transmitted is obtained in roughly the following way. An image of the picture to be transmitted is produced by lenses on the photosensitive screen of a tube known as an *image orthicon tube*. Each element of the sensitive surface emits photoelectrons, the number of which depends upon the intensity of the incident light. By a complex process the number of photoelectrons emitted from each of the tiny sensitive elements is determined and is converted into information which can be used to modulate the carrier wave. At the receiver this information is used to control the grid potential at the instant the corresponding element of the picture is to be reproduced on the receiving screen.

47.12 RADAR

The wavelengths of ordinary radio waves are of the order of a thousand feet. Such waves bend readily around obstacles such as buildings or airplanes. If radio waves of much higher frequency, and therefore shorter wavelength, are produced, the waves are readily reflected by aircraft and other objects of comparable size. Further, these waves can be focused into sharp beams by a reasonably small antenna. If a sharply focused beam of high-frequency waves is used to scan the sky, it is reflected by airplanes or other objects in the sky. By measuring the time lapse between the transmission of a pulse and the reception of the reflected pulse, one can determine the distance of the object from the transmitter, since radio waves are transmitted with the speed of light.

It is possible to scan a relatively large area by means of a radar transmitter and show on an oscilloscope screen the areas from which there is substantial reflection and those from which the reflection is much smaller. In this way one can make a radar map of an area which is not visible because of clouds, fog, or darkness.

One of the earliest and simplest examples of the use of radio waves to measure distance is found in the radio altimeter. If a radio wave is sent down from an airplane, it is reflected by the earth. If this radiation is sent in pulses, and if it is possible to measure the time it takes the wave to travel to the earth and return, the altitude of the plane above the terrain can be calculated.

QUESTIONS

1 How can a simple diode be used as a rectifier to charge a battery? Draw an appropriate circuit.

2 A simple diode rectifier provides a varying unidirectional current. How can one obtain an almost constant direct current from such a rectifier by adding suitable components to the circuit?

3 How does a triode act as an oscillator? From what source does the energy come to maintain the oscillations?

4 How can a triode be used to amplify a small signal? Explain with the aid of suitable curves.

5 When a triode is used as an amplifier, it is customary to operate in a region of the characteristic curve (Fig. 45.11) which is linear (nearly a straight line). Why is this desirable? What happens if a nonlinear region is used?

6 Draw a diagram of a full-wave rectifier using a transformer as a power source to charge a battery.

7 How can a triode be used as a detector? Why is the grid of a triode that is used as a detector biased more negatively than the same grid would be if the triode were used as an amplifier?

8 What is "space charge"? Discuss the importance of space charge in electron tubes.

9 How does one change stations in an ordinary radio? In a television receiver? Why the difference?

10 How does a cathode-ray oscilloscope work? What are some of its uses?

11 How is the picture drawn on the screen of a television tube?

PROBLEMS

1 The current in the plate circuit of a vacuum tube is 50 mA. How many electrons are arriving at the plate each second? *Ans.* 3.12×10^{17}

2 Three filaments of different materials each emit a saturation current of 30 mA. It requires input power of 0.8 W for an oxide-coated filament, 12 W from a thoriated-tungsten filament, and 55 W from a tungsten filament. Calculate the power requirement per milliampere of emission current for each type of filament.

3 A vacuum tube has an amplification factor of 15. If the grid voltage is decreased 0.24 V, how much must the plate potential be raised for the plate current to remain constant? *Ans.* 3.6 V

4 For a certain triode the plate current changes 5 mA for an 8-V change in plate potential. If it requires a 0.5-V change in grid potential to restore the plate current to its original value, calculate the amplification factor.

5 The plate current of a vacuum tube is 5 mA when the grid potential is −3.2 V and the plate potential is 150 V. It is also 5 mA when the grid potential is −2.8 V and the plate potential is 141 V. What is the amplification factor? *Ans.* 22.5

6 The signal available from an antenna has an amplitude of 5 μV. If this signal is passed through two stages of amplification, each with a voltage gain of 30, what is the amplitude of the potential fluctuations in the output circuit? Through how many more stages of the same voltage gain would the signal have to be passed to obtain an output voltage in excess of 3 V?

7 A thoriated-tungsten filament at 1500 K has a total emission current of 2 mA. Find the emission current at 1800 K if the constant b in Eq. (47.1) is 28,000 K. It can be shown that b is equal to the work function W_{min} divided by the Boltzmann constant k. Find the work function W_{min} for this filament.   *Ans.* 65 mA; 2.41 eV

8 When an ac signal of 0.5 V amplitude is impressed on the grid of a triode which is biased at -2.5 V, the plate current varies from 5 to 8 mA. If the resistance R_L in the plate circuit is 9,000 Ω, find the voltage amplification of the triode.

9 A triode is used as an amplifier. If the grid is biased at -3 V and a plate resistor R_L of 12,000 Ω is used, the plate current is 20 mA. If the grid potential is varied from -3.2 to -2.8 V, the plate current varies from 19 to 21 mA. Find the maximum and minimum potential difference across the plate resistor and the voltage gain.   *Ans.* 252 V; 228 V; 60

10 A triode amplifier uses a plate resistor of 15,000 Ω. When no signal is applied to the grid, the plate current is 8 mA. If the voltage gain is 80, find the amplitude of the potential fluctuations across the plate resistor when a signal of amplitude 0.005 V is applied to the grid. Between what limits does the plate current fluctuate?

11 An electron is projected along the axis of a cathode-ray tube midway between two parallel plates with a velocity of 1.5×10^7 m/sec. The plates are 1.2 cm apart and 3 cm long and have a potential difference of 120 V between them. Find (*a*) the angle the electron makes with the axis as it leaves the plates, (*b*) the distance of the electron from the axis as it leaves the plates, and (*c*) the distance from the axis at which the electron strikes a fluorescent screen 20 cm beyond the plates.

    Ans. (*a*) 0.23 rad; (*b*) 0.35 cm; (*c*) 5.0 cm

12 A cathode-ray tube has electrostatic deflection plates 2 cm square with a separation of 0.5 cm located 25 cm from the fluorescent screen of the tube. The electrons are accelerated through a potential difference of 1,500 V between emitter and anode. Calculate the approximate deflection of the beam on the screen for a 15-V potential applied to the plates.

13 If the cathode-ray tube of Prob. 12 uses magnetic rather than electrostatic deflection, find the deflection produced by a magnetic field of 0.002 T which is assumed uniform over a length of 2 cm and zero elsewhere.   *Ans.* 8 cm

14 Electrons in a television tube are accelerated through a potential difference of 10 kV. Find the speed of the electrons using the appropriate relativistic equations. How much smaller is this than the erroneous value calculated by using $\frac{1}{2}mv^2$ for the kinetic energy of the electrons?

IN EARLIER CHAPTERS we discussed many characteristics of solids—elastic constants, indices of refraction, thermal and electric conductivities, and densities—as unrelated phenomena and without any consideration of the contributions of individual atoms to these properties. In solid-state physics we seek to understand properties of solids in terms of their structure and organization as societies of atoms. In recent years solid-state physics has been an area of extraordinary activity. Advances have come rapidly, enlarging our understanding of the behavior of materials. Research has brought a host of new and improved materials—high-temperature alloys for jet engines; superior phosphors for fluorescent lights and television screens; new materials for construction and for magnets; transistors replacing vacuum tubes; molecular circuits replacing large, heavy, unreliable circuit elements in specialized fields. In this chapter we introduce a few fundamental ideas about solids and discuss qualitatively the p-n junction which lies at the heart of solid-state electronics.

CHAPTER FORTY-EIGHT
SOLID-STATE ELECTRONICS

48.1 BINDING IN SOLIDS

As we saw in Sec. 15.1, at a sufficiently low temperature the attractive forces between gas molecules typically are great enough for liquefaction to occur; upon further cooling fluid properties disappear, and eventually there remains a block of substance which maintains both its shape and its volume. Such a mass is commonly known as a *solid*, although some scientists prefer to reserve this term for what we shall call a *crystalline solid*. Of a piece of glass, a bar of gold, and a strip of wood, only the gold is a crystalline solid. The distinction is made on the basis of the kind of order which exists in the material, i.e., on the type of geometrical arrangement in which the atoms fall. By means of x-rays we can often establish the relative positions of various kinds of atoms in a material. Suppose for a moment that we could actually see the individual atoms in a piece of glass. If we locate a silicon atom and examine its neighbors, we find that they are not distributed in a random fashion. There is a definite geometrical order. However, if we look at other silicon atoms we find that their neighbors form a somewhat different pattern. For glass there is no orderly arrangement which persists throughout the material; there is *short-range order* (Fig. 15.1*b*), but not *long-range order*. A material which exhibits short-range order and retains its shape is said to be an *amorphous solid* or sometimes a supercooled liquid.

The geometrical arrangement of atoms in any one region of a crystalline solid is the same as in any other region. The ordering extends in uninterrupted fashion over distances of thousands of atomic diameters. For crystalline solids there exist 14 different possible arrays for three-dimensional lattices. The structure of a given crystal, including the locations of specific kinds of atoms within the lattice, can be studied by x-ray diffraction.

There are many examples of the same kinds of atoms arranging themselves in more than one kind of crystal lattice. In these cases the properties of the solid depend on the crystal structure. This is well illustrated by carbon, which forms crystals both in the diamond lattice and in the graphite lattice. Graphite is opaque and soft, rubbing off on paper, while diamond is translucent and hard enough to scratch glass. Carbon crystallizes in a hexagonal lattice as graphite and in a type of cubic lattice as diamond.

The forces of attraction which hold the atoms of a solid in their regular geometrical array determine not only the type of crystal lattice which is formed but also other physical properties of the material such as elastic moduli, optical constants, and conductivities, both electric and thermal. The origin of these forces differs from crystal to crystal; among the many kinds are the following.

Ionic Binding An ionic crystal is composed of positive and negative ions. For example, a crystal of sodium chloride (NaCl) consists of a regular array of Na$^+$ and Cl$^-$ ions (Figs. 48.1 and 43.9). The binding forces, which come essentially from the coulomb attraction between these charged ions, are relatively large. Once a molecule of NaCl becomes part of the crystal, it is no longer possible to identify a particular chlorine ion as belonging to a particular sodium ion. Instead, each ion has a number of equidistant neighbors. Ordinarily, ionic crystals are relatively transparent to visible and ultraviolet light but show strong absorption in the infrared. At low temperatures they are excellent electric and thermal insulators; at high temperatures they become electric conductors by virtue of the motion of ions through the crystal. A crystal formed from ions exhibits *ionic* or *heteropolar* binding.

Covalent Binding In covalent bonds electrons are exchanged between atoms; the average charge density between the atoms may be relatively high. Examples of crystals exhibiting covalent binding are germanium, silicon, and diamond. In a diamond crystal a single carbon atom shares electrons with four other carbon atoms at alternate corners of a cube. At low temperature, covalent crystals are ordinarily hard and often brittle and have high electric resistance. As the temperature is

Figure 48.1
NaCl crystal; the Na$^+$ ions (dark) have an effective radius of 0.98 Å compared with a radius of 1.81 Å for the Cl$^-$ ions (light).

increased, there is a marked decrease in resistance. Practical semi-conductors (Sec. 48.6) are based on covalent crystals.

Metallic Binding In metals the valence electrons are not tied to any particular atom or pair of atoms but are free to wander through the crystal. A metallic crystal may be thought of as an array of closely packed positive ions immersed in a sea of uniformly distributed electrons. This type of binding leads to high electric and thermal conductivity and to the optical and mechanical properties characteristic of metals.

Hydrogen Binding The hydrogen atom has a single electron, so one might expect it to form a bond with only one other atom. However, frequently hydrogen forms a strong bond between two atoms. We may think of this bond as primarily ionic in character, since it is formed only with strongly electronegative atoms. The hydrogen atom loses or transfers its electron to one of these atoms, and the proton binds two negative ions, nestling between these very much larger ions. Only hydrogen forms such bonds; no other ion has the very small size of the proton—so small that there are only two nearest neighbors. The hydrogen bond occurs in ice $(H_2O)_n$, solid hydrogen fluoride, and a wide variety of proteins and other organic compounds. Therefore, it is of great interest in biophysics and biochemistry.

Molecular Binding Atoms of argon, helium, and other inert gases are bound in the solid phase by relatively small electric-polarization forces, known as *van der Waals' forces*. Because of the weakness of these forces, the binding energies are low. The crystals have low melting points and are highly compressible and mechanically weak.

48.2 THERMAL PROPERTIES OF SOLIDS

Quantum mechanics predicts (and experiment confirms) that at 0 K there still exists in an atom some vibrational energy, known as the *zero-point energy*. At any higher temperature the energy associated with atomic vibrations is greater. The atoms of a solid do not vibrate independently, since they are joined to one another by elastic forces. The motion of any one atom is influenced by, and in turn influences, the motion of its neighbors. Indeed, the vibration of any individual atom can be thought of as a portion of a wave moving through the crystal. In the crystal only those waves exist which satisfy the boundary conditions at the surface of the material. Calculations of the allowed frequencies show that they are great in number for a solid which contains billions of atoms. Nevertheless, it has been possible to confirm by experiment many of the predictions of the theory.

We have seen that electromagnetic waves come in photons of energy $h\nu$. Similarly, vibrational waves in crystals are quantized, with the minimum energy associated with a frequency given by the same relation $h\nu$. The quanta of vibrational waves are called *phonons*. Near 0 K only low-energy phonons are present; as the temperature is raised, phonons of higher frequency come into existence. When we heat a crystal, we excite more phonons. These lattice vibrations make by far the largest contribution to the specific heat, except near 0 K, where the small contribution to the specific heat from the free electrons must be taken into account for a metal.

The transfer of heat by conduction from one point to another in a solid has been discussed phenomenologically in Sec. 17.2. For the solid as a geometrical assembly of atoms, we can calculate the thermal conductivity in terms of lattice vibrations which transfer energy from atom to atom as waves progress through the crystal. When we raise the temperature in one region of the crystal, phonon waves spreading out from this region of increased amplitude augment the amplitudes at points some distance away. This is equivalent to our phenomenological observation that heat is transferred from one region to another. We can compute the thermal conductivities of crystals with considerable accuracy by the phonon description so long as the crystals are not metallic. We do not get valid predictions for metallic conductors because in metals the primary conduction of heat is by free electrons.

48.3 IMPERFECTIONS IN CRYSTALS

A perfect single crystal is one in which a pattern is repeated without variation over the entire volume. Such a crystal is an idealization; all crystals have *imperfections*. Here the term *imperfection* does not imply anything undesirable. For example, we make steel from iron by deliberately introducing imperfections. Many properties of solids—specific heat, thermal conductivity, thermal expansion, and density, for example—are essentially the same for perfect crystals and for those with moderate impurities. On the other hand, some properties of great engineering significance are sensitive to relatively minute imperfections; among these are mechanical strength, resistance to rust, corrosion, and mechanical abrasion, and electric conductivity in semiconductors. Among the common types of crystal imperfections are the following.

Substitutional Impurities No crystal is ever chemically pure. A substitutional impurity is one in which a foreign atom occupies the site of one of the atoms of the host crystal and plays its role. For example, in a *p*-type semiconductor an aluminum atom may replace a germanium atom in the crystal. If the impurity atom has roughly the same size as the host atom, the regularity of the crystal at the lattice site is disturbed only slightly and the mechanical properties of the crystal are unchanged. The change in electrical properties is associated primarily with the fact that the impurity atom brings an electron configuration different from that of a host atom. The impurity atom also introduces additional electron energy levels into the crystal. These may play important roles in changing the color and radiation-absorption properties of the crystal.

Interstitials and Vacancies A defect in which an atom lies in a crystal lattice at a site which is not ordinarily occupied by atoms is known as an *interstitial* (Fig. 48.2) if the atom is simply a misplaced one of the host material, or as an *interstitial impurity* if it is an impurity atom. A defect in which a lattice site normally occupied by an atom is unoccupied is called a *vacancy*. Interstitial atoms and vacancies behave

Figure 48.2
The imperfection is *interstitial* if it is of the host material and an *interstitial impurity* if it is an atom of some other element.

much like chemical impurities in that they too bring new energy levels into the crystal. They also scatter electrons moving through the crystal and increase the resistivity.

Dislocations   A crystal with no chemical impurities may contain a region, known as a *dislocation*, in which the atoms are not arranged in the perfect lattice structure. Dislocations, which play an important role in determining the strength of ductile materials, are produced during solidification from the liquid phase. Ordinarily the crystal does not grow at a single point as the liquid solidifies. Rather there are many nuclei from which grow a collection of small crystals oriented more or less at random. The resulting solid is said to be *polycrystalline*. At the *grain boundaries* of the small crystals the matching of the crystal pattern is not perfect. The result is a series of dislocations.

48.4   ENERGY BANDS IN SOLIDS

When a group of atoms is brought together to form a solid, the energy levels associated with each atom undergo shifts, as a result of the presence of the neighboring atoms, which change the potential energy of an electron at any point. In particular, the potential energies of the outer (or valence) electrons are shifted significantly. Quantum mechanics reveals that as a group of atoms is assembled to form a solid, the energy levels which were sharp for individual atoms become broadened to become *energy bands*. The broadening is particularly great for the higher energy levels (Fig. 48.3).

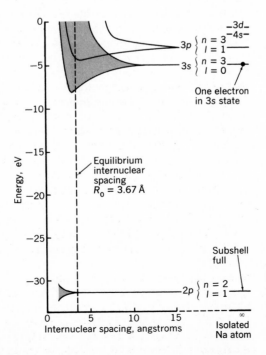

Figure 48.3
Energy levels are sharp for gaseous sodium atoms, but the higher energy levels broaden and overlap when sodium atoms crystallize to form a metal.

Figure 48.4

Comparison of energy bands of (a) a metallic conductor, (b) a semiconductor, and (c) an insulator. At 0 K electrons occupy all states indicated in black, while energy states in gray are allowed but unoccupied. Some occupied states (such as K and L levels) lie too low to show on this figure.

In a crystal at 0 K the individual atoms are as near to rest as they can ever be, and complications associated with thermal vibrations are minimized. Every electron is in its lowest possible energy state, so the lowest energy levels are filled and no electron is in an excited state. Now we can distinguish between metallic conductors, semiconductors, and insulators by the situations which pertain in the highest energy band as follows.

Metallic Conductors If the energy-level characteristics of the least tightly bound electrons are such that the energy state permits electrons to move freely about the potential hills, the material is a *metallic conductor*. The energy band contains some electrons but is not completely filled (Fig. 48.4a). An electric field can produce a flow of these electrons, and the material is an excellent electric conductor at all temperatures.

The Pauli exclusion principle requires that free electrons in a metal have a broad range of kinetic energies. Even at 0 K the most energetic electrons have a kinetic energy of a few electron volts; this maximum is called the *Fermi kinetic energy*. The energy distributions of the conduction electrons in a metal are indicated in Fig. 48.5. In metals (but not in other solids!) the energy E_F of the highest state occupied by electrons at 0 K is called the *Fermi energy* of the metal. When the temperature is increased, a few electrons are raised to slightly higher energies, leaving vacancies in levels just below the Fermi level. Even

Figure 48.5

Fermi distribution showing the relative number of conduction electrons having a given kinetic energy at 0 K (dashed line) and at room temperature (solid line).

Figure 48.6
Approximate energy-level diagram for conduction electrons in a metal. At 0 K all states up to the Fermi level E_F are filled; an energy W_{min} is needed to remove one of the most energetic electrons. The solid curve at the right is the approximate potential energy of an electron just outside the surface of the metal.

when the temperature is brought up to 300 K, the average kinetic energy of the electrons is increased very little; only a small fraction of the heat supplied to warm a metal goes into augmenting the kinetic energies of the electrons.

In order to remove one of the least tightly bound electrons from a metal at 0 K, the electron must be supplied enough energy to take it from the Fermi level to the potential just outside the metal (Fig. 48.6). In Sec. 44.1 we found evidence for this statement in the photoelectric effect. We identify the energy W_{min} in Fig. 48.6 as the minimum required by a photon to eject a photoelectron from the Fermi level. Similarly in order to achieve thermionic emission (Sec. 47.1), the temperature must be raised high enough for electrons to escape via thermal excitation.

Insulators If the gap between the highest filled band and the next permitted (but empty) band is large (Fig. 48.4c), modest temperatures are not sufficient to lift electrons to the conduction bands. Electrons in the filled bands are not free, and the material serves as an excellent insulator. Of course, if the temperature is raised sufficiently high, electrons can be excited to conduction levels, so *no material is an insulator at extremely high temperatures*. However, if the energy gap is reasonably large, practically no electrons are lifted to the conduction band at room temperature.

Semiconductors When the energy band associated with the valence electrons is completely filled but within a small energy range there is another band in which electrons could exist and move through the crystal, the material is a good insulator at 0 K (Fig. 48.4b). However, as the temperature is raised, thermal agitation gives some electrons enough energy to shift to the higher band. The electrons so excited serve as *conduction electrons*. Further, the *holes* left by the electrons when they move to the higher band also serve as charge carriers. Such an arrangement of energy levels is characteristic of semiconductors, which are insulators at 0 K but become poor conductors at somewhat higher temperature. Semiconductors become steadily better conductors as the temperature is increased, because more electrons are transferred to the conduction band. Each electron raised to the conduction band leaves a vacancy, or *hole*, in the valence band.

An energy band in which all allowed states are filled by electrons does not give rise to electric conductivity. However, when there are not quite enough electrons to fill the band completely, an applied electric field does produce a current. The holes (electron vacancies) move through the material, behaving like positive charges. In 1879, well before the electron was discovered, Hall devised an experimental arrangement by which the sign of the charge carriers in a conductor could be determined. Following Hall, let us consider a thin sheet of conductor (Fig. 48.7) with a pair of contacts C_1 and C_2 attached at opposite sides of the central cross section. For a uniform conductor carrying a current in the $+x$ direction C_1 and C_2 are at the same potential if there is no magnetic field. However, when a magnetic field B_z is applied normal to the sheet, a potential difference develops between C_1 and C_2; C_1 becomes positive relative to C_2 if the charge carriers are negative (electrons) and negative if the carriers are positive (holes).

To see how this potential difference arises consider the case of positive carriers, which drift to the right (Fig. 48.7) if the conventional current is in that direction. The force exerted by the magnetic field on the moving charges deflects them downward (Sec. 36.4), producing a positive layer at the bottom of the strip and a negative one at the top. These charge layers build up until the net upward force on the carriers from the resulting electric field E_y is equal to the average downward force due to the magnetic induction so that

$$qE_y = qv_xB_z \qquad (48.1)$$

where q is the charge and v_x the average drift speed of the carriers.

Now suppose that the carriers are negative, moving to the left for conventional current to the right. Again the force exerted on the carriers by the magnetic field deflects them downward, but now the bottom becomes negative and the top positive. Thus the electric field is now in the $-y$ direction so C_2 is negative relative to C_1, opposite to the case for positive carriers.

We define the *Hall coefficient* R_H for the arrangement of Fig. 48.7 to be

$$R_H = \frac{E_y}{J_xB_z} \qquad (48.2)$$

where J_x is the current density (or ratio of current I to cross-sectional

Figure 48.7
The Hall effect tells the sign of the charge carriers in a conductor. With a magnetic field out of the plane of the paper and a conventional current in the x direction the charge carriers are deflected downward by the qvB force. If the charge carriers are positive, C_2 becomes positive relative to C_1; for electrons C_2 becomes negative.

area A of the strip). As we saw in Sec. 36.4, $I = qnAv_x$ so $J_x = qnv_x$, where n is the number of carriers per unit volume. Using this relation and Eq. (48.1) in Eq. (48.2) leads to

$$R_H = \frac{v_x B_z}{qnv_x B_z} = \frac{1}{nq} \qquad (48.3)$$

Thus the experimental determination of the Hall coefficient of a conductor tells us both the sign of the charge carrier and the number of carriers per unit volume since $|q|$ is 1.6×10^{-19} C. The Hall coefficient is negative for electron conduction and positive for holes. For conductors such as sodium, copper, silver, and aluminum, the Hall coefficient is negative and has the right order of magnitude to correspond to one free electron per atom for Na, Cu, and Ag and to three electrons per atom for Al. Valence-2 elements (such as beryllium, zinc, and cadmium) have positive Hall coefficients, showing that conduction is primarily by holes. Studies of the Hall coefficients have been invaluable in gaining an understanding of semiconductor physics, which is fundamental to transistors and other solid-state devices.

48.6 SEMICONDUCTORS

Crystals of pure germanium are well known as a semiconductor material. Each atom of germanium has four valence electrons, which form four covalent bonds with neighboring atoms. This structure has the semiconductor arrangement of energy states, a completely filled energy band separated from an empty higher-lying conduction level by a small energy gap. Such a crystal is called an *intrinsic semiconductor*.

At any temperature above 0 K a few electrons in the semiconductor are thermally excited to the conduction band. The minimum energy required to move an electron from one of the covalent bonds into the conduction band is called the *energy gap E_g*. For silicon E_g is 1.09 eV; for germanium, 0.72 eV. These are small enough so that at room temperature a significant number of electrons are excited to the conduction band. The resulting holes are available for conduction in the lower band. In this case the Fermi level (Fig. 48.8) is located midway between the top of the valence band and the bottom of the conduction band. Although electrons and holes are separated on an energy diagram, in the crystal itself they exist in the same space. A dynamic equilibrium is achieved between the thermal production of holes and conduction electrons and the destruction of these carrier pairs by electrons dropping to the lower energy state (accessible because of the holes).

Now consider the effects of adding a small amount of arsenic to a pure germanium crystal. Addition of foreign atoms to an intrinsic semiconductor produces an *impurity semiconductor* (or *extrinsic* or *doped* semiconductor). Most of the important applications of semiconductors call for doped crystals, in which the electrical properties are drastically changed by the impurity atoms. Arsenic impurity atoms occupy sites in the crystal normally filled by germanium atoms (Fig. 48.9a). Each arsenic atom brings a total of five valence electons, of which four are used in forming covalent bonds with the four nearest germanium neighbors. The fifth electron finds no convenient neighbor with which

Figure 48.8

In an intrinsic semiconductor at room temperature some electrons are excited to the conduction band, leaving vacancies, or holes, in the valence band. Both the conduction electrons and the holes are available for electric conduction. Energies in white are not allowed to the electrons. The Fermi energy lies at the middle of the forbidden band.

(a)

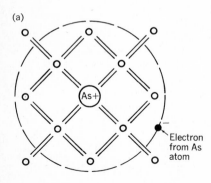

Electron
from As
atom

(b)

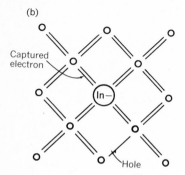

Captured
electron

Hole

Figure 48.9

(*a*) In a doped germanium crystal an arsenic impurity atom brings five electrons where only four are required to complete the covalent bonds with neighboring germanium atoms; the fifth electron is very loosely bound and readily excited to the conduction band.

(*b*) An indium acceptor atom in a germanium crystal provides only three of the four valence electrons needed to complete the covalent bonds. Much of the time, an electron from some other part of the crystal is captured by the indium atom, leaving a hole in the crystal.

to form a bond and is very loosely tied to the arsenic ion, even at 0 K. Only a small amount of energy is needed to free this electron and permit it to wander about in the lattice. Room temperature is enough to supply this energy. Almost every arsenic atom introduced into the crystal provides one electron which is available for conduction. The arsenic atoms are called *donors*, since each atom donates an electron to the crystal. An intrinsic semiconductor doped with a donor impurity is called an *n*-type semiconductor (*n* stands for negative) because conduction is primarily by electrons provided by the donor atoms. Arsenic is by no means the only donor element for *n*-type semiconductors; other elements from the same group (V_A) of the periodic table (such as phosphorus and antimony) are good donor impurities. Adding a donor impurity to an intrinsic semiconductor raises the position of the Fermi level (Fig. 48.10*a*) as a consequence of the increase in the number of electrons in the conduction band and the decrease in the number of holes.

Next consider what happens when we add to a germanium crystal atoms of valence 3, such as indium (or gallium, aluminum, boron). Each indium atom (Fig. 48.8*b*) brings only three electrons to form the four covalent bonds expected from each atom in the germanium crystal, producing a site at which one electron is missing from the four bonds. Such a site readily captures a wandering electron and utilizes it to complete the bonding structure, leaving a vacancy, or *hole*, somewhere in the valence band. Atoms of valence-3 elements are called *acceptor* impurities. They form *p*-type semiconductors, since in a crystal with such impurities there are sites at which an electron is missing and there are positive *holes* available for conduction. Adding an acceptor impurity to an intrinsic semiconductor lowers the Fermi level (Fig. 48.10*b*), since electrons are captured from both the conduction band and the valence band to fill the acceptor levels. This reduces the number of conduction electrons and increases the number of holes. As expected, *p*-type semiconductors exhibit positive Hall coefficients, while *n*-type ones have negative values for R_H.

48.7 *p-n* JUNCTIONS

If a region of *n*-type material is in intimate contact with a region of *p*-type material, they form a *p-n junction*. Such junctions are of great importance, since they are the building blocks for semiconductor devices. One cannot form a *p-n* junction by pressing *n*-type material against *p*-type material, since the act of pressing would bring such a tremendous concentration of impurities and imperfections to the interface that the junction would not have the required properties. One way of making a *p-n* junction is to melt an indium pellet which has been placed on top of an *n*-type germanium crystal. Where the germanium has been saturated with molten indium, *p*-type material is produced. A *p-n* junction is formed at the interface between the indium-saturated *p* region and the original *n* material.

As we saw in Sec. 30.2, the Fermi level of a material is of great importance. When two isolated materials are placed in contact, electrons move from the one in which the Fermi level is greater to the other. This lowers the Fermi level in the first and raises it in the second.

Figure 48.10

Isolated energy levels are introduced in the forbidden energy band of a semiconductor by doping it with impurity atoms. (*a*) Donor atoms give levels just below the conduction band and raise the Fermi energy above the middle of the forbidden band.
(*b*) Acceptor atoms introduce levels just above the valence band and lower the Fermi energy below the middle of the forbidden band.

Charges flow until the Fermi levels become equal in the two materials. In an isolated *n*-type semiconductor the Fermi level lies above the center of the forbidden energy region (Fig. 48.10), while for a *p*-type material the Fermi level lies below the center of the energy gap. At a *p-n* junction, where *n* and *p* materials are in intimate contact, electrons from the donor atoms near the junction fill the nearby acceptor states just across the junction (Fig. 48.11). This makes the *p*-material negative and raises the Fermi level there, while the loss of electrons in the *n*-region leaves it positively charged. In the immediate vicinity of the junction there is a scarcity of free carriers of either sign, because the electrons from the *n* side have filled acceptor states and holes on the *p* side. Thus at the junction we have a *depletion layer* perhaps 20 nm ($\approx$50 atom layers) thick which is a relatively poor conductor together with a double layer of bound charges, negative on the *p* side and positive on the *n* side.

When the Fermi level becomes the same on both sides, equilibrium is established. This does not mean that there is no charge transfer across the junction but rather that the charge transfers in both directions are equal. At the junction in equilibrium there are still many more electrons per unit volume in the conduction band in the *n* material at the right of Fig. 48.11 than in the *p* material at the left; consequently, there is an electron current associated with the diffusion of these electrons against the electric field of the double layer. At the same time there is an equal and opposite electron current from electrons which are accelerated "down the hill" by this field. This latter current has a magnitude which depends on the concentration of electrons in the *p*-type material and on the low rate at which the free electrons appear at the junction. Hence this current has a *saturation* value I_s which is ordinarily small. (The saturation current does not depend on the height

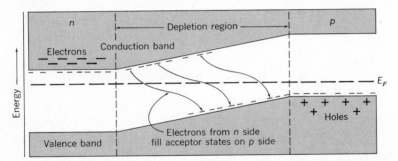

Figure 48.11

Energy-level diagram for an unbiased *n-p* junction. Electrons from donor atoms on the *n* side fill acceptor levels on the *p* side, making the *p* material negatively charged and the *n* material positively charged. A scarcity of charge carriers in the transition region increases the resistivity there. At equilibrium electrons diffuse from the *n* side to the *p* side at the same rate as electrons from the *p* side move "downhill" to the *n* side.

Figure 48.12
How current varies with bias potential
difference across a *p-n* junction.
Application of bias voltage shifts the
Fermi levels in the two sides of the
junction as shown.

of the potential hill any more than the water which goes over a waterfall depends on the height of the fall.) Holes are also exchanged across the boundary; their behavior is directly analogous to that of electrons.

When we apply a potential difference across the *p-n* junction, making the *p* material positive relative to the *n* material, the hill down which the electrons slide is reduced. This has essentially no effect on the saturation current I_s due to electrons going from the *p* to the *n* side. However, many more electrons from the *n* material diffuse to the *p* material, so the electron flow from the *n* to *p* side is greatly increased by this bias. The current in what is known as the *forward direction* rises rapidly with increasing forward bias (or potential difference), as shown in Fig. 48.12. On the other hand, if we make the *n*-type material positive relative to the *p*-type, the hill is increased, and electron flow from *n* to *p* is substantially reduced, while the current I_s remains the same. Consequently, with negative or reverse bias there is a very small current essentially independent of bias. Thus a *p-n* junction acts as a rectifier—not quite a perfect one, since there is some current in the reverse direction for reverse bias, but nevertheless an effective one, since we get a substantial current in the forward direction for a modest potential difference and a very small reverse current for the same (magnitude) reverse bias. The *p-n* junction rectifier performs essentially the same functions as the vacuum-tube diode (Sec. 47.2).

An alternative explanation for the rectifying action of a *p-n* junction can be made in terms of the depletion layer. Reverse bias removes electrons from the *n*-side and adds them to the *p*-side, thus increasing the width of the transition region in which there is a deficiency of carriers. This leads to a small current under reverse bias. On the other hand, forward bias adds electrons to the *n* side and removes them (leaving holes) in the *p* region. This reduces the width of the transition region and provides carriers for a substantial forward current.

48.8 THE *n-p-n* TRANSISTOR

An extremely important electronic device which is capable of amplifying signals is the *n-p-n* junction transistor (Fig. 48.13). This device consists of a thin layer of weakly *p*-type material known as the *base* sandwiched between strongly n-type material called the *emitter* and relatively weak n-type material which comprises the *collector*. Typically, the base is only of the order of a few micrometers in thickness, and the entire transistor may occupy a volume of only a few thousandths of a cubic centimeter. The base must be sufficiently thin for the probability of an electron's combining with a hole within the base to be small compared with the probability of the electron's passing completely through the base. Essentially all electrons which leave the emitter pass through the base to the collector, so the electron current to the base is small compared with the electron current to the collector.

The emitter is operated at a small forward bias ($\approx$0.5 V) relative to the base, while the collector is reverse-biased by a potential difference of several volts. These potential differences appear almost entirely at the junctions themselves, since both the n regions are relatively good conductors. In the emitter, thermal agitation provides a generous and continuous supply of conduction electrons, many of which reach the emitter-base interface and diffuse into and through the base. The electrons lost by the emitter are restored through the emitter electrode.

Since the emitter is forward-biased, a small change in emitter potential results in a substantial change in the electron current from emitter to base to collector and hence to a substantial change in the potential difference across the load. Thus a small voltage signal at the emitter leads to a larger change in the potential difference across the load, and the transistor operates as a voltage amplifier with a high impedance load.

Transistors can perform the same general functions as triodes (Secs. 47.3 to 47.5) and other vacuum tubes. While we have discussed here only an n-p-n transistor, it should be pointed out that p-n-p transistors are of comparable importance; their characteristics are similar except that the carriers are holes rather than electrons.

Among the major advantages of the transistor are its extremely small volume, the small potential differences it requires for operation, its extremely long life, and its ruggedness. There are disadvantages as well, among them the fact that at sufficiently high temperatures there are so many free electrons in all parts of the transistor that it fails to function properly.

Figure 48.13
Energy-level diagram for an n-p-n transistor used as an amplifier.

QUESTIONS

1 In what various ways can the "free" electrons be removed from a metal? Discuss.

2 When two materials are rubbed together, what determines which material gains electrons and which loses them?

3 What happens to the Fermi level of a material when electrons are added to it? Why?

4 What is the role of donor impurities in a semiconductor? What valence do donor atoms for a germanium crystal usually have?

5 Why do acceptor atoms lower the Fermi level in a crystal?

6 How does the number of conduction electrons compare with the number of holes in (a) an ideal intrinsic semiconductor, (b) a p-type semiconductor, and (c) an n-type semiconductor? Explain.

7 Why do transistors malfunction or fail if the temperature becomes too great?

8 Does an ordinary semiconductor at constant temperature obey Ohm's law? Does a p-n junction? Explain.

9 What is the Hall effect? Why is it of great importance in connection with semiconductors? How could it be used to measure magnetic fields?

10 Why should the disordered region at an edge dislocation be etched away much faster by an acid than a perfectly ordered region?

11 Draw a p-n-p transistor circuit analogous to the n-p-n transistor circuit of Fig. 48.13. Explain the operation of the p-n-p transistor, in which the bias voltages are opposite those of the n-p-n transistor.

PROBLEMS

1 An experiment on the threshold wavelength for photoelectric emission from tungsten yields the value 270 nm. Find the energy of the Fermi surface (relative to the configuration in which the electron is at rest an infinite distance from the metal) in joules and in electron volts. Compute the maximum energy of photoelectrons ejected by photons of $\lambda = 150$ nm in both joules and electron volts.
$Ans.$ 7.36×10^{-19} J; -4.6 eV; 5.89×10^{-19} J; 3.67 eV

2 The photocurrent of a cell can be cut to zero by a minimum retarding potential of 2.4 V when monochromatic light of 200 nm is incident. (a) What is the energy of the Fermi surface relative to the configuration when the electron is held an infinite distance from the material? (b) What is the maximum kinetic energy of photoelectrons for light of 250-nm wavelength? (c) What is the longest wavelength which can produce photoemission?

3 In a Hall experiment on silver a potential difference of 59 μV is developed across a foil 1.5 cm wide and 0.05 mm thick when the foil bears a current of 28 A in a magnetic field of strength 1.25 T in the direction of the thickness. If C_1 (Fig. 48.7) is positive, calculate the Hall coefficient of silver. $Ans.$ -8.4×10^{-11} m^3/C

4 The Hall coefficient of sodium is -2.5×10^{-10} m^3/C, and the density of sodium is 971 kg/m^3. From these data estimate the average number of free electrons per sodium atom.

5 In a NaCl crystal the ions are 2.82 Å from their nearest neighbors. Assuming that each ion is attracted only by the nearest ion in the adjacent layer, calculate the stress (force per unit area) required to pull the crystal apart. Do you expect your answer to be larger or smaller than an experimental determination? Why?
$Ans.$ 3.65×10^{10} N/m^2; larger; next nearest neighbors repel

6 In a potassium chloride crystal the ions are 3.14 Å apart. Assuming that each ion is attracted only by the nearest ion in the adjacent layer, find the stress (force per unit area) required to pull the crystal apart. Why is the answer smaller than for NaCl (Prob. 5)?

7 Find the average volume occupied by (a) a copper atom in a copper bar of density 8,930 kg/m^3, (b) an osmium atom in a crystal of osmium, the densest of the metals at 22,500 kg/m^3, and (c) a lithium atom in the least dense of metallic crystals at 530 kg/m^3.
$Ans.$ (a) 11.8×10^{-30} m^3; (b) 14×10^{-30} m^3; (c) 21.7×10^{-30} m^3

THE NUCLEI OF ATOMS are small indeed, but they contain most of the particles which make up the atom and almost all the atomic mass. Since mass is a form of energy, it is in nuclei that most of the energy of atoms is concentrated. The possibility of utilizing some of this energy is a continuing challenge to physicists. Quite apart from this incentive, the structure of the nucleus and its reactions are of great interest.

CHAPTER FORTY-NINE
NUCLEI AND NUCLEAR ENERGY

49.1 BUILDING BLOCKS OF NUCLEI

In our discussion of atoms thus far, we have been concerned primarily with the energy states of the electrons which are bound to the nuclei by coulomb attraction. We have attributed to each nucleus a charge Z and a mass number A, which is the integer closest to the atomic weight of the nuclear species to which the atom belongs. In this chapter we consider the structure and properties of the nuclei.

The simplest of all nuclei is that of ordinary hydrogen, an atom of which we designate by $_1^1H$. Note that we shall indicate the atomic number Z of an atomic species with a subscript at the lower left and the mass number A with a superscript at the upper left.† The nucleus of the hydrogen atom, called the *proton* (p), bears a positive charge equal to the negative charge on the electron. It has a mass of 1.6725×10^{-27} kg (1,836 times the mass of the electron) or 1.00728 atomic mass units (u). *One unified atomic mass unit is one-twelfth the mass of a carbon 12 ($_6^{12}C$) atom.*

The proton has an intrinsic spin angular momentum such that the component in the direction of an applied external field is either $\frac{1}{2}h/2\pi$ or $-\frac{1}{2}h/2\pi$. Associated with this spin is a small magnetic moment (Sec. 36.5) of the order of magnitude one would expect from a spinning positive charge the size of the proton, which has a radius of about 10^{-15} m.

Since all nuclei have charges which are integral multiples of the protonic charge, it is reasonable and fruitful to assume that protons are fundamental building blocks of all nuclei. A nucleus of atomic number Z contains Z protons. However, for a typical nucleus only about one-half the mass can be attributed to the protons. It was suggested by Rutherford that there might be a second building block of roughly the same mass as the proton but having no charge. This particle, the *neutron*, was discovered in 1932 (Sec. 50.2). The neutron mass is 1.00866 u,

---

† We follow the recommendation of the International Union of Pure and Applied Physics. Many books and papers indicate the mass number A by a superscript at the right—$_1H^1$, $_6C^{12}$—a form which requires less composition time.

slightly greater than that of the proton. It has the same intrinsic angular momentum, and, although chargeless, it does have a magnetic moment about seven-tenths that of the proton and of opposite sign (as though negative charge were rotating). There is evidence that the neutron has a positive core with enough negative charge in its outer reaches to make it electrically neutral. A spinning neutral system of this kind would have a negative magnetic moment.

Protons and neutrons are the building blocks with which we may construct any nucleus. The number of protons in a nucleus determines the atomic number Z, which is the positive charge on the nucleus in elementary units and the number of electrons associated with the neutral atom. The sum of the number of protons and the number of neutrons is the *mass number* A. Protons and neutrons in a nucleus are collectively referred to as *nucleons*. The mass number A is then the number of nucleons in the nucleus. In the quantum-mechanical treatment of nuclear structure, the strong binding of nuclei can be explained in terms of the exchange of charge between nucleons. A nucleon may be a proton at one instant and become a neutron the next instant by transferring its charge to another nucleon, which transmutes from neutron to proton. In addition to mass, each nucleon contributes angular momentum and magnetic moment to the nucleus. However, these are vector quantities, and their resultant is usually not large—indeed, it is typically zero for nuclei with even numbers of both protons and neutrons. This can be understood by applying the Pauli exclusion principle to nuclei.

The chemical behavior of any atom is determined by its electronic configuration and hence by its atomic number Z. All atoms with $Z = 1$ are hydrogen, all with $Z = 8$ are oxygen, and all with $Z = 92$ are uranium. However, atoms with the same number of protons may have different numbers of neutrons. Atoms with the same atomic number, but different mass number, are *isotopes* (Sec. 42.6). Most elements have at least two stable isotopes; tin has ten, the largest number for any element. In addition to stable isotopes, one or more radioactive isotopes are known for every element.

49.2 THE SIZE OF NUCLEI

If a typical atom were expanded to a radius of 10 m, its nucleus would have a radius of less than 1 mm. The volume of a nucleus is only about 10^{-12} of the volume of the atom. To put it another way, a nucleus viewed from an outer electron subtends an angle of 1 percent of the angle subtended by the sun at the earth ($0.5°$). In a sense an atom is a very open structure largely comprising empty space through which electrons are passing.

Although nuclei are exceedingly small, several types of measurements have been devised to determine nuclear radii. While there are minor differences in the empirical values, all methods agree on the order of magnitude. The volumes of all nuclei are roughly proportional to the number of nucleons A. If we assume a roughly spherical shape, we find that the radius (in meters) of a nucleus of mass number A is given approximately by

$$r = 1.2 \times 10^{-15} \sqrt[3]{A} \qquad\qquad (49.1)$$

The heaviest of all common nuclei is that of uranium 238, for which the nuclear radius is somewhat less than 10^{-14} m, as compared with the typical atomic radius of about 10^{-10} m.

The densities of nuclei of all kinds are roughly equal, about 2×10^{17} kg/m³. A pint of packed nucleons would weigh 100 billion tons!

49.3 THE DEUTERON

In 1932 Urey and his collaborators discovered a stable isotope of hydrogen with mass number 2. Such a hydrogen atom is known as *deuterium* or *heavy hydrogen*. Its nucleus, the *deuteron*, is composed of one proton and one neutron and has a mass of 2.01355 u. When a proton and a neutron combine to form a deuteron, a mass of 0.00239 u (1.00728 + 1.00866 − 2.01355) disappears and is transformed to a gamma photon.

According to Einstein's theory of relativity, the mass-energy equivalence is expressed (Sec. 41.9) by

$$\text{Energy} = mc^2 \qquad\qquad (49.2)$$

By Eq. (49.2),

$$(1 \text{ u})c^2 = (1.66 \times 10^{-27} \text{ kg})(3 \times 10^8 \text{ m/sec})^2 = 1.49 \times 10^{-10} \text{ J}$$

Since 1 eV = 1.60×10^{-19} J,

$$1 \text{ u} = \frac{931.5 \text{ MeV}}{c^2}$$

The mass which disappears when a deuteron is formed corresponds to 2.225 MeV. In order to break a deuteron into a proton and a neutron, we must supply this energy. Indeed, one of our best methods of measuring the neutron mass involves finding the lowest-energy photon which can split the deuteron into a proton and a neutron, a process called *photodisintegration*.

The fact that energy is released in the reaction $^1_1\text{H} + ^1_0 n \rightarrow {}^2_1\text{H}$ and is required to break up the deuteron is analogous to the situation in chemical reactions. When carbon and oxygen unite to form carbon dioxide, energy is set free. If we wish to decompose carbon dioxide into carbon and oxygen, we must add this amount of energy to the system. The total energy which would be released if we could perform a series of operations to build a nucleus from neutrons and protons is called the *binding energy* of the nucleus. Alternatively, the binding energy is the energy required to tear the nucleus apart into protons and neutrons.

A proton and a neutron are bound together only if their spins are essentially aligned in the same direction. As a result the spin angular momentum of the deuteron is one unit of $h/2\pi$, and the magnetic moment of the deuteron is only about one-third that of the proton, because the neutron contributes negative magnetic moment to the system.

There is a third isotopic form of hydrogen, called *tritium*, which has mass number 3. Its nucleus, the *triton*, is composed of one proton and two neutrons. It is unstable, half a given sample transforming to helium 3 in 12 years.

49.4 THE BINDING ENERGY OF NUCLEI

Nuclear forces are so great and the amount of energy involved when nucleons are brought together to form a new nucleus is so tremendous that we can observe the difference in masses before and after. We have called the difference between the mass of the component nucleons from which a nucleus is composed and the actual mass of the nucleus the nuclear *binding energy* c^2. In practice, mass spectrometers determine the masses of heavy atoms (or ions) rather than the masses of nuclei directly. It is convenient to think of any neutral atom of mass number A and atomic number Z as being composed of Z hydrogen atoms and $A - Z$ neutrons. The hydrogen atoms bring the protons for the nucleus and the electrons for the shells. Then the binding energy is given by the difference between the mass of Z hydrogen atoms plus $A - Z$ neutrons and the mass of the neutral atom:

$$\text{Binding energy} = [Zm_\text{H} + (A - Z)m_n - M]c^2 \qquad (49.3)$$

where m_H, m_n, and M are the masses of the hydrogen atom, the neutron, and the neutral atom, respectively. Table 49.1 lists the masses of a number of common atomic species.

To obtain the mass of the nucleus from the known atomic mass, one subtracts the mass of the Z electrons.

$$\text{One electron mass} = 0.0005486 \text{ u} = \frac{0.511 \text{ MeV}}{c^2}$$

The binding energy increases with the mass number A. For purposes of comparison it is convenient to divide the total binding energy by the number of nucleons to obtain the *binding energy per nucleon*. A plot of the binding energy per nucleon for many stable nuclei is shown in Fig. 49.1.

Table 49.1
Masses of Atoms
(Based on C^{12} as 12.00000)

Element	Z	A	Mass, u
(Neutron)	0	1	1.008663
Hydrogen	1	1	1.007825
(Deuterium)		2	2.01410
(Tritium)		3	3.01605
Helium	2	3	3.01603
		4	4.002603
Lithium	3	6	6.01513
		7	7.01600
Carbon	6	12	12.000000
Nitrogen	7	14	14.00307
Oxygen	8	16	15.99491
Sodium	11	23	22.98977
Sulfur	16	32	31.97207
Nickel	28	58	57.93534
Copper	29	63	62.92959
Tin	50	120	119.90220
Lead	82	208	207.97665
Uranium	92	238	238.05077

Example

Find the binding energy and the binding energy per nucleon for ^{7}Li, which is composed of three hydrogen atoms and four neutrons.

$$3 \times 1.007825 = 3.023475$$
$$4 \times 1.008663 = \underline{4.034652}$$
$$7.058127$$
$$^7\text{Li} \qquad \underline{7.01600}$$
$$\text{Binding energy} = 0.0421 \quad \text{u} \times c^2$$
$$0.0421 \times 931.5 = 39.3 \text{ MeV binding energy}$$
$$\text{Binding energy per particle} = \frac{39.3}{7} = 5.6 \text{ MeV/nucleon} \qquad ////$$

49.5 STABLE NUCLEI

We do not know the exact nature of the forces which hold nucleons together, but we do know what stable nuclei exist in nature and some of the conditions which must be satisfied if a nucleus is to be stable. Among the light elements we never find stable nuclei which have radically

Figure 49.1
Binding energy per nucleon as a function of mass number for stable nuclei.

different numbers of protons and neutrons. For example, oxygen has three stable isotopes with masses 16, 17, and 18. The 8 protons of the oxygen nucleus form stable configurations with 8, 9, or 10 neutrons, but not with 4 neutrons or with 20 neutrons. As we go up the periodic table, the number of neutrons increases more rapidly than the number of protons, until in ^{238}U there are almost 1.6 neutrons for every proton.

The increase in binding energy per nucleon for light elements (Fig. 49.1) is associated in part with the fact that on the average each nucleon has more neighbors to which it is bound. However, the rise is not steady; for example, the binding energy per nucleon is greater for $^{4}_{2}$He and $^{16}_{8}$O than for nuclei with one or two more nucleons. This can be explained in terms of a *shell model* of nuclei, which predicts that 2, 8, 20, 50, 82, and 126 nucleons of either type form particularly stable arrays. Throughout the realm of nuclei there is strong evidence of the pairing of two protons of opposite spin and of two neutrons of opposite spin to form very stable nuclei. Except for hydrogen, elements of even atomic number are several times as abundant as those of odd atomic number. Elements of odd Z have at most two stable isotopes, both of which ordinarily have an even number of neutrons (prominent exceptions: $^{2}_{1}$H, $^{6}_{3}$Li, $^{10}_{5}$B, and $^{14}_{7}$N). On the other hand, elements of even Z often have many isotopes, most of which have also an even number of neutrons.

For the heavier elements the mutual electrostatic repulsion of the protons increases rapidly with Z, and consequently the net binding energy per nucleon falls off. To this coulomb repulsion we attribute the fact that the number of neutrons exceeds the number of protons for stable nuclei heavier than $^{40}_{20}$Ca. It is also the reason for the instability of the very heavy nuclei. Whenever the arrangement of the nucleons is unstable, the nucleus is radioactive and undergoes one of several kinds of transition. It was in the observation of such transitions that nuclear physics had its birth.

49.6 ALPHA PARTICLES

In Sec. 43.5 we learned that naturally radioactive nuclei emit three types of radiation, named *alpha*, *beta*, and *gamma* rays. A few nuclear species, for example, $^{227}_{89}$Ac and $^{218}_{84}$Po, emit all three, but most emit gamma rays and either alpha particles or beta particles but not both.

Figure 49.2

(*a*) Graph of range of 7.68-MeV alpha particles from $^{214}_{84}$Po in air, as measured in apparatus (*b*).

The alpha particle is the nucleus of the helium atom. This fact was established by separating an alpha emitter from an evacuated region by means of a thin wall through which alpha particles could pass. After a few hours enough gas collected in the evacuated region so that its spectrum could be excited. This showed that the gas was helium. Alpha particles have been found to bear a double elementary positive charge, substantiating the fact that they are helium nuclei.

Alpha particles are easily absorbed by metal foils or by a few centimeters of air. They affect a photographic plate, cause many materials of fluoresce brilliantly, and ionize the air through which they pass. When alpha particles strike screens of fluorescent material, they produce tiny flashes of light called *scintillations*. Many of the early researches on alpha particles were carried out in darkened rooms by patient observers who sat for hours counting these flashes. If an alpha emitter is placed at R in Fig. 49.2*b*, a fluorescent screen S receives many scintillations each second when it is close to the source. As the screen is moved away from the alpha emitter, the number of scintillations per second remains roughly constant over a considerable distance. Then the number drops off sharply (Fig. 49.2*a*). As the particles move through the gas, they lose energy by ionizing the gas. In passing through the first few centimeters of air, the alpha particles are slowed down significantly, but few are lost. This type of range curve is characteristic of particles which lose energy gradually as they pass through matter. A curve such as that of Fig. 49.2*a* suggests that all the alpha particles from the source had the same energy. For any given alpha emitter we find that either all the alpha particles have the same energy or that there are a few groups of alpha particles emitted, each group with its own discrete energy.

In general, a nucleus is unstable against alpha decay if its mass exceeds that of an alpha particle plus that of the residual nucleus. This is equivalent to saying that a neutral atom is unstable against alpha decay if its mass exceeds that of a helium atom plus that of the remaining atom. In the reaction

$$^{226}_{88}\text{Ra} \rightarrow {}^{222}_{86}\text{Rn} + {}^{4}_{2}\text{He} + \text{energy}$$

the energy released is the difference between the mass of the radium (226.02536 u) and the sum of the masses of the radon (222.01753) and the helium (4.00260). This is 0.00523 u, or 4.87 MeV.

49.7 BETA EMISSION

Beta particles are electrons ejected from nuclei with speeds which may exceed $0.9c$, where c is the speed of light. These electrons have a penetrating power far greater than that of alpha particles.

The energies of the beta rays from a given radioactive nucleus vary continuously from very low energy to a maximum (Fig. 49.3). There is a *continuous energy spectrum of the beta rays*. This is in sharp contrast with the energy spectrum of the alpha particles, which has at most a few discrete energies. If beta rays were the only particles coming out of a given kind of nucleus, one would expect the energies of the beta rays to be discrete. However, if some other particle were emitted

Figure 49.3

Beta-ray spectrum of $^{210}_{83}$Bi showing the relative number of electrons emitted at various energies.

simultaneously with the electron, this particle would share the total energy available and the electron energy spectrum would be continuous. There is convincing evidence that another particle is indeed ejected from beta emitters along with the electron. It has no charge or rest mass but has a spin equal in magnitude to that of the electron. This particle was originally called the *neutrino*, meaning "little neutral one," but for reasons suggested in Sec. 50.9 it seems preferable to call it the *antineutrino* (represented by $\tilde{\nu}$).

When a nucleus contains too many neutrons for the number of protons present, the unbalance may be corrected by the transformation of a neutron into a proton with the emission of a beta ray and an antineutrino:

$$_0^1 n \rightarrow {}_1^1 H + {}_{-1}^0 \beta + \tilde{\nu}$$

An isolated neutron is not a stable particle but decays by the reaction above with a half-life of about 12 min, i.e., in 12 min one-half of any sample of neutrons not bound in a nucleus decays.

The beta-decay process results in a product nucleus of the same mass number A but atomic number Z higher by 1. Beta decay is expected whenever the mass of the product nucleus is less than that of the original nucleus. As an example, the mass of $_6^{14}C$ exceeds that of $_7^{14}N$, and

$$_6^{14}C \rightarrow {}_7^{14}N + {}_{-1}^0 \beta + \tilde{\nu}$$

When an element of low to moderate mass has too many protons for the number of neutrons, a transmutation occurs which results in the changing of a proton into a neutron. This may happen in one of two ways: (1) a positron may be emitted or (2) an orbital electron may be captured. In either case a neutrino is also involved. An example of a positron emitter is carbon 11, which decays by the reaction

$$_6^{11}C \rightarrow {}_5^{11}B + {}_1^0 \beta + \nu$$

where $_1^0 \beta$ represents a positron and ν a neutrino.

49.8 GAMMA RAYS

High-energy electromagnetic rays are called gamma rays when they are emitted by radioactive nuclei. The properties of gamma rays are identical with those of x-rays of the same wavelength.

When a radioactive nucleus emits an alpha or beta particle, it is likely that the residual nucleus will be left not in its most stable arrangement but in an excited state. Ordinarily, when the nucleus goes to a more stable configuration, the energy released is radiated as a gamma ray. Like the alpha particles, the gamma rays from a given radioactive species have discrete energies.

49.9 THE URANIUM-RADIUM SERIES

The most abundant radioactive nucleus in nature is that of the uranium isotope of mass 238 u, which we abbreviate ^{238}U. This material decays by the emission of an alpha particle with a half-life of 4.51×10^9 years.

The half-life is the time required for one-half the nuclei in a sample to disintegrate. For example, the half-life of radon is 3.85 days; one-half of a sample of radon disappears in 3.85 days. At the end of 7.7 days three-quarters of it has disappeared, and one-fourth remains. After 15.4 days, only one-sixteenth of the original radon is present.

The reaction in which ^{238}U emits an alpha particle may be written

$$^{238}_{92}\text{U} \rightarrow {}^{4}_{2}\text{He} + {}^{234}_{90}\text{Th}$$

Note that the emission of an alpha particle reduces the atomic number (left subscript) by 2 and the mass number (left superscript) by 4. The ^{234}Th decays by beta emission to protactinium 234, which in turn emits a beta ray and becomes ^{234}U:

$$^{234}_{90}\text{Th} \rightarrow {}^{0}_{-1}\beta + {}^{234}_{91}\text{Pa}$$
$$^{234}_{91}\text{Pa} \rightarrow {}^{0}_{-1}\beta + {}^{234}_{92}\text{U}$$

^{234}U is an alpha emitter with the daughter ^{230}Th. Ultimately the decay series ends with lead 206. The detailed list of reactions and half-lives for the uranium-radium series is presented in Table 49.2.

In addition to the uranium-radium series, three other natural radioactive series exist. The longest-lived member of the *thorium series* is ^{232}Th, and the final stable nucleus formed is again lead, but this time ^{208}Pb. The *neptunium series* has ^{237}Np as its longest-lived member, and the radioactive chain ends with bismuth 209, while the *actinium series* has ^{235}U as its longest-lived member and ^{207}Pb as its final stable nucleus.

Table 49.2
The Uranium-Radium Series

Nuclide	Element	Early name	Half-life	Energy of rays, MeV Alpha	Beta	Gamma
$^{238}_{92}$U	Uranium	Uranium I	4.5×10^9 yr	4.18		
$^{234}_{90}$Th	Thorium	Uranium X$_1$	24.5 days		0.103	0.09
$^{234}_{91}$Pa	Protactinium	Uranium X$_2$	1.14 min		2.32	0.8
$^{234}_{92}$U	Uranium	Uranium II	2.33×10^5 yr	4.76		
$^{230}_{90}$Th	Thorium	Ionium	8.3×10^4 yr	4.66		
$^{226}_{88}$Ra	Radium	Radium	1,620 yr	4.79		0.19
$^{222}_{86}$Rn	Radon	Ra emanation	3.8 days	5.49		
$^{218}_{84}$Po	Polonium	Radium A	3.05 min	6.00		
$^{214}_{82}$Pb	Lead	Radium B	26.8 min		0.65	0.29
$^{214}_{83}$Bi	Bismuth	Radium C	19.7 min	5.5	3.15	1.8
$^{214}_{84}$Po or	Polonium	Radium C′	1.5×10^{-4} sec	7.68		
$^{210}_{81}$Tl	Thallium	Radium C″	1.32 min		1.80	
$^{210}_{82}$Pb	Lead	Radium D	22 yr		0.026	0.047
$^{210}_{83}$Bi	Bismuth	Radium E	5.0 days		1.17	
$^{210}_{84}$Po	Polonium	Radium F	138 days	5.3		0.8
$^{206}_{82}$Pb	Lead	Radium G	Stable			

Since each alpha particle is a helium nucleus, helium is in the process of formation in all minerals containing alpha-emitting radioactive substances. The number of alpha particles given out by 1 g of uranium in equilibrium with its radioactive products has been found to be 9.7×10^4 per second. If this helium were all occluded and retained by the mineral, the ratio of the amount of helium to the amount of uranium would give an estimate of the age of the mineral. Without doubt some of the helium escapes, and this estimate of the age of the mineral would be too low. A determination of the uranium and helium in different kinds of rocks has shown that the ratio of the amount of helium to the amount of uranium is largest in those formations which, from geological considerations, are known to be the oldest. Studies of many kinds of radioactive decay point to an age for the earth of the order of 5×10^9 years.

For determining the ages of artifacts or archaeological interest, radioactive carbon 14 has proved to be a highly useful tool. Radiocarbon decays by beta emission with a half-life of 5,570 years. The $^{14}_{6}C$, produced by the bombardment of atmospheric nitrogen with neutrons, is in equilibrium in the earth's atmosphere and is present in all living matter in equilibrium concentration through exchange with the atmosphere. However, once the matter ceases to live, exchange stops, and the $^{14}_{6}C$ concentration decreases steadily. Libby has established the dates when life ceased in a wide variety of organic materials. This development is an important tool for geology and archaeology.

49.11 FUSION

The alpha particle is composed of four nucleons, two protons, and two neutrons. The mass of a ^4He atom is 4.00260 u, while the sum of the masses of two hydrogen atoms and two neutrons is 4.03298 u. The binding energy is 0.0304 u, or 28.2 MeV.

If we could build helium nuclei from protons and neutrons, we would have a source of tremendous energy. Actually, the probability of getting two protons and two neutrons simultaneously in so small a volume that they would interact to form an alpha particle is small. However, we might well carry out the process by first having a proton and a neutron combine to form a deuteron. Then two deuterons might be brought close enough together to interact. When we do this, so much energy is available that the four particles do not ordinarily stick together. Rather, either a high-energy proton is emitted, leaving us a triton, or a neutron is emitted, leaving us with a ^3He nucleus. The reactions are

$$^2_1H + {}^2_1H \Big\langle {{}^3_2He + {}^1_0n + 3.28 \text{ MeV energy} \atop {}^3_1H + {}^1_1H + 4.04 \text{ MeV energy}}$$

The energy released in a nuclear reaction such as this is calculated by subtracting the mass of the reaction products for the mass of the interacting particles. One of the many ways in which we can get an alpha particle from these products is to bring a deuteron and a triton together. The resulting reaction is

$$_1^2H + _1^3H \rightarrow {}_2^4He + {}_0^1n + 17.6 \text{ MeV energy}$$

The process of building helium from hydrogen is called *fusion*. As we have seen, 28.2 MeV is released for each helium nucleus built from the building blocks. If we produce fusion reactions on a large scale, we release a colossal amount of energy in a very short time and thus create a great explosion. The hydrogen bomb is an example of a fusion device. If deuterons and tritons are to come in contact with each other in spite of the repulsion between the positive charges, they must have high speeds. In a hydrogen bomb these speeds may be achieved by bringing the particles to an exceedingly high temperature, thereby producing *thermonuclear reactions*. One way to produce temperatures of this magnitude is by a nuclear explosion utilizing fission. An alternative is to pour vast amounts of energy into a plasma of ionized deuterium and tritium. In achieving the temperature needed (tens of millions of kelvins) in a laboratory fusion experiment, the hot plasma is confined by magnetic fields, since any type of material container would be vaporized by contact with the plasma.

The release of energy by means of the combination of nucleons into heavier particles (fusion) is directly analogous to the release of chemical energy by the combination of atoms. In the nuclear case one may release energy by combining a proton and a neutron to form a deuteron. In the analogous chemical situation one may combine an atom of sodium with an atom of chlorine to form sodium chloride. One great difference between these two reactions lies in the amount of energy released per interacting particle. In the case of proton and neutron the energy resulting from the combination is 2.2 MeV per proton, while in the case of the sodium chlorine reaction the energy is about 4 eV per sodium atom.

49.12 FISSION

One can obtain energy from chemical reactions not only by combining atoms of various substances to form molecules but also by breaking very large and complex molecules, such as trinitrotoluene molecules, into smaller and more tightly bound units. Similarly, heavy and complex nuclei can be made to break up with the release of energy.

When any heavy nucleus is highly excited, it may break up into two roughly equal parts. This process is called *fission*. Some heavy nuclei fission readily; others do not. The excitation can be achieved in several ways. For example, gamma rays may induce fission. The most familiar method is by introducing a neutron into a nucleus of uranium 235. When a slow neutron collides with a ^{235}U nucleus, it may be captured. The resulting ^{236}U nucleus breaks up into two roughly equal parts, plus a few additional neutrons. A typical reaction is

$$_{92}^{235}U + {}_0^1n \rightarrow {}_{92}^{236}U \rightarrow {}_{56}^{145}Ba + {}_{36}^{88}Kr + 3{}_0^1n + \text{energy}$$

The mass of the products is less than that of the fissioning nucleus; thus energy is released, on the average about 200 MeV per fission. The excited ^{236}U nucleus may break up in many different ways. Figure

49.4 shows the relative fission yields of the many mass numbers which may be created as fission fragments.

When a heavy nucleus fissions, the product nuclei usually contain several too many neutrons for stability. For example, one common fission product is strontium 95. The heaviest stable isotope of strontium is ^{88}Sr. The radioactive ^{95}Sr nucleus undergoes three successive beta decays and finally reaches stability as niobium 95. Some fission products transmute through beta decays of long half-life and produce serious contamination after a nuclear explosion.

49.13 CHAIN REACTIONS AND THE REACTOR

Since two to three neutrons are released in a typical fission, two or three other atoms may be fissioned if these neutrons are captured by ^{235}U nuclei. The neutrons released in these fissions can produce four or five more fissions, and a *chain reaction* is set off if a sufficient supply of ^{235}U atoms is present in a small volume. Such a chain reaction does not occur in ordinary uranium because a large fraction of the neutrons are captured by ^{238}U nuclei, which are 140 times as abundant in natural uranium as ^{235}U nuclei. Few of the ^{238}U nuclei fission.

To produce the first nuclear weapons, ^{235}U was separated from ^{238}U so a large fraction of the neutrons emitted by fissioning nuclei would be captured by ^{235}U nuclei and thereby produce additional fissions. However, even pure ^{235}U does not support a chain reaction unless an amount known as a *critical mass* is present, because too large a fraction of the neutrons escapes from the surface of a subcritical mass. If two almost critical masses are brought together, the system becomes critical, and an explosion occurs.

The neutrons released in fission are *fast*, with energies of several MeV. They are quickly slowed down by collisions with nuclei. ^{235}U does not capture fast neutrons readily, but it captures slow neutrons (energy less than 1 eV) very effectively. On the other hand, ^{238}U captures neutrons of intermediate energies more readily than slow ones. If the neutrons from fissions in natural uranium are slowed down in some other material so the ^{238}U does not capture them, a sustained reaction using natural uranium can be maintained in a nuclear *reactor* (or *pile*). Figure 49.5 shows a reactor in which carbon is used to slow down or *moderate* the neutrons. (Heavy water, containing deuterium, is another material which is an excellent *moderator*.) Fissions produce fast neutrons which enter the carbon, where they have many collisions and slow down to thermal velocities before they reach another uranium slug. The probability that a slow neutron will be captured by a ^{235}U nucleus is much greater than the probability it will be captured by a ^{238}U nucleus, even though there are 140 times as many of the latter. Cadmium control rods are moved into or out of the reactor to regulate the power level of the pile and keep it from blowing up. Cadmium 113 is an excellent capturer of slow neutrons. A small fraction ($<$1 percent) of the neutrons are not emitted until several seconds have elapsed after a fission has occurred. These *delayed neutrons* make the orderly control of a reactor feasible.

Each fission in a reactor releases about 200 MeV of energy, and

Figure 49.4
Fission yield of ^{236}U as it varies with the mass numbers of the fragments.

Figure 49.5
Uranium-graphite reactor in which uranium slugs are canned in aluminum; cadmium rods are used to control the power level at which the reactor operates.

the total energy transformed from mass to thermal energy in a large reactor is great. Nuclear energy is being used for generating electric power, for heating, for propelling submarines, and for nuclear weapons. Nuclear power plants for aircraft promise to extend the range and revolutionize the operation of airplanes. In the next century it is likely that an ever-increasing fraction of our total energy requirements will be supplied from nuclear sources.

49.14 PLUTONIUM

When neutrons are captured in ^{238}U, fission sometimes occurs, but usually the ^{239}U formed does not split up. However, it does emit a beta particle (half-life 23 min) to become neptunium 239 (Fig. 49.6), a beta emitter with a half-life of 2.3 days which becomes plutonium 239. ^{239}Pu is an alpha emitter with a half-life of 24,400 years. This plutonium fissions readily. Practically no plutonium exists in natural materials. Substantial quantities have been produced for use in nuclear weapons. It is produced in reactors which use natural uranium as fuel.

Figure 49.6
Production of plutonium from ^{238}U.

1 How many protons are associated with the nucleus $^{238}_{92}U$? How many neutrons?

2 The two stable isotopes of neon have mass numbers of 20 and 22. How many neutrons does each isotope have?

3 Why do the atomic weights of many elements differ greatly from integer values? Are the binding energies of the nucleons sufficient to explain this fact?

4 What are the properties of alpha, beta, and gamma rays?

5 How do atomic number and atomic weight change when a nucleus emits (*a*) an alpha particle, (*b*) a beta particle, (*c*) a gamma ray, and (*d*) a positron?

6 Why did early observations of beta decay seem to be in conflict with one of the conservation laws? How has this conflict been resolved?

7 What is meant by the binding energy of a nucleus? How do the binding energies of nuclei vary with atomic number? The binding energy per particle?

8 Which of the following properties of neutrinos and antineutrinos are not zero? (*a*) Kinetic energy; (*b*) rest mass; (*c*) linear momentum; (*d*) angular momentum.

9 A neutron is not charged, but it does have a magnetic dipole moment associated with its spin. How can this be explained on the basis of classical electromagnetism?

10 (*a*) A nucleus of thorium 232 ($^{232}_{90}Th$) eventually becomes lead 208 ($^{208}_{82}Pb$). In the series of alpha and beta decays by which this occurs, how many alpha and beta particles are emitted? (*b*) After a series of alpha and beta decays plutonium 239 ($^{239}_{94}Pu$) becomes lead 207 ($^{207}_{82}Pb$). How many alpha and beta particles are emitted in the complete decay scheme?

11 The half-life of radium 226 is 1,622 years. There is considerable evidence that the age of the earth may be 3 million times as great. If this is true, how can you explain the natural occurrence of radium in moderate abundance?

12 What is the function of graphite in a typical reactor?

13 Why are fission products ordinarily radioactive while fusion products are not?

14 In a typical fission reaction a neutron is captured by a $^{235}_{92}U$ nucleus, and three neutrons and two fission fragments are formed. If one of the fragments is a $^{90}_{36}Kr$ nucleus, what must the other fragment be?

PROBLEMS

1 How much energy would be released if a carbon 12 atom were assembled from six hydrogen atoms and six neutrons? What is the average binding energy per nucleon?
Ans. 92.2 MeV; 7.7 MeV/nucleon

2 Find the binding energy and the binding energy per nucleon for $^{40}_{20}Ca$, which has a mass of 39.9627 u.

3 Find the binding energy and the binding energy per nucleon for tin 120.
Ans. 1.02 GeV; 8.5 MeV/nucleon

4 In Sec. 49.11 three fusion reactions involving interactions of deuterium with deuterium and tritium are written. Check the energy released in each case by finding the mass converted to energy in the reaction.

5 Find the minimum energy required for the photodisintegration of an alpha particle into (*a*) $^3He + ^1_0n$, (*b*) $^3H + ^1H$, and (*c*) $^2H + ^2H$.
Ans. (*a*) 20.6 MeV; (*b*) 19.8 MeV; (*c*) 23.8 MeV

6 Find the binding energy and the binding energy per nucleon for sulfur 32.

7 Find the maximum energy of the beta particles from the reaction $^1_0n \rightarrow ^1_1p + ^0_{-1}\beta + \tilde{\nu}$. If the half-life of free neutrons is 11.7 min, what fraction of a sample of free neutrons remains after 46.8 min?
Ans. 780 KeV; one-sixteenth

8 Find the binding energy and the binding energy per nucleon for tritium. The maximum energy of beta rays emitted in the decay of tritium to ^3_2He is 18 keV. Find the mass difference between ^3_1H and ^3_2He.

9 (*a*) A radioactive nucleus has a half-life of 18 min. What fraction of an initial sample has decayed after 1.5 hr? (*b*) At the end of 15 days one-eighth of a sample of radioactive material remains. What is the half-life?
Ans. (*a*) $\frac{31}{32}$; (*b*) 5 days

10 When a ^{238}U nucleus captures a slow neutron, uranium 239, with a half-life of 23 min, is produced. If a sample of radioactive material from a nuclear reactor contains 10 g of ^{239}U, what mass of this isotope remains at the end of 46 min? 69 min?

11 Carbon 14 decays by beta emission to nitrogen 14 by beta decay with a half-life of 5,570 years. A gram of carbon from a living tree undergoes 720 disintegrations per hour. If 1 g of carbon from a wooden relic undergoes 180 disintegrations per hour, approximately what is the age of the relic? (Assume that when the tree was cut, its radioactivity corresponded to that of a living tree.)   *Ans.* 11,000 to 12,000 years

12 (*a*) The half-life of sodium 24 is 15 hr. What fraction of a sample of ^{24}Na remains after 45 hr? 60 hr? (*b*) The half-life of iodine 131 is 8 days. What fraction of a given sample of ^{131}I remains after 16 days? After 40 days?

13 Polonium 212 emits 8.78-MeV alpha particles. Find the speed of the alpha particles and the mass of the $^{212}_{84}$Po atom. (The mass of $^{208}_{82}$Pb is given in Table 49.1.)

    Ans. 2.06×10^7 m/sec; 211.98867 u

14 Uranium 238 decays to thorium 234 by the emission of an alpha particle of energy 4.18 MeV. Assuming that essentially all the energy released in this decay goes to the alpha particle, calculate the mass of thorium 234. What is the speed of the alpha particle?

15 It is estimated that the energy released in the atomic-bomb explosion at Hiroshima was about 7.6×10^{13} J, equivalent to 20,000 tons of TNT. If an average of 200 MeV was released per fission, and if all fissions were by neutron capture in ^{235}U, find the number of ^{235}U atoms fissioned

and the mass of ^{235}U consumed. If 20 percent of the ^{235}U atoms fissioned, what mass of ^{235}U was needed for the bomb?

    Ans. 2.37×10^{24}; 0.93 kg; 4.63 kg

16 A ^{226}Ra nucleus emits an alpha particle with an energy of 4.79 MeV. Find the speed of the alpha particle and the recoil speed of the ^{222}Rn nucleus.

17 How close can a 4.84-MeV alpha particle approach a lead nucleus assuming the latter remains at rest?   *Ans.* 48.9 fm

18 How close can a 8.78-MeV alpha particle approach a uranium nucleus assuming the latter remains at rest in its crystal?

19 A nuclear reactor converts 2 mg of mass to energy each day. Compute the power output in megawatts. If the average fission yields 200 MeV, how many nuclei fission each second?

    Ans. 2.08 MW; 6.5×10^{16}

20 When a slow neutron is captured by a ^{235}U nucleus, a fission releasing 200 MeV results. How much mass disappears? What fraction of the mass of the reacting particles is converted into energy?

21 The heat of combustion of coal (assume pure carbon) is about 3×10^7 J/kg. (*a*) When 1 kg of coal burns, what mass is converted to energy? (*b*) How many electron volts are released per atom of carbon burned? (*c*) If one-tenth the mass of carbon were annihilated, how many joules could be obtained per kilogram?

    Ans. (*a*) 3.3×10^{-10} kg; (*b*) 3.7 eV/atom; (*c*) 9×10^{15} J

22 How many fissions per day are required for a reactor to develop electric power at the rate of 30 MW if 12 percent of the energy released in fission is converted to electric energy?

23 A fission fragment of 88-u mass has an energy of 120 MeV. Find its speed and the temperature at which molecules of this mass would have an average energy of 120 MeV.

    Ans. 1.62×10^7 m/sec; 9.28×10^{11} K

IN CHAP. 49 we presented many facts concerning nuclei and the particles of which they are formed. But how were these facts established? In this chapter we discuss how physicists have induced reactions by bombarding various nuclear species with particles and how they determine the mechanisms involved from information provided by a wide variety of detectors.

CHAPTER 50
NUCLEAR REACTIONS

50.1 THE FIRST ARTIFICIAL TRANSMUTATION

The discovery of radioactivity in 1896 stimulated research activity which led to an understanding of both the decay schemes of the heavy elements such as uranium and the properties of the alpha, beta, and gamma rays arising from these natural transmutations. In 1919 Rutherford found that when alpha particles passed through air, a penetrating charged particle was produced which had a range as great as 40 cm, although the range of the incident alpha particles was only 7 cm. The more penetrating particles were not produced when the alpha particles passed through oxygen or carbon dioxide, but they were observed whenever nitrogen was present. Rutherford showed that the penetrating particles were high-energy protons, produced by the capture of alpha particles by nitrogen nuclei (Fig. 50.1).

When an alpha particle is captured by a nitrogen nucleus, a highly excited fluorine 18 nucleus is produced. This excited nucleus has too much energy to remain together; it breaks up by ejecting one of the protons, leaving a residual $^{17}_{8}O$ nucleus. We write the reaction as

$$^{4}_{2}He + \,^{14}_{7}N \rightarrow \,^{18}_{9}F^{*} \rightarrow \,^{17}_{8}O + \,^{1}_{1}H + energy$$

Here the asterisk indicates that the ^{18}F nucleus is excited.

Rutherford's discovery of artificial transmutation suggested that physicists could change one element into another by shooting high-speed particles into nuclei. In Rutherford's day alpha particles from radioactive materials were the only convenient high-energy projectiles available. Only high-energy charged particles can enter nuclei, because positive charges repel one another; these repulsive forces become exceedingly great before the two particles are close enough for the attractive nuclear forces to become dominant. Rutherford's discovery was a great stimulus toward the development of accelerators to produce high-energy charged particles. Until such high-voltage machines could be developed, there was always the possibility of using alpha particles from radioactive materials as bombarding projectiles. Here again Rutherford's discovery provided a major stimulus.

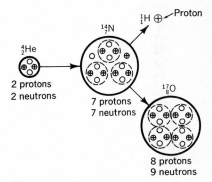

Figure 50.1
A nitrogen 14 nucleus captures a bombarding alpha particle to yield a proton and an oxygen 17 nucleus.

50.2   THE DISCOVERY OF THE NEUTRON

Figure 50.2

A neutron may be ejected when an alpha particle is captured by a beryllium 9 nucleus, leaving a carbon 12 nucleus.

In 1930 Bothe and Becker observed that several of the lighter elements emitted a very penetrating radiation when bombarded by alpha particles from polonium. They thought this radiation consisted of gamma rays, since it was not affected by magnetic fields and was therefore uncharged. The radiation was so penetrating that when alpha particles bombarded beryllium, it required 2 cm of lead to cut the intensity in half.

Two years later the Joliots found that these penetrating rays were fairly readily absorbed in paraffin, water, or cellophane. They concluded that the penetrating radiation was absorbed through some reaction which resulted in the ejection of protons from hydrogenous materials.

Later that same year Chadwick repeated the Joliot experiment and measured the energy of the ejected protons. He concluded that this energy was much too high to have been received in any sort of gamma-ray reaction. Twelve years earlier Rutherford had suggested that there might be a neutral particle with approximately the mass of the proton. If the penetrating radiation consisted of high-speed particles of this kind, they should produce not only recoil protons but also recoils in other nuclei. It turned out that Feather had already published data on the recoil velocities of nitrogen nuclei when they were bombarded by the mysterious particles. By applying the laws of conservation of energy and of momentum to what he assumed were elastic collisions between these particles and target nuclei, Chadwick was able to show that both his own data on proton recoils and those of Feather were consistent with the idea that the penetrating radiation consisted of neutral particles of mass slightly greater than the mass of the proton. These particles are called *neutrons*. The reaction by which neutrons are produced when beryllium is bombarded with alpha particles (Fig. 50.2) is

$$\,^{9}_{4}\mathrm{Be} + \,^{4}_{2}\mathrm{He} \rightarrow \,^{13}_{6}\mathrm{O}^{*} \rightarrow \,^{12}_{6}\mathrm{C} + \,^{1}_{0}n + Q$$

where Q represents the energy released through the conversion of mass.

50.3   REACTIONS INDUCED BY ACCELERATED PARTICLES

Figure 50.3

A lithium 7 nucleus captures a proton to produce beryllium 8 nucleus, which promptly breaks into two alpha particles.

Soon after Rutherford's discovery of artificial transmutation, physicists in several countries began to build accelerators to produce high-energy protons. The first device which was successful in accelerating charged particles artificially to produce nuclear reactions was put into operation by Cockcroft and Walton in England in 1930. Their accelerator produced a beam of protons of 0.5 MeV energy by utilizing high-voltage transformers, rectifiers, and capacitors.

Cockcroft and Walton found that when lithium was bombarded with protons of 150 keV energy, alpha particles were emitted with energies of about 8.6 MeV (Fig. 50.3). The following reaction occurred:

$$\,^{7}_{3}\mathrm{Li} + \,^{1}_{1}\mathrm{H} \rightarrow \,^{8}_{4}\mathrm{Be}^{*} \rightarrow \,^{4}_{2}\mathrm{He} + \,^{4}_{2}\mathrm{He} + 17.3 \text{ MeV energy}$$

The energy released is shared equally by the alpha particles, since this is the only way in which momentum can be conserved.

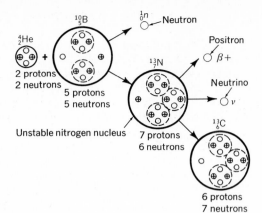

Figure 50.4
A boron 10 nucleus captures an alpha particle, producing a neutron and a nitrogen 13 nucleus; the $^{13}_{7}$N nucleus is unstable and decays to carbon 13 by the emission of a positron and a neutrino.

The source of the energy released in this and other nuclear reactions is the conversion of mass into energy. The masses of the lithium and hydrogen atoms add to 8.02381 u (Table 49.1), while the two helium atoms have a total mass of 8.00520 u. The decrease in mass is 0.0186 u, which corresponds to 17.3 MeV.

Shortly after Cockcroft and Walton made their discovery, other types of accelerators were put into operation, some of which are described in Secs. 31.8 and 50.5.

When a beam of high-energy particles (electrons, protons, deuterons, alpha particles, or other ions) bombards nuclei, a wide variety of reactions take place in which the bombarding particle is captured to produce a radioactive "compound" nucleus, which then emits one or more particles before it reaches a stable state. So long as the bombarding particles have energies less than about 100 MeV, the reactions can be described in terms of protons, neutrons, electrons, positrons, neutrinos, and antineutrinos. However, when the bombarding particles have energies of hundreds and thousands of MeV, new and strange particles are created—a whole new science of particle physics is born. Before introducing these particles, we consider some of the reactions induced by nuclear projectiles of moderately high energy and describe some of the detection devices which physicists use to study reactions.

50.4 ARTIFICIAL TRANSMUTATIONS

The development of high-voltage accelerators made it possible to bombard atomic nuclei with a variety of swiftly moving projectiles. When an accelerated particle is captured by a nucleus, the resulting system of nucleons is initially in a state of high excitation. It may get rid of its excess energy by emitting one or more particles. Hundreds of artificial transmutations have been studied. Table 50.1 lists some typical reactions. Figures 50.4 and 50.5 show two reactions in schematic form. A reaction in which a proton is captured and a neutron is emitted is called a (p,n) reaction, etc. In this shorthand the entering particle appears first, and the emitted particle after the comma.

Table 50.1
Some Typical Nuclear Transmutations

Kind	Example
(α,p)	$^{27}_{13}\text{Al} + ^{4}_{2}\text{He} \rightarrow ^{1}_{1}\text{H} + ^{30}_{14}\text{Si}$
(α,n)	$^{27}_{13}\text{Al} + ^{4}_{2}\text{He} \rightarrow ^{1}_{0}n + ^{30}_{15}\text{P}$
(p,α)	$^{19}_{9}\text{F} + ^{1}_{1}\text{H} \rightarrow ^{4}_{2}\text{He} + ^{16}_{8}\text{O}$
(p,n)	$^{11}_{5}\text{B} + ^{1}_{1}\text{H} \rightarrow ^{1}_{0}n + ^{11}_{6}\text{C}$
(p,γ)	$^{31}_{15}\text{P} + ^{1}_{1}\text{H} \rightarrow ^{32}_{16}\text{S} + \gamma$
(d,α)	$^{27}_{13}\text{Al} + ^{2}_{1}\text{H} \rightarrow ^{4}_{2}\text{He} + ^{25}_{12}\text{Mg}$
(d,p)	$^{14}_{7}\text{N} + ^{2}_{1}\text{H} \rightarrow ^{1}_{1}\text{H} + ^{15}_{7}\text{N}$
(d,n)	$^{25}_{12}\text{Mg} + ^{2}_{1}\text{H} \rightarrow ^{1}_{0}n + ^{26}_{13}\text{Al}$
$(d,2n)$	$^{3}_{1}\text{H} + ^{2}_{1}\text{H} \rightarrow 2^{1}_{0}n + ^{3}_{2}\text{He}$
(n,α)	$^{27}_{13}\text{Al} + ^{1}_{0}n \rightarrow ^{4}_{2}\text{He} + ^{24}_{11}\text{Na}$
(n,p)	$^{106}_{48}\text{Cd} + ^{1}_{0}n \rightarrow ^{1}_{1}\text{H} + ^{106}_{47}\text{Ag}$
$(n,2n)$	$^{107}_{47}\text{Ag} + ^{1}_{0}n \rightarrow 2^{1}_{0}n + ^{106}_{47}\text{Ag}$
(n,γ)	$^{113}_{48}\text{Cd} + ^{1}_{0}n \rightarrow ^{114}_{48}\text{Cd} + \gamma$
(γ,n)	$^{9}_{4}\text{Be} + \gamma \rightarrow ^{1}_{0}n + ^{8}_{4}\text{Be}$
(γ,p)	$^{27}_{13}\text{Al} + \gamma \rightarrow ^{1}_{1}\text{H} + ^{26}_{12}\text{Mg}$

Figure 50.5
A sodium 23 nucleus captures a deuteron, emitting a proton and leaving an unstable sodium 24 nucleus which decays to magnesium 24 by the emission of an electron and an antineutrino; the $^{24}_{12}$Mg nucleus is left in an excited state and emits a gamma photon.

If a target is bombarded by protons (or some other particles), it is common to find that protons of one energy are strongly captured while protons of slightly higher or lower energy are rejected. We say that such capture reactions show *resonances*. A proton is likely to be captured if it brings in just the right energy and angular momentum to form a compound nucleus in one of its allowed energy states. A proton with more or less energy than this required amount is not captured.

Many nuclei created by transmutations are unstable and eventually decay. In some cases these unstable nuclei have relatively long half-lives. When a proton is captured by carbon 12, nitrogen 13 is formed. This nitrogen nucleus is unstable, with a half-life of 10.1 min. It decays to ^{13}C by the emission of a positron. When tellurium 130 is bombarded with neutrons, ^{131}Te is formed but decays by electron emission to iodine 131. ^{131}I is also a beta emitter; its half-life is 8 days. This radioactive isotope is used in treating hyperthyroid conditions.

50.5 CHARGED-PARTICLE ACCELERATORS

In both nuclear and atomic physics we use beams of high-energy ions (in particular, protons and deuterons) for studying the reactions induced in various targets. A variety of particle accelerators have been developed, many of which depend on magnetic fields to constrain the particle to move in a circular orbit while electric fields produce the accelerations. Although particle accelerators are basic tools of nuclear physics, the underlying principles are those which govern the interactions of ions with electric and magnetic fields.

The Cyclotron The first major accelerator based on circular orbits in a magnetic field was the cyclotron, developed in 1932 by Lawrence and Livingston. The key parts of the cyclotron are a large electromagnet which provides a constant magnetic field over a large area and an oscillating electric field between two semicircular hollow elements, called *dees* because they have the shape of a D. Ions are produced at the center of the magnetic field between the two dees, as shown in Fig.

50.6. A charged particle which is accelerated toward the lower dee gains energy as it crosses the gap between the dees. Once in the lower dee its path is circular, since

$$qvB = \frac{mv^2}{r} \tag{50.1}$$

Soon the particle finds itself in the region between the two dees once more. If now the upper dee is negative, and the lower dee positive, the particle is accelerated across the gap and enters the upper dee with a higher velocity, where it traverses a semicircular arc of greater radius. When it again reaches the gap between the dees, the electric field has reversed, and the particle is accelerated downward. At each passing of the gap it gains energy, and the radius of its path in the magnetic field becomes greater, until eventually the particle has a very high energy and its orbit approaches the circumference of the magnetic poles. It is then removed from the cyclotron and directed against a target. The time $T_{1/2}$ required for the particle to make the half circle in one of the dees is

$$T_{1/2} = \frac{\pi r}{v} \tag{50.2}$$

In this time the accelerating field should be reversed; therefore this is the time for one-half cycle of the alternating potential difference applied to the dees. The number of complete revolutions made per second is

$$\nu = \frac{v}{2\pi r} \tag{50.3}$$

Substituting the ratio v/r from Eq. (50.1), we obtain

$$\nu = \frac{Bq}{2\pi m} \tag{50.4}$$

If we apply the constant frequency given by Eq. (50.4) to the dees, ions are accelerated each time they pass the gap.

The time required for a charged particle to make half a revolution in the cyclotron is independent of radius only so long as the mass of the particle is constant. As we saw in Sec. 41.9, the mass of any particle

Figure 50.6
Cyclotron, showing the path of a positively charged particle in the vacuum chamber. There is a magnetic field into the plane of the paper.

varies rapidly with velocity when v approaches c, the speed of light. If the cyclotron is used to accelerate a particle to a velocity which is more than a few percent of the speed of light, the mass of the particle increases with time. By Eq. (50.4), the accelerating field can be kept in step with the particles only by decreasing the frequency as the mass increases. In very large cyclotrons the speed of the particles becomes great enough so that relativistic effects are important. Then the frequency of the cyclotron must be varied during an acceleration cycle. A cyclotron operated in such a way is said to be *frequency modulated* and is sometimes called a *synchrocyclotron*.

The Betatron   The betatron is used to produce high-speed electrons. It consists essentially of a large electromagnet between the poles of which is placed an evacuated tube in the form of a doughnut (Fig. 50.7). Alternating current is applied to the coils of the electromagnet, and the accelerating potential difference is induced by the changing magnetic field, which acts on the electrons in the vacuum tube as if the tube were the secondary of a transformer. The electrons previously injected into the tube are whirled around in circular paths. As the electrons gain energy, the magnetic field is increased at such a rate that it is just strong enough to keep the electrons rotating in the same circular path. While the magnetic field builds up to its maximum value, the electrons travel around the vacuum tube many thousand times and acquire a velocity which closely approaches the speed of light.

The Synchrotron   In a synchrotron particles are accelerated at essentially constant radius in a doughnut-shaped evacuated tube. The acceleration is produced by a radio-frequency electric field applied across one or more gaps, as in the cyclotron. The magnetic field serves to keep the particle traversing the same circular path as it goes around. As the speed of the particle increases, the strength of the magnetic field must be increased so that the particle does not hit the walls of the doughnut. The frequency applied across the gap is increased, so that each time the particle crosses the gap it is properly accelerated. Synchrotrons ordinarily are designed to accelerate either protons or elec-

Figure 50.7
Doughnut-shaped vacuum tube of a betatron with magnetic field into the plane of the paper.

Figure 50.8
Tunnel of the 33-GeV
alternating-gradient synchrotron where
it joins the 50-MeV linear accelerator
which injects protons into the
"doughnut." At the right is a small
portion of the magnet ring providing
the guide field for the
$\frac{1}{2}$-mile-circumference ring.
[*Brookhaven National Laboratory.*]

trons. Protons with energies of many gigaelectron volts (GeV) are currently available from synchrotrons (Fig. 50.8).

50.6 THE DETECTION OF NUCLEAR REACTIONS

In nuclear reactions the target nuclei, the bombarding particle, and the resulting products are far too small to see and too light to weigh on the most sensitive balance. The problem of studying a reaction is a difficult one, since all judgments as to what has happened must be based on indirect evidence. However, scientists have developed a number of ingenious methods for studying nuclear reactions and detecting the products of the reactions.

Nuclear Emulsions The first detector of nuclear reactions was the photographic plate. Gamma rays striking such a plate produce blackening, just as visible light and x-rays do. By using an exceedingly fine-grained emulsion, the path of a high-energy proton, alpha particle, or electron may be studied under a microscope.

The Cloud Chamber The Wilson cloud chamber consists of a cylinder with a glass top and a movable piston at the bottom (Fig. 50.9). The volume of the chamber is filled with air and saturated water vapor or alcohol. When the piston is moved downward quickly, the mixture expands adiabatically, the temperature falls, and the formerly saturated

Figure 50.9
Wilson cloud chamber.

(a)

(b)

Figure 50.10
Cloud chamber photographs of (*a*) carbon and oxygen atoms breaking apart when struck by high-energy neutrons and (*b*) oxygen stars produced by 90-MeV neutrons in a 1.4-T magnetic field. [*The University of California.*]

vapor becomes supersaturated. Vapor molecules condense on any tiny particle which is present to serve as nucleus for a droplet. In particular, droplets form readily on charged ions. If a charged particle is moving through the chamber when the expansion occurs, droplets form on the ions produced by the passage of this charged particle, and a *track* is produced. Figure 50.10 shows cloud-chamber pictures of the breakup of nuclei struck by high-energy neutrons. The ionized fragments leave visible tracks.

When a strong magnetic field is impressed on a cloud chamber, charged particles moving across the field are deflected, producing curved paths. From the radius of curvature of a path, its length, and the number of ion pairs formed per unit path length, it is often possible to determine the mass and charge of the particle responsible for the track. It is not practical to build cloud chambers large enough to show the entire path of an extremely high-energy particle. To observe the paths and the reactions induced by such particles, photographic emulsions or bubble chambers are used.

The Bubble Chamber   In 1952 Glaser invented the bubble chamber, in which a charged particle moves through a superheated liquid rather than a supersaturated gas. The liquid is maintained at a temperature just under its boiling point at an elevated pressure. When the path of a particle is desired, the pressure is suddenly reduced. Ions produced by the passing particles offer nuclei on which bubbles form preferentially before general boiling begins. The bubble tracks are photographed (Fig. 50.11). Bubble chambers containing liquid hydrogen have proven to be peerless detectors of many high-energy reactions for which the particle tracks are too long to be shown completely in a cloud chamber of reasonable size. A charged particle loses its energy in a much shorter distance in a liquid than in a gas, because it interacts with many more atoms per unit path length.

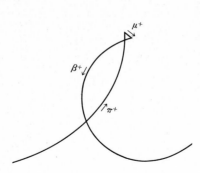

(a)   (b)

Figure 50.11
(*a*) Bubble-chamber photogram
showing the path of a π^+ meson
entering at the lower left and curving to
top center, where it decays into a
muon (and a mu neutrino which leaves
no track). The μ^+ moves a short
distance to the right and down before
decaying to a positron, a neutrino, and
a mu antineutrino. The positron leaves
an almost circular path, while the other
particles leave no tracks. There is a
magnetic field B perpendicular to the
plane of the paper. (*b*) Explanatory
sketch. [*Lawrence Radiation
Laboratory.*]

The Geiger-Müller Counter   A common detector of beta and gamma
rays is the Geiger-Müller counter. An elementary counter can be made
by stretching an insulated wire along the axis of a conducting metal
cylinder (Fig. 50.12). This chamber is evacuated and then filled to a
pressure of about 10 cm Hg with a mixture of argon and methyl alcohol.
The central wire is made about 1,000 V positive relative to the metal
cylinder. When an ionizing particle passes through this counter, electrons are ejected from atoms. The electrons move to the central wire,
while the positive ions move outward. The electric field near the wire
is sufficiently great for the gas to break down, and an avalanche results
during which a large instantaneous current is drawn. As the current
rises, the potential difference across the Geiger counter falls, while that
across the resistance R (Fig. 50.12) rises. Soon the potential difference
across the Geiger counter becomes too low to maintain the discharge.
The current drops to zero, and the potential difference across the counter
again rises to 1,000 V, so that the counter is once more ready to detect
an incident particle. Each time the gas breaks down in the Geiger
counter, the potential difference across the resistor R rises sharply; this
rise can be made to trigger a scaling circuit, which automatically counts
the pulses.

The Scintillation Detector   A detector which is the modern de-
scendent of the scintillation screen of Sec. 49.6 is the scintillation

Thin glass envelope   Metal cylinder

To detector

R

Figure 50.12
Geiger-Müller counter.

Ultraviolet and
visible light

Crystal

+200 V   +300 V   +600 V   +700 V   Anode
+1,000 V

Gamma ray

Photocathode

+100 V   +400 V   +500 V   +800 V   +900 V

Each electron produces several secondaries

Figure 50.13
Scintillation detector with
photomultiplier tube.

counter. When gamma rays fall on a crystal of sodium iodide with a little thallium impurity, absorption of the gamma rays results in the fluorescent emission of ultraviolet and visible light. In the scintillation detector shown in Fig. 50.13, the gamma-ray energy is absorbed in the crystal, producing light which falls on the photosensitive surface of the cathode of an *electron multiplier tube*. Electrons are ejected by the photoelectric effect and accelerated by a potential difference of the order of 200 V to electrode 1, which is made of a material which has the property of copious secondary emission. When they strike this electrode, several secondary electrons are ejected for each incident electron. The electrons from electrode 1 are accelerated to electrode 2, where a further multiplication is produced, and so forth. At the final collector an electron pulse is collected which may be 1 million times as great as that from the photocathode. Each time a gamma ray strikes and is absorbed in the crystal, a charge pulse is delivered to the anode. The charge collected is a measure of the energy of the gamma ray. The scintillation counter has the great advantage over the Geiger counter of not only telling when a gamma ray has been absorbed in the crystal but also yielding a charge pulse which is a measure of the energy.

Cerenkov Detectors When a charged particle passes through a medium with a speed greater than the speed of light in the medium, an electromagnetic "shock wave" of light is produced which is closely analogous to acoustic shock waves (Sec. 22.10). For example, electrons with a speed of $0.9c$ produce a bluish glow as they pass through water in which the speed of light is $c/1.33 = 0.75c$. Just as in the acoustic case, the sine of the angle between the path of the particle and the wavefront generated is the ratio of the speed of the electromagnetic wave c/n to the speed of the particle. (Here n is, of course, the index of refraction of the material through which the particle is moving.) Measurement of this angle permits the physicist to determine the speed of the particle generating the radiation, called *Cerenkov radiation* in honor of the physicist who made an early investigation of the phenomenon.

50.7 COSMIC RAYS

There are many accelerators which give charged particles (chiefly electrons and protons) energies in the range from 300 MeV to 35 GeV (billion electron volts in U.S. usage). In addition to these accelerators, there is another source of high-energy particles which is of great importance in studying nuclear reactions. This source is *cosmic radiation*. The primary cosmic rays consist of charged particles, mostly (85 percent) protons, with energies as great as 10^{20} eV. Also present are alpha particles, as well as some primary (and many secondary) electrons and positive nuclei with atomic numbers from 3 to 26. Few primary cosmic rays reach the surface of the earth. As they pass through the upper atmosphere, they collide with atmospheric nuclei and blast them apart. At the earth's surface the progeny of the primaries are high-energy photons, electrons, positrons, and other particles (Sec. 50.8). We call these particles *secondary cosmic rays*. A single proton may initiate a "shower" containing thousands of these secondaries.

The cosmic radiation which reaches the earth changes drastically in time, both in intensity and in energy spectrum. The major variations, which occur chiefly in particles with energies below 50 GeV, are closely related to solar magnetic activity. The particles are accelerated by magnetic fields in interstellar space and by accelerating fields associated with disturbances such as solar flares on the sun and other stars.

50.8 PIONS AND MUONS

When cosmic-ray particles collide with nuclei in the upper atmosphere, particles called *pions* (or π mesons) are often produced. Pions, which may be produced equally well by beams of accelerators, may be positive, negative, or neutral. A charged pion has a mass 273 times that of an electron, while the neutral pion is less massive—about 264 electron masses. Neutral π mesons disappear with the emission of two photons, which share the 134 MeV of energy released. Negative pions are strongly attracted to nuclei and are captured quickly. Positive pions are repelled by nuclei and are less likely to be captured. A π^+ decays (Fig. 50.11) in a very short time to a positive *muon* (μ^+) and a *mu neutrino* (ν_μ), while a negative pion decays according to the reaction

$$\pi^- \rightarrow \mu^- + \tilde{\nu}_\mu$$

where the tilde means an antineutrino. The mu neutrino is similar to, but distinguishable from, the neutrino of beta decay. Both the μ^+ and μ^- particles have a mass of 207 electron masses. They are short-lived, decaying by the reactions

$$\mu^- \rightarrow \beta^- + \tilde{\nu} + \nu_\mu \qquad \mu^+ \rightarrow \beta^+ + \nu + \tilde{\nu}_\mu$$

Pions are intimately linked to the attractive force between nucleons, each of which is sometimes imagined to be surrounded by a virtual pion cloud. In terms of this picture, the binding forces in nuclei are attributed to the exchange of pions much as we attribute covalent binding between hydrogen atoms to electron-exchange forces.

When a beam of protons with energies up to tens of gigaelectron volts is incident on a target, many other kinds of particles may be produced; indeed, any of the particles listed in Table 50.2 may be generated, either directly or indirectly, in this way.

50.9 PARTICLES AND ANTIPARTICLES

In 1955 a particle identical with the proton except for a negative charge was discovered by Segré, Chamberlain, and their collaborators at the University of California at Berkeley. The negative proton, or *antiproton*, was created by bombarding protons in a target with 6-GeV protons, thereby inducing the reaction

$$p + p + \text{energy} \rightarrow p + p + \tilde{p} + p$$

where $\tilde{p}$ indicates an antiproton. The energy of the bombarding proton is converted to a proton-antiproton pair plus the kinetic energy of the four residual particles. As soon as the antiproton is slowed down, it is annihilated by a proton; in a typical annihilation reaction the rest mass of the annihilating pair may appear as five pions and their kinetic energy (Fig. 50.14):

$$p + \tilde{p} \rightarrow \pi^+ + \pi^- + \pi^+ + \pi^- + \pi^0$$

One year later the antineutron $\tilde{n}$ was discovered at the same laboratory. Since the neutron bears no charge, the antineutron is also neutral. It is quickly annihilated, either by a proton or a neutron, usually

Table 50.2
Particles of Modern Physics

Name	*Symbol (charge)*	*Anti- particle*	*Rest mass, MeV/c²*	*Spin, h/2π*	*Mean life, second*	*Dominant decay modes*
Photon	γ		0	1		
Leptons:						
Neutrinos	ν, ν_μ	$\tilde{\nu}, \tilde{\nu}_\mu$	0	$\frac{1}{2}$		Stable
Electron	β^-	β^+	0.511	$\frac{1}{2}$		Stable
Muon	μ^-	μ^+	105.7	$\frac{1}{2}$	2.2×10^{-6}	$\beta^- + \nu_\mu + \tilde{\nu}$
Mesons:						
Pion	π^0	(π^0)	135	0	1×10^{-16}	Two γ rays
	π^+	π^-	139.6	0	2.6×10^{-8}	$\mu^+ + \nu_\mu$
Kaon	K^+	K^-	494	0	1.2×10^{-8}	$\mu^+ + \nu_\mu; \mu^+ + \nu_\mu + \pi^0$
						$\beta^+ + \nu + \pi^0; \pi^+ + \pi^0$
						$2\pi^+ + \pi^-; \pi^+ + 2\pi^0$
	K_S^0		498	0	9×10^{-11}	$\pi^+ + \pi^-; 2\pi^0$
	K_L^0		498	0	5×10^{-8}	$3\pi^0; \pi^+\pi^-\pi^0; \pi\beta\nu$
Eta	η^0		549	0		$\gamma\gamma; 3\pi^0; \pi^+\pi^-\pi^0$
Baryons:						
Proton	p^+	p^-	938.3	$\frac{1}{2}$		Stable
Neutron	n	$\tilde{n}$	939.5	$\frac{1}{2}$	1×10^3	$p + \beta^- + \tilde{\nu}$
Lambda hyperon	Λ	Λ	1,115	$\frac{1}{2}$	2.5×10^{-10}	$p + \pi^-; n + \pi^0$
Sigma hyperons	Σ^+	$\tilde{\Sigma}^+$	1,189	$\frac{1}{2}$	8×10^{-11}	$p + \pi^0; n + \pi^+$
	Σ^0	$\tilde{\Sigma}^0$	1,192	$\frac{1}{2}$	$(<10^{-14})$	$\Lambda + \gamma$
	Σ^-	$\tilde{\Sigma}^-$	1,196	$\frac{1}{2}$	1.5×10^{-10}	$n + \pi^-$
Xi hyperons	Ξ^0	$\tilde{\Xi}^0$	1,311	$\frac{1}{2}$	3×10^{-10}	$\Lambda + \pi^0$
	Ξ^-	$\tilde{\Xi}^-$	1,318	$\frac{1}{2}$	1.7×10^{-10}	$\Lambda + \pi^-$
Omega	Ω^-		1,672	$\frac{3}{2}$	1×10^{-10}	$\Xi^0\pi^-; \Xi^-\pi^0; \Lambda K^-$

with the production of several pions. If an antineutron is not annihilated by a nucleon, it decays by the reaction $\tilde{n} \rightarrow \tilde{p} + \beta^+ + \nu$.

We note a remarkable similarity here to the behavior of the electron and positron, ordinarily created together and disappearing by the annihilation reaction $\beta^- + \beta^+ \rightarrow 2\gamma$. We therefore identify the positron as the *antielectron*. Many of the particles of modern physics (Table 50.2) have antiparticles by which they can be annihilated. Of course, any antiparticle has a charge opposite to that of the corresponding particle, or else charge would not be conserved in annihilations. Many elementary particles decay in such a short time that they rarely find an antiparticle to annihilate. For example, the π^+ is antiparticle to the π^-, and vice versa. However, as we saw in Sec. 50.8 the most common fates for a π^- are (1) capture by a nucleus and (2) decay to a μ^-. If we know a decay mode of any particle, we can often write a corresponding possible decay mode for its antiparticle by changing each particle in the reaction to its antiparticle (just as was done for the muon decays in Sec. 50.8).

There is every reason to believe that a positron and a negative proton could form an atom of *antihydrogen*, which would have a spectrum similar to that of ordinary hydrogen. Indeed, from a collection of antiprotons, antineutrons, and positrons a world of antimatter could be constructed which would be indistinguishable from our world so long as everything were made of antiparticles. However, if some of this *contraterrene* matter were to come in contact with terrene matter, particle-antiparticle annihilation would occur, and almost the entire mass of the annihilating particles would be converted to energy of other forms. Contrast this to a fission reaction in which only about 0.1 percent of the mass is transformed.

When we organize the observed particles of physics, we classify neutrons, electrons, and muons as *leptons* (light particles). All have a spin quantum number of $\frac{1}{2}$ (in units of $h/2\pi$). When we assign a lepton number of $+1$ to the electron and the μ^-, ν, and ν_μ and a lepton number of -1 to the positron, μ^+, $\tilde{\nu}$, and $\tilde{\nu}_\mu$, we find that there is a conservation of lepton number in all observed reactions. When a neutron decays according to $n \rightarrow p + \beta^- + \tilde{\nu}$, we note that one lepton (β^-) and one antilepton ($\tilde{\nu}$) appear, for a net of zero leptons. In the μ^- decay we find one electron, one μ neutrino, and one antineutrino—one lepton before the decay and a net one lepton after the decay. The principle of *conservation of leptons* is a key conservation law in modern particle physics in the sense that any time a lepton appears or disappears there is the corresponding creation or annihilation of an antilepton.

Similarly, particles of half-integer spin and mass equal to or greater than that of the proton are called *baryons* (heavy particles). If to each ordinary baryon we assign the baryon number $+1$ and to the corresponding antibaryon -1, we find again that in all observed reactions the net baryon number does not change. Thus it appears that the *conservation of baryons* is also a valid law of particle physics. On the other hand, the particles classified as *mesons* have zero spin. A glance at the decay modes in Table 50.2 shows that there is no conservation of mesons, just as there is no conservation of photons.

A study of elementary particles is an area of intense activity in contemporary physics and a number of "new" mesons and baryons have

(a)

(b)

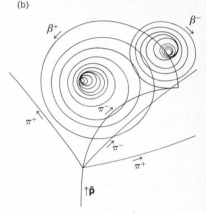

Figure 50.14

(a) Bubble-chamber photograph showing the path of an antiproton entering from below. Near the bottom of the picture the antiproton annihilated a proton, forming two π^+, two π^-, and a π^0, which leaves no track. In the upper quadrant of the picture a π^- interacted with a proton. Among the resulting products were a positron, which formed the counterclockwise spiral at top center, and an electron, which produced the smaller clockwise spiral. (b) Explanatory sketch. [*Courtesy of Lawrence Radiation Laboratory.*]

been discovered. What is the role of these strange new entities in nature? How many particles remain undiscovered? Will important uses be found for the new particles? Are there regions in the universe where stars and planets are composed of antiparticles? All these questions have been speculated upon; definitive answers await further research.

QUESTIONS

1   Why are low-energy neutrons much more effective in producing nuclear reactions than low-energy protons or alpha particles?

2   In Fig. 50.14 why are the electron and positron tracks spirals?

3   How can nuclear events be detected? Describe as many methods as you can.

4   How does a cyclotron give protons millions of electron volts of energy even though the highest potential difference involved is only a few kilovolts?

5   Relativity theory imposes a practical limit to the maximum energy which a proton can acquire from a simple cyclotron. Why does this limit arise? How is this limitation overcome in the synchrocyclotron and the synchrotron?

6   How are tracks produced in a Wilson cloud chamber?

7   Neutrons incident on the upper atmosphere collide with and are captured by ^{14}N nuclei with the subsequent emission in many cases of protons or alpha particles. What nuclei remain in each case?

8   What are cosmic rays? What is the distinction between primary and secondary cosmic rays?

9   What is a meson? A lepton? A baryon?

10   Roughly in what direction do you suppose the $\pi°$ meson produced in the p-p annihilation of Fig. 50.14 went? Why?

11   When a π^+ meson at rest decays to a μ^+ lepton, the muon always has the same energy, but when a μ^+ lepton decays to a positron, the positron may have any of a broad range of energies. Why?

PROBLEMS

1   Find the Q (energy released) in the ^9_4Be (α,n) $^{12}_6C$ reaction if the mass of ^9_4Be is 9.012186 u.
                                            Ans. 5.70 MeV

2   The first observed artificial transmutation was reported by Rutherford in 1910. He bombarded $^{14}_7N$ with alpha particles, and protons came out of the reaction. Find the Q (energy released) in the reaction if the mass of $^{17}_8O$ is 16.99913 u. (In this case Q is negative, meaning that energy must be supplied to the system.)

3   Calculate the Q (energy released) in the ^6_3Li (n,α) ^3_1H reaction.   *Ans.* 4.79 MeV

4   Complete each of the following reactions:
^{27}Al (n,α)? ^{31}P (p,α)? ^3H (γ,n)?
^{44}Ca (p,n)? $^{58}Cu \rightarrow$ ^{58}Ni + ? + ? ^{16}O (α,p)?
^7Li (γ,α)? ^{109}Ag (n,γ)? ^{24}Mg (d,α)?

5   Find the Q (energy released) in the reaction $^{19}_9F$ (p,α) $^{16}_8O$ if the mass of $^{19}_9F$ is 18.99841 u.
                                            Ans. 8.1 MeV

6   From the masses of ^6Li and ^7Li compute the binding energy of a neutron captured by a ^6Li nucleus. What is the wavelength of the gamma ray emitted in the ^6Li (n,γ) ^7Li reaction?

7   A cyclotron has pole faces with a diameter of 1.5 m and operates at a frequency of 20 MHz. Find the time required for a proton to traverse a semicircle in one dee, the magnetic field required to accelerate protons, and the maximum kinetic energy which protons can receive under these conditions.
               Ans. 2.5×10^{-8} sec; 1.31 T; 46.3 MeV

8   A synchrocyclotron has a magnetic field of 1.40 T. Find the appropriate frequency for accelerating protons when the speed of the protons is (a) $0.0100c$ and (b) $0.500c$.

9   Calculate the maximum energy of protons obtainable from a cyclotron of 1.2-m dee diameter and 1.5-T magnetic induction. At what frequency must the cyclotron be operated? If the average energy gain per dee passage is 40 keV, how many revolutions do the protons make?
                          Ans. 39 MeV; 22.9 MHz, 488

10   Calculate the maximum energy which the cyclotron of Prob. 9 can give alpha particles (He^{++} ions). At what frequency must the cyclotron be operated?

11   An alpha particle (He^{++} ion) with a speed of

4×10^7 m/sec is moving perpendicular to a uniform magnetic field of 1.25 T in a cyclotron. Find the force on the alpha particle and the radius of the circular path which it follows. What frequency must be applied to the dees to accelerate this alpha particle? If the radius of the cyclotron is 1.5 m, what is the maximum energy an alpha particle can be given in this cyclotron? *Ans.* 1.60×10^{-11} N; 0.663 m; 9.6 MHz; 169 MeV

12 Show that in a cyclotron the maximum kinetic energy which can be given an ion is $q^2B^2r^2/2m$, where B is the magnetic intensity and r the maximum allowable radius.

13 A proton of rest mass 938 MeV (1.67×10^{-27} kg) is given a kinetic energy of 9.38×10^9 eV in a proton synchrotron. Find the mass of the proton at this speed. What must be the radius of the orbit if the vertical magnetic field is 1.3 T? *Ans.* 1.84×10^{-26} kg; 26.3 m

14 A 5-MeV alpha particle (mass 4 u) collides elastically with an oxygen atom (mass 16 u) at rest. Calculate the maximum kinetic energy which can be transferred to the oxygen atom. (Use non-relativistic equations.)

15 A π° meson at rest decays into two photons which must have equal and opposite momenta. Find the energy and wavelength of one of the photons. *Ans.* 67.5 MeV; 1.84×10^{-14} m

16 Find the energy released in the decay of a positive muon to a positron (plus neutrino and antineutrino).

17 Find the energy released in the decay of a pion to a muon. When a pion at rest decays, the muon receives about 4.1 MeV of kinetic energy. Find the momentum carried off by the neutrino. *Ans.* 33.9 MeV; 1.58×10^{-20} kg-m/sec

18 A neutron of mass m and initial speed v_i collides head on with a nucleus of mass M at rest. Show that if the collision is elastic, the energy W transferred to the nucleus is $2Mm^2v_i^2/(M + m)^2$. Find the fraction of the energy of the neutron transferred to (a) a proton of mass m, (b) an alpha particle of mass $4m$, and (c) a uranium atom of mass $238m$.

19 A negative muon may be captured by a nucleus and occupy a "Bohr orbit" until it decays or interacts with a proton. Find the radius of the first Bohr orbit of a muon captured by lead, and show that this orbit lies inside the nucleus. *Ans.* 3.1×10^{-15} m; nuclear radius is about 7×10^{-15} m

20 A nucleus of charge Z may temporarily capture a pion which may occupy a hydrogenlike level of very low energy. If a negative pion with a mass of 273 electron masses is captured by an aluminum atom, find (a) the Bohr radius and binding energy for a pion in the ground ($n = 1$) state, and (b) the energy of the photon emitted when the pion makes a transition from the $n = 2$ to the $n = 1$ state.

APPENDIX

APPENDIX A
MATHEMATICAL FORMULAS

A.1 TRIGONOMETRIC RELATIONS

Figure A.1

Consider angle θ in the right triangle ABC (Fig. A.1) which has sides of lengths a, b, and c. By definition, the sine of θ is the ratio of the side opposite θ to the hypotenuse, the cosine of θ is the ratio of the side adjacent to θ to the hypotenuse, and the tangent of θ is the ratio of side opposite θ to the side adjacent to θ. The values of these functions are listed in the Table A.1. The cotangent of θ is the reciprocal of $\tan \theta$.

$$\sin \theta = \frac{a}{c} \qquad \cos \theta = \frac{b}{c} \qquad \tan \theta = \frac{a}{b} \qquad \cot \theta = \frac{b}{a}$$

Angle ϕ is the complement of θ. Clearly,

$$\sin \phi = \frac{b}{c} = \cos \theta = \cos (90° - \phi)$$

and

$$\cos \phi = \frac{a}{c} = \sin \theta = \sin (90° - \phi)$$

By the pythagorean theorem, $c^2 = b^2 + a^2$. From this it follows immediately that $\sin^2 \theta + \cos^2 \theta = 1$. Other trigonometric identities which are useful are

$$\sin (\theta \pm \phi) = \sin \theta \cos \phi \pm \cos \theta \sin \phi$$
$$\cos (\theta \pm \phi) = \cos \theta \cos \phi \mp \sin \theta \sin \phi$$

In any plane triangle having angles α, β, and γ with opposite sides a, b, and c, respectively (Fig. A.2),

$$\frac{\sin \alpha}{a} = \frac{\sin \beta}{b} = \frac{\sin \gamma}{c} \qquad \text{law of sines}$$

and

$$\cos \alpha = \frac{b^2 + c^2 - a^2}{2bc} \qquad \text{law of cosines}$$

If an angle exceeds 90°,

$$\sin \beta = \sin (180° - \beta)$$

and

$$\cos \beta = -\cos (180° - \beta)$$

If an angle θ is expressed in radians,

$$\sin \theta = \theta - \frac{\theta^3}{3!} + \frac{\theta^5}{5!} - \frac{\theta^7}{7!} + \cdots$$

$$\cos \theta = 1 - \frac{\theta^2}{2!} + \frac{\theta^4}{4!} - \frac{\theta^6}{6!} + \cdots$$

Figure A.2

Table A.1
Trigonometric Functions

Angle	Sine	Cosine	Tangent	Angle	Sine	Cosine	Tangent
1°	.01745	.9998	.01746	46°	.7193	.6947	1.0355
2°	.03490	.9994	.03492	47°	.7314	.6820	1.0724
3°	.05234	.9986	.05241	48°	.7431	.6691	1.1106
4°	.06976	.9976	.06993	49°	.7547	.6561	1.1504
5°	.08716	.9962	.08749	50°	.7660	.6428	1.1918
6°	.1045	.9945	.1051	51°	.7771	.6293	1.2349
7°	.1219	.9925	.1228	52°	.7880	.6157	1.2799
8°	.1392	.9903	.1405	53°	.7986	.6018	1.3270
9°	.1564	.9877	.1584	54°	.8090	.5878	1.3764
10°	.1736	.9848	.1763	55°	.8192	.5736	1.4281
11°	.1908	.9816	.1944	56°	.8290	.5592	1.4826
12°	.2079	.9781	.2126	57°	.8387	.5446	1.5399
13°	.2250	.9744	.2309	58°	.8480	.5299	1.6003
14°	.2419	.9703	.2493	59°	.8572	.5150	1.6643
15°	.2588	.9659	.2679	60°	.8660	.5000	1.7321
16°	.2756	.9613	.2867	61°	.8746	.4848	1.8040
17°	.2924	.9563	.3057	62°	.8829	.4695	1.8807
18°	.3090	.9511	.3249	63°	.8910	.4540	1.9626
19°	.3256	.9455	.3443	64°	.8988	.4384	2.0503
20°	.3420	.9397	.3640	65°	.9063	.4226	2.1445
21°	.3584	.9336	.3839	66°	.9135	.4067	2.2460
22°	.3746	.9272	.4040	67°	.9205	.3907	2.3559
23°	.3907	.9205	.4245	68°	.9272	.3746	2.4751
24°	.4067	.9135	.4452	69°	.9336	.3584	2.6051
25°	.4226	.9063	.4663	70°	.9397	.3420	2.7475
26°	.4384	.8988	.4877	71°	.9455	.3256	2.9042
27°	.4540	.8910	.5095	72°	.9511	.3090	3.0777
28°	.4695	.8829	.5317	73°	.9563	.2924	3.2709
29°	.4848	.8746	.5543	74°	.9613	.2756	3.4874
30°	.5000	.8660	.5774	75°	.9659	.2588	3.7321
31°	.5150	.8572	.6009	76°	.9703	.2419	4.0108
32°	.5299	.8480	.6249	77°	.9744	.2250	4.3315
33°	.5446	.8387	.6494	78°	.9781	.2079	4.7046
34°	.5592	.8290	.6745	79°	.9816	.1908	5.1446
35°	.5736	.8192	.7002	80°	.9848	.1736	5.6713
36°	.5878	.8090	.7265	81°	.9877	.1564	6.3138
37°	.6018	.7986	.7536	82°	.9903	.1392	7.1154
38°	.6157	.7880	.7813	83°	.9925	.1219	8.1443
39°	.6293	.7771	.8098	84°	.9945	.1045	9.5144
40°	.6428	.7660	.8391	85°	.9962	.08716	11.4301
41°	.6561	.7547	.8693	86°	.9976	.06976	14.3007
42°	.6691	.7431	.9004	87°	.9986	.05234	19.0811
43°	.6820	.7314	.9325	88°	.9994	.03490	28.6363
44°	.6947	.7193	.9657	89°	.9998	.01745	57.2900
45°	.7071	.7071	1.0000	90°	1.0000	.00000	∞

A.2 QUADRATIC EQUATIONS

Any quadratic equation can be put in the form $ax^2 + bx + c = 0$, for which the solutions are

$$x = \frac{-b \pm \sqrt{b^2 - 4ac}}{2a}$$

A.3 MENSURATION FORMULAS

The circumference of a circle of radius r is $2\pi r$ ($\pi = 3.1416$).

The area of a circle of radius r is πr^2.

The area of an ellipse with semiaxes a and b is πab.

The surface area of a sphere of radius r is $4\pi r^2$, and its volume is $\frac{4}{3}\pi r^3$.

TABLES OF DATA

Table B.1
Heat Constants of Solids

Substance	Melting point, °C	Coefficient of linear expansion per Celsius degree	Specific heat, kcal/(kg)(C°)	Heat of fusion	
				kcal/kg	Btu/lb
Aluminum	657	0.0000255	0.22	76.8	140
Bismuth	268	0.0000157	0.030	12.6	22.7
Brass		0.0000193	0.090		
Copper	1084	0.0000167	0.093	43	77
Glass		0.0000083	0.20		
Gold	1063	0.0000139	0.030		
Ice	0	0.000051	0.50	79.8	144
Iron	1503	0.0000119	0.11	30	54
Lead	327	0.0000276	0.030	5.4	9.7
Mercury	−38.8		0.033	2.8	5.4
Nickel	1452	0.0000128	0.109	4.6	8.3
Platinum	1756	0.0000089	0.032	27	48.6
Silver	960	0.0000188	0.056	22	39
Steel		0.000012	0.11		
Tungsten	3360	0.0000044	0.034		
Zinc	418	0.0000263	0.092	28.1	50.6

Table B.2
Heat Constants of Liquids

Substance	Boiling point, °C	Volume expansion per Celsius degree	Specific heat, kcal/(kg)(C°)	Heat of vaporization	
				kcal/kg	Btu/lb
Ammonia	−34			294	529
Aniline	184		0.514	110	198
Alcohol (ethyl)	78.1	0.0011	0.55	205	369
Benzine	80.3	0.00124	0.34	94.4	170
Chloroform	61	0.00126	0.232	58	106
Ether (ethyl)	34.5	0.00163	0.56	88.4	159
Gasoline	70–90	0.0012		71–81	128–146
Glycerin	290	0.00053	0.58		
Mercury	358	0.000182	0.0332	68	122
Turpentine	159	0.00094	0.42	70	126
Water	100	0.00030	1.00	540	970

Note: Densities are tabulated in Sec. 12.8.

Table B.3
Boiling Point of Water
(Boiling points of water at pressures near standard atmospheric pressure; the pressures are given in millimeters of mercury at 0°C)

Pressure, mm Hg	Temperature, °C	Pressure, mm Hg	Temperature, °C	Pressure, mm Hg	Temperature, °C
733	98.99	749	99.59	765	100.18
735	99.07	751	99.67	767	100.26
737	99.14	753	99.74	769	100.33
739	99.22	755	99.82	771	100.40
741	99.29	757	99.89	773	100.47
743	99.37	759	99.96	775	100.55
745	99.44	761	100.04	777	100.62
747	99.52	763	100.11	779	100.69

Table B.4
Properties of Saturated Water Vapor

Temperature, °C	Pressure		Volume		Heat units per unit mass, kcal/kg		
	N/m^2	$lb/in.^2$	m^3/kg	ft^3/lb	Of water	Latent heat	Total heat of vapor
0	6.2×10^2	0.089	204.970	3283.00	0	594.7	594.7
10	1.23×10^3	0.178	106.620	1707.60	10	589.4	599.4
20	2.31	0.336	58.150	931.48	20	584.1	604.1
30	4.20	0.61	33.132	530.72	30	578.8	608.8
40	7.32	1.06	19.670	314.77	40.1	573.4	613.5
50	1.23×10^4	1.78	12.091	193.68	50.1	567.9	618
60	1.98×10^4	2.88	7.695	123.26	60.1	562.4	622.6
70	3.11	4.51	5.050	80.89	70.2	556.8	627
80	4.72	6.86	3.4085	54.60	80.3	551	631.5
90	7.00	10.16	2.3592	37.79	90.4	545.2	635.6
100	1.013×10^5	14.70	1.6702	26.754	100.5	539.6	639.7
110	1.433	20.79	1.2073	19.339	110.7	532.9	643.6
120	1.986×10^5	28.83	0.8894	14.247	120.9	526.6	647.4
130	2.705	39.26	0.6664	10.675	131.1	520	651
140	3.621	52.56	0.5071	8.123	141.3	513.2	654.5
150	4.771	69.24	0.3917	6.274	151.6	506.2	657.8
160	6.197	89.93	0.3065	4.91	161.9	498.9	660.8
170	7.942	115.27	0.2429	3.891	172.2	491.4	663.7
180	1.005×10^6	145.90	0.1945	3.116	182.6	483.7	666.3
190	1.258	182.56	0.1575	2.523	193.1	475.7	668.8
200	1.557	226.00	0.1288	2.063	203.6	467.5	671.1
210	1.910	277.20	0.1063	1.703	214.1	459.1	673.2

C.1 ENERGY STORED IN AN INDUCTIVE CIRCUIT

The energy stored in the magnetic field associated with a circuit for which the coefficient of self-induction is L is equal to the work required to establish the current in the circuit or the energy which will be released when the current is eliminated. The back emf due to the self-inductance in the circuit is

$$e = -L \frac{di}{dt}$$

The work done against this back emf in the time dt is

$$ei \, dt = Li \frac{di}{dt} \, dt = Li \, di$$

The whole work is

$$W = \int_0^I ei \, dt = \int_0^I Li \, di = \tfrac{1}{2}LI^2$$

C.2 EFFECTIVE VALUE OF AN ALTERNATING CURRENT

By definition the effective value of an alternating current is the steady current which dissipates heat in a resistor at the same average rate as the alternating current. Let the alternating current be

$$i = I_{\max} \sin 2\pi\nu t$$

The power developed instantaneously in a resistor R is

$$I_{\max}^2 R \sin^2 2\pi\nu t$$

The heat developed during one cycle is

$$I_{\text{eff}}^2 RT = \int_0^T I_{\max}^2 R \sin^2 2\pi\nu t \, dt$$

where T is the period ($T = 1/\nu$). Therefore,

$$I_{\text{eff}}^2 T = I_{\max}^2 \int_0^T \sin^2 \frac{2\pi}{T} t \, dt = \frac{I_{\max}^2 T}{2}$$

from which

$$I_{\text{eff}} = \frac{I_{\max}}{\sqrt{2}} = 0.707 I_{\max}$$

Application of Kirchhoff's second law to the circuit of Fig. 40.3 yields

$$v = \mathcal{E}_{max} \sin 2\pi\nu t = L\frac{di}{dt}$$

from which

$$i = \int \frac{\mathcal{E}_{max} \sin 2\pi\nu t \, dt}{L} = -\frac{\mathcal{E}_{max} \cos 2\pi\nu t}{2\pi\nu L}$$

and

$$I_{max} = \frac{\mathcal{E}_{max}}{2\pi\nu L}$$

Similarly, when an alternating potential difference $\mathcal{E}_{max} \sin 2\pi\nu t$ is applied across a capacitor,

$$\mathcal{E}_{max} \sin 2\pi\nu t = \frac{q}{C}$$

Since $i = dq/dt$,

$$\frac{i}{C} = \frac{d(\mathcal{E}_{max} \sin 2\pi\nu t)}{dt} = 2\pi\nu\mathcal{E}_{max} \cos 2\pi\nu t$$

Therefore,

$$i = \frac{\mathcal{E}_{max} \cos 2\pi\nu t}{1/2\pi\nu C} \qquad \text{and} \qquad I_{max} = \frac{\mathcal{E}_{max}}{1/2\pi\nu C}$$

For an ac series circuit containing a source of emf $e = \mathcal{E}_{max} \sin 2\pi\nu t$ and resistance, inductance, and capacitance (Fig. 40.7), application of Kirchhoff's second law at any instant leads to

$$e = \mathcal{E}_{max} \sin 2\pi\nu t = Ri + L\frac{di}{dt} + \frac{q}{C}$$

If we differentiate with respect to time,

$$2\pi\nu\mathcal{E}_{max} \cos 2\pi\nu t = R\frac{di}{dt} + L\frac{d^2i}{dt^2} + \frac{i}{C}$$

The solution of this differential equation is

$$i = \frac{\mathcal{E}_{max} \sin(2\pi\nu t - \theta)}{\sqrt{R^2 + [2\pi\nu L - (1/2\pi\nu C)]^2}}$$

where

$$\tan\theta = \frac{2\pi\nu L - 1/2\pi\nu C}{R}$$

and

$$I = \frac{\mathcal{E}}{\sqrt{R^2 + [2\pi\nu L - (1/2\pi\nu C)]^2}}$$

If either the inductor or the capacitor is removed from the circuit and the circuit is completed once more, the appropriate term vanishes from the equations above. This leads to Eq. (40.4) if the capacitor is removed and to Eq. (40.6) if there is no inductance.

The instantaneous power supplied to an ac circuit is given by $p = ei$ [Eq. (40.9)]. The average power over a complete cycle is

$$P = \frac{1}{T} \int_0^T p \, dt = \frac{1}{T} \int_0^T ei \, dt$$

$$= \frac{1}{T} \int_0^T (\mathcal{E}_{max} \sin 2\pi\nu t) \, I_{max} \sin (2\pi\nu t - \theta) \, dt$$

where T is the period ($T = 1/\nu$). Since

$$\sin (2\pi\nu t - \theta) = \sin 2\pi\nu t \cos \theta - \cos 2\pi\nu t \sin \theta$$

P becomes

$$P = \frac{\mathcal{E}_{max} I_{max}}{T} \int_0^T \left(\sin^2 \frac{2\pi t}{T} \cos \theta + \sin \frac{2\pi t}{T} \cos \frac{2\pi t}{T} \sin \theta \right) dt$$

$$= \frac{\mathcal{E}_{max} I_{max} \cos \theta}{2} = \mathcal{E}I \cos \theta$$

which is Eq. (40.10).

INDEX

PERIODIC TABLE OF THE ELEMENTS

(Based on Carbon 12 as 12.00000)

Orbital blocks: $s\,(l=0)$ · $d\,(l=2)$ · $p\,(l=1)$ · $f\,(l=3)$

n	I_A	II_A	III_B	IV_B	V_B	VI_B	VII_B	VIII	VIII	VIII	I_B	II_B	III_A	IV_A	V_A	VI_A	VII_A	Inert gases
1	1 H 1.00797	2 He 4.0026																2 He 4.0026
2	3 Li 6.939	4 Be 9.0122											5 B 10.811	6 C 12.01115	7 N 14.0067	8 O 15.9994	9 F 18.9984	10 Ne 20.183
3	11 Na 22.9898	12 Mg 24.312											13 Al 26.9815	14 Si 28.086	15 P 30.9738	16 S 32.064	17 Cl 35.453	18 Ar 39.948
4	19 K 39.102	20 Ca 40.08	21 Sc 44.956	22 Ti 47.90	23 V 50.942	24 Cr 51.996	25 Mn 54.9380	26 Fe 55.847	27 Co 58.9332	28 Ni 58.71	29 Cu 63.546	30 Zn 65.37	31 Ga 69.72	32 Ge 72.59	33 As 74.9216	34 Se 78.96	35 Br 79.904	36 Kr 83.80
5	37 Rb 85.47	38 Sr 87.62	39 Y 88.905	40 Zr 91.22	41 Nb 92.906	42 Mo 95.94	43 Tc (97)	44 Ru 101.07	45 Rh 102.905	46 Pd 106.4	47 Ag 107.868	48 Cd 112.40	49 In 114.82	50 Sn 118.69	51 Sb 121.75	52 Te 127.60	53 I 126.904	54 Xe 131.30
6	55 Cs 132.905	56 Ba 137.34	*	72 Hf 178.49	73 Ta 180.948	74 W 183.85	75 Re 186.2	76 Os 190.2	77 Ir 192.2	78 Pt 195.09	79 Au 196.967	80 Hg 200.59	81 Tl 204.37	82 Pb 207.19	83 Bi 208.980	84 Po (209)	85 At (210)	86 Rn (222)
7	87 Fr (223)	88 Ra (226)	†															

Transition elements

$f\,(l=3)$

* Lanthanide series (rare earths) — d

57 La 138.91	58 Ce 140.12	59 Pr 140.907	60 Nd 144.24	61 Pm (145)	62 Sm 150.35	63 Eu 151.96	64 Gd 157.25	65 Tb 158.924	66 Dy 162.50	67 Ho 164.930	68 Er 167.26	69 Tm 168.934	70 Yb 173.04	71 Lu 174.97

† Actinide series

89 Ac (227)	90 Th (232)	91 Pa (231)	92 U (238)	93 Np (237)	94 Pu (244)	95 Am (243)	96 Cm (247)	97 Bk (247)	98 Cf (251)	99 Es (254)	100 Fm (257)	101 Md (256)	102 No (254)	103 Lw (254)